Solar Magnetic Fields

André Balogh • Edward Cliver • Gordon Petrie •
Sami Solanki • Michael Thompson •
Rudolf von Steiger
Editors

Solar Magnetic Fields

From Measurements Towards Understanding

Chapters 2–14 were previously published in *Space Science Reviews* Volume 210, Issues 1–4, 2017

Springer

Editors
André Balogh
Imperial College London
London, UK

Edward Cliver
Advanced Technology Solar Telescope
National Solar Observatory
Sunspot, NM, USA

Gordon Petrie
National Solar Observatory
Tucson, AZ, USA

Sami Solanki
Max-Planck-Institut für
Sonnensystemforschung
Göttingen, Niedersachsen, Germany

Michael Thompson
High Altitude Observatory
University Corporation for
Atmospheric Research
Boulder, CO, USA

Rudolf von Steiger
International Space Science Institute
Bern, Switzerland

ISSN 1385-7525 Space Sciences Series of ISSI
ISBN 978-94-024-1520-9

Library of Congress Control Number: 2018942944

Cover Image: Inversion of a spectropolarimetric raster scan of a sunspot in AR 12049, observed at 1.568 μm with GRIS at GREGOR in May 2014. From left to right: temperature, magnetic field strength, azimuth and inclination of the magnetic field. Courtesy of GREGOR under the leadership of the Kiepenheuer-Institut für Sonnenphysik (KIS) with the German partners AIP, IAG and MPS, the Spanish IAC and with contributions by the Czech ASCR

Printed on acid-free paper

This Springer imprint is published by the registered company Springer Science+Business Media B.V. part of Springer Nature.
The registered company address is: Van Godewijckstraat 30, 3311 GX Dordrecht, The Netherlands

Contents

The Life Cycle of Active Region Magnetic Fields
M.C.M. Cheung, L. van Driel-Gesztelyi, V. Martínez Pillet and M.J. Thompson: Space Science Reviews 2016, 2017:317–349 (24 August 2016)
https://doi.org/10.1007/s11214-016-0259-y

Surface Flux Transport and the Evolution of the Sun's Polar Fields
Y.-M. Wang: Space Science Reviews 2016, 2017:351–365 (26 April 2016)
https://doi.org/10.1007/s11214-016-0257-0

The Global Solar Dynamo
R.H. Cameron, M. Dikpati and A. Brandenburg: Space Science Reviews 2016, 2017:367–395 (11 February 2016) https://doi.org/10.1007/s11214-015-0230-3

Prospects of Solar Magnetometry—From Ground and in Space
L. Kleint and A. Gandorfer: Space Science Reviews 2015, 2017:397–426 (17 December 2015) https://doi.org/10.1007/s11214-015-0208-1

Preface

André Balogh[1] · Rudolf von Steiger[2]

The variability in solar activity and its causes have remained the subject of theoretical and experimental investigations for several decades. The objective is to understand the causal chains that shape the near and far solar environment, from the interior of the Sun all the way to interplanetary space. The observational means as well as the theoretical and modelling resources devoted to the topic have been formidable. Built on a broad range of terrestrial and space-based observational resources, solar science has progressed to the point where there are vast quantities of observational data and very detailed models of phenomena that describe the processes in the interior of the Sun, its surface and atmosphere, as well as in the heliosphere that extends to the interstellar medium at a distance of more than 100 AU.

In parallel with the ever-increasing volume of observations and constantly refined and renewed models, there is a flow of publications, scientific papers and books, and there are numerous conferences and workshops yearly that are devoted to aspects of solar physics. Yet, despite the proliferation of phenomenological descriptions, there are many open questions concerning the physical processes underlying the causal chains of solar variability. Some of these have been enumerated and discussed, among others, by Parker in this journal (Vol. 144, 2009).

There is, however, general agreement on the key role played by magnetic fields in all solar phenomena and throughout the sun and its close and distant environment. Reviews of magnetic fields in the sun's interior, its atmosphere and in the heliosphere in that volume and elsewhere provide an overview of the place of magnetic fields in solar physics. The key variability manifests itself in the many phenomena related to the 11-year activity cycle, which was also reviewed in this journal (Vol. 186, 2014).

While the central role of magnetic fields is acknowledged, understanding the many processes that involve magnetic fields is hampered by limitations of their quantitative knowledge. Measuring magnetic fields remotely in the solar environment is difficult, even though

✉ R. von Steiger
vsteiger@issibern.ch

1 Imperial College, London, UK

2 International Space Science Institute, Bern, Switzerland

solar observers have always taken advantage of newly discovered techniques in their instruments. The first discovery of strong magnetic fields in sunspots by Hale in the early 20th century came from applying the then recently described Zeeman effect of magnetic fields on the splitting of spectral lines. Progress in solar physics has come about from increasing observational capabilities, outlined in the historical sequence of some of the reviews setting out progress in the field by Babcock and others. A detailed personal account of observing solar magnetic fields, with a fifty-year perspective, is given by Stenflo (this volume).

Helioseismology, the study of solar oscillations is a powerful tool for the study of the solar interior. Since the start of its use three decades ago, great progress has been made in measuring by inference solar magnetic fields in the interior of the sun, in the near-surface layers and in the photosphere. In particular, it has provided substantial observational support for refining global models of the solar dynamo—even though the debate concerning the "best" model remains open. This is one clear example of observational data prompting major advances in understanding the mechanisms at the origin of the sun's magnetic field and setting constraints on theoretical models, yet not being sufficiently comprehensive to enable a firm choice to be made among competing dynamo models.

Whole sun magnetographs were the first instruments, following the original observing instruments used by Hale, that allowed progressing beyond measurements of the strong magnetic fields associated with sunspots. The pioneering magnetograph constructed at the Mount Wilson Observatory by Babcock, carefully designed incorporating the best available technology, allowed an estimate of about 1 G as the general surface magnetic field of the sun. He was also able to observe the polarity reversal of the sun associated with the sunspot cycle and propose, on the basis of his observations, a phenomenological model of the 22-year magnetic cycle that forms the first element of what is now known as the Babcock-Leighton model; for an up-to-date assessment of the status of that model in the current context, see Charbonneau in Living Reviews in Solar Physics.

Since the early days of magnetograph measurements in the 1950s, new generations of magnetographs and spectropolarimeters have considerably increased the resolution and general capabilities of solar surface magnetic measurements from ground-based observatories. It is now well established that the surface field is very highly non-uniform, with magnetic flux concentrated in narrow structures that represent the complex emergence of magnetic flux from below the photosphere.

The ability to fly a high-performance solar magnetograph on a space platform, such as the Solar Orbiter to be launched in 2018 will further increase the observational data base for improving the resolution and comprehensive coverage of solar surface magnetic measurements (see Kleint and Gandorfer, this volume). The resolution capability of the most recent observations of solar surface fields, now makes it possible to combine direct observations of photospheric fields to match the observed chromospheric fields. Such observations, made in this case with the Ultraviolet Filter Imager (SuFI) onboard the Sunrise balloon-borne observatory, challenge the modelling capabilities by providing very tight constraints on the determination of very small scale magnetic field structures and their dynamics in the solar atmosphere. A broader based discussion of these observational capabilities and their interpretation is given by Borrero et al. (this volume).

In the solar photosphere, chromosphere and corona, the magnetic field measurements are generally very difficult. There are several measurement possibilities: Zeeman and Hanle spectropolarimetry of atomic line profiles, the structure and dynamics of visible plasma entrained in the magnetic field, and radio techniques. The principal challenge of spectropolarimetry is to devise an optimum strategy of sampling the spectropolarimetric hypercube, which spans the five dimensions x, y, λ, S, and t (space, wavelength, Stokes vector, and time). Different strategies of sampling that hypercube and their relative merits and disadvantages are discussed by Lagg et al. (this volume). Our ability to make solar magnetic measurements critically depends on technology availability and understanding of radiative transfer in the solar atmosphere.

The in situ measurement of the magnetic field, where it can be performed in space, while accurate, suffers from the limitation of spatial coverage. Heliospheric magnetic field measurements have confirmed the solar origin of magnetic fields (in the heliosphere) and have narrowed the links between the Sun and the heliospheric medium. The first key results came quickly, in the 1960s—the relatively crude magnetometers (and spacecraft) performed well (if not always reliably). In particular, confirmation of the basic truth of Parker's model of coronal expansion and the consequent large-scale geometry of the magnetic field in the solar wind was achieved by the first few missions—Mariner 2, IMP 1 (and followers), and Mariners 4 & 5. The basic target of measuring fields of $\sim$5 nT with accuracies better than 1 nT was quickly achieved. New results since the early days have been achieved by a combination of the following factors: new trajectories through the heliosphere (in latitude and radial distance); improvements in magnetometers, but mostly in their data acquisition capabilities; an increased attention to achieve magnetically clean spacecraft in the 1970s, 1980s and 1990s (although nowadays agencies try workaround solutions that remain to be trusted and proved). The upcoming solar missions of ESA and NASA, Solar Orbiter and Solar Probe Plus (now aptly renamed Parker Solar Probe), respectively, both to be launched in 2018, will once again extend our knowledge vastly in the inner heliosphere down to ten solar radii.

As models of solar phenomena increase in complexity, thanks to improved understanding, a vastly increased computing capability and more refined observations covering broader parameter ranges, it is now more difficult to transfer measurement uncertainties and errors from the observables to the finer details of phenomenological models. Given also the increasing range of measurements and the need to combine observations and measurements from different sources, a good understanding of instrumental capabilities and limitations as well as quantitative error assessments have become necessary for confident modelling of complex solar magnetic phenomena.

Recognising this challenge, a Scientific Forum held at the International Space Science Institute in Bern, Switzerland on 20–21 November 2012 identified the potential benefits of assessing solar magnetic field measurements and how they can lead to a better understanding of solar phenomena. A Workshop was held in ISSI in February 2015 with the title *Solar Magnetic Fields: From Measurements Towards Understanding*. The review papers in this volume of the Space Science Series of ISSI, Volume 57, has been reprinted from Space Science Reviews (Volume 210, Issues 1–4 published in September 2017). This collection provides a highly topical overview of the achievements of solar magnetic field measurements in improving the quantitative observational data base for making further progress in solar physics. In the same spirit, the ISSI Forum in 2012 had also recommended a broad review of the observables and parameters used to describe and explain the 11-year solar activity cycle. The collection of reviews arising from the Workshop held in ISSI in November 2013 was published in Space Science Reviews (Volume 186, 2014) and in Volume 53 of the Space Science Series of ISSI: *The Solar Activity Cycle—Physical Causes and Consequences*, published in April 2015. The present volume thus completes the objectives elaborated in 2012 in the ISSI Scientific Forum on solar activity and solar magnetism.

Space Sci Rev (2017) 210:5–35
DOI 10.1007/s11214-015-0198-z

History of Solar Magnetic Fields Since George Ellery Hale

J.O. Stenflo[1,2]

Received: 20 May 2015 / Accepted: 11 August 2015 / Published online: 7 September 2015

Abstract As my own work on the Sun's magnetic field started exactly 50 years ago at Crimea in the USSR, I have been a participant in the field during nearly half the time span since Hale's discovery in 1908 of magnetic fields in sunspots. The present historical account is accompanied by photos from my personal slide collection, which show a number of the leading personalities who advanced the field in different areas: measurement techniques, from photographic to photoelectric and imaging methods in spectro-polarimetry; theoretical foundations of MHD and the origin of cosmic magnetic fields (birth of dynamo theory); the quest for increased angular resolution from national projects to international consortia (for instruments both on ground and in space); introduction of the Hanle effect in astrophysics and the Second Solar Spectrum as its playground; small-scale nature of the field, the fundamental resolution limit, and transcending it by resolution-independent diagnostics.

Keywords Sun: atmosphere · Magnetic fields · Polarization · Dynamo · Magnetohydrodynamics (MHD)

1 Sunspots as a Window to Cosmic Magnetism

Astrophysics can be seen as remote sensing: the information on the physical conditions in the universe reaches us encoded in the spectra that we record with our telescopes. Our task is to decode that information through spectral analysis.

With the discovery by Zeeman (1897) of the Zeeman effect, it became known that magnetic fields leave their encoded "fingerprints" in the splitting and polarization of spectral lines. Through spectro-polarimetry we can therefore extract information on magnetic fields in remote objects. George Ellery Hale was the first to make use of this opportunity in astrophysics (cf. Fig. 1). He had noticed how the shape of the Sun's corona suggests that the Sun

✉ J.O. Stenflo
stenflo@astro.phys.ethz.ch

1 Institute of Astronomy, ETH Zurich, 8093 Zurich, Switzerland

2 Istituto Ricerche Solari Locarno, Via Patocchi, 6605 Locarno-Monti, Switzerland

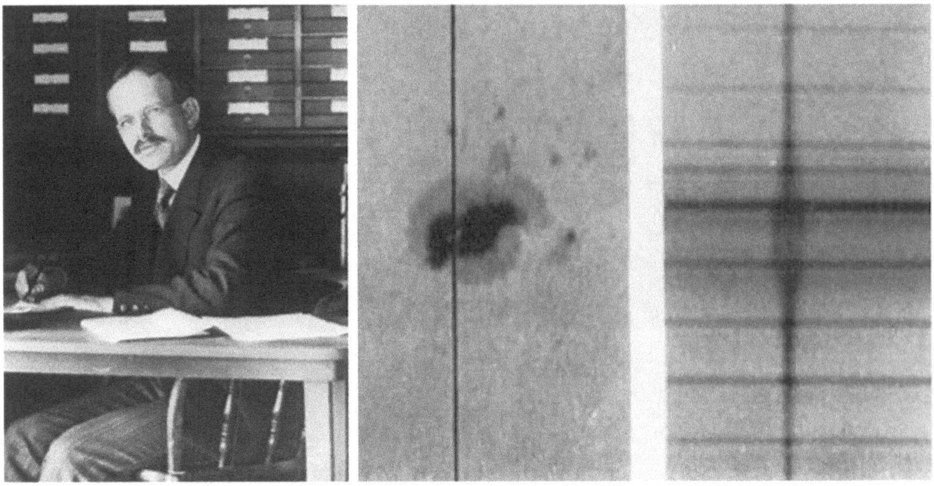

Fig. 1 George Ellery Hale and his discovery of magnetic fields in sunspots through observations of the Zeeman effect. At the place where the spectrograph slit crosses a sunspot the spectral lines get split in polarized components. The magnitude of the splitting is proportional to the magnetic field strength, while the polarization state (circular or linear) reveals the orientation of the field

is a magnetized sphere with a global dipole-like field, and that the filamentary vortex structure of the Hα fibrils around sunspots indicate that sunspots are the seats of strong magnetic fields and vertical electric currents. This led him to look for and find pronounced Zeeman splitting in sunspots (Hale 1908).

Sunspots thus became the gateway to the exploration of cosmic magnetic fields, including the origin of the fields and their role in astrophysical plasmas. We now know that magnetic fields govern most cosmic variability on intermediate time scales and pervade all cosmic plasmas, generating structuring, thermodynamic effects, and instabilities. Measurements of the Sun's magnetic field has guided the development of magnetohydrodynamics and dynamo theories.

The pioneering work by Hale and his team led to the discovery of Hale's polarity law (the law that governs the E-W orientation of the sunspot polarities with respect to solar hemisphere and 11-yr cycle), which showed that the magnetic cycle is 22 yr (the Hale cycle), twice the length of the sunspot cycle (Hale et al. 1919). Another fundamental discovery in the same paper is Joy's law, which tells that the polarity orientation of bipolar magnetic regions deviates systematically (in a statistical sense) from the E-W direction: the orientation is tilted so that the leading part of the region (with respect to the direction of solar rotation) is closer to the equator than the following part. Hale's and Joy's laws serve as observational cornerstones of solar dynamo theory. The Sun's dynamo can be seen as a prototype for all cosmic dynamos.

2 The Enigmatic General Magnetic Field of the Sun

Soon after his discovery of magnetic fields in sunspots Hale wanted to explore the non-spot "background field", the so-called "general" magnetic field of the Sun, which he believed to be a global dipole based on the appearance of the solar corona at eclipses. The observational technique was to place a grid of mica strips across the spectrograph slit, so that one gets an alternating sequence of left- and right-handed circularly polarized spectra. In the presence of a magnetic field there would be a relative line displacement between the opposite polarization states.

The spectra were recorded on photographic plates, and the line positions were determined manually with a micrometer. With modern standards this was a crude technique not adequate for detection of the weak background fields. Aware of the marginal nature of this undertaking and the danger of subjectivity in the results, Hale let many of his assistants independently reduce the same plate material, and he claimed in the publications that the persons reducing the data were not informed about which heliographic latitude a given plate referred to. This was supposed to eliminate possible bias from the results.

2.1 An Example of Personal Bias

All of the assistants except two got null results (consistent with what is to be expected in view of the measurement uncertainties). One of these two was a young postdoc who had recently arrived to Pasadena from Holland, Van Maanen. He immediately impressed Hale by getting positive results, and Hale gave most weight to his findings. This resulted in claims that the strength of the polar field was about an order of magnitude stronger than we find with modern, far more accurate techniques (Hale et al. 1918). In addition the polarity of the polar fields was opposite to what one would expect according the much later discovered model of the solar cycle by Babcock (1961).

In 1968, shortly after completion of my Doctoral Thesis with the title "The Sun's Magnetic Field", I spent seven months in Pasadena with the Mt Wilson Observatory and decided to use this opportunity to explore the Hale archives in search for an explanation of his enigmatic results on the Sun's general magnetic field. I was able to locate 400 photographic plates recorded in 1914, together with the meticulous notebook of Van Maanen with his micrometer measurements of the Zeeman shifts for each of these plates. After having the plates carefully cleaned I took them in my car and drove to the Sacramento Peak Observatory (Sunspot, New Mexico), where they had a digitized microphotometer with which I could (with the help of Jacques Beckers) have all the plates scanned. I then brought the big magnetic tapes with the scanned data to Lund, Sweden, where I analyzed them with a CDC mainframe computer. My computer reduction gave null results (within the standard error) for the general magnetic field, while Van Maanen had found for the same set of plates a field much stronger than the standard error, which varied with heliographic latitude as a global dipole field with the wrong polarity (Stenflo 1970).

Around the time of Hale's pioneering solar work the astronomical community was involved in an intense debate on the nature of spiral nebulae: Were they "island universes" (what we now call galaxies), or were they nearby objects inside our Milky Way system? Van Maanen joined this debate and provided what at that time seemed to be very convincing evidence against the island universe idea. Around 1922–1923 he measured on photographic plates substantial proper motions along the spiral arms of M33, which implied that the rotation period of M33 was smaller than that of the Milky Way by a factor of one thousand and that M33 could therefore not be very distant and large but had to be inside our galaxy (cf. Lundmark 1926). We now know that M33 is a galaxy and that whatever Van Maanen measured was not a real property.

2.2 The Photoelectric Magnetograph

The general, background magnetic fields, which we now refer to as quiet-sun magnetic fields, gave too weak Zeeman-effect signatures to be measurable with the technology that was available to Hale. Further progress had to wait until the breakthrough in sensitivity provided by the photoelectric magnetograph, introduced by Babcock (1953). Instead of manually measuring wavelength positions of spectral lines on photographic plates one used photomultipliers as detectors and electro-optical polarization modulation with an ADP-type Pockels cell. Two exit windows in the spectral focal plane selected the light from the blue and red wing of a chosen Zeeman-sensitive spectral line. A servo-controlled tilting glass plate near the spectral focal plane kept the spectral line centered with respect to the two exit slits. The tilt angle of the plate represented the measured Doppler shift. The modulated polarization signals from the two exit slits were combined in a difference amplifier (since the polarization signals from the two line wings have opposite signs—today we say that the Stokes V profile is anti-symmetric) and then demodulated.

Since photomultipliers are "1-pixel detectors", a circular-polarization image (which represents a "magnetogram") could only be built up through raster scanning of the solar image. The initial observations were crude, with a spatial resolution of order one arcmin, but they allowed the recording of full-disk magnetograms with a polarimetric sensitivity that was sufficient for mapping the weak background fields. Thus Babcock could initiate a synoptic full-disk magnetogram program at the Mt Wilson Observatory.

This synoptic program led Babcock to the discovery that the general magnetic field, in particular the field in the polar caps of the Sun, reverses sign every 11 years. Thus not only the sunspot polarities but also the global magnetic field varies with the 22-yr Hale cycle. The photoelectric magnetograph gave us a comprehensive empirical picture of solar magnetism. The underlying mechanism that governs the Hale cycle must physically link the behavior of sunspots with the behavior of the global field.

Babcock (1961) could integrate the various empirical results into a phenomenological theory for the solar cycle, which explicitly clarified how the bits and pieces were connected. The Hale cycle represents an oscillation between a poloidal and a toroidal configuration of the magnetic field. The frozen-in field lines of the poloidal field are wound up by the latitudinal shear of the Sun's differential rotation, so that a toroidal field is built up. When the strength of the amplified toroidal field exceeds a certain threshold, magnetic tension can no longer prevent buoyant sections of the subsurface flux ropes to float up to the surface to form bipolar magnetic regions with polarity orientations in the general E-W direction, with opposite orientations in the N and S hemispheres, as required by Hale's polarity law. During the rise to the surface the flux loop is rotated by the Coriolis force, resulting in a systematic tilt of the bipolar orientation with respect to the E-W direction, in qualitative agreement with Joy's law. The tilt is such that the poleward (following) part of the bipolar region has a polarity that is opposite to that of the polar fields in the same hemisphere. When bipolar regions age, they fragment and spread by turbulent diffusion to form the general background field. Because of the initial tilt of the bipolar regions it is the following polarity that will dominate the new background field at high latitudes, thereby canceling and reversing the polar field. Coriolis forces acting on buoyant toroidal flux combined with turbulent diffusion thus lead to a regeneration of a poloidal field with reversed polarity.

While Babcock's model was phenomenological, qualitative, and descriptive, it contained the physical ingredients that are central to modern dynamo theories. Leighton (1969) significantly extended Babcock's model by giving it a formulation in terms of a set of semi-empirical equations that could be used for quantitative modeling of the evolution of the pattern of surface magnetic fields. The Babcock-Leighton framework has been used since the early 1980s by Neil Sheeley and coworkers at NRL (e.g. Sheeley et al. 1985) for systematic modeling of the field pattern to understand its role for coronal holes, solar wind streams, and other large-scale phenomena on the Sun.

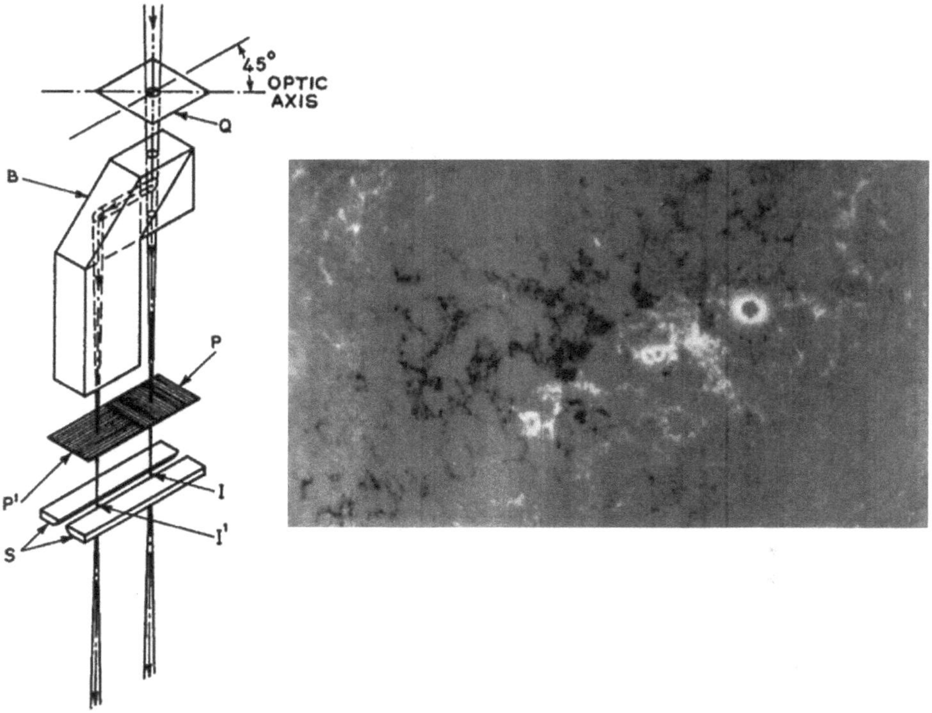

Fig. 2 The principles of the Leighton photographic magnetograph are illustrated to the left, from Leighton (1959). With a polarizing beam splitter two images in opposite circular polarization states of the chosen solar region are formed at the entrance slit of the spectrograph. The photographic plate is located behind the exit slit in the spectral focus. The entrance and exit slits are scanned in unison to build up monochromatic images (spectroheliograms) on the photographic plate. After the plate has been developed, the two orthogonal-polarization images are subtracted, to give a magnetogram like the one in the right part of the figure (from Vrabec 1971). It reveals the network structure of the magnetic-field pattern

2.3 The Magnetic Network

With the superior sensitivity of the photoelectric magnetograph one might think that the days of doing magnetometry with photographic plates were over. However, photographic plates represented 2-D detectors with an enormous number of "pixels", which for mapping purposes with high angular resolution represented a huge advantage over photomultipliers, which are 1-pixel devices.

Robert Leighton developed a photographic magnetograph technique based on the photographic recording of two simultaneous spectroheliograms in orthogonal circular polarization states (Leighton 1959), as illustrated in the left portion of Fig. 2. Subsequent photographic subtraction of the two images then results in a polarization map (line-of-sight magnetogram). Such photographic subtraction is of course a tricky business due to the non-linear response of photographic plates, but Leighton and his followers developed it into an art.

One place in particular where the Leighton technique was perfected to produce not only high-resolution maps but also magnetic-field movies of outstanding quality was the San Fernando Observatory. The right part of Fig. 2 gives an example, from Vrabec (1971), of such a magnetogram, which clearly reveals that the magnetic pattern has a network structure.

Fig. 3 Neil Sheeley in his Tucson office in 1968. He perfected the art of recording Kitt Peak spectroheliograms to explore the photospheric network

Remarkable movies of the moat structure around sunspots were made, showing how flux fragments of opposite polarities stream away from the spots.

Leighton also used his subtractive spectroheliogram technique to make full-disk Doppler maps of the Sun, which led to his discovery of the supergranulation (Leighton et al. 1962). The largely horizontal flows in the supergranulation were much more effective than the granular flows for the redistribution of the fields through "turbulent diffusion". They came to play a major role in Leighton's kinematic model of the solar cycle, in which he treated the redistribution as a random walk on the solar surface of the frozen-in field lines.

Spectroheliograms had since many decades been an established tool to make narrow-band monochromatic images of the Sun for any chosen wavelength in the visible solar spectrum. They had revealed that chromospheric lines like the H and K lines of ionized calcium show a network-like emission pattern. Leighton's Doppler and magnetic-field maps showed that both the supergranular velocity network and the magnetic-field network were closely correlated with the emission network, they seemed to coincide with each other. The statistical correlations between these three types of network were intensely explored by many solar physicists around the end of the 1960s, in particular by recoding Doppler and magnetic data through raster scans with a photoelectric magnetograph, which generally allowed the analysis to be more quantitative than with the photographic method.

In 1967 Neil Sheeley (cf. Fig. 3) published a seminal paper in which he showed that many photospheric lines had line gaps (very significant line weakenings) at the locations of the quiet-sun network, and that these were places with strong, small-scale flux concentrations, with field strengths reaching up to hundreds of G (Sheeley 1967). In spectroheliograms the line gaps show up as a photospheric emission network. Subsequently Chapman and Sheeley (1968) made an extensive spectroheliogram survey of the behavior of different photospheric spectral lines with various excitation potentials, to demonstrate that the line gaps were the result of increased photospheric temperature. The observations allowed the temperature structure of the photospheric network to be modeled.

Sheeley's hundred G quiet-sun field strengths were shockingly high for that time, but they still represented a lower limit determined by the spatial resolution of his instrument. Since the intrinsic sizes of the network elements were not known and therefore not their filling factors f, the intrinsic field strength would be $1/f$ times the apparent field strength.

To find the intrinsic field strength without achieving infinite angular resolution one had to come up with a resolution-independent approach, which occurred soon afterwards with the introduction of the magnetic line-ratio method (cf. Sect. 5.2).

3 Mapping the Vector Magnetic Fields and Electric Currents

Already Hale in his early work on sunspot magnetic fields knew how to combine the measured circular and linear polarizations of the longitudinal and transverse Zeeman effects to derive the magnetic-field vector (strength and orientation of the field). Formally such a Zeeman-effect "inversion" is very straightforward. In practice the inversion often leads to false results because of a combination of factors, in particular non-linear response, non-Gaussian noise, and unknown subresolution structure of the field. In contrast to the transverse Zeeman effect, the polarization of the longitudinal Zeeman effect has a nearly linear dependence on the field (via its line-of-sight component) with a Gaussian noise distribution, and also has high field sensitivity. This is the reason why most observational explorations of the magnetic field has been of its line-of-sight component. What we call "magnetogram" is always a circular-polarization map; in the more rare cases when also the Zeeman effect in the linear polarization is being recorded we use the term "vector magnetogram".

Soon after Babcock's introduction of the photoelectric magnetograph his concept was being extended at various places, including Crimea, IZMIRAN (near Moscow), Pulkovo by Leningrad, and Ondřejov by Prague, to record vector magnetic fields. Both the circular and linear polarizations in a selected spectral line were modulated by Pockels cells. The demodulated signals were recorded with ink pen on strip-chart recorders and combined to give the magnetic-field vector. The pioneering work in vector magnetometry was done at the Crimean Astrophysical Observatory, where Stepanov and Severny (1962) published the first vector-field maps of the Sun. Figure 4 from Severny (1966) shows an example of maps of the vector magnetic field around sunspots (see also Severny 1964).

Vector magnetometry was a hot topic in solar physics of the 1960s, because it was seen as a tool for the exploration of the full MHD structure of the solar atmosphere. Maxwell's equations were being used to derive maps of the vertical electric current density from the horizontal gradients of the observed transverse magnetic fields. The pattern of vertical electric currents was found to correlate with the locations where flares occurred, as expected in the current-interruption theory of solar flares by Alfvén and Carlqvist (1967). Around that time Alfvén was finding it more revealing to describe MHD environments in terms of the electric current structure instead of magnetic fields, since currents are responsible for the pinch effect and could cause instabilities with the development of double layers, in which the MHD approximations break down.

The upper panel of Fig. 4 indicates how the apparent transverse field rotates with height in the atmosphere. Superposing the results obtained with four spectral lines formed at different heights one reveals how the azimuth angle varies. This behavior suggests that the field lines are twisted, as expected in the presence of vertical electric currents. However, the interpretation has to also account for the apparent twist induced by magneto-optical effects.

The Crimean Astrophysical Observatory was not only the world-leading institute in vector magnetometry, it was also the place where the theoretical foundations for the interpretation of magnetograph measurements were developed. While Wasaburo Unno in Japan had

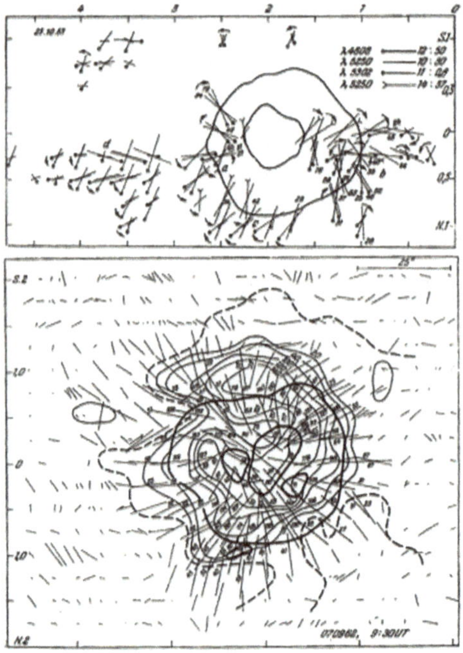

Fig. 4 Examples of maps of vector magnetic fields in regions around sunspots, from Severny (1966). The top panel shows how the field azimuth rotates as observed with four spectral lines formed at different heights. The contours represent the sunspot umbra and penumbra. In the bottom panel the thin solid contours represent the longitudinal magnetic field, superposed on the background of short lines that represent the strength and orientation of the transverse magnetic field

Fig. 5 Dimitry Nikolaevich Rachkovsky (1928–2013), who gave us a comprehensive formulation of LTE radiative transfer of the full Stokes vector in magnetized media, including magneto-optical effects. His formulation is used in all current Stokes inversion work. The photo was taken in 1989 in Kiev

published a phenomenological but correct derivation of the radiative transfer equation for the Stokes vector (Unno 1956), it was the work of Stepanov and Rachkovsky at Crimea that brought us a self-consistent mathematical derivation that also included the magneto-optical effects. While Stepanov formulated this approach (Stepanov 1958a, 1958b), Rachkovsky (cf. Fig. 5) completed the theory to give us a mathematically stringent and comprehensive (within the assumption of LTE) formulation of the Stokes vector radiative transfer problem (Rachkovsky 1962a, 1962b). Rachkovsky's theory was used to identify observational signatures of magneto-optical effects in the recordings with the Crimean magnetograph.

In 1965 the academies of sciences in Sweden and the Soviet Union signed an exchange agreement. It happened, thanks to the contacts and support of my mentor Hannes Alfvén, that I was the first from the Swedish side to be sent over to work in the USSR under this

Fig. 6 At the Crimean Astrophysical Observatory in 1965: Jan Stenflo and the observatory director, Andrei Borisovich Severny, the leading Soviet solar physicist at the time

agreement. Alfvén believed that the Sun's magnetic field had a filamentary fine structure induced by the pinch effect of electric currents, and that this fine structure invalidated Babcock's observations since it was not resolved. He conjectured that if the field consisted of "microspots" with less visibility because of their different thermodynamic properties, then the spatial averaging process would lead to the wrong interpretation of the measurements (Alfvén 1967).

Alfvén had a strong incentive to see Babcock's results invalidated, since a general magnetic field that reversed sign every 11 years contradicted Alfvén's own theory of the solar cycle. He sent me to Crimea with the mission to confirm his conjecture by exploring the fine structure with the Crimean magnetograph. My mission failed, since my work rather helped to validate Babcock's results. However, Alfvén's ideas put me on a path to explore the nature of the unresolved magnetic fine structure and how it can be properly diagnosed. This path took me in productive directions that none of us could foresee at that time.

Thus at age 22 I spent three summer months in 1965 to study the fine structure of quiet-sun magnetic fields with the Crimean magnetograph, and I returned for another four months in the summer of 1966. This work gave the observational basis for my Doctoral Thesis, which I completed in the spring of 1968 (at the University of Lund, Sweden). Figure 6 shows me during the visit in 1965 together with the Director of the Crimean Astrophysical Observatory, Andrei Borisovich Severny, who was the most prominent Soviet solar physicist at the time.

The Crimean Astrophysical Observatory was a world leader in solar physics in the 1960s, it was an exciting time to be there. However, towards the end of the decade began a period of political and economic stagnation and decline, which finally ended in the collapse of the Soviet Union. Figure 7 contains photos that I took of the solar magnetograph when I visited Crimea again in 2011. The electronic equipment, control panels, and strip-chart recorders were much the same as when I worked there 45 years before. It was like going back in time, not much had changed. In the right panel is Valery Kotov, who still uses this equipment to study the 160-min oscillations of the Sun. I became befriended with him in 1965, when he

Fig. 7 Control panels and strip-chart recorders of the Crimean magnetograph system in photos taken in 2011. The equipment has not changed much since the mid 1960s. In the right panel is Valery Kotov, who used the instrumentation in the 1960s to determine the vector electric current structure of active regions, and who is now using it for continued studies of the Sun's 160-min oscillations

was a PhD student at Crimea. In his thesis on the electromagnetic structure of active regions he used vector magnetic-field mapping not only to determine the vertical electric-current density, but also the horizontal current density and therefore the *vector electric currents*. The horizontal electric currents were found by recording the transverse magnetic field at different heights, and then from the height gradients of the horizontal field components use Maxwell's equations to obtain the horizontal components of the current.

Since there has been a revival in the mapping of vector magnetic fields in recent years in the context of Stokes inversions, one may wonder how reliable all the pioneering results from the 1960s about the vector magnetic fields were. My general answer is that the maps of the relative distributions of azimuth orientations of the transverse fields are reliable (as long as one accounts for the magneto-optical effects in the interpretation). Likewise the maps of the line-of-sight flux density (regular magnetograms) are also reliable. However, the magnitudes of the transverse flux density have been greatly overestimated, with the result that the determined field inclinations are much too large, the fields appear much more transverse than they really are.

The main source of the problem is that the relation between the measured linear polarization and the transverse field strength is highly non-linear (in contrast to the case for the circular polarization). This has two consequences for observations that do not have high S/N ratio: (1) The spatial averaging over unresolved subpixel structures depends on the nature of this structuring. Ignoring this problem always leads to errors in the direction of artificially enhancing the transverse fields. (2) The noise distribution of the transverse fields is highly non-Gaussian, with an extended non-Gaussian tail that makes the fields look much more transverse than they are. Since in addition the polarization amplitudes that are produced by a given field strength are much smaller (by typically a factor of 25) for the linear polarization relative to the circular polarization, the noise infiltrates the transverse fields much more than the longitudinal fields, always in the direction of making the fields appear much too transverse.

This problem does not go away by using improved Stokes inversion techniques, because it is not a technical problem. The only known way to fully eliminate the transverse-field bias (in the weak-field quiet-sun case) is to use a statistical approach with ensemble averages (cf. Sect. 5.5), averaging the Stokes Q linear polarization as observed away from disk center (to break the symmetry), without trying to convert Q to transverse field strength, since such conversion is the source of the non-linearities and the non-Gaussian noise. Because the ensemble averages are independent of the spatial resolution of the instrument (since Q is proportional to the number of photons received), and because the noise distribution in Q is Gaussian and thus symmetric, the ensemble averages of Stokes Q give unbiased results that neither favors horizontal nor vertical. Application of this technique has shown that quiet-sun magnetic fields are preferentially vertical (with respect to the isotropic angular distribution) in the low to middle photosphere (as observed in photospheric lines over most over the solar disk) but become preferentially horizontal in the upper photosphere (as observed close to the solar limb, cf. Stenflo 2013a).

4 Origin of the Sun's Magnetic Cycle

We have seen how Babcock integrated the various empirical findings into a phenomenological model of the solar cycle, which contained the main physical ingredients that govern the activity cycle of the Sun in terms of an interplay between turbulence and magnetic fields in a rotating magnetized plasma. We are now convinced that these ingredients are responsible for the generation of large-scale magnetic fields all over the universe, not only in stars and planets, but also on galactic scales. The Sun's 22-yr magnetic cycle is a prototype for all cosmic dynamos.

4.1 An Incorrect Theory Worthy of a Nobel Prize

In a Letter to Nature in 1942 (my birth year) my mentor Hannes Alfvén (cf. Fig. 8) presented (two decades before Babcock) an alternative explanation for the origin of the sunspot cycle (Alfvén 1942). In his scenario the Sun possessed a global dipole-like field that was a remnant from the formation of the solar system. This field therefore penetrated the core of the Sun, so it could not reverse (in contradiction to Babcock's later findings). At that time it was believed that the CNO cycle rather than the p-p reaction chain was responsible for the thermonuclear energy production. This had the consequence that the central temperature was considered to be as high as 25 million K, and that a steep temperature gradient would cause the solar core to be in a convective state. Alfvén therefore postulated that the sunspots originated from disturbances in the central regions of the Sun, and that these disturbances traveled as waves (later called Alfvén waves) along the global dipole field, emerging at the surface as sunspots.

Fig. 8 Jan Stenflo with his mentor Hannes Alfvén on the Pacific beach in La Jolla in 1968. Two years later Alfvén received the Nobel Prize in Physics for his discovery of what we now refer to as Alfvén waves

We now know this explanation of the solar cycle to be incorrect. However, in this short paper Alfvén for the first time introduced the concept of MHD waves, explaining why disturbances in a magnetized plasma would propagate as transverse waves along the field lines with a speed that we now refer to as the Alfvén velocity. It was for the discovery of these waves that he received the 1970 Nobel Prize in Physcis. Although he introduced this wave concept within the context of an incorrect theory for the solar cycle, it was a Nobel-prize worthy discovery that has since played a fundamental role in all of plasma physics. It is an example of how wrong theories can contain elements of great value!

4.2 Beginnings of Dynamo Theory

The concept of explaining the origin of macroscopic magnetic fields in terms of dynamo processes goes back to Larmor (1919), but subsequent progress towards a viable theory was relatively slow. One serious obstacle was the "anti-dynamo theorem" of Cowling (1933) (cf. Fig. 9), which was a proof that an axially symmetric velocity field, like that of a rotating medium, was incapable of generating a dynamo. Such idealized geometries were insufficient, the symmetry had to be broken for dynamo action to occur.

While Walter Elsasser from the University of Utah had overcome this obstacle and established the mathematical framework of dynamo theory, and in particular applied it to explain terrestrial magnetism (Elsasser 1946, 1955), Eugene Parker of Chicago extended the ideas to the solar case and highlighted how Coriolis forces in a rotating and turbulent medium make the turbulence cyclonic and thereby can regenerate the poloidal field from the toroidal one (Parker 1955). Babcock made use of these ideas when he built his phenomenological model of the solar cycle. Parker was the leading theoretical solar physicist for several decades, with fundamental contributions over a broad range of topics, including the origin of the solar wind, the cause of intermittency in the photospheric magnetic flux, and the mechanism for coronal heating. Figure 10 shows him at the COSPAR meeting in Tokyo in 1968.

Fig. 9 T.G. Cowling from Leeds, UK, in conversation with R.F. Howard from Mt Wilson Observatory, Pasadena, during the First Solar Wind Conference at Asilomar, California, in 1971. Cowling is best known for his anti-dynamo theorem (Cowling 1933). Bob Howard took over and modernized Babcock's work on recording full-disk solar magnetograms and developed the successful synoptic program of Mt Wilson magnetograms that has for many decades provided fundamental data on the global evolution of the magnetic field

Fig. 10 Eugene Parker (left), here at the 1968 COSPAR meeting in Tokyo, made seminal contributions in theoretical solar physics over a wide range of topics

4.3 Mean-Field MHD and the α–ω Dynamo

Alfvén had heard that Max Steenbeck in Jena, DDR, had developed a new and comprehensive formulation of the dynamo problem, and therefore sent me to Jena in 1969 to find out how the theory worked. Max Steenbeck was a leading person in the East German establishment, head of the country's research council with the rank like that of a government minis-

Fig. 11 Max Steenbeck in his office in Jena in 1969. He was the "father" of mean-field electrodynamics and α–ω dynamo theory

ter. However, he was foremost an innovative scientist, who had built up a leading group in plasma physics in Jena. Soon after my arrival he took me into his office, called in his young collaborators Fritz Krause and Karl-Heinz Rädler, and described to me on the blackboard the ideas behind the α–ω dynamo. Figure 11 shows Steenbeck at his desk on that occasion in 1969.

The key to a comprehensive mathematical formulation of the dynamo problem was the development of mean-field electrodynamics (Steenbeck and Krause 1969). Assuming that the spatial and temporal scales of the turbulence are small in comparison with the global scales to allow ensemble averaging, one could express the statistical effect of cyclonic turbulence as an electromotive force along the field lines with α as the proportionality constant. A second-order term in the ensemble average of the turbulence represented turbulent diffusion with coefficient β. The poloidal field lines get wound up by the shear (radial and latitudinal) in the Sun's rotational angular velocity ω. The evolution of the global solar magnetic field can then be formulated in terms of two coupled equations describing the interplay between the α, β, and ω effects, one equation for the poloidal component, one for the toroidal. Decomposition of these components in their spherical harmonics leads to a coupled system of equations for the eigenvalue problem, which allows very explicit and consistent modeling of the solar cycle. This type of modeling is very general and can be applied to account for the magnetic fields in various types of other stars, planets, and galaxies.

I returned to Jena in 1970 at the time when the institute was being moved from Jena to Potsdam, and I participated in the move. The Potsdam institute established itself as a leading center for dynamo research during the following decades.

4.4 Synoptic Programs

In the caption to Fig. 9 we mentioned how Bob Howard took over the pioneering work of Horace Babcock and developed the first synoptic program for full-disk line-of-sight component magnetograms recorded daily at the Mount Wilson Observatory. Later (in the early 1970s) similar synoptic programs were started at Stanford by John Wilcox and at Kitt Peak by Bill Livingston and Jack Harvey. The Stanford program continues to this day with well defined but low spatial resolution. The Kitt Peak program has had the highest spatial resolution of the ground-based programs and was upgraded in the late 1990s by Jack Harvey and Christoph Keller as part of the SOLIS synoptic program.

Fig. 12 John Wilcox (left) with Jan Stenflo in 1967 at the Swedish solar station on Capri, Italy. Wilcox was a leading solar wind scientist and founded the synoptic full-disk magnetogram program at the Stanford University

The most advanced synoptic full-disk magnetogram projects have been the space programs *SOHO*/MDI (1995–2011) and its superior successor, *SDO*/HMI. Both have been managed by the Stanford group.

John Wilcox had close contacts with Hannes Alfvén and was in the audience in Stockholm when I in 1966 gave a colloquium at Alfvén's institute on my observations of the Sun's magnetic fine structure at the Crimean Astrophysical Observatory. His main scientific focus at the time was on the physical properties of the solar wind and its sources on the Sun. He visited me at the Swedish solar station on Capri in 1967 (see Fig. 12). When I for the first time set foot on US soil (in 1968 after the defense of my Doctoral dissertation in Lund) Wilcox had arranged that I spent the first four nights at the Faculty Club of the Berkeley campus (later Wilcox moved from Berkeley to Stanford). I had made a world-around trip, so I entered the US from the west. It was Berkeley June 1968 at the height of the Vietnam war, and it was my first impression of what American society was like!

After Wilcox' tragic accidental death in 1984 by drowning while swimming in Hawaii, the synoptic Stanford Solar Observatory was renamed Wilcox Solar Observatory.

5 The Quest for Angular Resolution

When the Sun is observed with improved angular resolution we discover previously unknown structures which deepen our physical understanding, like the granular convection pattern, the chromospheric filamentary structures, network and internetwork bright points, etc. Photospheric magnetic flux is highly intermittent with basic kG flux elements having typical sizes of order 100 km or below. Most of the size distribution is beyond reach of current telescopes, although flux tubes in the larger-scale tail of the distribution have been resolved. The magnetic-field pattern has a fractal-like structure, with a structuring that continues far beyond the resolution limit.

Fig. 13 JOSO meeting 1969 in Freiburg, Germany. From left to right: Karl-Otto Kiepenheuer (Freiburg), Jean Rösch (Pic du Midi), Peter Brandt (Freiburg), Kees de Jager (Utrecht), Guglielmo Righini (Arcetri)

It has therefore always been a high priority for solar physicists to push for ever higher angular resolution, since it is widely believed that the fundamental physical mechanisms to a large extent are connected to processes at the smallest scales.

5.1 JOSO and LEST

Half a century ago the leading solar physics facilities were in the US (Mount Wilson in California, Kitt Peak with the McMath telescope in Arizona, Sacramento Peak in New Mexico) and the USSR (Crimean Astrophysical Observatory). The European solar physics community felt a strong need to overcome the political fragmentation after two devastating world wars and create a joint European solar facility that would again give Europe a leading role in science, as had been done by CERN for particle physics and by ESO for night-time astronomy. The driving force behind this pan-European solar physics movement was Karl-Otto Kiepenheuer, Director of the Fraunhofer Institute (after his death renamed Kiepenheuer Institute for Solar Physics) in Freiburg, Germany. In the late 1960s Kiepenheuer set up an informal European organization with the name JOSO, Joint Organization for Solar Observations (Kiepenheuer 1973), which had the goal of constructing a major European solar telescope, LEST (Large European Solar Telescope). This goal was to be achieved in two steps: (1) Search by systematic site testing campaigns for the optimum location of LEST. (2) Establishment of a new organization with legal structure and adequate financial means to design and construct LEST at the chosen site.

Figure 13 shows Kiepenheuer at one of the first meetings of JOSO, in Freiburg 1969. Beside him is Jean Rösch, Director of the Pic du Midi Observatory, representing France. Holland was represented by Kees de Jager, who founded the journal Solar Physics and later served as president of COSPAR, Italy by Guglielmo Righini, Director of the Arcetri Observatory in Florence, while I was representing Sweden.

After having done site testing of a number of coastal sites in Portugal, Sicily, and Greece it was concluded that high-altitude volcanic island sites would be better, and that therefore the site testing should concentrate on Tenerife (Izaña) and La Palma (Roque de los Muchachos) in the Canary Islands. Towards the end of the 1970s these site testing campaigns came to the conclusion that both Izaña and Roque de los Muchachos were indeed outstanding sites of comparable quality. There was no clear recommendation which of the two was better. Both are at the same elevation, just above the well-defined inversion layer where the clouds form. At these sites the wind direction is most of the time such that the line of sight from the telescope to the Sun is in a laminar air flow if the telescope sits on top of a tower high enough to avoid ground turbulence. This leads to excellent seeing, while the high altitude causes scattered light from the sky to be low.

Rather than waiting for a LEST to be realized, various countries quickly decided to set up their national facilities on these sites. Germany chose Izaña, where the Kiepenheuer Institute in Freiburg built the VTT (Vacuum Tower Telescope) while the University Observatory of Göttingen set up the Gregory-Coudé telescope that they had transferred from their solar station in Locarno, Switzerland. Similarly the French built their THEMIS 90 cm aperture polarization-free telescope at Izaña. The Swedes (through their observatory director Arne Wyller, who had succeeded Yngve Öhman, founder of the Swedish Capri observatory) transferred their solar station in Anacapri to Roque de los Muchachos, where also the Dutch established themselves with the DOT (Dutch Open Telescope), designed by Rob Hammerschlag. Soon after, the telescope at the Swedish La Palma observatory was completely rebuilt with an innovative design by Göran Scharmer, who later took over as observatory director. Under Scharmer the La Palma facility became the unrivaled world leader in solar imaging with the highest angular resolution. No other solar facility in the world has yet consistently achieved such resolution.

In parallel with these national activities the efforts towards the realization of the LEST dream made rapid progress. A new organization, the LEST Foundation, was established in 1983 with legal seat at the Royal Swedish Academy of Sciences. Its task was to execute the intentions of JOSO, to obtain the financial means, develop the detailed design, and to construct and operate LEST at the chosen site. The scientific aim was focused on recording magnetic fields with the highest possible spatial resolution (Stenflo 1985). Throughout its existence the LEST Foundation was led by a "triumvirate", with me as President, Oddbjørn Engvold of Oslo as Project Director, and Kai-Inge Hillerud of the Royal Swedish Academy in Stockholm as Executive Secretary. The innovative design was led by Engvold with crucial input from Richard Dunn, US master optical designer, who made the Sacramento Peak Observatory (New Mexico) become the leading high-resolution facility in the US, and from Torben Andersen, who was a Danish telescope designer at ESO. Since other non-European countries like the US and China joined the LEST Foundation as members, the acronym was redefined as Large Earth-based Solar Telescope, to emphasize that it was not just a European but a truly international undertaking.

Around the end of the 1980s the detailed design for the entire facility had been completed, the definite site had been selected (on Roque de los Muchachos near the Swedish La Palma observatory), we were ready to start the construction phase, and it appeared that the needed funding support was forthcoming (Engvold 1991). Dramatic external events changed the situation. When the Soviet Union collapsed, Germany turned all its priorities and financial resources toward the enormous task (and historical opportunity) of German reunification, with the consequence that the expected substantial German contributions to the construction phase of LEST would not be available in the foreseeable future. The LEST project was put on hold for another decade (although during this time many design studies for various focal-plane instrumentations were carried out), until it was finally decided to dissolve the LEST Foundation as of June 30, 2002.

As the LEST project lost momentum in the 1990s, the idea of a next-general type international solar facility was being developed in the US in the form of ATST (Advanced Technology Solar Telescope), with an innovative 4 m aperture telescope design by Jacques Beckers. This ambitious project, renamed DKIST (Daniel K. Inouye Solar Telescope) is now in construction at the Haleakala site on Maui and is expected to be in operation by 2020.

Before the LEST Foundation was dissolved, Torben Andersen and Oddbjørn Engvold designed a scaled down version of LEST, which has since been implemented in the form of the Chinese 1-m solar telescope at the Fuxian lake near Kunming and the 1.5-m German GREGOR telescope on Tenerife (which has replaced the 45 cm Göttingen Gregory-Coudé telescope). The original LEST idea of a joint European solar facility has been resurrected within Europe in the form of EST (European Solar Telescope), aiming for a telescope in the DKIST class, and it is progressing well with the help of EU funding.

5.2 The McMath-Pierce Facility as a Center of Innovation

In the late 1960s and the following decades the leading US facility for solar observations with high spatial resolution was the tower telescope at the Sacramento Peak Observatory in New Mexico, which had been designed by Richard Dunn and later given the name DST (Dunn Solar Telescope). However, the main innovations and advances in the exploration of solar magnetic fields took place at the McMath telescope at Kitt Peak because of the many contributions of in particular William Livingston, Jack and Karen Harvey (cf. Fig. 14), Jim Brault, and Neil Sheeley. This giant telescope had been designed by Robert McMath but was renamed the McMath-Pierce facility after the death of the head of the Kitt Peak solar department, Keith Pierce.

Harvey and Livingston implemented innovative techniques for magnetic-field mapping, which much advanced our insight and understanding of the small-scale nature of the fields (cf. Livingston et al. 1976; Harvey 1977). Around 1974 they initiated the synoptic program of daily Kitt Peak full-disk magnetograms and synoptic maps that resolved the field in much greater detail than anywhere else. Jack's wife Karen discovered together with Sara Martin of the San Fernando Observatory the ephemeral active regions (Harvey and Martin 1973). Jim Brault developed a Fourier transform spectrometer (FTS) for the visible part of the spectrum (Brault 1978, 1985), which with the help of Jack Harvey could be converted into an FTS polarimeter.

My own main observational work has been done with the McMath-Pierce facility. In September 1971 I applied the magnetic line-ratio technique (with the 5250–5247 Å line pair) to discover that more than 90 % of the magnetic flux that had been recorded in solar magnetograms (with a resolution of a few arcsec) on the quiet Sun actually is in kG form but appears weak since it is unresolved with a small filling factor (Stenflo 1973). In 1978–1979 I used the FTS polarimeter to make atlases of the Stokes V spectra in plages and the network and surveys of the linearly polarized spectrum near the solar limb. The Stokes V spectra (Stenflo et al. 1984) led to detailed physical validation of the line-ratio technique, revealed the nature of the Stokes V asymmetries, and could be used by Sami Solanki, my first PhD student at ETH Zurich, to apply multi-line techniques for the empirical modeling of solar magnetic flux tubes. The linear polarization survey led to the discovery of the Second Solar Spectrum (cf. Sect. 6.2), which is formed exclusively by coherent scattering processes and is the playground for the Hanle effect. When the ZIMPOL technology had been developed

Fig. 14 Jack and Karen Harvey with Jan Stenflo at Stonehenge, 1970. Jack did ground-breaking work in the 1970s on the small-scale structure of solar magnetic fields, and Karen codiscovered the ephemeral active regions

at ETH Zurich it was used from 1994 onwards at the McMath-Pierce facility for systematic explorations of the Second Solar Spectrum with a polarimetric precision of 10^{-5}. The main Hanle-effect constraints on the properties of the hidden turbulent magnetic fields have come from ZIMPOL observations at Kitt Peak.

The main reason why the McMath-Pierce facility has served as the premier place for innovative, experimental solar science is, besides its high light-gathering capacity (1.5 m entrance aperture) the large, open experimental environment in the spatious observing room, with free access to the large solar image and unrestricted access to the beam in front of the entrance slit of the spectrograph, which makes it simple to insert novel experimental equipment. Other large astronomical telescopes generally do not allow experimental freedom of this kind. It will be greatly missed with the closure of the McMath-Pierce facility.

5.3 FTS Polarimetry

The FTS polarimeter allowed unprecedented insight into the physical nature of the spatially unresolved magnetic fields. It allowed the simultaneous recording of the spectra of Stokes I (the ordinary intensity) and one of Q, U, or V with full spectral resolution (absence of significant instrumental spectral smearing), no spectral stray light, and with simultaneous wavelength coverage over the full spectral range admitted by the broad-band prefilter used (typically 1000 Å). This was far beyond what one could dream of achieving with any spectrograph system. The trade-off was low angular resolution (5 arcsec at the time) and temporal resolution (tens of minutes). Since the vast majority of solar physicists had their focus on high spatial resolution, few took interest in the use of the FTS polarimeter, I became almost its only user.

As an example Fig. 15 shows a 4.5 Å wide portion of the FTS spectral atlas of a weak plage recorded in 1979. In particular it illustrates the exact behavior of the Stokes V profiles

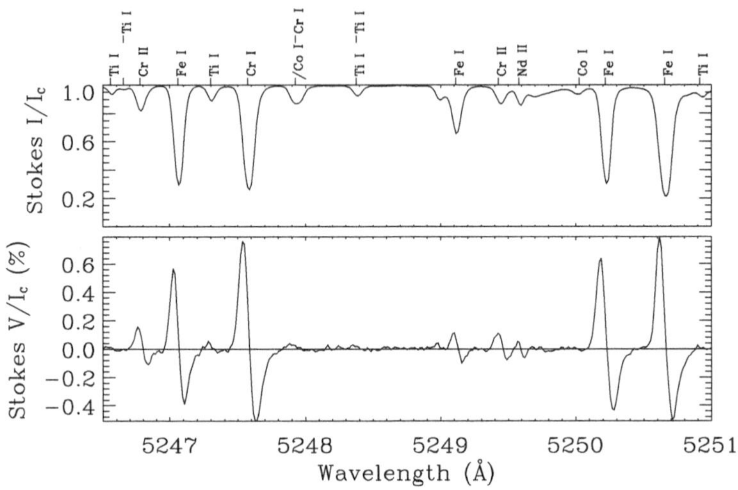

Fig. 15 Illustration of a small spectral window from the Stokes *I* and *V* atlas of a weak plage, recorded with the FTS polarimeter in 1979 (Stenflo et al. 1984; Stenflo 2013b)

of the lines Fe I 5247.06 and 5250.22 Å, which belong the magnetic line-ratio pair that was used with the McMath Babcock-type magnetograph system to reveal that most of the photospheric magnetic flux that had been recorded in magnetograms has its origin in strong, 1–2 kG bundled fields with small filling factors. Analysis of the observed Stokes *V* line shapes allows a detailed validation of this interpretation. The FTS Stokes *V* atlases are available both as pdf and data files at http://www.irsol.ch/data_archive/#ftsv.

The Stokes *V* spectral atlases and their center-to-limb variations represented a gold mine for the construction of empirical models of the spatially unresolved kG flux tubes, which was exploited by Sami Solanki in his PhD thesis project at ETH Zurich. Sami went on to become director at the Max-Panck-Institute for Solar System Research in Göttingen, where he led the successful *Sunrise* project to observe solar magnetic fields with unprecedented spatial resolution from a stratospheric balloon (Solanki et al. 2010).

The FTS polarimeter represented a remarkable advance in solar instrumentation, but it has now been decommissioned, and no similar instrument exists anywhere else. There are not even plans within any of the main solar telescope projects to build another instrument of this type. Like the 1970s represented a heroic era when we could fly people to the moon but are not able to do it any more (or have lost interest in it), FTS polarimetry represents a lost art, without sufficient motivation anywhere to resurrect it.

It is true that the FTS polarimeter in the 1970s had low spatial and temporal resolution and used 1-pixel detectors, but it was in principle possible to improve the resolutions and to use multi-pixel detector arrays, although it is technically difficult. However, these improvements were never done, because they were not requested by the broader solar community.

5.4 The Endless Debate on the Existence of kG Fields

Application of the magnetic 5250/5247 line ratio method through my 1971 Kitt Peak observations allowed me to conclude that most of the measured quiet-sun magnetic flux is in kG form (Stenflo 1973). For me the evidence was so clear that I considered the case as closed. Since however this conclusion was not reached by resolving the kG fields, but indirectly, through the identification of a telltale spectral signature that could not have any other interpretation, it took many years before my result found broad acceptance within the community, even after the interpretation got validated with the independent FTS polarimeter data. Over the years various people presented their own conclusions about the existence of quiet-sun kG fields as if it were a new discovery.

While as time went by ever fewer people questioned the existence of kG flux on the quiet Sun, two leading scientists within the solar community never accepted it: Andrei Severny, director of the largest astrophysical observatory in the Soviet Union, and Harold Zirin of Caltech, who founded the Big Bear Solar Observatory (which presently has the largest-aperture solar telescope in the world). After Severny passed away in 1987, Hal Zirin remained the stronghold of resistance.

Hal seemed to enjoy this controversy. In 1992 he organized an IAU Colloquium in Beijing and invited me to have a debate with him there on the reality of the kG fields (cf. Stenflo 1993; Zirin 1993). The debate between the two of us was set up to represent a highlight of the conference. The lively event did not change any opinions, but I remember the atmosphere as friendly and entertaining. Hal and I respected each other in spite of our opposite views.

Long after, it seems that some remember this event as a debate about whether the quiet-sun fields are weak or strong. This is not at all what it was about. There has always been agreement that most of the photospheric volume is filled with weak fields. The intermittent kG flux picture that resulted from application of the line-ratio technique implied that only about 1 % of the photospheric volume is occupied by kG fields, while the remaining 99 % has much weaker fields. The discovery of the kG fields was followed by a focus on finding the properties of the remaining 99 % weak fields. These efforts led to the discovery that one can use the symmetry properties of the Hanle effect to constrain the field strengths of these weak fields (Stenflo 1982).

So the Beijing debate in 1992 was not about whether the fields are strong or weak, it was whether there exist kG-type magnetic fields on the quiet Sun, which carry much of the magnetic flux that we see in solar magnetograms. Zirin always denied the existence of such fields, and he never gave up this position.

5.5 Ensemble Averages and Angular Distributions

While the initial observations of quiet-sun magnetic fields by Babcock was with arcmin angular resolution, the observing techniques advanced rapidly in subsequent years, so that by the end of the 1960s an angular resolution of better than a few arcsec could be reached. With each new resolution breakthrough there was a tendency to proclaim that the magnetic structures had finally been resolved. We now know that such proclamations were wildly premature, since the magnetic structuring continues to far smaller scales (cf. Fig. 21 below).

With the realization that the fields remain largely unresolved came the need to find resolution-independent spectral signatures that could reveal the intrinsic properties of the fields as they would appear with infinite resolution. Since the field morphology is not resolved, such techniques must be statistical in nature, making use of ensemble averages of field elements to find the intrinsic statistical properties of the elements that make up the ensemble. The magnetic line-ratio method is an example of such a technique. Since the field elements are far from resolved, the angular-resolution window of the instrument averages over an ensemble of fields with a vast range of field strengths, both weak and strong. The *differential* effects between the ensemble-averaged Stokes V profiles of the 5250–5247 Å line pair allow us to conclude that most of the resolved Stokes V signal gets its contribution from intrinsically kG flux elements. The 5250–5247 pair is the only known line combination for which the kG interpretation is unique, because thermodynamic and radiative-transfer effects are unable to produce differential effects of the observed kind (in contrast to the case for all other considered line combinations).

Two other types of spectral signatures from ensemble averages of Stokes parameters have led to profound insights into the hidden small-scale nature of solar magnetic fields: (1) Depolarization signature in the linear scattering polarization, caused by the Hanle effect of an ensemble of microturbulent (optically thin) magnetic fields with random orientations of their field vectors (Stenflo 1982). These fields represent most of the weaker (non-kG) fields that occupy approximately 99 % of the photospheric volume. (2) Sign pattern of the ensemble-average of Stokes Q from the transverse Zeeman effect as observed away from disk center (which is needed to break the symmetry to give non-zero signatures). This sign pattern is an unbiased and resolution-independent signature of the vertical or horizontal preference (with respect to the isotropic case) of the angular distribution of the field vectors of the ensemble (Stenflo 1987, cf. the discussion at end of Sect. 3).

6 Hanle Effect and the Second Solar Spectrum

Since Hale the Zeeman effect has been the main tool for the diagnostics of cosmic magnetic fields. In recent decades, however, the Hanle effect has emerged as a complementary tool that is sensitive to fields in parameter domains that are nearly inaccessible to the Zeeman effect. The main reason why it has not yet been so widely used but rather been an area for a smaller community of specialists is that the underlying physics is complicated and the applications are not so straightforward. The observations require higher polarimetric sensitivity, and the interpretations generally need vector radiative transfer with scattering and partial frequency redistribution in magnetized media, a subject area that is still not fully developed. The interpretations are also sensitive to the assumptions concerning the 3-D geometry of the atmosphere. Due to such complications much of the astrophysical exploitation of the Hanle effect still lies in the future.

The Hanle effect was discovered by Wilhelm Hanle (cf. Fig. 16) in 1923 during his PhD work in Göttingen (Hanle 1924). It played an important role in the early development of quantum mechanics because it represented an explicit expression of the concept of linear superposition of quantum states and the decoherence that is caused by an external magnetic field. It soon found applications over a wide range of fields in physics before it was realized that it could also have important applications in astrophysics (cf. Moruzzi and Strumia 1991). Hanle himself was enthusiastic when he learnt about the astrophysical applications.

6.1 Coherent Scattering and Hanle Effect on the Sun

The blue sky is polarized due to Rayleigh scattering at molecules in the Earth's atmosphere. Similarly the Sun's spectrum gets linearly polarized because coherent scattering processes in the continuum and the various spectral lines contribute to the formation of the spectrum. Magnetic fields are not the source of the scattering polarization, but they modify it and thereby generate observable spectral signatures. It is this magnetically-induced modification that is referred to when we speak of the Hanle effect.

It was realized already in the 1920s by Öhman (1929) that such scattering polarization should exist on the Sun. Yngve Öhman was the founder of the Swedish solar station on Capri and was one of my mentors (since I started my solar work on Capri at age 20 in 1963). The first reliable observational evidence came when Günter Brückner recorded the scattering

Fig. 16 Wilhelm Hanle visiting me in my ETH office in 1983 on the occasion of the 60th anniversary of his effect

Fig. 17 Jean-Louis Leroy (right) at Pic du Midi, 1970

polarization near the Sun's limb in the Ca I 4227 Å line with the telescope at Göttingen's solar station in Locarno, Switzerland (Brückner 1963). In the mid 1980s the ownership of the observatory was transferred to Switzerland and it was renamed IRSOL (Istituto Ricerche Solari Locarno). Its scientific focus is the exploration of the scattering polarization and the Hanle effect with the powerful imaging polarimetry system ZIMPOL that was developed at ETH Zurich (see Sect. 6.2).

The first consistent interpretation of the non-magnetic 4227 Å observations in terms of polarized radiative transfer was made by Dumont et al. (1973) in Paris. The first observation of the Hanle effect in an astrophysical context was done in solar prominences by Leroy et al. (1977) at the Pic du Midi Observatory. Figure 17 shows Jean-Louis Leroy when I visited him at Pic du Midi in 1970.

The personal contacts developed during my extended periods of work in the Soviet Union in the 1960s led to the opportunity for Sweden to build instruments to be flown on Soviet Intercosmos satellites. Due to the severe limitations in space, weight, and telemetry we had to come up with an idea of doing something qualitatively different, not attempted with the much larger resources available elsewhere. Our choice was to build a Swedish spectro-polarimeter for the solar vacuum ultraviolet, to search for signatures of polarization due to coherent scattering in H Lyman α and in emission lines formed in the chromosphere-corona transition region. This project represented my own entry into the field of scattering polarization physics.

During the first attempt to launch the Swedish experiment, in 1975, the Soviet rocket exploded and all the experiments were destroyed. However, we had a spare model that could be successfully put in orbit the following year on the satellite Intercosmos 16 (the satellite that exploded was never assigned a number in the Intercosmos series—officially it did not exist). I participated in the launch preparations during three weeks at the super-secret military launch site of Kapustin Yar near Volgograd. It was the first time that anyone from a western country was admitted to a Soviet launch site.

Our UV spectro-polarimeter functioned as it was supposed to, but contamination of the UV optics due to outgassing in orbit from other parts of the spacecraft (because the integration had not been performed in a cleanroom but in a dirty garage environment) caused the optical transmission to quickly degrade by a factor of one hundred, increasing the polarimetric noise by an order of magnitude. It was therefore only possible to set an upper limit of 1 % for the H Lyman α scattering polarization near the limb as observed with one arcmin spatial resolution (Stenflo et al. 1980). Although this meager scientific output was a disappointment, the experiment demonstrated the feasibility of doing spectro-polarimetry in the vacuum ultraviolet.

There is considerable potential for using the scattering polarization and its magnetic modification via the Hanle effect in this part of the spectrum as a tool to explore the elusive magnetic fields in the chromosphere-corona transition region, where the physical processes responsible for coronal heating take place. Still, four decades later, this potential has not yet been exploited, but finally there is hope for some progress: the *CLASP* rocket experiment for recording the Hanle effect in H Lyman α is ready to be launched (cf. Ishikawa et al. 2011).

6.2 Second Solar Spectrum

The UV space experiment led to my interest in a systematic exploration of the physics of scattering polarization and its signatures in the Sun's spectrum. In 1978 I used the grating spectrograph and the FTS at the Kitt Peak McMath telescope to make a spectral survey of the linear polarization inside the solar limb, from 3165 to 9950 Å (Stenflo et al. 1983a, 1983b). This led to the discovery of the so-called Second Solar Spectrum, which is the linearly polarized spectrum that is exclusively caused by coherent scattering processes on the Sun. For more convenience it is now referred to with the acronym SS2, implying that the ordinary intensity spectrum is SS1.

There is a need to refer to SS2 as a totally different spectrum because the fractional linear polarization Stokes Q/I is as richly spectrally structured as the ordinary intensity spectrum, but with structures that look entirely different and which are formed by largely different physical processes. Its discovery led to intense theoretical work trying to identify all the unfamiliar spectral features and develop the quantum-mechanical foundations for their interpretations. SS2 is a playground for the Hanle effect, but the underlying physics is complex and still not fully understood. A modern atlas of SS2 has been produced by Gandorfer (2000, 2002, 2005) and is now available both in pdf and data format at http://www.irsol.ch/data_archive/#ss2.

Figure 18 gives an example of SS2 around the K and H lines of ionized calcium, to indicate the nature of some of the phenomena that we encounter in SS2. The nearly 200 Å

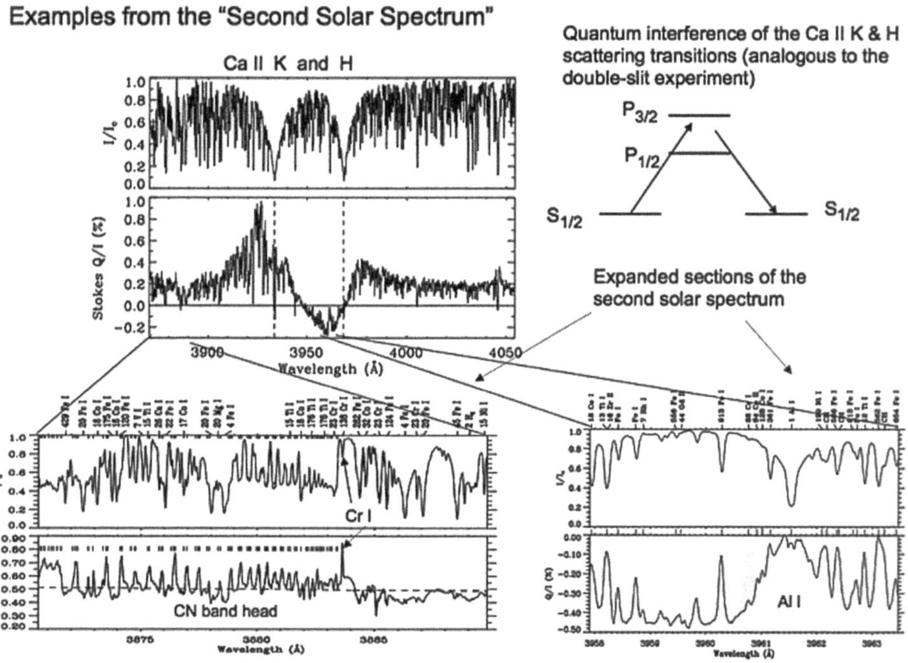

Fig. 18 Examples of various structures in the Second Solar Spectrum (SS2). The panels to the upper left show a nearly 200 Å wide portion around the strong Ca II K and H resonance lines (marked by the *vertical dashed lines*). The upper panel shows the intensity spectrum (SS1), the panel below SS2 with its remarkable polarization sign reversal, which has its explanation in terms of quantum interference between the K and H line scattering amplitudes, as indicated to the upper right. With the imaging polarimeter ZIMPOL it became possible in the late 1990s to zoom in on small sections of the spectrum, to uncover the enormous wealth of new kinds of structures in SS2, as indicated in the sets of panels in the lower half of the figure

wide section shown in the double panel to the upper left was recorded in 1978 (Stenflo 1980). The upper panel shows SS1 (the intensity spectrum) with the two dominating Ca II resonance lines, whose extended damping wings form a quasi-continuum that is populated by a multitude of blend lines. The strange shape of the SS2 spectrum in the panel below is a signature of quantum-mechanical interference between the $J = 1/2$ and $3/2$ upper states of the H and K lines (Stenflo 1980), as indicated in the diagram to the upper right. The intermediate state of the coherent scattering process represents a "Schrödinger cat state", a coherent superposition of states having different total angular momenta. The resulting quantum interference is responsible for the sign reversal of SS2 in Fig. 18.

Although major SS2 features like those of the H and K lines in the upper part of Fig. 18 could be uncovered with the Kitt Peak observations in 1978, the polarimetric precision that could be achieved was not much better than 0.1 %, far insufficient for most of the SS2. It was only with the introduction of the ZIMPOL (Zurich Imaging Polarimeter) technology in the mid 1990s that the full wealth of spectral structures in SS2 became accessible to exploration. With ZIMPOL at Kitt Peak we could routinely reach a polarimetric precision better than 10^{-5} in combination with high spectral resolution. The lower half of Fig. 18 illustrates how this allowed us to zoom in on any small spectral section to expose the new world of polarization phenomena.

The ZIMPOL technology, which was invented by Hanspeter Povel (cf. Fig. 19) at ETH Zurich (Povel 1995, 2001), revolutionized imaging polarimetry by achieving a polarimetric

Fig. 19 Hanspeter Povel, who at ETH Zurich invented the ZIMPOL technology for CCD-based imaging polarimetry with a precision of better than 10^{-5} in the fractional polarization

precision nearly two orders of magnitude better than other imaging systems. This opened a window to polarization physics that had not been accessible before. The problem had been that the available 2-D electronic detectors of CCD type have slow readout, which seemed incompatible with the fast (kHz) polarization modulation needed to eliminate seeing noise. If instead of modulation one would form the difference between two simultaneous images in orthogonal polarization states, the accuracy would be limited by gain-table noise, which is caused by inaccuracies in the flat-field calibration. Povel's solution to eliminate these two dominating noise sources was a way to create hidden fast buffer storage areas within the CCD sensor, between which the photocharges could be cycled in synchrony with the kHz polarization modulation. The second and later generations of the ZIMPOL system have four sets of fast hidden buffers, implying that the photocharges can be cycled between four simultaneous image planes within a single CCD sensor. Linear combinations between these four images give us the simultaneous images of the four Stokes parameters, free from seeing and gain table noise. The technology has now been successfully used in solar research for two decades.

The interpretation of the wealth of new phenomena in SS2 and the application of the Hanle effect for the diagnostics of solar magnetic fields require that the fundamental quantum processes that govern the interaction between matter and radiation in magnetized media are sufficiently understood. Figure 20 shows two of the main pioneers in the work to establish a foundation based on general principles in quantum field theory: Egidio Landi Degl'Innocenti of Firenze, Italy, and Veronique Bommier of Paris. They developed the theory in terms of the density matrix formalism with irreducible tensors, treated optical pumping by solving the statistical equilibrium problem for multi-level systems with atomic polarization, and applied the theory for the diagnostics of prominence magnetic fields (cf. Bommier 1980; Landi Degl'Innocenti 1983). The monograph by Landi Degl'Innocenti and Landolfi (2004) provides a comprehensive account of this theory.

An alternative approach to the scattering theory in terms of the Kramers-Heisenberg scattering amplitudes is presented in the monograph on solar magnetic fields by Stenflo (1994). It is particularly suited for the inclusion of the important effects of partial frequency redistribution effects in polarized radiative transfer with scattering in magnetized media. It has been the approach used by the groups in Bangalore (headed by K.N. Nagendra) and Nice (led by Marianne Faurobert and Helène Frisch), who have developed the theoretical tools needed to interpret the SS2.

The physics of polarized radiation with the Zeeman and Hanle effects in magnetized stellar atmospheres has been the subject of a series of Solar Polarization Workshops that

Fig. 20 Egidio Landi Degl'Innocenti of Firenze and Veronique Bommier of Paris in 2013 during Solar Polarization Workshop No. 7 in Kunming, China. They helped establish the quantum-mechanical foundation for the interaction between radiation and matter in magnetized media

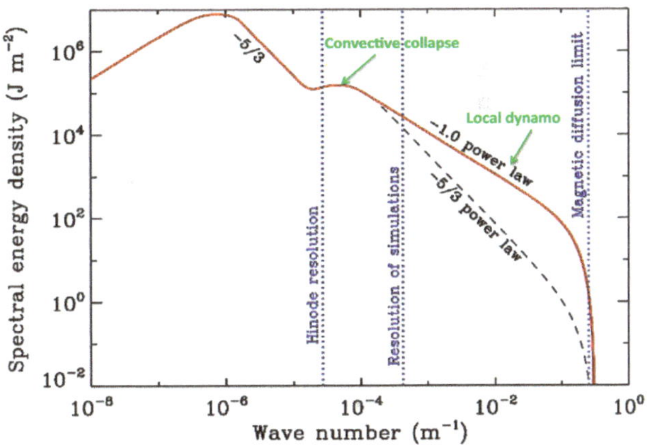

Fig. 21 Empirically determined magnetic energy spectrum extending over seven orders of magnitude in scale (*red line*), from the global scales to the magnetic diffusion limit, where the field lines decouple from the plasma. The spectrum needs to be elevated in the optically thin inertial range to a power law of -1.0 to be compatible with the constraints from the observed Hanle depolarization effect. From Stenflo (2014)

have taken place every three years in different parts of the world since 1995. The first of them was organized in St. Petersburg, Russia, by V.V. Ivanov, who coined the expression "Second Solar Spectrum", when referring to the linearly polarized solar spectrum that is formed by coherent scattering processes. The proceedings of these Workshops provide an updated account of the field. The latest is from Solar Polarization 7, which was held in Kunming in 2013 (Nagendra et al. 2014).

6.3 Scale Spectrum of Magnetic Structures: Empirical View

The ongoing quest to diagnose the smallest scales has led to the understanding that the magnetic structuring continues on scales orders of magnitude smaller than the resolved ones, down to the magnetic diffusion limit (of order 25 m), where the ohmic diffusion time scale becomes smaller than the convective time scale and the field lines therefore cease to be frozen-in and decouple from the plasma. Since most scales are still unresolved, indirect methods based on the observations of ensemble averages (see Sect. 5.5) have had to be used, with guidance from numerical simulations of magneto-convection, which have become increasingly advanced with the rapid growth in computing power.

Figure 21 gives a global overview of the empirically determined magnetic energy spectrum of the quiet Sun over seven orders of magnitude in scale size, from Stenflo (2014). The portion to the left of the vertical dotted line marked "Hinode resolution" (at a scale of about 200 km) is based on resolved observations with *Hinode* SOT/SP. The spectral bump immediately to the right of this line, at scales 10–100 km, follows from diagnostics with the line-ratio method, from which a population of intermittent kG flux elements with sizes in this range can be inferred. Theoretically the existence of such a flux population can be understood in terms of the convective collapse mechanism (Parker 1978; Spruit and Zweibel 1979).

One of the main applications of the Hanle effect has been to diagnose the properties of magnetic flux which is hidden to the Zeeman effect, because the magnetic elements are optically thin with random orientations of their field vectors. When this method was first introduced (Stenflo 1982) it revealed the existence of a seething ocean of microturbulent magnetic flux with field strengths in the range 10–100 G. The most detailed radiative-transfer modeling of this Hanle depolarization effect with the use of 3-D atmospheres produced by MHD numerical simulations has led to a turbulent field strength of 60 G if a single-valued field is assumed (δ function PDF). With a more realistic PDF an average field strength about twice as large is obtained (Trujillo Bueno et al. 2004). The 60 G value should therefore be considered as a lower limit.

So much volume-filling flux in the optically thin scale range is not compatible with the standard Kolmogorov-type $-5/3$ power law (dashed line at high wave numbers in Fig. 21). The energy spectrum needs to be substantially elevated and flatter to be compatible with the Hanle observational constraints, which are best satisfied with the -1.0 power law that is illustrated in Fig. 21. Such an elevated spectrum cannot be explained in terms of the ordinary turbulent cascade from larger to smaller scales, there is a need for a new source of flux at small scales. This source could be provided by a local dynamo. While observations at resolved scales rule out a significant contribution from a local dynamo (Stenflo 2012), such dynamo action appears to be needed for consistency with the Hanle constraints.

7 Concluding Remarks

Current research on solar magnetic fields is very diverse, with an interplay between theory (e.g. plasma physics, quantum field theory, polarized radiative transfer), observations (ground-based or space-based, polarimetric imaging with spectrographs or narrow-band filter systems, etc.), technological developments (detector and modulation systems, telescope technology, enhanced resolutions, computing power), and numerical simulations of magneto-convection. The present rather sketchy and personal overview has focused on some highlights that have advanced our empirical view of solar magnetic fields since the time of Hale. It is unavoidable that many of the important contributions to the field have not been covered here, because it is far beyond the scope of the present overview to aim for a comprehensive historical account.

Hale opened a new window to the universe by applying the physical effect discovered by Pieter Zeeman a decade before (Zeeman 1897). A new polarimetric window has now been opened, although with much delay, by applying the physical effect discovered by Wilhelm Hanle at the time when quantum mechanics was being formulated in Göttingen (Hanle 1924). The complexity of all the coherence phenomena has led to a return to an exploration of the fundamental quantum nature of matter-radiation interactions in magnetized media. This work is still ongoing.

In contrast the non-coherent Zeeman-effect physics is well understood. The problems when applying it to the Sun arise because we sample an inhomogeneous medium in which the physical parameters vary over regions that are not resolved. Even if we could observe with infinite angular resolution, the spatial sampling along the line of sight remains about 100 km, since this is the typical depth of line formation. This is more than three orders of magnitude larger than the small-scale end of the magnetic energy spectrum. One may in principle improve the resolution along the line of sight by using combinations of spectral lines, whose contribution functions are slightly shifted relative to each other, but this improvement can hardly exceed one order of magnitude. Therefore the observed quantities will always represent ensemble averages over distributions of unresolved structures. The diagnostics of the intrinsic properties of the elements that make up such ensembles is a formidable challenge. It is however a challenge that non-solar astronomers are well familiar with, since for much more distant objects nobody can be misled to believe that the physical structures are resolved.

It is therefore obvious that the history of solar magnetic field observations will not come to an end point any time soon. As much else in science it is an open-ended enterprise. The past history has seen many spectacular advances, which have given us insights about the role played by magnetic fields not only in the Sun but throughout the universe. The deepened understanding has led to new questions that could not be contemplated at the time of Hale. The future exploration of these questions on the Sun will continue to fertilize the rest of astrophysics.

References

H. Alfvén, Existence of electromagnetic-hydrodynamic waves. Nature **150**, 405–406 (1942)

H. Alfvén, Solar magnetic fields, in *Magnetism and the Cosmos*, ed. by W.R. Hindmarsh, F.J. Lowes, P.H. Roberts, F.R. Runcorn (Oliver & Boyd, Edinburgh, 1967), pp. 246–261

H. Alfvén, P. Carlqvist, Currents in the solar atmosphere and a theory of solar flares. Sol. Phys. **1**, 220–228 (1967)

H.W. Babcock, The solar magnetograph. Astrophys. J. **118**, 387 (1953)

H.W. Babcock, The topology of the Sun's magnetic field and the 22-year cycle. Astrophys. J. **133**, 572–587 (1961)

V. Bommier, Quantum theory of the Hanle effect. II – Effect of level-crossings and anti-level-crossings on the polarization of the D3 helium line of solar prominences. Astron. Astrophys. **87**, 109–120 (1980)

J.W. Brault, Solar Fourier transform spectroscopy. Oss. Mem. Oss. Astrofis. Arcetri **106**, 33 (1978)

J.W. Brault, Fourier transform spectroscopy, in *High Resolution in Astronomy. Fifteenth Advanced Course of the Swiss Society of Astronomy and Astrophysics*, ed. by A.O. Benz, M. Huber, M. Mayor (Geneva Obs., Sauverny, 1985), pp. 3–61

G. Brückner, Photoelektrische Polarisationsmessungen an Resonanzlinien im Sonnenspektrum. Mit 6 Textabbildungen. Z. Astrophys. **58**, 73 (1963)

G.A. Chapman, N.R. Sheeley Jr., The photospheric network. Sol. Phys. **5**, 442–461 (1968)

T.G. Cowling, The magnetic field of sunspots. Mon. Not. R. Astron. Soc. **94**, 39–48 (1933)

S. Dumont, J.-C. Pecker, A. Omont, Theoretical study of the Fraunhofer lines polarization: the case of Ca I 4227. Sol. Phys. **28**, 271–288 (1973)

W.M. Elsasser, Induction effects in terrestrial magnetism part I. Theory. Phys. Rev. **69**, 106–116 (1946)

W.M. Elsasser, Hydromagnetism. I. A review. Am. J. Phys. **23**, 590–609 (1955)

O. Engvold, Large Earth-based Solar Telescope – LEST. Adv. Space Res. **11**, 157–168 (1991)

A. Gandorfer, *The Second Solar Spectrum: A High Spectral Resolution Polarimetric Survey of Scattering Polarization at the Solar Limb in Graphical Representation. Volume I: 4625 Å to 6995 Å* (VdF, Zurich, 2000)

A. Gandorfer, *The Second Solar Spectrum: A High Spectral Resolution Polarimetric Survey of Scattering Polarization at the Solar Limb in Graphical Representation. Volume II: 3910 Å to 4630 Å* (VdF, Zurich, 2002)

A. Gandorfer, *The Second Solar Spectrum: A High Spectral Resolution Polarimetric Survey of Scattering Polarization at the Solar Limb in Graphical Representation. Volume III: 3160 Å to 3915 Å* (VdF, Zurich, 2005)
G.E. Hale, On the probable existence of a magnetic field in Sun-spots. Astrophys. J. **28**, 315 (1908)
G.E. Hale, F.H. Seares, A. van Maanen, F. Ellerman, The general magnetic field of the Sun. Apparent variation of field-strength with level in the solar atmosphere. Astrophys. J. **47**, 206–254 (1918)
G.E. Hale, F. Ellerman, S.B. Nicholson, A.H. Joy, The magnetic polarity of Sun-spots. Astrophys. J. **49**, 153–178 (1919)
W. Hanle, Über magnetische Beeinflussung der Polarisation der Resonanzfluoreszenz. Z. Phys. **30**, 93–105 (1924)
J.W. Harvey, Observations of small-scale photospheric magnetic fields. Highlights Astron. **4**, 223 (1977)
K.L. Harvey, S.F. Martin, Ephemeral active regions. Sol. Phys. **32**, 389–402 (1973)
R. Ishikawa, T. Bando, D. Fujimura, H. Hara, R. Kano, T. Kobiki, N. Narukage, S. Tsuneta, K. Ueda, H. Wantanabe, K. Kobayashi, J. Trujillo Bueno, R. Manso Sainz, J. Stepan, B. de Pontieu, M. Carlsson, R. Casini, A sounding rocket experiment for spectropolarimetric observations with the Ly_α line at 121.6 nm (CLASP), in *Solar Polarization 6*, ed. by J.R. Kuhn, D.M. Harrington, H. Lin, S.V. Berdyugina, J. Trujillo-Bueno, S.L. Keil, T. Rimmele. Astronomical Society of the Pacific Conference Series, vol. 437 (2011), pp. 287–295
K.O. Kiepenheuer, Joint Organization for Solar Observations (JOSO), in *Proceedings of the First European Astronomical Meeting, Solar Activity and Related Interplanetary and Terrestrial Phenomena*, ed. by J. Xanthakis (Springer, Heidelberg, 1973), p. 165
E. Landi Degl'Innocenti, Polarization in spectral lines. I – A unifying theoretical approach. Sol. Phys. **85**, 3–31 (1983)
E. Landi Degl'Innocenti, M. Landolfi, *Polarization in Spectral Lines*. Astrophysics and Space Science Library, vol. 307 (Kluwer, Dordrecht, 2004)
J. Larmor, How could a rotating body such as the sun become a magnet? Rep. - Br. Assoc. Adv. Sci. **87**, 159–160 (1919)
R.B. Leighton, Observations of solar magnetic fields in plage regions. Astrophys. J. **130**, 366 (1959)
R.B. Leighton, A magneto-kinematic model of the solar cycle. Astrophys. J. **156**, 1–26 (1969)
R.B. Leighton, R.W. Noyes, G.W. Simon, Velocity fields in the solar atmosphere. I. Preliminary report. Astrophys. J. **135**, 474 (1962)
J.L. Leroy, G. Ratier, V. Bommier, The polarization of the D3 emission line in prominences. Astron. Astrophys. **54**, 811–816 (1977)
W.C. Livingston, J. Harvey, C. Slaughter, D. Trumbo, Solar magnetograph employing integrated diode arrays. Appl. Opt. **15**, 40–52 (1976)
K. Lundmark, Internal motions of Messier 33. Astrophys. J. **63**, 67–71 (1926)
G. Moruzzi, F. Strumia (eds.), *The Hanle Effect and Level-Crossing Spectroscopy* (Plenum Press, New York, 1991)
K.N. Nagendra, J.O. Stenflo, Z.Q. Qu, M. Sampoorna (eds.), *Solar Polarization 7*. Astronomical Society of the Pacific Conference Series, vol. 489 (2014)
Y. Öhman, Astronomical consequences of the polarization of fluorescence. Mon. Not. R. Astron. Soc. **89**, 479–482 (1929)
E.N. Parker, Hydromagnetic dynamo models. Astrophys. J. **122**, 293 (1955)
E.N. Parker, Hydraulic concentration of magnetic fields in the solar photosphere. VI – Adiabatic cooling and concentration in downdrafts. Astrophys. J. **221**, 368–377 (1978)
H. Povel, Imaging Stokes polarimetry with piezoelastic modulators and charge-coupled-device image sensors. Opt. Eng. **34**, 1870–1878 (1995)
H.P. Povel, Ground-based instrumentation for solar magnetic field studies, with special emphasis on the Zurich imaging polarimeters ZIMPOL-I and II, in *Magnetic Fields Across the Hertzsprung-Russell Diagram*, ed. by G. Mathys, S.K. Solanki, D.T. Wickramasinghe. Astronomical Society of the Pacific Conference Series, vol. 248 (2001), pp. 543–552
D.N. Rachkovsky, Magneto-optical effects in spectral lines of sunspots. Izv. Krym. Astrofiz. Obs. **27**, 148–161 (1962a)
D.N. Rachkovsky, Magnetic rotation effects in spectral lines. Izv. Krym. Astrofiz. Obs. **28**, 259–270 (1962b)
A. Severny, Solar magnetic fields. Space Sci. Rev. **3**, 451–486 (1964)
A.B. Severny, Magnetic fields of the Sun and stars (in Russian). Usp. Fiz. Nauk **88**, 3–50 (1966)
N.R. Sheeley Jr., Observations of small-scale solar magnetic fields. Sol. Phys. **1**, 171–179 (1967)
N.R. Sheeley Jr., C.R. DeVore, J.P. Boris, Simulations of the mean solar magnetic field during sunspot cycle 21. Sol. Phys. **98**, 219–239 (1985)
S.K. Solanki, P. Barthol, S. Danilovic, A. Feller, A. Gandorfer, J. Hirzberger, T.L. Riethmüller, M. Schüssler, J.A. Bonet, V. Martínez Pillet, J.C. del Toro Iniesta, V. Domingo, J. Palacios, M. Knölker, N. Bello

González, T. Berkefeld, M. Franz, W. Schmidt, A.M. Title, SUNRISE: instrument, mission, data, and first results. Astrophys. J. **723**, 127–133 (2010)
H.C. Spruit, E.G. Zweibel, Convective instability of thin flux tubes. Sol. Phys. **62**, 15–22 (1979)
M. Steenbeck, F. Krause, Zur Dynamotheorie stellarer und planetarer Magnetfelder I. Berechnung sonnenähnlicher Wechselfeldgeneratoren. Astron. Nachr. **291**, 49–84 (1969)
J.O. Stenflo, Hale's attempts to determine the Sun's general magnetic field. Sol. Phys. **14**, 263–273 (1970)
J.O. Stenflo, Magnetic-field structure of the photospheric network. Sol. Phys. **32**, 41–63 (1973)
J.O. Stenflo, Resonance-line polarization. V – Quantum-mechanical interference between states of different total angular momentum. Astron. Astrophys. **84**, 68–74 (1980)
J.O. Stenflo, The Hanle effect and the diagnostics of turbulent magnetic fields in the solar atmosphere. Sol. Phys. **80**, 209–226 (1982)
J.O. Stenflo, LEST – a large international solar telescope for the 1990s. Vistas Astron. **28**, 571–576 (1985)
J.O. Stenflo, Observational constraints on a 'hidden', turbulent magnetic field of the Sun. Sol. Phys. **114**, 1–19 (1987)
J.O. Stenflo, Strong and weak magnetic fields: nature of the small-scale flux elements, in *IAU Colloq. 141: The Magnetic and Velocity Fields of Solar Active Regions*, ed. by H. Zirin, G. Ai, H. Wang. Astronomical Society of the Pacific Conference Series, vol. 46, (1993), pp. 205–214
J.O. Stenflo, *Solar Magnetic Fields – Polarized Radiation Diagnostics* (Kluwer, Dordrecht, 1994)
J.O. Stenflo, Basal magnetic flux and the local solar dynamo. Astron. Astrophys. **547**, 93 (2012)
J.O. Stenflo, Horizontal or vertical magnetic fields on the quiet Sun. Angular distributions and their height variations. Astron. Astrophys. **555**, 132 (2013a)
J.O. Stenflo, Solar magnetic fields as revealed by Stokes polarimetry. Astron. Astrophys. Rev. **21**, 66 (2013b)
J.O. Stenflo, Nature of quiet-Sun magnetic fields, in *Solar Polarization 7*, ed. by K.N. Nagendra, J.O. Stenflo, Z.Q. Qu, M. Sampoorna. Astronomical Society of the Pacific Conference Series, vol. 489 (2014), pp. 3–19
J.O. Stenflo, D. Dravins, N. Wihlborg, A. Bruns, V.K. Prokofev, I.A. Zhitnik, H. Biverot, L. Stenmark, Search for spectral line polarization in the solar vacuum ultraviolet. Sol. Phys. **66**, 13–19 (1980)
J.O. Stenflo, D. Twerenbold, J.W. Harvey, Coherent scattering in the solar spectrum – survey of linear polarization in the range 3165–4230 Å. Astron. Astrophys. Suppl. Ser. **52**, 161–180 (1983a)
J.O. Stenflo, D. Twerenbold, J.W. Harvey, J.W. Brault, Coherent scattering in the solar spectrum – survey of linear polarization in the range 4200–9950 Å. Astron. Astrophys. Suppl. Ser. **54**, 505–514 (1983b)
J.O. Stenflo, S. Solanki, J.W. Harvey, J.W. Brault, Diagnostics of solar magnetic fluxtubes using a Fourier transform spectrometer. Astron. Astrophys. **131**, 333–346 (1984)
V.E. Stepanov, The absorption coefficient of atoms in the case of reverse Zeeman effect for an arbitrary directed magnetic field. Izv. Krym. Astrofiz. Obs. **18**, 136–150 (1958a)
V.E. Stepanov, On the theory of the formation of absorption lines in a magnetic field and the profile of Fe λ 6173 Å in the solar sunspot spectrum. Izv. Krym. Astrofiz. Obs. **19**, 20–45 (1958b)
V.E. Stepanov, A.B. Severny, A photoelectric method for measurements of the magnitude and direction of the solar magnetic field. Izv. Krym. Astrofiz. Obs. **28**, 166–193 (1962)
J. Trujillo Bueno, N. Shchukina, A. Asensio Ramos, A substantial amount of hidden magnetic energy in the quiet Sun. Nature **430**, 326–329 (2004)
W. Unno, Line formation of a normal Zeeman triplet. Publ. Astron. Soc. Jpn. **8**, 108 (1956)
D. Vrabec, Magnetic fields spectroheliograms from the San Fernando observatory, in *Solar Magnetic Fields*, ed. by R. Howard. IAU Symposium, vol. 43 (1971), pp. 329–339
P. Zeeman, On the influence of magnetism on the nature of the light emitted by a substance. Philos. Mag. **43**, 226 (1897)
H. Zirin, The interaction of weak and strong magnetic fields on the Sun (invited), in *IAU Colloq. 141: The Magnetic and Velocity Fields of Solar Active Regions*, ed. by H. Zirin, G. Ai, H. Wang. Astronomical Society of the Pacific Conference Series, vol. 46 (1993), pp. 215–221

Space Sci Rev (2017) 210:37–76
DOI 10.1007/s11214-015-0219-y

Measurements of Photospheric and Chromospheric Magnetic Fields

Andreas Lagg[1] · Bruce Lites[2] · Jack Harvey[3] · Sanjay Gosain[3] · Rebecca Centeno[2]

Received: 31 July 2015 / Accepted: 23 October 2015 / Published online: 17 December 2015

Abstract The Sun is replete with magnetic fields, with sunspots, pores and plage regions being their most prominent representatives on the solar surface. But even far away from these active regions, magnetic fields are ubiquitous. To a large extent, their importance for the thermodynamics in the solar photosphere is determined by the total magnetic flux. Whereas in low-flux quiet Sun regions, magnetic structures are shuffled around by the motion of granules, the high-flux areas like sunspots or pores effectively suppress convection, leading to a temperature decrease of up to 3000 K. The importance of magnetic fields to the conditions in higher atmospheric layers, the chromosphere and corona, is indisputable. Magnetic fields in both active and quiet regions are the main coupling agent between the outer layers of the solar atmosphere, and are therefore not only involved in the structuring of these layers, but also for the transport of energy from the solar surface through the corona to the interplanetary space.

Consequently, inference of magnetic fields in the photosphere, and especially in the chromosphere, is crucial to deepen our understanding not only for solar phenomena such as chromospheric and coronal heating, flares or coronal mass ejections, but also for fundamental physical topics like dynamo theory or atomic physics. In this review, we present an overview

✉ A. Lagg
lagg@mps.mpg.de

B. Lites
lites@ucar.edu

J. Harvey
jharvey@nso.edu

S. Gosain
sgosain@nso.edu

R. Centeno
rce@ucar.edu

1 Max Planck Institute for Solar System Research, Justus-von-Liebig-Weg 3, 37077 Göttingen, Germany

2 High Altitude Observatory, National Center for Atmospheric Research, Boulder, USA

3 National Solar Observatory, 950 N. Cherry Ave., Tucson, AZ 85719, USA

of significant advances during the last decades in measurement techniques, analysis methods, and the availability of observatories, together with some selected results. We discuss the problems of determining magnetic fields at smallest spatial scales, connected with increasing demands on polarimetric sensitivity and temporal resolution, and highlight some promising future developments for their solution.

Keywords Sun · Magnetic field · Chromosphere · Photosphere · Measurement · Observations · Spectro-polarimetry

1 Introduction

The energy produced by nuclear fusion in the core of our closest star, the Sun, is transported into space almost exclusively by photons emitted from the solar surface. In the absence of solar magnetic fields, this energy flux would likely be nearly constant. It is the solar activity cycle, manifested in the reversal of the solar magnetic field with a 22-year periodicity, which is the main contributor to the variation of the solar energy flux impinging on earth. This variation is only at the order of less than one percent (Solanki et al. 2013), but this is sufficient to affect our natural and technical environment in a significant manner. Spectacular eruptions of high energy particles in so-called coronal mass ejections also have the solar magnetic field as their driver. Any reliable space weather forecast, important for example to avoid damage to Earth orbiting satellites, requires the understanding of the energetics in the magnetized solar atmosphere. A robust determination of the solar magnetic field in all atmospheric layers is therefore mandatory.

The transition from the regime dominated by convective gas and plasma motions to the one driven by magnetic forces takes place between the photosphere and the chromosphere. In this layer, the plasma β (i.e., the ratio of gas pressure to magnetic pressure) becomes unity. Extremely fast physical processes on size scales at and well below the resolution limit of current solar telescopes, like reconnection or wave dissipation, occur ubiquitously in active regions but also in places of low solar activity. As a logical consequence, solar instrumentation development aims for the determination of the magnetic field at highest spatial and temporal resolution directly above, in and below the $\beta = 1$ layer (Kleint and Gandorfer 2015, this issue).

Since in situ measurements of the field using magnetometers in these deep layers of the solar atmosphere are impossible, we have to rely on remote-sensing observations. The information about the magnetic field must be extracted from the radiation emitted from the solar surface and the layers above it. Instruments in the near ultra-violet, the visible regions (where the energy flux from the photosphere and the chromosphere is highest), and the infrared regions of the solar spectrum record the intensity and polarization profiles of Fraunhofer lines. The conditions in the solar atmosphere encoded in these spectral lines are deciphered by the combination of advanced instrumentation and sophisticated analysis techniques.

For more than a century the Zeeman effect has been used to reliably determine the magnetic field in the photosphere (Hale 1908; Stenflo 2015, this issue). Enhanced instrumental capabilities allowed one to extend the measurements from regions with strong Zeeman signals, like sunspots or pores (Rempel and Scharmer 2015, this issue), down to small-scale events and also towards higher atmospheric layers: the chromosphere and the corona (Trujillo Bueno and Landi degl'Innocenti 2015, this issue; Wiegelmann et al. 2015, this issue).

Especially in these weak field environments, the Hanle effect has gained outstanding importance during the last two decades. In some special spectral lines the advantages of both effects are combined, and allow for continuous measurements from weak, unresolved magnetic fields in the sub-gauss regime up to strong, kilo-gauss field in sunspots.

The unambiguous interpretation of observations using these two effects requires the exact treatment of the physical mechanisms behind them. The radiative transfer equation must be solved, the population of the energy sub-levels in the relevant atoms of the solar atmosphere must be computed, and the effect of radiation and collisions must be considered (de la Cruz Rodriguez et al. 2015, this issue). In addition, the tiny imprints of weak, small scale magnetic fields in the polarization signals of spectral lines require advanced instrumentation with cutting-edge technology in, e.g., optical engineering and detector design.

2 Overview of Solar Magnetic Field Measurement

Several methods have been used to estimate the properties of the magnetic field in various regions of the solar atmosphere. The assumption that density structures and their dynamics are entrained along or guided by the magnetic field has long been used to infer magnetic properties from observations of the solar corona, prominences and chromospheric fibrils (e.g. Bigelow 1889). Frequencies of oscillation of features thought to be tied to magnetic fields have been used to estimate magnetic field strengths (e.g. Hyder 1966; Ballester 2014). Various properties of solar radio emission are used to deduce magnetic field characteristics in the chromosphere and corona (see review by Kundu 1990). In addition to these remote sensing observations, direct measurement of heliospheric magnetic fields has been done for decades, and by 2024 we can expect in situ magnetic field measurements as close as 8.5 solar radii above the solar surface (Solar Probe Plus 2015).

By far the majority of solar magnetic field remote sensing measurements have been made by observations of the Zeeman and Hanle polarization effects on atomic spectral lines. Some reviews of solar spectropolarimetry methods include Stenflo (1994, 2013) and Keller (2002). The vastly larger general field of polarimetry measurements and techniques is frequently reviewed. Some recent reviews of special note include Trippe (2014), Snik et al. (2014), and Rodenhuis et al. (2014).

It is useful to consider the key elements of solar spectropolarimetry as a sequence from the source to the inferred measurement: source magnetic field → radiative transfer → atmosphere → telescope → calibration → imaging optics → polarization modulator → wavelength selector → detector → data recording → calculation of source polarization → inference of source magnetic field properties. Our task is to determine the source magnetic field as functions of 3D spatial position and time from observations of the Stokes vector elements that convey intensity and polarization information as functions of 2D spatial position, spectral wavelength, and time. This is a big challenge. In the future, it may be possible to derive more robust 3D information by using simultaneous stereoscopic observations from space. One could discuss each of the key elements of solar spectropolarimetry in detail but space constraints prevent that. Here we give a simple overview with focused attention on some topics of current interest.

2.1 Contemporary Observational and Research Areas

At present, one may identify four major domains of solar polarimetry research driven by technical tradeoff considerations and scientific interests.

1. High spatial resolution, with the goal of exploring small-scale magnetoconvection.
2. High polarimetric sensitivity, for studies of weak magnetic fields including the chromosphere and corona.
3. Full solar disk synoptic sequences, for investigations of large-scale magnetic field evolution, the solar dynamo and solar cycle, and for space weather applications.
4. Fast cadence, to better understand solar activity such as flares, CMEs, spicules, filaments, etc.

There is not enough light available to meet all these needs simultaneously using available instrumentation or methods. Tradeoffs are required, and to guide these tradeoffs various system considerations are useful. A spectropolarimetric solar observation can be represented as an 5D hypercube with time (t) as one dimension and Stokes vector elements, which represent polarization information (I, Q, U, V) as one additional dimension. The remaining three dimensions are two plane-of-the-sky coordinates (x, y) and wavelength (λ). Usually this datacube is drawn at one instant of time with edges x, y, λ and cube volume elements containing one of the four Stokes vector elements. Ideally we would like to completely fill such a datacube with a single, brief snapshot in order to obtain observations with the most efficient use of available light. Hagen and Kudenov (2013) have very usefully reviewed progress toward snapshot spectral imaging technology and techniques. In practice, current measurements of solar magnetic fields are mostly made using spatially-scanning slit spectrometers or wavelength-scanning narrow band filters. In the former case observing time speed is compromised in order to maintain spectrum line profile integrity at the expense of non-simultaneous spatial sampling. In the latter case observing speed is emphasized but the profile integrity is compromised by the need to observe spectra at different wavelengths and polarizations at different times. The history of magnetic field estimation is full of attempts to mitigate these tradeoffs and their compromises.

2.2 Time Constraints

The duration of an observation is important for two main reasons. First, the Stokes vector must be measured accurately at each datacube element before any observation conditions change. Second, the acquisition of a spectropolarimetric data cube must be made sufficiently fast compared to the evolutionary time scale of the magnetic structure observed. These requirements conflict with both the acquisition of a sufficient number of photons for low-noise Stokes polarimetry and construction of the spatial/spectral images quickly enough to suppress noise from solar feature evolution and ground-based seeing or space-based pointing jitter. Thus, progress in solar magnetic feature measurement is strongly influenced by such tradeoffs, but it is also strongly linked to technology improvements that enhance the efficiency of observations. In current practice, intensity measurements in at least four different polarization modulation states are needed to fully define a Stokes vector measurement. A polarization modulator device sequentially produces the different states. Measurement noise arises irreducibly not only from variations in the number of photons detected (shot-noise), but also spuriously from any image intensity or geometry changes during the modulation sequence. A goal is to not add significant spurious noise in excess of the shot noise. For ground-based observations, atmospheric seeing is a major source of noise. Measurements of the frequency spectrum of seeing effects on solar images (Rimmele 2000) show large power at low frequencies but also that significant noise is present at frequencies in excess of 100 Hz.

Practical experience shows that polarization noise produced by atmospheric seeing may be reduced by rapid acquisition of modulated solar images. Spatial resolution and site conditions are important factors, but completing a polarization modulation sequence in 10 ms

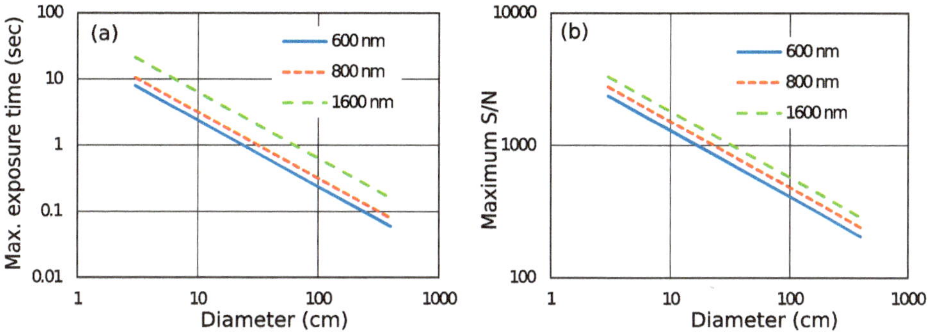

Fig. 1 Left (**a**): Maximum exposure time for diffraction-limited disk-center photospheric observation constrained by <0.05 pixel motion of target at 5 $\mathrm{km\,s^{-1}}$. Right (**b**): Maximum signal to noise ratio per exposure for diffraction-limited photospheric observations at disk center resulting from the number of photons per pixel impinging the detector within the maximum exposure time

(from the ground) or a few seconds (from space—depending on the frequency spectrum of pointing jitter) is highly desired for minimizing spurious noise. Ground-based slit spectrograph polarimeters using fast polarization modulation can reach noise levels (uncertainty of Q, U, or V relative to I) of order 10^{-5} while ground-based filter polarimeters are usually limited to 10^{-3} due to an additional time penalty of acquiring many images at different wavelengths and polarization states during which observation conditions change. Adaptive optics and reconstruction techniques for ground-based filter images can reduce the spurious noise caused by seeing. For both ground and space filter observations the evolving pattern of solar intensity and polarization produce polarization noise that depends on the properties of the instrument and the acquisition sequence of the various wavelength and polarization state images.

2.3 Solar Polarimetry is Light-Starved

To determine the basic structure of strong magnetic features like sunspots, noise levels of 10^{-2} of the continuum intensity are acceptable. Some high sensitivity work requires noise of the order 10^{-5}, or even smaller. Considering just shot noise, 10^4 to 10^{10} photons must be collected to reach these noise levels. Suppose we want to make a photospheric observation at disk center using two spatial pixels per diffraction limit with an overall instrument efficiency of 10 %. The temporal modulation typically used for polarization measurements requires us to accept a displacement of no more than 0.05 pixel on the imaging device during the observation, otherwise spurious noise is introduced to the data. Under these conditions and assuming a typical photospheric velocity of 5 $\mathrm{km\,s^{-1}}$, Fig. 1a shows the maximum exposure time allowed at three different wavelengths and a range of telescope apertures. At the small aperture size limit shown (GONG) exposure times of about 10 seconds can be used, whilst at the large size limit (DKIST), 60–160 ms are the longest exposures allowed.

If we use two spectral samples per spectral resolution element of $\lambda/\Delta\lambda = 90\,000$ and assume an overall system transmission of 10 %, then Fig. 1b shows the maximum signal-to-noise ratio (S/N) for Stokes I that can be obtained in the continuum computed using the maximum available exposure times. Note that the S/N ($I/\Delta I$ or $I/\Delta p$) will be lower for Q, U, V ($\Delta p \approx \sqrt{3}/\Delta I$) and in the darker parts of spectral lines. It is obvious that achieving high polarimetric sensitivity in diffraction limited observations is very difficult.

2.4 Strategies to Improve Efficiency

The grim outlook shown in Fig. 1b has motivated many tradeoffs to be able to secure useful high-resolution, high-precision measurements of solar magnetic fields. Several of these were used in the first solar magnetographs. One may, for example, use larger spatial and spectral samples, combine signals from more than one spectrum line, use sparse sampling of the spectrum, slice the incoming solar image into a format better suited to the spectrometer, and/or select a wavelength compatible with high overall instrumental efficiency. Dual-beam polarimetry was introduced so that all the photons are used all the time. As detector technology advanced, better quantum efficiency and higher read-out speeds were achieved all the time (in addition to greatly reducing crosstalk among polarization states arising from residual image motion and blurring).

There are only a few chromospheric and coronal spectral lines that are useful for magnetography, but many lines forming in the photosphere are available. The photospheric photon flux per spectrum line Doppler width peaks at near-IR wavelengths around 1 μm so this should be an optimum wavelength for observations. However, IR spectrum lines tend to be weak compared to those at shorter wavelengths (though this is counteracted by stronger Zeeman splitting at long wavelengths), and spatial resolution for a given telescope aperture is smaller at longer wavelengths. Historically, lines in the visible part of the spectrum have been favored mainly owing to the availability of lower-cost detectors with high quantum efficiency. At present, the optimum wavelength for photospheric magnetic field Zeeman effect measurements ranges from the yellow to the near infrared.

One long-used strategy is to observe only circular polarization (Stokes V), from which the line-of-sight component of the magnetic field can be estimated. Provided the magnetic elements are spatially resolved, to first order the strength of the circular polarization signal depends linearly on the source magnetic field strength, while the linear polarization signal (Stokes Q and U) has a quadratic dependence, and for modest field strengths it is rather weak.

The availability of detectors with large numbers of pixels that can be read out fairly quickly has stimulated clever techniques to fill the image/polarization/wavelength hypercube in single snapshots (Hagen and Kudenov 2013). The basic idea is to divide the large space on the detector in ways that allow recording spectral, spatial, and polarization information simultaneously. One older example first used photographically (Martin et al. 1974) is as follows. Instead of using a small format detector in a spectrograph having one long entrance slit, a large format detector allows one to use many entrance slits each producing a spectrum of a different part of the solar image. Figure 2 is an early example of what such an observation would look like. Overlapping of the output spectra is mitigated by using a blocking filter with a narrow pass band. So instead of scanning a solar image with one slit, using N slits reduces the required scan time to 1/N, a significant efficiency improvement. There are dozens of other clever ideas to improve efficiency based on large format detectors. Many of the ideas use fiber optics, microlens, micropolarizer, and microfilter arrays and other modern optical technologies. It is almost inevitable that a new generation of solar magnetographs will utilize some of these efficiency improvements.

Detector performance is key to magnetic field measurement. Fortunately, detector technology continues to improve rapidly. High efficiency sometimes requires conflicting properties. These include: high quantum efficiency over a large range of wavelengths, large pixel count, large dynamic range, excellent linearity, fast readout, no image smear, excellent modulation transfer function, no image lag (signal retention between successive readouts), no etalon fringing, and, if possible, pixel level on-chip demodulation capability. Manufacturers

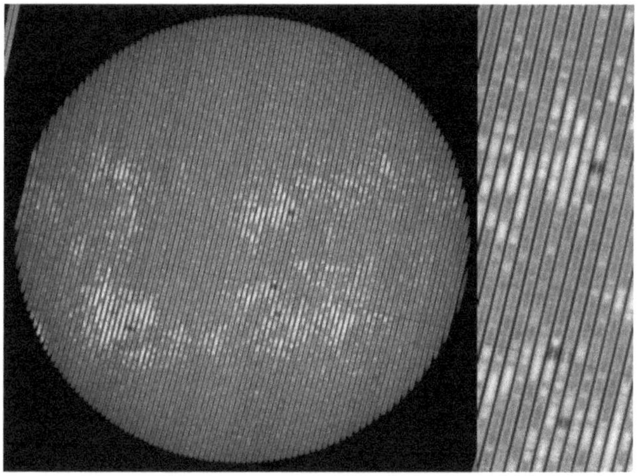

Fig. 2 An early example of trading sparse spatial sampling to gain spectral coverage. Spectroheliogram and zoom showing individual K-line spectra at numerous positions of the entrance slit (Meudon, 1938 August 8)

have devised various architectures based on converting photons to charges in a detection layer followed by conversion of the charges to a measurable voltage. In some devices the detection layer is stacked above a layer of readout electronics. In others all the components are essentially in one layer. In some devices (e.g. CCD, most CMOS) the readout process destroys the original charge while in others (e.g. CID, DEPFET, ZIMPOL, sCMOS) the charge is shuffled and not necessarily destroyed immediately by the readout process. To obtain high speed, the sensor area is divided into smaller regions and readout is accomplished simultaneously through multiple channels.

2.5 State of the Art

Figure 3 summarizes the major domains of solar spectropolarimetry, and some telescopes/instruments in use (or planned in parentheses) placed close to the domains for which they were designed. This is not a full listing (for example, a few years ago there were more than 20 instruments capable of measuring the chromospheric magnetic field). The figure is intended to suggest the large range of capabilities available for measuring the solar magnetic field limited primarily by financial and personnel resources and the advance of technology.

3 Ground-Based Techniques

Ground based observations have several advantages over space- or balloon-borne observatories. Their large aperture allows for highest spatial resolution, they can be equipped with complex instrumentation with a diverse selection of observing modes in a multitude of spectral lines, elaborate calibration schemes can be applied, and virtually no limits on data rates do exist. The only impediments to observing from ground in past few decades were the atmospheric seeing effects and the diurnal cycle. The latter has been very successfully mitigated by establishing globally distributed observing networks with identical instrumentation such as GONG, which can be run for extremely long time periods by constant maintenance and upgrades. On the other hand the techniques that are used to overcome seeing effects has seen tremendous progress. In the following we will give examples of the current state-of-the-art in high resolution imaging techniques, narrow band two-dimensional spectrometers, and high precision polarimeters.

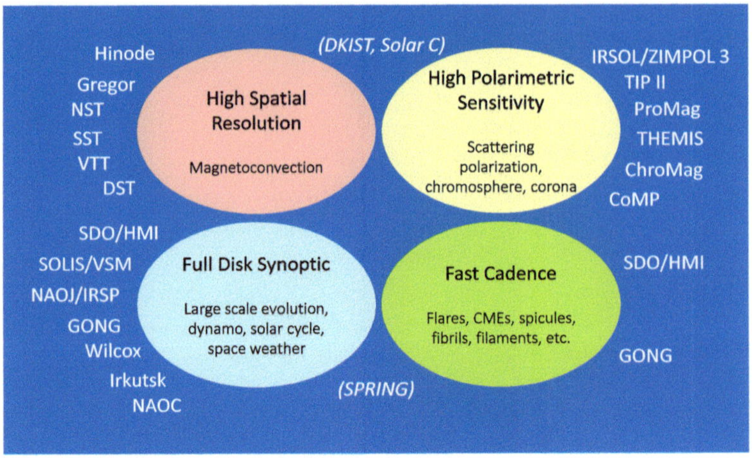

Fig. 3 Domains of current solar spectropolarimetry and a sampling of telescope/instrument names clustered to the domain for which they are best suited

3.1 Longitudinal Magnetometry: Still a Viable Tool

The term magnetometry has been loosely used in solar physics for the measurement of magnetic fields in the solar atmosphere. However, in reality we can only remotely sense the solar magnetic field by analyzing the photons emitted from the solar atmosphere and applying diagnostic techniques such as Zeeman and Hanle effect to the analyzed photons. Depending on the extent of analysis of the collected photons one can infer either complete vector or just longitudinal component of the magnetic field. The strength of longitudinal magnetometry is its simplicity in interpreting the measured signal, resulting from the almost linear dependence of the longitudinal Zeeman effect to the line-of-sight component of the magnetic field (Stenflo 2015, this issue). Complex, model dependent techniques, like inversions, can often be avoided, and still the fine structure of the solar atmosphere can be retrieved (Stenflo 1994). Babcock (1953) used the polarization properties of the Zeeman components of the spectral line to measure the projection of the magnetic field vector along the line-of-sight. The instruments that use this method to map the longitudinal magnetic field of the Sun are popularly known as Babcock type magnetographs. Longitudinal magnetographs have been used extensively for studying the distribution and evolution of magnetic flux on the Sun at a variety of spatial and temporal scales. Modern longitudinal magnetographs based on two dimensional image sensors exist in various flavors (e.g., SOLIS, *Hinode* SP/NFI, GONG, MDI), and are still very popular for variety of studies such as quiet sun magnetism, flux emergence, active region evolution, magnetic helicity flux estimation, flare related changes, polar field strengths, coronal field extrapolations and so forth.

3.2 The Age of Vector Magnetic Field Measurements

In contrast to longitudinal measurements the transverse fields are more challenging to measure and, even more, to interpret. The Zeeman effect induced linear polarization signal is proportional to the square of the transverse field strength and is therefore weaker compared to the circular polarization signal which is proportional to the longitudinal field strength. This different response of the measured signals to the orientation of the magnetic field results often in ambiguous interpretations. For example, the distribution of the magnetic field

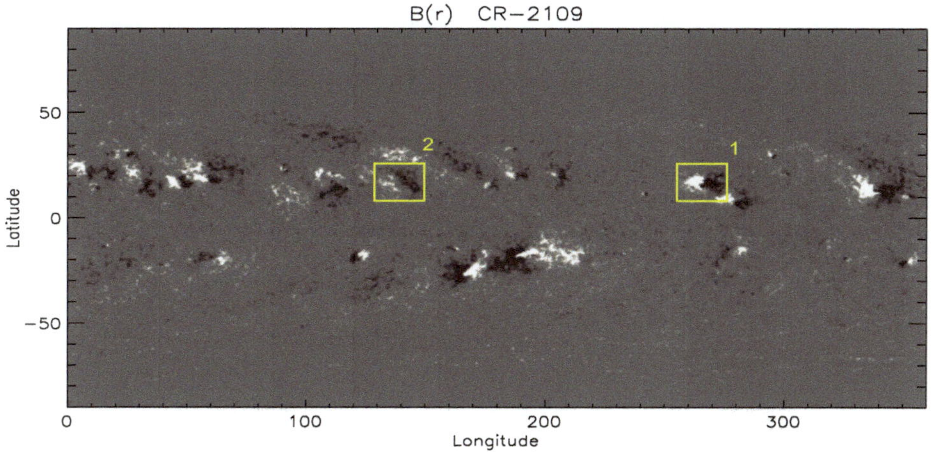

Fig. 4 Synoptic Carrington map of the vector magnetic field component B_r synthesized using full-disk SOLIS/VSM vector magnetograms is shown for CR-2109 (the B_θ and B_ϕ components are not shown). The map is scaled between ±100 G with positive (*white*) values of B_r pointing upward. The *yellow square boxes* '1' and '2' show examples of a typical compact active region and a diffuse decaying region, respectively

orientation in weak field regions is still under debate as to whether the weaker fields are predominantly horizontal or vertical (Borrero and Kobel 2013). In active regions, especially near the polarity inversion line, the transverse component of the field becomes strong, often accompanied with a highly non-potential or sheared magnetic field topology. The presence of free or excess magnetic energy above the potential field energy is a necessary condition for the occurrence of energetic phenomena such as flares and coronal mass ejections (CMEs). It is therefore necessary to measure the full magnetic field vector in solar active regions in order to estimate the non-potentiality of the magnetic field and its evolution leading to flare and or CME. Further, it is now well known that solar active regions show a weak hemispheric bias in their magnetic field twist (Pevtsov et al. 1994). The twist of the magnetic field can be derived by using various proxies such as force-free parameter α, current helicity, magnetic helicity, chirality of filament barbs, sense of rotation of chromospheric superpenumbral whirls around sunspots.

Most of the above mentioned proxies require the measurement of the full magnetic vector field. The origin of the helicity in active regions and their hemispheric bias is still an open question. Various mechanisms have been proposed to explain it, for example its origin in the solar dynamo itself, or during the rise of buoyant flux tubes via their interaction with turbulent convection, or after the flux emergence due to reconnection near the polarity inversion line. Some authors think that the solar dynamo itself is responsible for the generation of helicity, others believe that the turbulent convection interacting with the buoyant flux tubes as they rise towards the surface is the culprit, and a third group thinks that helicity is generated via reconnection processes when the emerged flux interacts with the ambient magnetic properties of the solar corona.

A systematic monitoring of vector fields on the solar disk is therefore needed to understand the relation between magnetic twist and other observables, as well as to build up statistics. Synoptic observations of the magnetic field vector over the full disk derived from full Stokes polarimetry in the photospheric Fe I 630.1 nm line are routinely done by the SOLIS/VSM instrument. Figure 4 shows the synoptic Carrington map of the photospheric field (radial component B_r) generated using the daily fulldisk measurements by SOLIS/VSM (Gosain et al. 2013). Such maps are useful, for example, for monitoring the hemispheric helicity trends (Gosain et al. 2013), their relation with the kinetic helicity of subsurface flows derived from helioseismology (Komm et al. 2014; Komm and Gosain 2015), global nonlinear force-free field (NLFFF) extrapolations (Tadesse et al. 2014a, 2014b) and so on. In addition to synoptic vector field measurements of the full disk it is also desirable to have highly sensitive magnetic field measurements in certain areas of the solar disk such as polar regions and beneath non-active region filaments that also can erupt and be associated with a CME.

3.2.1 Filter-Based or Imaging Instruments

As mentioned in Sect. 2 the magnetic field is determined from multi-dimensional spectropolarimetric observations: for every pixel of a 2-dimensional map the spectrum and the temporal evolution should be recorded, resulting in a 4-dimensional data cube for each of the four Stokes components (I, Q, U, V). Since detectors can only record two-dimensional images, a choice of the order of the sampling must be made. Two different approaches are commonly taken, the filter-based and spectrograph-based measurement. Both are eventually leading to same output but are fundamentally different in that the imaging instruments sample two spatial dimensions simultaneously and spectral samples are obtained sequentially, while in the case of spectrograph-based instruments one spatial dimension (along the slit) and the spectral dimension are obtained simultaneously and the second spatial dimension is obtained sequentially by scanning the slit laterally. The fourth dimension of the data cube, time, is achieved by the repetition of these measurements.

Examples of current filter type instruments are the Imaging Vector Magnetograph (IVM, Mickey et al. 1996), the Interferometric BIdimensional Spectrometer (IBIS, Cavallini 2006), the Solar Vector Magnetograph (SVM, Gosain et al. 2006), the CRisp Imaging SpectroPolarimeter (CRISP) at the SST (Scharmer et al. 2008), and the GREGOR Fabry-Pérot Instrument (GFPI, Denker et al. 2010). Tunable imaging spectrometers come in various flavors. For example, HMI uses a Michelson Interferometer as a filter, and CRISP, IBIS, GFPI, IVM, and SVM use tunable air-gap Fabry-Pérot etalons in single or tandem configuration. The rapid tunability of piezo-mounted air-gap etalons and the high throughput are main advantages of these devices. Further, Fabry-Pérot etalons have the advantage that, unlike for example birefringent filters, they do not contain linear polarizers. This makes dual beam polarimetry possible (see Sect. 3.3) by placing the polarizing beam splitter close to the focal plane of the cameras. Alternatively, one can also choose solid electro-optic crystal based etalons such as lithium niobate ($LiNbO_3$) crystal wafer based etalons (Mathew et al. 1998; Debi Prasad and Gosain 2002; Martínez Pillet et al. 2011) where the highly polished thin wafer is coated with highly reflective coatings. Here, the wavelength tuning is based on the change of the refractive index and the thickness of the $LiNbO_3$ crystals by applying different voltages. The advantage of these solid etalons compared to normal, air-spaced etalons is their high refractive index ($\approx$2.28), allowing for smaller etalon diameters for a given field-of-view. For instance, the largest air-gap etalons (15 cm aperture) ever made and used in the Improved Solar Observing Optical Network (ISOON, Neidig et al. 1998) could be replaced by 6.5 cm aperture $LiNbO_3$ etalons of equivalent throughput. On the other hand their limitations are that they are fragile (specially with larger apertures), have higher absorption losses, and are susceptible to damage if tuned faster than 1000 Volts/s.

Under the assumption that the acquisition of a spectropolarimetric data-cube is faster than the evolution timescale of the solar structures and that of the atmospheric seeing, filter-based instruments provide a snapshot of the magnetic field over the field-of-view. Thus, such

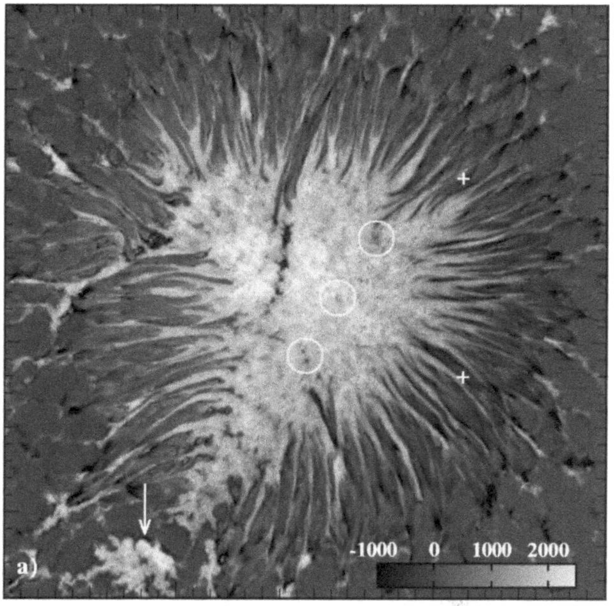

Fig. 5 Longitudinal magnetic field for a $35'' \times 35''$ large FoV at $\tau_c = 0.95$ derived from NICOLE inversions show the presence of opposite polarity fields in most parts of the penumbra. Such opposite polarity fields are absent in higher layers (e.g., at $\tau_c = 0.09, 0.01$). The *long white arrow* points towards the disk center. Figure reproduced from Scharmer et al. (2013), Fig. 4(a)

measurements are useful for studies of the temporal evolution of magnetic fields at high cadence, ideal for studies about, e.g., flares or waves. Until the beginning of the new millennium, the magnetic field inference from these observations had some limitations in terms of spectral resolution, sparse wavelength sampling, or spectral purity (Lites et al. 1994). However, significant progress has been made since then in improving the performance of these instruments in various aspects, for example by combining two or more etalons in tandem configurations to increase the spectral resolution. Instruments like CRISP, IBIS, TESOS and more recently GFPI, therefore now allow to obtain depth dependent, high spatial resolution vector magnetograms. As an example, Fig. 5 shows the longitudinal field obtained from SST/CRISP observations with twice the angular resolution of Hinode/SOT (reproduced from Scharmer et al. 2013). Such observations are possible by combining several techniques, like adaptive optics, post-facto image reconstruction algorithms and the high spectral resolution and throughput of the CRISP instrument. Due to short acquisition time of the spectropolarimetric dataset from such instruments it is possible to acquire high quality observations even during brief moments of good seeing.

3.2.2 Stokes Spectrographic Observations

The spectrograph based instruments sample the spectral dimension simultaneously. Spatial scanning by moving the slit over the field-of-view is needed to build two-dimensional maps. Prominent examples of spectrograph based imaging polarimeters are ASP (Lites 1996), SOLIS/VSM (Keller and Solis Team 2001), SPINOR (Socas-Navarro et al. 2006), TIP (Collados et al. 2007), GRIS (Collados et al. 2013). In order to expedite the scanning time, a few variants have been developed, such as the multi-slit spectrograph, that uses multiple slits separated far enough so that the usable range of spectra can be recorded without overlap (e.g., FIRS, Jaeggli et al. 2010). Here, an order-sorting pre-filter must be used to isolate the wavelength range of interest. Other variants can be grouped into two classes: (1) integral field spectrographs (IFS), where the 2D field-of-view is converted to a one dimensional arrangement feeding a regular slit spectrograph, using either an image slicer (Bowen 1938) or a bundle of optical fibres (e.g. DLNIRSP, Elmore et al. 2014), and (2) microlens arrays producing multiple point images out of an extended field-of-view, again feeding a spectrograph with the multiple point sources, oriented such that the dispersion avoids overlapping of the individual spectra (microlens fed spectrograph, M. van Noort, private communication).

Simultaneous wavelength information together with typically higher spectral resolution, higher signal-to-noise ratios and finer wavelength sampling are the main benefits of these instruments. This is of particular advantage when inferring the accurate field strength and orientation in the pixel, including the height stratification of the physical parameters in the line forming region of the solar atmosphere. Sit-and-stare observing modes cater well to the studies of dynamic phenomena, ideally complemented with high temporal resolution slit-jaw imaging. With an IFS the dynamical studies can be extended to a small but 2-dimensional field-of-view. The disadvantage of spectrograph based instruments is that they are not well suited for observations of a large field-of-view such as a full AR on time-scales of minutes, which is required by some studies such as flare research.

The ideal solution, therefore, would be to take hybrid approach where the high-quality reflective slit-jaw projection of the spectrograph is used to feed a filter based spectropolarimeter. This allows for information over whole field-of-view with high cadence and limited spectral resolution while the smaller field-of-view sampled by slit (or an IFS) has more detailed spectral information. Real-time processing of the imaging data can help to rapidly re-position the slit to an interesting region, e.g., a flaring kernel or an emerging flux region within the slit-jaw field-of-view.

3.3 Overcoming Seeing

One of the biggest challenges for any ground-based measurements is the atmospheric seeing, which changes rapidly within a few tens of milliseconds. The atmospheric seeing can be one of the major sources of spurious polarization signals in sequential measurements (Lites 1987; Leka and Rangarajan 2001). Dual beam measurements, where the orthogonal polarization states are separated by means of a polarizing beam splitter and recorded simultaneously, are essential for reducing the seeing effects in measurements. Unlike single beam setup, the dual beam setup fully utilizes 100 percent of the photons available for polarimetry. However, different optical paths for the measurement of the two polarization states in a dual beam setup might introduce a systematic differential gain/aberrations error, this puts strict constraints on the optical quality of the two optical paths in such a system. Alternatively, if single beam setup is preferred, a very fast modulation scheme with a matching fast camera readout must be used, in order to allow a complete modulation cycle before the seeing changes. However, polarization measurements require accumulation of various modulation cycles to build up the desired S/N. This usually results in a loss of spatial resolution, since frames taken under variable seeing conditions must be combined to acquire the desired S/N ratio. In order to preserve or restore the spatial resolution in such observations we then resort to the high angular resolution techniques as described below.

3.3.1 Image Correction in Real Time

The optical systems which can compensate for distorting effects of atmospheric seeing in real-time are known as adaptive optics (AO) systems. Such systems typically have a wavefront sensor (WFS), which provides information on the wavefront distortions. This information is used to control in real-time a corrector optics, typically a combination of tip-tilt mirror and a deformable mirror (Rimmele and Marino 2011; Scharmer et al. 2000). The advantage of such systems is that it (i) enhances the angular resolution in the images, and (ii) allows us to make longer exposures to attain high S/N ratio in polarimetric measurements. The latter benefit relieves the requirement for extremely fast camera readouts in order to freeze the atmospheric seeing. However, due to the highly variable nature of seeing and due to time constraints in AO systems, only a limited number of higher order modes can be corrected. Thus, the residual seeing effects still remain, introducing seeing-induced cross-talk (Krishnappa and Feller 2012). Higher modulation frequencies help to reduce the seeing-induced crosstalk (Judge et al. 2004; Gandorfer et al. 2004; Ramelli et al. 2010; Casini et al. 2012; Krishnappa and Feller 2012).

The seeing is generally expressed with the Fried parameter, r_0, which is a measure of the spatial scale over which the wavefront can be considered approximately distortion-free (rms distortion of less than 1 radian). The performance of AO systems depends heavily upon the basic seeing conditions (r_0), the performance being better when r_0 is larger. Another property of AO systems is that the quality of real-time corrections is good near the center of isoplanatic patch (typically $\approx$1–2″) in the FoV, the patch which is used for the wave front sensing. A correction over a field-of-view much larger then the isoplanatic patch can be achieved with so called multi-conjugate adaptive optics (MCAO) systems (Rimmele et al. 2010). First prototypes of MCAO systems are currently being developed (e.g., Schmidt et al. 2014).

Observations in infrared wavelengths, such as the Fe I 1.56 μm lines, are best suited for magnetometry, since the magnetic sensitivity scales linearly with the wavelength. A further benefit of observing in this wavelength region is that the atmospheric seeing varies with wavelength such that r_0 is proportional to $\lambda^{6/5}$, i.e., the isoplanatic size is much larger at longer wavelengths.

3.3.2 Image Reconstruction: Post Processing Techniques

To further improve the quality of AO corrected data, offline image reconstruction methods, like speckle imaging (von der Luehe 1993), phase diversity (Lofdahl and Scharmer 1994), and multi-object multi-frame blind deconvolution (MOMFBD, van Noort et al. 2005) are applied. These techniques are not limited by isoplantism and can be used to improve image quality over a large field-of-view. Typically they require separate broadband channel images, recorded at a similar wavelength and strictly simultaneously with the narrowband channel images. The high S/N ratios in these broadband channel images deliver the additional information required for the reconstruction of the narrowband channel images.

Such reconstruction techniques are essential when narrow band imaging polarimetry is performed as the seeing introduces spatial to spectral crosstalk, and as a consequence spurious polarization signals. Reconstruction methods achieve a good alignment between the pixels of sequentially taken, narrowband images (van Noort et al. 2005), which helps avoid spurious polarization signals. The reconstruction techniques are, however, prone to errors or lead to spurious features when the noise in the data is not properly accounted for, and therefore require accurate calibration of parameters. For further details on speckle polarimetry reader is referred to Keller and von der Luehe (1992) and references therein, and for MOMFBD techniques we refer to van Noort et al. (2005) and Löfdahl and Scharmer (1994).

3.4 Ground-Based Observing Networks

A network of identical instrumentation distributed geographically around the globe in order to achieve continuous observations of the Sun has been a very successful concept. The long-term operations of networks, such as Global Oscillations Network Group (GONG, Harvey et al. 1996) or the Birmingham Solar Observing Network (BiSON, Chaplin et al. 1996), have led to a continuous data series of uniform quality. The benefit of running such networks are (i) redundancy of sites (e.g., if one site fails there is still flow of data from other nodes), (ii) a possibility of cross calibration and merging of data, (iii) easy upgradeability and maintenance of instruments. Earlier networks were designed for helioseismology studies, which require long, uninterrupted time series of solar oscillation measurements to achieve high frequency resolution. Nowadays, emphasis is put on long term global solar magnetic field measurements for solar cycle studies and continuous monitoring of vector magnetic fields in solar active regions for space weather prediction research. A proposal to set up a new network of identical instruments designed for measuring fulldisk multi-height velocity and magnetic field vector measurements along with high resolution context imaging in different wavelength bands, such as white light, G-band, H-alpha and Calcium K, is currently funded by European Research Council (ERC). This new network is called Solar Physics Research Integrated Network Group (SPRING) (Hill et al. 2013).

3.5 Examples of the State of the Art and Current Status

The magnetic field observations got a big boost in the last decade owing to the space based *Hinode*/SP and *SDO*/HMI instruments. The current next-big-thing in solar magnetic field measurements is expected from the 4-meter aperture Daniel K. Inouye Solar Telescope (DKIST, Rimmele and ATST Team 2008; Tritschler et al. 2015), currently being built on Maui/Hawaii. Equipped with adaptive optics and very sensitive infrared detectors, DKIST will provide magnetic field measurements with unprecedented sensitivity and spatial resolution. However, since the number of photons per diffraction limit resolution elements remains independent of the telescope aperture size, the telescope will be very sensitive when used as a so-called "photon bucket", i.e., operated at a resolution coarser than its diffraction limit. Another series of large aperture solar telescopes have become fully operational in the last decade: the 1.6 m New Solar Telescope (NST, Goode et al. 2010) at Big Bear, the 1.5 m GREGOR telescope on Tenerife (Schmidt et al. 2012), and the 1 m SST on La Palma (Scharmer et al. 2003).

All of these telescopes are equipped with fully operational AO systems, while the development of MCAO systems is being carried out in parallel. In the near future, these MCAO systems will allow for magnetic field measurements at a higher spatial resolution over a much larger field-of-view than conventional AO systems deliver (Rimmele et al. 2010). Also, recent demonstration of ground-layer solar AO by Ren et al. (2015) looks promising.

Specialized detectors designed for high speed polarimetry are capable of reaching a polarization sensitivity of the order of 1×10^{-5} of continuum intensity. Examples of such detectors are as below:

- Charge shuffling cameras, e.g., the Zurich Imaging POLarimeter (ZIMPOL), contain hidden buffers for caching the charges on the chip thus allowing to buildup sufficient signal before the actual readout is performed (Povel et al. 1994). Further, since the same pixel is exposed for different modulation states, no differential gain variations are present. The latest version of the ZIMPOL system employs sensors with improved electronics and high sensitivity. New designs/versions of ZIMPOL based on CMOS detectors are also being planned (Ramelli et al. 2010).
- The Fast Solar Polarimeter (FSP, Feller et al. 2014) uses pn-CCD cameras with extremely low readout noise while delivering high frame rates. A polarization sensitivity of below 1×10^{-4} of the continuum intensity has already been achieved. The development of these

cameras for larger formats (1 megapixel) is a current project at the Max Planck Institute for Solar System Research.

4 Space-Based Techniques

4.1 Motivation for Measurement of Photospheric Magnetic Fields from Space

Owing to the circumstance that many aspects of photospheric magnetism have remained largely unresolved, the study of solar magnetism has resulted in an ongoing quest for observations of higher angular resolution. A fundamental scale of the solar atmosphere is the density scale height (h_ρ, of order 150 km in the photosphere). The scale height is the characteristic length for many of the most important hydrodynamic and radiative processes. Resolving photospheric magnetic fields on this scale has been a goal of solar physics for several decades, and the cleanest way to arrive to a spatial resolution at h_ρ would be observations from a space-based platform. Unfortunately, the cost of deploying a telescope of the necessary size (diameter of order 1 m) in space has been prohibitive to realization of this goal. Parallel efforts from ground-based observatories—larger telescopes, adaptive optics, advanced image restoration techniques—are pushing observations closer to this long-sought goal. As a result, the quest for the highest angular resolution from space might now be relegated to a secondary status relative to the benefits of uninterrupted observations with uniformly *very high* angular resolution. Provided that continuity of observations is achieved, many extremely important science problems of solar physics may be addressed effectively without completely resolving the magnetic fields with resolution equivalent to h_ρ.

Another motivation for space-based observations is access to the hemisphere of the Sun not visible from Earth (or near-Earth orbit), including detailed study of magnetic fields at the solar poles. Imaging instrumentation has been flown (e.g., the *STEREO* mission), but so far no observations of photospheric magnetic fields have been carried out from such a vantage point, although one such mission is in development (see Sect. 4.5).

A less obvious motivation for space observations of solar magnetism is support of coordinated observing campaigns. Science objectives that require diverse observations from multiple observatories (both space- and ground-based) have a much higher chance of success because the space-based components of the campaign have a predictably high probability of success.

4.2 Challenges of Space-Based Magnetic Field Measurements

The development of any space mission is a costly and lengthy process, often necessitating international collaboration in order to make the program possible within budgets of the individual participating countries. International collaborations increase the complexity of management of a space instrumentation program, but they also broaden the pool of available scientific and engineering talent. Many of the missions providing measures of photospheric magnetic fields (as summarized below) result from significant international collaboration.

The technical challenges of space instrumentation are significant in comparison to ground-based observing. Weight and power are always at a premium, as is physical size. Space instrumentation must be resilient under extremes of temperature and radiation exposure never encountered on the ground. The vacuum environment can lead to contamination of optical surfaces, so careful attention must be paid to the outgassing properties of materials for space flight. Furthermore, complex instrumentation must be ultra-reliable and multiply redundant to reduce the possibility of a single-point failure that would bring an end to the mission. These challenges of space instrumentation drive the high development cost and result in significant compromises in design. As a rule, the simpler the design and the fewer the moving parts, the more reliable the instrument becomes. As a result of simplicity, space instrumentation usually has far less flexibility of operation than comparable ground-based instrumentation.

To date, space instrumentation for measurement of photospheric magnetism has carried out measurements of the Zeeman effect in the visible spectrum only. There is little motivation for observing photospheric fields in ultraviolet lines because the ratio of magnetic splitting to the Doppler width decreases linearly with decreasing wavelength. Furthermore, several other factors render observations at ultraviolet wavelengths more difficult, from both scientific and technical standpoints. Conversely, the advantages of observing in the infrared are countered by the cost of deploying proportionately larger telescopes in order to achieve spatial resolution equivalent to that of an instrument operating in the visible.

Solar observations require high angular resolution, so the measurement of magnetic fields over the spatial extent of even a single active region demands very high data rates. Full Stokes polarimetry at multiple wavelengths is necessary for measurement of the vector magnetic field, hence, at least four times the data volume of intensity-only measurements are required. All of this data must be transmitted to the ground at a cadence appropriate for monitoring the evolution of the target solar features. In order to confront the data requirements in the face of limited telemetry, the polarization data are usually compressed on-board. The compression leads to some loss of information that would not be present in equivalent ground-based measurements. Even with compression, limitations imposed by telemetry remain a limitation to the quantity of data that may be acquired during any given time period.

4.3 Past and Ongoing Space Missions

To date there have been only three successful space missions capable of measurements of the photospheric magnetic field. Two of those, *SoHO* and *SDO*, carried synoptic instruments with full-disk capability intended for precision Doppler velocity measurements for helioseismology in addition to sensing the magnetic field. The third mission, *Hinode*, has instrumentation optimized for high-resolution study of magnetic fields as a primary objective. These three missions are described in more detail below, and an example of one significant scientific accomplishment is provided for each.

4.3.1 Solar and Heliospheric Observatory/Michelson Doppler Imager (SoHO/MDI)

The *SoHO*/MDI instrument (Scherrer et al. 1995) was primarily intended for helioseismology, but fortunately the capability for measurement of circular polarization, and hence the line-of-sight component of the magnetic apparent flux density $\mathrm{B}^{L}{}_{\mathrm{app}}$, was retained in spite of pressure to "de-scope" the instrument owing to budgetary considerations. *SoHO*/MDI operated from December 1995 to April 2011 and therefore provided coverage of the magnetic field spanning more than one complete solar cycle.

The optical layout of *SoHO*/MDI is shown in Fig. 6. The heart of the instrument, the Michelson interferometers, had heritage from the Fourier Tachometer (Brown 1981), a ground-based instrument for helioseismology. The rotatable waveplates between the interferometers allowed the device to tune over the 0.5 Å bandpass of the Lyot filter centered on the Ni I line at 6768 Å. The full-width half-maximum (FWHM) of the combined system was

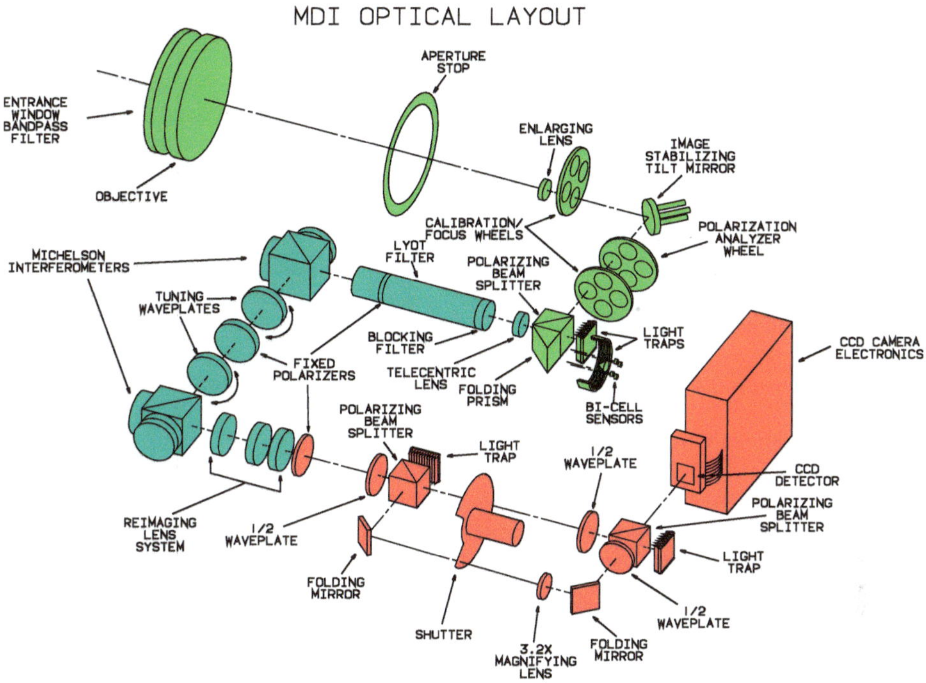

Fig. 6 The optical layout of the *SoHO*/MDI instrument is shown. The wavelength isolation was accomplished by a pair of Michelson interferometers preceded by a Lyot filter. Circular polarization analysis was accomplished by alternately rotating quarter-wave retarders into the beam with the polarization analyzer wheel. Diagram taken from Scherrer et al. (1995)

listed at 94 mÅ, and in typical operation *SoHO*/MDI sampled the line at five wavelengths separated by 75 mÅ.

The emphasis on helioseismology led to several aspects of this instrument that were not optimal for precision measures of photospheric magnetic fields: performing the polarization analysis in by alternately inserting separate quarter-wave plates, the use of a spectrum line with less than optimal magnetic sensitivity, a spectral resolution a factor of 2–3 times greater than the typical Doppler width of the Ni I line in the photosphere, and limited angular resolution (4 arcsec in full-disk mode, 1.2″ in high resolution mode). That being said, owing to its station at the first Earth-Sun Lagrangian point L1, *SoHO*/MDI was the first instrument to provide an uninterrupted sequence of longitudinal magnetograms of the Sun at a regular cadence and uniform quality. This instrument facilitated the study of the evolution of active regions in a way never before realized.

The magnetogram data from *SoHO*/MDI were (and are still) being used extensively to understand solar phenomena, as witnessed by the more than 800 abstracts citing the MDI instrument between 1996 and 2015. The numerous applications of the data from this instrument facilitated such diverse studies as extrapolation of photospheric fields to compare with structures in the chromosphere and corona, studies of active region evolution including measures of injection of helicity into the upper solar atmosphere, polar fields, flare-induced magnetic field changes, the decay of active regions, and many more.

The science made possible uniquely by this space experiment is illustrated by the statistical study of the orientation axis of bipolar regions (Stenflo and Kosovichev 2012). They examined over 73 000 active regions from 1995–2011. This selection included regions with bipolar strength (or flux) ranging from just above the typical flux of ephemeral active regions to the largest observed regions. They found that the distribution of tilt angles (Joy's law) with solar latitude obeys the same relationship independently of the strength of the dipole (or net flux) of the region. This behavior led the authors to conclude that the tilt of active regions is established at the source of the dynamo giving rise to the active regions, not in the buoyant rise through the convection zone as was previously hypothesized. Furthermore, the authors showed examples (representing just a few percent of the total number of active regions) that did not obey the Hale hemispheric polarity law. The authors then conclude that these exceptions "...rule out the possibility of well-defined, coherent toroidal flux systems as a source of all active regions, even the large ones".

4.3.2 Solar Dynamics Observatory/Helioseismic and Magnetic Imager (SDO/HMI)

The Solar Dynamics Observatory (*SDO*), launched in February 2011, carries as one of its complement of instruments the Helioseismic and Magnetic Imager (HMI, Scherrer et al. 2012; Schou et al. 2012). This instrument is so similar in design to the *SoHO*/MDI that no optical layout of the system is provided herein, but it incorporates a number of features that represent significant upgrades from *SoHO*/MDI:

- A different spectral line is observed: Fe I 6173 Å. Not only is this line isolated within a fairly clean region of the spectrum to enable precision helioseismology, but it is also a normal Zeeman triplet with high magnetic sensitivity.
- Polarization analysis is carried out with rotating retarders mounted permanently in the beam. This arrangement allows for a higher precision polarimetry than *SoHO*/MDI. Additionally, these waveplates allow for measurement of the full Stokes vector, so the instrument is capable of vector magnetometry.
- The angular resolution of the instrument is $1''$ ($0.5''$ pixels), with the field-of-view covering entire solar disk.
- Two CCD cameras are used, one is dedicated to helioseismology, the other is dedicated to polarimetry. The data system operates at a higher cadence than that of *SoHO*/MDI.

As the name of the instrument implies, this instrument has a much greater scientific emphasis on magnetometry than the *SoHO*/MDI did. Besides continuing the helioseismic function beyond the end of the *SoHO*/MDI program, *SDO*/HMI brings a new dimension in continuous *vector* magnetometry of the full disk. This instrument makes it possible, for the first time, to perform synoptic studies of the vector magnetic field of every active region on the disk. The work of Liu et al. (2014) serves as one example of a study involving the *SDO*/HMI vector magnetogram data of many active regions. In that work they sought to validate, with data of higher resolution and uniform quality, the rather weak hemispheric preference for sign of the twist of active region fields as reported by several authors who used ground-based data. As a measure of twist they adopted a $B_z{}^2$-weighted average of the force-free parameter α (obeying $\nabla \times \mathbf{B} = \alpha \mathbf{B}$) over the active region. Figure 7 shows the result of their analysis of 151 active regions appearing during the first 3.5 years of the current activity cycle. This study demonstrates a very clear hemispheric preference for sign of the magnetic helicity. The high cadence of *SDO*/HMI vector magnetic field measurements allowed them to select for study observations of active regions as they traversed the solar central meridian, thereby avoiding any bias that could arise as a result of viewing angle with respect to solar longitude.

Being a very similar cousin to the *SoHO*/MDI instrument, the *SDO*/HMI instrument is optimized for helioseismology rather than precision polarimetry. As a result, it has some limitations for measurement of the vector magnetic field. Firstly, the instrument has a rather low

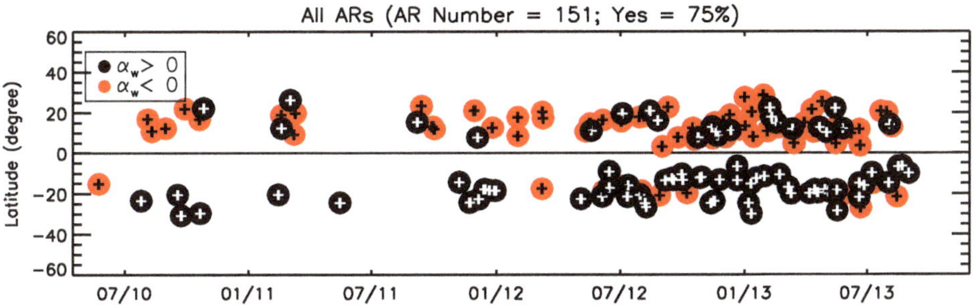

Fig. 7 The latitudinal distribution of the sign of the $B_z{}^2$-weighted average α_w of the force-free parameter α is indicated as a function of solar latitude and time during the rise phase of Solar Cycle 24. The figure results from analysis of vector magnetograms obtained with *SoHO*/MDI. Sign of this proxy of magnetic helicity is indicated by color. For this distribution of 151 active regions, 75 % show the hemispheric preference for helicity suggested by previous studies of ground-based data. Diagram adapted from Liu et al. (2014)

duty cycle and therefore it does not use the available photons efficiently. The signal-to-noise ratio of typical polarimetric observations is such that studies of the vector magnetic field in quiet regions are extremely difficult at best. Second, the instrument is a single-beam polarimeter with a slow polarization modulation cycle. This means that there will be crosstalk among the Stokes parameters arising from uncompensated image motion and evolution of the solar scene. Third, only one spectral line is sampled, and that with six samples of the line profile. Each wavelength sample is subject to a filter bandpass of 76 mÅ. Instruments optimized for Stokes polarimetry sample the complete spectral profile of two or more lines having different sensitivities to the Zeeman effect, and do so with a spectral resolution comparable to the Doppler width of the line in question (typically 30–40 mÅ for photospheric lines). The compromises of the *SDO*/HMI spectral sampling and resolution result in greatly reduced sensitivity to subtle features of the Stokes profiles that would otherwise assist in precision measures of the magnetic field vector. They also prevent the accurate extraction of magnetic filling factors in the data reduction process, so the standard *SDO*/HMI data inversions assume unit filling factor. For regions outside of sunspots, the extracted field strength is then actually a measure of the average field strength over the resolution element—a measure that is lower than the actual field for the usual case where intense magnetic elements are not resolved spatially. Furthermore, the inferred field inclination will be more transverse to the line-of-sight than the actual field. Thus one must anticipate significant systematic errors in the inferred magnetic field vector for regions outside of sunspots. These errors may or may not be significant depending upon the nature of the science question being addressed. For studies that are highly sensitive to such errors, one should use data from instruments specifically optimized for Stokes polarimetry, such as the Solar Optical Telescope on the *Hinode* mission described in the following. Furthermore, limited spectral sampling hinders the extraction of vertical gradients of physical parameters that might otherwise be available to analysis in finely-sampled Stokes profiles, and the limited wavelength range of these samples precludes measurement of strongly Doppler-shifted fields or very strong Zeeman splitting in sunspot umbrae.

4.3.3 Hinode SOT/FG and SOT/SP

The *Hinode* mission (Kosugi et al. 2007) is the first space mission carrying instrumentation specifically optimized for precision Stokes polarimetry. For this purpose it carries the 50 cm

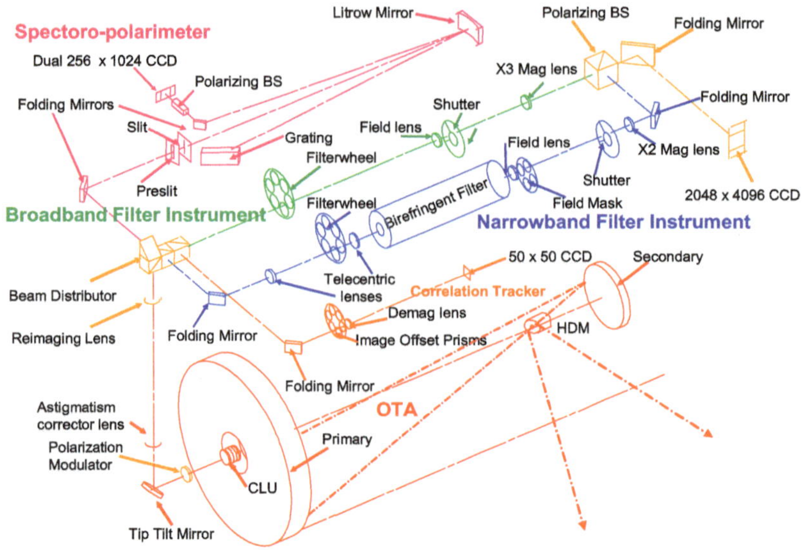

Fig. 8 The optical layout of the *Hinode* SOT identifies the five major optical systems with different colors. Of particular interest for this paper are the Narrowband Filter Instrument (*blue*) and the Spectro-Polarimeter (*magenta*), both of which are specifically designed to perform precision polarimetry for measurement of solar magnetic fields. For a detailed description of the Spectro-Polarimeter see Lites et al. (2013). Diagram taken from Tsuneta et al. (2008)

Solar Optical Telescope (SOT: Tsuneta et al. 2008) with capability of both spectrographic observations with the Spectro-Polarimeter (SP: Lites et al. 2013) and filtergraphic observations with the Narrowband Filter Instrument (NFI). The spectrographic observations have the advantages that they provide measures of the spectral Stokes profiles simultaneously at all wavelengths, so that the integrity of the profiles in wavelength is retained, and also that solar events with high Doppler shift do not fall outside of the sampled wavelength range. These advantages are countered by the non-simultaneity of images of the solar scene that must be built up by stepping the narrow spectrograph slit across the image. One may have the best of both worlds with simultaneous filtergraphic and spectrographic Stokes polarimetry as implemented in the SOT/SP/NFI instrumentation.

The optical layout of the SOT Focal Plane Package (FPP) is shown in Fig. 8. Polarization modulation is accomplished by a rotating bi-crystalline retarder (quartz and sapphire) located in the symmetric part of the optical system just after the Optical Telescope Assembly (OTA). After the reflection by the tip-tilt mirror the beam is divided among four optical paths: the SP, the Broadband Filter Instrument (BFI), the NFI, and the Correlation Tracker. The NFI and BFI share a common focal plane and detector. The SP and NFI (or BFI) can operate simultaneously because the SP has its own detector. The polarization modulator rotates at 0.625 Hz such that the frame transfer detectors sample continuously at 10 Hz: a sampling speed fast enough to "freeze" moving features in the solar atmosphere for most polarimetric measurements. The SOT/SP operates in dual-beam mode by imaging both orthogonal linear polarizations exiting the polarizing beam splitter near its focal plane. Owing to continuous integration/readout of the detector, the instrument operates at close to 100 % duty cycle. Several modes of polarimetry are available to the NFI, including frame-transfer modes using an externally positioned mask of the focal plane. So both SOT/NFI and SOT/SP are capable of high precision polarimetry at high angular resolution.

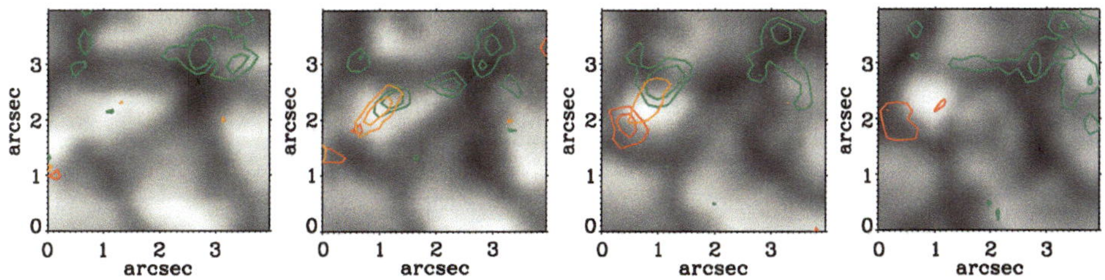

Fig. 9 This sequence of four continuum intensity images derived from repeated, short maps of the *Hinode*/SOT Spectro-Polarimeter reveal the emergence of a small dipole loop in a granule of the quiet Sun. *Red* and *green contours* represent positive and negative circular polarization (upward- and downward-directed magnetic field), and *orange contours* indicate linear polarization (horizontal field component). A loop structure is seen to form in the second frame with horizontal fields between the two vertically-oriented fields of opposite polarity. As the top of the loop rises above the photosphere, only vertical fields remain in the photosphere, each centered over a dark intergranular lane. Diagram taken from Centeno et al. (2007)

Freedom from the disturbing effects of the Earth's atmosphere allows the *Hinode* polarimeters to continuously achieve angular resolution of the 50 cm SOT. This has permitted the detailed study of small-scale magnetism in the solar atmosphere. One area of intense study by *Hinode* researchers has been the magnetism of the quiet solar atmosphere, as illustrated by the example in Fig. 9. Prior to *Hinode*, glimpses from ground-based instrumentation revealed the presence of small-scale horizontal fields in the solar photosphere (Lites et al. 1996), and it was suspected that they might correspond to emergence of small magnetic loops into the photosphere. With the advent of *Hinode* the widespread occurrence of these features was revealed for the first time (Lites et al. 2008) owing to the excellent image quality combined with high polarimetric precision. Using repeated short maps of quiet regions with SOT/SP, Centeno et al. (2007) were able to demonstrate that the horizontal fields are indeed the tops of tiny magnetic loops emerging within granules, as shown in Fig. 9.

As a consequence of the seeing-free space environment, the point spread function (PSF) does not vary significantly during the time of an observation. This attribute allowed the development of inversion codes that take into account the PSF self-consistently during the minimization process, with the result that maps having diffraction-limited resolution of the SOT may be obtained for the physical parameters in the solar atmosphere. Such techniques are described in van Noort (2012), Asensio Ramos and de la Cruz Rodríguez (2015) and also in de la Cruz Rodriguez et al. (2015, this issue). They have already resulted in numerous publications (e.g., Lagg et al. 2014; Tiwari et al. 2013; Buehler et al. 2015).

The *Hinode* mission was launched in September 2006 and still continues to operate providing measures of the photospheric magnetic field reliably and on demand. It is the only such space mission operating now, and comparable or better instrumentation is only on the horizon, being at least 6–10 years away. The SOT/SP is operating nominally, having experienced only a 25 % drop in throughput over the past nine years on orbit. The SOT/NFI instrument continues to provide limited magnetogram data in the wing of the Na D-line only, owing to a progressive drift off-band of the pre-filters to the birefringent filter. Also, about a year into the mission, *Hinode* suffered a failure of its high-rate X-band telemetry. Fortunately, the lower rate S-band telemetry continued to function well, and conservative management of the data volume from SOT combined with a greatly increased number of downlink passes has allowed most of the SOT science objectives to continue to be addressed. This history of *Hinode* telemetry issues is yet another testament to the need for redundancy in space mission hardware.

4.4 High-Altitude Balloon Missions

Although they are not strictly space missions, a brief discussion of high-altitude balloon missions carrying instrumentation to measure the photospheric magnetic field is provided here. Balloon platforms reproduce some desirable conditions of space: (1) the rarified atmosphere at high altitude nearly completely eliminates the disruptive effect of atmospheric seeing, (2) during the high-latitude mid-summer extended periods of solar observing without night interruption are possible, and (3) some ultraviolet wavelengths become nearly transparent.

The cost of a balloon experiment is much smaller than that of an equivalent experiment in space because of relaxed restrictions on weight, power, contamination, vibration/acoustic tolerance, documentation, and other factors. Cost issues aside, development of a balloon mission is still a major undertaking. Taking the two prior solar magnetographic balloon missions as typical, the development time for a major long-duration balloon experiment is comparable to that of a space mission. Reliability constraints of a balloon mission are similar to those of a space experiment since high altitude balloon launches are expensive and infrequent. One usually has the opportunity to recover the flight hardware after the end of a balloon mission, but that hardware must be able to survive a crash landing if it is to be flown again. In most respects, pointing and image stabilization of a balloon-borne high resolution telescope is more difficult than a space platform because one has to contend with variable winds at float altitude, and variable wind gradients between the balloon itself and its payload drive a pendulum motion.

Solar long-duration balloon experiments may be flown at mid-summer in the polar regions in order to have continuous observing throughout the flight. Two solar magnetographic experiments have been flown: the Flare Genesis Experiment (FGE, Bernasconi et al. 2000, 2001, 2002), and the successful Sunrise program as described in the following.

4.4.1 Sunrise/IMaX

The Sunrise program (Barthol et al. 2011) is a high-altitude, long-duration balloon experiment intended to explore solar phenomena at very high angular resolution (see also Kleint and Gandorfer 2015, this issue). It was a joint effort of Germany, Spain, and the USA. It consisted of a 1.0 m telescope carrying the Imaging Magnetograph eXperiment (IMaX—Fig. 10, Martínez Pillet et al. 2011) and an ultraviolet imager (Sunrise Filter Imager (SuFI), Gandorfer et al. 2011). The former operated in the highly Zeeman-sensitive Fe I line at 5250 Å. Also like FGE, IMaX used a pair of liquid crystal variable retarders to accomplish polarization modulation, but as a polarimeter IMaX had many features that qualified it as a quantitative solar polarimeter. IMaX measured combinations of four states that uniquely define the Stokes 4-vector. Compared to a polarimeter measuring pure states with six measurements, the IMaX scheme has significantly higher polarimetric efficiency (del Toro Iniesta and Collados 2000). Furthermore, the instrument had a nearly continuous expose/readout cycle resulting in a duty cycle of near unity. These features allowed IMaX to perform polarimetry at a S/N appropriate for quantitative determination of the magnetic field vector. IMaX also used a tunable lithium niobate Fabry-Pérot filter in double pass so that it achieved a spectral resolution of 85 mÅ at 5250 Å. In typical full-Stokes polarimetry mode it sampled five wavelengths throughout the line profile, each with a S/N of about 1000.

Two flights of Sunrise have been realized. The first flight in the north polar region lasted six days in June 2009, and produced outstanding results. High resolution sequences of data of duration up to 30 minutes were obtained of quiet Sun conditions. The Sunrise/IMaX system achieved close to its design angular resolution of about 0.15″, i.e. a factor of two better than that of *Hinode* SOT/SP.

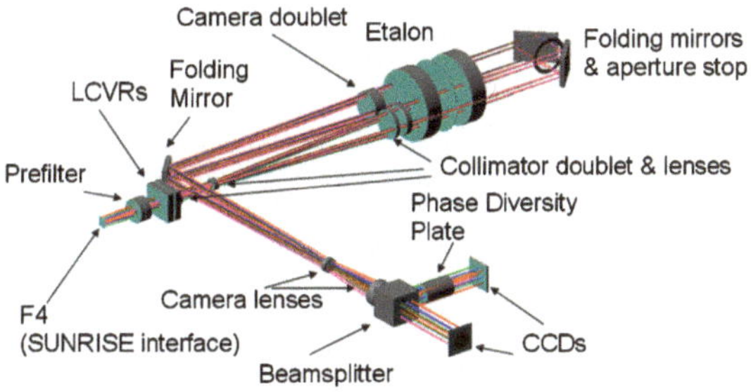

Fig. 10 The optical layout of the IMaX instrument on the Sunrise balloon program illustrates the simplicity of its design. Having two focal planes, it may operate as either a dual-beam polarimeter or, by applying a slight de-focus to one of the beams, as a phase-diversity imager. The wavelength isolation is accomplished by a tunable lithium niobate crystalline Fabry-Pérot etalon in double pass. Diagram taken from Martínez Pillet et al. (2011)

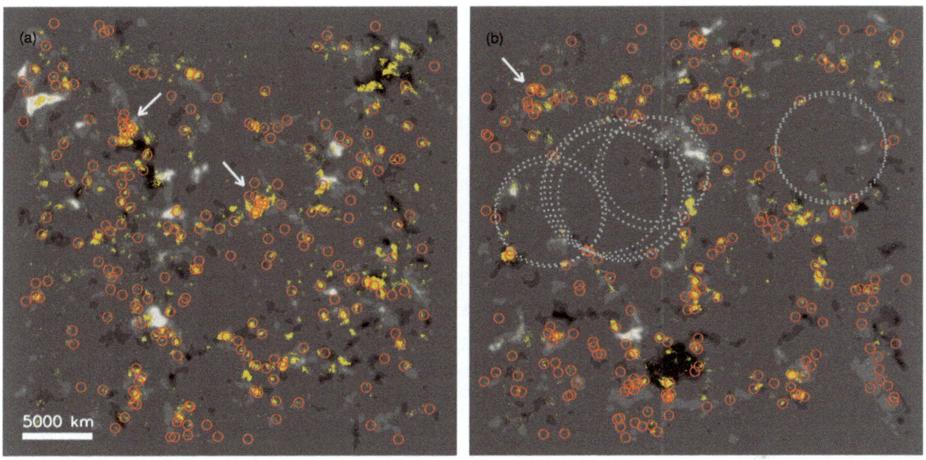

Fig. 11 Two sequences of polarimetric data from Sunrise/IMaX of durations 22 and 31 minutes are presented. Average longitudinal magnetograms are shown in greyscale. During these sequences, small-scale loops in the photosphere (*red circles*) appear as evidenced by linear polarization (*yellow patches*) separating two patches of opposite circular polarization. *Dotted circles* ("dead calm") in the image at right delimit areas where there is very little or no emergence of these small-scale loops. *Arrows* point to regions where several loops appear at the same location in the time series. Figure taken from Martínez González et al. (2012)

An example of the science that was made possible by Sunrise/IMaX is illustrated by Fig. 11. In that work, Martínez González et al. (2012) identify the presence of small-scale magnetic loops in the photosphere as evidenced by opposite circular polarization patches (upward- and downwardly-directed magnetic fields) connected by a detectable linear polarization patch between them (horizontal component of the field). Such loops were identified in short maps with the *Hinode* SOT/SP (see Fig. 9), but the high-resolution and high-precision imaging polarimetry of Sunrise/IMaX for the first time permitted this view of their spatial/temporal organization. This work reveals persistent areas where none of these small-scale loops appear: "dead calm" regions of the quiet Sun. The authors interpret this result—that the emergence of small-scale magnetic loops in the quiet Sun is not isotropic—as an important observed property of quiet Sun magnetism that must be explained by proposed mechanisms for its generation (such as a small-scale dynamo).

4.5 Future Space Missions to Measure Solar Photospheric Magnetic Fields

Perhaps there remain some hidden surprises regarding photospheric magnetic fields that are lurking below our current capability of resolving them because we have certainly not resolved the fine-scale structure of magnetism in the internetwork, and especially in the intergranular lanes. Both theory and simulations (e.g., Vögler and Schüssler 2007; Rempel 2014) and interpretation of observations (e.g., Pietarila Graham et al. 2009) suggest that observation of magnetic structure at the dominant scale of photospheric magnetism may be beyond reach with current diagnostic techniques, in recognition of the scattering properties of spectral lines in the photosphere when sampling scales at or below the photon mean free path. In any event, improvement of adaptive optical systems for the next generation of large ground-based telescopes is proceeding at a rapid pace, so future exploration of photospheric magnetism at very small scales would appear to be most aptly addressed using those facilities.

How then could future space observations contribute uniquely to further the understanding of photospheric magnetism? Little is still known regarding details of the photospheric field at the extreme solar poles. The polar regions are important in that it is believed that they play a crucial role in the progression of the large-scale solar dynamo. Magnetism of the polar regions has been studied extensively using *Hinode* (Ito et al. 2010; Shiota et al. 2012, and Petrie and Ettinger 2015, this issue), but spatial resolution of the extreme poles is limited. Vertical fields are detected mainly by the transverse Zeeman effect and is therefore not very sensitive to weaker fields, and one is detecting fields well above the physical height corresponding to the $\tau_c = 1$ level at disk center because of foreshortening. The Solar Orbiter mission will fly the Polarimetric and Helioseismic Imager (PHI) in an inclined orbit that will pass within 0.28 AU of the Sun (Kleint and Gandorfer 2015, this issue). PHI will perform precision polarimetry of the polar regions near closest approach, so magnetic fields of the polar regions will be attained from a more favorable viewpoint, albeit only for a brief period.

A new frontier for observational studies of solar magnetism is the chromosphere (see Sect. 5). Solar-C is an initiative for a major space mission lead by Japan but, like *Hinode*, having significant contributions from NASA and ESA. The major scientific thrust for Solar-C is the magnetism of the chromosphere, but interpretation of chromospheric field measurements will require accompanying measurements of fields in the photosphere. The Solar-C concept provides for simultaneous measurements of the field vector using spectral lines forming at several heights from the photosphere through the chromosphere.

In the near term, it is extremely important for the discipline to continue to exploit its excellent resources for photospheric field measurement that are still on orbit. The community should strive to ensure that *SDO*/HMI and *Hinode* SOT/SP are operational as long as possible in order to build a synoptic data base of vector magnetic fields through a full solar cycle. One important motivation to sustain these observations is to understand the long-term behavior of non-AR fields: the remnant fields of active regions and also fields apparently not associated with the global dynamo. This is especially important if the most recent extended solar minimum and the present weak maximum heralds a major shift in the dynamo activity, for example the beginning of a new Maunder minimum.

5 Chromospheric Magnetic Field Measurements

5.1 The Importance of the Chromospheric Field Measurement to the Current Status of Solar Physics

The previous Sects. 3 and 4 demonstrated the high level of sophistication reached especially for the measurement[1] of the magnetic field in the photosphere, achieved by using advanced instrumentation in ground-based, balloon-borne and space-based observatories. The spatial and temporal resolution of the measurements in this deepest layer of the solar atmosphere directly accessible by observations has been increased by an order of magnitude within the last three decades. Similarly, recent space observatories operating at wavelengths not accessible from the ground permitted exploitation of the structure of the coronal magnetic field by analyzing the plasma emission of ions trapped in the coronal magnetic field with unprecedented precision.

The chromosphere is the coupling element between the photosphere and the corona. This extremely dynamic layer is located only one convection-cell size above the solar photosphere (≈1000 km). Exposed to the cold environment of space, the radiative energy losses effectively cool down the chromosphere. To maintain the typical temperature of approximately 10 000 K a substantial amount of energy is required, about 15 times higher (Aschwanden et al. 2007) than the energy necessary to heat the solar corona. Whereas the "coronal heating problem" has attracted the attention of the solar physics community for decades, the equally important and interesting problem for the chromosphere has only very recently been emphasized. Similarly to the corona, the candidates for the chromospheric heating are wave dissipation, magnetic reconnection and Ohmic dissipation. Not only are the details of the underlying heating mechanisms poorly understood, but even the relative importance of the possible mechanisms in various structures remains poorly determined. The investigation of the chromospheric magnetism is crucial for both the understanding of the Sun-Earth connection with all its consequences to our technical and natural environment, and the detailed study of fundamental plasma physical processes in density and temperature regimes difficult to achieve in laboratories.

Progress in this field of research has been hindered by several aspects of inferring the properties of the chromosphere. The low gas and plasma densities result in large mean free paths for photons. In this radiatively-dominated regime the population of atomic levels deviates strongly from thermodynamic equilibrium, requiring complex and computationally expensive three-dimensional radiative transfer modeling to interpret the observations. Of similar complexity are magneto-hydrodynamic simulations, which for the photosphere have in the meantime become a reliable and robust tool to understand and analyze the fundamental physical processes. The basis to shed light into this complex field of research are reliable measurements of the chromospheric magnetic field at high spatial and temporal resolution.

5.2 Challenges to High-Resolution Measurements of Chromospheric Fields

The chromosphere "hides" itself between the observationally easier accessible layers photosphere and corona. With an average height of 500 to 2000 km above the solar surface it may be observed almost exclusively on the solar disk, with the consequence that the chromospheric information contained in the solar spectrum is to a large extent overwritten by the

[1] In this section, the term measurement refers to the determination of the magnetic field from remote sensing instruments by interpreting signals directly influenced by the solar magnetic field.

photospheric emission. The pressure scale height in the solar atmosphere is approximately 150 km, leading to typical densities in the chromosphere of about four orders of magnitude lower than the photospheric ones. This decrease of gas pressure leads to an expansion and subsequently weakening of the magnetic field, rooted in often sub-arcsecond sized magnetic patches in the quiet-Sun photosphere.

The main tool for diagnosing the chromospheric magnetic field is spectropolarimetry in Fraunhofer lines formed under chromospheric conditions. With a few exceptions, the chromospheric signature in these spectral lines is only present in a narrow range around the line core. Weak fields and the low densities require highly sensitive spectropolarimetric measurements to detect these signals. Additionally, these low densities result in an increase of the typical velocities in the chromosphere: the Alfén speed reaches values of 100 $\mathrm{km\,s^{-1}}$, the sound speed is around 20 $\mathrm{km\,s^{-1}}$. Gas, plasma and wave motions can easily achieve these velocities, shock waves can form and reconnection of magnetic field lines can occur, making measurements at high temporal and spatial resolution mandatory.

The resulting dilemma was already discussed in Sect. 2 for photospheric conditions. There, the typical velocities are on the order of 5 $\mathrm{km\,s^{-1}}$, but the velocities in the chromosphere can easily be an order of magnitude higher. As a consequence, the maximum permissible exposure time to freeze in the solar evolution for a large aperture solar telescope ($\gtrsim$1 m) lies in the ten millisecond range (see Fig. 1a). On the other hand, the photon flux in the line core, i.e., the part of the spectrum containing the chromospheric information, is on the average only 10–20 % of the continuum intensity. Additionally, to characterize the weak chromospheric fields, a high S/N ratio in the range of a few times 10^{-4} is desirable. To achieve these goals, highly photon efficient instrumentation on large aperture solar telescopes, not necessarily operated at their diffraction limit, is mandatory. The Daniel K. Inouye Solar Telescope (DKIST, Rimmele and ATST Team 2008; Tritschler et al. 2015), currently under construction on Maui/Hawaii, and the European Solar Telescope (EST, Collados et al. 2013) are the logical steps in this direction.

Besides these technical challenges, chromospheric conditions introduce a high level of complexity to the interpretation of the polarization signals. Simplifying assumptions, like the local thermodynamic equilibrium (LTE) conditions prevailing in the photosphere, do not hold anymore. Chromospheric absorption and emission processes need to be modeled under non-LTE conditions involving on average 6–12 atomic levels in the most prominent atoms used for chromospheric diagnostics (H, Ca, Mg, and He). The population of the levels in these atoms additionally strongly depends on the incident coronal radiation field. Its anisotropy in the illumination of the chromospheric layer is responsible for creating imbalances in the population of the sub-levels of the atoms, leading to atomic polarization. Long time scales of up to several hours for ionization and recombination processes introduce a memory effect of the chromosphere to past events, complicating the determination of the initial conditions for theoretical models (de la Cruz Rodriguez et al. 2015, this issue).

5.3 Diagnostic Techniques

Our knowledge about the chromospheric magnetic field is mainly based on the interpretation of the polarimetric signal in chromospheric spectral lines. The above mentioned difficulties in measuring the polarimetric signal in these lines motivates the search for alternative diagnostic techniques. Two possible candidates are: extrapolations based on photospheric magnetic field maps and measurements in the millimeter and sub-millimeter wavelength range, soon available at high spatial resolution using the ALMA (Atacama Large Millimeter/submillimeter Array).

Extrapolations of the photospheric field rely on the robustness and simplicity of photospheric magnetic field measurements, which can be obtained on a routine basis at higher spatial and temporal resolution than chromospheric measurements. During the past decade a significant improvement on the reliability of these extrapolations has been achieved, for example by including Hα images in the pre-processing of the data (Wiegelmann et al. 2008). Unfortunately, the principle of extrapolations, based on minimizing the complexity of the magnetic field in the observed atmospheric volume, is not applicable for the most interesting and very dynamic conditions of, e.g., shocks or reconnection sites. To uncover the secrets at these locations, direct measurements are mandatory.

The linear relation between millimeter wavelength emissions and the chromospheric temperature makes ALMA an ideal thermometer for the solar chromosphere. In addition, in the near future it will perform magnetic field measurements in the chromosphere with a spatial resolution of down to $0.2''$ based on two physical processes: the Zeeman effect in high-n recombination lines of hydrogen and diatomic molecules, and the influence of the magnetic field to the temperature distribution by suppressing the power of propagating waves. Details about the usage of ALMA for chromospheric diagnostics can be found in Judge et al. (2015, this issue) and in Wedemeyer et al. (2015).

In the following sections we return to the current state-of-the-art measurement technique for chromospheric fields using spectropolarimetry in Zeeman and Hanle sensitive spectral lines in the visible and the near infrared regime. Here the improved observational capabilities have led to significant progress during the last couple of years, which will continue in the coming decade with the advent of large-aperture solar telescopes.

5.3.1 Diagnostic Spectral Lines

In the strictest sense the term chromosphere is defined as the region of Hα emission. Using the Hα line as the tool to investigate the chromospheric magnetism therefore seems to be an obvious choice. Unfortunately, the above-mentioned difficulties in modeling spectral lines are most prominent in this line. Hα originates from an excited level of the hydrogen atom ($n = 3$), whose population strongly depends on the temperature and radiative processes in the atmosphere, and therefore making the interpretation of the spectral profiles exceptionally complex (Leenaarts et al. 2012).

Spectral lines that are easier to interpret have been identified and are now observed on a routine basis. Figure 12 (snapshot from Bifrost chromospheric modeling code, taken from Carlsson et al. 2015) presents an overview of these lines which have emerged as especially well suited for diagnosing chromospheric magnetism. Sorted by their formation height from bottom to top, obtained from non-LTE modeling, these lines are: Mg I B_2 at 517 nm, the Na D line pair at 589 nm, the Ca II infrared triplet at 854 nm, the Ca II h&k lines at 390 nm, He I D3 at 588 nm, the He I triplet at 1083 nm, and the Mg II h&k lines at 279 nm. The latter has been observed on a routine basis only in spectroscopic and imaging mode by the IRIS mission (De Pontieu et al. 2014) and the second Sunrise stratospheric balloon flight (Barthol et al. 2011). Its potential to retrieve the magnetic field vector is very high, but so far limited to theoretical calculations (Trujillo Bueno 2014).

The quiet-Sun modeling underlying Fig. 12 demonstrates the high corrugation of especially the upper chromospheric layers (Carlsson et al. 2015). The corrugation becomes even stronger above active regions containing pores or sunspots. The height variation can easily span several Mm and must be taken into account for the correct interpretation of the spectropolarimetric signals.

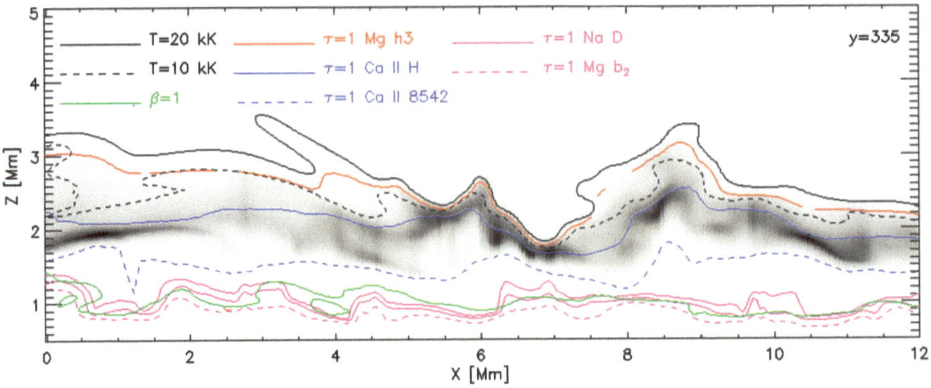

Fig. 12 Formation heights of various chromospheric spectral lines computed from the self-consistent chromospheric modeling code Bifrost (Gudiksen et al. 2011). The snapshot from a time series shows the iso-temperature surfaces (*black lines*), the plasma-$\beta = 1$ layer (*green*) and the optical depth $\tau = 1$ layers for the spectral lines as labeled in the figure. The layer of absorption in the He I 1083 nm triplet is indicated by the *gray-shaded area*. Taken from Carlsson et al. (2015)

The Ca II and He I infrared triplets are currently in the focus for the diagnostics of chromospheric magnetism. The analysis of the Ca II lines has benefited from improvements in the computationally expensive non-LTE modeling (e.g., Leenaarts et al. 2009; de la Cruz Rodríguez et al. 2012) and the availability of image reconstruction techniques, thereby permitting production of polarimetric maps at the diffraction limit of the modern solar telescopes at high signal-to-noise ratio (Scharmer 2006; Cavallini 2006; Puschmann et al. 2012). Inversions of the radiative transfer equation (RTE) to retrieve the physical conditions in the chromosphere from the measured Stokes spectra are on a good route to become a standard tool for the solar community, similarly to the inversion tools currently available for the photosphere. The progress in this field is summarized in de la Cruz Rodriguez et al. (2015, this issue).

5.3.2 The He I 1083 nm Triplet

This triplet has gained attraction due to two major improvements: developments in detector technology allow for spectropolarimetric observations at noise levels down to the 10^{-4} range (in units of the continuum), and the theoretical explanation for the linear polarization signal in the absence of magnetic fields (Trujillo Bueno et al. 2002). The latter led to the development of easy-to-use inversion codes (Asensio Ramos et al. 2008; Lagg et al. 2009). The biggest advantage of the He I 1083 nm line over other chromospheric lines is the absence of almost any photospheric contamination in the signal. The reason for this lies in the special formation process of this triplet. The He I 1083 nm line (as well as the He I D3 line) is a transition in the triplet state of the He atom, which is only sparsely populated under normal photospheric and chromospheric conditions. Coronal ultra-violet (UV) radiation with a wavelength shorter than 50.4 nm (i.e., 24.58 eV) is required to ionize the He atoms. The subsequent recombination occurs with equal probability either back to the singlet state, or to the triplet state. Since the chromosphere is highly opaque to coronal UV radiation, this ionization process occurs only in the top most layer of the chromosphere, resulting in a thin slab containing He I atoms in the triplet state. To first order, the physical conditions in this thin slab are height-invariant, an assumption validated by the simplicity of the observed Zeeman signals in this triplet. The flip side of this special formation process is the weak absorption signature above quiet Sun regions, where coronal UV illumination is very low.

5.3.3 He I *1083 nm and the Hanle Effect*

A further strength of the He I triplet is the coverage of a wide range of magnetic field strengths: The Landé factors of the He I triplet ($g_{\text{eff}} = 2.0, 1.75, 1.25$) allow for reliable measurements of the magnetic field vector above ≈50–100 G using the Zeeman effect. For higher field strengths the Paschen-Back effect sets in, shifting the Zeeman sub-levels and introducing an asymmetry in the Stokes profiles (Socas-Navarro et al. 2004, 2005; Sasso et al. 2006). The Hanle effect covers the region of lower field strengths (0.1–100 G, Trujillo Bueno et al. 2002). Anisotropic illumination of the thin slab by the photospheric radiation field introduces imbalances in the population of the lower and upper levels of the transition, producing a characteristic linear polarization signal. Weak magnetic fields of up to ≈8 G change the strength and the direction of polarization (Hanle sensitive regime), whereas the linear polarization signals for stronger fields are only sensitive to the field direction (Hanle saturated regime).

Despite the above-mentioned advantages of the He triplet, the correct interpretation of the signals poses major challenges for the near future. For the correct modeling of the strength of the line, the coronal environment must be taken into account as well as the importance of collisions for the population of the triplet state. The Hanle measurements require an exact determination of the anisotropy of the incident radiation field, influenced not only by the height of the slab, but also by the presence of brightness variations in the photosphere (e.g. sunspots), and the absorption of optically thick He I layers. Additionally, the ambiguities introduced by the Hanle effect must be treated carefully.

The near-infrared spectral region is particularly well suited for the large-aperture ground-based telescopes which were commissioned recently. The reasons are better and more stable seeing (Turon and Léna 1970), less scattered light (Staveland 1970) and less atmospheric extinction. The New Solar Telescope (NST, Goode et al. 2010) at the Big Bear Solar Observatory with an aperture of 1.6 m and the GREGOR Telescope on Tenerife (1.5 m, Schmidt et al. 2012) are both equipped with instrumentation for magnetic field diagnostics in the near infrared region, with one of the main foci being on imaging polarimetry and spectropolarimetry in the He I 1083 m triplet.

5.4 Examples of Current Chromospheric Field Measurement Results

5.4.1 He I *1083 nm Triplet*

Two examples of recent He I 1083 nm observations were selected to demonstrate recent advances in the interpretation of the spectropolarimetric data and the progress in instrumentation, allowing for observations at unprecedented spatial resolution and polarimetric sensitivity in this line. The first one shows data from the Facility Infrared Spectropolarimeter (FIRS, Jaeggli et al. 2010) at the National Solar Observatory's Dunn Solar Telescope (DST): Schad (2013) presents an observation of superpenumbral chromospheric fibrils spanning from the penumbra of NOAA AR 11408 to the internetwork regions outside the active region (see Fig. 13). Limited by the diffraction limit of the instrument of 0.6″, full Stokes maps in He I 1083 nm were recorded with an average noise level of 3–4 × 10^{-4} in units of the continuum intensity. The spectra were analyzed with the inversion code HAZEL

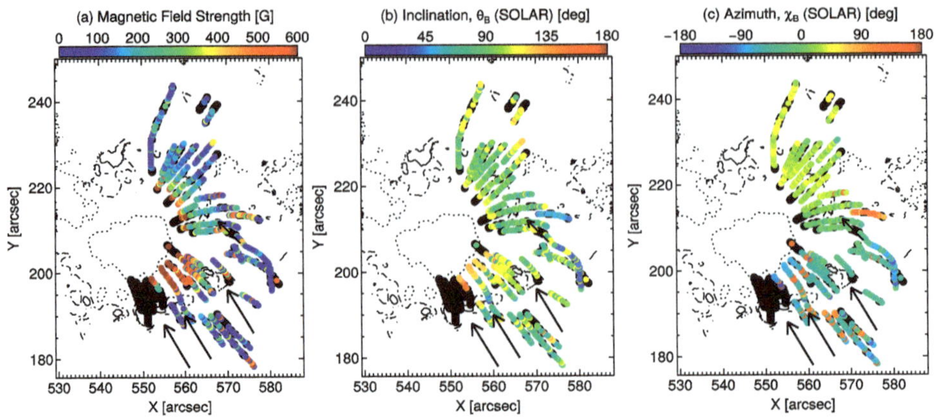

Fig. 13 Magnetic field strength, inclination and azimuth (in solar reference frame) of superpenumbral fibrils observed in the He I 1083 nm triplet with FIRS at the DST on January 29 2012. The central part of the filament is characterized by horizontal fields below 300 G, well aligned to the visible fibrils (from Schad 2013)

("Hanle and Zeeman Light", Asensio Ramos et al. 2008) which is based on the multi-term calculations of the orthohelium atom of Landi Degl'Innocenti and Landolfi (2004). Schad focussed their analysis on the determination of the magnetic field orientation in these fibrils to answer the question whether the magnetic field vector is aligned with the structures observed simultaneously in Hα and Ca II infrared intensity maps. This required a thorough analysis to select the correct solution out of up to 8 ambiguous solutions, a consequence of the combination of the well known 180° ambiguity of Zeeman effect with the so called Van Vleck ambiguity (see, e.g., Casini et al. 2005, 2009; Merenda et al. 2006; Orozco Suárez et al. 2015).

As a result of this analysis, Schad (2013) concluded that the projected direction of the field derived from He I inversions does not deviate by more than ±10° from the visible fibrils. For most of the fibrils observed in the Ca II 854.2 nm line, de la Cruz Rodríguez and Socas-Navarro (2011) obtained a similar alignment, but in some of the Ca II fibrils the magnetic field direction deviates significantly from the visible structures. A thorough investigation of more fibrils at high spatial resolution, ideally simultaneously in He I and Ca II, is required to resolve this discrepancy.

The largest European solar telescope, GREGOR, entered the scientific phase in 2014 (Schmidt et al. 2012). The GREGOR Infrared Spectrograph (GRIS, Collados et al. 2012) is producing spectropolarimetric He I 1083 nm data with an unprecedented spatial resolution close to the diffraction limit of the telescope at signal to noise ratios in the Stokes parameters of up to 3000. An example of such Stokes parameter maps, composed from the scan of the slit over AR 12096 on June 27 2014, is presented in Fig. 14. A detailed analysis of the fine structure of the chromospheric field is currently being performed at the partner institutions (Kiepenheuer Institut für Sonnenphysik, Freiburg; Max-Planck-Institut für Sonnensystemforschung, Göttingen; Leibniz-Institut für Astrophysik, Potsdam; Instituto de Astrofísica de Canarias, La Laguna) and will soon appear in a peer-reviewed journal.

5.4.2 Ca II *854.2 nm*

The highest spatial resolution for spectropolarimetric measurements is currently achievable only using filter-based instruments. As described in Sect. 3.2.1, these instruments allow to

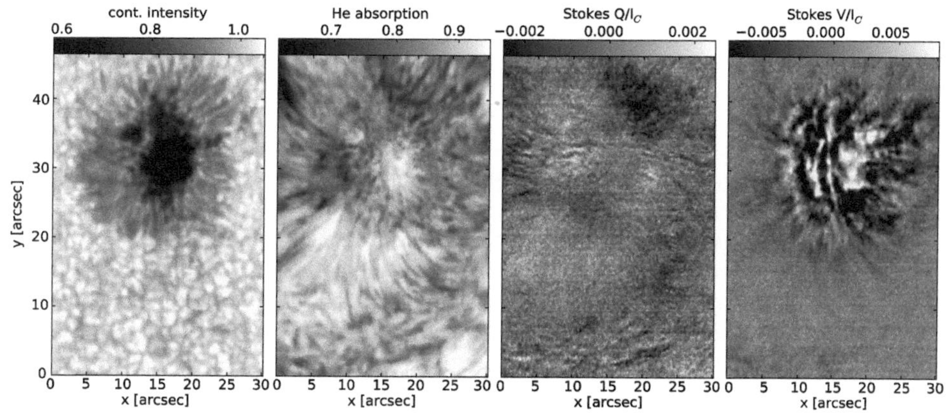

Fig. 14 Continuum image and He I 1083 nm Stokes I, Q, and V maps of AR 12096 obtained with GREGOR/GRIS on June 27 2014 close to disk center ($\mu = 0.98$). The achieved spatial resolution is $0.44''$ at a spectral resolution of 200 000. The total exposure time of 1.6 s per slit position for the full Stokes vector (without readout and processing time) resulted in a noise level of 7–10 $\times$ 10^{-4} of the continuum intensity

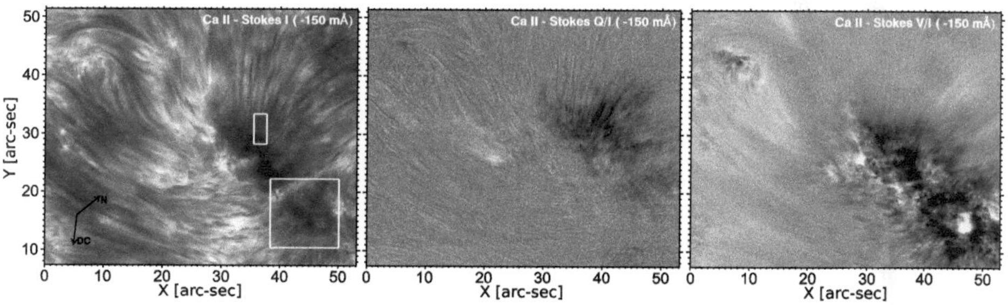

Fig. 15 CRISP Stokes I, Q/I, and V/I filtergrams at $\Delta\lambda = -150$ mÅ from the core of the Ca II 854.2 nm line (adapted from de la Cruz Rodríguez et al. 2013). The *boxes* mark the regions for which a detailed analysis of umbral flashes and running penumbral waves was performed

apply image reconstruction routines to fully exploit solar telescopes down to their theoretical resolution limits. The most successful instruments of this type are the CRisp Imaging SpectroPolarimeter at the 1-meter Swedish Solar Telescope (SST/CRISP, Scharmer et al. 2008) and the Interferometric BIdimensional Spectrometer (IBIS, Cavallini 2006).

An example of a chromospheric magnetic field measurement with CRISP is presented in Fig. 15, showing the Stokes I, Q/I, and V/I map of AR 11204 observed on May 4 2011. de la Cruz Rodríguez et al. (2013) applied the non-LTE inversion code NICOLE (Socas-Navarro et al. 2000) to this dataset and obtained the height dependence of temperature, line-of-sight velocity and magnetic field vector above the umbra and the penumbra of the sunspot at a spatial resolution of $\approx 0.15''$. The high temporal cadence of CRISP (16 s for a full-Stokes line scan) enabled the authors to perform a detailed analysis of umbral flashes, revealing their $\approx$1000 K hotter shock front when compared to the surrounding material, without a measurable influence on the magnetic field strength and direction. Additionally, a surprisingly large variation of the magnetic field strength (up to 200 G) in the chromosphere above the penumbra could be attributed to running penumbral waves. For more examples of high-quality measurements in the Ca II line using CRISP and IBIS data the reader is referred to Pietarila et al. (2007) and Kleint (2012).

6 Technology and Data Analysis Challenges

The previous chapters demonstrated that the determination of the magnetic field vector using remote sensing instruments requires a close intermeshing of observational techniques and analysis methods. Technology must be pushed to its limits to obtain data sets with high S/N ratios in all Stokes parameters. An accurate knowledge of all instrumental deficiencies, unavoidably resulting in data degradation, is necessary in order to treat them carefully in the data analysis pipelines. The most common tool for extracting the magnetic field from the observations is the inversion of the Stokes spectra: From an initial guess model atmosphere synthetic Stokes spectra are computed, then the parameters of this model atmosphere are adjusted iteratively until the synthetic Stokes spectra agree with the observed ones (see also de la Cruz Rodriguez et al. 2015, this issue). In this process it is extremely important that the synthetic Stokes spectra are degraded in exactly the same way as the instrument alters the spectra originating from the Sun. Only then the comparison between the observed and synthetic Stokes profile will deliver reliable information about the conditions in the solar atmosphere.

The aim is, of course, to keep the instrumental deficiencies as small as possible. This is where technological improvements come into play, which have allowed for a significant increase in the accuracy of spectro-polarimetric measurements at high spatial resolution during the last decade. The key ingredients here are advances in detector technology (e.g. FSP, see Sect. 3.5), massively multiplexed focal planes (2-dimensional spectropolarimeters, see Sect. 3.2.2), and the use of adaptive optic systems mandatory for operating large aperture solar telescopes at their diffraction limit (see Sect. 3.3.1).

In the near future, large-aperture solar telescopes will allow to observe solar features well below scales of 50 km. Another challenge for the future is to improve the height resolution by performing spectropolarimetric measurements in many spectral lines simultaneously. The difference in the formation height of the lines and in the response to changes of atmospheric parameters will help modern inversion tools to retrieve the stratification in the solar atmosphere with unprecedented accuracy.

Despite all the efforts to avoid instrumental deficiencies, they will always be present in the data. The proper treatment of these instrumental effects is a big challenge on the data analysis side. This has been done successfully during the last years in terms of image reconstruction techniques based on phase-diversity, speckle, or MOMFBD techniques (Lofdahl and Scharmer 1994; von der Luehe 1993; van Noort et al. 2005). The next evolutionary steps in this direction are novel inversion techniques, like the spatially-coupled inversions by van Noort (2012) or the sparse inversion of Stokes profiles by Asensio Ramos and de la Cruz Rodríguez (2015), which take into account the point-spread-function (PSF) of the telescope self-consistently during the inversion process. In a next development step, inversion programs must take into account the natural 2D-coupling between the single-column atmospheres.

But even the cleanest data are subject to ambiguous solutions in the determination of the physical conditions in the solar atmosphere. Such ambiguities are caused by the physical process producing the measured signal (e.g., the well-known 180° ambiguity of the Zeeman effect, or the Van Vleck ambiguity in the Hanle diagnostics), or by the mere presence of a multitude of model atmospheres producing the same Stokes vector within the noise level of the data (degeneracy of the fit parameters).

The resolution of ambiguities is a complex and very often not unique process, involving assumptions about the physical conditions in the solar atmosphere. A successful technique to solve for these ambiguities is to prescribe certain conditions to the solar atmosphere, for

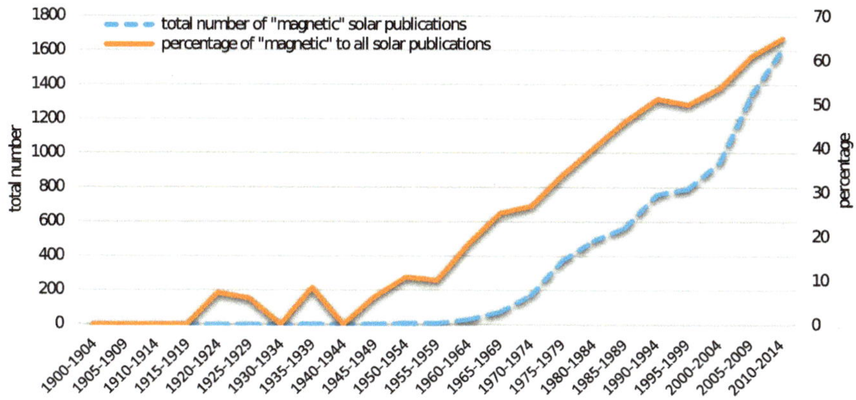

Fig. 16 5-year totals of papers published in refereed journals including the words "(solar OR sun) AND (photosphere OR chromosphere) AND magnetic" in the title or abstract according to ADS (*dashed blue curve*). The *solid orange curve* indicates the percentage of these publications relative to the ones without the search tag "magnetic" (right ordinate axis)

example minimizing the divergence of the magnetic field or the vertical current density. A description about available tools for removing the 180° Zeeman ambiguity was given by Metcalf et al. (2006). The solution of the so-called Van Vleck ambiguity, present in Hanle measurements, often requires to choose the "more likely" solution, i.e., the solution matching a realistic, physical model of the observed structure (e.g., Orozco Suárez et al. 2014; Schad et al. 2013).

The identification of the degeneracy in the solution is another important step towards the reliable interpretation of spectropolarimetric data. It helps to identify the simplest model atmosphere compatible with the observed Stokes spectra, which is usually the one with the lowest number of free parameters. Bayesian inversion techniques (Asensio Ramos et al. 2007) or Markov chain Monte Carlo (MCMC) methods are promising approaches to detect degeneracies.

Ideally, ambiguities and degeneracies in the solutions should be addressed directly by the observations: multi-line spectropolarimetry can provide the complementary information needed to find the unique solution. Similarly, observations from different vantage points are beneficial. The latter will be achieved for the first time by the out-of-ecliptic mission Solar Orbiter (see Sect. 4), which will provide high-resolution Stokes maps of the photosphere with the Polarimetric and Helioseismic Imager (PHI).

7 Conclusions

As mentioned in the introduction, the Sun would be quietly evolving along the main sequence if it were not for its magnetic field. Solar magnetism is responsible for solar activity and space weather events, some of which affect our modern civilization. Magnetic processes appear to be ubiquitous throughout the universe. So for both astrophysical and practical reasons there has been increasing interest in observing and understanding the solar magnetic field. This increase is suggested by Fig. 16 which shows 5-year totals of papers containing "solar or sun" and "photosphere or chromosphere" and "magnetic" in either the title or the abstract of refereed journals (based on NASA ADS). Even more impressive is the increase in the percentage of the publications relative to the ones without the search tag "magnetic":

during the last 5 years, more than 60 % of all papers on the solar photosphere and chromosphere deal with the subject magnetism.

This impressive growth of interest in solar magnetism stems to a large extent from the recognition of its importance to almost all fundamental physical processes on the Sun and its relevance to space weather, including the efforts to predict space weather events. The rapid development of measurement techniques for solar magnetic fields is a logical consequence, additionally supported by progress in technology, especially in the field of detectors and optical instrumentation. Together with the recent and forthcoming introduction of new telescopes we expect a steady increase of new measurements and understanding of the solar magnetic field. Recent excitement in exoplanet discovery and research has awakened interest in new techniques of spectropolarimetry of stars that should benefit solar observational methods. In the entire two-century history of solar polarimetry, the next decade should be the most exciting and scientifically productive.

Acknowledgements *Hinode* is a Japanese mission developed and launched by ISAS/JAXA, collaborating with NAOJ as a domestic partner, NASA and STFC (UK) as international partners. Scientific operation of the *Hinode* mission is conducted by the *Hinode* science team organized at ISAS/JAXA. This team mainly consists of scientists from institutes in the partner countries. Support for the post-launch operation is provided by JAXA and NAOJ (Japan), STFC (UK), NASA, ESA, and NSC (Norway). B. Lites was supported in part by the FPP project at LMSAL and HAO under NASA contract NNM07AA01C. The National Center for Atmospheric Research is sponsored by the National Science Foundation. The National Solar Observatory is operated by the Association of Universities for Research in Astronomy (AURA, Inc.) under a cooperative agreement with the National Science Foundation. The 1.5-meter GREGOR solar telescope was built by a German consortium under the leadership of the Kiepenheuer-Institut für Sonnenphysik in Freiburg with the Leibniz-Institut für Astrophysik Potsdam, the Institut für Astrophysik Göttingen, and the Max-Planck-Institut für Sonnensystemforschung in Göttingen as partners, and with contributions by the Instituto de Astrofísica de Canarias and the Astronomical Institute of the Academy of Sciences of the Czech Republic. This study is supported by the European Commission's FP7 Capacities Programme under the Grant Agreement number 312495.

References

M.J. Aschwanden, A. Winebarger, D. Tsiklauri, H. Peter, The coronal heating paradox. Astrophys. J. **659**, 1673–1681 (2007). doi:10.1086/513070

A. Asensio Ramos, J. de la Cruz Rodríguez, Sparse inversion of Stokes profiles. I. Two-dimensional Milne-Eddington inversions. Astron. Astrophys. **577**, 140 (2015). doi:10.1051/0004-6361/201425508

A. Asensio Ramos, M.J. Martínez González, J.A. Rubiño-Martín, Bayesian inversion of Stokes profiles. Astron. Astrophys. **476**, 959–970 (2007). doi:10.1051/0004-6361:20078107

A. Asensio Ramos, J. Trujillo Bueno, E. Landi Degl'Innocenti, Advanced forward modeling and inversion of Stokes profiles resulting from the joint action of the Hanle and Zeeman effects. Astrophys. J. **683**, 542–565 (2008). doi:10.1086/589433

H.W. Babcock, The solar magnetograph. Astrophys. J. **118**, 387 (1953). doi:10.1086/145767

J.L. Ballester, Prominence seismology, in *IAU Symposium*, ed. by B. Schmieder, J.-M. Malherbe, S.T. Wu. IAU Symposium, vol. 300 (2014), pp. 30–39. doi:10.1017/S1743921313010703

P. Barthol, A. Gandorfer, S.K. Solanki, M. Schüssler, B. Chares, W. Curdt, W. Deutsch, A. Feller, D. Germerott, B. Grauf, K. Heerlein, J. Hirzberger, M. Kolleck, R. Meller, R. Müller, T.L. Riethmüller, G. Tomasch, M. Knölker, B.W. Lites, G. Card, D. Elmore, J. Fox, A. Lecinski, P. Nelson, R. Summers, A. Watt, V. Martínez Pillet, J.A. Bonet, W. Schmidt, T. Berkefeld, A.M. Title, V. Domingo, J.L. Gasent Blesa, J.C. Del Toro Iniesta, A. López Jiménez, A. Álvarez-Herrero, L. Sabau-Graziati, C. Widani, P. Haberler, K. Härtel, D. Kampf, T. Levin, I. Pérez Grande, A. Sanz-Andrés, E. Schmidt, The sunrise mission. Sol. Phys. **268**, 1–34 (2011). doi:10.1007/s11207-010-9662-9

P.N. Bernasconi, D.M. Rust, H.A.C. Eaton, High resolution vector magnetograms with the flare genesis vector polarimeter, in *Advanced Solar Polarimetry—Theory, Observation, and Instrumentation*, ed. by M. Sigwarth. Astronomical Society of the Pacific Conference Series, vol. 236 (2001), p. 399

P.N. Bernasconi, D.M. Rust, H.A. Eaton, G.A. Murphy, Balloon-borne telescope for high-resolution solar imaging and polarimetry, in *Airborne Telescope Systems*, ed. by R.K. Melugin, H.-P. Röser. Society of Photo-Optical Instrumentation Engineers (SPIE) Conference Series, vol. 4014 (2000), pp. 214–225
P.N. Bernasconi, D.M. Rust, M.K. Georgoulis, B.J. Labonte, Moving dipolar features in an emerging flux region. Sol. Phys. **209**, 119–139 (2002). doi:10.1023/A:1020943816174
F.H. Bigelow, The solar corona discussed by spherical harmonics. Contrib. Smithsonian Inst. **691**, 22 (1889)
J.M. Borrero, P. Kobel, Inferring the magnetic field vector in the quiet Sun. III. Disk variation of the Stokes profiles and isotropism of the magnetic field. Astron. Astrophys. **550**, 98 (2013). doi:10.1051/0004-6361/201118239
I.S. Bowen, The image-slicer a device for reducing loss of light at slit of stellar spectrograph. Astrophys. J. **88**, 113 (1938). doi:10.1086/143964
T. Brown, The Fourier tachometer: principles of operation and current status, in *Solar Instrumentation: What's Next?*, ed. by R.B. Dunn (1981), p. 150
D. Buehler, A. Lagg, S.K. Solanki, M. van Noort, Properties of solar plage from a spatially coupled inversion of Hinode SP data. Astron. Astrophys. **576**, 27 (2015). doi:10.1051/0004-6361/201424970
M. Carlsson, V.H. Hansteen, B.V. Gudiksen, J. Leenaarts, B. De Pontieu, A publicly available simulation of an enhanced network region of the Sun. ArXiv e-prints (2015, in press). ArXiv:1510.07581
R. Casini, R. Bevilacqua, A. López Ariste, Principal component analysis of the He I D_3 polarization profiles from solar prominences. Astrophys. J. **622**, 1265–1274 (2005). doi:10.1086/428283
R. Casini, A.G. de Wijn, P.G. Judge, Analysis of seeing-induced polarization cross-talk and modulation scheme performance. Astrophys. J. **757**, 45 (2012). doi:10.1088/0004-637X/757/1/45
R. Casini, A. López Ariste, F. Paletou, L. Léger, Multi-line Stokes inversion for prominence magnetic-field diagnostics. Astrophys. J. **703**, 114–120 (2009). doi:10.1088/0004-637X/703/1/114
F. Cavallini, IBIS: a new post-focus instrument for solar imaging spectroscopy. Sol. Phys. **236**, 415–439 (2006). doi:10.1007/s11207-006-0103-8
R. Centeno, H. Socas-Navarro, B. Lites, M. Kubo, Z. Frank, R. Shine, T. Tarbell, A. Title, K. Ichimoto, S. Tsuneta, Y. Katsukawa, Y. Suematsu, T. Shimizu, S. Nagata, Emergence of small-scale magnetic loops in the quiet-Sun internetwork. Astrophys. J. Lett. **666**, 137–140 (2007). doi:10.1086/521726
W.C. Chaplin, Y. Elsworth, G.R. Isaak, R. Lines, C.P. McLeod, B.A. Miller, R. New, H.B. van der Raay, Observing the Sun with the Birmingham Solar-Oscillations Network (BISON). Observatory **116**, 32–33 (1996)
M. Collados, A. Lagg, J.J. Díaz Garcí A, E. Hernández Suárez, R. López López, E. Páez Mañá, S.K. Solanki, Tenerife infrared polarimeter II, in *The Physics of Chromospheric Plasmas*, ed. by P. Heinzel, I. Dorotovič, R.J. Rutten. Astronomical Society of the Pacific Conference Series, vol. 368 (2007), p. 611
M. Collados, R. López, E. Páez, E. Hernández, M. Reyes, A. Calcines, E. Ballesteros, J.J. Díaz, C. Denker, A. Lagg, R. Schlichenmaier, W. Schmidt, S.K. Solanki, K.G. Strassmeier, O. von der Lühe, R. Volkmer, GRIS: the GREGOR infrared spectrograph. Astron. Nachr. **333**, 872 (2012). doi:10.1002/asna.201211738
M. Collados, F. Bettonvil, L. Cavaller, I. Ermolli, B. Gelly, A. Pérez, H. Socas-Navarro, D. Soltau, R. Volkmer, EST Team, The European Solar Telescope. Mem. Soc. Astron. Ital. **84**, 379 (2013)
J. de la Cruz Rodríguez, H. Socas-Navarro, Are solar chromospheric fibrils tracing the magnetic field? Astron. Astrophys. **527**, 8 (2011). doi:10.1051/0004-6361/201016018
J. de la Cruz Rodriguez, M. van Noort, H. Uitenbroek, Radiative diagnostics of the solar photosphere and chromosphere (2015, this issue)
J. de la Cruz Rodríguez, H. Socas-Navarro, M. Carlsson, J. Leenaarts, Non-local thermodynamic equilibrium inversions from a 3D magnetohydrodynamic chromospheric model. Astron. Astrophys. **543**, 34 (2012). doi:10.1051/0004-6361/201218825
J. de la Cruz Rodríguez, L. Rouppe van der Voort, H. Socas-Navarro, M. van Noort, Physical properties of a sunspot chromosphere with umbral flashes. Astron. Astrophys. **556**, 115 (2013). doi:10.1051/0004-6361/201321629
B. De Pontieu, A.M. Title, J.R. Lemen, G.D. Kushner, D.J. Akin, B. Allard, T. Berger, P. Boerner, M. Cheung, C. Chou, J.F. Drake, D.W. Duncan, S. Freeland, G.F. Heyman, C. Hoffman, N.E. Hurlburt, R.W. Lindgren, D. Mathur, R. Rehse, D. Sabolish, R. Seguin, C.J. Schrijver, T.D. Tarbell, J.-P. Wülser, C.J. Wolfson, C. Yanari, J. Mudge, N. Nguyen-Phuc, R. Timmons, R. van Bezooijen, I. Weingrod, R. Brookner, G. Butcher, B. Dougherty, J. Eder, V. Knagenhjelm, S. Larsen, D. Mansir, L. Phan, P. Boyle, P.N. Cheimets, E.E. DeLuca, L. Golub, R. Gates, E. Hertz, S. McKillop, S. Park, T. Perry, W.A. Podgorski, K. Reeves, S. Saar, P. Testa, H. Tian, M. Weber, C. Dunn, S. Eccles, S.A. Jaeggli, C.C. Kankelborg, K. Mashburn, N. Pust, L. Springer, R. Carvalho, L. Kleint, J. Marmie, E. Mazmanian, T.M.D. Pereira, S. Sawyer, J. Strong, S.P. Worden, M. Carlsson, V.H. Hansteen, J. Leenaarts, M. Wiesmann, J. Aloise, K.-C. Chu, R.I. Bush, P.H. Scherrer, P. Brekke, J. Martinez-Sykora, B.W. Lites, S.W. McIntosh, H. Uitenbroek, T.J. Okamoto, M.A. Gummin, G. Auker, P. Jerram,

P. Pool, N. Waltham, The Interface Region Imaging Spectrograph (IRIS). Sol. Phys. **289**, 2733–2779 (2014). doi:10.1007/s11207-014-0485-y

C. Debi Prasad, S. Gosain, Comparative study of $LiNbO_3$ and servo controlled air gap Fabry-Perot etalons for solar application. Exp. Astron. **13**, 153–158 (2002). doi:10.1023/A:1025564731907

J.C. del Toro Iniesta, M. Collados, Optimum modulation and demodulation matrices for solar polarimetry. Appl. Opt. **39**, 1637–1642 (2000). doi:10.1364/AO.39.001637

C. Denker, H. Balthasar, A. Hofmann, N. Bello González, R. Volkmer, 2010, The GREGOR Fabry-Perot interferometer: a new instrument for high-resolution solar observations. doi:10.1117/12.856830

D.F. Elmore, T. Rimmele, R. Casini, S. Hegwer, J. Kuhn, H. Lin, J.P. McMullin, K. Reardon, W. Schmidt, A. Tritschler, F. Wöger, The Daniel K. Inouye solar telescope first light instruments and critical science plan, in *Society of Photo-Optical Instrumentation Engineers (SPIE) Conference Series*. Society of Photo-Optical Instrumentation Engineers (SPIE) Conference Series, vol. 9147 (2014), p. 7. doi:10.1117/12.2057038

A. Feller, F.A. Iglesias, K. Nagaraju, S.K. Solanki, S. Ihle, Fast solar polarimeter: description and first results, in *Solar Polarization 7*, ed. by K.N. Nagendra, J.O. Stenflo, Q. Qu, M. Samooprna. Astronomical Society of the Pacific Conference Series, vol. 489 (2014), p. 271

A.M. Gandorfer, H.P.P.P. Steiner, F. Aebersold, U. Egger, A. Feller, D. Gisler, S. Hagenbuch, J.O. Stenflo, Solar polarimetry in the near UV with the Zurich Imaging Polarimeter ZIMPOL II. Astron. Astrophys. **422**, 703–708 (2004). doi:10.1051/0004-6361:20040254

A. Gandorfer, B. Grauf, P. Barthol, T.L. Riethmüller, S.K. Solanki, B. Chares, W. Deutsch, S. Ebert, A. Feller, D. Germerott, K. Heerlein, J. Heinrichs, D. Hirche, J. Hirzberger, M. Kolleck, R. Meller, R. Müller, R. Schäfer, G. Tomasch, M. Knölker, V. Martínez Pillet, J.A. Bonet, W. Schmidt, T. Berkefeld, B. Feger, F. Heidecke, D. Soltau, A. Tischenberg, A. Fischer, A. Title, H. Anwand, E. Schmidt, The filter imager SuFI and the image stabilization and light distribution system ISLiD of the sunrise Balloon-Borne observatory: instrument description. Sol. Phys. **268**, 35–55 (2011). doi:10.1007/s11207-010-9636-y

P.R. Goode, R. Coulter, N. Gorceix, V. Yurchyshyn, W. Cao, The NST: first results and some lessons for ATST and EST. Astron. Nachr. **331**, 620 (2010). doi:10.1002/asna.201011387

S. Gosain, P. Venkatakrishnan, K. Venugopalan, Design and status of Solar Vector Magnetograph (SVM-I) at Udaipur solar observatory. J. Astrophys. Astron. **27**, 285–292 (2006). doi:10.1007/BF02702531

S. Gosain, A.A. Pevtsov, G.V. Rudenko, S.A. Anfinogentov, First synoptic maps of photospheric vector magnetic field from SOLIS/VSM: non-radial magnetic fields and hemispheric pattern of helicity. Astrophys. J. **772**, 52 (2013). doi:10.1088/0004-637X/772/1/52

B.V. Gudiksen, M. Carlsson, V.H. Hansteen, W. Hayek, J. Leenaarts, J. Martínez-Sykora, The stellar atmosphere simulation code bifrost. Code description and validation. Astron. Astrophys. **531**, 154 (2011). doi:10.1051/0004-6361/201116520

N. Hagen, M.W. Kudenov, Review of snapshot spectral imaging technologies. Opt. Eng. **52**(9), 090901 (2013)

G.E. Hale, On the probable existence of a magnetic field in Sun-spots. Astrophys. J. **28**, 315 (1908). doi:10.1086/141602

J.W. Harvey, F. Hill, R.P. Hubbard, J.R. Kennedy, J.W. Leibacher, J.A. Pintar, P.A. Gilman, R.W. Noyes, A.M. Title, J. Toomre, R.K. Ulrich, A. Bhatnagar, J.A. Kennewell, W. Marquette, J. Patron, O. Saa, E. Yasukawa, The Global Oscillation Network Group (GONG) project. Science **272**, 1284–1286 (1996). doi:10.1126/science.272.5266.1284

F. Hill, M.J. Thompson, M. Roth, Workshop report: a new synoptic solar observing network. Space Weather **11**, 392–393 (2013). doi:10.1002/swe.20068

C.L. Hyder, Winking filaments and prominence and coronal magnetic fields. Z. Astrophys. **63**, 78 (1966)

H. Ito, S. Tsuneta, D. Shiota, M. Tokumaru, K. Fujiki, Is the polar region different from the quiet region of the Sun? Astrophys. J. **719**, 131–142 (2010). doi:10.1088/0004-637X/719/1/131

S.A. Jaeggli, H. Lin, D.L. Mickey, J.R. Kuhn, S.L. Hegwer, T.R. Rimmele, M.J. Penn, FIRS: a new instrument for photospheric and chromospheric studies at the DST. Mem. Soc. Astron. Ital. **81**, 763 (2010)

P. Judge, S. White, R. Casini, Coronal measurements (2015, this issue)

P.G. Judge, D.F. Elmore, B.W. Lites, C.U. Keller, T. Rimmele, Evaluation of seeing-induced cross talk in tip-tilt-corrected solar polarimetry. Appl. Opt. **43**, 3817–3828 (2004). doi:10.1364/AO.43.003817

C.U. Keller, Instrumentation for astrophysical spectropolarimetry, in *Astrophysical Spectropolarimetry*, ed. by J. Trujillo-Bueno, F. Moreno-Insertis, F. Sánchez (2002), pp. 303–354

C.U. Keller, Solis Team, The SOLIS Vector-Spectromagnetograph (VSM), in *Advanced Solar Polarimetry—Theory, Observation, and Instrumentation*, ed. by M. Sigwarth. Astronomical Society of the Pacific Conference Series, vol. 236 (2001), p. 16

C.U. Keller, O. von der Luehe, Solar speckle polarimetry. Astron. Astrophys. **261**, 321–328 (1992)

L. Kleint, Spectropolarimetry of C-class flare footpoints. Astrophys. J. **748**, 138 (2012). doi:10.1088/0004-637X/748/2/138

L. Kleint, A. Gandorfer, Prospects of solar magnetometry—from the ground and in space. Space Sci. Rev. (2015, this issue). doi:10.1007/s11214-015-0208-1

R. Komm, S. Gosain, Current and kinetic helicity of long-lived activity complexes. Astrophys. J. **798**, 20 (2015). doi:10.1088/0004-637X/798/1/20

R. Komm, S. Gosain, A.A. Pevtsov, Hemispheric distribution of subsurface kinetic helicity and its variation with magnetic activity. Sol. Phys. **289**, 2399–2418 (2014). doi:10.1007/s11207-014-0477-y

T. Kosugi, K. Matsuzaki, T. Sakao, T. Shimizu, Y. Sone, S. Tachikawa, T. Hashimoto, K. Minesugi, A. Ohnishi, T. Yamada, S. Tsuneta, H. Hara, K. Ichimoto, Y. Suematsu, M. Shimojo, T. Watanabe, S. Shimada, J.M. Davis, L.D. Hill, J.K. Owens, A.M. Title, J.L. Culhane, L.K. Harra, G.A. Doschek, L. Golub, The Hinode (Solar-B) mission: an overview. Sol. Phys. **243**, 3–17 (2007). doi:10.1007/s11207-007-9014-6

N. Krishnappa, A. Feller, Precision in ground-based solar polarimetry: simulating the role of adaptive optics. Appl. Opt. **51**, 7953 (2012). doi:10.1364/AO.51.007953

M.R. Kundu, Measurement of solar magnetic fields from radio observations. Mem. Soc. Astron. Ital. **61**, 431–455 (1990). ISSN 0037-8720

A. Lagg, R. Ishikawa, L. Merenda, T. Wiegelmann, S. Tsuneta, S.K. Solanki, Internetwork horizontal magnetic fields in the quiet Sun chromosphere: results from a joint Hinode/VTT study, in *The Second Hinode Science Meeting: Beyond Discovery-Toward Understanding*, ed. by B. Lites, M. Cheung, T. Magara, J. Mariska, K. Reeves. Astronomical Society of the Pacific Conference Series, vol. 415 (2009), p. 327

A. Lagg, S.K. Solanki, M. van Noort, S. Danilovic, Vigorous convection in a sunspot granular light bridge. Astron. Astrophys. **568**, 60 (2014). doi:10.1051/0004-6361/201424071

E. Landi Degl'Innocenti, M. Landolfi (eds.), Polarization in spectral lines, in *Astrophysics and Space Science Library*, vol. 307 (2004)

J. Leenaarts, M. Carlsson, L. Rouppe van der Voort, The formation of the Hα line in the solar chromosphere. Astrophys. J. **749**, 136 (2012). doi:10.1088/0004-637X/749/2/136

J. Leenaarts, M. Carlsson, V. Hansteen, L. Rouppe van der Voort, Three-dimensional non-LTE radiative transfer computation of the CA 8542 infrared line from a radiation-MHD simulation. Astrophys. J. Lett. **694**, 128–131 (2009). doi:10.1088/0004-637X/694/2/L128

K.D. Leka, K.E. Rangarajan, Effects of 'Seeing' on vector magnetograph measurements. Sol. Phys. **203**, 239–254 (2001). doi:10.1023/A:1013373424510

B.W. Lites, Rotating waveplates as polarization modulators for Stokes polarimetry of the Sun—evaluation of seeing-induced crosstalk errors. Appl. Opt. **26**, 3838–3845 (1987). doi:10.1364/AO.26.003838

B.W. Lites, Performance characteristics of the advanced Stokes polarimeter. Sol. Phys. **163**, 223–230 (1996). doi:10.1007/BF00147996

B.W. Lites, V. Martinez Pillet, A. Skumanich, A quantitative comparison of vector magnetic field measurement and analysis techniques. Sol. Phys. **155**, 1–27 (1994). doi:10.1007/BF00670727

B.W. Lites, K.D. Leka, A. Skumanich, V. Martinez Pillet, T. Shimizu, Small-scale horizontal magnetic fields in the solar photosphere. Astrophys. J. **460**, 1019 (1996). doi:10.1086/177028

B.W. Lites, M. Kubo, H. Socas-Navarro, T. Berger, Z. Frank, R. Shine, T. Tarbell, A. Title, K. Ichimoto, Y. Katsukawa, S. Tsuneta, Y. Suematsu, T. Shimizu, S. Nagata, The horizontal magnetic flux of the quiet-Sun internetwork as observed with the Hinode spectro-polarimeter. Astrophys. J. **672**, 1237–1253 (2008). doi:10.1086/522922

B.W. Lites, D.L. Akin, G. Card, T. Cruz, D.W. Duncan, C.G. Edwards, D.F. Elmore, C. Hoffmann, Y. Katsukawa, N. Katz, M. Kubo, K. Ichimoto, T. Shimizu, R.A. Shine, K.V. Streander, A. Suematsu, T.D. Tarbell, A.M. Title, S. Tsuneta, The Hinode spectro-polarimeter. Sol. Phys. **283**, 579–599 (2013). doi:10.1007/s11207-012-0206-3

Y. Liu, J.T. Hoeksema, X. Sun, Test of the hemispheric rule of magnetic helicity in the Sun using the Helioseismic and Magnetic Imager (HMI) data. Astrophys. J. Lett. **783**, 1 (2014). doi:10.1088/2041-8205/783/1/L1

M.G. Lofdahl, G.B. Scharmer, Application of phase-diversity to solar images, in *Image Reconstruction and Restoration*, ed. by T.J. Schulz, D.L. Snyder. Society of Photo-Optical Instrumentation Engineers (SPIE) Conference Series, vol. 2302 (1994), pp. 254–267

M.G. Löfdahl, G.B. Scharmer, Wavefront sensing and image restoration from focused and defocused solar images. Astron. Astrophys. Suppl. Ser. **107**, 243–264 (1994)

S.F. Martin, H.E. Ramsey, G.A. Carroll, D.C. Martin, Multi-slit spectrograph and H alpha Doppler system. Sol. Phys. **37**, 343–350 (1974). doi:10.1007/BF00152493

M.J. Martínez González, R. Manso Sainz, A. Asensio Ramos, E. Hijano, Dead calm areas in the very quiet Sun. Astrophys. J. **755**, 175 (2012). doi:10.1088/0004-637X/755/2/175

V. Martínez Pillet, J.C. Del Toro Iniesta, A. Álvarez-Herrero, V. Domingo, J.A. Bonet, L. González Fernández, A. López Jiménez, C. Pastor, J.L. Gasent Blesa, P. Mellado, J. Piqueras, B. Aparicio, M. Balaguer, E. Ballesteros, T. Belenguer, L.R. Bellot Rubio, T. Berkefeld, M. Collados, W. Deutsch, A. Feller,

F. Girela, B. Grauf, R.L. Heredero, M. Herranz, J.M. Jerónimo, H. Laguna, R. Meller, M. Menéndez, R. Morales, D. Orozco Suárez, G. Ramos, M. Reina, J.L. Ramos, P. Rodríguez, A. Sánchez, N. Uribe-Patarroyo, P. Barthol, A. Gandorfer, M. Knoelker, W. Schmidt, S.K. Solanki, S. Vargas Domínguez, The Imaging Magnetograph eXperiment (IMaX) for the sunrise Balloon-Borne solar observatory. Sol. Phys. **268**, 57–102 (2011). doi:10.1007/s11207-010-9644-y

S.K. Mathew, A. Bhatnagar, C.D. Prasad, A. Ambastha, Fabry-Perot filter based solar video magnetograph. Astron. Astrophys. Suppl. Ser. **133**, 285–292 (1998). doi:10.1051/aas:1998312

L. Merenda, J. Trujillo Bueno, E. Landi Degl'Innocenti, M. Collados, Determination of the magnetic field vector via the Hanle and Zeeman effects in the He I λ10830 multiplet: evidence for nearly vertical magnetic fields in a polar crown prominence. Astrophys. J. **642**, 554–561 (2006). doi:10.1086/501038

T.R. Metcalf, K.D. Leka, G. Barnes, B.W. Lites, M.K. Georgoulis, A.A. Pevtsov, K.S. Balasubramaniam, G.A. Gary, J. Jing, J. Li, Y. Liu, H.N. Wang, V. Abramenko, V. Yurchyshyn, Y.-J. Moon, An overview of existing algorithms for resolving the 180° ambiguity in vector magnetic fields: quantitative tests with synthetic data. Sol. Phys. **237**, 267–296 (2006). doi:10.1007/s11207-006-0170-x

D.L. Mickey, R.C. Canfield, B.J. Labonte, K.D. Leka, M.F. Waterson, H.M. Weber, The imaging vector magnetograph at Haleakala. Sol. Phys. **168**, 229–250 (1996). doi:10.1007/BF00148052

D. Neidig, P. Wiborg, M. Confer, B. Haas, R. Dunn, K.S. Balasubramaniam, C. Gullixson, D. Craig, M. Kaufman, W. Hull, R. McGraw, T. Henry, R. Rentschler, C. Keller, H. Jones, R. Coulter, S. Gregory, R. Schimming, B. Smaga, The USAF Improved Solar Observing Optical Network (ISOON) and its impact on solar synoptic data bases, in *Synoptic Solar Physics*, ed. by K.S. Balasubramaniam, J. Harvey, D. Rabin. Astronomical Society of the Pacific Conference Series, vol. 140 (1998), p. 519

D. Orozco Suárez, A. Asensio Ramos, J. Trujillo Bueno, The magnetic field configuration of a solar prominence inferred from spectropolarimetric observations in the He I 10830 Å triplet. Astron. Astrophys. **566**, 46 (2014). doi:10.1051/0004-6361/201322903

D. Orozco Suárez, A. Asensio Ramos, J. Trujillo Bueno, Height variation of the vector magnetic field in solar spicules. Astrophys. J. Lett. **803**, 18 (2015). doi:10.1088/2041-8205/803/2/L18

G. Petrie, S. Ettinger, Polar field reversals and active region decay. Space Sci. Rev. (2015, this issue). doi:10.1007/s11214-015-0189-0

A.A. Pevtsov, R.C. Canfield, T.R. Metcalf, Patterns of helicity in solar active regions Astrophys. J. **425**, L117–L119 (1994). doi:10.1086/187324

A. Pietarila, H. Socas-Navarro, T. Bogdan, Spectropolarimetric inversions of the Ca II 8498 and 8542 Å lines in the quiet Sun. Astrophys. J. **670**, 885–902 (2007). doi:10.1086/521271

J. Pietarila Graham, S. Danilovic, M. Schüssler, Turbulent magnetic fields in the quiet Sun: implications of Hinode observations and small-scale dynamo simulations. Astrophys. J. **693**, 1728–1735 (2009). doi:10.1088/0004-637X/693/2/1728

H.P. Povel, C.U. Keller, I.-A. Yadigaroglu, Two-dimensional polarimeter with a charge-coupled-device image sensor and a piezoelastic modulator. Appl. Opt. **33**, 4254–4260 (1994). doi:10.1364/AO.33.004254

K.G. Puschmann, C. Denker, F. Kneer, N. Al Erdogan, H. Balthasar, S.M. Bauer, C. Beck, N. Bello González, M. Collados, T. Hahn, J. Hirzberger, A. Hofmann, R.E. Louis, H. Nicklas, O. Okunev, V. Martínez Pillet, E. Popow, T. Seelemann, R. Volkmer, A.D. Wittmann, M. Woche, The GREGOR Fabry-Pérot interferometer. Astron. Nachr. **333**, 880 (2012). doi:10.1002/asna.201211734

R. Ramelli, S. Balemi, M. Bianda, I. Defilippis, L. Gamma, S. Hagenbuch, M. Rogantini, P. Steiner, J.O. Stenflo, ZIMPOL-3: a powerful solar polarimeter, in *Society of Photo-Optical Instrumentation Engineers (SPIE) Conference Series*, vol. 7735 (2010), p. 1. doi:10.1117/12.857120

M. Rempel, Numerical simulations of quiet Sun magnetism: on the contribution from a small-scale dynamo. Astrophys. J. **789**, 132 (2014). doi:10.1088/0004-637X/789/2/132

M. Rempel, G. Scharmer, Sunspots (2015, this issue)

D. Ren, L. Jolissaint, X. Zhang, J. Dou, R. Chen, G. Zhao, Y. Zhu, Solar ground-layer adaptive optics. Publ. Astron. Soc. Pac. **127**, 469–478 (2015). doi:10.1086/681672

T.R. Rimmele, Solar adaptive optics, in *Adaptive Optical Systems Technology*, ed. by P.L. Wizinowich Society of Photo-Optical Instrumentation Engineers (SPIE) Conference Series, vol. 4007 (2000), pp. 218–231

T.R. Rimmele, ATST Team, The unique scientific capabilities of the Advanced Technology Solar Telescope. Adv. Space Res. **42**, 78–85 (2008). doi:10.1016/j.asr.2008.02.024

T.R. Rimmele, J. Marino, Solar adaptive optics. Living Rev. Sol. Phys. **8**, 2 (2011). doi:10.12942/lrsp-2011-2

T.R. Rimmele, F. Woeger, J. Marino, K. Richards, S. Hegwer, T. Berkefeld, D. Soltau, D. Schmidt, T. Waldmann, Solar multiconjugate adaptive optics at the Dunn Solar Telescope, in *Society of Photo-Optical Instrumentation Engineers (SPIE) Conference Series*, vol. 7736 (2010), p. 31. doi:10.1117/12.857485

M. Rodenhuis, F. Snik, G. van Harten, J. Hoeijmakers, C.U. Keller, Five-dimensional optical instrumentation: combining polarimetry with time-resolved integral-field spectroscopy, in *Polarization: Measurement, Analysis, and Remote Sensing XI*, ed. by D.B. Chenault, D.H. Goldstein. Society of Photo-Optical Instrumentation Engineers (SPIE) Conference Series, vol. 9099 (2014). doi:10.1117/12.2053241

C. Sasso, A. Lagg, S.K. Solanki, Milne-Eddington inversions of the He I 10830 Å Stokes profiles: influence of the Paschen-Back effect. Astron. Astrophys. **456**, 367–371 (2006). doi:10.1051/0004-6361:20065257

T.A. Schad, Spectropolarimetry of fine magnetized structures in the upper solar atmosphere, PhD thesis, The University of Arizona (2013)

T.A. Schad, M.J. Penn, H. Lin, He I vector magnetometry of field-aligned superpenumbral fibrils. Astrophys. J. **768**, 111 (2013). doi:10.1088/0004-637X/768/2/111

G.B. Scharmer, Comments on the optimization of high resolution Fabry-Pérot filtergraphs. Astron. Astrophys. **447**, 1111–1120 (2006). doi:10.1051/0004-6361:20052981

G.B. Scharmer, M. Shand, M.G. Lofdahl, P.M. Dettori, W. Wei, Workstation-based solar/stellar adaptive optics system, in *Adaptive Optical Systems Technology*, ed. by P.L. Wizinowich. Society of Photo-Optical Instrumentation Engineers (SPIE) Conference Series, vol. 4007 (2000), pp. 239–250

G.B. Scharmer, K. Bjelksjo, T.K. Korhonen, B. Lindberg, B. Petterson, The 1-meter Swedish solar telescope, in *Innovative Telescopes and Instrumentation for Solar Astrophysics*, ed. by S.L. Keil, S.V. Avakyan. Society of Photo-Optical Instrumentation Engineers (SPIE) Conference Series, vol. 4853 (2003), pp. 341–350

G.B. Scharmer, G. Narayan, T. Hillberg, J. de la Cruz Rodríguez, M.G. Löfdahl, D. Kiselman, P. Sütterlin, M. van Noort, A. Lagg, CRISP spectropolarimetric imaging of penumbral fine structure. Astrophys. J. Lett. **689**, 69–72 (2008). doi:10.1086/595744

G.B. Scharmer, J. de la Cruz Rodriguez, P. Sütterlin, V.M.J. Henriques, Opposite polarity field with convective downflow and its relation to magnetic spines in a sunspot penumbra. Astron. Astrophys. **553**, 63 (2013). doi:10.1051/0004-6361/201220899

P.H. Scherrer, R.S. Bogart, R.I. Bush, J.T. Hoeksema, A.G. Kosovichev, J. Schou, W. Rosenberg, L. Springer, T.D. Tarbell, A. Title, C.J. Wolfson, I. Zayer, MDI Engineering Team, The solar oscillations investigation—Michelson Doppler imager. Sol. Phys. **162**, 129–188 (1995)

P.H. Scherrer, J. Schou, R.I. Bush, A.G. Kosovichev, R.S. Bogart, J.T. Hoeksema, Y. Liu, T.L. Duvall, J. Zhao, A.M. Title, C.J. Schrijver, T.D. Tarbell, S. Tomczyk, The Helioseismic and Magnetic Imager (HMI) investigation for the Solar Dynamics Observatory (SDO). Sol. Phys. **275**, 207–227 (2012). doi:10.1007/s11207-011-9834-2

D. Schmidt, T. Berkefeld, F. Heidecke, A. Fischer, O. von der Lühe, D. Soltau, GREGOR MCAO looking at the Sun, in *Society of Photo-Optical Instrumentation Engineers (SPIE) Conference Series*, vol. 9148 (2014), p. 1. doi:10.1117/12.2055154

W. Schmidt, O. von der Lühe, R. Volkmer, C. Denker, S.K. Solanki, H. Balthasar, N. Bello Gonzalez, T. Berkefeld, M. Collados, A. Fischer, C. Halbgewachs, F. Heidecke, A. Hofmann, F. Kneer, A. Lagg, H. Nicklas, E. Popow, K.G. Puschmann, D. Schmidt, M. Sigwarth, M. Sobotka, D. Soltau, J. Staude, K.G. Strassmeier, T.A. Waldmann, The 1.5 meter solar telescope GREGOR. Astron. Nachr. **333**, 796 (2012). doi:10.1002/asna.201211725

J. Schou, P.H. Scherrer, R.I. Bush, R. Wachter, S. Couvidat, M.C. Rabello-Soares, R.S. Bogart, J.T. Hoeksema, Y. Liu, T.L. Duvall, D.J. Akin, B.A. Allard, J.W. Miles, R. Rairden, R.A. Shine, T.D. Tarbell, A.M. Title, C.J. Wolfson, D.F. Elmore, A.A. Norton, S. Tomczyk, Design and ground calibration of the Helioseismic and Magnetic Imager (HMI) instrument on the Solar Dynamics Observatory (SDO). Sol. Phys. **275**, 229–259 (2012). doi:10.1007/s11207-011-9842-2

D. Shiota, S. Tsuneta, M. Shimojo, N. Sako, D. Orozco Suárez, R. Ishikawa, Polar field reversal observations with Hinode. Astrophys. J. **753**, 157 (2012). doi:10.1088/0004-637X/753/2/157

F. Snik, J. Craven-Jones, M. Escuti, S. Fineschi, D. Harrington, A.D. Martino, D. Mawet, J. Riedi, J.S. Tyo, An overview of polarimetric sensing techniques and technology with applications to different research fields, in *Polarization: Measurement, Analysis, and Remote Sensing XI*, ed. by D.B. Chenault, D.H. Goldstein. Society of Photo-Optical Instrumentation Engineers (SPIE) Conference Series, vol. 9099 (2014). doi:10.1117/12.2053245

H. Socas-Navarro, J. Trujillo Bueno, E. Landi Degl'Innocenti, Signatures of incomplete Paschen-back splitting in the polarization profiles of the He I λ10830 multiplet. Astrophys. J. **612**, 1175–1180 (2004)

H. Socas-Navarro, J. Trujillo Bueno, E. Landi Degl'Innocenti, Polynomial approximants for the calculation of polarization profiles in the He I 10830 Å multiplet. Astrophys. J. Suppl. Ser. **160**, 312–317 (2005). doi:10.1086/431934

H. Socas-Navarro, J. Trujillo Bueno, B. Ruiz Cobo, Non-LTE inversion of Stokes profiles induced by the Zeeman effect. Astrophys. J. **530**, 977–993 (2000). doi:10.1086/308414

H. Socas-Navarro, D. Elmore, A. Pietarila, A. Darnell, B.W. Lites, S. Tomczyk, S. Hegwer, Spinor: visible and infrared spectro-polarimetry at the National Solar Observatory. Sol. Phys. **235**, 55–73 (2006). doi:10.1007/s11207-006-0020-x

S.K. Solanki, N.A. Krivova, J.D. Haigh, Solar irradiance variability and climate. Annu. Rev. Astron. Astrophys. **51**, 311–351 (2013). doi:10.1146/annurev-astro-082812-141007

L. Staveland, Determination of the spread function for solar stray light. Sol. Phys. **12**, 328–331 (1970). doi:10.1007/BF00227125
J. Stenflo (ed.), Solar magnetic fields: polarized radiation diagnostics, in *Astrophysics and Space Science Library*, vol. 189 (1994)
J. Stenflo, History of solar magnetic fields since George Ellery Hale. Space Sci. Rev. (2015, this issue). doi:10.1007/s11214-015-0198-z
J.O. Stenflo, Solar magnetic fields as revealed by Stokes polarimetry. Astron. Astrophys. Rev. **21**, 66 (2013). doi:10.1007/s00159-013-0066-3
J.O. Stenflo, A.G. Kosovichev, Bipolar magnetic regions on the Sun: global analysis of the SOHO/MDI data set. Astrophys. J. **745**, 129 (2012). doi:10.1088/0004-637X/745/2/129
T. Tadesse, T. Wiegelmann, S. Gosain, P. MacNeice, A.A. Pevtsov, First use of synoptic vector magnetograms for global nonlinear, force-free coronal magnetic field models. Astron. Astrophys. **562**, 105 (2014a). doi:10.1051/0004-6361/201322418
T. Tadesse, A.A. Pevtsov, T. Wiegelmann, P.J. MacNeice, S. Gosain, Global solar free magnetic energy and electric current density distribution of Carrington rotation 2124. Sol. Phys. **289**, 4031–4045 (2014b). doi:10.1007/s11207-014-0581-z
S.K. Tiwari, M. van Noort, A. Lagg, S.K. Solanki, Structure of sunspot penumbral filaments: a remarkable uniformity of properties. Astron. Astrophys. **557**, 25 (2013). doi:10.1051/0004-6361/201321391
S. Trippe, [Review] polarization and polarimetry. J. Korean Astron. Soc. **47**, 15–39 (2014). doi:10.5303/JKAS.2014.47.1.15
A. Tritschler, T.R. Rimmele, S. Berukoff, R. Casini, S.C. Craig, D.F. Elmore, R.P. Hubbard, J.R. Kuhn, H. Lin, J.P. McMullin, K.P. Reardon, W. Schmidt, M. Warner, F. Woger, DKIST: observing the Sun at high resolution, in *18th Cambridge Workshop on Cool Stars, Stellar Systems, and the Sun*, ed. by G.T. van Belle, H.C. Harris. Cambridge Workshop on Cool Stars, Stellar Systems, and the Sun, vol. 18 (2015), pp. 933–944
J. Trujillo Bueno, Polarized radiation observables for probing the magnetism of the outer solar atmosphere, in *Solar Polarization 7*, ed. by K.N. Nagendra, J.O. Stenflo, Q. Qu, M. Samooprna. Astronomical Society of the Pacific Conference Series, vol. 489 (2014), p. 137
J. Trujillo Bueno, E. Landi degl'Innocenti, UV polarized radiation observables for probing the upper chromosphere (2015, this issue)
J. Trujillo Bueno, E. Landi Degl'Innocenti, M. Collados, L. Merenda, R. Manso Sainz, Selective absorption processes as the origin of puzzling spectral line polarization from the Sun. Nature **415**, 403–406 (2002)
S. Tsuneta, K. Ichimoto, Y. Katsukawa, S. Nagata, M. Otsubo, T. Shimizu, Y. Suematsu, M. Nakagiri, M. Noguchi, T. Tarbell, A. Title, R. Shine, W. Rosenberg, C. Hoffmann, B. Jurcevich, G. Kushner, M. Levay, B. Lites, D. Elmore, T. Matsushita, N. Kawaguchi, H. Saito, I. Mikami, L.D. Hill, J.K. Owens, The solar optical telescope for the Hinode mission: an overview. Sol. Phys. **249**, 167–196 (2008). doi:10.1007/s11207-008-9174-z
P.J. Turon, P.J. Léna, High resolution solar images at 10 microns: sunspot details and photometry. Sol. Phys. **14**, 112–124 (1970). doi:10.1007/BF00240164
M. van Noort, Spatially coupled inversion of spectro-polarimetric image data. I. Method and first results. Astron. Astrophys. **548**, 5 (2012). doi:10.1051/0004-6361/201220220
M. van Noort, L. Rouppe van der Voort, M.G. Löfdahl, Solar image restoration by use of multi-frame blind de-convolution with multiple objects and phase diversity. Sol. Phys. **228**, 191–215 (2005). doi:10.1007/s11207-005-5782-z
A. Vögler, M. Schüssler, A solar surface dynamo. Astron. Astrophys. **465**, 43–46 (2007). doi:10.1051/0004-6361:20077253
O. von der Luehe, Speckle imaging of solar small scale structure. I—methods. Astron. Astrophys. **268**, 374–390 (1993)
S. Wedemeyer, T. Bastian, R. Brajsa, M. Barta, H. Hudson, G. Fleishman, M. Loukitcheva, B. Fleck, E.P. Kontar, B. De Pontieu, S.K. Tiwari, Y. Kato, R. Soler, P. Yagoubov, J.H. Black, P. Antolin, E. Scullion, S. Gun'ar, N. Labrosse, A.O. Benz, H.-G. Ludwig, P. Hauschildt, J.G. Doyle, V.M. Nakariakov, S.K. Solanki, S.M. White, T. Ayres, P. Heinzel, M. Karlicky, T. Van Doorsselaere, D. Gary, C.E. Alissandrakis, A. Nindos, L. Rouppe van der Voort, M. Shimojo, T. Zaqarashvili, E. Perez, Solar science with the Atacama large millimeter/submillimeter array—a revolutionizing new view of our Sun. ArXiv e-prints (2015)
T. Wiegelmann, G.J.D. Petrie, P. Riley, Coronal magnetic field models. Space Sci. Rev. (2015, this issue). doi:10.1007/s11214-015-0178-3
T. Wiegelmann, J.K. Thalmann, C.J. Schrijver, M.L. Derosa, T.R. Metcalf, Can we improve the preprocessing of photospheric vector magnetograms by the inclusion of chromospheric observations? Sol. Phys. **247**, 249–267 (2008). doi:10.1007/s11207-008-9130-y

Space Sci Rev (2017) 210:77–108
DOI 10.1007/s11214-015-0189-0

Polar Field Reversals and Active Region Decay

Gordon Petrie[1] · **Sophie Ettinger**[2]

Received: 19 May 2015 / Accepted: 15 July 2015 / Published online: 29 July 2015

Abstract We study the relationship between polar field reversals and decayed active region magnetic flux. Photospheric active region flux is dispersed by differential rotation and turbulent diffusion, and is transported poleward by meridional flows and diffusion. We summarize the published evidence from observation and modeling of the influence of meridional flow variations and decaying active region flux's spatial distribution, such as the Joy's law tilt angle. Using NSO Kitt Peak synoptic magnetograms covering cycles 21–24, we investigate in detail the relationship between the transport of decayed active region flux to high latitudes and changes in the polar field strength, including reversals in the magnetic polarity at the poles. By means of stack plots of low- and high-latitude slices of the synoptic magnetograms, the dispersal of flux from low to high latitudes is tracked, and the timing of this dispersal is compared to the polar field changes. In the most abrupt cases of polar field reversal, a few activity complexes (systems of active regions) are identified as the main cause. The poleward transport of large quantities of decayed trailing-polarity flux from these complexes is found to correlate well in time with the abrupt polar field changes. In each case, significant latitudinal displacements were found between the positive and negative flux centroids of the complexes, consistent with Joy's law bipole tilt with trailing-polarity flux located poleward of leading-polarity flux. The activity complexes of the cycle 21 and 22 maxima were larger and longer-lived than those of the cycle 23 and 24 maxima, and the poleward surges were stronger and more unipolar and the polar field changes larger and faster. The cycle 21 and 22 polar reversals were dominated by only a few long-lived complexes whereas the cycle 23 and 24 reversals were the cumulative effects of more numerous, shorter-lived regions. We conclude that sizes and lifetimes of activity complexes are key to understanding the diversity of polar reversals.

✉ G. Petrie
gpetrie@nso.edu

S. Ettinger
saettinger@uchicago.edu

1 National Solar Observatory, 950 N. Cherry Ave., Tucson, AZ 85719, USA

2 University of Chicago, 5801 South Ellis Avenue, Chicago, IL 60637, USA

Keywords Solar magnetic fields · Activity cycle · Solar poles · Photosphere · Chromosphere · Corona · Solar interior · Magnetic flux-transport models

1 Introduction

The Sun's polar fields are globally influential. They dominate the large-scale structure of the solar corona and the heliosphere (Hoeksema 1984), they supply the majority of the interplanetary mean field, and channel the fast solar wind (Wang et al. 2009). They are also believed to form the seed fields of new solar cycles through subsurface equator-ward meridional flow and differential rotation (Charbonneau 2010). In this review we are interested in their cyclical evolution and their formation from the decay and transport of strong fields from low latitudes.

We focus on observations of three classes of photospheric magnetic field: active region fields, which are compact, intense, flux-balanced and usually bipolar in structure, located equatorward of about ±30°; polar fields, which are located poleward of ±60° and are relatively weak, almost unipolar and have large spatial scale, so that over most of the solar cycle they appear dipole-like; and high-latitude surges of field forming from decaying active regions, observed to stream poleward between about ±30° and ±60° from active latitudes in plume-like formations. These fields can all be seen in Fig. 1, which shows a butterfly diagram (time-latitude) plot of the radial photospheric flux density distribution over the past 40 years. This plot was constructed using line-of-sight full-disk photospheric magnetograms. Assuming that the photospheric field is approximately radial (Svalgaard et al. 1978; Wang and Sheeley 1992; Petrie and Patrikeeva 2009), the line-of-sight flux density measurements were converted to radial flux density by dividing by $\cos\rho$, where ρ is the heliocentric angle between the line-of-sight vector and the local vertical. Then the measurements were averaged in longitude over a longitudinal range (here within ±30° of central meridian). Finally the two-dimensional butterfly plot was formed by locating all longitudinal image averages along the time axis. The solar rotation axis is tilted at an angle of 7.25° with respect to the ecliptic plane, and so the fields near the solar poles are either observed with very large viewing angles or, for six months at a time, not observed at all. The locations in the butterfly diagram nearest the poles were filled or corrected using values for these fields estimated using spatial and temporal interpolation as described in Petrie (2012). The active regions, located equatorward of ±40°, have much larger flux density (a few kG) than the high-latitude fields. However, because the active regions are generally bipolar and approximately flux-balanced, and the butterfly diagram shows longitudinally-averaged measurements, the diagram emphasizes both the latitudinal flux biases of the active latitudes and the more unipolar fields of the high latitudes. Thus the butterfly diagram is an excellent aid to the study of flux transport across the solar surface.

These three classes of field, the active regions, poleward surges and polar fields, are observed to interact over the solar cycle. Solar cycles wax and wane with an approximately 11-year period: maxima and minima of activity appear approximately every 11 years, but the magnetic polarity patterns of the bipolar active regions and the polar fields swap every 11-year cycle, so that the magnetic cycle period is about 22 years. During activity maxima, sunspots, flares and coronal mass ejections (CMEs) are strongest and most numerous. Solar minima are characterized by a lack of magnetic activity, but at these times the polar fields are strongest. The polar fields wax and wane cyclically with a phase lag of approximately 90° relative to the 22-year magnetic activity cycle. They maximum and

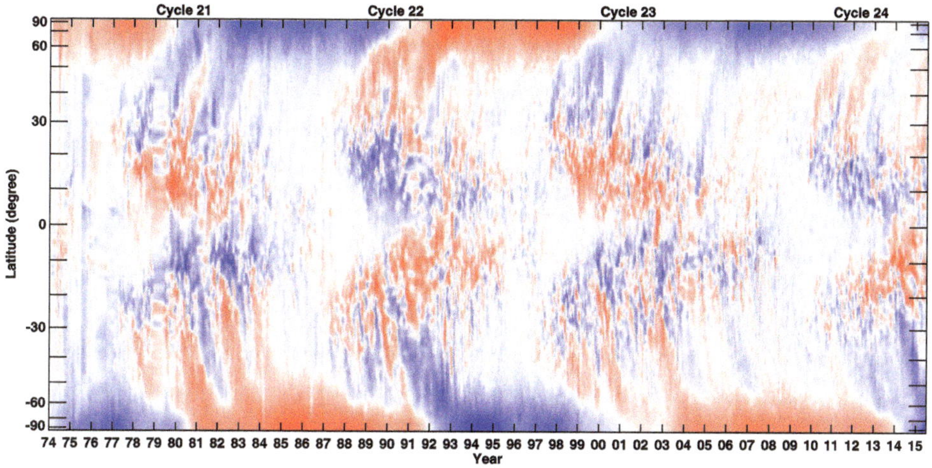

Fig. 1 Butterfly diagram based on Kitt Peak magnetogram data, summarizing the photospheric radial field distributions derived from the longitudinal photospheric field measurements. Each pixel is colored to represent the average field strength at each time and latitude. *Red/blue* represents positive/negative flux, with the color scale saturated at 15 G. Updated from Petrie (2012, 2013)

minimum strength in anti-phase with the activity maxima and minima and their polarity reversals occur during activity maxima when the active regions are most numerous and influential.

While the lifetime of an individual active region is only a few months, the evolution of the polar fields is well organized over much longer timescales—see Fig. 2. The curves in Fig. 2 represent longitudinal averages of measurements within a range of polar latitudes, and are averaged also in time, but their coherent behavior over long timescales is clear in the figure. This implies that the decayed active region flux that reaches the poles has systematic biases over these long timescales. It is well known that, according to Hale's law (often called Hale's polarity rule, Hale and Nicholson 1925), the active region fields usually appear in bipolar formations, whose leading and trailing polarities generally match in each hemisphere, are opposite in the north and south hemispheres, and swap in each hemisphere every 11-year cycle. According to Joy's law (Hale et al. 1919) these bipoles have an average tilt such that the leading polarity is generally closer to the equator than the trailing polarity. The leading polarity also tends to be more compact and intense than the trailing polarity. When the dipole decays and its flux is dispersed by global near-surface flows, the decayed flux that travels poleward therefore tends to have a net bias of trailing polarity. Because the polar field polarity matches the leading dipole polarity in each hemisphere at the beginning of each cycle, the decayed flux reaching the polar latitudes has opposite net polarity to the polar fields, and reverses them (Babcock 1961).

Figure 1 clearly shows that, while the activity cycles share the same overall latitude asymmetries, with leading-/trailing-polarity flux located equatorward/poleward on average, with poleward surges dominated by trailing-polarity flux and polar fields changing accordingly in a cyclic fashion, the character of the activity emergence patterns, poleward surges and polar field variations change significantly from cycle to cycle. In particular, the sizes and strengths of the poleward surges of cycles 21 and 22 are generally greater than those of cycles 23 and 24. Strong, giant surges only tend to be found in cycles 21 and 22, with the cycle 22 maximum southern high-latitude fields being dominated by a single especially

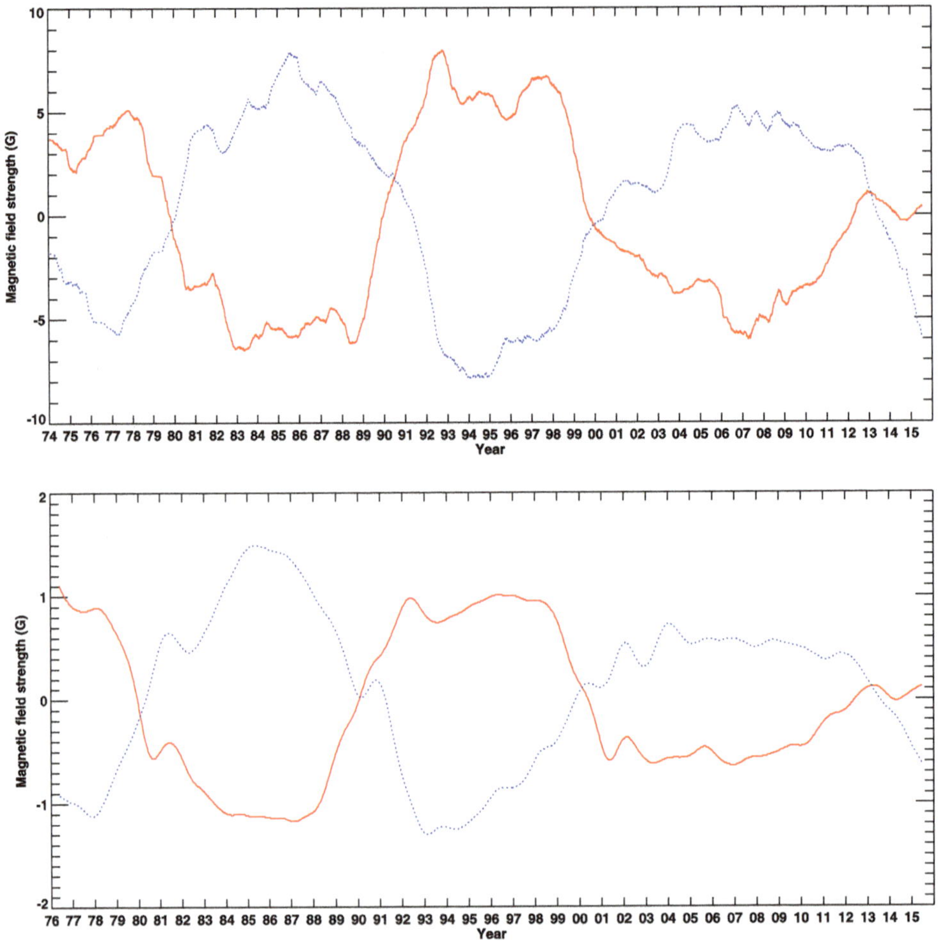

Fig. 2 10-day averages of the north (*red solid lines*) and south (*blue dotted lines*) polar fields measured at Kitt Peak (*top*) and Wilcox (*bottom*) are plotted against time. The Kitt Peak data are for the radial field component and derive from heliographic latitudes $\pm 65^\circ$. The Wilcox measurements are for the line-of-sight field component and come from the $3'$ apertures nearest the poles, which cover between about $\pm 55^\circ$ and the poles. Updated from Petrie (2013)

intense negative-polarity surge. The cycle 23 and 24 high-latitude fields tend to appear in the form of multiple, episodic surges of mixed polarity. The polar field changes shown in Fig. 2 are correspondingly larger and more abrupt in general during cycles 21 and 22 than during cycles 23 and 24. Figure 2 shows that the cycle 24 polar reversal is well advanced in the south at the time of writing. This is not the case in the north, where multiple sign reversals have occurred since late 2012. The WSO polar data, which include lower latitudes, between about $\pm 55^\circ$ and the poles, indicate a recent change of sign to positive field. The NSO data, representing $\pm 65^\circ$ latitude, also shows this latest sign reversal, with a time lag compared to the WSO data. The latest positive poleward surge in the north shown in Fig. 1 indicates that this reversal is on its way to the pole. In Sect. 8 we will explore the photospheric field measurements in more detail, covering both low and high latitudes during the periods of polar field reversal.

The photosphere and interior are fluid-dominated media, and the surface magnetic field is advected by the plasma flows. The large-scale cyclical organization of the global photospheric field is therefore closely related to the large-scale photospheric flow patterns: the turbulent diffusion, meridional flow, and differential rotation. Active region fields are weakened and dispersed by the turbulent granulation of the photosphere, whose large-scale effects resemble those of diffusion, and have usually been modeled as such since the landmark work of Leighton (1964). The differential rotation also expands the field structures longitudinally. Though this has no direct effect on the latitude motion of the field, the weakening of the field may influence its advection by other flows. The slow, systematic poleward meridional flow plays a major role in transporting the decayed active region flux poleward. The interplay of the meridional flow and the Joy's law tilt will be a major focus of this paper.

Petrie et al. (2014) recently reviewed observations and models of the interactions between active regions and the polar field, focusing in particular on the interrelated phenomena that are observed to migrate in both directions across the high-latitude corridor between the active and polar latitudes, then Petrie (2015) surveyed the observed properties of the polar fields and the flux transport modeling of their evolution, before reviewing atmospheric phenomena that clearly exhibit signs of the polar fields' global influence. In this review we will examine the dependence of the polar field variations on the flux and flow patterns at low latitudes, focusing on the effects of Joy's law dipole tilt and meridional flow variations, as well as summarizing photospheric field observations at low and high latitudes during the past four (cycles 21–24) polar reversals. The review is organized as follows. Section 2 recaps the global features of photospheric flux transport processes, and Sect. 3 summarizes evidence that the Babcock-Leighton model for the cycle is essentially correct. Sections 4 and 5 focus on the properties and effects of background meridional flows and local perturbations to these flows, and Sect. 6 explores the influence of Joy's law bipole tilt variations on polar fields. Section 7 presents evidence of varying flux transfer between the north and south hemispheres. Section 8 goes into more detail regarding the relationship between individual active regions' decaying field and changes in the high-latitude net flux, and shows how the proper ties of these fields can vary from cycle to cycle, producing polar changes of varying size and abruptness. We summarize in Sect. 9.

2 Solar Flux Transport

Flux transport dynamo models have been the main theoretical tool for understanding the origins of the solar cycle and solar activity, and have produced a complex theoretical picture of how the solar dynamo may operate and how the solar convection zone and tachocline may behave. The global solar magnetic flux transport processes of interest to us here are summarized in Fig. 3, with the different stages of the cycle represented by schematic diagrams labeled (*a–i*). The figure is based on the phenomenological model of Babcock (1961) from which Leighton (1969) developed his numerical kinematic model for the solar cycle. This phenomenological model is usually referred to as the Babcock-Leighton model. The sequence begins with a dipole-like field characteristic of solar activity minimum. Because low latitudes rotate faster than high latitudes (Snodgrass 1983), this dipole field is sheared (Fig. 3, panel *a*) such that strong toroidal fields are created, of opposite signs in the two hemispheres (panel *b*). When the toroidal field becomes strong enough the flux ropes become buoyant, instabilities develop, and loop structures grow and rise to the surface (Parker 1955, panel *c*). They twist and tilt as they rise because of the Coriolis force. Tilted bipoles

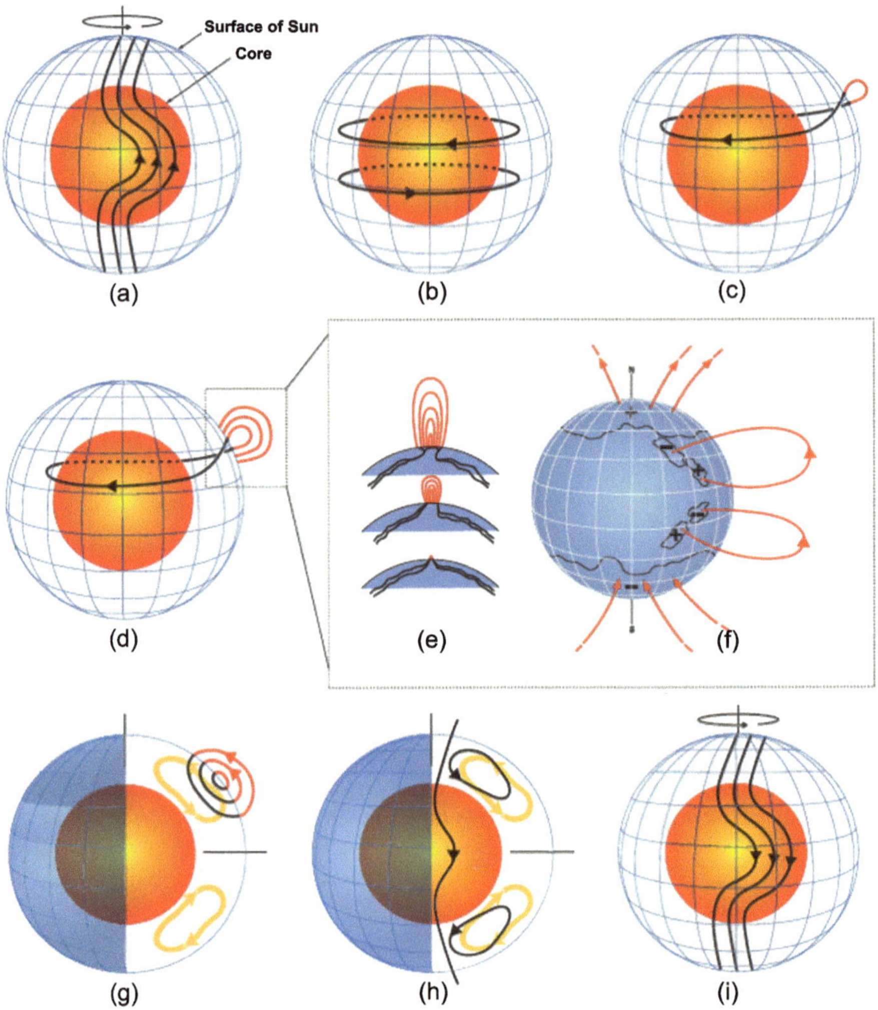

Fig. 3 Schematic diagram of the major global photospheric flux-transport dynamo processes. The *red inner sphere* represents the Sun's radiative core and the *blue mesh* the solar surface. In between is the solar convection zone. (**a**) Shearing of poloidal field by the Sun's differential rotation near the bottom of the convection zone. (**b**) The toroidal field resulting from this shearing by differential rotation. (**c**) When the toroidal field is strong enough, buoyant loops rise to the surface, twisting as they rise due to the Coriolis force. Tilted bipoles are formed from these loops. (**d–f**) Additional flux emerges (**d**, **e**) and spreads (**f**) in latitude and longitude from decaying spots. (**g**) Meridional flow (*yellow circulation* with *arrows*) carries surface magnetic flux poleward, causing the polar fields to reverse. (**h**) Some of this flux is transported downward to the bottom and towards the equator. These poloidal fields have sign opposite to those at the beginning of the sequence, in (**a**). (**i**) This reversed poloidal flux is sheared by the differential rotation to produce the new toroidal field opposite in sign to that shown in (**b**). From Dikpati and Gilman (2007). Babcock (1961) did not explicitly include meridional circulation in his model, though he speculated about its existence

are formed at the photospheric surface by these emerging loop fields (panels *d–f*). As the loops continue to emerge and expand the bipoles spread in longitude and latitude. Meridional flow carries the photospheric flux poleward (panel *g*). Because of the Joy's law tilt the flux reaching the pole tends to be mostly of trailing polarity, causing the polar fields to

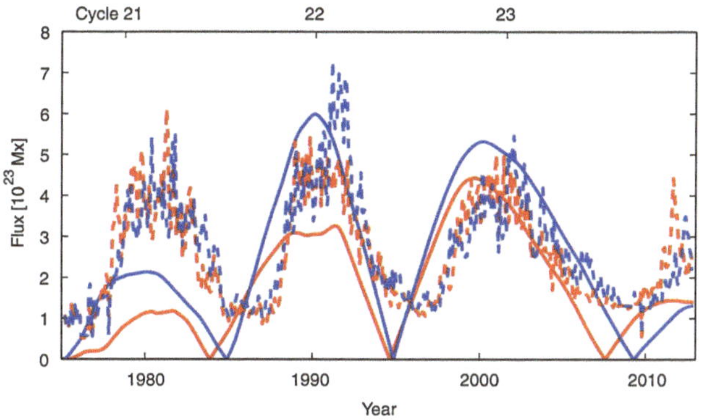

Fig. 4 Hemispheric total unsigned surface flux (*dashed lines*) from synoptic magnetograms and the modulus of the net toroidal flux (*solid lines*) are given for the last three solar cycles. *Red lines* refer to the northern and *blue lines* to the southern hemisphere. From Cameron and Schüssler (2015)

reverse polarity (panel *h*). Some of this polar flux is transported into the interior and equatorward by the meridional flow. The resulting poloidal field has polarity opposite to that of the dipole-like field at the beginning of the cycle, and a new cycle is created from this dipole (panel *i*).

The Babcock-Leighton phenomenological model is one important example of a flux transport dynamo models, in which the Joy's law tilt of the active regions is the source of the poloidal field component that results in the polar fields. This model was the first attempt to unite the observed cyclical patterns of bipolar active regions and polar fields into a coherent picture of the solar cycle, and Babcock-Leighton modeling work continues to the present day. In this paper we focus on this phenomenological picture. Charbonneau (2010) offers much more general discussion of solar dynamo models.

3 Polar Field as Source Field of Succeeding Cycle

In this section we present evidence that the description of the solar cycle provided by the Babcock-Leighton model is essentially correct. In the last steps of the cycle shown in Fig. 3 the activity of the new cycle is produced by the interior differential rotation acting on the poloidal field associated with the polar fields of the previous cycle minimum. This suggests a relationship between the strength of one cycle's polar field and the strength of the succeeding activity cycle. Schatten et al. (1978) first suggested the use of polar magnetic field strength around sunspot minimum as a predictor of the amplitude of the next activity cycle. Using various proxies including polar faculae and corona shape at eclipses, they predicted a value between ~ 120 and 180 for the maximum smoothed monthly sunspot number in cycle 21. This forecast proved to be correct (the actual value being 165). A later forecast for cycle 22 using WSO polar field measurements was similarly successful (Schatten and Sofia 1987).

A limitation of the polar precursor method is that the prediction becomes available at the time of solar minimum, only 3–4 years ahead of the maximum. In order to remedy this situation, Schatten and Pesnell (1993) introduced a combined predictor, called the Solar Dynamo Amplitude (SoDA) index, exploiting the fact that the poloidal (polar) and toroidal (active-region) fields appear oscillate between maxima and minima, like sine and cosine functions such that the sum of their squared amplitudes may be introduced as a phase-independent variable with only long-term amplitude variations. Around minimum, the SoDA index is equivalent to the polar precursor and its value yields the prediction for the next cycle activity maximum (Schatten et al. 1996). Its gradual change, however, makes it possible to use it for prediction purposes well before the actual minimum. In the case of cycle 23, a prediction within ±10 % or so of the actual peak was possible as early as 1993. This success strengthens the case that the phenomenological model described in Sect. 2 is essentially correct. More recently, Muñoz-Jaramillo et al. (2013) analyzed a MWO data set of photospheric polar faculae (1917–1985) and found reasonable correlation between the proxy polar field strength (facular count) and the cycle amplitude (maximum sunspot area) of the next cycle. Their result, like the success of polar precursor cycle prediction methods and the success of the SoDA index, strongly suggest that the polar field of a given cycle acts as the seed field for the next cycle, as in the Babcock-Leighton model.

Cameron and Schüssler (2015) argued that the net toroidal flux generated by differential rotation in the convection zone in each hemisphere can be observed from the observed surface magnetic flux distribution. The main assumptions are that the differential rotation angular velocity is independent of radius in the equatorial plane, the magnetic field does not penetrate the bottom of the convection zone, and the angular velocity is zero at the rotation axis. In a reference frame rotating with the angular velocity at the equator, the kinematic induction equation can be reduced using Stokes' theorem to an expression for the toroidal flux time derivative in terms of the surface magnetic flux density, the differential rotation profile, and the toroidal flux, which is difficult to estimate from first principles. However, assuming zero net toroidal flux in February 1975, near cycle 20 solar minimum when they assume the net toroidal field to be close to sign-reversal, Cameron and Schüssler (2015) integrated the inductive term of their equation in time. The resulting net toroidal flux integral, which includes the flux diffusion term after the first timestep, is compared in Fig. 4 to the unsigned magnetic flux for each hemisphere. The figure shows that the toroidal flux estimate is of the same order of magnitude (around 1–5 $\times 10^{23}$ Mx) as the unsigned surface flux, and it takes peak values around activity maximum, with a phase shift due to the toroidal flux diffusion term. The toroidal flux cycle peaks earlier than the activity cycle because of the ongoing decay of the previous cycle's flux, but because the new cycle flux is decaying at the same rate, the net toroidal flux amplitude is only weakly dependent on the decay rate. The main effect of the decay term is the phase shift. The authors argued that the toroidal flux generated from the polar fields is sufficient to explain the observed surface flux emergence over time, thus reinforcing the conjecture of Babcock (1961) the observed polar fields represent the poloidal field source for the subsurface toroidal field.

In the following sections we will discuss the possible impact of specific flux-transport processes on the character of the polar fields. We will focus in particular on the Joy's law tilt that the active regions are observed to emerge with, the meridional flows that help to carry the decayed active region fields towards the poles, and the interplay of these phenomena with the diffusive processes that weaken and disperse the active region fields.

4 Meridional Flow

Many modeling studies have focused on the influence of the meridional flow on the global transport of photospheric flux. Although the possible existence of poleward near-surface meridional flow was discussed by Babcock (1961), such flows were not included in his phenomenological model, nor in the subsequent numerical kinematic modeling of Leighton (1969), although interior equator ward flows did play a role in these models. When reliable measurements of meridional flows became available, in Doppler measurements by Duvall (1979) and Labonte and Howard (1982), followed by photospheric feature-tracking (Komm et al. 1993), Doppler (Ulrich 1993) and helioseismic observations (Hathaway et al. 1996) of poleward surface meridional flows of 20 $\mathrm{m\,s^{-1}}$, meridional flows began to appear explicitly

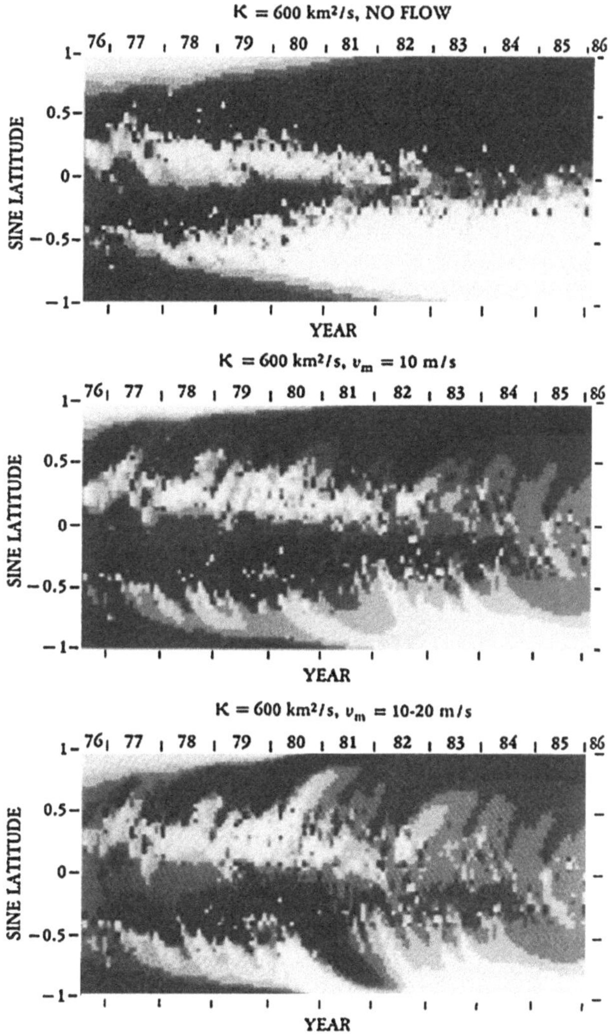

Fig. 5 Gray-scale maps showing the latitudinal field distributions as a function of time, representing the simulated evolution of photospheric fields in the presence of diffusion at a rate $\kappa = 600\ \mathrm{km^2\,s^{-1}}$ and meridional flows of different speeds as follows. *Top*: $v = 0\ \mathrm{m\,s^{-1}}$. *Middle*: $v = 10\ \mathrm{m\,s^{-1}}$. *Bottom*: $v = 20\ \mathrm{m\,s^{-1}}$. These plots show that the poleward transport of flux by diffusion alone does not produce the observed surges, and yields a polar field that is too weak and too broadly spread over latitude. Poleward flow and equatorward diffusion combine to establish strong, concentrated fields, as observed; however, the simulation with $v = 10\ \mathrm{m\,s^{-1}}$ (*middle*) does not reproduce the observed giant surges and the accompanying fluctuations of the polar field strengths during 1980–1982. The simulation with $v = 20\ \mathrm{m\,s^{-1}}$ (*bottom*) reproduces better the giant surges observed near sunspot maximum and the corresponding fluctuations in the polar field strength. From Wang et al. (1989)

in models. Sheeley et al. (1985) extended the photospheric flux transport work of Leighton (1969) by adding a background meridional flow explicitly to the model, and by modeling the flux transport in the two surface spatial dimensions over an entire solar cycle. The main effect of the meridional flow was to weaken the mean photospheric field by disrupting long-lived patterns of strong photospheric flux, and by removing active region flux from active to polar latitudes.

The important influence of meridional flow in polar field reversal is demonstrated in Figs. 5 and 6. The top plot in Fig. 5 shows a butterfly diagram of a kinematic photospheric flux transport simulation that does not include the effects of meridional flow. The polar fields reverse in this simulation, but much more slowly and smoothly than observed (compare this plot with Fig. 1). The decayed active region flux reaches polar latitudes due to the diffusion model alone, and so the poleward surges are very smooth and diffuse. The addition of meridional flows causes the poleward surges to be more structured with alternating polarity, as seen in Fig. 1, and also leads to a fluctuating polar field with multiple polarity reversals

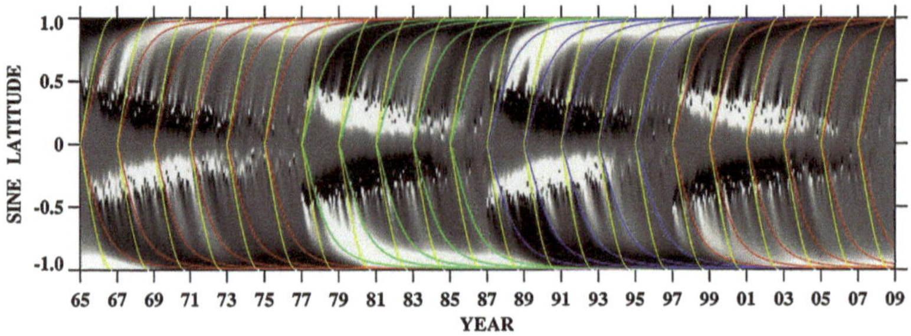

Fig. 6 Latitude-time plot of the simulated photospheric field. Each column of pixels represents a 27.3-day Carrington map of the photospheric field averaged over longitude. The *gray scale* ranges from -4 G (*black*) to $+4$ G (*white*). Overlaid *color curves* indicate the trajectories that flux elements would follow if their poleward transport were determined solely by the underlying meridional flow field $v(L) = \pm v_m \sin 0.1|L| \cos 1.8L$ (*red*: $v_m = 17\ \mathrm{m\,s^{-1}}$; *green*: $v_m = 15.5\ \mathrm{m\,s^{-1}}$; *blue*: $v_m = 14.5\ \mathrm{m\,s^{-1}}$). These values for v_m are optimized for each solar cycle, so as to approximately match the measured polar field strengths during 1976, 1986, 1996, and 2009. Also over-plotted (*yellow curves*) are the trajectories corresponding to a latitude-independent poleward flow $v(L) = \pm 20\ \mathrm{m\,s^{-1}}$. The slopes of the poleward surges are closer to those of the constant flows because of the contribution of supergranular diffusion to the flux transport. From Wang et al. (2009)

that are seen in Fig. 2, where even the spatially- and temporally-averaged polar measurements can change multiple times during reversal, e.g., during the cycle 24 reversal at the north pole, which is still in progress at the time of writing. Only when a fast meridional flow profile, with amplitude around 20 $\mathrm{m\,s^{-1}}$ is included does the model produce the giant surges observed during some solar maxima, e.g., during cycles 21 and 22 in Fig. 6. As well as producing more structured polar changes, meridional flows also affect the sizes of the polar changes. Whereas a diffusion-like process increases the latitude separation of the two polarities in a tilted dipole, the effect of the meridional flow on the bipole's flux distribution is highly dependent on the amplitude and shape of the meridional flow profile. For example, a fast meridional flow will sweep a dipole poleward before the diffusion has a chance to separate the polarities significantly, which inhibits the lead polarity from interacting with opposite-polarity from the opposite hemisphere. The net effect of the dipole flux on the polar field would be small in this case. If, on the other hand, the meridional flow is slow enough that the dipole flux can expand significantly then more of the leading-polarity flux can remain at low latitudes, allowing the leading-polarity flux to dominate the high latitudes and to have a greater net effect on the polar fields.

This behavior is illustrated in Fig. 6, which shows a flux-transport simulation representing cycles 20–23. This model is designed to match the polar field measurements from MWO covering these cycles (1965–2009). In this model, identical bipoles were introduced at a rate proportional to the sunspot number, with fixed Joy's law tilt and latitude distribution. The meridional flow speed was allowed to vary from cycle to cycle but was kept constant during each cycle. The meridional flow amplitudes assigned to cycles 20–23 were 17, 15.5, 14.5 and 17 $\mathrm{m\,s^{-1}}$. The two fasted meridional flow profiles produced the weakest polar fields and poleward surges of more mixed polarity, and the two slowest meridional flow profiles produced giant, nearly unipolar surges at solar maximum and the strings polar fields.

It is therefore clear that the meridional flows play a key role in defining the character of a solar cycle. Schrijver and Liu (2008) and Dikpati (2011) have argued that not only the amplitude but the shape of the meridional flow profile can have a profound effect on the polar field changes. Comparing the effects of profiles with different flow gradients near the equator,

Schrijver and Liu (2008) obtained weak/strong axial dipole components (i.e., polar fields) in models with strong/weak flow gradients near the equator. This is because strong/weak low-latitude poleward flows inhibit/promote interaction and cancelation of northern- and southern-hemispheric lead-polarity flux across the equator, producing small/large net fluxes in the two hemispheres, and small/large polar field changes. Dikpati (2011) compared the effects on models of changing the latitude of the peak meridional flow speed. Flows peaking at low latitudes tend to cause leading polarities to catch up with trailing polarities on their way poleward, enhancing cancellation between them, and weakening their effect on the polar fields. Flows peaking at high latitudes help the trailing polarities to escape polewards faster than leading polarities, inhibiting cancellation between them and producing large polar field changes.

It has been suggested that an individual active region straddling the equator, with large net flux on each side of the equator, might create large polar field changes (Cameron et al. 2013). Because a single active region can contain more magnetic flux than the polar regions, and the meridional flow transports flux poleward in each hemisphere, if a bipolar active region can be separated by the diverging meridional flow pattern at the equator such that a significant flux imbalance is created in each hemisphere, then the large net fluxes must ultimately reach the poles and produce large flux changes there.

Wang and Sheeley (1991) modeled the effects of the large-scale photospheric flows on individual emerged bipoles. Differential rotation has no effect on the axisymmetric component of the field, including the axial dipole moment. The discrepancy between the dipole moments that the bipoles emerge with and the subsequent dipole moment of the photospheric field is therefore due to diffusion and meridional flow. The cases with only diffusion and only meridional flow are simple: Leighton (1964) showed that with diffusion alone the magnetic multipoles would decay such that the low-order multipoles decay slowest and the lowest-order multipole, the dipole, would decay on a timescale comparable to a solar cycle. On the other hand, Wang and Sheeley (1991) argued that, for a poleward meridional flow in the absence of diffusion, a bipole located entirely within the north or south hemisphere would be transported to the nearest pole, and its two polarities would merge there or in transit. Thus, either diffusion or meridional flow acting alone would reduce the dipole moment of an emerged bipole.

Wang and Sheeley (1991) then modeled the behavior of bipoles situated at a variety of latitudes and subjected to diffusion and meridional flows of different speeds. At most latitudes meridional flow and diffusion acting together on a bipole will decrease its dipole moment or the reasons given above, except in some cases of bipoles at low latitudes. The dipole moment can increase if some of the bipole's flux manages to diffuse across the equator and a significant flux imbalance is created in each hemisphere. If this happens the meridional flows can carry these unbalanced fluxes to the two poles and concentrate them there, producing an enhanced dipole moment. However, this happens to a small minority of active regions on the real Sun, in which case the effect on the hemispheric net flux balance could be significant (Cameron et al. 2013).

5 Local Meridional Flow Perturbations

In the Babcock-Leighton phenomenological model the flux emerges from the interior to the surface as bipoles and is dispersed by the differential rotation, meridional flow and super granular diffusion. Since the leading-polarity flux is generally closer to the equator than the trailing-polarity flux, more leading- than trailing-polarity flux diffuses towards or across the equator and cancels with the leading-polarity flux in the opposite hemisphere leaving a net flux of trailing polarity to be transported poleward. It is this net flux that changes the polar field. This chain of events may be modified so that the trailing-polarity flux, being poleward of the leading-polarity flux, dominates the high latitude, resulting in a net trailing/leading-polarity flux at high/low latitudes. The high-latitude trailing-polarity net flux then results in a polar field change, and the leading-polarity low-latitude net flux is eventually canceled through interactions with lead-polarity flux in the opposite hemisphere. Since the poleward surge patterns in the two hemispheres are independent of each other, the relationship between the two hemispheres is not simple.

Higher active region flux leads to slower poleward transport of the decayed flux, and the surges associated with large regions tend to have trailing-polarity flux. This may explain the connection between trailing/leading polarity and slower/faster poleward meridional flow. This explanation would relegate the meridional flow in the flux transport to a more passive role, with its properties determined by the magnetic properties of the magnetic regions themselves. Higher active region flux leads to greater surge flux but also stronger converging flows at active latitudes, which tend to reduce the Joy's law tilts of active regions in flux-transport models (Jiang et al. 2010). Including flux-dependent local meridional flow perturbations improves the modeling of cycle amplitudes (Cameron and Schüssler 2010), and prescribing faster overall flow speeds to stronger cycles improves the modeling of polar fields and open flux (Wang et al. 2002, 2009). Thus the surge field originates from the poloidal component of the active region field and is modulated by the meridional flow and the inflows towards the activity belts.

Hathaway and Rightmire (2010) studied the time-dependence of the meridional flow using magnetic elements as tracers, whose bulk motion includes the effects of both the plasma bulk flows and the diffusion-like motions associated with the granulation. They found lower flow speeds during cycle 23 activity maximum and higher speeds during the cycle 22 and 23 minima. Basu and Antia (2010) analyzed helioseismic measurements of meridional flows and found similar patterns of overall flow speed variations, but the latitude profiles and amplitudes that they found were quite different from those of Hathaway and Rightmire (2010). Furthermore, Basu and Antia (2010) discovered that this flow speed variation was associated with a flow pattern that migrated equator ward with the magnetic activity belts. When this pattern was removed from their data, the speed variations vanished.

The distinction between global meridional flow speed variations, presumably extending to the depths of the convection zone, and local variations associated with the activity, has profound consequences for the role of the flow variations in the global flux transport and the formation and reversal of the polar fields. Cameron and Schüssler (2010) showed that the observed meridional flow variations could be qualitatively reproduced by a flux-transport model that includes inflows towards the activity belts.

The Babcock-Leighton flux transport model has usually been cast in linear terms (but see Schrijver et al. 2002, for a model including nonlinear, field-strength-dependent flux dispersal). Cameron and Schüssler (2012) discussed the possible back-reaction of the magnetic field on differential rotation, the tilt angles of the bipoles and the trans-equatorial diffusion of preferentially leading-polarity magnetic flux. The evidence concerning back-reaction to the meridional flow was considered inconclusive. Dasi-Espuig et al. (2010) found variations of Joy's law tilt angles that are anti-correlated with cycle amplitude. Since smaller tilt angles should reduce the source of the poloidal field, they discussed this anti-correlation as a possible saturation mechanism. This may be the result of a change in the Coriolis force that twists the magnetic flux tubes as they rise through the convection zone. Alternatively they may be caused by near-surface flow perturbations associated with the presence of the field, such as

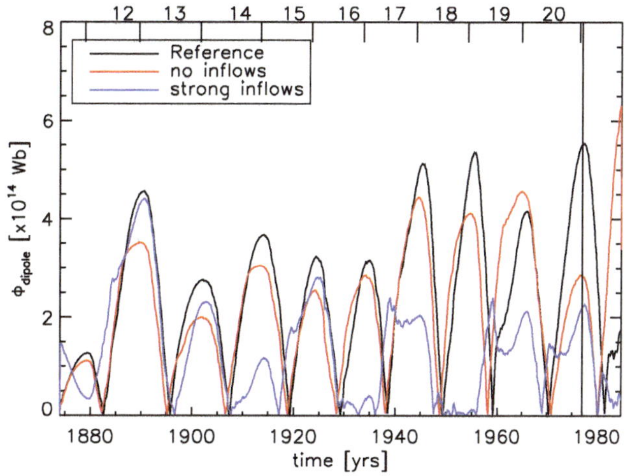

Fig. 7 Magnetic flux corresponding to the axial dipole, from the flux-transport simulation with inflows (reference case, *black curve*), without inflows (*red curve*), and with strong inflows (amplitude multiplied by 1.5 with respect to the reference case, *blue curve*). The *black vertical line* indicates the termination of the Royal Greenwich Observatory sunspot data in 1976. From Cameron and Schüssler (2012)

a field-dependent inflow towards active regions. Being more observable, the latter candidate mechanism has received more attention. Collectively, active region inflows produce mean meridional flow perturbations towards the activity belts whose amplitudes depend on the magnetic flux of the regions.

Jiang et al. (2010) and Cameron and Schüssler (2012) found from flux-transport modeling that the inflows tend to reduce the centroid latitude separation of the positive and negative active region fluxes, leading to weaker high-latitude surge fields and smaller polar field changes, as well as less trans-equatorial diffusion of leading-polarity flux.

In the models of Cameron and Schüssler (2012) the presence of the active region inflows tended to produce stronger polar fields (represented by the axial dipole strength of the model field). Their results are shown in Fig. 7. The authors identified two opposing effects of the inflows on the trans-equatorial diffusion of leading-polarity flux: 1) for low-latitude regions the inflows can cross the equator and enhance the transport of leading-polarity flux from the opposite hemisphere, increasing the low-latitude flux imbalance in each hemisphere, leading to stronger polar fields; and 2) the convergence of the inflows tends to reduce the Joy's law tilts of the bipoles, counteracting the diffusive separation of the regions' polarities and weakening their contribution to the polar fields. For weaker cycles the first effect was found to be dominant, giving stronger polar fields, whereas for stronger cycles the first effect was suppressed, producing weaker polar fields. This explanation was based on the Waldmeier effect, where stronger cycles tend to peak earlier than weaker cycles: in stronger cycles the activity peaks before it has migrated close enough to the equator for the inflows to produce significant transport of leading flux across the equator.

6 Joy's Law Tilt

Since the transport of net magnetic flux depends sensitively on the latitude separation of active regions' positive and negative fluxes, a systematic change in the bipoles' Joy's law tilt distribution could have an impact on the development and reversal of the polar fields. There is evidence that the average Joy's law tilt is not constant in time, and that Joy's law tilt variations play a role in determining polar field changes. Figure 8 shows the annual average latitude centroids of the positive and negative active region flux in the northern and

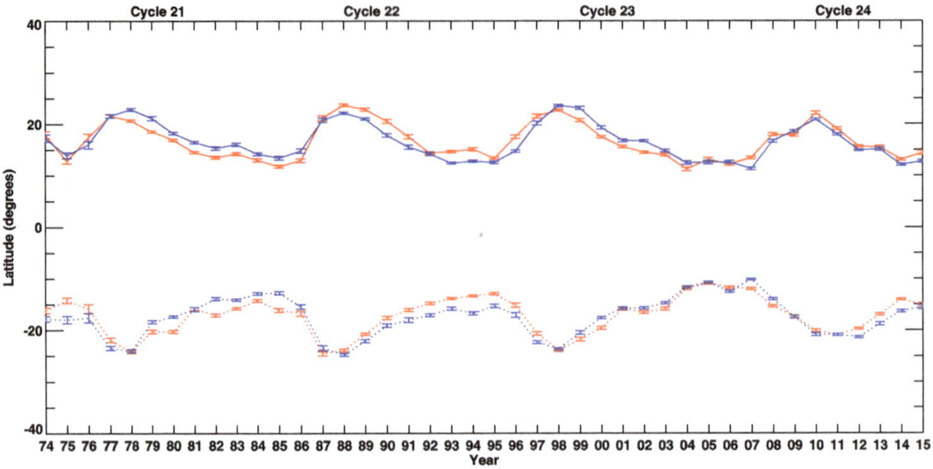

Fig. 8 The latitudinal centroids of the active region fields as functions of time for positive (*red*) and negative (*blue*) polarities in the northern (*solid lines*) and southern (*dotted lines*) hemispheres for the Kitt Peak data summarized in Fig. 1. The *error bars* indicate the standard deviations of the annual means. The displacements between the positive and negative centroids are a consequence of and consistent with Joy's law. Updated from Petrie (2012)

southern hemispheres. The equator-ward migration of the flux emergence can be clearly seen in these curves: the centroids begin each cycle at high latitudes, poleward of $\pm 20°$, and migrate slowly equator-ward over the cycle. The systematic displacement between the positive and negative centroids in each hemisphere is also clear, and the associated dipole components have matching sign in the two hemispheres. These produce a net dipole component which at the beginning of each cycle opposes the dipole component of the polar fields. As the cycle progresses the polar fields weaken, reverse and strengthen with dipole component of sign matching the dipole of the activity. Figure 8 also shows that the size of the centroid displacement is not constant in time. In particular, during the decline of cycle 23, the south-hemisphere centroids after 2001 and the north-hemisphere centroids after 2003 became very close together. Cycle 24 also shows signs of this phenomenon: in the past few years (2012–2015) the northern-hemisphere centroids have been very close together while the southern-hemispheric centroids have been well separated. This behavior coincided with episodes of stalled polar field evolution: during the decline of cycle 23 the polar field reversal slowed down, resulting in the weak polar fields that characterized that cycle and the ensuing minimum. The cycle 24 northern polar field reversal has been stalled since late 2012. Petrie (2012) demonstrated the close relationship between the centroid separation and the flux density of the high-latitude surges. When multiplied by the total active region flux, the centroid separation closely matches the high-latitude surge flux density. This result is independent of meridional flow speed.

Muñoz-Jaramillo et al. (2013) studied the effects of Joy's law tilt on the polar field strength over a longer timeframe: 1905 until the end of cycle 23. They used proxy data for the polar field strength: facular counts from MWO and MDI, calibrated using polar field measurements from WSO. They used Joy's law tilt measurements derived from MWO images by Dasi-Espuig et al. (2010). These measurements have since been challenged by Ivanov (2012) and corrected by Dasi-Espuig et al. (2013). Although, as Fig. 9 shows, Muñoz-Jaramillo et al. (2013) found no significant correlation between the total polar flux during cycle minimum and the maximum or total cycle sunspot area during the preceding maximum, signif-

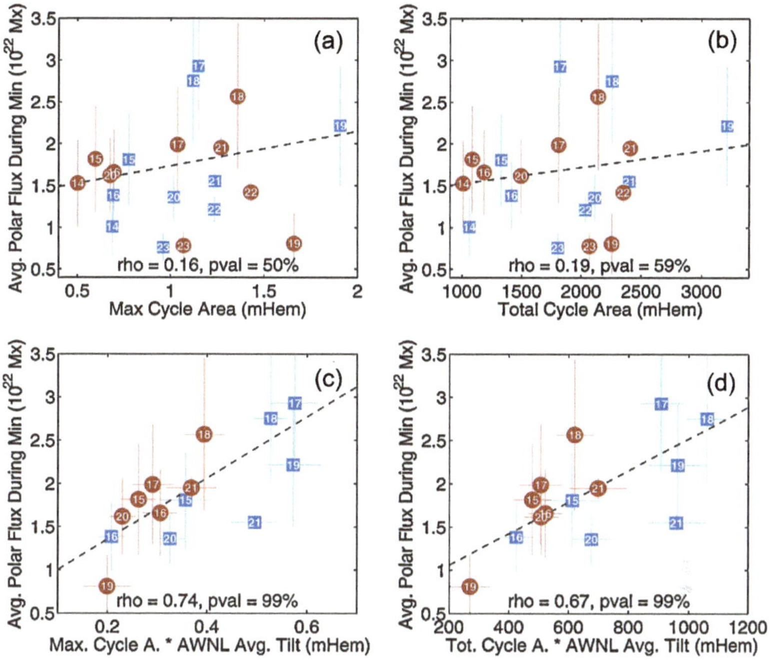

Fig. 9 Scatter plots of proxy polar magnetic flux (based on polar facular counts) against maximum (*left plots*) and average (*right plots*) sunspot area (*top plots*) and sunspot area multiplied by average Joy's law tilt (*bottom plots*). No correlation is found between the maximum (**a**), or total (**b**), sunspot area of a cycle and polar flux during the subsequent minimum. However, both maximum (**c**) and total (**d**) sunspot area become correlated with polar flux if multiplied by the area-weighted average tilt normalized by latitude of emergence (AWNL). *Square* (*circular*) *markers* represent data for the northern (southern) hemispheres. *Markers* are numbered using cycle amplitude as reference. The *dashed line* corresponds to a linear fit using the least absolute residuals method. The text inside the figure panels indicates the Pearson's correlation coefficients and their statistical significance. From Muñoz-Jaramillo et al. (2013)

icant correlations appeared when they multiplied the total or maximum area by the average Joy's law sunspot tilt (normalized by latitude). This provides further evidence that the Joy's law tilt is an important factor in determining the polar fields (though this experiment should perhaps be repeated using corrected MWO (or KK) Joy's law tilt data).

Sun et al. (2015) compared the flux density of high-latitude poleward surges to both of the quantities discussed above that are directly relevant to flux-transport processes: the separation between the active regions' positive and negative flux latitude centroids, the active region flux itself, and the meridional flow speed. Their calculations used line-of-sight magnetograms and Doppler images from HMI and MDI (2010–2014). The high-latitude flux density was averaged over a 10° latitude band poleward of the activity. As Fig. 10 shows, significant correlations were found between high-latitude flux density and both the active region flux centroid separation and the meridional flow speed. As might be expected, trailing-polarity surges were found in cases with positive centroid separation (positive Joy's law dipole tilt)—a moderate correlation with Pearson linear correlation coefficient around 0.6 was found. More surprisingly, the meridional flow velocity was anti-correlated with the high-latitude flux density in both hemispheres: slower/faster-than-average flow speeds corresponded to trailing/leading-polarity surges. This result remains unexplained, but preferentially fast transport of leading-polarity flux would have a dampening effect on polar field

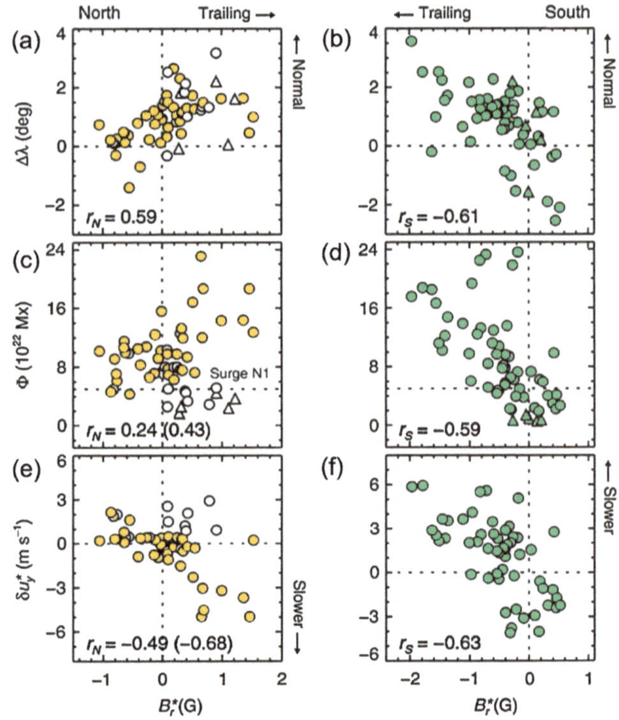

Fig. 10 Scatter plots showing correlation between selected variables of solar flux transport: mean active region tilt (separation $\Delta\lambda$ between positive and negative flux centroids), active region flux Φ, surge field B_r^*, and residual meridional flow velocity δu_y^* for the northern (*left plots*) and southern (*right plots*) hemispheres. *Circles* show Carrington rotation means from HMI, *triangles* from MDI. *Open symbols* indicate a surge named N1 by the authors, which deviates from the overall trend in (**c**) and (**e**). B_r^* and δu_y^* are from observations of 10° latitude bands poleward of the activity. In each plot the Pearson's correlation coefficient r is indicated (HMI only); values in parentheses have the outlier surge N1 excluded. The probability of chance occurrence $p = 0.068$ for plot (**c**), and $p < 10^{-4}$ for the other plots. From Sun et al. (2015)

changes. Such behavior may be related to local meridional flow perturbations which we will discuss in the following section(s). In Fig. 1 the high-latitude poleward transport velocity of the flux, represented by the slopes of the streaks of color in the high-latitude regions between about ±30° and ±65°, appear to be independent of the polarity of the flux. As we discussed in Sect. 4 regarding Fig. 6, the transport velocity combines the effects of both the meridional flow and the diffusion.

Cameron et al. (2010) studied the effect of Joy's law tilt variation on flux-transport models. They used the Joy's law tilt results from MWO data covering 1905–1986, derived by Dasi-Espuig et al. (2010) who found an anti-correlation between the amplitude of an activity cycle and the average Joy's law tilt angle during that cycle. In previous models the reversal of the polar fields after a strong cycle often failed to take place. Additional decay processes were needed, such as radial diffusion (Schrijver) or large overall meridional flow changes (Wang et al. 2002), to achieve polar field reversals every cycle, including after strong cycles. Cameron et al. (2010) found that the reversal of the polar fields after strong cycles could be explained instead by this anti-correlation between cycle amplitudes and Joy's law tilt angles: the polar field changes during strong cycles can be dampened by the small tilt angles during these cycles.

7 Equatorial Fields and Trans-equatorial Diffusion

One of the central processes of the Babcock-Leighton model described in Sect. 2 is the interaction of leading-polarity flux across the equator, which allows a net trailing-polarity flux to develop in each hemisphere which in turn changes the polar field. The previous sections

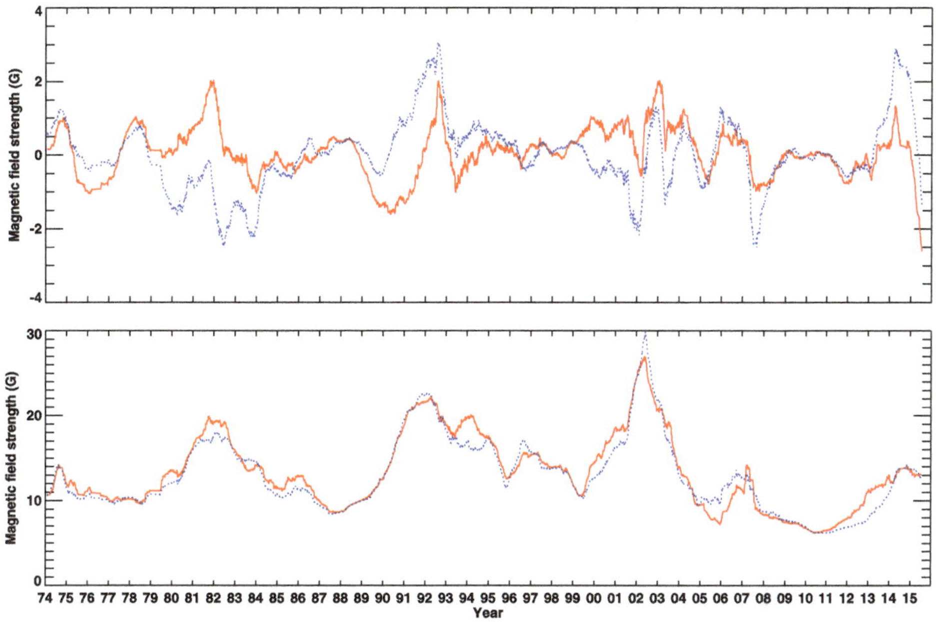

Fig. 11 Plots of the average near-equatorial net (*top*) and absolute (*bottom*) magnetic field strength in the northern (between -2° and 0° latitude, *red*) and southern (between 0° and 2° latitude, *blue*) hemispheres. The calculations were based on the time-latitude magnetic flux density data shown in Fig. 1

have described how both the Joy's law tilt and the meridional flows have much influence over this process. As discussed in Sect. 4, meridional flows could influence polar field changes by inhibiting leading-polarity flux interactions across the equator. In this section we summarize the behavior of the fields located close to the equator over the duration of the NSO magnetogram data set, 1974–present.

Figure 11 summarizes the evolution of the fields close to the equator over the time covered by the NSO magneto grams, 1974–2015. The figure emphasizes the diversity of behavior of these fields on both sides of the equator. During the cycle 21 maximum the leading polarity in each hemisphere determined the dominant polarity of the net flux density close to the equator, with positive peak north of the equator and two negative peaks south of the equator. This matches the two-step pattern of positive flux transport into southern high latitudes shown in the middle right plot of Fig. 14, and the two episodes of surge formation shown in Fig. 13. Cycle 22 was more asymmetric near the equator: a large negative equatorial flux density on the north side was quickly followed by a larger positive peak on the south side. This positive peak corresponds to the giant poleward surge of negative flux in the early 1990s evident in Fig. 1. Towards the end of the southern equatorial flux density peak in Fig. 11, the flux on the north side also briefly develops a net positive bias corresponding to a brief negative north poleward surge during the decline of cycle 22, but both the north equatorial flux and poleward surge polarities revert back to normal before the end of the cycle.

The equatorial flux patterns for cycle 23 are much more complex. There are multiple changes of polarity on both sides of the equator. Though the cycle 23 net flux density on the south side has a clear negative bias overall, and some of the peaks have amplitude as large as those of cycles 21 and 22, the number of changes of sign contrasts strongly with cycles 21

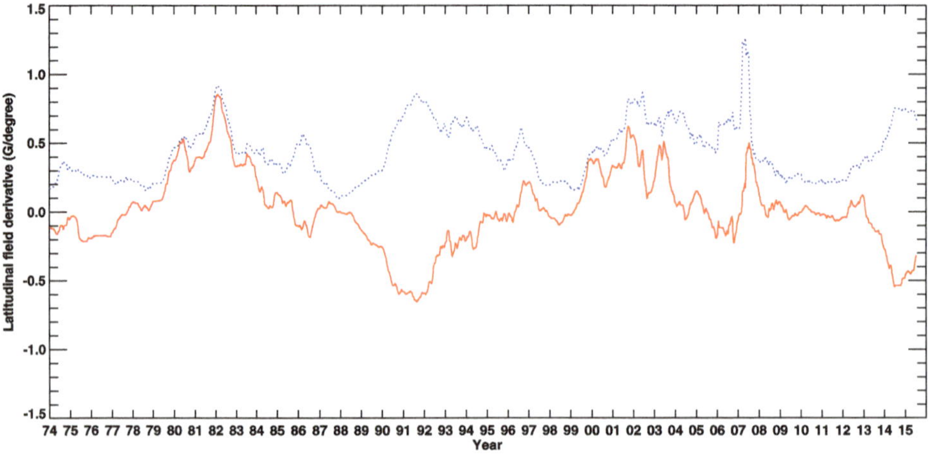

Fig. 12 Plot of the average net (*red*) and absolute (*blue*) latitudinal field derivative across the equator as a function of time. The calculations were based on the time-latitude magnetic flux density data shown in Fig. 1

and 22. This contrast matches the pattern of small, relatively weak, mixed-polarity poleward surges at high latitudes in Fig. 1, and the tortuously slow polar field reversal (especially in the south) in Fig. 2 during cycle 23 compared to cycles 21 and 22. There is a similar pattern in the north hemisphere, but with positive bias and shallower peaks and troughs. The data indicate that there was a similar quantity of magnetic flux near the equator during the cycle 23 maximum compared to the cycle 21 and 22 maxima, but the flux density tended to be of more changeable and mixed polarity during cycle 23. The bottom panel of Fig. 11 shows the average absolute equatorial flux density on each side of the equator over time. Here the 11-year cyclical variations are clear and the deep cycle 23 minimum is emphasized. The plot also shows that there was at least as much magnetic flux near the equator during the cycle 23 maximum as during cycles 21 and 22. This serves as evidence in support of the idea that the Joy's law tilt angle, or latitudinal separation of the positive and negative active-region flux centroids, was significantly smaller on average during cycle 23 compared to cycles 21 and 22, as Fig. 8 shows. It argues against the theory that enhanced low-latitude meridional flows during cycle 23 produced the weak polar fields of that cycle side fast meridional flows would have inhibited the presence and interaction of flux across the equator, as discussed in Sect. 4.

The cycle 24 equatorial flux density patterns in Fig. 11 have so far resembled those of cycle 23, with multiple peaks and troughs and changes of sign. However, the southern hemisphere had a large positive peak lasting between 2012 and 2014, corresponding to the persistent southern active region latitude flux centroid separation (Fig. 8), the consistently negative southern poleward surges (Fig. 1), and the monotonic negative change in the polar field (Fig. 2) over this time. The average absolute equatorial flux densities on the north and south sides of the equator have each exhibited mostly a nearly monotonic increase during cycle 24 so far, though the north equatorial absolute flux density began decreasing slightly in late 2014 while the southern flux density leveled off around the same time. Thus, while the sign of the cycle 24 equatorial absolute flux density has risen and fallen with multiple changes of sign north and south of the equator, the total equatorial flux increased steadily until late 2014. The cycle 24 peaks of these curves are modest compared to those of the previous three cycles, though the peak net values have been comparable as the top plot of Fig. 11 shows.

A more directly relevant quantity with respect to kinematic photospheric flux transport modeling is the trans-equatorial field derivative. The latitude derivative of the radial field at the equator is directly proportional to the field diffusion across the equator in the model (Durrant et al. 2004; Cameron et al. 2013). Figure 12 shows the average derivative as a function of time. This derivative has a clear positive/negative peak during the cycle 21/22 maximum, consistent with much interaction and cancelation between the leading polarities in the two hemispheres during these two cycle maxima. The cycle 23 pattern is more complex, with multiple peaks and a few changes of sign. While the derivative is usually positive during cycle 23, the peaks are significantly smaller and less well defined than those of cycles 21 and 22.

The average absolute value of the latitudinal derivative of the field at the equator is also plotted in Fig. 12. Unlike the average net derivative, the average absolute derivative has a broad peak of approximately equal amplitude for cycle 23 compared to cycles 21 and 22. This curve also has a second tall, narrow peak in late cycle 23 corresponding to the asymmetric decline of the cycle. Thus just as much field diffusion between the hemispheres appears to have occurred during the maximum of cycle 23 as during cycles 21 and 22, but this diffusion changed sign much more during cycle 23, corresponding to the smaller, weaker, more mixed poleward surges and the smaller, slower polar field changes during cycle 23.

The equatorial field derivative for cycle 24 is (so far) much simpler than for cycle 23, with a monotonic change towards ever larger negative derivatives since late 2012, when the persistent southern active region latitude flux centroid separation began (Fig. 8), the southern poleward surges became decisively negative (Fig. 1), and the south polar field began its negative monotonic change and reversal (Fig. 2). The simplicity of this field derivative pattern reflects that the southern flux density curve in the upper plot of Fig. 11 has remained above the northern curve since late 2012. The size of the derivative during cycle 24 is comparable to the peak values of the previous three cycles.

8 Observations of Photospheric Fields During the Polar Reversals of Cycles 21–24

We now summarize NSO KPVT (512-channel and SPMG) and SOLIS/VSM observations of active- and high-latitude flux densities during times of polar field reversal. The object of this exercise is to identify and characterize the particular active regions that supply the large net fluxes to the high latitudes that in turn produce the largest polar field changes. We adopt a stack plot approach first used by Gaizauskas et al. (1983, 2001). Examples are shown in Fig. 13, which shows stackplots of synoptic map latitude slices covering Carrington rotations 1684–1702. The left stackplot shows the south active latitudes with color scale saturated at ± 50 G, the right stackplot the southern high latitudes saturated at ± 10 G. Red/blue is positive/negative. In the following sections we will use these to trace major high-latitude surges of decayed active region flux to the activity complexes.

8.1 Cycle 21

Figure 13 summarizes the low- and high-latitude magnetic flux distributions in the southern hemisphere over the period of the south polar field reversal (rotations 1684–1701). The left stack plot shows the active latitudes of the southern hemisphere with greyscale saturated at ± 50 G, and the right stack plot the southern high latitudes with greyscale saturated at

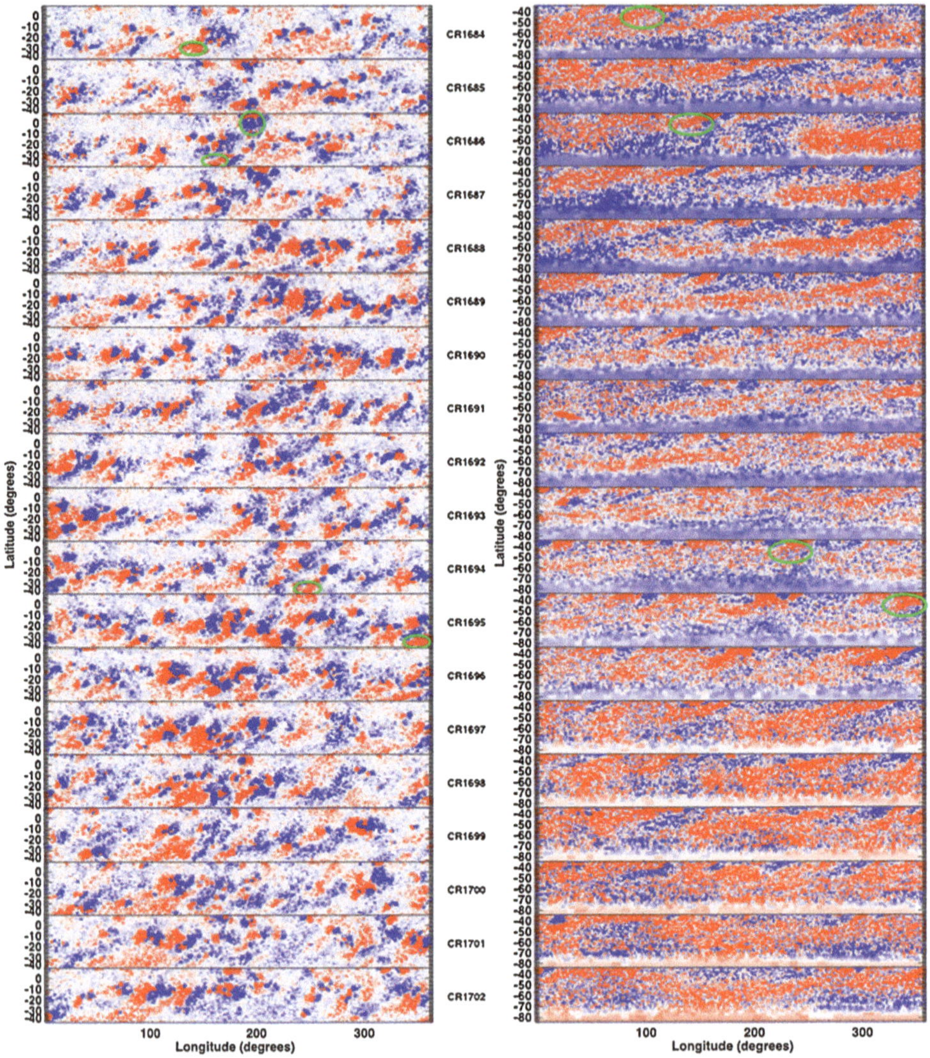

Fig. 13 Above are stackplots of synoptic map latitude slices covering Carrington rotations 1684–1702, corresponding to the cycle 21 south polar reversal. The *left* stackplot shows the south active latitudes with color scale saturated at ±50 G, the *right* stackplot the southern high latitudes saturated at ±10 G. *Red/blue* coloring represents positive/negative flux. The *ovals* in the *left* stackplot indicate activity complex trailing polarities that are associated with influential poleward surges of trailing-polarity flux. These surges are indicated by *ovals* in the *right* stackplot. The *green circle* in the *left* stackplot (in the panel for rotation 1686) indicates a magnetic dipole whose polarities lie on opposite sides of the equator

±10 G. The figure therefore allows us to relate the major high-latitude plumes of decayed active region flux in the right stack plot to the activity complexes emphasized in the left stack plot, and thereby trace the evolving relationship of each complex and poleward surge from rotation to rotation. The low-latitude stack plot demonstrates the large scale of the active region clusters, referred to as activity complexes by Gaizauskas et al. (2001), and their long duration of a year or more. There appear to be two major episodes of change in the flux

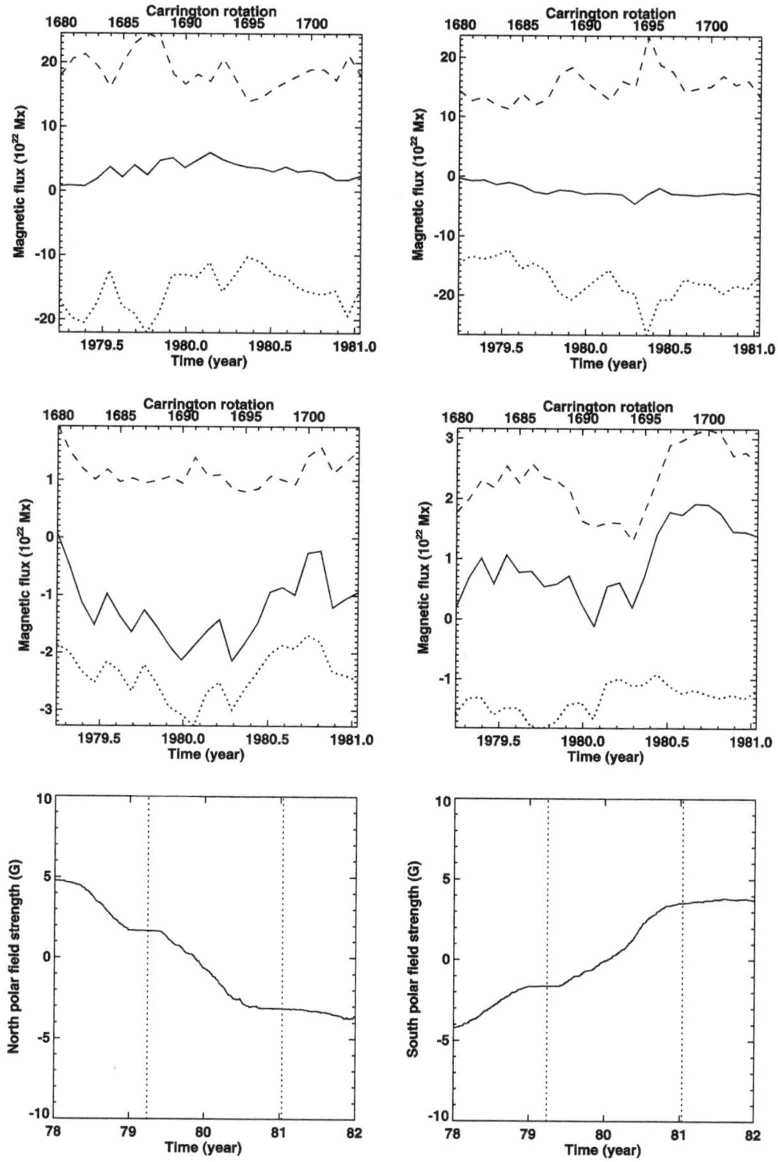

Fig. 14 Shown above are the low-latitude (−40° to 0°, *top*) and high-latitude (−65° to −40°, *middle*) fluxes and the polar field strength (*bottom*), in the northern (*left* plots) and southern (*right* plots) hemispheres, against time for Carrington rotations 1680–1704, corresponding to the cycle 21 polar reversal. The *vertical dotted lines* in the *bottom* plots indicate the timespans covered by the *top* and *middle* plots

balance at high latitudes, apparently associated with major poleward surges of positive flux indicated in the panels for Carrington rotations 1684 and 1686 in the right plot in the figure, producing the first episode, and 1694 and 1695, producing the second episode. These surges emanate from the trailing polarities of the activity complexes indicated in the left stackplot. The influence of the two surge episodes on the net high-latitude flux is visible in the right stackplot as a change from an almost equal mix of blue and red before rotation 1695 to a strong red bias after rotation 1695.

Section 4 discussed how a single active region straddling the equator, with large net magnetic fluxes in each hemisphere, could produce large polar field changes if these net fluxes are transported by meridional flows to the poles. This phenomenon does not seem to play a major role in the polar field reversals studied here. One of the example equatorial active regions pointed out by Cameron et al. (2013) is shown in Fig. 13. The region in question is indicated by the circle in the low-latitude panel for rotation 1686. The main neutral line of this region nearly coincides with the equator, so that most of its flux north/south-hemispheric flux is positive/negative. The region might therefore be expected to produce a positive/negative change in the north/south polar field, but this is evidently not what occurred. Instead, the region's positive and negative fluxes appear to have dispersed and canceled with each other and with neighboring flux systems. The region had no significant effect on the northern and southern high-latitude net fluxes, which in fact became negative and positive, respectively. Instead, the cumulative effect of multiple instances of diffuse, trailing-polarity flux concentrations located poleward of their lead-polarity counterparts lead to trailing-polarity poleward surges that clearly changed the high-latitude flux balance, including the indicated examples in rotations 1684, 1686, 1694 and 1695. We therefore agree with Cameron et al. (2013) that polar field reversals can be dominated by only a few active regions, but we find that large, long-lived complexes are responsible rather than regions straddling the equator.

This southern high-latitude flux change is shown also in Fig. 14 as a steep positive change around rotation 1695, corresponding to the change of color in the right plot of Fig. 13, and which is well correlated in time with the polar field reversal (recall that the polar field data are smoothed). The low-latitude flux developed a negative bias that does not appear to have been immediately fully compensated by cross-equatorial diffusion of positive lead-polarity flux. There is some evidence of negative-polarity preference in the flux shown immediately south of the equator in Fig. 13, consistent with Fig. 11. The northern hemisphere produced a near-mirror-image of the behavior of the southern hemisphere, and its polar field reversal occurred at almost exactly the same time. At the end of this synchronized polar reversal the low and high latitudes had significant flux biases in both hemispheres: positive and low northern and high southern latitudes, and negative at high northern and low southern latitudes. This kind of near-lock-step hemispheric anti-symmetry does not appear to be typical. The size and duration of the activity complexes in Fig. 13 corresponds to the scale of the giant poleward surges during cycle 21 as shown in Fig. 1.

8.2 Cycle 22

Among the simplest of the polar reversals covered by the data set is the southern reversal of cycle 22. The polar field seems to be reversed by a single giant surge of negative flux in Fig. 1. Figure 15 shows a stackplot of latitude slices of synoptic maps covering Carrington rotations 1837–49. The figure demonstrates the bipolar form of active regions, and the prevailing asymmetry of these bipoles with trailing flux generally more diffuse and more poleward than leading flux, and the polarity bias of the corresponding high-latitude surges. The longitudinal drift of the high-latitude features over time is due to the differential rotation: high latitudes rotate more slowly than low latitudes, and Carrington rotation periods are defined by the surface rotation period at about $\pm 26°$ latitude.

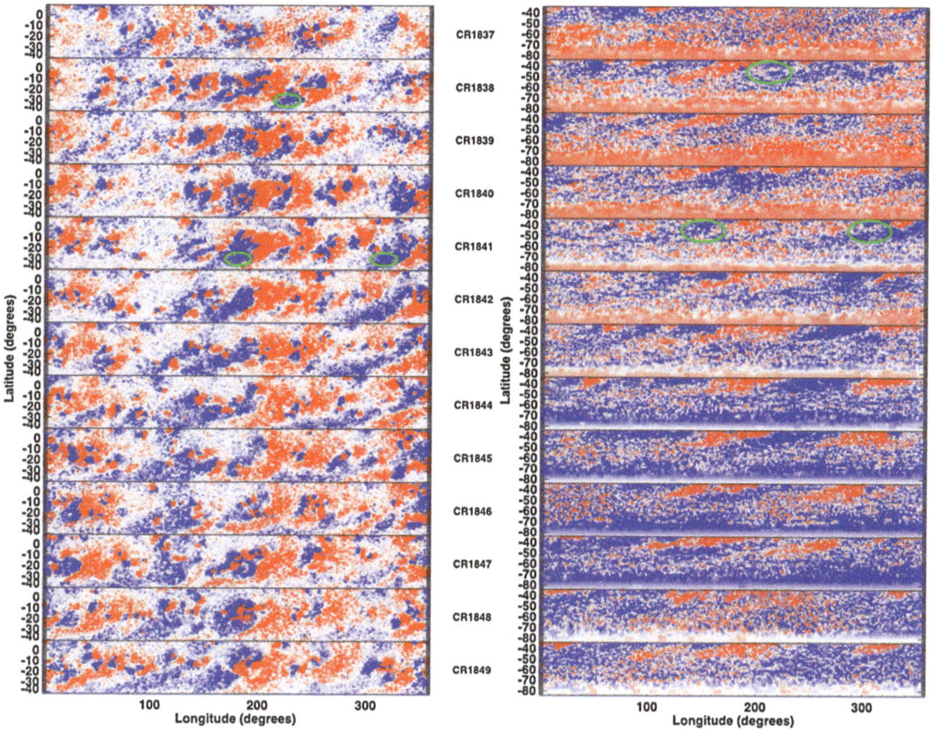

Fig. 15 Above are stackplots of synoptic map latitude slices covering Carrington rotations 1837–1849, corresponding to the cycle 22 south polar reversal. The *left* stackplot shows the south active latitudes with color scale saturated at ±50 G, the *right* stackplot the southern high latitudes saturated at ±10 G. *Red/blue* coloring represents positive/negative flux. The *ovals* in the *left* stackplot indicate activity complex trailing polarities that are associated with influential poleward surges of trailing-polarity flux. These surges are indicated by *ovals* in the *right* stackplot

According to the stackplots in Fig. 15, there were two major activity complexes at low latitudes during this period: one centered around ±0° longitude, and one centered around ±200°. Both complexes were present throughout this year-long period, though significant temporal evolution is evident in the plot. Plumes of positive and negative flux are to be seen early in the sequence, as are regions of mixed polarity. There is some evidence of negative (trailing) polarity dominating the southern sides of the complexes, particularly at the locations indicated in the left panels for rotations 1838, 1839 and 1841. These correspond to the strong plumes of negative flux at high latitudes indicated in the right panels for the same rotations. Although positive plumes remain present throughout, it is clear that the negative plumes dominate the high latitudes during the rotations following 1839, indicated by the coloring of the high latitudes becoming progressively bluer after this time.

Figure 16 confirms this high-latitude flux change in the south. The net high-latitude flux is almost balanced during rotation 1831, and steadily develops a strong negative bias over the next 10–20 rotations. The development of this bias is well correlated in time with the reversal of the polar field. Over the same period, the low-latitude flux develops a compensating bias which is eventually balanced over time. This is consistent with the net transfer of negative flux from low to high latitudes, after which a net positive flux is diffused across

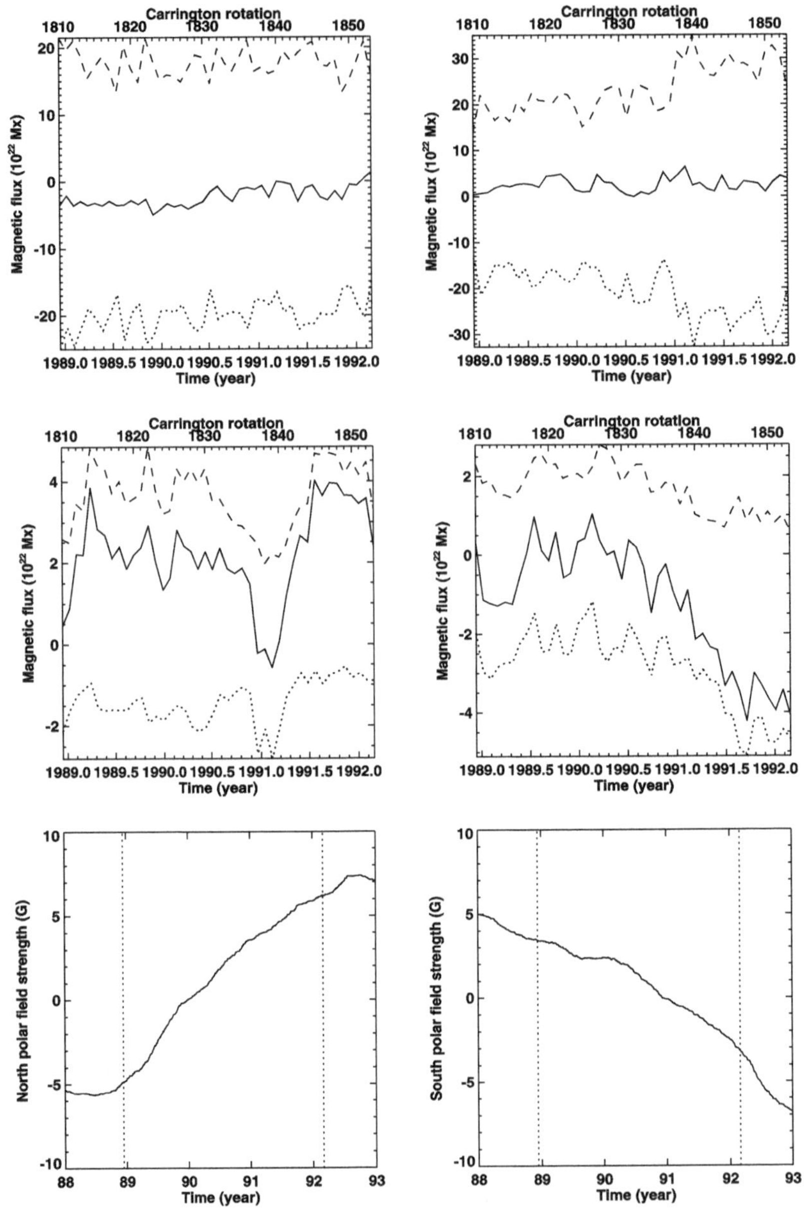

Fig. 16 Shown above are the low-latitude (−40° to 0°, *top*) and high-latitude (−65° to -40°, *middle*) fluxes and the polar field strength (*bottom*), in the northern (*left* plots) and southern (*right* plots) hemispheres, against time for Carrington rotations 1810–1853, corresponding to the cycle 22 polar reversal. The *vertical dotted lines* in the *bottom* plots indicate the timespans covered by the *top* and *middle* plots

the equator, as also implied by the plots in Figs. 11 and 12. In Fig. 15 the polarity of the flux immediately south of the equator becomes predominantly positive after rotation 1840. It is noteworthy that the poleward transport of trailing flux occurs before the compensating removal of leading flux in this case.

In the northern hemisphere there were three distinct episodes of poleward surges associated with three activity complexes. These occurred between rotations 1810–20, 1820–30 and 1840–50. The middle-left plot in Fig. 16 shows the more protracted, complex and intermittent nature of the northern high-latitude net flux changes compared to the south, reflecting the influence of three activity complexes over different time intervals in the north, compared to two simultaneous complexes in the south. The size and duration of the two northern activity complexes in Fig. 15 corresponds to the scale of the southern giant poleward surge during cycle 22 as shown in Fig. 1. The several northern giant poleward surges during cycle 22 visible in Fig. 1 correspond to the more complex, non-monotonic change in the high-latitude flux balance seen in Fig. 16. The flux change in the north seems to have been unusually quick and simple. The size and duration of the activity complexes in Fig. 15 corresponds to the scale of the giant poleward surges during cycle 21 as shown in Fig. 1.

8.3 Cycle 23

The polar reversals of cycle 23 were not as simple as those of cycles 21 or 22. In both hemispheres the high-latitudes were much less unipolar and the polar changes were shallower, particularly at the south pole, more intermittent. Figure 17 shows stackplots latitude slices of synoptic maps covering Carrington rotations 1938–1956, covering the duration of the cycle 23 northern polar field reversal. Unlike in Fig. 15 for the cycle 22 southern reversal where the activity was arranged in two large complexes which persisted over a year, here the complexes are smaller, are scattered over various longitudes, show more temporal change and have shorter durations of 5–10 rotations. Nevertheless, the same pattern of trailing-polarity flux tending to be located at higher latitudes and to be more diffuse than leading-polarity flux is again clear in the figure. Some clear examples are indicated in the left panels of Fig. 17, with corresponding trailing-polarity high-latitude surges indicated in the right panels of the figure. There are two major episodes of such occurrences: one during rotations 1940–43 and the other during rotations 1947–49. Before rotation 1940 the high latitudes appear almost flux-balanced, whereas after rotation 1940 the strong negative surges seem to emerge, and the high-latitudes have acquired a significant negative net flux. With the introduction of new negative surges in rotations 1947–49 the high-latitude flux appears negative at almost all longitudes. In this case there is evidence that the flux just north of the equator after rotation 1940 is mostly of positive polarity, though this pattern is not as strong in Fig. 17 as in Fig. 15 for cycle 22. As discussed in Sect. 7 the flux polarities on each side of the equator were much more changeable during the cycle 23 maximum than during the cycle 21 and 22 maxima.

Figure 18 confirms the steadily-developing negative high-latitude flux bias after rotation 1940, and shows that this development is well correlated in time with the north polar field reversal. Figure 18 also shows that a compensating positive flux bias developed at low latitudes around the same time. This time, the low-latitude bias does not seem to be balanced by cross-equatorial diffusion over the time interval covered by the plots. This is unlike the situation in Fig. 15 for cycle 22. This difference may be because the activity was located at higher latitudes during cycle 23 polar reversal than during the cycle 22 reversal. On the other hand, this displacement of the activity from the equator does not appear to have compromised the regions' ability to change the high-latitude and polar fluxes, in contrast to the modeling result of Cameron and Schüssler (2012) mentioned in Sect. 5. The cycle 23 northern high-latitude net flux became only half as strong, about -2×10^{22} Mx as the southern high-latitude net flux, -4×10^{22} Mx, and the cycle 23 flux was of more mixed polarity. The cycle 23 polar field change was correspondingly smaller and slower. The smaller size and

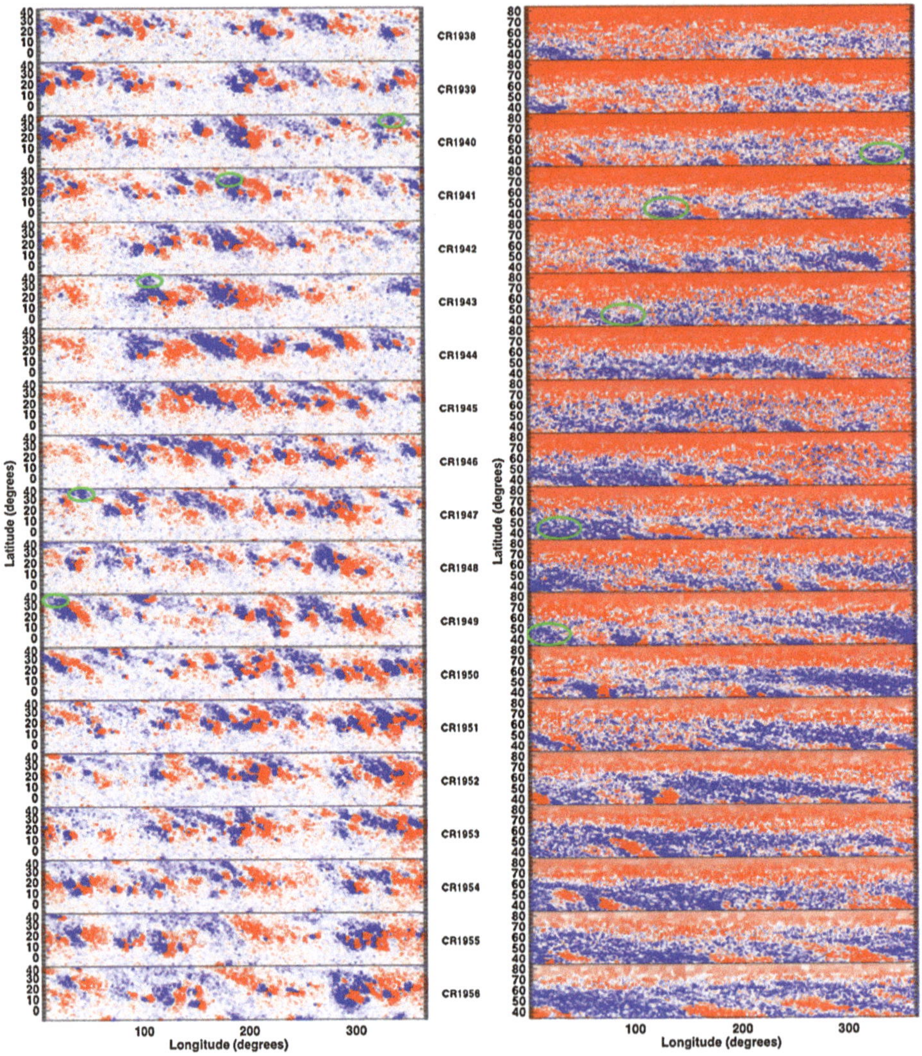

Fig. 17 Above are stackplots of synoptic map latitude slices covering Carrington rotations 1938–1956, corresponding to the cycle 23 north polar reversal. The *left* stackplot shows the south active latitudes with color scale saturated at ±50 G, the *right* stackplot the southern high latitudes saturated at ±10 G. *Red/blue* coloring represents positive/negative flux. The *ovals* in the *left* stackplot indicate activity complex trailing polarities that are associated with influential poleward surges of trailing-polarity flux. These surges are indicated by *ovals* in the *right* stackplot

shorter duration of the activity complexes in Fig. 17 compared to Fig. 13 and Fig. 15 corresponds to the smaller scale of the poleward surges during cycle 23 compared to the giant surges of cycles 21 and 22 shown in Fig. 1.

8.4 Cycle 24

Figure 19 shows stackplots of the low- and high-latitude fluxes during Carrington rotations 2119–35, covering the duration of the cycle 24 southern polar field reversal. The field pat-

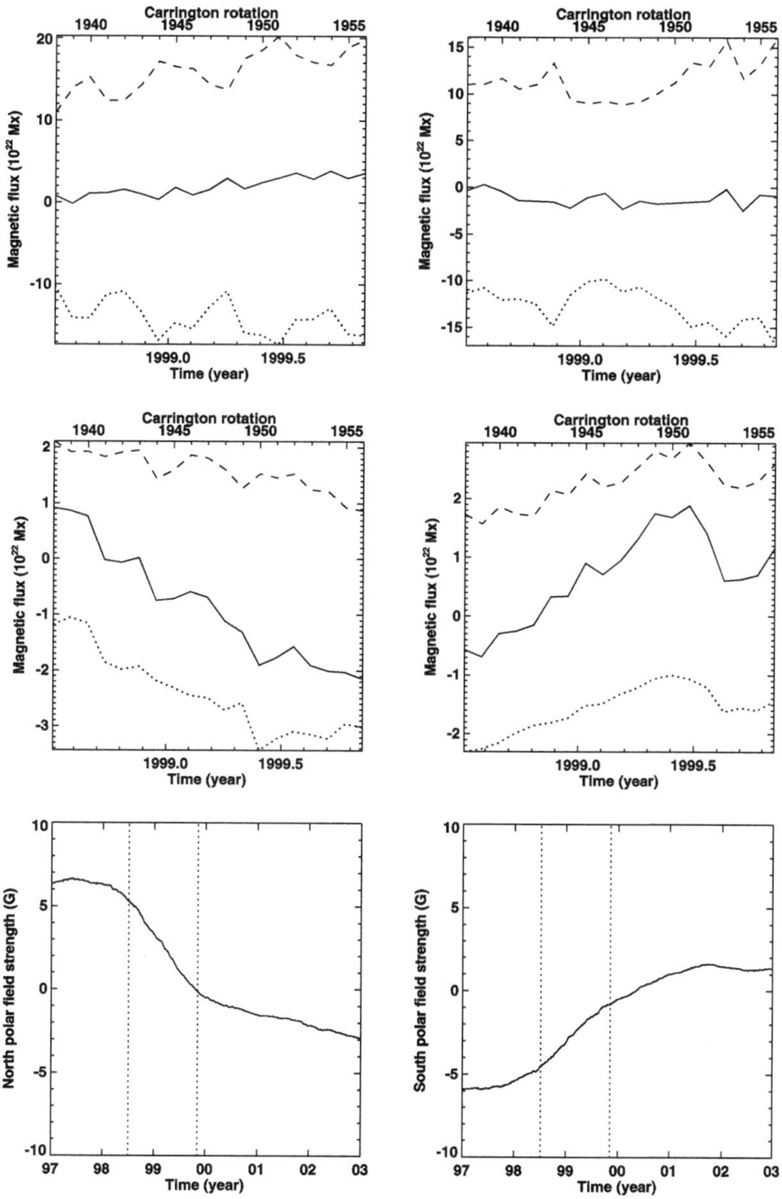

Fig. 18 Shown above are the low-latitude (0° to 40°, *top*) and high-latitude (40° to 65°, *middle*) fluxes and the polar field strength (*bottom*), in the northern (*left* plots) and southern (*right* plots) hemispheres, against time for Carrington rotations 1938–1956, corresponding to the cycle 23 polar reversal. The *vertical dotted lines* in the *bottom* plots indicate the timespans covered by the *top* and *middle* plots

terns at this time more closely resemble those of cycle 23 than cycle 22, with smaller activity complexes lasting between 5 and 10 rotations and exhibiting no strong longitudinal preferences. The tendency of trailing-polarity flux to be more diffuse and further from the equator than the leading-polarity flux is again evident, and three specific instances of such trailing-

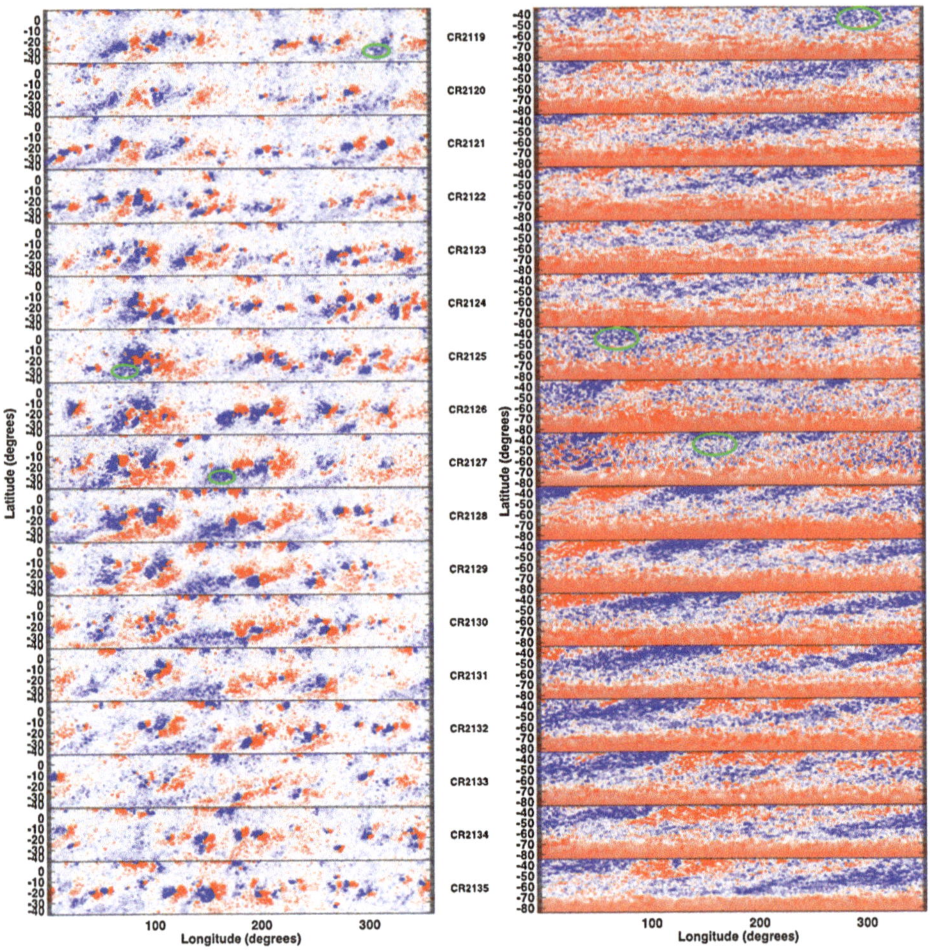

Fig. 19 Above are stackplots of synoptic map latitude slices covering Carrington rotations 2119–2135, corresponding to the cycle 24 south polar reversal. The *left* stackplot shows the south active latitudes with color scale saturated at ±50 G, the *right* stackplot the southern high latitudes saturated at ±10 G. *Red/blue* coloring represents positive/negative flux. The *ovals* in the *left* stackplot indicate activity complex trailing polarities that are associated with influential poleward surges of trailing-polarity flux. These surges are indicated by *ovals* in the *right* stackplot

polarity flux patterns leading to major poleward surges are indicated in the figure. The high latitudes do not acquire a significant flux bias until all three plumes have contributed, after rotation 2127. This can be seen more clearly in Fig. 20. Even after rotation 2127 the net flux is weaker than during the cycle 22 and 23 reversals, less than 10^{22} Mx. The development of this high-latitude flux bias is well correlated in time with the southern polar field reversal, which is slower, taking > 2 yr, and results in a smaller field change than the cycle 22 and 23 reversals. As with the cycle 23 northern reversal there was a positive low-latitude net flux during the polar reversal, and the flux close to the equator did not show an obvious polarity bias. This may again be because the activity was displaced from the equator in the southern hemisphere at the time of the reversal (the north-hemispheric activity was much closer to the equator at this time).

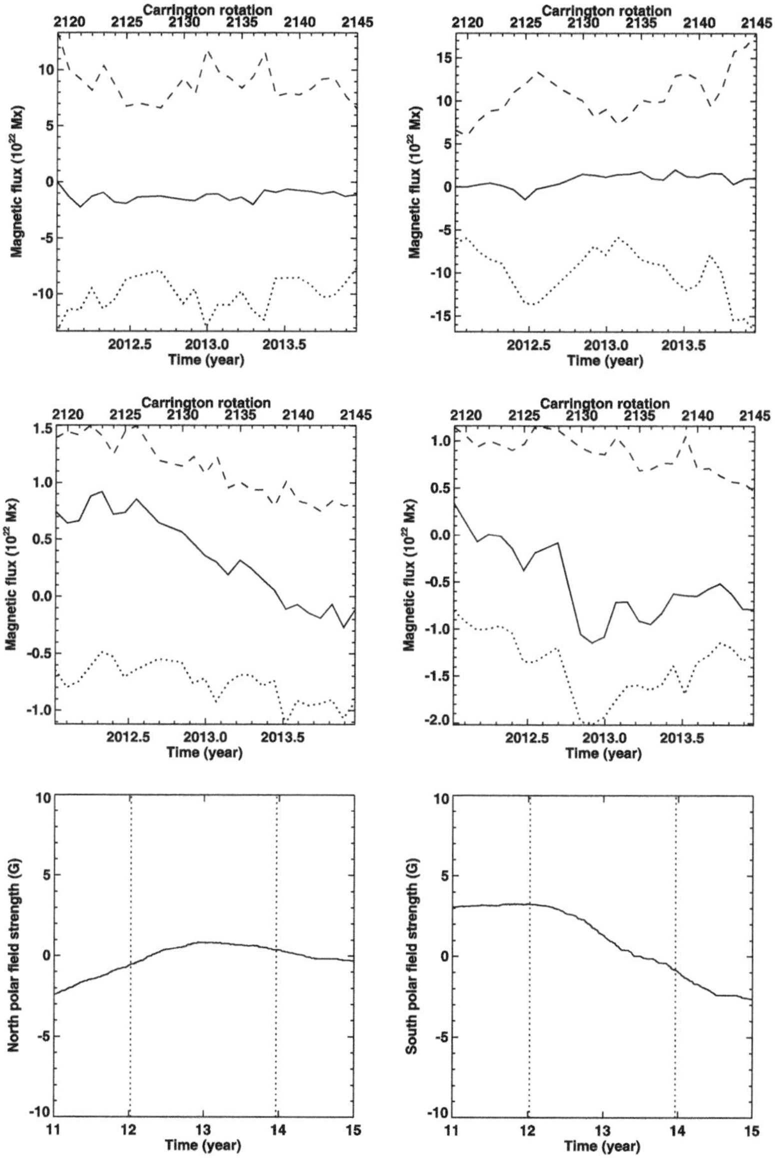

Fig. 20 Shown above are the low-latitude ($-40°$ to $0°$, *top*) and high-latitude ($-65°$ to $-40°$, *middle*) fluxes and the polar field strength (*bottom*), in the northern (*left* plots) and southern (*right* plots) hemispheres, against time for Carrington rotations 2119–2145, corresponding to the cycle 24 polar reversal. The *vertical dotted lines* in the *bottom* plots indicate the timespans covered by the *top* and *middle* plots

9 Conclusion

We have reviewed the evidence from observations and models that polar field changes are governed by near-surface flows, in particular meridional flows, and by the flux distribution that active regions emerge with, especially the Joy's law bipole tilt. While changes in the amplitude and shape of meridional flow profiles can theoretically have an enormous impact on the polar fields, observational evidence of systematic global meridional flow changes is inconclusive. Local meridional flow perturbations seem to be caused by the presence of active regions. If these perturbations systematically reduce the Joy's law tilt of all active regions, as suggested by the flux-transport models, then they do not reduce the correlation between bipole tilts (or latitude separations of positive and negative active region flux centroids) and polar field changes. Furthermore we found no signatures of enhanced low-latitude meridional flows inhibiting the interaction of leading-polarity fields of the two hemispheres across the equator. Instead, periods of reduced polar field change, such as the decline of cycle 23, are characterized by just as much flux interaction across the equator, but this interaction tends to involve much more mixed polarity each side of the equator, producing much less net flux transfer between the two hemispheres. Times of mixed-polarity trans-equatorial flux transfer correspond to poleward surges also of mixed polarity in each hemisphere, which in turn correlate to times of reduced Joy's law bipole tilt (flux latitude centroid separation) in each hemisphere.

The sizes and polarity biases of the poleward surges are clearly related to the sizes, lifetimes and latitude flux separations of the activity complexes at active latitudes. Whereas the polar field reversals of cycles 21 and 22 were dominated by only a few large, long-lived activity complexes producing large surges of trailing-polarity flux and large, abrupt polar field changes/reversals, the cycle 23 and 24 reversals did not feature such enormous activity complexes. These latter reversals were instead the cumulative result of numerous smaller, shorter-lived regions producing numerous weak poleward surges. We therefore conclude that understanding the creation and reversal of the polar field will require us to answer the question of why large, long-lived complexes with large latitudinal flux separations are produced during some cycles and not during others.

Acknowledgements This work utilizes data obtained by the NSO Integrated Synoptic Program (NISP), managed by the National Solar Observatory, which is operated by the Association of Universities for Research in Astronomy (AURA), Inc. under a cooperative agreement with the National Science Foundation. SE's contribution to this work was carried out through the National Solar Observatory Summer Research Assistantship (SRA) Program.

References

H.W. Babcock, The topology of the sun's magnetic field and the 22-year cycle. Astrophys. J. **133**, 572 (1961). doi:10.1086/147060

S. Basu, H.M. Antia, Characteristics of solar meridional flows during Solar cycle 23. Astrophys. J. **717**, 488–495 (2010). arXiv:1005.3031. doi:10.1088/0004-637X/717/1/488

R.H. Cameron, M. Schüssler, Changes of the solar meridional velocity profile during cycle 23 explained by flows toward the activity belts. Astrophys. J. **720**, 1030–1032 (2010). arXiv:1007.2548. doi:10.1088/0004-637X/720/2/1030

R.H. Cameron, M. Schüssler, Are the strengths of solar cycles determined by converging flows towards the activity belts? Astron. Astrophys. **548**, A57 (2012). arXiv:1210.7644. doi:10.1051/0004-6361/201219914

R. Cameron, M. Schüssler, The crucial role of surface magnetic fields for the solar dynamo. Science **347**, 1333–1335 (2015). 1503.08469. doi:10.1126/science.1261470

R.H. Cameron, J. Jiang, D. Schmitt, M. Schüssler, Surface flux transport modeling for Solar cycles 15–21: Effects of cycle-dependent tilt angles of sunspot groups. Astrophys. J. **719**, 264–270 (2010). arXiv:1006.3061. doi:10.1088/0004-637X/719/1/264

R.H. Cameron, M. Dasi-Espuig, J. Jiang, E. Işık, D. Schmitt, M. Schüssler, Limits to solar cycle predictability: Cross-equatorial flux plumes. Astron. Astrophys. **557**, A141 (2013). arXiv:1308.2827. doi:10.1051/0004-6361/201321981

P. Charbonneau, Dynamo models of the solar cycle. Living Rev. Sol. Phys. **7**, 3 (2010). doi:10.12942/lrsp-2010-3

M. Dasi-Espuig, S.K. Solanki, N.A. Krivova, R. Cameron, T. Peñuela, Sunspot group tilt angles and the strength of the solar cycle. Astron. Astrophys. **518**, A7 (2010). arXiv:1005.1774. doi:10.1051/0004-6361/201014301

M. Dasi-Espuig, S.K. Solanki, N.A. Krivova, R. Cameron, T. Peñuela, Sunspot group tilt angles and the strength of the solar cycle (Corrigendum). Astron. Astrophys. **556**, C3 (2013). doi:10.1051/0004-6361/201014301e

M. Dikpati, Polar field puzzle: Solutions from flux-transport dynamo and surface-transport models. Astrophys. J. **733**, 90 (2011). arXiv:1104.0269. doi:10.1088/0004-637X/733/2/90

M. Dikpati, P.A. Gilman, Global solar dynamo models: Simulations and predictions of cyclic photospheric fields and long-term non-reversing interior fields. New J. Phys. **9**, 297 (2007). doi:10.1088/1367-2630/9/8/297

C.J. Durrant, J.P.R. Turner, P.R. Wilson, The mechanism involved in the reversals of the sun's polar magnetic fields. Sol. Phys. **222**, 345–362 (2004). doi:10.1023/B:SOLA.0000043577.33961.82

T.L. Duvall Jr, Large-scale solar velocity fields. Sol. Phys. **63**, 3–15 (1979). doi:10.1007/BF00155690

V. Gaizauskas, K.L. Harvey, J.W. Harvey, C. Zwaan, Large-scale patterns formed by solar active regions during the ascending phase of cycle 21. Astrophys. J. **265**, 1056–1065 (1983). doi:10.1086/160747

V. Gaizauskas, D.H. Mackay, K.L. Harvey, Evolution of solar filament channels observed during a major poleward surge of photospheric magnetic flux. Astrophys. J. **558**, 888–902 (2001). doi:10.1086/322298

G.E. Hale, S.B. Nicholson, The law of sun-spot polarity. Astrophys. J. **62**, 270 (1925). doi:10.1086/142933

G.E. Hale, F. Ellerman, S.B. Nicholson, A.H. Joy, The magnetic polarity of sun-spots. Astrophys. J. **49**, 153 (1919). doi:10.1086/142452

D.H. Hathaway, L. Rightmire, Variations in the Sun's meridional flow over a solar cycle. Science **327**, 1350 (2010). doi:10.1126/science.1181990

D.H. Hathaway, P.A. Gilman, J.W. Harvey, F. Hill, R.F. Howard, H.P. Jones, J.C. Kasher, J.W. Leibacher, J.A. Pintar, G.W. Simon, GONG observations of solar surface flows. Science **272**, 1306–1309 (1996). doi:10.1126/science.272.5266.1306

J.T. Hoeksema, Structure and evolution of the large scale solar and heliospheric magnetic fields. PhD thesis, Stanford University, CA (1984)

V.G. Ivanov, Joy's law and its features according to the data of three sunspot catalogs. Geomagn. Aeron. **52**, 999–1004 (2012). doi:10.1134/S0016793212080130

J. Jiang, E. Işik, R.H. Cameron, D. Schmitt, M. Schüssler, The effect of activity-related meridional flow modulation on the strength of the solar polar magnetic field. Astrophys. J. **717**, 597–602 (2010). arXiv:1005.5317. doi:10.1088/0004-637X/717/1/597

R.W. Komm, R.F. Howard, J.W. Harvey, Meridional flow of small photospheric magnetic features. Sol. Phys. **147**, 207–223 (1993). doi:10.1007/BF00690713

B.J. Labonte, R. Howard, Solar rotation measurements at Mount Wilson. III—Meridional flow and limbshift. Sol. Phys. **80**, 361–372 (1982). doi:10.1007/BF00147982

R.B. Leighton, Transport of magnetic fields on the Sun. Astrophys. J. **140**, 1547 (1964). doi:10.1086/148058

R.B. Leighton, A magneto-kinematic model of the solar cycle. Astrophys. J. **156**, 1 (1969). doi:10.1086/149943

A. Muñoz-Jaramillo, M. Dasi-Espuig, L.A. Balmaceda, E.E. DeLuca, Solar cycle propagation, memory, and prediction: Insights from a century of magnetic proxies. Astrophys. J. Lett. **767**, L25 (2013). arXiv:1304.3151. doi:10.1088/2041-8205/767/2/L25

E.N. Parker, The formation of sunspots from the solar toroidal field. Astrophys. J. **121**, 491 (1955). doi:10.1086/146010

G.J.D. Petrie, Evolution of active and polar photospheric magnetic fields during the rise of cycle 24 compared to previous cycles. Sol. Phys. **281**, 577–598 (2012). arXiv:1204.1048. doi:10.1007/s11207-012-0117-3

G.J.D. Petrie, Solar magnetic activity cycles, coronal potential field models and eruption rates. Astrophys. J. **768**, 162 (2013). arXiv:1303.1218. doi:10.1088/0004-637X/768/2/162

G.J.D. Petrie, Solar magnetism in the polar regions. Living Rev. Sol. Phys. (2015)

G.J.D. Petrie, I. Patrikeeva, A comparative study of magnetic fields in the solar photosphere and chromosphere at equatorial and polar latitudes. Astrophys. J. **699**, 871–884 (2009). arXiv:1010.6041. doi:10.1088/0004-637X/699/1/871

G.J.D. Petrie, K. Petrovay, K. Schatten, Solar polar fields and the 22-year activity cycle: Observations and models. Space Sci. Rev. **186**, 325–357 (2014). doi:10.1007/s11214-014-0064-4

K.H. Schatten, W.D. Pesnell, An early solar dynamo prediction: Cycle 23 is approximately cycle 22. Geophys. Res. Lett. **20**, 2275–2278 (1993). doi:10.1029/93GL02431

K.H. Schatten, S. Sofia, Forecast of an exceptionally large even-numbered solar cycle. Geophys. Res. Lett. **14**, 632–635 (1987). doi:10.1029/GL014i006p00632

K.H. Schatten, P.H. Scherrer, L. Svalgaard, J.M. Wilcox, Using dynamo theory to predict the sunspot number during solar cycle 21. Geophys. Res. Lett. **5**, 411–414 (1978). doi:10.1029/GL005i005p00411

K. Schatten, D.J. Myers, S. Sofia, Solar activity forecast for solar cycle 23. Geophys. Res. Lett. **23**, 605–608 (1996). doi:10.1029/96GL00451

C.J. Schrijver, Y. Liu, The global solar magnetic field through a full sunspot cycle: Observations and model results. Sol. Phys. **252**, 19–31 (2008). doi:10.1007/s11207-008-9240-6

C.J. Schrijver, M.L. De Rosa, A.M. Title, What is missing from our understanding of long-term solar and heliospheric activity? Astrophys. J. **577**, 1006–1012 (2002). doi:10.1086/342247

N.R. Sheeley Jr., C.R. DeVore, J.P. Boris, Simulations of the mean solar magnetic field during sunspot cycle 21. Sol. Phys. **98**, 219–239 (1985). doi:10.1007/BF00152457

H.B. Snodgrass, Magnetic rotation of the solar photosphere. Astrophys. J. **270**, 288–299 (1983). doi:10.1086/161121

X. Sun, J.T. Hoeksema, Y. Liu, J. Zhao, On polar magnetic field reversal and surface flux transport during solar cycle 24. Astrophys. J. **798**, 114 (2015). 1410.8867. doi:10.1088/0004-637X/798/2/114

L. Svalgaard, T.L. Duvall Jr., P.H. Scherrer, The strength of the sun's polar fields. Sol. Phys. **58**, 225–239 (1978). doi:10.1007/BF00157268

R.K. Ulrich, The controversial sun, in *IAU Colloq. 137: Inside the Stars*, ed. by W.W. Weiss, A. Baglin. Astronomical Society of the Pacific Conference Series, vol. 40 (Astronomical Society of the Pacific, San Francisco, 1993), pp. 25–42

Y.M. Wang, N.R. Sheeley Jr., Magnetic flux transport and the sun's dipole moment—new twists to the Babcock-Leighton model. Astrophys. J. **375**, 761–770 (1991). doi:10.1086/170240

Y.M. Wang, N.R. Sheeley Jr., On potential field models of the solar corona. Astrophys. J. **392**, 310–319 (1992). doi:10.1086/171430

Y.M. Wang, A.G. Nash, N.R. Sheeley Jr., Evolution of the sun's polar fields during sunspot cycle 21—poleward surges and long-term behavior. Astrophys. J. **347**, 529–539 (1989). doi:10.1086/168143

Y.M. Wang, J. Lean, N.R. Sheeley Jr., Role of a variable meridional flow in the secular evolution of the sun's polar fields and open flux. Astrophys. J. Lett. **577**, L53–L57 (2002). doi:10.1086/344196

Y.M. Wang, E. Robbrecht, N.R. Sheeley Jr., On the weakening of the polar magnetic fields during solar cycle 23. Astrophys. J. **707**, 1372 (2009). doi:10.1088/0004-637X/707/2/1372

Space Sci Rev (2017) 210:109–143
DOI 10.1007/s11214-016-0294-8

Radiative Diagnostics in the Solar Photosphere and Chromosphere

J. de la Cruz Rodríguez[1] · M. van Noort[2]

Received: 14 December 2015 / Accepted: 27 September 2016 / Published online: 15 November 2016

Abstract Magnetic fields on the surface of the Sun and stars in general imprint or modify the polarization state of the electromagnetic radiation that is leaving from the star. The inference of solar/stellar magnetic fields is performed by detecting, studying and modeling polarized light from the target star. In this review we present an overview of techniques that are used to study the atmosphere of the Sun, and particularly those that allow to infer magnetic fields. We have combined a small selection of theory on polarized radiative transfer, inversion techniques and we discuss a number of results from chromospheric inversions.

Keywords Radiative transfer · Magnetic fields · Polarimetry · Photosphere · Chromosphere

1 Introduction

Stellar spectra encode information of the composition and of the physical conditions present in the hosting star where radiation originates. Compared to other research fields in physics, astronomy is mostly a *remote sensing* science. Solar and stellar physics have developed through the study of radiation and polarization, attempting to infer the underlying physical state of the atmospheric plasma. While most stellar observations are spatially unresolved,[1] our proximity to the Sun provides a high-resolution insight into the outer layers of its atmosphere. The solar atmosphere poses a wonderful laboratory to study stellar atmospheres and

[1] See latest advances in full-Stokes Zeeman Doppler Imaging techniques (e.g., Rosén et al. 2015).

✉ J. de la Cruz Rodríguez
jaime@astro.su.se

M. van Noort
vannoort@mps.mpg.de

1 Institute for Solar Physics, Stockholm University, Albanova University Center, 10691, Stockholm, Sweden

2 Max Planck Institute for Solar System Research, Justus-von-Liebig-Weg 3, 37077 Göttingen, Germany

the processes that are responsible for energy transport, and *forward modeling* and *inversion* have been particularly successful for understanding line formation and in the interpretation of spectropolarimetric and spectroscopic observations.

Forward modeling techniques usually involve magnetohydrodynamical simulations of a stellar atmosphere that may include waves, magnetic fields, flows and a number of dynamic events. Synthetic spectra from those model atmospheres can be qualitatively compared with observations to study plasmas or to understand how spectral lines react/form under different physical conditions. Recent developments in 3D MHD simulations have made calculations that simultaneously include a photosphere, a chromosphere and a corona possible, and these have been extensively used to study the elusive solar interface region.

Inversion techniques rely upon algorithms that systematically modify the parameters of a *guessed* atmospheric model, to minimize the difference between the emerging spectra from that model and the observations. Therefore, inversions can infer quantitative information from an observation, assuming that the proposed model and radiative transfer can *realistically* describe the conditions in the atmosphere. Inversions have been extensively used to analyze photospheric observations and, more recently, to study the chromosphere.

In this chapter we review the state-of-the-art of radiative diagnostics in solar physics. In Sect. 2 we introduce radiative transfer and response functions. In Sects. 3 and 4 we describe inversion techniques and the main spectral lines used in chromospheric studies. In Sect. 5, some recent inversion results are presented. Finally, future perspectives of radiative diagnostics are discussed in Sect. 6.

2 Radiative Transfer and Response Functions

In the field of stellar and solar physics, almost all knowledge that has been accumulated has been acquired through the analysis and interpretation of radiation, combined with the known laws of physics. The field is thus crucially dependent on the correct understanding of the information contained in the light that we receive.

The field of radiative transfer has a rich history, dating back to the 19th century, when the study of stellar and solar light led to the discovery of many new elements, some of which, such as Helium, were even discovered first on the Sun, before they were found also on Earth. Particularly the first half of the 20th century saw a revolution in our understanding of the structure of atoms and molecules, and the interaction of light with them, which layed the foundations for the inference of the composition and conditions of stellar atmospheres.

2.1 Radiative Transfer in Stellar Atmospheres

In stellar atmospheres, the structure of the atmosphere is dominated by the transport of the energy, generated in the core of the star, to the outside. In the part of the atmosphere that is visible to an observer, the conversion of thermal energy to radiation is thus of crucial importance for the vertical structure, and more importantly is, at least in part, determined by the radiation field itself. The radiation thus carries the only information we receive from any stellar atmosphere, and at the same time is an essential constituent of that same atmosphere.

For the correct interpretation of the light that we receive, it is thus necessary to self-consistently model the radiation field and the atmospheric structure, by solving the radiative transfer equation and the equations of statistical equilibrium, along with hydrostatic equilibrium.

The monochromatic radiative transfer equation (RTE) for a single ray of polarized light can be expressed as

$$\mu \frac{d\boldsymbol{I}}{ds} = -\mathbf{K}\boldsymbol{I} + \boldsymbol{j}, \tag{1}$$

where $\boldsymbol{I} = (I, Q, U, V)^T$ is the Stokes parameter vector, $\boldsymbol{j} = (j_I, j_Q, j_U, j_V)^T$ is the total emission vector, $\mathbf{K}$ is the total absorption matrix, and μ is the angle of the ray relative to the surface. There are seven independent terms in the absorption matrix: k_I is related to the line opacity; k_Q, k_U and k_V describe the coupling of the intensity I with Q, U and V; and f_Q, f_U and f_V are conversion terms between Q, U and V due to magneto-optical effects (Landi Degl'Innocenti and Landolfi 2004):

$$\mathbf{K} = k_c \begin{pmatrix} 1 & 0 & 0 & 0 \\ 0 & 1 & 0 & 0 \\ 0 & 0 & 1 & 0 \\ 0 & 0 & 0 & 1 \end{pmatrix} + k_l \begin{pmatrix} k_I & k_Q & k_U & k_V \\ k_Q & k_I & f_V & -f_U \\ k_U & -f_V & k_I & f_Q \\ k_V & f_U & -f_Q & k_I \end{pmatrix}, \tag{2}$$

where k_c is the background opacity and k_l is the line opacity. In general, the line opacity depends on the populations of the lower and upper level of the transition (n_l and n_u), the oscillator strength of the transition (f), the statistical weight of the atomic levels (g_l and g_u) and the Doppler width of the line ($\Delta\nu_D$),

$$k_l = \frac{\pi e_0^2}{mc} \frac{n_l f}{\Delta\nu_D} \left(1 - \frac{n_u g_l}{n_l g_u}\right),$$

where the Doppler width of the line is given by

$$\Delta\nu_D = \frac{\nu_0}{c} \sqrt{\frac{2T K_B}{M} + \mathrm{v}_{turb}^2}, \tag{3}$$

where K_B is the Boltzmann constant, T is the temperature, M is the atomic mass and v_{turb} is the microturbulent Doppler broadening.

The terms involved in the RTE are normally derived from the physical parameters of the model atmosphere: temperature, macroscopic line-of-sight velocity, turbulent motions and the magnetic field vector. In practice, the parameters of the model atmosphere are usually known on a discrete grid of depth points along the ray. In most cases, the depth-stratification of $\mathbf{K}$ and $\boldsymbol{j}$ are not known analytically and the RTE must be integrated numerically assuming some boundary condition. Over the years, many methods have been developed to integrate the RTE to obtain the emerging full-Stokes intensity. In the early 70s, it was common to use a Runge-Kutta solver (Landi Degl'Innocenti 1976), but this type of solver is very inefficient and shows poor accuracy on coarse depth grids. The *diagonal element lambda operator* method (DELO, Rees et al. 1989) reformulates the RTE in terms of the *source vector* ($\boldsymbol{S} = \mathbf{j}/k_I)^T$ by dividing all terms in Eq. (1) by the absorption coefficient k_I. This change leaves all diagonal terms in $\mathbf{K}$ equal to unity, which allows for the substitution

$$\bar{\mathbf{K}} = \frac{\mathbf{K}}{\eta_I} - \mathbb{1},$$

where $\mathbb{1}$ is the 4×4 identity matrix. The RTE becomes:

$$\mu \frac{d\boldsymbol{I}}{d\tau} = \boldsymbol{I} - \mathscr{S}, \tag{4}$$

with the monochromatic optical depth defined as $d\tau = k_I ds$ and $\mathscr{S} = \boldsymbol{S} - \bar{\mathbf{K}}\boldsymbol{I}$. On a discrete grid of depth points the solution is normally expressed as an integration starting at the *upwind* point, towards the next point in the direction of propagation (usually referred as the *central* point), which we denote with the sub-indexes u and 0 respectively:

$$\boldsymbol{I}(\tau_0) = \boldsymbol{I}(\tau_u)\mathbf{O}(\delta\tau_0) + \int_{\tau_u}^{\tau_0} \mathbf{O}(t)\mathscr{S}(t)dt.$$

$\mathbf{O}$ is the evolution operator, a 4×4 real matrix that applied to the Stokes vector at point u yields the Stokes vector at point 0. The evolution operator can have a complicated analytical form, but assuming that the absorption matrix is constant between two consecutive grid points and equal to the value in the middle of the interval, the evolution operator can be expressed in terms of exponential terms.

$$\boldsymbol{I}(\tau_0) = \boldsymbol{I}(\tau_u)e^{-\delta\tau_0} + \int_{\tau_u}^{\tau_0} e^{-(t-\tau_u)}\mathscr{S}(t)dt. \tag{5}$$

An underlying assumption of the DELO method is that the diagonal terms of $\mathbf{K}$ are much smaller than the non-diagonal terms that are included in the effective source vector $\mathscr{S}$. This way of writing the transfer equation has allowed us to use the same evolution operator as in the unpolarized case, but we have conveniently hidden some dirt under the effective source vector. Strictly speaking the evolution operator is exact only when the non-diagonal terms of $\mathbf{K}$ are zero, but this approximation seems to hold rather well when compared against other methods for solving the PRT equation.

Equation (5) can be integrated *analytically* if the equivalent source vector is assumed to have a given dependence with depth within the interval (τ_u, τ_0). The solution can be written as a function of the ensuing intensity and the upwind, center and (sometimes) downwind values of the modified source function. Rees et al. (1989) assumed a linear dependence giving rise to the DELO-Linear solution. Trujillo Bueno (2003) showed that the linear approximation used in the DELO-Linear yields a poor absolute error in the computation of the atomic populations and developed the DELO-Parabolic method by splitting the equivalent source vector in the integral term of Eq. (5) and assuming a parabolic dependence for $\boldsymbol{S}$ and linear dependence for $\bar{\mathbf{K}}\boldsymbol{I}$. A problem of the parabolic dependence is that parabolae can easily overshoot if the monochromatic depth scale is not equidistant. Therefore, de la Cruz Rodríguez and Piskunov (2013) introduced two integration strategies based on non-overshooting Bezier splines that solve the issues of DELO-parabolic while keeping quadratic and cubic accuracy, usually referred as DELO-Bezier. The DELO solvers can be computed with relatively low computational cost and usually provide good accuracy in most situations, making them especially valuable so compute non-LTE problems and inversions where the RTE must be solved accurately many times. An alternative Hermitian method proposed by Bellot Rubio et al. (1998) has also been extensively used in solar inversion codes.

A different approach to solve the PRT equation has been proposed by Steiner et al. (2016). Instead of searching for a smooth interpolant, they allow for a discontinuous solution at the boundary of each grid cell in the atmosphere. Within each cell, the source function is allowed to change linearly and the PRT equation is integrated using the Milne-Eddington solution. This approach shows very promising results, and is especially accurate when coarse grids are used to describe the model atmosphere, a property that is especially relevant for inversion methods. However, the convergence stability of this approach when used to solve the statistical equilibrium equations with approximate lambda operators remains to be explored.

2.2 Coupling Between Atmospheric and Radiative Quantities

We now consider the dependencies of the RTE on the parameters of the model atmosphere. Assuming a known set of physical quantities, such as temperature, line of sight velocity, turbulent velocities, the three components of the magnetic field vector, and a depth scale, it is possible to compute the radiative properties that appear in Eq. (4). For reasons of simplicity, we first consider here the case of non-polarized radiation, in which case the radiative transfer equation can be written as:

$$\mu \frac{dI_{\mu\nu}}{ds} = \eta_{\mu\nu} - \chi_{\mu\nu} I_{\mu\nu}, \tag{6}$$

where $\eta_{\mu\nu}$ and $\chi_{\mu\nu}$ are the total emissivity and opacity, including continuum and line contributions. Eq. (6) can be re-written using the source function $S_{\mu\nu} = \eta_{\mu\nu}/\chi_{\mu\nu}$ as

$$\mu \frac{dI_{\mu\nu}}{-\chi_{\mu\nu} ds} = I_{\mu\nu} - S_{\mu\nu} \Rightarrow \mu \frac{dI_{\mu\nu}}{d\tau_{\mu\nu}} = I_{\mu\nu} - S_{\mu\nu},$$

where $d\tau_{\mu\nu} = -\chi_{\mu\nu} ds$ is the optical depth along the ray.

The line emissivity (η_ν) and opacity (χ_ν) characterize the line radiative properties as a function of the atomic level populations n and the Einstein coefficients ($B_{ll'}$, $B_{l'l}$ and $A_{ll'}$). In the unpolarized case and assuming complete redistribution (CRD) of scattered photons, the expressions are simplified to the following when the energy of level l' is smaller than that of level l ($l \succ l'$, using the same notation as Rybicki and Hummer 1991):

$$\eta_{ll'}(\mu, \nu) = \frac{h\nu}{4\pi} n_u A_{ll'} \varphi(\mu, \nu), \tag{7}$$

$$\chi_{ll'}(\mu, \nu) = \frac{h\nu}{4\pi} (n_l B_{l'l} - n_u B_{ll'}) \varphi(\mu, \nu), \tag{8}$$

where φ is the profile, which we have imposed to be the same when a photon is absorbed or emitted (CRD). The total emissivity and opacity can be expressed as the sum of all contributions:

$$\eta_{\mu\nu} = \sum_{l \succ l'} \eta_{ll'}(\mu, \nu) + \eta_c(\nu),$$

$$\chi_{\mu\nu} = \sum_{l \succ l'} \chi_{ll'}(\mu, \nu) + \chi_c(\nu),$$

where $\eta_c(\nu)$ and $\chi_c(\nu)$ are the continuum contribution to the total emissivity and opacity. For time-independent equilibrium situations, the populations of the atomic levels n can be obtained by solving the statistical equilibrium equations:

$$\sum_{l' \prec l} \left[n_l A_{ll'} - (n_{l'} B_{l'l} - n_l B_{ll'}) \bar{J} \right] - \sum_{l' \succ l} \left[n_{l'} A_{l'l} - (n_l B_{ll'} - n_{l'} B_{l'l}) \bar{J} \right] = - \sum_{l'} (n_l C_{ll'} + n_{l'} C_{l'l}), \tag{9}$$

where $C_{ll'}$ is the collisional rate between level l and l', which is generally a function of temperature and density only, and $\hat{J}$ is the integrated mean intensity (CRD) given by

$$\bar{J} = \frac{1}{4\pi} \int d\Omega \int d\nu \varphi(\mu, \nu) I_{\mu,\nu},$$

which depends on temperature, density and LOS velocity, as well as any microturbulent velocity.

In the unpolarized case then, the dependence of all elements in Eq. (9) on the atmospheric quantities in most cases not especially complicated. However, calculating the radiative quantities α and η for a given atmosphere is by no means a straightforward task. The principal difficulty in solving (9) is that we need to know the atomic level populations to calculate the intensities using Eq. (7) and (8), but the atomic populations depend on the intensity through $\bar{J}$. Therefore, the problem is not only non-local, it also highly non-linear and must therefore be solved iteratively, using an appropriate approximation to describe the dependence of $\bar{J}$ on the atomic level populations, as pioneered by Cannon (1973), Scharmer (1981), Olson and Kunasz (1987) and Rybicki and Hummer (1991) among others. For every set of temperature, density and velocity, however, a solution can be calculated.

In the general polarized case, Eq. (9) contains the population densities of all substates of every energy level, which increases both the size and the difficulty of calculating the coefficients of the linear system to solve significantly. More specifically, the energy of the various sub-states within each energy level is now dependent on the magnetic field strength and direction. Although this is what makes it possible to quantify the magnetic field strength and direction in the solar atmosphere, the associated computational cost is significant. Many synthesis codes therefore resort to the use of the *polarization-free* approximation, that first solves for the populations using the unpolarized approximation, and distributes the populations of the unpolarized levels across the sub-levels according to their statistical weight, thus allowing for a single full-Stokes solution to be computed (Trujillo Bueno and Landi Degl'Innocenti 1996).

Although the polarization-free approximation offers a great simplification, in many cases the density is sufficiently high that the collisional rates dominate Eq. (9), and the solution can be accurately described by the approximation of LTE. Under these conditions, the source function is directly given by the Planck function, and in practice we only need to compute explicitly the terms in the absorption matrix. Assuming that only the Zeeman effect is at work hereafter, the terms of the absorption matrix in Eq. (2) are given by the following expressions (Landi Degl'Innocenti and Landolfi 2004):

$$
\begin{aligned}
k_I &= \frac{1}{2}\left[h_p \sin^2\theta + \frac{h_b + h_r}{2}\left(1+\cos^2\theta\right)\right],\\
k_Q &= \frac{1}{2}\left[h_p - \frac{h_b + h_r}{2}\right]\sin^2\theta\cos 2\chi,\\
k_U &= \frac{1}{2}\left[h_p - \frac{h_b + h_r}{2}\right]\sin^2\theta\sin 2\chi,\\
k_V &= \frac{1}{2}[h_r - h_b]\cos\theta,\\
f_Q &= \frac{1}{2}\left[g_p - \frac{g_b + g_r}{2}\right]\sin^2\theta\cos 2\chi,\\
f_U &= \frac{1}{2}\left[g_p - \frac{g_b + g_r}{2}\right]\sin^2\theta\sin 2\chi,\\
f_V &= \frac{1}{2}[g_r - g_b]\cos\theta,
\end{aligned}
$$

where θ and χ are the inclination and the azimuth of the magnetic field vector and the h and g components are computed from the Voigt-Faraday functions including the effect of Doppler motions and the Zeeman splitting. The latter is normally characterized by a wavelength shift of the polarized components and their relative strength.

The wavelength shift of the Zeeman components ($\Delta\lambda_B$) can be expressed in terms of the quantum state M according to the selection rules:

$$\Delta M = M_u - M_l = \begin{cases} +1 \equiv b, \\ 0 \equiv p, \\ -1 \equiv r. \end{cases}$$

The computation of the profiles is directly given by the following expressions (for $q = b, p, r$):

$$h_q = \sum_{M_l M_u} S_q^{J_l J_u} \frac{1}{\Delta\nu_D\sqrt{\pi}} H\big(a, v - v_a + v_B(g_u M_u - g_l M_l)\big),$$

$$f_q = \sum_{M_l M_u} S_q^{J_l J_u} \frac{1}{\Delta\nu_D\sqrt{\pi}} L\big(a, v - v_a + v_B(g_u M_u - g_l M_l)\big),$$

where S_q is the strength of the Zeeman component, H and L are the absorption and dispersion profiles respectively, a is the damping and v, v_a and v_B are the frequency, Doppler velocity ($V_{l.o.s.}$) and Zeeman splitting respectively expressed in Doppler width units,

$$v = \frac{\nu_0 - \nu}{\Delta\nu_D}, \qquad v_a = \frac{\nu_0 V_{l.o.s}}{c\Delta\nu_D}, \qquad v_B = \frac{e_0}{4\pi m_e c}\frac{B}{\Delta\nu_D}, \qquad a = \frac{\Gamma_{stark} + \Gamma_{vdW} + \Gamma_{rad}}{\Delta\nu_D}. \tag{10}$$

We have chosen to work with frequencies instead of wavelength, but v_B can also be expressed in terms of the central wavelength of the line λ_0 (Eq. 11). This form of v_B explicitly shows that the Zeeman splitting scales quadratically with λ and linearly with B (with $\Delta\lambda_D$ the Doppler width in units of wavelength),

$$v_B = \frac{e_0}{4\pi m_e c}\frac{B\lambda_0^2}{\Delta\lambda_D}. \tag{11}$$

The damping parameters originate from collisions with charged particles (Γ_{stark}), collisions with neutral particles (Γ_{vdW}) and from the finite lifetime of atomic levels (Γ_{rad}). These parameters have traditionally been computed theoretically or measured and they are typically considered part of the input atomic data that can be retrieved from an atomic database like VALD (Piskunov et al. 1995; Barklem et al. 1998). The damping parameters are typically functions of the electron density, the hydrogen and helium densities and the temperature.

Finally, the strength of the Zeeman components can be computed in general using $3-j$ symbols ($S_q^{J_l J_u}$), but in the case of the Zeeman effect they are given in Table 1 as a function of the J and M quantum numbers of the sub-levels of the transition.

Although the above outline is far from complete, it is possible to conclude that there is a limited number of atmospheric quantities that control the radiative properties of the solar atmosphere. While some of the dependencies on these quantities can be approximated in a fairly simple way, others enter the expressions in many different places, some of them both directly and indirectly. Perhaps the most convoluted dependency is that on the temperature,

Table 1 Strength of the Zeeman components as a function of the quantum numbers J and M. For clarity, we note that $\Delta M = M_u - M_l$. Source: Landi Degl'Innocenti and Landolfi (2004), Table 3.1

	$J_u = J_l + 1$	$J_u = J_l$	$J_u = J_l - 1$
$\Delta M = +1 \equiv b$	$\frac{3(J_l+M_l+1)(J_l+M_l+2)}{2(J_l+1)(2J_l+1)(2J_l+3)}$	$\frac{3(J_l-M_l)(J_l+M_l+1)}{2J_l(J_l+1)(2J_l+1)}$	$\frac{3(J_l-M_l)(J_l-M_l-1)}{2J_l(2J_l-1)(2J_l+1)}$
$\Delta M = 0 \equiv p$	$\frac{3(J_l-M_l+1)(J_l+M_l+1)}{(J_l+1)(2J_l+1)(2J_l+3)}$	$\frac{3M_l^2}{J_l(J_l+1)(2J_l+1)}$	$\frac{3(J_l-M_l)(J_l+M_l)}{J_l(2J_l-1)(2J_l+1)}$
$\Delta M = -1 \equiv r$	$\frac{3(J_l-M_l+1)(J_l-M_l+2)}{2(J_l+1)(2J_l+1)(2J_l+3)}$	$\frac{3(J_l+M_l)(J_l-M_l+1)}{2J_l(J_l+1)(2J_l+1)}$	$\frac{3(J_l+M_l)(J_l+M_l-1)}{2J_l(2J_l-1)(2J_l+1)}$

which appears in the expressions for the Doppler width, the line opacity, the source function and usually the damping constants. The densities and the ionization balance mostly influence the damping and the line opacity. The line of sight velocities and the magnetic field influence the absorption matrix, which encodes the Doppler shifts and Zeeman splitting at each depth of the atmospheric model. In Sect. 2.3 we formalize and illustrate these dependencies using so-called *response functions.*

2.3 Response Functions

In order to change a given atmosphere in such a way that the emerging spectrum matches an observed spectrum, information on how the spectrum changes when the atmosphere is perturbed must be obtained. Unfortunately, since the radiation influences the state of the atmosphere in a non-local and non-linear way, the response of the spectrum to the atmospheric perturbation is also not linear, so that a perfect correction generally cannot be produced at once. In most cases the response, although not linear, will reduce the difference, so that an iterative approach can be used.

The response of the spectrum to an atmospheric perturbation can be calculated *numerically* using finite differences, by simply calculating the spectrum, perturbing the atmosphere and calculating the spectrum again. The difference is then divided by the perturbation, yielding the desired response function. Alternatively, a *centered* derivative (in the nominal value of the parameter) is likely to be more accurate but it will be twice as slow:

$$R_\lambda(x, \tau_k) \approx \underbrace{\frac{S_\lambda(x+\delta x, \tau_k) - S_\lambda(x-\delta x, \tau_k)}{2\delta x \Delta\tau_k}}_{\text{centered}} \approx \underbrace{\frac{S_\lambda(x+\delta x, \tau_k) - S_\lambda(x, \tau_k)}{\delta x \Delta\tau_k}}_{\text{non-centered}}, \tag{12}$$

where R_λ is the monochromatic response function to perturbations to a parameter x and S_λ is the monochromatic emerging Stokes vector. *Response functions* for polarized light were proposed by Landi Degl'Innocenti and Landi Degl'Innocenti (1977), who derived an analytical expression by perturbing the RTE. However, those early studies did not couple directly the response of the spectra to changes in typical parameters of model atmospheres like, e.g., temperature or density but rather to perturbations of the parameters of the RTE. For depth-stratified atmospheres, analytical response functions have been derived in the LTE case (Ruiz Cobo and del Toro Iniesta 1992), but it is not trivial to write them as the response of all parameters must be also be propagated through all the calculations involved in the computation of the terms of the RTE and into the equation of state of the gas. Analytical response function in Milne-Eddington codes are also commonly used and somewhat less convoluted to derive (e.g., Orozco Suárez and Del Toro Iniesta 2007). The advantage of analytical response functions is that they are usually more precise and much more efficient

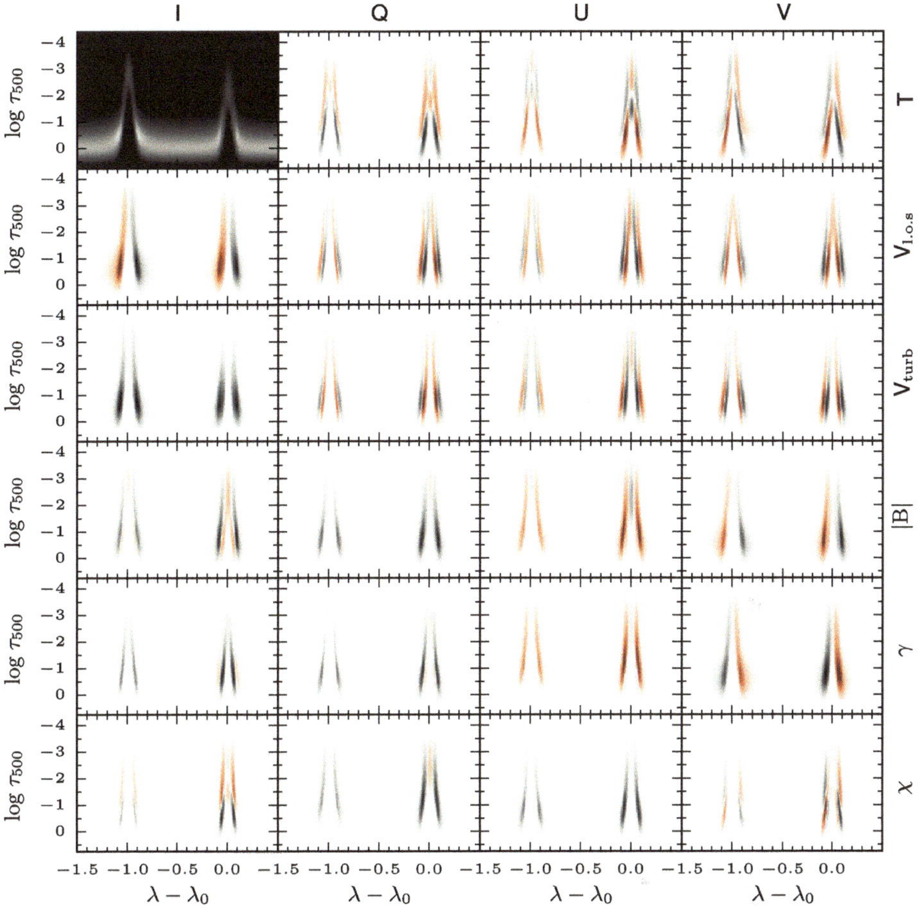

Fig. 1 Numerical response functions of Stokes I, Q, U and V (from left to right) to the temperature, line of sight velocity, turbulent motions, magnetic field strength, magnetic field inclination (γ) and magnetic field azimuth (χ) from top to bottom. The calculations have been performed in LTE using the FAL-C atmosphere with an ad-hoc magnetic field vector of 500 G

to compute than the finite differences approach, as the overhead compared to synthesizing a single profile is small.

Figure 1 illustrates the numerical response functions of the Fe I λ6301.5 and λ6302.5 lines to temperature, line of sight velocity, turbulent broadening, magnetic field strength, inclination and azimuth along the optical depth scale of the FAL-C model atmosphere, including a constant magnetic field strength of 500 G inclined at 45 deg with and azimuth of 20 deg. It is beyond the scope of this review to analyze in detail each of the panels, since response functions are model dependent, but we note some basic properties that can be illustrated using these panels:

- Spectral lines sample a range of optical depths, in this case increasing in height from the continuum towards line center.
- Of the 6 considered parameters, only the temperature influences the continuum.

- The coupling effect of magneto-optical terms in the absorption matrix is clearly visible for example in the response of all Stokes parameters to changes in the azimuth ((Landi Degl'Innocenti 1979)).
- In most parameters the shape of the response function in Stokes Q, U and V can appear to be quite convoluted as an effect of the shape of the profiles.

3 Inversion Methods for Solar Observations

Unlike what is suggested by the name, most inversion codes do not provide an inverse mapping from the spectra to an atmosphere, but instead *infer* the atmosphere from the data in a semi-automated way. Since the mapping from atmosphere to spectrum is not reversible in general, simplifying assumptions generally need to be made to ensure the solution remains physically plausible. These assumptions have a significant impact on the reliability of the result, in a way that is not evident from either the precision with which the data are reproduced, nor from the physical plausibility of the inverted result.

The restrictions imposed are often ad-hoc, because the amount of constrainable information in the spectrum is generally difficult to estimate, but sometimes some a-priori restrictions can be applied, based on physical arguments, that can add stability, allowing for less restrictive assumptions to be made elsewhere.

The basic ingredients of any inversion code are the ability to generate a synthetic spectrum through a forward calculation on an atmospheric model, knowledge of the important instrumental effects involved in mapping the synthetic spectrum to the observed data, and an automated minimization engine, which can be based on a number of techniques.

3.1 Atmospheric Model

Most inversion codes are fundamentally based on a physical formation model of a varying degree of complexity. There are two main areas in which the model can vary in complexity: the atmospheric structure and the physics assumed to be involved in generating the emerging spectrum.

3.1.1 Model Atmospheres

The sophistication of the atmosphere that can be fitted to an observation depends to a large extent on the number and type of spectral lines that it is required to reproduce the profile of. Where a single spectral line contains already a large amount of information about the atmosphere in which it originated, the exact shape of several different line profiles can significantly restrict the possibilities.

For the inversion of a single spectral line, a commonly made choice is that of a Milne-Eddington (ME) atmosphere (Harvey et al. 1972; Auer et al. 1977), which is characterized by a constant opacity, linear source function, and a host of arbitrary fit parameters that are needed to fit the profile, but do not really correspond to useful physical information. The solution can be integrated analytically, however, leading to a very efficient calculation of the line profile, which allows for a fast inversion. This type of atmosphere is able to reproduce simple spectral line profiles. Many codes exist, that are based on this type of atmosphere (e.g. HELIX, Lagg et al. 2004; MERLIN, Skumanich and Lites 1987; HMI VFISV, Borrero et al. 2011, to name just a few), each of them using different minimization strategies to find the best fit to the observed spectrum (see comparison by Borrero et al. 2014).

In those cases where more sophistication is needed to fit the observed spectra, either due to their large formation range, or due to their specific sensitivity to gradients in some physical quantity, a fully stratified, physical model of the atmosphere must be used. The advantage of this approach is that the fitted atmosphere has a firm physical basis, and can thus be used to study the magneto-hydrodynamic structure underlying an observation (Ruiz Cobo and del Toro Iniesta 1992).

A number of codes exists that allows for a height dependent stratification (SIR, Ruiz Cobo and del Toro Iniesta 1992; SPINOR, Frutiger et al. 2000; NICOLE, Socas-Navarro et al. 1998, etc.), all of which make use of a gradient search based optimization method, using response functions or finite differencing to calculate the gradient. However, for all of them *it is crucial that the level of stratification that is required to produce the spectrum is met, but not exceeded, or the inversion problem is no longer well posed.* This last problem in particular gives rise to a level of arbitrariness that is very difficult to quantify.

More recently, a new generation of methods has begun development that tries to reduce some of the degrees of freedom of the inversions, by introducing additional constraints on the solution, that are either inspired by physics, or by other arguments. One of these is the sparse inversion code (Asensio Ramos and de la Cruz Rodríguez 2015), that requires the solution to be sparse, a quality that is inferred from the sparseness of the data (see Fig. 2). This method is able to condense the information from many pixels in a reduced number of fit parameters, yielding a large reduction in the work needed to invert the data, and better constrained results.

3.1.2 Physics

The physics that is assumed to be of importance in the formation of the observed spectral lines can vary enormously, depending on the nature of the transition of interest, the spectral resolution, the signal to noise and the atmospheric heights of interest. Many inversion codes make specific assumptions about the atmosphere, but also about the physics believed to be important for the spectral line that was of primary interest to the author of the code at the time it was written. Here, we give a brief overview of the most important physical effects in which most of the major codes differ.

LTE/NLTE: The most commonly employed assumptions are those of Local Thermodynamic Equilibrium (LTE) and plane-parallellity (1D). Both of these are never strictly applicable, since the mere observability of the radiation field implies the escape of energy and thus by definition non-local equilibrium conditions, and no object in the known universe is infinitely large and flat. Nonetheless, both approximations appear to be remarkably accurate in many cases of interest, and continue to enjoy popularity in the inversion community in particular, owing to the huge simplifications that can be made by using them.
The formation of many photospheric lines is well described using LTE, in which case the dimensionality of the problem can be disregarded altogether, and the spectrum can be directly and efficiently calculated for each image pixel from the magneto-hydrodynamic structure. For lines forming higher up in the solar atmosphere, however, this approximation often does not hold, and the assumption of LTE must be abandoned. Normally, this happens when collisional rates are very low and radiation is weakly coupled to the local physical conditions (e.g., in the chromosphere), invalidating the convenient assumption of Local Thermodynamical Equilibrium, that is commonly used in the photosphere. This is particularly important for lines in which the scattering of radiation plays an important role, but comes at a hugely increased computational cost. Such costs must often be compensated

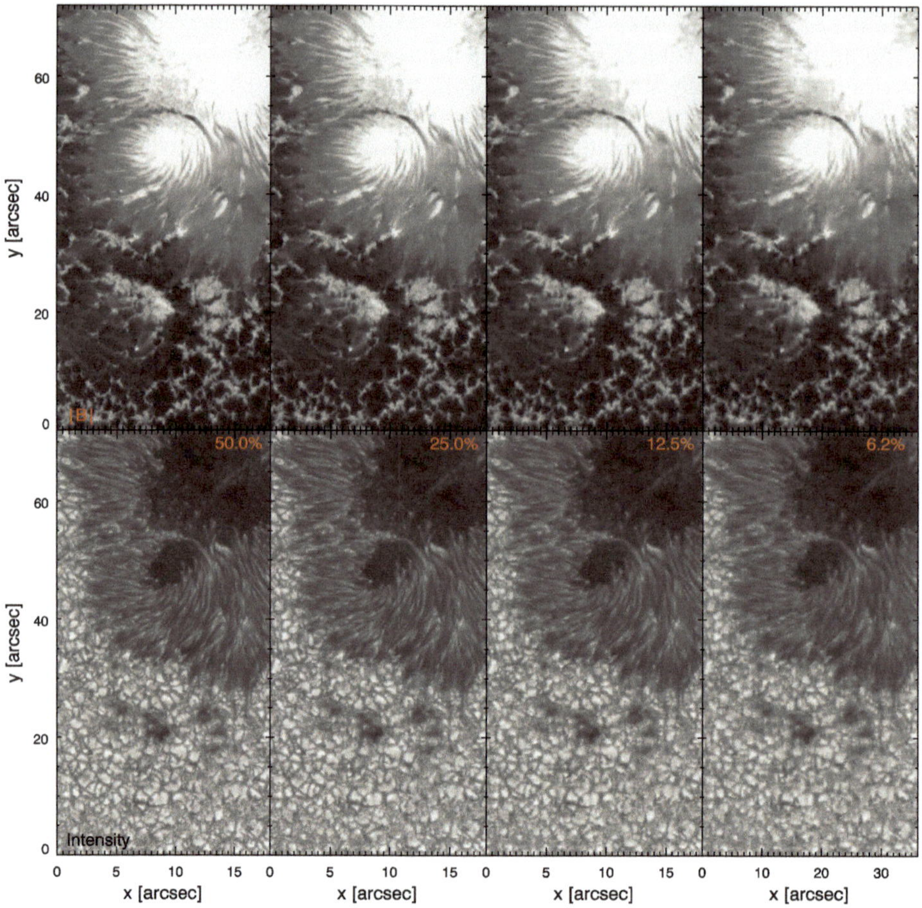

Fig. 2 Compressibility of solar data from the inversion of a Hinode/SOT observation. *Top*: Inferred magnetic field from a sparse inversion assuming a sparsity level of 50 % (*left*). The other maps have then been compressed assuming [25 %, 12.5 %, 6.25 %] sparsity. *Bottom*: The emerging intensity in the wing of the Fe I λ6302 line, computed from the sparse models above

for by simplifying the atmospheric model drastically, so that the overall sophistication of the formation model may not actually increase by much.

Geometry/Dimensionality: Once the assumption of LTE is abandoned, the dimensionality of the atmosphere comes into play, since the non-locality of the radiation field manifests itself in the vertical, as well as in the horizontal dimensions. Although several forward solvers exist that can calculate the emergent spectrum in full 3D (e.g. RH, Uitenbroek 2001; MULTI3D, Leenaarts and Carlsson 2009), no inversion code currently exists that can invert a dataset using full 3D radiative transfer, largely due to the huge computational effort this would require.

Hanle vs. Zeeman effect: The calculation of the imprint that the magnetic field leaves in the emergent radiation, in the form of polarization, can depend on many details in the atomic physics, that influence the formation of the line profiles. Two different effects exist that can be used to extract the strength and orientation of the magnetic field in the line forming region: the Zeeman effect and the Hanle effect. While the Zeeman effect is the most widely used, owing to its simplicity and relatively high signal level, the Hanle effect can have a much higher sensitivity to weak magnetic fields, and it is an important tool for the characterization of magnetic fields in the low-density chromosphere.

The Zeeman effect (Zeeman 1897) has been known for over a century and the theoretical description of it is relatively straightforward. Interpretation of Zeeman induced polarization has therefore been the cornerstone of photospheric inversions for many decades, in particular for the inversion of more complex, height-dependent atmospheres. However, its sensitivity to the magnetic field has intrinsic limitations: in regions where the field strength is very weak and collisional rates are low, scattering polarization can dominate over Zeeman induced polarization.

The use of the Hanle effect (Hanle 1924), is much less widespread, largely due to the complex physics and the large uncertainties in some of the physical quantities that are important for the interpretation, but it is able to return information on much weaker magnetic fields, especially in very low density regions. This effect has been the workhorse for measurements of the very weak magnetic field in chromospheric structures, such as loops and filaments.

Molecules: While most inversions are based on the interpretation of a single atomic transition, clearly the use of a larger number of transitions, where both effects are present, would constitute a much better constrained problem. Especially in cool regions, such as sunspot umbrae, the presence of molecular lines provides an opportunity to make use of very different line transition properties to constrain the atmospheric parameters. Significant progress has been made in that direction by a number of groups (Berdyugina et al. 2000; Asensio Ramos et al. 2005), but the complexity of the physics and the numerical cost have thus far remained substantial and few good inversions with molecules currently exist.

While this list is not intended to be exhaustive, it covers the most important differences in the physics between codes.

3.2 Data Formation

There are many ways in which an instrument can modify the true spectra while converting them to digital form. Any of them, when neglected, can alter the fitted atmospheric properties in various ways, which must be considered when interpreting the result.

3.2.1 Spectral Degradation

Traditionally, the formation of the observed data from a synthetic, undegraded spectrum, has involved mainly consideration of the spectral response of the instrument used to record the data. This is a simple effect to take into account, since it involves only the convolution of the synthetic spectra with the instrumental spectral response function, after which a direct comparison with the observed data can be made.

The spectral resolution of the observation is one of the most important factors in the ability of an inversion code to recover information from the spectral line, and determines to a large extent how complicated an atmosphere can be recovered from the data. There is such a direct link between the complexity of the atmosphere and the spectral resolution, that several instruments were designed and built, with the specific intent to produce data suitable for ME inversion codes, that have specifically optimized their spectral resolution to achieve the best result. Figure 3 is an example of a study of this effect for the Solar Orbiter PHI instrument, and is the result of the inversion of sythesized observations. The error in

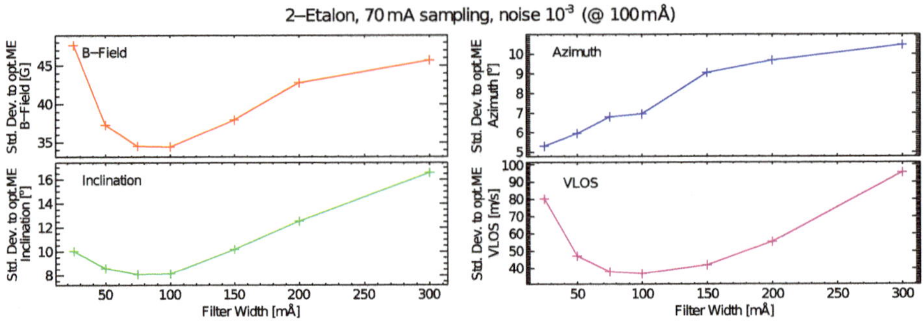

Fig. 3 Effect of spectral resolution in the accuracy of a Milne-Eddington inversion, using 5 spectral points. Courtesy: A. Lagg

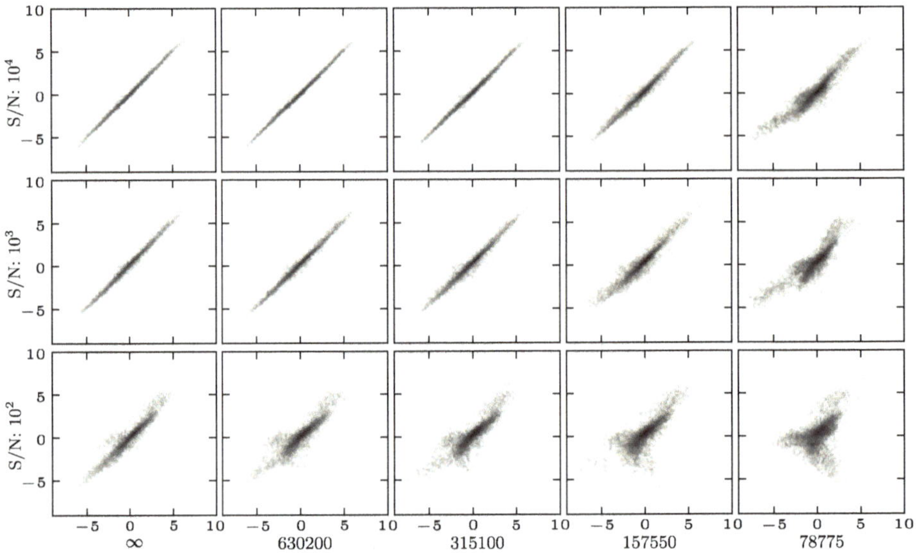

Fig. 4 Effect of spectral resolution vs. S/N on the recovered LOS velocity of a photospheric inversion based on simulated spectra of the FeI lines at 6301.5 and 6302.5 Å

the recovered inversion results decreases with increasing spectral resolution for almost all inverted quantities, but then increases again below a critical value, where aliasing of high-frequency spectral information starts to occur. This result suggests that since a ME inversion needs only 5 points to be adequately constrained, and the result is most accurate when the degraded spectrum is critically sampled, there is a maximum spectral resolution, beyond which the accuracy is not increased. Obviously, better wavelength coverage and spectral resolution are usually desired, as long as the observations are sampled accordingly. This example is particularly relevant for observations acquired with Fabry-Perot interferometers and Lyot filters, where wavelength coverage can be sacrificed in favor of temporal cadence for a given integration time.

If a height dependent atmosphere is desired, however, the spectral resolution needs to be sufficiently high to obtain a reliable result. An illustration of this effect is given in Fig. 4, where the sythesized line profiles of the photospheric FeI lines at 6301.5 and 6302.5 Å of a quiet Sun MHD simulation, calculated using the MURaM code, were inverted using SPINOR assuming a height dependent atmospheric model, for several different amounts of noise and degraded using different spectral resolutions. The scatter, an indication of the accuracy with which the "true" value was recovered, although not negligible for noisless, full-resolution data, clearly increases with increasing noise level. It also increases rapidly with decrease in the spectral resolution, where a factor 2 in spectral resolution corresponds roughly to an order of magnitude in S/N. Most likely, the subtle variations in the line profile are reduced by a low spectral resolution, and are subsequently more easily overpowered by the noise.

3.2.2 Spatial Degradation

Traditionally, the contamination of spectra with spectra from other parts of the field-of-view (FOV) has been considered as part of the formation process, most commonly in the form of a stray-light contribution of some sort, that is added to the undegraded spectra before or after degradation by the instrument. Typically, since the actual stray-light contribution depends on many factors, such as the observing conditions, time of day, etc., the stray-light profile is calculated from the average profile over the FOV or is modeled using a "typical" average quiet Sun atmosphere, and the amount of stray-light is fitted for as a free parameter (e.g. Stenflo 1968; Skumanich and Lites 1991; Bernasconi and Solanki 1996, and many more).

However, in most cases, the true nature of such *spatial* contamination is much more complicated, because there is a, more or less compact, mapping from the source (the undegraded spectra), to the data. This implies that to calculate the stray-light profile for a particular pixel, the undegraded spectra around it are required, which are not available, as they are themselves contaminated by an unknown amount of stray-light.

Orozco Suárez et al. (2007) deal with this problem by using the data itself, and subtracting it from the original data by fitting. While this so-called "local-straylight" method indeed removes a lot of the unwanted stray-light, it has the significant disadvantage that it approximates a convolution of the real profiles with the convolution of the data, which has already been convolved.

If spectra are available over an extended FOV, deconvolution can be considered to remove the spatial degradations directly. While it is in principle possible to recover the undegraded spectra by deconvolution, this procedure will amplify any noise in the data, preferentially at the highest possible spatial frequencies. One way to reduce this problem is to first accumulate the signal in the spectra in a reduced number of quantities, for instance by using a principal component analysis (PCA, Rees et al. 2000), and then deconvolve the maps (Ruiz Cobo and Asensio Ramos 2013). This method relies on the PCA basis functions of the data to be a good basis for the undegraded profiles, which for Hinode SP data seems to be a good approximation. Regularized spatial deconvolution has also been used to compensate the effect of straylight from atmospheric high-order aberrations (Scharmer et al. 2013).

Another approach to calculate the stray-light is to use the actual undegraded profiles and apply the spatial degradation to them. Since the undegraded spectra are not know, we must self-consistently calculate the degraded and the non-degraded spectra, so that the line profiles calculated from the inversion, degraded by telescope diffraction, produces the observed data. This method is known as a coupled inversion (van Noort 2012), and requires the simultaneous fitting of all profiles in the FOV. While it is mathematically the most consistent way to calculate the inverted physical atmosphere, it introduces additional degrees of freedom in the solution, because the sum of a large number of spectra, and not each individual spectrum is fitted to the data. This implies that a change in the spectrum in one pixel can be effectively

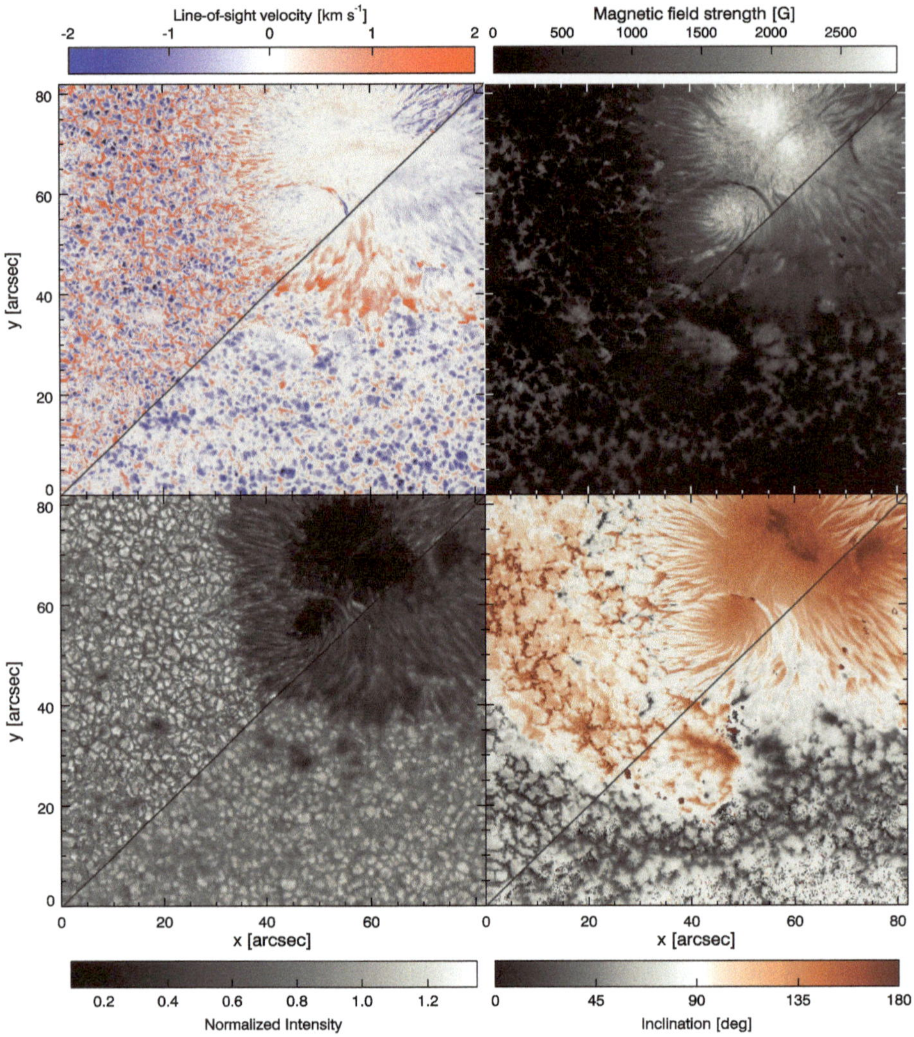

Fig. 5 Coupled inversion of a sunspot (*top left of each panel*) observed with Hinode SOT/SP on the 5th of June 2007, compared with a "traditional" inversion, where stray-light is ignored (*bottom right of each panel*). Observation acquired with Hinode/SOT (Kosugi et al. 2007)

"hidden" in the degraded data by an opposite change in the spectrum of the adjacent pixel. In addition, the current implementation is unable to deal with non-uniform instrumental properties, currently limiting its application to data from space-based instruments.

A selection of inverted quantities obtained using this method is shown in Fig. 5, where a comparison is made between a coupled inversion and a traditional "1D" inversion, where instrumental effects are ignored. Despite the additional degrees of freedom, the coupled inversion maps look much more consistent, and show more detail than the 1D result.

Further treatment of instrumental degradation will need to include the capability to deal with changing degradation conditions, to be able to invert ground based data. The framework for this already exists, but still requires impractical numerical resources to implement.

3.3 Optimization

At present, there are a number of basic methods in use for obtaining the most probable atmospheric structure, responsible for producing the observed spectra: classical downhill minimization, genetic algorithms and Bayesian methods.

3.3.1 Genetic

This method is loosely based on the principle of evolution by survival of the fittest, over a given number of generations, and is implemented using the PIKAIA algorithm (Charbonneau 1995) as the optimization method for the HELIX code. The optimization strategy consists of setting up a number of populations of trial solutions, each with an internal variation of the fit parameters. A given fraction of the populations is then eliminated, with a chance of survival that is mostly determined by the quality of the agreement with the observed data, usually based on a least squares metric. The remaining solutions are then duplicated, and random changes are introduced, thus giving rise to a new generation.

There are a number of advantages, such as a guarantee that the global minimum will be found if enough generations are allowed to pass, and automatic detection of degeneracies in parameter space. It also only requires the forward calculation of the synthetic line profiles, and not the derivative of the merit function to the atmospheric fit parameters, which makes it very flexible.

One major disadvantage is that the number of trials per generation that is necessary to probe all fit parameters independently is dependent on the number of atmospheric fit parameters, so that the numerical cost grows very rapidly with the complexity of the fitted atmosphere. This has limited the use of this method to relatively simple slab and Milne-Eddington atmospheres.

3.3.2 Bayesian/MCMC

Bayesian inversions (Asensio Ramos et al. 2007) are a probabilistic approach to inversion, in the sense that they determine for a distribution of trial solutions the posterior, that is: a distribution of the probability that any of the trial solutions is in agreement with the observed data. The most probable value of the solution is the one with the largest probability, but the uncertainty in the parameter values can be extracted from the posterior as well.

The key ingredient for this type of inversion is a method that generates a set of trial solutions that efficiently samples the posterior. A brute force grid computation will obviously do the job, but scales poorly with the number of fit parameters. The Markov Chain Monte Carlo (MCMC, f.i. Metropolis et al. 1953) method is one way to reduce this effort, and has been used successfully applied in recent years (Asensio Ramos 2011; Asensio Ramos and Martínez González 2014). MCMC is not unlike the genetic method, with the difference that new generations are based on old ones in a different way, that most efficiently samples the posterior, for which an estimate of the posterior is used.

The main limitation of this method is that to sample the posterior, many forward calculations are needed, which limits the applicability of the method to relatively simple and fast atmospheric models.

3.3.3 Database/PCA

This method is the most direct, but therefore also the most thorough. The aim is to quantify the spectra in some way that can be used to categorize them. This can be done by using the intensity values in each wavelength bin of the data, but also by using the value of the inner product of the data with some basis of spectral basis functions. An optimal set of such basis functions can be produced using Principal Components (PCA), a small number of which is usually sufficient to describe each spectral profile to the noise limit (Rees et al. 2000; Socas-Navarro et al. 2001). The objective is now to build an exhaustive database of spectra of all possible atmospheres, which are all described by the coefficients used to categorize the spectra. This database then can be used as a look-up table for an observed spectrum, for which the database lookup then returns all possible atmospheres that yield a sufficiently similar spectrum.

The result is thus a distribution of atmospheric values, not a specific value. If the distribution is normal, an error may be specified, as is often done for gradient search minimizations. However, frequently the distributions are far from normal, and there can be co-dependencies between parameters, that indicate there are degeneracies in the model.

Clearly, these methods return a lot of information that can be used to evaluate the result, but this comes at the cost of having to calculate a database that is complete. This limits the applicability of such methods to relatively simple atmospheres, since the dimensionality of the database quickly grows beyond the capabilities of available computational resources.

3.3.4 Gradient Search

This is a classic method for finding the optimum solution to a given problem, for which a large body of mathematical work exists (see for instance Press et al. 1992). The method is really a collection of methods, most of which are very efficient, in that they require only a few trial calculations of the synthetic spectra to converge, but they suffer from the problem that convergence to the global minimum can only be guaranteed if the merit function is monotonic. In addition, the gradient of the merit function to the atmospheric fit parameters is required, which can be difficult or costly to compute. In most inversion codes, the gradient is not explicitly calculated, but instead the response functions of the atmosphere are used to drive the solution in the direction of the minimum, by eliminating the remaining difference between the forward calculated spectrum and the observed one. It is relatively easy to show that when a ℓ_2-norm is used, the gradient of the merit function depends on the response function. Assuming that the parameters of our model atmosphere are encoded in a vector $\boldsymbol{x} = x_1, x_2, ..., x_i$, for an observation $\boldsymbol{O}(\lambda)$ consisting of N_λ wavelengths in the four Stokes parameters, the merit function (χ^2) and the gradient relative to a parameter x_i are given by

$$\chi^2 = \frac{1}{4N_\lambda} \sum_{j<4N_\lambda} \frac{(S_j(\boldsymbol{x}) - O_j)^2}{\sigma_j^2}, \tag{13}$$

$$\frac{\partial \chi^2}{\partial x_i} = \frac{1}{2N_\lambda} \sum_{j<4N_\lambda} \frac{(S_j(\boldsymbol{x}) - O_j)}{\sigma_j^2} \underbrace{\frac{\partial S_j(\boldsymbol{x})}{\partial x_i}}_{\boldsymbol{R}(x_i)}, \tag{14}$$

where $\boldsymbol{S}(\boldsymbol{x})$ are the synthetic profiles and σ is the noise of the observations. The last term in Eq. (14) is the response function of $\boldsymbol{S}$ to perturbations in x_i.

The most common gradient search minimization algorithm in use is the Levenberg-Marquardt method (Levenberg 1944; Marquardt 1968), which is a mix of the minimization of the linearized problem and a local gradient search that kicks in when the forward problem is very non-linear. It is at the heart of all commonly used height dependent atmospheric inversion codes, and has a robust convergence, as long as the derivatives of the synthetic

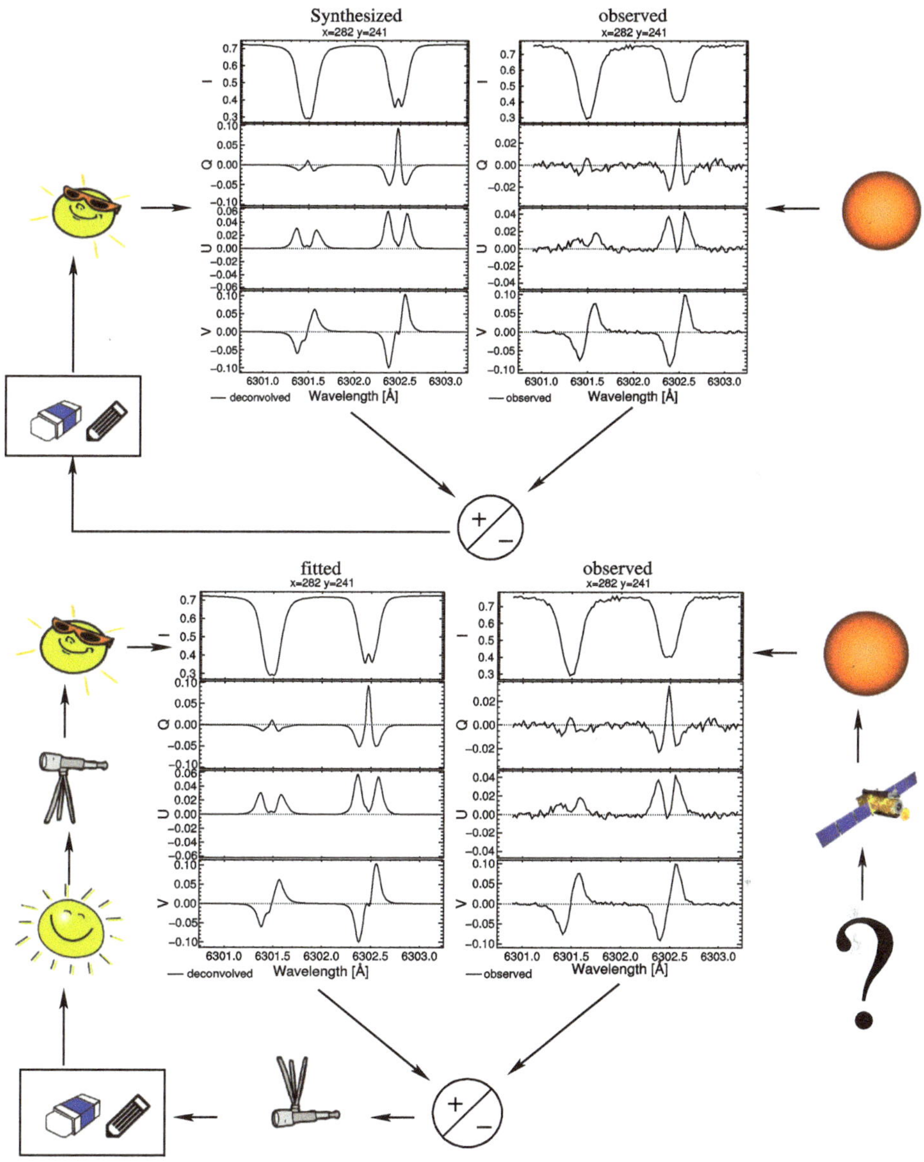

Fig. 6 Gradient search optimization. *Top*: single spectrum, without instrumental effects. *Bottom*: coupled inversion, with instrumental effects considered

spectrum with respect to the atmospheric fit parameters can be efficiently obtained. Figure 6 (left) shows the basic scheme, where the difference between the observed and synthetic spectrum are used to drive corrections to the current estimate of the atmosphere.

It is possible to adapt the gradient search algorithm to search for the minimum of the difference between the spatially degraded synthetic profiles and the data, instead of using the synthetic profiles directly, by applying the effect that instruments have on the spectra also to the response functions, before the correction is calculated. Although this requires some additional effort in the computation of the correction, the gradient search method is currently the only method that has been successfully applied to simultaneously invert more than 10^6 atmospheric variables (van Noort 2012). Figure 6 illustrates gradient search algorithm for spatially coupled inversions and traditional pixel-by-pixel inversions.

4 Main Diagnostics in the Chromosphere

Observing the chromosphere and the transition region poses a challenge for solar physicists. There is a very limited selection of spectral lines that have sufficient opacity to sample the chromosphere.

Ground based observations are limited to visible and infrared wavelengths, and the most common chromospheric diagnostics are perhaps the Ca II H & K lines (λ3934 & λ3968), the H I 6563 line (Hα), the Ca II infrared triplet lines (IR, λ8489, λ8542 & λ8662) and the He I 5876 (D_3) and λ10830 lines.

The ultra-violet includes a list of interesting lines that must be observed from space. These diagnostics are highly unexplored compared with those in the visible and infrared: the H I λ1216 line (Lα), Mg II h & k lines (λ2796 & λ2804), the He I 584 line, the He II 304 line, Si IV λ1394 and λ1403, and many more. In Table 2 we summarize modelling requirements for some chromospheric diagnostics, and in Fig. 7 the formation height of the core of some chromospheric lines is illustrated as computed from a 3D MHD simulation.

Table 2 Main chromospheric diagnostics and the corresponding modelling requirements

Line(s)	Scattered photons	Zeeman/Hanle	Geometry	Ionization
Ca II H & K	PRD	Zeeman (AR), Hanle (QS)	1.5D	stat. equilibrium
Hα	CRD	Zeeman (AR), Hanle (QS)	3D	non-equilibrium
Ca II IR triplet	CRD	Zeeman (AR), Hanle (QS)	1.5D	stat. equilibrium
Mg II h & k	PRD	Zeeman, Hanle (k line)	1.5D	stat. equilibrium
He I D_3 & λ10830	CRD (?)	Zeeman + Hanle	3D (?)	non-equilibrium

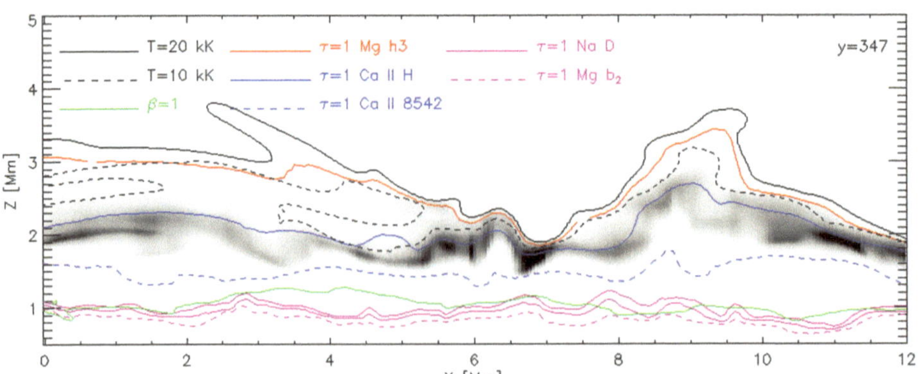

Fig. 7 Slice from a 3D MHD simulation indicating the $\tau = 1$ surface at line center for Mg II h_3 & b_2, Ca II H & λ8542, and Na I D. The contribution function of the He I λ10830 line is rendered in grey-scale because it is optically thin. Courtesy of M. Carlsson (Institute for theoretical astrophysics, University of Oslo)

The Ca II H & K lines are particularly sensitive to temperature because the Planck function is very steep at those wavelengths. Thus, their broad wings can act as remarkable thermometers for the photosphere and the lower chromosphere. These lines have slightly higher Landé factor than the IR triplet lines ($g_{3933} = 1.16$, $g_{3968} = 1.33$) but the lower photon count in the UV and the dependency of the Zeeman splitting with λ^2 makes them effectively less sensitive to magnetic fields (see Eq. (11)). Partial redistribution effects (PRD) must be included to reproduce the intensity profile close to line center (Uitenbroek 1989). The real advantage of these lines is that they can be observed from the ground at the highest resolution that a telescope can deliver as the diffraction limit of a telescope scales linearly with λ. The importance of 3D non-LTE effects remains to be assessed in a similar way as de la Cruz Rodríguez et al. (2012) did for the λ8542 line.

The Mg II h & k lines sample the transition region, providing valuable information of the physical coupling between the chromosphere and the corona (Uitenbroek 1997). With the launch in 2013 of NASA's Interface Region Imaging Spectrograph (IRIS, De Pontieu et al. 2014), these lines have become particularly interesting for the solar community (Leenaarts et al. 2013; Pereira et al. 2013), although without measuring polarization. However, these lines could be combined with the Ca II IR triplet lines to constrain the temperature stratification of the solar chromosphere and to measure the magnetic field in the chromosphere. PRD of scattered photons must be used to model these lines.

The Hα line has been considered for a very long time one of the best chromospheric diagnostics (see comments by Rutten 2007). It is well suited for the measurement of chromospheric velocities because the line profile is observed in absorption in most chromospheric conditions (not always though). Cauzzi et al. (2009) showed that the width of the Gaussian core of the Hα line is strongly affected by the chromospheric temperature because the hydrogen atom is very light. However, this line is not well suited for current implementations of non-LTE inversions because to reproduce the observed intensities, the radiation field must be evaluated in 3D dimensions (Leenaarts et al. 2012a). Hα is not an ideal diagnostic for chromospheric magnetic fields because its response to the Zeeman effect is originated in the photosphere (Socas-Navarro and Uitenbroek 2004), although the line is sensitive to the Hanle effect close to the line center (e.g., Štěpán and Trujillo Bueno 2010).

The Ca II IR triplet lines have a relatively high sensitivity to magnetic fields (with Landé factors $g_{8498} = 1.07$, $g_{8542} = 1.10$, $g_{8662} = 0.87$), and they can be modelled assuming CRD (see Uitenbroek 1989) and statistical-equilibrium (Wedemeyer-Böhm and Carlsson 2011). Furthermore, these lines are very sensitive to the temperature stratification of the atmosphere and encode information about velocity gradients. Therefore, most, if not all, depth-stratified non-LTE inversions in the chromosphere have been carried out using the Ca II IR triplet lines in active regions (). In the quiet-Sun, scattering polarization and the Hanle effect must be included to describe the linear polarization signals (Manso Sainz and Trujillo Bueno 2010; Carlin et al. 2013). Recently, Leenaarts et al. (2014) showed that the effect of isotopic splitting must be included in the inversions to properly retrieve the velocity stratification.

The physical processes involved in the formation of the He I D_3 and λ10830 lines are significantly different than those of the Hα line and of the aforementioned Ca II lines (see Goldberg 1939; Andretta and Jones 1997; Trujillo Bueno and Asensio Ramos 2007; Centeno et al. 2008; Golding et al. 2014; Leenaarts et al. 2016). The He I atom consists of two spectral series of singlets (para-helium) and triplets (ortho-helium). The ground state of the atom is a singlet level, but the D_3 and the λ10830 transitions take place between two levels of the triplet system. Since transitions between the para and the ortho levels are forbidden by the dipole selection rules and collisional rates in the chromosphere are very low, the only way to populate the ortho-levels is through photo-ionization from the para-helium system

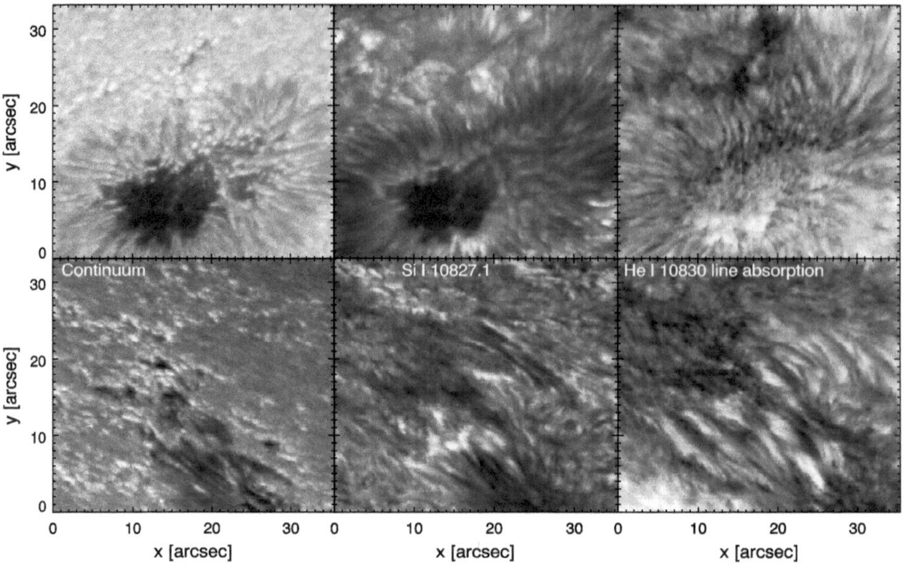

Fig. 8 Raster-scan observation of AR12394 (*top*) and AR12390 (*bottom*) on 2014-08-01 at 1083 nm with the TRIPPEL spectrograph at the Swedish 1-m Solar Telescope. From left to right the *panels* show images in the continuum, in Si I λ10827 line center and in He I λ10830 line absorption. Courtesy of J. Joshi, T. Libbrecht, G. Sliepen and D. Kiselman (Institute for Solar Physics, Stockholm University)

followed by a recombination into the ortho-system. The electrons cascade down through the levels of the ortho-system until the ground meta-stable level of the ortho-helium system. The ionization potential of the para-system is 24.6 eV, which means that the photoionization of the He I atom can only be effectively done with UV photons ($\lambda < 504$ Å). Such radiation can only originate in the solar corona, which makes these lines particularly interesting because they are sensitive to both chromospheric and coronal physical conditions; see observations with the TRIPPEL spectrograph (Kiselman et al. 2011) at the Swedish 1-m Solar Telescope (SST, Scharmer et al. 2003) in Fig. 8 and with Gregor/GRIS (Schmidt et al. 2012; Collados et al. 2007) in Lagg et al. (2015). Since UV photons can only penetrate a thin layer of the chromosphere before they are absorbed, these lines are assumed to form in a relatively thin and localized range of heights, making the constant slab or Milne-Eddington descriptions applicable.

5 Inversions in the Chromosphere: a Selection of Results

This volume includes dedicated chapters about sunspots (Rempel & Scharmer, same issue) and about photospheric magnetic fields in the quiet-Sun (Schussler et al., same issue). Therefore, we have chosen to focus on a small selection of results obtained from chromospheric studies, and more specifically on those concerning inference of magnetic fields.

The first attempts to *invert* chromospheric data can be tracked back to the VAL and FAL atmospheres (see Vernazza et al. 1981 and Fontenla et al. 1993). Those studies used spatially-averaged spectra, including many chromospheric lines from different species (atoms and ions), to derive models that could reproduce the observations. Nowadays, the VAL and FAL atmospheres are still used to derive radiative fluxes in the chromosphere

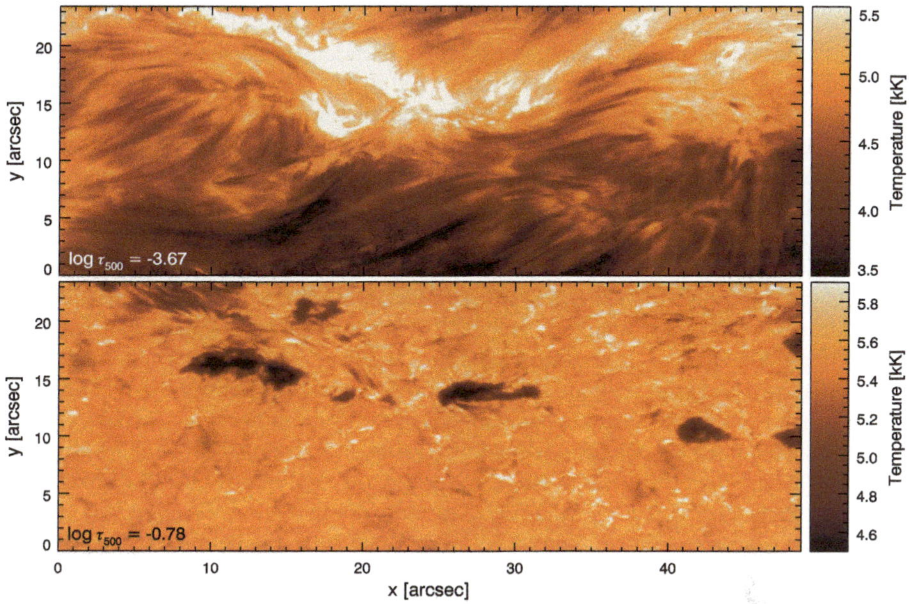

Fig. 9 Inferred chromospheric (*top*) and photospheric (*bottom*) temperatures from a non-LTE inversion using observations in the Ca II λ8542 line acquired with CRISP (Scharmer 2006; Scharmer et al. 2008) at the SST. The temperature maps are given at fixed optical depths of $\log\tau_{500} = -3.67$ (*top*) and $\log\tau_{500} = -0.78$ (*bottom*). A small patch from this inversions was used by de la Cruz Rodríguez et al. (2015) to study small scale magnetic flux emergence

or as reference chromospheric models. However, the chromosphere is highly dynamic and finely structured. These 1D models cannot catch those two inherent properties of the chromosphere, and it is now widely accepted that spatio-temporal models must be used to characterize this region of the Sun.

The following techniques have been particularly successful to derive magnetic fields in the chromosphere:

1. Depth-stratified non-LTE inversions, assuming statistical equilibrium and complete redistribution in angle and frequency of scattered photons (CRD). Socas-Navarro et al. (2000b) developed inversion methods that are used nowadays (Socas-Navarro et al. 2015) to derive spatio-temporal model atmospheres including a photosphere and a chromosphere (see detailed temperature maps in Fig. 9).
2. Milne-Eddington atmosphere or constant physical slab inversions. These inversions work directly with the parameters of the radiative transfer equation, and therefore only provide *direct* information about the line-of-sight velocity and magnetic field vector.
3. The weak-field approximation can be used in most chromospheric situations to infer the magnetic field vector. Most chromospheric lines have a relatively low Landé factor and a relatively large Doppler width, making them suitable candidates for this simple approximation.

5.1 Depth-Stratified Non-LTE Inversions

We focus now in approach 1. The statistical-equilibrium equations (Eq. (9)) have an explicit dependency on the mean intensity $\bar{J}$, which in current implementations is computed for each pixel assuming plane-parallel geometry. In reality, if radiation is not coupled to the local conditions of the plasma, horizontal radiative transfer can become important, and the evaluation of the radiation field must be performed in 3D. Therefore, this method is only suited for lines that can be modelled assuming plane-parallel geometry.

A number of studies including the inversion of Ca II IR data can be found in the literature:

Sunspots: Non-LTE inversions have allowed to identify the origin of umbral flashes (UF) in sunspots and to derive the chromospheric temperature increase during the flash phase (Socas-Navarro et al. 2000a). The polarity reversals observed in Stokes V during UF are induced by emission in the intensity profile, not to actual changes in the polarity of the magnetic field (de la Cruz Rodríguez et al. 2013b). In fact, most of the variability that is observed in monochromatic Stokes V images is produced by the thermal properties of the plasma that leave an imprint in Stokes V, like Doppler motions and Doppler broadening. Similar conclusions were drawn from the study of Ca II H & K data in plage (Martinez Pillet et al. 1990). The 3D structure of the magnetic field vector has been retrieved by Socas-Navarro (2005b), who found a complex field configuration with opposite-sign torsion that could be interpreted as opposite helicity flux ropes inside the same spot. In a second study Socas-Navarro (2005a) analyzed chromospheric heating induced by currents dissipation in the upper layers of the same sunspot. One aspect that remains to be investigated is the full depth stratification of the penumbra of sunspots, and how the Evershed effect and the moat flow relate to the inverse Evershed flow and the superpenumbra observed in the chromosphere.

Plage: Shine and Linsky (1974) discussed ad-hoc 1D models that could reproduce some observed properties in plage profiles. However, the first and probably only attempt to apply field-dependent non-LTE inversions to a patch of network was carried out by Pietarila et al. (2007a), using slit-spectrograph observations in the Ca II λ8498 and λ8542 lines, recorded with SPINOR (Socas-Navarro et al. 2006). One of their main results is the appearance of a field that expands with height, and becomes more horizontal in the chromosphere, suggesting a canopy effect (see flux-tube models by Solanki et al. 1991, or measurements of increasing magnetic field gradients in plage by Sanchez Almeida et al. 1989). However, their conclusions are based in the visual appearance of the inverted field topology because they did not have enough signal in Stokes Q and U to derive the inclination of the magnetic field. This canopy effect has also been found in 3D MHD simulations (de la Cruz Rodríguez et al. 2013a), which has a clear diagnostic associated: the core of the Ca II IR lines is observed in emission or with greatly enhanced intensity (see Fig. 10). It remains unclear what process heats the chromosphere in this canopy, above the *quiet* photosphere below, but it must have a magnetic origin.

Quiet-Sun: In quiet-Sun observations, the amplitudes of Zeeman induced polarization (in most chromospheric lines) are similar to those originated from scattering polarization and from the Hanle effect. These amplitudes are expected to be very low, of the order of 10^{-4} relative to the continuum intensity. Therefore, the challenge to measure chromospheric magnetic fields is twofold: first it is very hard (probably impossible at the moment) to reach the required sensitivity with high/medium spatial resolution, but even if such observations were available, modelling the observed polarization profiles would be extremely hard. One of the best estimates of the field strength in the chromosphere of quiet-Sun is given by Trujillo Bueno et al. (2004).

5.2 Constant Slab Inversions and Milne-Eddington Inversions

The first realistic inversions of the D_3 line, including a full treatment of the Hanle effect and scattering polarization, were originally introduced by López Ariste and Casini (2002). To

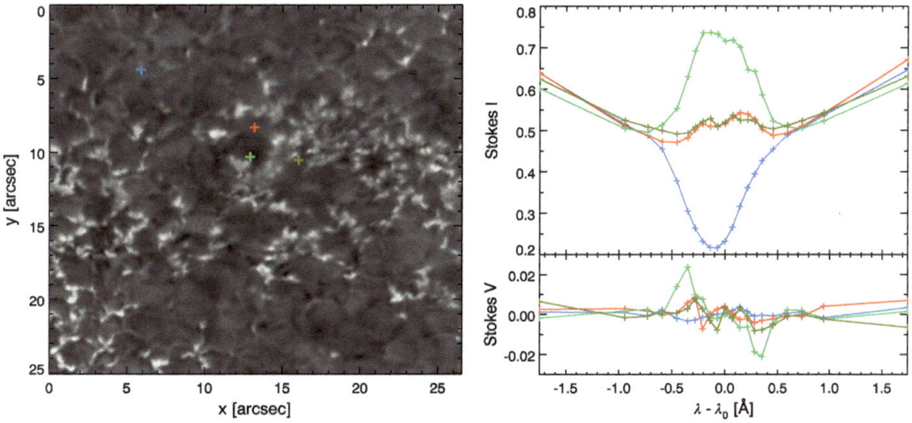

Fig. 10 Observation of plage in the Ca II λ8542 line recorded with the CRISP instrument at Swedish 1 m Solar Telescope. *Left*: Monochromatic image acquired in the wing of the line, showing bright-points embedded in photospheric granulation. The *color markers* indicate the locations of the spectra plotted on the *right panel*, with the same color coding

overcome the computational challenge (at that time) of performing a full least-squares-fit for this non-linear problem, they used a method based on PCA decomposition and precomputed synthetic profiles (Rees et al. 2000). Solanki et al. (2003) studied the λ10830 line in an active region, including only Zeeman induced polarization in their inversions. HAZEL was the first code that included full spectral synthesis with Zeeman induced polarization, scattering polarization and the Hanle effect (Asensio Ramos et al. 2008), and Helix+ is now also capable of similar calculations (Lagg et al. 2009).

The D_3 and λ10830 lines have been used to study prominences and filaments, spicules, fibrils, and flares. Here we summarize some important results from the literature:

Fibrils: To our knowledge, most of the studies carried out in fibrils (using He I lines) have focused on measuring the direction of the magnetic field and on assessing whether the magnetic field is aligned along the direction defined by chromospheric fibrils. In particular, the fibrils observed in the superpenumbra of sunspots seem to be aligned with the measured direction of the magnetic field (Schad et al. 2013), although de la Cruz Rodríguez and Socas-Navarro (2011) found a number of cases with significant miss-alignment using the λ8542 line. Further studies of fibrils in the Hα line using a 3D MHD simulation, showed that fibrils are usually aligned with the magnetic field in the horizontal direction, but that is not always the case vertically (Leenaarts et al. 2015).

Prominences: Theoretical models of solar prominences predict that the magnetic field must be highly inclined to support the material inside the prominence. Observationally, most measurements of the magnetic field topology in prominences have come from observations in He I lines (λ5876 and λ10830). A difficulty that arises from these observations is the presence of ambiguities that are compatible with several orientations of the magnetic field. The validity of a more inclined versus a more vertical topology of the magnetic field has been greatly debated by the community (see latest results by Casini et al. 2005; Orozco Suárez et al. 2014; Schmieder et al. 2014; Martínez González et al. 2015).

Filaments: Magnetic fields in filaments have been recently studied in the photosphere and in the chromosphere separately (Lites 2005; López Ariste et al. 2006; Merenda et al. 2006). For the first time, Kuckein et al. (2012) carried out a detailed analysis of simultaneous photospheric and chromospheric spectropolarimetric observations, and they were capable of

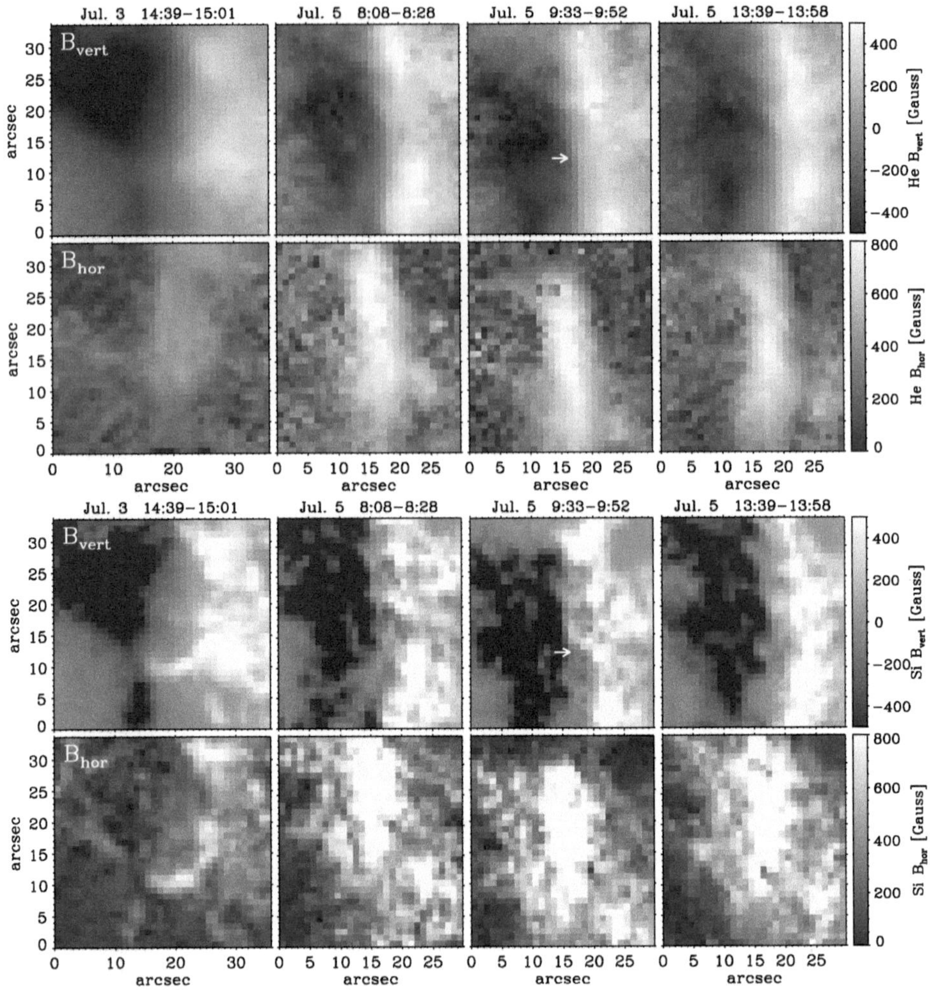

Fig. 11 Magnetic field in a filament as a function of time. The upper two rows correspond to chromospheric inversions of the He I 10830 line, wheres the lower two rows correspond to photospheric inversions of the Si I λ10827. For each set of inversions, the upper row corresponds to the vertical component, and the lower panel corresponds to the horizontal component. Reproduced from Kuckein et al. (2012), Figs. 11 and 12 combined

inferring the magnetic field vector in the photosphere (using the Si I λ10827 line and in the chromosphere (using the He I λ10830 line). These studies have shown that the field in the chromosphere has a strong horizontal component along the spine of the filament (up to 800 G), whereas the vertical component of the field has typical values of 400–500 G and are placed on both sides of the spine (see Fig. 11).

Spicules: Polarimetric observations of spicules outside the limb have been used to derive the magnetic field orientation (López Ariste and Casini 2005; Centeno et al. 2010), including the variation of the magnetic field with height (Orozco Suárez et al. 2015). However, to achieve a high enough signal to noise ratio, these observations have usually been integrated for periods of 10 to 45 minutes. Unfortunately, the chromosphere is highly dynamic and those measurements reflect the average magnetic field vector over that integration period.

5.3 The Weak-Field Approximation in the Chromosphere

The weak field approximation (WFA hereafter) poses a very fast method to estimate the magnetic field vector from full-Stokes observations of magnetically sensitive lines (Rayrole 1967). It is applicable when the Doppler width of the line ($\Delta\lambda_D$) is much larger than the Zeeman splitting of the line ($\Delta\lambda_B$) and if the magnetic field strength, field inclination and line-of-sight velocity are close to constant as a function of depth in the atmosphere. Under those conditions, Stokes Q, U and V can be written as a function of constant magnetic field vector components (with depth) and the first and second derivatives of the intensity profile (see summary in Landi Degl'Innocenti and Landolfi 2004).

Most chromospheric lines have a relatively low Landé factor and are usually broader than photospheric lines. Therefore, it is relatively easy to satisfy the conditions mentioned above, except that many chromospheric spectral lines are sensitive to a large range of regimes, and can sample photospheric and chromospheric conditions simultaneously. However, this inconvenience can be easily overcome by isolating a small wavelength range where the line can be assumed to sample one atmospheric regime. We note that Wöger et al. (2009) proposed a hybrid *center of gravity* method to derive depth information of the magnetic field. Jennerholm Hammar (2015) used synthetic observations in the λ8542 line, computed from a 3D MHD simulation to assess the usability of the WFA. Their results show good agreement between the inferred components of the magnetic field and that in the simulation.

The weak field approximation has been used to study the evolution of magnetic fields in flares (Harvey 2012), in a sunspot umbra (de la Cruz Rodríguez et al. 2013b) and in plage (Pietarila et al. 2007b), from observations in the Ca II 8542 Å line.

6 Discussion and Future Developments

6.1 Structuring of Magnetic Fields in the Chromosphere

In Sects. 5.1 and 5.2 we have summarized a number of studies that have inferred magnetic fields, in most cases targeting particular features or phenomena in the chromosphere. The question of understanding how is the magnetic field topology in the chromosphere seems to persist. Recent developments in solar instrumentation have allowed to acquire spectropolarimetric observations with very high-spatial resolution (see summary by Kleint and Gandorfer 2015), and monochromatic images in all the Stokes parameters can appear very finely structured. Particularly in the chromosphere, this fine structures may be (wrongly) assumed to indicate a similar topology of the magnetic fields, although magnetic fields may govern the intricate force balance in the chromosphere. This fine structure is likely to contain strong *imprints* of Doppler motions, line broadening and opacity variations and corrugations, and it should not be assumed to resemble the magnetic field topology (similarly discussed by Judge 2006). In the chromosphere gas pressure is several orders of magnitude lower than in the photosphere, whereas magnetic fields are usually one order of magnitude lower. While the photosphere is dominated by gas pressure gradients, that dictate the structuring of magnetic fields, in the chromosphere this is not clearly the case.

Figure 12 illustrates a highly structured Stokes I image close to line center in the λ8542 line (left panel), but the inferred l.o.s. component of the magnetic field is much smoother, and similar results are obtained, e.g., in filaments (see Fig. 11 in Kuckein et al. 2012). Similarly, de la Cruz Rodríguez et al. (2013b) found that most of the variability observed as a function of time in umbral flashes can be explained with an almost constant magnetic field

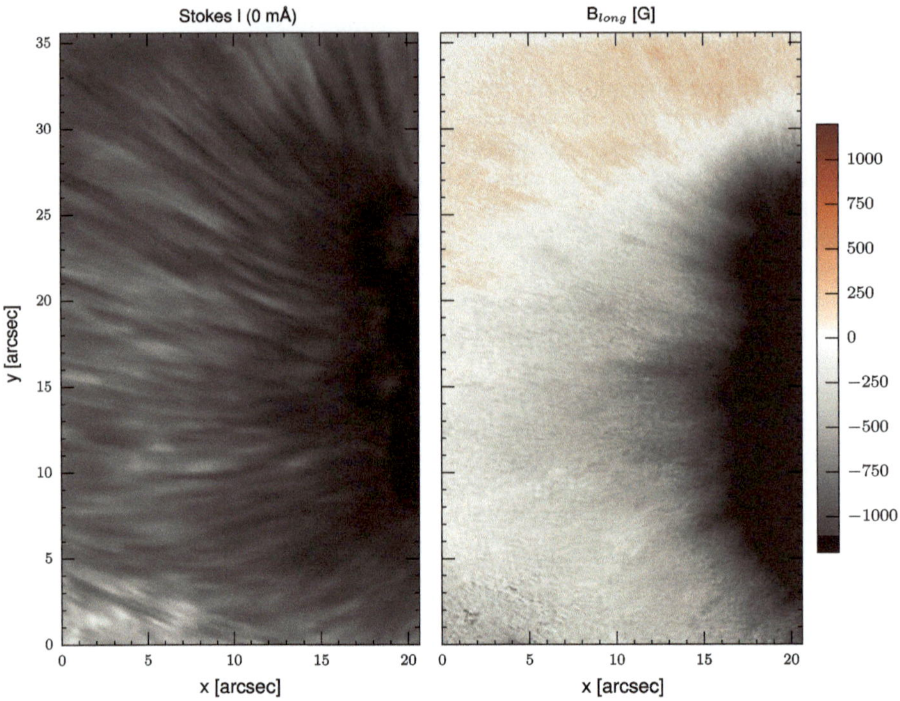

Fig. 12 SST/CRISP spectropolarimetric observation of AR1793 on 2013-07-22 (08:51:55UT) in the Ca II 8542 Å line. *Left*: umbra and penumbra of a sunspot (AR1793) at the line center. *Right*: the corresponding line of sight component of the magnetic field vector, computed with the WFA

(at each location). LTE inversions in the photosphere in sunspots and plage regions also reveal that the magnetic field rapidly expands as a function of height and becomes relatively smooth (e.g., Scharmer et al. 2013; Buehler et al. 2015), supporting this scenario.

Therefore, magnetic fields in the chromosphere must be smoother than the visual appearance of many high-resolution observations suggests, especially in active regions. Schad et al. (2015) draw similar conclusion from the analysis of Milne-Eddington inversions of a He I λ10830 dataset.

Very recently, results from the inversion of high-resolution GREGOR/GRIS observations in the Si I λ10827 and the He I λ10830 lines have revealed that the magnetic field strength is very smooth in the upper chromosphere, but the field inclination contains fine structure that is correlated with the inclination of the photospheric magnetic field (see Fig. 13, reproduced from Joshi et al. 2016).

6.2 An Outline for the Future

The future of radiative diagnostics continues to build upon forward modelling and inversions. Nowadays, RH (Uitenbroek 2001), MULTI3D (Leenaarts and Carlsson 2009) and NICOLE (Socas-Navarro et al. 2015) (among others) are well developed and available to the community. Hanle calculations may be possible with PORTA (Štěpán and Trujillo Bueno 2013, closed-source) and with TRAVIATA (Carlin and Asensio Ramos 2015, under development). These codes shall be used to develop new techniques to accelerate the computation

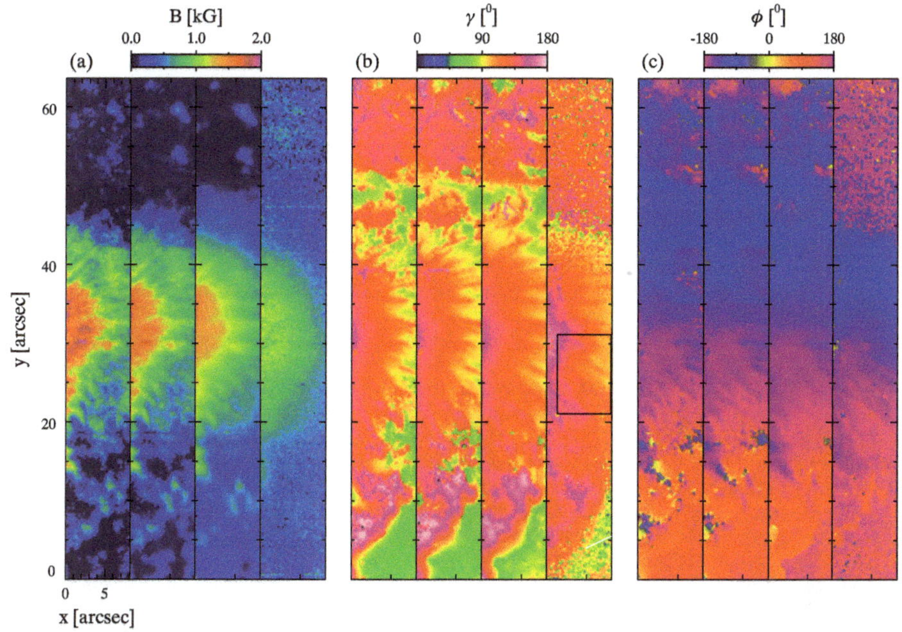

Fig. 13 Reconstruction of the magnetic field vector from GREGOR/GRIS observations in the Si I λ10827 and the He I λ10830 lines. From left to right there are three groups of panels showing the magnetic field strength, inclination and azimuth. Each *group of panels* contains the reconstructed quantity as a function of decreasing optical depth from left to right. The *first three panels* are reconstructed from LTE inversions in the Si I line. The *fourth panel* corresponds for a Milne-Eddington inversion of the He I line. Reproduced from Joshi et al. (2016), Fig. 3

of the NLTE problem (e.g., Fabiani Bendicho et al. 1997; Leenaarts et al. 2012b) to implement missing microphysics in the radiative transfer calculations (e.g., Osorio et al. 2012), and to study line formation with theoretical/numerical model atmospheres.

The methods presented by van Noort (2012) and just recently by Asensio Ramos and de la Cruz Rodríguez (2015) pose the latest development in photospheric inversions. These methods couple the parameters of the model atmosphere spatially and they properly account for the smearing of the telescope PSF. The application of these methods in the chromosphere will be a major step forward in the near future, although we have barely started to exploit the potential of simpler 1.5D inversions in non-LTE. Another point to improve upon is related to the nature of the spatial coupling, which has so far been instrumentally motivated or mathematically imposed, but the usability of the MHD equations remains, to our knowledge, to be proven.

There is also great potential in the exploration of interesting chromospheric lines like the Ca II H & K, or the Mg II h & k, although PRD effects must be included to study these lines. To assess the potential of inversions in Hα, the radiative transfer problem must be solved in 3D which largely complicates the inversion and the description of the model atmosphere.

Acknowledgements J. de la Cruz Rodríguez acknowledges financial support from Vetenskapsrådet (VR, the Swedish Research Council) and the Swedish National Space Board (SNSB).

References

V. Andretta, H.P. Jones, On the role of the solar corona and transition region in the excitation of the spectrum of neutral helium. Astrophys. J. **489**, 375–394 (1997)

A. Asensio Ramos, Bayesian inference of solar and stellar magnetic fields in the weak-field approximation. Astrophys. J. **731**, 27 (2011). doi:10.1088/0004-637X/731/1/27

A. Asensio Ramos, J. de la Cruz Rodríguez, Sparse inversion of Stokes profiles, I: two-dimensional Milne-Eddington inversions. Astron. Astrophys. **577**, 140 (2015). doi:10.1051/0004-6361/201425508

A. Asensio Ramos, M.J. Martínez González, Hierarchical analysis of the quiet-Sun magnetism. Astron. Astrophys. **572**, 98 (2014). doi:10.1051/0004-6361/201423860

A. Asensio Ramos, J. Trujillo Bueno, M. Collados, Observation and modeling of anomalous CN polarization profiles produced by the molecular paschen-back effect in sunspots. Astrophys. J. Lett. **623**, 57–61 (2005). doi:10.1086/430060

A. Asensio Ramos, M.J. Martínez González, J.A. Rubiño-Martín, Bayesian inversion of Stokes profiles. Astron. Astrophys. **476**, 959–970 (2007). doi:10.1051/0004-6361:20078107

A. Asensio Ramos, J. Trujillo Bueno, E. Landi Degl'Innocenti, Advanced forward modeling and inversion of Stokes profiles resulting from the joint action of the Hanle and Zeeman effects. Astrophys. J. **683**, 542–565 (2008). doi:10.1086/589433

L.H. Auer, L.L. House, J.N. Heasley, The determination of vector magnetic fields from Stokes profiles. Sol. Phys. **55**, 47–61 (1977). doi:10.1007/BF00150873

P.S. Barklem, S.D. Anstee, B.J. O'Mara, Line broadening cross sections for the broadening of transitions of neutral atoms by collisions with neutral hydrogen. Publ. Astron. Soc. Aust. **15**, 336–338 (1998). doi:10.1071/AS98336

L.R. Bellot Rubio, B. Ruiz Cobo, M. Collados, An Hermitian method for the solution of polarized radiative transfer problems. Astrophys. J. **506**, 805–817 (1998). doi:10.1086/306260

S.V. Berdyugina, C. Frutiger, S.K. Solanki, W. Livingstone, Successful spectral synthesis of Zeeman-split molecular bands in sunspot spectra. Astron. Astrophys. **364**, 101–104 (2000)

P.N. Bernasconi, S.K. Solanki, Inversion of Stokes vector profiles in terms of a 3-component model. Sol. Phys. **164**, 277–290 (1996). doi:10.1007/BF00146640

J.M. Borrero, S. Tomczyk, M. Kubo, H. Socas-Navarro, J. Schou, S. Couvidat, R. Bogart, VFISV: very fast inversion of the Stokes vector for the helioseismic and magnetic imager. Sol. Phys. **273**, 267–293 (2011). doi:10.1007/s11207-010-9515-6

J.M. Borrero, B.W. Lites, A. Lagg, R. Rezaei, M. Rempel, Comparison of inversion codes for polarized line formation in MHD simulations, I: Milne-Eddington codes. Astron. Astrophys. **572**, 54 (2014). doi:10.1051/0004-6361/201424584

D. Buehler, A. Lagg, S.K. Solanki, M. van Noort, Properties of solar plage from a spatially coupled inversion of Hinode SP data. Astron. Astrophys. **576**, 27 (2015). doi:10.1051/0004-6361/201424970

C.J. Cannon, Angular quadrature perturbations in radiative transfer theory. J. Quant. Spectrosc. Radiat. Transf. **13**, 627–633 (1973). doi:10.1016/0022-4073(73)90021-6

E.S. Carlin, A. Asensio Ramos, Chromospheric diagnosis with Ca II lines: forward modeling in forward scattering, I. Astrophys. J. **801**, 16 (2015). doi:10.1088/0004-637X/801/1/16

E.S. Carlin, A. Asensio Ramos, J. Trujillo Bueno, Temporal evolution of the scattering polarization of the Ca II IR triplet in hydrodynamical models of the solar chromosphere. Astrophys. J. **764**, 40 (2013). doi:10.1088/0004-637X/764/1/40

R. Casini, R. Bevilacqua, A. López Ariste, Principal component analysis of the He I D_3 polarization profiles from solar prominences. Astrophys. J. **622**, 1265–1274 (2005). doi:10.1086/428283

G. Cauzzi, K. Reardon, R.J. Rutten, A. Tritschler, H. Uitenbroek, The solar chromosphere at high resolution with IBIS, IV: dual-line evidence of heating in chromospheric network. Astron. Astrophys. **503**, 577–587 (2009). doi:10.1051/0004-6361/200811595

R. Centeno, J. Trujillo Bueno, H. Uitenbroek, M. Collados, The influence of coronal EUV Irradiance on the emission in the He I 10830 Å and D_3 multiplets. Astrophys. J. **677**, 742–750 (2008). doi:10.1086/528680

R. Centeno, J. Trujillo Bueno, A. Asensio Ramos, On the magnetic field of off-limb spicules. Astrophys. J. **708**, 1579–1584 (2010). doi:10.1088/0004-637X/708/2/1579

P. Charbonneau, Genetic algorithms in astronomy and astrophysics. Astrophys. J. Suppl. Ser. **101**, 309 (1995). doi:10.1086/192242

M. Collados, A. Lagg, J.J. Dıaz Garcıa, E. Hernández Suárez, R. López López, E. Páez Mañá, S.K. Solanki, Tenerife infrared polarimeter II, in *The Physics of Chromospheric Plasmas*, ed. by P. Heinzel, I. Dorotovic, R.J. Rutten. ASP Conference Series, vol. 368 (Astronomical Society of the Pacific, San Francisco, 2007), p. 611

J. de la Cruz Rodríguez, N. Piskunov, DELO-Bezier formal solutions of the polarized radiative transfer equation. Astrophys. J. **764**, 33 (2013). doi:10.1088/0004-637X/764/1/33

J. de la Cruz Rodríguez, H. Socas-Navarro, Are solar chromospheric fibrils tracing the magnetic field? Astron. Astrophys. **527**, 8 (2011). doi:10.1051/0004-6361/201016018

J. de la Cruz Rodríguez, H. Socas-Navarro, M. Carlsson, J. Leenaarts, Non-local thermodynamic equilibrium inversions from a 3D magnetohydrodynamic chromospheric model. Astron. Astrophys. **543**, 34 (2012). doi:10.1051/0004-6361/201218825

J. de la Cruz Rodríguez, B. De Pontieu, M. Carlsson, L.H.M. Rouppe van der Voort, Heating of the magnetic chromosphere: observational constraints from Ca II λ8542 spectra. Astrophys. J. Lett. **764**, 11 (2013a). doi:10.1088/2041-8205/764/1/L11

J. de la Cruz Rodríguez, L. Rouppe van der Voort, H. Socas-Navarro, M. van Noort, Physical properties of a sunspot chromosphere with umbral flashes. Astron. Astrophys. **556**, 115 (2013b). doi:10.1051/0004-6361/201321629

J. de la Cruz Rodríguez, V. Hansteen, L. Bellot-Rubio, A. Ortiz, Emergence of granular-sized magnetic bubbles through the solar atmosphere, II: non-LTE chromospheric diagnostics and inversions. Astrophys. J. **810**, 145 (2015). doi:10.1088/0004-637X/810/2/145

B. De Pontieu, A.M. Title, J.R. Lemen, G.D. Kushner, D.J. Akin, B. Allard, T. Berger, P. Boerner, M. Cheung, C. Chou, J.F. Drake, D.W. Duncan, S. Freeland, G.F. Heyman, C. Hoffman, N.E. Hurlburt, R.W. Lindgren, D. Mathur, R. Rehse, D. Sabolish, R. Seguin, C.J. Schrijver, T.D. Tarbell, J.-P. Wülser, C.J. Wolfson, C. Yanari, J. Mudge, N. Nguyen-Phuc, R. Timmons, R. van Bezooijen, I. Weingrod, R. Brookner, G. Butcher, B. Dougherty, J. Eder, V. Knagenhjelm, S. Larsen, D. Mansir, L. Phan, P. Boyle, P.N. Cheimets, E.E. DeLuca, L. Golub, R. Gates, E. Hertz, S. McKillop, S. Park, T. Perry, W.A. Podgorski, K. Reeves, S. Saar, P. Testa, H. Tian, M. Weber, C. Dunn, S. Eccles, S.A. Jaeggli, C.C. Kankelborg, K. Mashburn, N. Pust, L. Springer, R. Carvalho, L. Kleint, J. Marmie, E. Mazmanian, T.M.D. Pereira, S. Sawyer, J. Strong, S.P. Worden, M. Carlsson, V.H. Hansteen, J. Leenaarts, M. Wiesmann, J. Aloise, K.-C. Chu, R.I. Bush, P.H. Scherrer, P. Brekke, J. Martinez-Sykora, B.W. Lites, S.W. McIntosh, H. Uitenbroek, T.J. Okamoto, M.A. Gummin, G. Auker, P. Jerram, P. Pool, N. Waltham, The Interface Region Imaging Spectrograph (IRIS). Sol. Phys. **289**, 2733–2779 (2014). doi:10.1007/s11207-014-0485-y

P. Fabiani Bendicho, J. Trujillo Bueno, L. Auer, Multidimensional radiative transfer with multilevel atoms, II: the non-linear multigrid method. Astron. Astrophys. **324**, 161–176 (1997)

J.M. Fontenla, E.H. Avrett, R. Loeser, Energy balance in the solar transition region, III: helium emission in hydrostatic, constant-abundance models with diffusion. Astrophys. J. **406**, 319–345 (1993). doi:10.1086/172443

C. Frutiger, S.K. Solanki, M. Fligge, J.H.M.J. Bruls, Properties of the solar granulation obtained from the inversion of low spatial resolution spectra. Astron. Astrophys. **358**, 1109–1121 (2000)

L. Goldberg, The temperature of the solar chromosphere. Astrophys. J. **89**, 673 (1939). doi:10.1086/144092

T.P. Golding, M. Carlsson, J. Leenaarts, Detailed and simplified nonequilibrium helium ionization in the solar atmosphere. Astrophys. J. **784**, 30 (2014). doi:10.1088/0004-637X/784/1/30

W. Hanle, Über magnetische Beeinflussung der Polarisation der Resonanzfluoreszenz. Z. Phys. **30**, 95–105 (1924). doi:10.1086/140355

J.W. Harvey, Chromospheric magnetic field measurements in a flare and an active region filament. Sol. Phys. **280**, 69–81 (2012). doi:10.1007/s11207-012-0067-9

J. Harvey, W. Livingston, C. Slaughter, *Line Formation in the Presence of Magnetic Fields* (National Center for Atmospheric Research, Boulder, 1972), p. 227

F. Jennerholm Hammar, Inference of chromospheric magnetic fields with the Ca II 8542 line, Master's thesis, Department of physics and astronomy, Uppsala University, 2015. http://urn.kb.se/resolve?urn=urn:nbn:se:uu:diva-238863

J. Joshi, A. Lagg, S.K. Solanki, A. Feller, M. Collados, D. Orozco Suárez, R. Schlichenmaier, M. Franz, H. Balthasar, C. Denker, T. Berkefeld, A. Hofmann, C. Kiess, H. Nicklas, A. Pastor Yabar, R. Rezaei, D. Schmidt, W. Schmidt, M. Sobotka, D. Soltau, J. Staude, K.G. Strassmeier, R. Volkmer, O. von der Lühe, T. Waldmann, Upper chromospheric magnetic field of a sunspot penumbra: observations of fine structure. ArXiv:1608.01988 (2016). doi:10.1051/0004-6361/201629214

P. Judge, Observations of the solar chromosphere, in *Solar MHD Theory and Observations: A High Spatial Resolution Perspective*, ed. by J. Leibacher, R.F. Stein, H. Uitenbroek Astronomical Society of the Pacific Conference Series, vol. 354, 2006, p. 259

D. Kiselman, T.M.D. Pereira, B. Gustafsson, M. Asplund, J. Meléndez, K. Langhans, Is the solar spectrum latitude-dependent? An investigation with SST/TRIPPEL. Astron. Astrophys. **535**, A14 (2011). doi:10.1051/0004-6361/201117553

L. Kleint, A. Gandorfer, Prospects of solar magnetometry—from ground and in space. Space Sci. Rev. (2015). doi:10.1007/s11214-015-0208-1

T. Kosugi, K. Matsuzaki, T. Sakao, T. Shimizu, Y. Sone, S. Tachikawa, T. Hashimoto, K. Minesugi, A. Ohnishi, T. Yamada, S. Tsuneta, H. Hara, K. Ichimoto, Y. Suematsu, M. Shimojo, T. Watanabe, S. Shimada, J.M. Davis, L.D. Hill, J.K. Owens, A.M. Title, J.L. Culhane, L.K. Harra, G.A. Doschek, L. Golub, The Hinode (Solar-B) mission: an overview. Sol. Phys. **243**(1), 3–17 (2007)

C. Kuckein, V. Martínez Pillet, R. Centeno, An active region filament studied simultaneously in the chromosphere and photosphere, I: magnetic structure. Astron. Astrophys. **539**, 131 (2012). doi:10.1051/0004-6361/201117675

A. Lagg, J. Woch, N. Krupp, S.K. Solanki, Retrieval of the full magnetic vector with the He I multiplet at 1083 nm. Maps of an emerging flux region. Astron. Astrophys. **414**, 1109–1120 (2004). doi:10.1051/0004-6361:20031643

A. Lagg, R. Ishikawa, L. Merenda, T. Wiegelmann, S. Tsuneta, S.K. Solanki, Internetwork horizontal magnetic fields in the quiet Sun chromosphere: results from a joint Hinode/VTT study, in *The Second Hinode Science Meeting: Beyond Discovery-Toward Understanding*, ed. by B. Lites, M. Cheung, T. Magara, J. Mariska, K. Reeves Astronomical Society of the Pacific Conference Series, vol. 415, 2009, p. 327

A. Lagg, B. Lites, J. Harvey, S. Gosain, R. Centeno, Measurements of photospheric and chromospheric magnetic fields. Space Sci. Rev. (2015). doi:10.1007/s11214-015-0219-y

E. Landi Degl'Innocenti, MALIP – a programme to calculate the Stokes parameters profiles of magnetoactive Fraunhofer lines. Astron. Astrophys. Suppl. Ser. **25**, 379 (1976)

E. Landi Degl'Innocenti, Magneto-optical effects and the interpretation of linearly polarized intensity distributions observed with a vector magnetograph. Sol. Phys. **63**, 237–245 (1979). doi:10.1007/BF00174529

E. Landi Degl'Innocenti, M. Landi Degl'Innocenti, Response functions for magnetic lines. Astron. Astrophys. **56**, 111–115 (1977)

E. Landi Degl'Innocenti, M. Landolfi, *Polarization in Spectral Lines*. Astrophysics and Space Science Library: A Series of Books on the Recent Developments of Space Science and of General Geophysics and Astrophysics, vol. 307 (Springer, New York, 2004). 9781402024146

J. Leenaarts, M. Carlsson, MULTI3D: a domain-decomposed 3D radiative transfer code, in *The Second Hinode Science Meeting: Beyond Discovery-Toward Understanding*, ed. by B. Lites, M. Cheung, T. Magara, J. Mariska, K. Reeves Astronomical Society of the Pacific Conference Series, vol. 415, 2009, p. 87

J. Leenaarts, M. Carlsson, L. Rouppe van der Voort, The formation of the Hα line in the solar chromosphere. Astrophys. J. **749**, 136 (2012). doi:10.1088/0004-637X/749/2/136

J. Leenaarts, T. Pereira, H. Uitenbroek, Fast approximation of angle-dependent partial redistribution in moving atmospheres. Astron. Astrophys. **543**, 109 (2012b). doi:10.1051/0004-6361/201219394

J. Leenaarts, T.M.D. Pereira, M. Carlsson, H. Uitenbroek, B. De Pontieu, The formation of IRIS diagnostics, II: the formation of the Mg II h & k lines in the solar atmosphere. Astrophys. J. **772**, 90 (2013). doi:10.1088/0004-637X/772/2/90

J. Leenaarts, J. de la Cruz Rodríguez, O. Kochukhov, M. Carlsson, The effect of isotopic splitting on the bisector and inversions of the solar Ca II 854.2 nm line. Astrophys. J. Lett. **784**, 17 (2014). doi:10.1088/2041-8205/784/1/L17

J. Leenaarts, M. Carlsson, L. Rouppe van der Voort, On fibrils and field lines: the nature of Hα fibrils in the solar chromosphere. Astrophys. J. **802**, 136 (2015). doi:10.1088/0004-637X/802/2/136

J. Leenaarts, T. Golding, M. Carlsson, T. Libbrecht, J. Joshi, The cause of spatial structure in solar He I 1083 nm multiplet images. ArXiv:1608.00838 (2016)

K. Levenberg, A method for the solution of certain non-linear problems in least squares. Q. Appl. Math. **2**, 164–168 (1944)

B.W. Lites, Magnetic flux ropes in the solar photosphere: the vector magnetic field under active region filaments. Astrophys. J. **622**, 1275–1291 (2005). doi:10.1086/428080

A. López Ariste, R. Casini, Magnetic fields in prominences: inversion techniques for spectropolarimetric data of the He I D_3 line. Astrophys. J. **575**, 529–541 (2002). doi:10.1086/341260

A. López Ariste, R. Casini, Inference of the magnetic field in spicules from spectropolarimetry of He I D3. Astron. Astrophys. **436**, 325–331 (2005). doi:10.1051/0004-6361:20042214

A. López Ariste, G. Aulanier, B. Schmieder, A. Sainz Dalda, First observation of bald patches in a filament channel and at a barb endpoint. Astron. Astrophys. **456**, 725–735 (2006). doi:10.1051/0004-6361:20064923

R. Manso Sainz, J. Trujillo Bueno, Scattering polarization of the Ca II IR triplet for probing the quiet solar chromosphere. Astrophys. J. **722**, 1416–1436 (2010). doi:10.1088/0004-637X/722/2/1416

D. Marquardt, An algorithm for least-squares estimation of nonlinear parameters. SIAM J. Appl. Math. **11**(2), 431–441 (1968)

M.J. Martínez González, R. Manso Sainz, A. Asensio Ramos, C. Beck, J. de la Cruz Rodríguez, A.J. Díaz, Spectro-polarimetric imaging reveals helical magnetic fields in solar prominence feet. Astrophys. J. **802**, 3 (2015). doi:10.1088/0004-637X/802/1/3

V. Martinez Pillet, R.J. Garcia Lopez, J.C. del Toro Iniesta, R. Rebolo, M. Vazquez, J.E. Beckman, S. Char, Circular polarization of the CA II H and K lines in solar quiet and active regions. Astrophys. J. Lett. **361**, 81–85 (1990). doi:10.1086/185832

L. Merenda, J. Trujillo Bueno, E. Landi Degl'Innocenti, M. Collados, Determination of the magnetic field vector via the Hanle and Zeeman effects in the He I λ10830 multiplet: evidence for nearly vertical magnetic fields in a polar crown prominence. Astrophys. J. **642**, 554–561 (2006). doi:10.1086/501038

N. Metropolis, A.W. Rosenbluth, M.N. Rosenbluth, A.H. Teller, E. Teller, Equation of state calculations by fast computing machines. J. Chem. Phys. **21**(6) (1953)

G.L. Olson, P.B. Kunasz, Short characteristic solution of the non-LTE transfer problem by operator perturbation, I: the one-dimensional planar slab. J. Quant. Spectrosc. Radiat. Transf. **38**, 325–336 (1987). doi:10.1016/0022-4073(87)90027-6

D. Orozco Suárez, J.C. Del Toro Iniesta, The usefulness of analytic response functions. Astron. Astrophys. **462**, 1137–1145 (2007). doi:10.1051/0004-6361:20066201

D. Orozco Suárez, L.R. Bellot Rubio, J.C. Del Toro Iniesta, S. Tsuneta, B. Lites, K. Ichimoto, Y. Katsukawa, S. Nagata, T. Shimizu, R.A. Shine, Y. Suematsu, T.D. Tarbell, A.M. Title, Strategy for the inversion of Hinode spectropolarimetric measurements in the quiet Sun. Publ. Astron. Soc. Jpn. **59**, 837 (2007). doi:10.1093/pasj/59.sp3.S837

D. Orozco Suárez, A. Asensio Ramos, J. Trujillo Bueno, The magnetic field configuration of a solar prominence inferred from spectropolarimetric observations in the He I 10 830 Å triplet. Astron. Astrophys. **566**, 46 (2014). doi:10.1051/0004-6361/201322903

D. Orozco Suárez, A. Asensio Ramos, J. Trujillo Bueno, Height variation of the vector magnetic field in solar spicules. Astrophys. J. Lett. **803**, 18 (2015). doi:10.1088/2041-8205/803/2/L18

Y. Osorio, P. Barklem, K. Lind, M. Asplund, The influence of electron collisions on non-LTE Li line formation in stellar atmospheres. J. Phys. Conf. Ser. **388**(4), 042018 (2012). doi:10.1088/1742-6596/388/4/042018

T.M.D. Pereira, J. Leenaarts, B. De Pontieu, M. Carlsson, H. Uitenbroek, The formation of IRIS diagnostics, III: near-ultraviolet spectra and images. Astrophys. J. **778**, 143 (2013). doi:10.1088/0004-637X/778/2/143

A. Pietarila, H. Socas-Navarro, T. Bogdan, Spectropolarimetric inversions of the Ca II 8498 and 8542 Å lines in the quiet Sun. Astrophys. J. **670**, 885–902 (2007a). doi:10.1086/521271

A. Pietarila, H. Socas-Navarro, T. Bogdan, Spectropolarimetric observations of the Ca II λ8498 and λ8542 in the quiet Sun. Astrophys. J. **663**, 1386–1405 (2007b). doi:10.1086/518714

N.E. Piskunov, F. Kupka, T.A. Ryabchikova, W.W. Weiss, C.S. Jeffery, VALD: the Vienna atomic line data base. Astron. Astrophys. Suppl. Ser. **112**, 525 (1995)

W.H. Press, S.A. Teukolsky, W.T. Vetterling, B.P. Flannery, *Numerical Recipes in C* (Cambridge Univ. Press, Cambridge, 1992)

J. Rayrole, Contribution à l'étude de la structure du champ magnétique dans les taches solaires. Ann. Astrophys. **30**, 257 (1967)

D.E. Rees, C.J. Durrant, G.A. Murphy, Stokes profile analysis and vector magnetic fields, II: formal numerical solutions of the Stokes transfer equations. Astrophys. J. **339**, 1093–1106 (1989). doi:10.1086/167364

D.E. Rees, A. López Ariste, J. Thatcher, M. Semel, Fast inversion of spectral lines using principal component analysis, I: fundamentals. Astron. Astrophys. **355**, 759–768 (2000)

L. Rosén, O. Kochukhov, G.A. Wade, First Zeeman doppler imaging of a cool star using all four Stokes parameters. Astrophys. J. **805**, 169 (2015). doi:10.1088/0004-637X/805/2/169

B. Ruiz Cobo, A. Asensio Ramos, Returning magnetic flux in sunspot penumbrae. Astron. Astrophys. **549**, 4 (2013). doi:10.1051/0004-6361/201220373

B. Ruiz Cobo, J.C. del Toro Iniesta, Inversion of Stokes profiles. Astrophys. J. **398**, 375–385 (1992). doi:10.1086/171862

R.J. Rutten, Observing the solar chromosphere, in *The Physics of Chromospheric Plasmas*, ed. by P. Heinzel, I. Dorotovič, R.J. Rutten Astronomical Society of the Pacific Conference Series, vol. 368, 2007, p. 27

G.B. Rybicki, D.G. Hummer, An accelerated lambda iteration method for multilevel radiative transfer, I: non-overlapping lines with background continuum. Astron. Astrophys. **245**, 171–181 (1991)

J. Sanchez Almeida, M. Collados, J.C. del Toro Iniesta, On the generation of the net circular polarization observed in solar faculae. Astron. Astrophys. **222**, 311–315 (1989)

T.A. Schad, M.J. Penn, H. Lin, He I vector magnetometry of field-aligned superpenumbral fibrils. Astrophys. J. **768**, 111 (2013). doi:10.1088/0004-637X/768/2/111

T.A. Schad, M.J. Penn, H. Lin, A. Tritschler, He I vector magnetic field maps of a sunspot and its superpenumbral fine-structure. Sol. Phys. **290**, 1607–1626 (2015). doi:10.1007/s11207-015-0706-z

G.B. Scharmer, Solutions to radiative transfer problems using approximate lambda operators. Astrophys. J. **249**, 720–730 (1981). doi:10.1086/159333

G.B. Scharmer, Comments on the optimization of high resolution Fabry-Pérot filtergraphs. Astron. Astrophys. **447**(3), 1111–1120 (2006)

G.B. Scharmer, K. Bjelksjo, T.K. Korhonen, B. Lindberg, B. Petterson, The 1-meter Swedish solar telescope, in *Innovative Telescopes and Instrumentation for Solar Astrophysics*, ed. by S.L. Keil, S.V. Avakyan. Proceedings of the SPIE, vol. 4853 (2003), pp. 341–350

G.B. Scharmer, G. Narayan, T. Hillberg, J. de la Cruz Rodriguez, M.G. Löfdahl, D. Kiselman, P. Sütterlin, M. van Noort, A. Lagg, CRISP spectropolarimetric imaging of penumbral fine structure. Astrophys. J. Lett. **689**(1), L69 (2008)

G.B. Scharmer, J. de la Cruz Rodríguez, P. Sütterlin, V.M.J. Henriques, Opposite polarity field with convective downflow and its relation to magnetic spines in a sunspot penumbra. Astron. Astrophys. **553**, 63 (2013). doi:10.1051/0004-6361/201220899

W. Schmidt, O. von der Lühe, R. Volkmer, C. Denker, S.K. Solanki, H. Balthasar, N. Bello González, T. Berkefeld, M. Collados Vera, A. Hofmann, F. Kneer, A. Lagg, K.G. Puschmann, D. Schmidt, M. Sobotka, D. Soltau, K.G. Strassmeier, The GREGOR solar telescope on Tenerife, in *The Second ATST-EAST Meeting: Magnetic Fields from the Photosphere to the Corona*, ed. by T. Rimmele, A. Tritschler, F. Wöger, V. Collados, H. Socos-Navarro, R. Schlichenmaier, M. Carlsson, T. Berger, A. Cadavid, P. Gilbert, P. Goode, M. Knölker. ASP Conference Proceedings, vol. 463 (Astronomical Society of the Pacific, San Francisco, 2012), p. 365

B. Schmieder, H. Tian, T. Kucera, A. López Ariste, N. Mein, P. Mein, K. Dalmasse, L. Golub, Open questions on prominences from coordinated observations by IRIS, Hinode, SDO/AIA, THEMIS, and the Meudon/MSDP. Astron. Astrophys. **569**, 85 (2014). doi:10.1051/0004-6361/201423922

R.A. Shine, J.L. Linsky, Physical properties of solar chromospheric plages, II: chromospheric plage models. Sol. Phys. **39**, 49–77 (1974). doi:10.1007/BF00154970

A. Skumanich, B.W. Lites, Stokes profile analysis and vector magnetic fields, I: inversion of photospheric lines. Astrophys. J. **322**, 473–482 (1987). doi:10.1086/165743

A. Skumanich, B. Lites, Velocity gradients across a flaring neutral line from Stokes II measurements, in *Solar Polarimetry*, ed. by L.J. November, 1991, pp. 307–317

H. Socas-Navarro, Are electric currents heating the magnetic chromosphere? Astrophys. J. Lett. **633**, 57–60 (2005a). doi:10.1086/498145

H. Socas-Navarro, The three-dimensional structure of a sunspot magnetic field. Astrophys. J. Lett. **631**, 167–170 (2005b). doi:10.1086/497334

H. Socas-Navarro, H. Uitenbroek, On the diagnostic potential of Hα for chromospheric magnetism. Astrophys. J. Lett. **603**, 129–132 (2004). doi:10.1086/383147

H. Socas-Navarro, B. Ruiz Cobo, J. Trujillo Bueno, Non-LTE inversion of line profiles. Astrophys. J. **507**, 470–481 (1998). doi:10.1086/306330

H. Socas-Navarro, J. Trujillo Bueno, B. Ruiz Cobo, Anomalous polarization profiles in sunspots: possible origin of umbral flashes. Science **288**, 1396–1398 (2000a). doi:10.1126/science.288.5470.1396

H. Socas-Navarro, J. Trujillo Bueno, B. Ruiz Cobo, Non-LTE inversion of Stokes profiles induced by the Zeeman effect. Astrophys. J. **530**, 977–993 (2000b). doi:10.1086/308414

H. Socas-Navarro, A. López Ariste, B.W. Lites, Fast inversion of spectral lines using principal components analysis, II: inversion of real Stokes data. Astrophys. J. **553**, 949–954 (2001). doi:10.1086/320984

H. Socas-Navarro, D. Elmore, A. Pietarila, A. Darnell, B.W. Lites, S. Tomczyk, S. Hegwer, Spinor: visible and infrared spectro-polarimetry at the national solar observatory. Sol. Phys. **235**, 55–73 (2006). doi:10.1007/s11207-006-0020-x

H. Socas-Navarro, J. de la Cruz Rodríguez, A. Asensio Ramos, J. Trujillo Bueno, B. Ruiz Cobo, An open-source, massively parallel code for non-LTE synthesis and inversion of spectral lines and Zeeman-induced Stokes profiles. Astron. Astrophys. **577**, 7 (2015). doi:10.1051/0004-6361/201424860

S.K. Solanki, O. Steiner, H. Uitenbroeck, Two-dimensional models of the solar chromosphere, I: the CA II K line as a diagnostic: 1.5-D radiative transfer. Astron. Astrophys. **250**, 220–234 (1991)

S.K. Solanki, A. Lagg, J. Woch, N. Krupp, M. Collados, Three-dimensional magnetic field topology in a region of solar coronal heating. Nature **425**, 692–695 (2003). doi:10.1038/nature02035

O. Steiner, F. Züger, L. Belluzzi, Polarized radiative transfer in discontinuous media. Astron. Astrophys. **586**, 42 (2016). doi:10.1051/0004-6361/201527158

J.O. Stenflo, On the interpretation of measurements with a solar magnetograph. Acta Univ. Lund. II **2** (1968)

J. Štěpán, J. Trujillo Bueno, On the probable existence of an abrupt magnetization in the upper chromosphere of the quiet Sun. Astrophys. J. Lett. **711**, 133–137 (2010). doi:10.1088/2041-8205/711/2/L133

J. Štěpán, J. Trujillo Bueno, PORTA: a three-dimensional multilevel radiative transfer code for modeling the intensity and polarization of spectral lines with massively parallel computers. Astron. Astrophys. **557**, 143 (2013). doi:10.1051/0004-6361/201321742

J. Trujillo Bueno, The generation and transfer of polarized radiation in stellar atmospheres, in *Stellar Atmosphere Modeling*, ed. by I. Hubeny, D. Mihalas, K. Werner Astronomical Society of the Pacific Conference Series, vol. 288, 2003, p. 551

J. Trujillo Bueno, A. Asensio Ramos, Influence of atomic polarization and horizontal illumination on the Stokes profiles of the He I 10830 Å multiplet. Astrophys. J. **655**, 642–650 (2007). doi:10.1086/509795

J. Trujillo Bueno, E. Landi Degl'Innocenti, The polarization-free approximation. Sol. Phys. **164**, 135–153 (1996). doi:10.1007/BF00146629

J. Trujillo Bueno, N. Shchukina, A. Asensio Ramos, A substantial amount of hidden magnetic energy in the quiet Sun. Nature **430**, 326–329 (2004). doi:10.1038/nature02669

H. Uitenbroek, An efficient method for the evaluation of general redistribution integration weights. Astron. Astrophys. **216**, 310–314 (1989)

H. Uitenbroek, The solar Mg II H and K lines – observations and radiative transfer modeling. Sol. Phys. **172**, 109–116 (1997). doi:10.1023/A:1004981412889

H. Uitenbroek, Multilevel radiative transfer with partial frequency redistribution. Astrophys. J. **557**, 389–398 (2001). doi:10.1086/321659

M. van Noort, Spatially coupled inversion of spectro-polarimetric image data, I: method and first results. Astron. Astrophys. **548**, 5 (2012). doi:10.1051/0004-6361/201220220

J.E. Vernazza, E.H. Avrett, R. Loeser, Structure of the solar chromosphere, III: models of the EUV brightness components of the quiet-sun. Astrophys. J. Suppl. Ser. **45**, 635–725 (1981). doi:10.1086/190731

S. Wedemeyer-Böhm, M. Carlsson, Non-equilibrium calcium ionisation in the solar atmosphere. Astron. Astrophys. **528**, 1 (2011). doi:10.1051/0004-6361/201016186

F. Wöger, S. Wedemeyer-Böhm, H. Uitenbroek, T.R. Rimmele, Morphology and dynamics of the low solar chromosphere. Astrophys. J. **706**, 148–157 (2009). doi:10.1088/0004-637X/706/1/148

P. Zeeman, On the influence of magnetism on the nature of the light emitted by a substance. Astrophys. J. **5**, 332–347 (1897). doi:10.1086/140355

Space Sci Rev (2017) 210:145–181
DOI 10.1007/s11214-017-0400-6

Magnetic Diagnostics of the Solar Corona: Synthesizing Optical and Radio Techniques

R. Casini[1] · S.M. White[1] · P.G. Judge[2]

Received: 17 November 2016 / Accepted: 8 July 2017 / Published online: 7 August 2017

Abstract In this contribution we review the current state-of-the-art of coronal magnetometry, in both optical and radio domains. We address the achievable objectives and the challenges of present measurement techniques and interpretation tools. In particular, we focus on the role that these observations can play for constraining and validating numerical models of the global coronal magnetic field. With regard to optical techniques, we mainly focus on the use of M1 diagnostics, further developing the theory of the formation of their polarization signatures in the magnetized corona.

Keywords Sun: corona · Sun: magnetic field · Polarization · Optical techniques · Radio techniques

1 Introduction

The observational evidence for the presence of magnetic fields in the solar corona can be dated back to the interpretation of historical eclipse data, which demonstrated the existence of loop-like structures that were reminiscent of the magnetic-field patterns that had been observed in laboratory experiments involving magnets and electric currents. Our current understanding of the nature of coronal magnetic fields indicates that these originate in the solar interior, through the dynamo action in the ionized gases (plasma) that constitute the convective layer of the solar atmosphere, and eventually make their way through the photosphere and into the outer layers of the solar atmosphere by buoyancy and hydro-magnetic instabilities. This mechanism of flux emergence is accompanied by processes of magnetic reconnection and generation of MHD and plasma waves, which ultimately control the dynamics and the heating of the solar corona. Because of the fundamental role played by

The National Center for Atmospheric Research is sponsored by the National Science Foundation.

✉ R. Casini
casini@ucar.edu

[1] High Altitude Observatory, National Center for Atmospheric Research, P.O. Box 3000, Boulder, CO 80307-3000, USA

[2] Space Vehicles Directorate, Air Force Research Lab., Kirtland AFB, NM, 87117, USA

magnetic fields in determining the thermal and magnetohydrodynamic characteristics of the solar corona, which in turn controls the fundamental processes of coronal heating, explosive energy release and particle acceleration, the measurement of the magnetic field vector in the corona (coronal magnetometry) has become a high priority goal for observational solar physics.

However, despite the successes and continuous advances of magnetic field measurements in the solar photosphere, reliable determinations of the vector magnetic field in the solar corona have remained elusive. This state of affairs can fundamentally be traced back to two reasons. First of all, coronal structures are typically 8–9 orders of magnitude less dense than the photospheric plasma, and 6 orders of magnitude dimmer than the solar disk in the optical region of the solar spectrum, making the detection of any optical signal from the corona so difficult that observations are usually restricted to specific observing conditions, only attainable through natural or artificial solar eclipses. Secondly, because of the much larger volume available to the coronal plasma and the significantly lower density, momentum balance forces the ambient magnetic field to be much weaker (by 2–3 orders of magnitude) than in the solar photosphere—with the exception of coronal regions immediately overlying strong magnetically active regions of the photosphere—and so the requirements on the sensitivity and accuracy of magnetic-field measurement instruments and techniques become extremely demanding and challenging.

In light of these problems, there has been a proliferation of numerical modeling efforts that attempt to simulate coronal magnetic structures through extrapolation of the observed photospheric field and other data. Schrijver et al. (2006) and De Rosa et al. (2009) review and compare methods that derive coronal fields by applying mathematical techniques to photospheric vector magnetograms. The addition of line-of-sight (Doppler) velocity measurements, horizontal flow data and observations of the evolution of the vector field allow the explicit determination of electric fields and the tracking of changes in the coronal field (e.g., Fisher et al. 2015; Cheung et al. 2015). As an alternative to using the photospheric vector field as a starting point, methods have been developed that use the line-of-sight photospheric field together with observations of the paths of coronal magnetic field lines, usually obtained from EUV images. The field lines themselves provide no information on magnetic field strengths in the corona, so these methods start with a potential-field solution derived from the line-of-sight (or radial) magnetograms and then add electric currents in the corona in order to match the observed field lines (e.g., Malanushenko et al. 2014; Aschwanden 2016). All such approaches face mathematical problems of defining proper boundary conditions, and the uniqueness of solutions, as well as ill-posedness. Thus, at least in the near term, numerical extrapolation techniques complement but cannot replace observational efforts to measure magnetic structures in the corona directly. The validation of any numerical technique for extrapolation of photospheric or chromospheric fields into the corona will require comparison with robust magnetic field measurements in the corona.

At present, there are two main types of observations that present the best opportunities to infer properties of the coronal magnetic field: 1) spectro-polarimetry of coronal emission lines from various ionized atomic species in the optical domain (by which we mean the wavelength range from the far UV through the visible to the near IR), probing different temperature and density regimes of coronal plasma, and 2) radio observations in various passbands, probing regions of the corona characterized by different plasma frequencies. In this review, we summarize the methodologies of these observations, their realm of applicability and limitations, and chances for success. We note that coronal seismology, i.e., the study of oscillations of coronal features such as loops, can also be used to determine magnetic field properties when the wave mode producing the oscillations can be identified and (in many cases) the density is also known (e.g., Nakariakov and Verwichte 2005). However, this application is limited to features that show such oscillations, and will not be discussed here.

2 Historical Overview

At optical wavelengths, the observation of magnetically sensitive, forbidden (i.e., magnetic dipole; M1) emission lines from highly ionized species (e.g., Fe XIII, Fe XIV, Ca XV) has historically had the precedence, because of the relative ease of detection of these lines, even from the ground with modest-size coronographic instruments, thanks to the relatively strong emissivity and absence of significant absorption features in the Earth's atmosphere at those wavelengths, and because of the relative simplicity of the associated line formation model. In fact, even for the very weak fields of the "quiet Sun" corona ($\lesssim 10$ G), the magnetic signature in the circular polarization profile is essentially determined by the *longitudinal* Zeeman effect, which provides a lower bound for the total strength of the ambient magnetic field. On the other hand, the linear polarization profiles are completely dominated by the scattering polarization induced by the anisotropic irradiation of the coronal plasma by the underlying photosphere and chromosphere. Because the ratio of the Zeeman splitting to the typical thermal Doppler width increases linearly with wavelength, observers have mostly concentrated their efforts in the observation of forbidden coronal lines that emit in the near IR (such as the Fe XIII lines at 1074.7 and 1079.8 nm, and the Si IX line at 3934 nm; e.g., Lin et al. 2004, Judge et al. 2002, 2013) in order to maximize the chances for a successful detection of the longitudinal Zeeman effect. The contribution of the transversal Zeeman effect to the linear polarization is so small, being only a second-order effect, that it is completely swamped by the dominant polarization signal of resonant scattering.

Because of the large nuclear charge acting on the optically active electron states, for the ionic species characteristic of the solar corona, permitted (i.e., electric dipole; E1) atomic transitions typically occur in the far UV region of the solar spectrum, and are therefore accessible only from space because of absorption in the Earth's atmosphere. This is the main reason why this type of diagnostic has not been extensively exploited in the past, but the situation has now changed because of the new opportunities offered by space-borne solar missions planned for the near future (e.g., Solar Orbiter, Solar-C). One of the advantages offered by this type of diagnostic is sensitivity to the Hanle effect for the typical magnetic strengths of the solar corona. In the Hanle effect (Hanle 1924), the scattering polarization that is induced by the anisotropic irradiation of the emitting atoms is very sensitive to changes in the ambient magnetic field around the so-called *Hanle critical magnetic strength*, as well as to the topology of the field. Therefore, *both* the linear and circular polarizations of the observed line profiles carry the full vector information about the ambient magnetic field.

E1 transitions in principle provide a more powerful magnetic diagnostic, because of their sensitivity to the Hanle effect. On the other hand the modeling of the line formation problem becomes more heavily dependent on the accurate description of the illumination and plasma conditions—such as the possible breaking of the cylindrical symmetry of the incident radiation field induced by plasma inhomogeneities in the photosphere and chromosphere, and the effects of collisional depolarization by charged perturbers in the corona (see, e.g., the review by Raouafi et al. 2016). While these conditions are obviously also important for a proper modeling of the M1 coronal lines, in the case of these E1 lines they critically affect the zero-field polarization of the model that sets the reference for the magnetic strength inference from the observed polarimetric signals. These modeling difficulties have been a deterrent for the solar physics community at large to fully engage in the necessary theoretical and numerical modeling efforts to tackle the problem of E1 coronal emission line polarization. Recent reviews that address these issues have been given by Trujillo Bueno et al. (2017), who present the state-of-the-art in modeling of UV spectropolarimetry for magnetic field diagnostics in the transition region and the corona, and by Raouafi et al. (2016), who use MHD simulations of coronal conditions to determine the magnitude of polarization signals likely to be detected in UV (H I Ly-α) and IR (He I 10830 Å) lines due to the Hanle effect.

Radio observations provide very contrasting diagnostics for coronal magnetic fields. Solar radio astronomy developed rapidly after 1945 when radar equipment leftover from the war, and scientists trained to use it, were applied to observing the Sun. As radio astronomy developed and polarization measurements became possible, it was found that solar radio emission is frequently polarized and that the polarization was a consequence of emission in a magnetized medium (e.g., Kundu 1965; McLean and Labrum 1985). Solar radio emission was found only to exhibit circular polarization, and this led to the understanding that Faraday rotation within the Sun's corona is so large that it wipes out any linear polarization across the observing bandwidths typical of radio telescopes. This simplifies the analysis since only circular polarization need be considered, but in any case the natural electromagnetic modes in the plasma that makes up the solar atmosphere are fully circularly polarized under most conditions. This fact makes radio measurements a fruitful technique for studying coronal magnetic fields in particular.

2.1 Magnetic Fields and Optical Emission

The study of the solar corona in the optical region of the solar spectrum has by necessity been driven by the brightest lines that could be detected during natural eclipses or through the use of coronographic instruments. The "green", "yellow", and "red" coronal lines are prime examples of such spectral diagnostics. All these three lines, respectively at 530, 569, and 637 nm, correspond to forbidden (M1) transitions of different ionic species (respectively, Fe XIV, Ca XV, and Fe X), mapping different thermal characteristics of the solar corona, and the problem of how they form rose to the attention of the solar community very shortly after their identification.

Pioneering work that led to the understanding of the polarization signals observed in the M1 emission lines of the solar corona started in the second half of the 1960s, most relevantly with the contribution of Charvin (1965), who was the first to point out that the observed direction of linear polarization in those lines can be related to the direction of the projection of the vector magnetic field on the plane of sky (POS). Successively, Harvey (1969) attempted the detection of the circular polarization signature of the component of the coronal magnetic field along the line of sight (LOS) in the spectrum of the "green line" of Fe XIV at 530.4 nm. These observations required long integration times and large binning areas ($\sim$ 1 arcmin^2) in order to extract the weak Zeeman signals, because of the limited sensitivity of the instrumentation. However, the measurements were affected by large error bars making the results often inconclusive, although in seeming agreement with potential field extrapolations. A few more precise measurements indicated values for the volume-averaged LOS component of the magnetic field of a few gauss.

The work of Charvin (1965) provided the driver for additional observational work, aimed at mapping the lines of force of the coronal magnetic field projected on the POS, through the observation of the position angle of the linear polarization of the M1 emission lines of Fe XIII at 1074.7 and 1079.8 nm (Arnaud 1982; Querfeld and Smartt 1984; Arnaud and Newkirk 1987).

At the same time, the theoretical description of the line polarization of M1 transitions was advanced by inclusion of collisional effects (Sahal-Brechot 1974, 1977; House 1977). These works were still exclusively focused on the modeling of the linear polarization signals, mainly because the observational challenges involved in detecting the weak circular polarization signals were insurmountable at the time. In addition, ongoing observational experiments were only looking at the broadband polarization signals integrated over the line profile, in order to increase the signal-to-noise ratio of the data, and because the circular polarization signal integrates out to zero in this case, the problem of modeling the full Stokes vector of the coronal emission lines was of secondary importance.

For this reason, the few attempts to observe the LOS components of the coronal magnetic field (Harvey 1969; Kuhn et al. 1994) had consistently assumed that the circular polarization profiles of the M1 coronal emission lines could be described through the well-known "magnetograph formula" (e.g., Landi Degl'Innocenti and Landolfi 2004). Consequently, the observational efforts to observe the linear and circular polarization profiles of the M1 lines had progressed independently.

Casini and Judge (1999) provided the first self-consistent derivation of the full-Stokes emission of the M1 lines to leading order in the Zeeman parameter (Zeeman splitting divided by the Doppler width of a given line), where it was shown that the amplitude of the circular polarization signals, while directly proportional to the LOS component of the magnetic field according to the longitudinal Zeeman effect, is nonetheless affected by the same mechanism of anisotropic irradiation that is responsible for the observed linear polarization signals (thus implying a modification of the magnetograph formula). This work provided a coherent framework to tackle the problem of the full-Stokes inversion of coronal emission line polarization, demonstrating that the full vector information of the ambient magnetic field can be inferred when the true height of the emitting plasma is known.

Successive applications of the M1 coronal line formation model to the synthesis of Stokes profiles of the Fe XIII lines in realistic configurations of coronal magnetic fields were initiated by Judge and collaborators (Judge and Casini 2001; Judge et al. 2006). Those studies demonstrated that significant characteristic polarization signatures of the coronal magnetic field survive the long integration path of these optical emission diagnostics, which can therefore be effectively used to compare with observations in order to validate proposed models of the global coronal magnetic field, as well as to validate and constrain numerical extrapolation in the solar corona of the magnetic field measured in the photosphere.

These studies motivated instrumental efforts aimed at developing synoptic observing programs of coronal full-Stokes data, such as the Coronal Multichannel Polarimeter (CoMP; Tomczyk et al. 2008) and its successor, the Coronal Solar Magnetism Observatory (CoSMO; Tomczyk 2012). On the modeling front, they triggered the development of tools for the synthesis of coronal polarimetric observables under different coronal model assumptions (e.g., the FORWARD package of Gibson et al. 2016).

The problem of full-Stokes inversion of coronal data has recently been revisited through an analysis combining both the 10747 and 10798 lines of the Fe XIII M1 multiplet (Plowman 2014), taking into account the effects of polarimetric noise. Later in this paper we provide a summary of the main findings of this recent study. Coronal tomography techniques (e.g., Kramar et al. 2013) could in theory improve the knowledge available for single-point inversions. However, the problem of the temporal evolution of the solar corona between sampling observations can result in large errors in the derived magnetic properties, and so it should be expected that coronal tomography is going to be helpful only if the polarization measurements are strictly simultaneous, such as with a STEREO-like mission armed with a coronagraph and polarimeter.

A significant effort has also been spent over the past two decades in studying the formation of permitted (E1) lines from ions emitting in the far UV of the solar spectrum. Using space-borne instruments, such as the Ultra-Violet Coronograph Spectrometer (UVCS) on board the Solar and Heliospheric Observatory (SOHO) spacecraft, spectroscopic observations of emission lines such as the H I Ly-α line at 121.6 nm and the lines of O VI at 103.2 and 103.7 nm have been conducted, which revealed the presence in the corona of a significant anisotropy in the velocity distribution of the emitting ions (Kohl et al. 1998). Because the excitation of these resonance lines in the corona is dominated by the relatively narrow emission features at the same wavelength coming from the underlying transition region, the large speeds and velocity anisotropy observed in organized plasma flows in the corona (e.g., in the solar wind) can be responsible for a significant modulation of the intensity of the incident radiation as seen from the moving ions, because of the condition of off-resonance irradiation that can be produced at such large speeds.

This has motivated the investigation of the effects of Doppler redistribution on the formation of these lines in a magnetized medium (*Doppler dimming*; Sahal-Brechot et al. 1998). Low-sensitivity linear polarization measurements of the O VI line at 103.2 nm were conducted using the SOHO/SUMER (Solar Ultraviolet Measurement of Emitted Radiation) instrument, exploiting the instrumental polarization of the spectrograph's optics to perform the polarization analysis, and the observed signals were successfully interpreted using a model of polarized line formation combining the magnetic Hanle effect and Doppler redistribution (Raouafi et al. 1999). Recently, this problem has been revisited through more advanced modeling of the emitted polarized signatures, particularly for the case of the H I Ly-α line (Khan et al. 2011; Khan and Landi Degl'Innocenti 2011, 2012; Khan 2012).

2.2 Magnetic Fields and Radio Emission

Essentially everything in the solar atmosphere from the chromosphere outwards is visible somewhere in the radio domain, and while the emission mechanism may be different in different frequency ranges, the measured circular polarization always contains some information about the magnetic field in the source region (e.g., articles in Gary and Keller 2004). In the case of plasma emission (Ginzburg and Zheleznyakov 1958), which is the primary mechanism for solar radio bursts at low frequencies, it is not always simple to derive anything more than the line-of-sight orientation of the magnetic field. In the case of gyromagnetic emission, where the spiralling motion of electrons in a magnetic field is an inherent aspect of the emission mechanism, quantitative information on the magnetic field is more readily available. Solar active regions with strong coronal magnetic fields radiate at low harmonics of the electron gyrofrequency (usually referred to as "gyroresonance" emission; Ginzburg and Zheleznyakov 1961; Zheleznyakov 1962; Kakinuma and Swarup 1962) and this can be used to produce direct measurements of coronal field strengths, while highly energetic electrons accelerated in solar flares radiate at higher harmonics (called "synchrotron" for highly relativistic electrons and "gyrosynchrotron" emission for mildly relativistic electrons) and their spectrum provides a measure of the field strength in the loops on which they are trapped. Even bremsstrahlung, which results from the simple mechanism of collisions between electrons and ions, can provide measurements of coronal magnetic fields.

3 Coronal Emission Lines: Polarization Diagnostics

In this section we present the basic aspects of the polarization of coronal emission lines, in order to clarify the limitations of such diagnostics in the study of coronal magnetism.

We attempt to provide a unifying approach to the problem of how the observed polarization signature is affected by the presence of the coronal magnetic field and by the excitation of the atomic transitions via anisotropic irradiation of the emitting plasma from the underlying atmospheric layers, regardless of whether the observed transitions are permitted (E1) or forbidden (M1).

Because of the large temperatures of coronal plasmas ($T > 10^5$ K), and the relatively low strength of the coronal magnetic field, even above strong active regions, at the typical coronal heights that are accessible with optical instrumentation ($B < 200$ G), we rely on the weak-field approximation for the emitted line Stokes profiles, i.e., $\omega_B/\Delta\omega_T \ll 1$, where $\omega_B = (\mu_0/\hbar)\, B$ is the Larmor angular frequency associated with the ambient magnetic field strength (μ_0 is Bohr's magneton, and $\hbar$ the reduced Planck constant), and $\Delta\omega_T$ is the Doppler width associated with the coronal plasma temperature T, in the same units.

We further rely on the fact that the coronal plasma is practically always optically thin at optical wavelengths, and so the transport of polarized radiation along the LOS is well approximated by a simple summation of the contributions along the LOS of the local polarized emissivity scattered toward the observer, plus a possible (unpolarized) contribution of the background radiation from the underlying atmosphere, in the case of on-disk observations.

In order to do this, we adopt the formalism presented by Casini (2002), where a polynomial expansion of the full Stokes emissivity for a transition between an upper state u and a lower state l in terms of the small parameter $\omega_B/\Delta\omega_T$ was derived. In particular, in the case of coronal lines, we can safely limit the Taylor expansion to first order only. For a transition at wavelength λ_0, we find

$$\frac{\omega_B}{\Delta\omega_T} = \lambda_0 \frac{\mu_0 B}{h} \sqrt{\frac{m_A}{2k_B T}}, \tag{1}$$

where m_A is the ion's mass, and k_B is Boltzmann's constant. For a spectral line of iron in the 1 μm range, and using the plasma constraints indicated above, we find $\omega_B/\Delta\omega_T \approx 0.05$. This value is thus representative of the worst case scenario for the applicability of a Taylor expansion in the small parameter $\omega_B/\Delta\omega_T$, as values significantly smaller than this should be expected in general, because of the typically larger temperatures and lower magnetic field strengths than assumed for the above estimate.

We anticipate here that by neglecting the second-order term in the polynomial expansion of the Stokes profiles in the small parameter $\omega_B/\Delta\omega_T$, we are essentially neglecting the contribution of the transverse Zeeman effect to the linear polarization of the scattered radiation. The justification for this is that other sources of linear polarization (such as atomic polarization induced by anisotropic excitation processes) are going to be dominant over the small quadratic contribution of the transverse Zeeman effect from the weak coronal magnetic field, with the exception of special geometric conditions, such as fields inclined at the so-called Van Vleck angle from the local vertical (see Sect. 3.3).

The formalism of Casini (2002) is based on the theory of polarized line formation described in Landi Degl'Innocenti and Landolfi (2004). That theory is applicable under the hypothesis that the frequency of the incident energy is completely redistributed before the re-emission process occurs, so that no memory of the frequency of the incident photon is preserved during radiation scattering. Accordingly, the absorption and emission profiles in the frame of reference of the scattering atom are identical. This hypothesis of *complete redistribution* (CRD) of the frequency of the incident radiation is typically valid in the denser

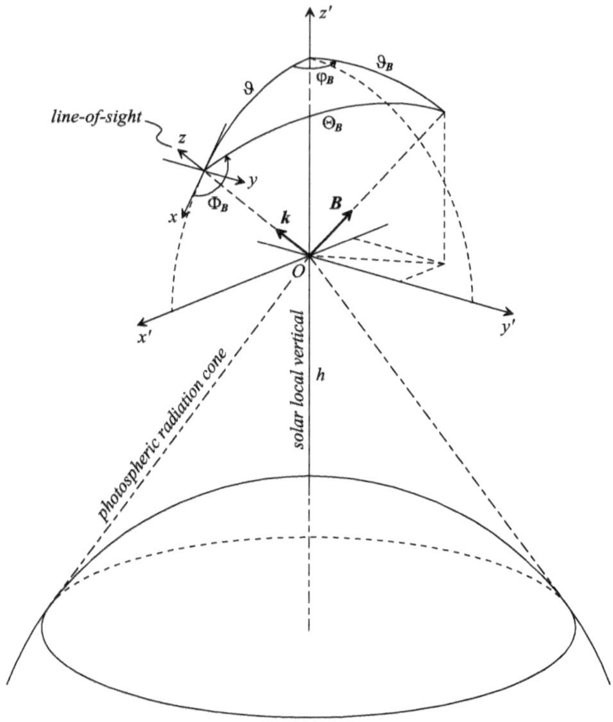

Fig. 1 Geometrical setting for the problem of coronal magnetic field diagnostics. The direction of $y \equiv y'$ identifies the parallel to the limb, which is assumed as the reference direction for positive Stokes Q polarization. The direction of the magnetic field, $\boldsymbol{B}$, is identified by the angles $\vartheta_{\boldsymbol{B}}$ and $\varphi_{\boldsymbol{B}}$ in the reference frame of the local vertical, and by $\Theta_{\boldsymbol{B}}$ and $\Phi_{\boldsymbol{B}}$ in the reference frame of the observer

plasmas of the solar atmosphere (e.g., in the photosphere), because of the shorter collisional times, which become comparable with the typical lifetimes of the atomic levels, even for permitted transitions. In the solar corona, as the gas density drops by 8–9 orders of magnitude with respect to the photosphere, the conditions for the collisional redistribution of the energy of the incident photon may no longer be valid for some of the permitted transitions, but they are still valid in the case of the forbidden M1 lines, because of the much longer lifetimes (by approximately the same 8–9 orders of magnitude) of the levels involved in those transitions, compared to the case of E1 lines.

Additionally, it can be shown (e.g., Heitler 1954) that the CRD hypothesis is formally equivalent to the assumption that the incident radiation has no spectral structure across the energy span of the atomic levels involved in a given transition, the so-called *flat-spectrum approximation*. In the weakly magnetized corona, this condition is normally satisfied, at least for transitions in the IR and in the visible, because of the small Zeeman splitting and the lack of strong, sharp spectral features in the spectrum of the solar photosphere that irradiates the corona at those wavelengths.

In contrast, because of the high density of spectral features in the solar UV spectrum, the *partial redistribution* (PRD) of the frequency of the incident radiation can in principle become important at shorter wavelengths. This is particularly true for important E1 resonance transitions at UV wavelengths, especially when the thermal motion of the atoms in the hot corona is taken into account (Doppler redistribution). However, for the ionized species considered here for the formation of the M1 coronal lines, the UV transitions within the atomic system are typically dominated by collisional excitation, and so the effects of PRD can be ignored when modeling the statistical equilibrium of those ions.

Table 1 Components of the irreducible spherical tensors $T_Q^K(i,\hat{\boldsymbol{k}})$, for $i=0,1,2,3$, and $Q\ge 0$, for the geometry of the observer given in Fig. 1. The components with negative Q are obtained from the conjugation property, $T_{-Q}^K(i,\hat{\boldsymbol{k}})=(-1)^Q\,T_Q^K(i,\hat{\boldsymbol{k}})^*$. The sign choice for $i=1,2$ corresponds to the case of E1 (+) or M1 (−) transitions

$T_0^0(0,\hat{\boldsymbol{k}})=1$
$T_0^1(0,\hat{\boldsymbol{k}})=0$
$T_1^1(0,\hat{\boldsymbol{k}})=0$
$T_0^2(0,\hat{\boldsymbol{k}})=\frac{1}{2\sqrt{2}}(3\cos^2\Theta_{\boldsymbol{B}}-1)$
$T_1^2(0,\hat{\boldsymbol{k}})=-\frac{\sqrt{3}}{2}\sin\Theta_{\boldsymbol{B}}\cos\Theta_{\boldsymbol{B}}$
$T_2^2(0,\hat{\boldsymbol{k}})=\frac{\sqrt{3}}{4}\sin^2\Theta_{\boldsymbol{B}}$
$T_0^0(1,\hat{\boldsymbol{k}})=0$
$T_0^1(1,\hat{\boldsymbol{k}})=0$
$T_1^1(1,\hat{\boldsymbol{k}})=0$
$T_0^2(1,\hat{\boldsymbol{k}})=\pm\frac{3}{2\sqrt{2}}\sin^2\Theta_{\boldsymbol{B}}\cos 2\Phi_B$
$T_1^2(1,\hat{\boldsymbol{k}})=\pm\frac{\sqrt{3}}{2}\sin\Theta_{\boldsymbol{B}}(\cos\Theta_{\boldsymbol{B}}\cos 2\Phi_B-\mathrm{i}\sin 2\Phi_B)$
$T_2^2(1,\hat{\boldsymbol{k}})=\pm\frac{\sqrt{3}}{2}[\frac{1}{2}(1+\cos^2\Theta_{\boldsymbol{B}})\cos 2\Phi_B-\mathrm{i}\cos\Theta_{\boldsymbol{B}}\sin 2\Phi_B]$
$T_0^0(2,\hat{\boldsymbol{k}})=0$
$T_0^1(2,\hat{\boldsymbol{k}})=0$
$T_1^1(2,\hat{\boldsymbol{k}})=0$
$T_0^2(2,\hat{\boldsymbol{k}})=\pm\frac{3}{2\sqrt{2}}\sin^2\Theta_{\boldsymbol{B}}\sin 2\Phi_B$
$T_1^2(2,\hat{\boldsymbol{k}})=\pm\frac{\sqrt{3}}{2}\sin\Theta_{\boldsymbol{B}}(\cos\Theta_{\boldsymbol{B}}\sin 2\Phi_B+\mathrm{i}\cos 2\Phi_B)$
$T_2^2(2,\hat{\boldsymbol{k}})=\pm\frac{\sqrt{3}}{2}[\frac{1}{2}(1+\cos^2\Theta_{\boldsymbol{B}})\sin 2\Phi_B+\mathrm{i}\cos\Theta_{\boldsymbol{B}}\cos 2\Phi_B]$
$T_0^0(3,\hat{\boldsymbol{k}})=0$
$T_0^1(3,\hat{\boldsymbol{k}})=\sqrt{\frac{3}{2}}\cos\Theta_{\boldsymbol{B}}$
$T_1^1(3,\hat{\boldsymbol{k}})=-\frac{\sqrt{3}}{2}\sin\Theta_{\boldsymbol{B}}$
$T_0^2(3,\hat{\boldsymbol{k}})=0$
$T_1^2(3,\hat{\boldsymbol{k}})=0$
$T_2^2(3,\hat{\boldsymbol{k}})=0$

3.1 Stokes Vector of Coronal Emission Lines in the Weak-Field Approximation

Following the formalism of Casini (2002), the polarization profiles of an emission line of Einstein coefficient A_{ul} corresponding to the transition $J_u\to J_l$ in a multi-level atom, formed in the optically thin corona under the CRD hypothesis, have the following general form for their emission coefficients (where $i=0,1,2,3$ refer to the Stokes parameters I, Q, U, V, respectively):

$$\begin{aligned}\varepsilon_i(\omega,\hat{\boldsymbol{k}})&=\frac{\hbar\omega}{4\pi}N_uA_{ul}\sum_{KQ}\sum_{K_uQ_u}\sqrt{3(2J_u+1)(2K+1)(2K_u+1)}\\&\quad\times\sum_{M_uM_u'M_l}\sum_{qq'}(-1)^{1+J_u-M_u+q'}\begin{pmatrix}J_u&J_l&1\\-M_u&M_l&-q\end{pmatrix}\begin{pmatrix}J_u&J_l&1\\-M_u'&M_l&-q'\end{pmatrix}\\&\quad\times\begin{pmatrix}J_u&J_u&K_u\\M_u'&-M_u&-Q_u\end{pmatrix}\begin{pmatrix}1&1&K\\q&-q'&-Q\end{pmatrix}\\&\quad\times\operatorname{Re}\bigl\{T_Q^K(i,\hat{\boldsymbol{k}})\,\sigma_{Q_u}^{K_u}(J_u)\,\Phi(\omega_{J_uM_u,J_lM_l}-\omega)\bigr\},\end{aligned}\tag{2}$$

where ω and $\hat{k}$ are the angular frequency and wave vector, respectively, of the emission, N_u is the number density of atoms in the upper level J_u at the scattering point, and $\sigma^{K_u}_{Q_u}(J_u)$ (with $K_u = 0, \ldots, 2J_u$, and $Q_u = -K_u, \ldots, K_u$) is the *fractional* atomic polarization of the upper level (subject to the normalization condition $\sigma^0_0(J_u) = 1$). The latter must be obtained from the solution of the statistical equilibrium problem for the atom subject to the ambient illumination and coronal plasma collisions, in the atomic reference frame where the quantization axis is defined by the direction of the magnetic field vector (hereafter, the *magnetic reference frame*). We will not enter into the specifics of the statistical equilibrium problem in this paper, for which we refer the reader to the monograph on polarized line formation by Landi Degl'Innocenti and Landolfi (2004; see also Casini and Landi Degl'Innocenti 2008). We point out, however, that Eq. (2) is only applicable when quantum interference between different J-levels of the atomic structure is negligible. Typically, this is true whenever the fine structure frequency of the ionic species of interest is much larger than both the inverse lifetime of the atomic levels involved and the Larmor frequency corresponding to the ambient magnetic field.

The irreducible spherical tensors $T^K_Q(i, \hat{\boldsymbol{k}})$ relate the observer's reference frame for polarization measurements to the magnetic reference frame. Their components for the four Stokes parameters I, Q, U, V are given in Table 1 (Landi Degl'Innocenti and Landolfi 2004), assuming the convention for the definition of positive Stokes Q as described in Fig. 1. We note the sign choice in the tensor components for the linear-polarization Stokes parameters ($i = 1, 2$), which is determined by whether the atomic transition is permitted (E1) or forbidden (M1) (Casini and Judge 1999, see Eq. (6) in that paper).

Finally, the (normalized) complex line profile $\Phi(\omega_{ul} - \omega)$ is in general given by the Voigt profile for the real part and the Faddeeva-Voigt function for its imaginary part. In the coronal case, because of the dominance of the Doppler component of the profile due to the large plasma temperature, this profile can be safely approximated by a gaussian $\phi(\omega_{ul} - \omega)$ for the real part, and the Dawson integral function $\psi(\omega_{ul} - \omega)$ for its imaginary part.

Relying on the first-order power expansion of the line profiles in the small parameter $\omega_B/\Delta\omega_T$, as discussed above, after some tedious Racah algebra, we find (cf. Casini 2002)

$$\begin{aligned}\varepsilon_i(\omega, \hat{\boldsymbol{k}}) \approx{}& \frac{\hbar\omega}{4\pi} N_u A_{ul} \\ &\times \sum_{KQ} \Bigg[(-1)^{1+J_u+J_l} \sqrt{3(2J_u+1)} \begin{Bmatrix} 1 & 1 & K \\ J_u & J_u & J_l \end{Bmatrix} T^K_Q(i, \hat{\boldsymbol{k}})\, \sigma^K_Q(J_u)\, \phi(\omega_{ul} - \omega) \\ &+ \omega_B \sum_{K_u} (-1)^Q \sqrt{(2K+1)(2K_u+1)} \begin{pmatrix} K & K_u & 1 \\ Q & -Q & 0 \end{pmatrix} \Gamma^K_{K_u}(J_u, J_l) \\ &\times \mathrm{Re}\big\{ T^K_Q(i, \hat{\boldsymbol{k}})\, \sigma^{K_u}_Q(J_u)\, \Phi'(\omega_{ul} - \omega) \big\} \Bigg],\end{aligned} \tag{3}$$

where we indicated with $\Phi'(\omega)$ the derivative of the complex profile $\Phi = \phi + \mathrm{i}\psi$ with respect to its argument. If we introduce the normalized frequency $v = \omega/\Delta\omega_T$ in the line profile, then $\Phi' = (d/d\omega)\Phi = (1/\Delta\omega_T)(d/dv)\Phi$, through which the small parameter $\omega_B/\Delta\omega_T$ can be made to appear explicitly in the first-order term of Eq. (3).

The quantity

$$\Gamma^K_{K_u}(J_u, J_l) = \sqrt{3(2J_u+1)} \Bigg[(-1)^{K+K_u+1} g_{J_l} \sqrt{J_l(J_l+1)(2J_l+1)} \begin{Bmatrix} K & K_u & 1 \\ 1 & J_u & J_l \\ 1 & J_u & J_l \end{Bmatrix}$$

$$+(-1)^{J_u-J_l+1}\, g_{J_u}\sqrt{J_u(J_u+1)(2J_u+1)}\begin{Bmatrix} K & K_u & 1 \\ J_u & J_u & J_u \end{Bmatrix}$$

$$\times\begin{Bmatrix} K & 1 & 1 \\ J_l & J_u & J_u \end{Bmatrix}\Bigg], \tag{4}$$

which enters the first-order term in Eq. (3), plays the role of a "generalized" effective Landé factor for the transition $J_u \to J_l$. In fact, for $K_u = 0$, $K = 1$, which corresponds to the contribution of the Zeeman effect to the circular polarization of the emitted line, we have exactly $\Gamma_0^1(J_u, J_l) = \sqrt{2/3}\, \bar{g}_{J_u J_l}$, where

$$\bar{g}_{J_u J_l} = \frac{1}{2}(g_{J_u} + g_{J_l}) + \frac{1}{4}(g_{J_u} - g_{J_l})\big[J_u(J_u+1) - J_l(J_l+1)\big], \tag{5}$$

is the usual effective Landé factor for the transition $J_u \to J_l$, g_J being the Landé factor for the level J.

Equation (3) describes the polarized light emitted by a weakly magnetized plasma (in the usual sense $\omega_B/\Delta\omega_T \ll 1$) under arbitrary conditions of illumination and collisional interaction with the plasma. These excitation conditions are fully captured by the fractional atomic polarization tensor $\sigma_{Q_u}^{K_u}(J_u)$ at statistical equilibrium. In this regard, the form of Eq. (3) is very interesting, because it shows that, in this weak-field limit, only the first few multipole orders K_u of the fractional atomic polarization contribute to the polarization of the scattered radiation, *regardless of the complexity of the atomic system* from which the emission line of interest originates. In particular, because the geometric tensors $\mathcal{T}_Q^K(i, \hat{\boldsymbol{k}})$ are at most of rank $K = 2$, we see from Eq. (3) that we must only take into account multipole orders of atomic polarization up to $K_u = \min(2J_u, 3)$. In the literature on atomic polarization, the odd multipole orders K_u describe the *orientation* of the atomic system under the effect of the anisotropic excitation mechanisms, whereas the even orders describe its *alignment*. In our case, the atomic alignment is principally caused by the anisotropic illumination of the scattering atom, due to the fact that the upper solar atmosphere is preferentially irradiated by the underlying photosphere. Atomic orientation is generally produced when the incident radiation carries some amount of net circular polarization as seen in the atom's frame of rest.

The components of the atomic polarization tensor with $Q_u \neq 0$ account for coherence effects in the scattered polarized radiation that are produced when the magnetic sublevels M_u and M'_u "interfere" in a quantum-mechanical sense (cf. Eq. (2), in particular the third 3-j symbol). When the magnetic splitting is much larger than the typical level width due to the finite lifetime of the atomic levels, these quantum interferences are completely relaxed, and only the (real) components with $Q_u = 0$ of the atomic polarization tensor survive. This regime of "relaxation of atomic coherence" is controlled by the dimensionless parameter

$$\xi = \frac{g_{J_u}\omega_B}{A_{ul}}. \tag{6}$$

When ξ is of order unity (typically, $0.1 \lesssim \xi \lesssim 10$) the ambient magnetic field is most effective at relaxing atomic coherence. This regime corresponds to the *Hanle effect* (Hanle 1924), which produces characteristic polarization signatures in the scattered radiation that are sensitive to both strength and topology of the ambient magnetic fields, and can therefore effectively be employed as diagnostics of plasma magnetism (Casini et al. 2003; Casini and Landi Degl'Innocenti 2008; Trujillo Bueno 2010).

The above description essentially holds for both cases of E1 and M1 coronal lines. The distinction between the formation of these two types of diagnostics in the magnetized corona is practically determined by the domain of the parameter ξ. Because of the smallness of the coronal magnetic field—at least in regions above the quiet Sun—the formation of E1 lines is typically realized in the regime of the Hanle effect ($\xi \sim 1$). The subject has been comprehensively discussed in the review of Raouafi et al. (2016), and will not be considered here. In this work, instead, we want to expand upon the current theory of the formation of M1 polarized diagnostics (e.g., Casini and Judge 1999), which pertains to the magnetic regime of the so-called "saturated" Hanle effect ($\xi \gg 1$).

In the limit $\xi \gg 1$, the components $\sigma_0^{K_u}(J_u)$ of the fractional atomic polarization tensor in the reference frame of the magnetic field attain asymptotic (real) values, which are no longer sensitive to the ambient magnetic field strength, but only to its topology. In contrast, the components with $Q \neq 0$ all vanish in such limit. Because of the very small values of A_{ul} in the case of forbidden lines, this regime characterizes the formation of all the M1 emission lines that are routinely observed in the corona (e.g., Fe X, Fe XIII, Fe XIV, Ca XV), for practically any measurable field strength. For these lines, and assuming $\omega_B > 0$, Eq. (3) becomes,

$$
\begin{aligned}
\varepsilon_i(\omega, \hat{\boldsymbol{k}}) \approx \frac{\hbar\omega}{4\pi} N_u A_{ul} \sum_K \mathcal{T}_0^K(i, \hat{\boldsymbol{k}}) \\
\times \Biggl[(-1)^{1+J_u+J_l} \sqrt{3(2J_u+1)} \begin{Bmatrix} 1 & 1 & K \\ J_u & J_u & J_l \end{Bmatrix} \sigma_0^K(J_u)\, \phi(\omega_{ul}-\omega) \\
+ \omega_B \sum_{K_u} \sqrt{(2K+1)(2K_u+1)} \begin{pmatrix} K & K_u & 1 \\ 0 & 0 & 0 \end{pmatrix} \\
\times \Gamma_{K_u}^K(J_u, J_l)\, \sigma_0^{K_u}(J_u)\, \phi'(\omega_{ul}-\omega) \Biggr], \qquad (7)
\end{aligned}
$$

for $i = 0, 1, 2, 3$, and Table 1 can conveniently be used to identify the various contributions to the polarization of the scattered radiation. We summarize here the nature of these contributions for each Stokes parameter:

1. The intensity profile, Stokes I ($i = 0$), is dominated by the isotropic spontaneous de-excitation from the upper level ($K_u = K = 0$), characterized by a profile of Gaussian shape, with a correction to the signal amplitude due to the possible presence of atomic alignment in the upper state ($K_u = K = 2$). In the presence of atomic orientation ($K = 0$, $K_u = 1$ and $K = 2$, $K_u = 1, 3$), the intensity profile also acquires a magnetic induced signal modulation, which is antisymmetric with respect to the line center.
2. Both linear polarization components (Stokes Q and U, i.e. $i = 1, 2$) are dominated by the atomic alignment ($K_u = K = 2$), with a profile shape that mimics the intensity profile. As in the case of the intensity profile, the presence of atomic orientation ($K = 2$, $K_u = 1, 3$) introduces an antisymmetric modulation of the linear polarization signal that is proportional to the strength of the ambient magnetic field.
3. The circular polarization signal, Stokes V ($i = 3$), is a combination of the longitudinal Zeeman effect ($K_u = 0$; plus a correction due to atomic alignment, $K_u = 2$), with its characteristic antisymmetric profile shape, and of the possible contribution of atomic orientation ($K_u = K = 1$), which adds a signal of Gaussian shape that is independent of the magnetic field.

We note that these general properties apply also to the case of permitted transitions, with the additional complication that the atomic coherence among magnetic sublevels is only partially relaxed by the ambient magnetic field, for magnetic strengths that fall within the regime of the Hanle effect.

For convenience, we explicitly derive the expressions for the four Stokes parameters in the case of M1 emission lines, for the lowest principal orders of atomic polarization ($K_u \leq 2$):

$$I(\omega, \hat{\boldsymbol{k}}) \approx \frac{\hbar\omega}{4\pi} N_u A_{ul} \left\{ \left[1 + \frac{1}{2}\big(3\cos^2\Theta_{\boldsymbol{B}} - 1\big) D(J_u, J_l)\, \sigma_0^2(J_u) \right] \phi(\omega_{ul} - \omega) \right.$$
$$- \omega_B \left[\Gamma_1^0(J_u, J_l) - \frac{1}{2}\big(3\cos^2\Theta_{\boldsymbol{B}} - 1\big) \Gamma_1^2(J_u, J_l) \right]$$
$$\left. \times\, \sigma_0^1(J_u)\, \phi'(\omega_{ul} - \omega) \right\}, \tag{8a}$$

$$\begin{Bmatrix} Q \\ U \end{Bmatrix} (\omega, \hat{\boldsymbol{k}}) \approx -\frac{\hbar\omega}{4\pi} N_u A_{ul} \frac{3}{2} \sin^2\Theta_{\boldsymbol{B}} \begin{Bmatrix} \cos 2\Phi_B \\ \sin 2\Phi_B \end{Bmatrix}$$
$$\times \big[D(J_u, J_l)\, \sigma_0^2(J_u)\, \phi(\omega_{ul} - \omega)$$
$$- \omega_B\, \Gamma_1^2(J_u, J_l)\, \sigma_0^1(J_u)\, \phi'(\omega_{ul} - \omega) \big], \tag{8b}$$

$$V(\omega, \hat{\boldsymbol{k}}) \approx \frac{\hbar\omega}{4\pi} N_u A_{ul} \cos\Theta_{\boldsymbol{B}} \big\{ E(J_u, J_l)\, \sigma_0^1(J_u)\, \phi(\omega_{ul} - \omega)$$
$$- \omega_B \big[\bar{g}_{J_u J_l} - \sqrt{3}\, \Gamma_2^1(J_u, J_l)\, \sigma_0^2(J_u) \big] \phi'(\omega_{ul} - \omega) \big\}, \tag{8c}$$

where we introduced the following parameters for the atomic transition $J_u \to J_l$,

$$D(J_u, J_l) = \frac{2}{\sqrt{5}} \frac{(-1)^{J_u - J_l + 1} (2J_u + 1)\sqrt{J_u(J_u+1)(2J_u-1)(2J_u+3)}}{(J_u + J_l)(J_u + J_l + 1)(J_u + J_l + 2)(J_u - J_l + 1)!(J_l - J_u + 1)!},$$

$$E(J_u, J_l) = 2\sqrt{3} \frac{(-1)^{J_u - J_l} (2J_u + 1)\sqrt{J_u(J_u+1)}\, [J_u(J_u+1) - J_l(J_l+1) - 1]}{(J_u + J_l)(J_u + J_l + 1)(J_u + J_l + 2)(J_u - J_l + 1)!(J_l - J_u + 1)!},$$

and explicitly wrote out the effective Landé factor (5) in the expression for Stokes V.

Equations (8a)–(8c) generalize previous results on the polarization of the forbidden coronal emission lines (Casini and Judge 1999; Casini and Landi Degl'Innocenti 2008) to the case when atomic orientation may be important. It can be expected that some contribution of atomic orientation may be present in coronal emission that is produced above strong active regions, as the incident radiation could be strongly circularly polarized, if it corresponds to a magnetically sensitive photospheric line. As an example, the Fe XIII coronal lines lie in the proximity of two strong photospheric lines of Si I, both of which are red-shifted with respect to the rest wavelengths of the Fe XIII lines. The velocity separation between the coronal Fe XIII line and these photospheric Si I lines is 72 km/s and 47 km/s, respectively, for Fe XIII 1074.7 nm and Fe XIII 1079.8 nm, which fall within the range of speeds that have been observed in coronal plasma downflows (e.g., Tripathi et al. 2006, and references therein). Thus, it cannot be excluded that the irradiation of the coronal Fe XIII ions from the underlying photosphere, in the presence of strongly magnetic active regions, may be affected by both the intensity and polarization of the Si I lines, possibly leading to Doppler dimming and the generation of atomic orientation in the Fe XIII ions (see Sect. 3.2).

Whether permitted or forbidden spectral lines are employed for coronal diagnostics, the importance of a proper modeling of the atomic polarization induced in the scattering atom, in order to infer the correct plasma properties, is fairly evident, because of the very nature of the polarization signals of the scattered radiation that we have outlined. This implies that the local conditions of illumination of the scattering plasma must be known sufficiently well, in order to set up and solve a statistical equilibrium problem for the irradiated atom that is a good representation of the observed phenomenon. Similarly, successful modeling of the collisional effects (both polarizing and depolarizing) on the scattered radiation under the local plasma conditions is also necessary.

Of course, such heavy reliance on a correct modeling of the photospheric boundary conditions and of the local collisional rates greatly exacerbates the difficulty of the inversion problem of coronal emission lines.

3.2 Models of Coronal Plasma Illumination

Often (see, e.g., Casini and Judge 1999; Casini and Landi Degl'Innocenti 2008) the simplifying assumption can be made that the incident radiation field is unpolarized and cylindrically symmetric around the local vertical to the Sun drawn from the scattering atom. While this can be a fairly good representation of the illumination conditions of the quiet-Sun corona, such an approximation may be inapplicable when strong active regions are present in the photosphere underlying the observed coronal region, as pointed out earlier.

To better illustrate this point, we consider the simplest case of a two-level atom with unpolarizable ground state. For this model atom, it can be shown (e.g., Casini and Landi Degl'Innocenti 2008) that the lowest orders of the fractional atomic polarization tensor, $\sigma_0^K(J_u)$, $K = 0, 1, 2$, which appear in Eqs. (8a)–(8c), are directly proportional to the tensor components $(J_0^K)_B$ of the radiation field expressed in the reference frame of the magnetic field. In turn, these tensor components are derived from the radiation tensor components $(J_Q^K)_S$ expressed in the solar reference frame via the tensor rotation

$$(J_0^K)_B = \sum_Q D_{Q0}^K(R_{SB})\,(J_Q^K)_S, \tag{9}$$

where $D_{QQ'}^K(R_{SB})$ is the quantum rotation matrix of order K, associated with the rotation operator R_{SB} that transforms the solar reference frame into the reference frame of the magnetic field.

The radiation tensors $(J_Q^K)_S$ can be expressed in terms of the Stokes vector of the incident radiation, through the relation

$$\bigl(J_Q^K(\omega)\bigr)_S = \oint \frac{d\hat{\boldsymbol{k}}}{4\pi} \sum_{i=0}^{3} T_Q^K(i,\hat{\boldsymbol{k}})_S\, S_i(\omega,\hat{\boldsymbol{k}}), \tag{10}$$

where the geometric tensors $T_Q^K(i,\hat{\boldsymbol{k}})_S$ are formally identical to the E1 type of the tensors $T_Q^K(i,\hat{\boldsymbol{k}})$ given in Table 1, with the substitution $(\Theta_B, \Phi_B) \to (\vartheta, \varphi)$, where $\hat{\boldsymbol{k}} \equiv (\vartheta, \varphi)$ is the direction of the incident light within the radiation cone subtended by the solar surface at the scattering center, as measured in the reference frame of the local vertical.

Thus, the evaluation of the tensor components $(J_Q^K)_S$ generally requires knowledge of the true height of the observed region over the solar surface, and a full description of the (polarized) radiation field from the underlying solar atmosphere that irradiates the ion. The distribution and polarization of the incident radiation will accordingly affect the various contributions to the atomic polarization produced in the scattering plasma, as exemplified by Eq. (9). In the simplest model of a non-polarized ($i = 0$) and cylindrically symmetric radiation field ($S_0(\omega, \hat{\boldsymbol{k}}) \neq f(\varphi)$), the only two nonvanishing components of $(J^K_Q)_S$ are $(J^0_0)_S$ and $(J^2_0)_S$, which describe, respectively, the mean intensity (independent of the reference frame) and the anisotropy of the incident radiation in the reference frame of the local vertical. If, on the other hand, the incident radiation is strongly circularly polarized ($i = 3$), it may be possible to generate nonvanishing components $(J^1_Q)_S$, leading to the appearance of atomic orientation in the scattering ions.

Typically, circular polarization in the incident radiation is produced by the Zeeman effect from strong photospheric magnetic fields, e.g., such as those found in sunspots. Then, the presence of bulk motions in the observed plasma region may shift the peak amplitude of the (antisymmetric) Stokes-V signal towards the central wavelength of the coronal line, so that the ions are illuminated by radiation with a broadband component of circular polarization, i.e., with non-vanishing $(J^1_Q)_S$ tensor components.

As a measure of this effect, we can estimate the ratio $(J^1_0)_B/(J^0_0)_B$ for a given diameter d of the polarizing region on the solar surface and the height h of the scattering plasma region in the corona, using simple geometric considerations (see, e.g., Sect. 3.3 of Casini and Judge 1999), and the expressions of the geometric tensors $\mathcal{T}^K_Q(i, \hat{\boldsymbol{k}})$ in Table 1. Neglecting for simplicity the effect of the center-to-limb variation (CLV) on $(J^0_0)_B$, we find

$$\frac{(J^1_0)_B}{(J^0_0)_B} \approx \sqrt{\frac{3}{8}}\, p_V \cos\vartheta_B \,\frac{d^2}{d^2+4h^2}\left[1-\sqrt{1-\left(\frac{R_\odot}{R_\odot+h}\right)^2}\,\right]^{-1}, \tag{11}$$

where we indicated with p_V the fractional circular polarization produced by the magnetic region, and with ϑ_B the inclination of the coronal magnetic field from the local vertical. As an example, for a medium-size sunspot (~ 20 Mm diameter) producing strongly circularly polarized radiation ($V/I \sim 25\%$), the maximum atomic orientation at a coronal height of $0.1\,R_\odot$ directly above the sunspot can be expected to be approximately 0.5%, which is further weighted by $\cos\vartheta_B$. This value can be compared with the expected maximum atomic alignment at the same height, which is approximately 20%, if we assume no CLV and neglect the additional effects of depolarizing collisions. By inspection of Eqs. (8a)–(8c), we see that the contribution of the atomic orientation $\sigma^1_0(J_u)$ to the intensity and the Stokes-Q and U profiles of a coronal line is additionally weighted by the strength of the weak coronal magnetic field. So it can be concluded that the effects of radiation-induced atomic orientation on the linearly polarized emission of coronal lines, under typical circumstances, only bring a negligible correction to the dominant contribution from the atomic alignment. In the case of Stokes V, instead, the contribution of atomic orientation to the signal may compete with the weak anti-symmetric signature of the longitudinal Zeeman effect, considering that under typical coronal conditions $\omega_B/\Delta\omega_T$ is only a small fraction of a percent (cf. discussion of (1)).

In conclusion, our estimate suggests that the circular polarization of the scattered radiation in coronal lines may carry a physical symmetric signature of the atomic orientation, which may be produced by spectrally neighboring magnetic sensitive lines formed in strong active regions of the underlying photosphere, and which may be brought within resonance with the coronal lines by the presence of plasma upflows and downflows in the corona.

It must be observed that other contributions to the statistical equilibrium of the emitting ions from various processes in the coronal plasma, such as depolarizing collisions, can affect and reduce the importance of atomic orientation to the Stokes-V contribution. In addition, the presence of mixed polarities on the solar surface will generally reduce the amount of circular polarization in the incoming light integrated over the photosphere. However, even enough residual signed magnetic flux capable to generate as little as 1% circular polarization in the radiation emitted from a medium-size active region, can trigger enough orientation in the atomic levels to produce detectable asymmetries in the scattered Stokes-V profiles. Such signals are of physical origin, and should therefore not be dismissed as instrumental effects of polarization cross-talk. A similar physical mechanism has indeed been invoked in recent work by Martínez González et al. (2012) to interpret the "anomalous" circular polarization signals in the chromospheric He I line at 1083 nm, observed in some solar prominences.

A different mechanism by which atomic orientation can be produced in the coronal plasma becomes available in the case of ions with complex atomic structures, where the weak coronal magnetic field can induce energy "crossing" between magnetic sublevels pertaining to different fine-structure levels. In this type of J–J' level interference, the atomic alignment generated by the anisotropy of the incident radiation field can be transformed internally into atomic orientation (Kemp et al. 1984). This *alignment-to-orientation* transfer mechanism is responsible, as an example, for the anomalous Stokes-V signals observed in the chromospheric lines of He I and Na I in the 588–590 nm spectral region (e.g., Casini and Landi Degl'Innocenti 2008). We point out that the formalism presented in this article cannot account for this mechanism, since the polarized emissivity described by the original Eq. (2) strictly applies to atomic structures where quantum interferences between different J-levels can be neglected.

The presence of bulk velocities in the coronal plasma may increase the complexity of the modeling and inversion of coronal spectro-polarimetric observables, even when the incident radiation is non-polarized. If the incident radiation field has no significant spectral structure across the coronal line width, then the photo-excitation of the scattering ions is largely independent of the plasma velocity, since the ions see the same radiation intensity from the Doppler shifted solar spectrum. This condition evidently breaks down when the incident radiation has instead a significant spectral modulation around the emission line. One possibility, which is commonly observed in the case of permitted UV transitions in the corona, is that the observed line in the corona is resonant with the corresponding emission line in the upper chromosphere and transition region, from which most of the photo-excitation of the coronal ions is produced. When the ions have a preferential velocity because of bulk motions in the coronal plasma, the incident radiation may drop out of resonance with the scattered radiation, if the Doppler shift is comparable with the width of the spectral line in the incident radiation. Accordingly, the ions see a "dimmed" source of radiation in the direction of the plasma velocity (see Fig. 2). This generally breaks the cylindrical symmetry of the incident radiation from the underlying atmosphere around the local vertical, even in the absence of strong inhomogeneities of the solar surface, such as a large active region. The higher complexity of the surface radiation field folds into Eq. (9), which is needed to set up the statistical equilibrium problem for the coronal ions embedded in the coronal magnetic field. For a more comprehensive description of Doppler dimming and related mechanisms, including the effect of anisotropic plasma temperature distributions, we refer the reader to the papers by Sahal-Brechot et al. (1998), Raouafi et al. (1999), Fineschi (2001), and Khan et al. (2011).

3.3 Inversion Problem for M1 Coronal Emission Lines

Introducing the frequency-integrated Stokes parameters, $\bar{S}_i = \int d\omega\, S_i(\omega, \hat{\boldsymbol{k}})$, and owing to the fact that $\int d\omega\, \phi'(\omega_{ul} - \omega) = 0$ for a Gaussian profile, from Eqs. (8a)–(8c) we easily see

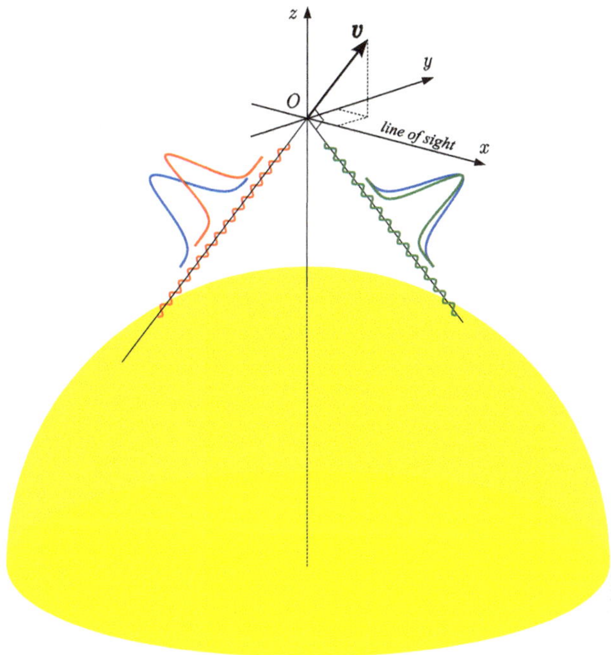

Fig. 2 Schematic diagram illustrating the origin of Doppler dimming. $(O; x, y, z)$ represents the reference frame of the solar local vertical through the scattering point O, with the x-axis directed towards the observer, and $\langle y, z \rangle$ defining the POS. The *blue bell curves* represent the absorption profile of the scatterer in the atomic rest frame for the coronal line of interest, whereas the *red* and *green bell curves* represent spectral line emission from the underlying atmosphere resonant with the coronal line. If the scatterer at O moves with velocity $\boldsymbol{v}$ due to bulk plasma motions, it sees a Doppler shifted emission from the underlying atmosphere in the direction of its motion (*red curve*), whereas any radiation received from the Sun in a plane normal to $\boldsymbol{v}$ is unshifted (*green curve*). Because of the lower absorption cross-section to Doppler shifted radiation, the illumination of the scatterer from the Sun cannot be assumed to be cylindrically symmetric around z, unless $\boldsymbol{v} \| z$

that

$$\bar{U}/\bar{Q} = \tan 2\Phi_{\boldsymbol{B}}, \tag{12a}$$

$$\sqrt{\bar{Q}^2 + \bar{U}^2} \propto \left|\sigma_0^2(J_u)\right| \sin^2\Theta_{\boldsymbol{B}}, \tag{12b}$$

$$\bar{V} \propto \sigma_0^1(J_u) \cos\Theta_{\boldsymbol{B}}. \tag{12c}$$

In the case when the atomic orientation $\sigma_0^1(J_u) = 0$, Eq. (12c) does not carry any information, but the Stokes V and I profiles are related via a generalized form of the well-known *magnetograph formula* (e.g., Landi Degl'Innocenti and Landolfi 2004)

$$V(\omega, \hat{\boldsymbol{k}}) = -\omega_B \cos\Theta_{\boldsymbol{B}}\, \bar{\Gamma}_{J_u J_l} \frac{dI(\omega, \hat{\boldsymbol{k}})}{d\omega}, \tag{13}$$

where

$$\bar{\Gamma}_{J_u J_l} = \frac{\bar{g}_{J_u J_l} - \sqrt{3}\,\Gamma_2^1(J_u, J_l)\,\sigma_0^2(J_u)}{1 + \frac{1}{2}(3\cos^2\Theta_{\boldsymbol{B}} - 1)\, D(J_u, J_l)\,\sigma_0^2(J_u)} \tag{14}$$

is a modified effective Landé factor, which differs from the usual version through a correction determined by the presence of atomic alignment (Casini and Judge 1999).

It is possible in principle to retrieve the full vector information for $\boldsymbol{B}$, if one takes into account both the linear and circular polarization signals of a line. However, one generally needs a preliminary knowledge of $\sigma_0^1(J_u)$ and $\sigma_0^2(J_u)$, which are determined by the illumination conditions of the emitting plasma, the rates of collisional polarization and depolarization of the upper level, and the inclination of the magnetic field from the local vertical, ϑ_B. As a consequence, one generally must rely on simultaneous polarimetric observations of multiple spectral lines, forming in the same plasma region, in order to collect a number of independent diagnostic measurements adequate for the number of model parameters that need to be inferred. In many cases, knowledge of the first few frequency moments of the observed Stokes profiles may be sufficient to set up the inversion problem. In the case of Gaussian line profiles, these moments are easily expressed analytically in terms of the Doppler shift and Doppler width of the line, which therefore can be included as two additional inversion unknowns (e.g., Plowman 2014).

The inversion problem for magnetically sensitive coronal lines is further complicated by the possible presence of polarization signal ambiguities that are intrinsic to the line formation process. The best known of these ambiguities corresponds to the degeneracy of the linear polarization signals, Stokes Q and U, between magnetic configurations where the orientation of the field on the POS differs by 180°. A more subtle ambiguity is instead associated with the fact that the sign of $\sigma_0^2(J_u)$ is generally unknown, and therefore the actual sign of $\cos 2\Phi_B$ and $\sin 2\Phi_B$ in the expressions of Stokes Q and U is undetermined. As a consequence, when the sign of $\sigma_0^2(J_u)$ is not known a priori, we can only determine the direction of the field on the POS with an ambiguity of 90° (Casini and Judge 1999). This is also immediately evident from considering Eq. (12a). For example, for the model of a two-level atom with non-polarized ground state mentioned earlier, if we can assume that the atomic alignment is originated only because of the anisotropy of a cylindrically symmetric radiation field around the solar local vertical (see Fig. 1), we have (see Eq. (9); also Casini and Judge 1999; Casini and Landi Degl'Innocenti 2008)

$$\sigma_0^2(J_u) \propto \frac{1}{2}\left(3\cos^2\vartheta_B - 1\right)\frac{(J_0^2)_S}{(J_0^0)_S} \tag{15}$$

where the geometric factor corresponds to the rotation matrix component $D_{00}^2(R_{SB})$. We then see that $\sigma_0^2(J_u)$ changes sign at the so-called *Van Vleck angle*, defined by $\vartheta_{\mathrm{VV}} \equiv \arccos(1/\sqrt{3}) \approx 54.7°$, and at its supplementary angle, $180° - \vartheta_{\mathrm{VV}}$. This is illustrated in Fig. 3, for a particular scattering geometry (see caption to the figure for more details). We remark that the use of all the information contained in the Stokes vector of the emitted radiation is a necessary condition (although not always sufficient) for the resolution of the 90° ambiguity in practical cases.

In the presence of Doppler dimming, Eq. (9) generally leads to a more involved dependence of atomic alignment on the geometry of the coronal magnetic field than Eq. (15). This implies that the null points of linear polarization may occur for magnetic inclinations significantly different from the Van Vleck angle, or even the absence of clear null points in the linear polarization pattern.

Plowman (2014) has investigated the intrinsic limitations of "single-point" inversions based on simultaneous observations of the Fe XIII lines at 1074.7 and 1079.9 nm. This study assumed an ideal situation where the observed signals originate from a well localized region of the corona, so that no LOS smearing of those signals is produced. The adopted model also

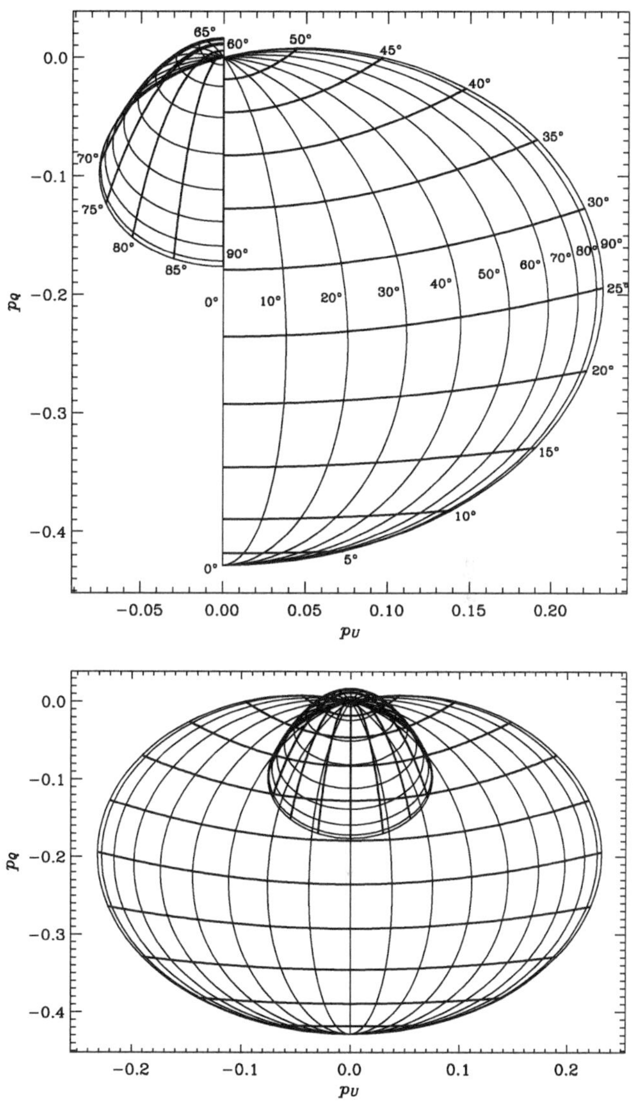

Fig. 3 Hanle diagrams of $p_U = \bar{U}/\bar{I}$ against $p_Q = \bar{Q}/\bar{I}$, for the Fe XIII λ1074.7 nm line in a 90° scattering event. The atom is illuminated by a radiation field with anisotropy factor $w(\omega_{10}) = 0.5$. The magnetic strength is $B = 1$ G, and for this line it already ensures the complete relaxation of the atomic coherences. The magnetic field inclination, $\vartheta_{\boldsymbol{B}}$, varies from 0° to 90°, and the azimuth, $\varphi_{\boldsymbol{B}}$, from −90° to 90° (see Fig. 1). The *thick lines* correspond to curves of constant $\vartheta_{\boldsymbol{B}}$, whereas the *thin lines* correspond to curves of constant $\varphi_{\boldsymbol{B}}$. In the *top* figure we show only the region of positive azimuth in greater detail. (The curve for $\vartheta_{\boldsymbol{B}} = 55°$ is not shown, because it practically reduces to the point of zero polarization, being very close to the Van Vleck angle.) The 90° ambiguity can only occur where a point (p_U, p_Q) in the diagram can belong to two different pairs of isocurves of $\vartheta_{\boldsymbol{B}}$ and $\varphi_{\boldsymbol{B}}$ (see *bottom* figure)

used the simplifying assumption of a nonpolarized and cylindrically symmetric illumination of the scattering atoms, so that the contribution from atomic orientation in Eqs. (8a)–(8c) can be dropped. That work relied on an analytic inversion of the wavelength-integrated polarization signals from the two iron lines, which, after taking into account information redundancy between the two lines, allows the inference of seven parameters characterizing the local properties of the emitting plasma region: the three components of the vector magnetic field, and the atomic populations and fractional alignment $\sigma_0^2(J_u)$ of the upper levels of the two transitions. In particular, the atomic alignment encodes the information of the anisotropy of the incident radiation, and of the local plasma conditions that determine the degree of depolarization due to plasma collisions. Despite the important simplifying assumptions of the adopted model, the results of Plowman (2014) show that a very high

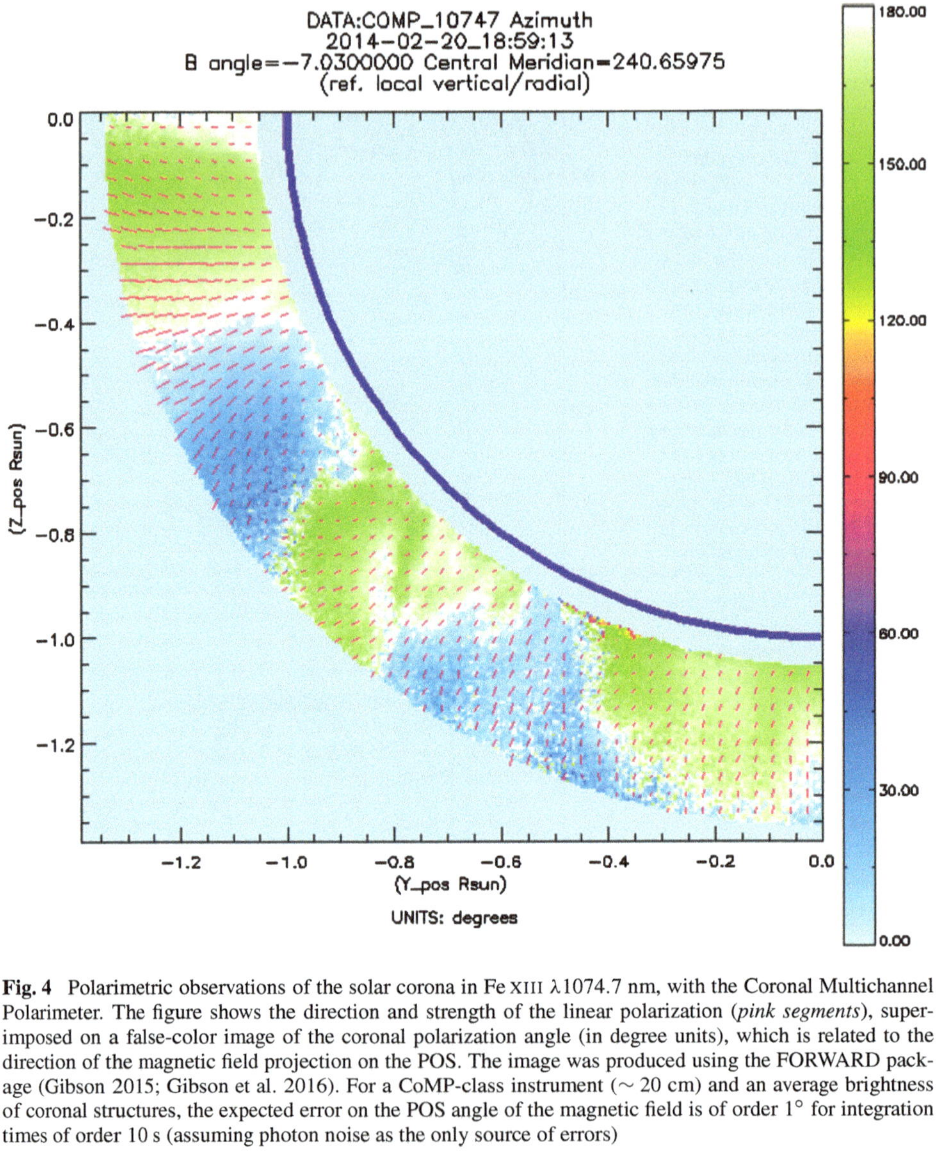

Fig. 4 Polarimetric observations of the solar corona in Fe XIII λ1074.7 nm, with the Coronal Multichannel Polarimeter. The figure shows the direction and strength of the linear polarization (*pink segments*), superimposed on a false-color image of the coronal polarization angle (in degree units), which is related to the direction of the magnetic field projection on the POS. The image was produced using the FORWARD package (Gibson 2015; Gibson et al. 2016). For a CoMP-class instrument (~ 20 cm) and an average brightness of coronal structures, the expected error on the POS angle of the magnetic field is of order 1° for integration times of order 10 s (assuming photon noise as the only source of errors)

S/N ratio of the observed signals (of the order of 10^5) may be necessary in order to reliably infer the magnetic properties of the observed corona, even in such a highly idealized case.

Figure 4 shows recent observations of the solar corona with the Coronal Multichannel Polarimeter (CoMP; Tomczyk et al. 2008), deployed at Mauna Loa Solar Observatory of the National Center of Atmospheric Research. The figure displays the direction and strength of the linear polarization at Fe XIII λ1074.7 nm (as inferred from Eqs. (12a) and (12b), respectively), superimposed on a false-color image of the polarization angle. We note the presence of regions of vanishing polarization, which indicate the possible occurrence of longitudinal fields ($\Theta_B \approx 0°$, $180°$) or of the Van Vleck effect ($\vartheta_B \approx \vartheta_{VV}$, $180° - \vartheta_{VV}$).

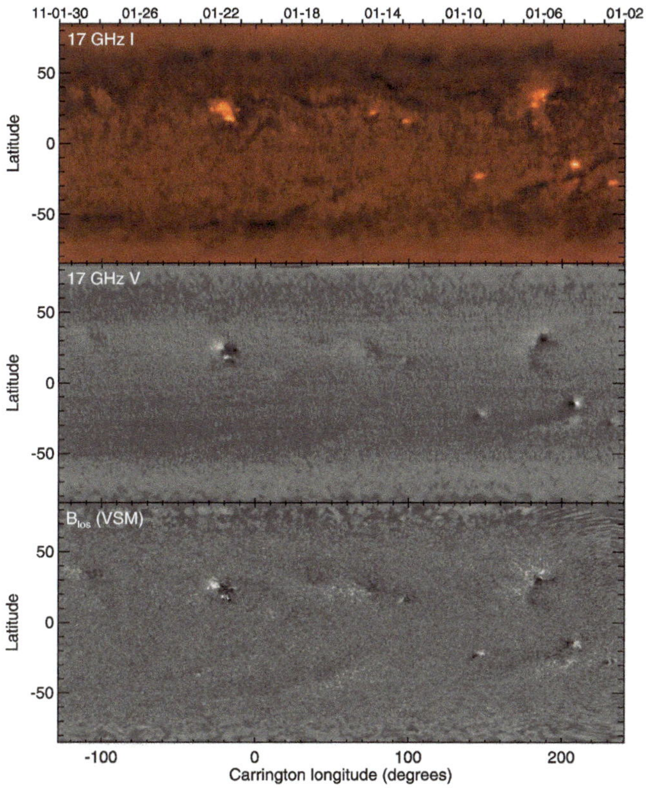

Fig. 5 A comparison of a radio Carrington plot with a line-of-sight photospheric magnetic field Carrington plot. The radio data are generated from sensitive daily maps from the Nobeyama Radioheliograph at 17 GHz: the *upper* plot shows total intensity, while the *middle* plot shows the circularly polarized flux. The *lower* panel shows the traditional magnetic Carrington plot using data from the VSM at NSO. The period (as labelled at the top of the plot) runs from 2011 Jan 2 to Jan 30. All three panels use a square-root transfer function to emphasize weaker features. The VSM map is regridded to a resolution element of size 0.2 heliographic degrees ($\sim 3''$), while the resolution in the 17 GHz maps is $12''$. Brightness temperatures in Stokes I range from -3 to 22×10^3 K (after the background disk level of 10^4 K has been subtracted), while the display range in Stokes V is -900 to 900 K and in B is -400 to $+400$ G

4 Radio Diagnostics

Any doubts that radio data contain information on solar magnetic fields should be erased by inspection of the lower two panels of Fig. 5. The panels compare a month of synoptic measurements of circularly polarized (Stokes V) radio flux at 17 GHz (middle panel, from the Nobeyama Radioheliograph) with photospheric line-of-sight magnetic measurements from the Vector Spectromagnetograph at the National Solar Observatory (lower panel), using the Carrington format of extracting data at central meridian passage. The radio data reproduce the pattern of the photospheric magnetic field image remarkably well, even down to weak plages regions. The radio data appear smoother both because the spatial resolution ($12''$) is poorer than in the VSM map and because more data have been averaged together in order to reduce the noise level in the V data. The radio emission at 17 GHz does not see below the 10^4 K layer in the chromosphere, so the 17 GHz circular polarization in this figure results from a combination of optically thick chromospheric and optically thin coronal contributions.

Radio emission cares about magnetic field both because the acceleration of electrons under the Lorentz force creates radiation, and because the presence of a magnetic field in a plasma breaks the degeneracy of the two propagating transverse electromagnetic modes. In magnetoionic theory (e.g., Melrose 1980), except when looking perpendicular to $\boldsymbol{B}$, the natural propagating electromagnetic radiation modes of a magnetized plasma are predominantly circularly polarized. Electrons gyrating in magnetic fields interact most strongly with the mode whose sense of circular polarization matches their sense of rotation, labelled the *extraordinary* or x mode. The other natural electromagnetic mode is the *ordinary* or o mode, and has the opposite sign of circular polarization. This difference between the two radiation modes is important irrespective of the specific radiation mechanism involved.

There are four such mechanisms that matter for solar radio emission:

1. *Gyroresonance emission* is the emission from nonrelativistic thermal plasma at low harmonics of the electron gyrofrequency, $f_B = 2.8B$ MHz, where B is expressed in gauss. This mechanism is strong wherever $B > 300$ G in the corona and produces optically thick emission in active regions which may be highly polarized. Typically the 3rd harmonic is responsible for emission at a given observing frequency.
2. *Bremsstrahlung* derives opacity from the acceleration of electrons colliding with ions. It occurs throughout the solar atmosphere and predominantly produces thermal emission. As demonstrated below, bremsstrahlung is strongest from cool dense material. Thus, active regions are typically optically thick in the corona (at million-degree brightness temperatures) up to frequencies of a few GHz (e.g., White 1999) due to this process.
3. *Gyrosynchrotron emission* is the same physical mechanism as gyroresonance emission, but it applies to the emission by mildly relativistic electrons at harmonics 10–100 of the gyrofrequency. The higher energy of the electrons results in a broadening of the cyclotron resonances that creates a continuum mechanism rather than narrow resonance lines. Generally only nonthermal electrons accelerated in flares have sufficient energy to radiate by this mechanism, which has a broad frequency response with a spectral peak typically in the 5–20 GHz range.
4. *Plasma emission* is produced by energetic electrons at low harmonics of the plasma frequency $f_p = 9000 {n_e}^{0.5}$. This is a multi-step process that starts with the emission of electrostatic Langmuir waves with frequency f_p which are subsequently converted to electromagnetic radiation at frequencies f_p and $2f_p$. This mechanism is most effective when $f_p \gg f_B$ and thermal absorption tends to suppress it at frequencies above a few GHz, so it is predominantly a low-frequency mechanism. It can produce very bright highly polarized bursts at low frequencies. In the presence of magnetic fields, fundamental plasma emission is highly polarized in the sense of the *ordinary* or o mode due to the fact that the x mode is cut-off at a frequency above f_p.

An important aspect of the interpretation of solar radio emission is that opacities may be significant and radiative transfer must be taken into account. A convenient simplification is to convert the radio flux S (units of power per area per Hertz) in any given spatial resolution element into a brightness temperature, T_B, as follows:

$$S = \frac{2k_B f^2 T_B}{c^2} \Delta\Omega \tag{16}$$

where f is the frequency, k_B is Boltzmann's constant, c is the speed of light and $\Delta\Omega$ is the solid angle of the resolution element. The factor of 2 applies when both polarizations are combined. Then radiative transfer for brightness temperature along a ray path through the solar atmosphere is governed by the equation

$$\frac{dT_B(s)}{ds} = \kappa(s)\left[T(s) - T_B(s)\right] \tag{17}$$

where s is distance along the path, κ is the absorption coefficient per unit length and T is the thermal temperature equivalent of the local energy distribution. For an isothermal corona at temperature $T_{\rm cor}$, we can write the solution of the radiative transfer through the atmosphere as

$$T_B = T_{\rm chr}\exp^{-\tau_{\rm cor}} + T_{\rm cor}\left(1 - \exp^{-\tau_{\rm cor}}\right) \tag{18}$$

where $T_{\rm chr}$ is the temperature of the layer of the chromosphere that is optically thick at the observing frequency and $\tau_{\rm cor}$ is the optical depth of the corona (i.e., the integral of $\kappa(s)$ through the corona): so when $\tau_{\rm cor} \ll 1$ we measure the chromospheric temperature with a coronal contribution added, but when $\tau_{\rm cor} \gg 1$ we measure the coronal temperature.

In addition to the fact that we understand the emission mechanisms fairly well, the utility of radio observations for the study of coronal magnetic fields results from the fact that the range of electron gyrofrequencies corresponding to magnetic field strengths in the corona exactly matches the range of frequencies accessible to radio astronomers from the ground: the field strength range 4–2000 G corresponds to the gyrofrequency range 10 MHz–6 GHz. We now discuss specific mechanisms in turn, spending most time on gyroresonance emission which holds the most promise as a useful diagnostic.

4.1 Gyroresonance Emission

The detailed form of gyroresonance opacity is derived in a number of places (e.g., Zheleznyakov 1970; Melrose 1980; Robinson and Melrose 1984) and its application to measuring coronal fields has been reviewed previously (e.g., Zlotnik 1968b; Gary and Hurford 1994; White and Kundu 1997; White 2004). At a given observing frequency f, magnetic field strengths whose corresponding gyrofrequencies $f_B = 2.8\times 10^6\,B$ (with B measured in gauss) correspond to f divided by an integer (the harmonic number), i.e., $f_B = f/s$ where $s = 1, 2, 3, 4$, can contribute opacity. Each such magnetic field strength will be located at a different height in the corona along an observer's line of sight, and each contributes opacity according to the following approximation:

$$\tau_{x,o}(s, f, \Theta_B) = 0.0133\,\frac{n_e\,L_B(\Theta_B)}{f}\,\frac{s^2}{s!}\left(\frac{s^2\sin^2\Theta_B}{2\mu}\right)^{s-1} F_{x,o}(\Theta_B) \tag{19}$$

where Θ_B is the angle between the magnetic field direction and the line of sight, and $L_B(\Theta_B)$ is the scale length of the magnetic field $(B/\frac{\partial B}{\partial l})$ evaluated along the line of sight. $F_{x,o}(\Theta_B)$ is a function of angle which is of order unity for the x mode near $\Theta_B = 90°$, but decreases sharply at smaller Θ_B, and is smaller in the o mode than in the x mode. At angles Θ_B away from 90° it is often approximated by

$$F_{x,o}(\Theta_B) \approx (1 - \sigma\cos\Theta_B)^2 \tag{20}$$

with $\sigma = -1$ for the x mode and $\sigma = 1$ for the o mode. However, this approximation is really only appropriate when the two natural electromagnetic modes are perfectly circularly polarized, and at the low harmonics ($s = 1, 2, 3, 4$) relevant to the corona this is rarely a good

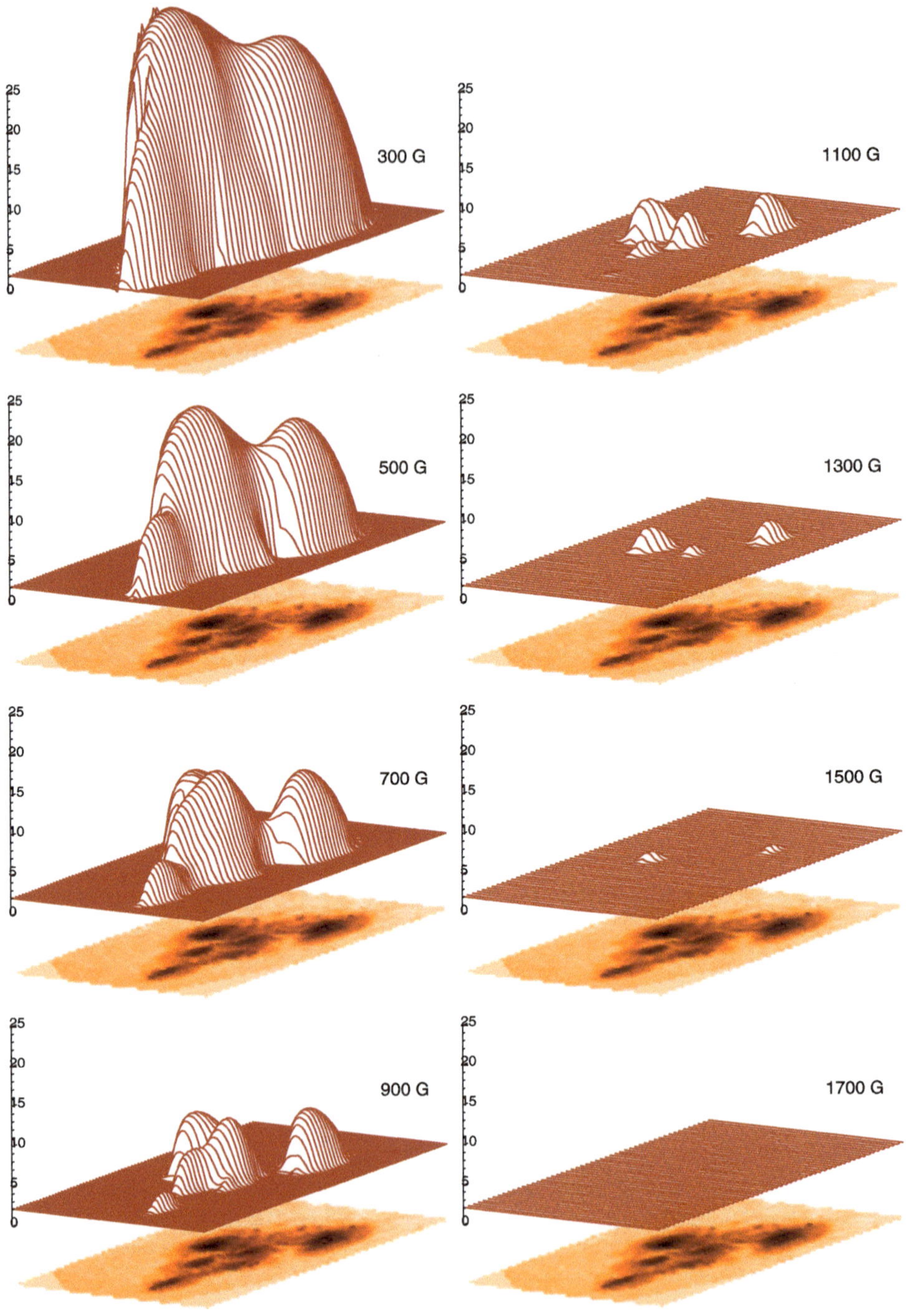

Fig. 6 Coronal isogauss surfaces for an active region on 1991 May 7. The white light image of the region, 160″ across, is shown under the extrapolated magnetic surface at the magnetic field strength labelled in each *panel*. The *vertical axis* shows height above the photosphere, and is exaggerated compared to the horizontal dimension and labelled in arcseconds. The surfaces are derived from a nonlinear force-free extrapolation of a Mees Solar Observatory vector magnetogram (Mikic and McClymont 1994; Lee et al. 1998)

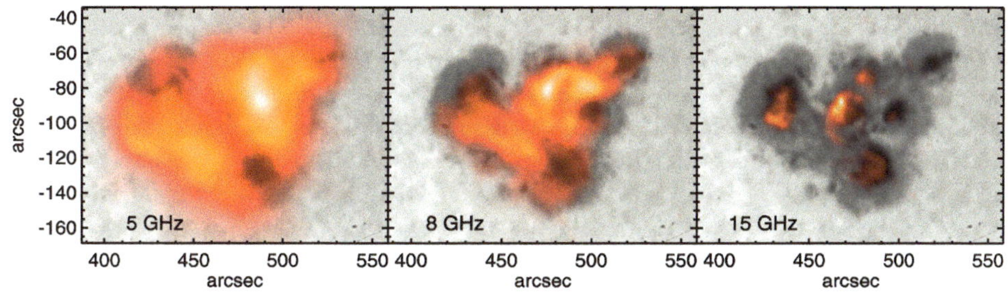

Fig. 7 Radio images of the solar active region shown in Fig. 6. The images were obtained with the Very Large Array at the frequencies as labelled: the radio emission using a red color table is overlaid on a white-light continuum image in greyscale. The red color table shows the variation of coronal temperatures across the corresponding isogauss surfaces. Maximum radio brightness temperatures are 4.5 MK, 5.1 MK and 2.4 MK at 5, 8 and 15 GHz, respectively: these temperatures are high enough to produce significant opacity in the 4th harmonic layer

assumption. Zlotnik (1968a) presents accurate expressions for $F_{x,o}(\Theta_B)$ which correctly handle the mode polarization properties, but we will not require them here.

The way to think about gyroresonance emission is to visualize the nested surfaces of constant B ("isogauss") corresponding to the harmonics that contribute at a given frequency. An example is shown in Fig. 6, derived from a fully nonlinear force-free extrapolation of a vector magnetogram (Lee et al. 1998). For a vigorous active region such as this, the entire corona above the active region has field strengths exceeding a few hundred gauss. The highest field strengths in the corona are not just found above the biggest sunspots where large magnetic fluxes reach the corona: the center of the active region contains highly sheared fields that produced strong coronal field strengths with high temperatures observed (Lee et al. 1997). Radio observations at a given frequency essentially measure the brightness temperature on the surface corresponding to the highest optically thick harmonic, and Fig. 7 shows observations of this same region by the Very Large Array at three frequencies. The radio observations matched the coronal magnetic field strength predictions of the nonlinear force-free model, and not those of potential or linear force-free extrapolations, thereby validating the extrapolation (Lee et al. 1998, 1999).

It is important to note that gyroresonance observations are sensitive to the absolute magnetic field strength B, not the line-of-sight component $B \cos \Theta_B$ that many other techniques measure. Since the frequency width of a given cyclotron resonance is proportional to $\mu^{-1/2} s f_B$, where $\mu = m_e c^2 / k_B T$, and typically $\mu \approx 3000$ in the corona, the cyclotron resonances are very narrow. Correspondingly, the layers in the corona that contribute at a given frequency are very thin: the width of the resonance means that B varies by less than 2% across a resonant layer, corresponding to a physical width of less than 200 km for typical coronal magnetic gradients (e.g., scale length $\sim 10^4$ km). The narrow physical thickness of the gyroresonant layers is an important feature of this mechanism: since they are much smaller than relevant gradients in n_e, B and T (except possibly in the vicinity of current sheets), these physical properties may be regarded as constant as one passes through any given gyroresonant layer, and by changing the observing frequency one is peeling away narrow shells that do not overlap in properties.

Applying typical coronal conditions to (19), we find that the x mode is optically thick ($\tau \geq 1$) in the $s = 2$ and $s = 3$ isogauss layers over a broad range of angles Θ_B, while the o mode is optically thick over most of the $s = 2$ layer, and may be at least marginally optically thick over a small portion of the $s = 3$ layer if Θ_B is large. $s = 4$ can contribute to the x mode if the corona is very hot (≈ 5 MK, e.g., Fig. 7 and Lee et al. 1997). The μ^{-s} dependence of (19) ensures that there is a large drop in opacity from one layer to the next, typically by a factor of over 100. This means that the observed brightness temperature image is dominated by the temperature variation across the highest optically thick isogauss layer, rather than optically thin contributions from higher layers (which, from radiative transfer theory, are proportional to the optical depth of the higher layer times the temperature difference between the two layers). At each harmonic the o-mode opacity is always at least an order of magnitude smaller than the x-mode opacity.

4.1.1 Gyroresonance Diagnostics for Active Regions

The way that these properties can be turned into coronal magnetograms is as follows. At frequencies above about 2 GHz, bremsstrahlung never seems to make coronal sources optically thick, and hence any source measured to have coronal brightness temperatures above this frequency must be optically thick due to gyroresonance emission. This means that the radio source seen in the images outlines the isogauss surface corresponding to a harmonic, usually $s = 2$ (o mode) or $s = 3$ (x mode). Since the background level on the solar disk above 3 GHz is below 60×10^3 K, there is a lot of contrast between the gyroresonance sources ($\gtrsim 10^6$ K) and other features, and they are easily identified. Ideally, along each line of sight we can measure the radio spectrum from 1 to 20 GHz in both x and o modes. At low frequencies an active region will be optically thick, but as frequency increases at some point the maximum magnetic field strength in the corona along that line of sight drops such that $3 f_B < f$, and at the point the x-mode brightness temperature spectrum will drop dramatically. We can measure the frequency at which this happens, and have thus determined the maximum coronal B along that line of sight. The o mode should show a drop above $2 f_B$, and the ratio of the frequencies of the sharp spectral drops in the x and o modes gives us confirmation that we have the right harmonic identification. In this way it is straightforward to measure B across the base of the corona, assuming that generally B will decrease with height so that the maximum value of B along a line of sight is always at the base of the corona.

The value of this technique has been demonstrated by Wang et al. (2015), who took the density, temperature and magnetic field structure of the physical active region model by Mok et al. (2005) and generated model images from 1 to 20 GHz using radiative transfer through the model, taking into account both bremsstrahlung and gyroresonance opacity. These model images were then embedded in a realistic frequency-dependent full solar disk model containing a number of active regions. The resulting images were then used to generate model telescope data for the 13-element Expanded Owens Valley Solar Array (EOVSA) array at 64 frequencies from 1 to 18 GHz assuming a full day's observation (and ignoring solar rotation), which were then processed through the standard EOVSA mapping and deconvolution steps. This produced images that looked very similar to the initial model images. The active-region images were then used in the method described above to determine the magnetic field at the base of the corona, and shown to reproduce the magnetic field at the 0.72 MK level in the Mok active-region model remarkably well. This study demonstrated that producing magnetograms of absolute field strength at the base of the corona is straightforward as long as suitable data are available.

Radio measurements such as these are also an important resource for testing extrapolation techniques (e.g., Lee et al. 1999), since they provide a measurement of magnetic field strength that is quite independent of other common techniques, and since the measurement only relies on our ability to measure radio frequencies, it is robust and reliable. Currently there are a number of efforts (e.g., Wiegelmann and Inhester 2006; De Rosa et al. 2009; Aschwanden and Malanushenko 2013; Aschwanden et al. 2014) to try to use field lines outlined by EUV emission to test extrapolations of surface fields into the corona, but the field line morphology provides no information on absolute field strengths. The radio measurements are largely insensitive to density, and therefore to the field line details that dominate EUV images, because the opacity of a gyroresonance layer is linearly proportional to density (whereas EUV brightness $\propto n_e{}^2$) and density variations are much smaller than the large changes in opacity between harmonic layers due to the μ^{-s}-dependence of (19).

An obvious limitation of the gyroresonance measurement is that, in common with most coronal diagnostics, it does not measure the height of the isogauss layers for regions on the disk without additional information such as stereoscopic measurements, which presently are not feasible. It can measure heights at the limb (Brosius and White 2006), but there the isogauss surfaces are severely foreshortened. Thus by observing at many radio frequencies covering the microwave range, we can build up a three-dimensional picture of the magnetic field in the corona with good *relative* height detail, but one without a measured *absolute* vertical scale. This suggests that radio measurements in combination with extrapolations of surface fields are likely to be necessary to develop a completely quantitative picture of coronal magnetic fields above active regions.

The increasing importance of bremsstrahlung emission (next subsection) at lower frequencies does place a lower limit to the field strengths that can usefully be measured by this technique: below 2 GHz the clear distinction between gyroresonance and bremsstrahlung sources does not occur, placing a lower limit of about 250 G (assuming 3rd harmonic emission) on the field strengths we can infer. Fields weaker than this must be measured by other techniques. This means that the gyroresonance technique essentially only works for active regions, since we do not see radio sources above 3 GHz with coronal temperatures (implying optically thick gyroresonance emission) outside active regions (e.g., White 1999).

4.2 Bremsstrahlung Polarization

The interpretation of the polarization of bremsstrahlung radio emission in terms of coronal magnetic fields has been reviewed by Gelfreikh (2004). Bremsstrahlung is simply the radiation excited by collisions between electrons and ions, and is always active in the solar corona. The absorption coefficient at frequency f due to collisions in a plasma at temperature T and density n_e is approximately

$$\kappa = 0.2 n_e{}^2 T^{-3/2} f^{-2} \text{ cm}^{-1} \tag{21}$$

It has the same density-squared dependence as EUV emission, and hence has the same bias towards high-density regions. In addition, opacity increases as frequency decreases, and is stronger in cool regions rather than hot regions. These dependencies are very different from those of gyroresonance emission, implying that the two mechanisms are usually diagnostics of different coronal conditions.

In the presence of a magnetic field, the magnetoionic mode properties result in different opacities for the x and o modes:

$$\kappa_{x,o} = \frac{0.2 n_e{}^2}{T^{3/2}(f + \sigma f_B \cos\Theta_B)^2} \text{ cm}^{-1} \tag{22}$$

where $\sigma = -1$ for the x mode and $\sigma = 1$ for the o mode, Θ_B is the angle between the magnetic field and the line of sight, and we assume $f \gg f_B$, f_p so that the refractive index is unity.

If the coronal radio emission is dominated by an optically thin source, then the degree of circular polarization r seen in the radio observation should be simply

$$r = \frac{\kappa_x - \kappa_o}{\kappa_x + \kappa_o} \approx 2\,\frac{f_B}{f}\cos\Theta_{\boldsymbol{B}} \tag{23}$$

and the magnetic field component along the line of sight in the source (weighted by the square of the density) is then

$$B\cos\Theta_{\boldsymbol{B}} \approx \frac{f}{5.6\times 10^6\ \mathrm{Hz}}\,r \tag{24}$$

e.g., a 10% circularly polarized bremsstrahlung source at 1 GHz would imply a line-of-sight field of just 18 G.

However, the simple expression in (24) does not apply when the polarization includes optically thick contributions as well, which will often be the case for bremsstrahlung. In particular, the weakly-polarized plages regions in Fig. 5 have brightness temperatures that obey (16): most of the solar disk at 17 GHz is at the chromospheric temperature $T_{\mathrm{chr}} = 10000$ K. The brightness temperature variations seen in the upper panel of Fig. 5 can have contributions from both temperature variations in the optically thick chromosphere as well as the density variations in optically thin coronal loops overlying the plages, and both components could contribute to the observed polarized flux. At lower frequencies, e.g., 1 GHz, active regions are often optically thick in the corona due to bremsstrahlung.

Grebinskij et al. (2000) demonstrated how one can determine the magnetic field from an optically thick bremsstrahlung source. In this case, the two polarizations are optically thick at slightly different heights in the atmosphere, and if the local temperatures at those heights are different, then the brightness temperatures in the polarizations will differ and thus produce a net polarization. Thus the degree of polarization in the optically thick case depends on the gradient of temperature with height in the source, which is, ab initio, unknown. Fortunately the height scale is not needed, since the actual heights of the layers in which the two polarizations are optically thick also depend on the local gradients along the line of sight (Grebinskij et al. 2000). Because of this, the degree of polarization r actually depends on the gradient of the temperature with respect to the frequency of the optically thick layer:

$$r = -\frac{f_B\cos\Theta_{\boldsymbol{B}}}{f}\,\frac{d\log T}{d\log f} \tag{25}$$

If we have sufficiently dense frequency coverage in the observations with reliable brightness temperature calibration, then we can actually measure $d\log T_B/d\log f$ along each line of sight from a cube of images versus frequency and thus (assuming $T_B = T$ for optically thick emission) determine the line-of-sight component of B as

$$B\cos\Theta_{\boldsymbol{B}} \approx \frac{f}{2.8\times 10^6\ \mathrm{Hz}}\,r\left(-\frac{d\log T_B}{d\log f}\right)^{-1} \tag{26}$$

Since this is an optically thick measurement, it determines the field strength in a relatively narrow layer of the corona; however, by virtue of (21), the thickness of the corresponding layer will be considerably larger than is the case for gyroresonance emission. The need to measure the gradient of T_B with frequency implies the need for well-calibrated, high-fidelity high-spatial-resolution imaging over a broad continuous frequency range, and since coronal sources are only optically thick due to bremsstrahlung below about 2 GHz where the spatial resolution of current telescopes is not high (typically tens of arcseconds), there has been little practical application of this technique so far.

4.3 Gyrosynchrotron Emission from Flares

Flares accelerate electrons to energies from 10 keV up to many MeV, and these energetic electrons radiate prolifically at radio wavelengths via gyrosynchrotron emission, producing solar microwave bursts (e.g., Bastian et al. 1998). As is true of the X-ray emission from flares, images often indicate that the radio emission is dominated by electrons on a confined set of magnetic loops in the solar corona. An example of such an event is shown in Fig. 8. The electrons responsible for the radio emission are believed to have a power law energy distribution above about 20 keV, consistent with measurements of hard X-ray spectra that can be well fit by power laws. For such an energy distribution, the radio emission spectrum can be calculated using gyrosynchrotron theory (Ginzburg and Syrovatskii 1965; Melrose 1980) and it is found that the frequency at which the radio flux spectrum peaks is almost linearly related to the magnetic field strength in the loop at that point (e.g., Dulk and Marsh 1982). Thus high-resolution radio imaging across a broad frequency range simultaneously can provide spatially resolved radio spectra along a loop and reveal the variation in magnetic field strength (Bastian et al. 1998). An example of radio spectra from a model flare loop is shown in Fig. 9, where the relationship between the frequency of the peak in the radio spectrum and the actual field strength in the emitting location is clear.

A number of papers have simulated the gyrosynchrotron emission from a flare loop, and Kuznetsov et al. (2011) have developed an IDL tool for such simulations, including the effects of thermal bremsstrahlung absorption as well as pitch angle anisotropy (Fleishman and Melnikov 2003). The simulator takes advantage of fast gyrosynchrotron calculation methods (Fleishman and Kuznetsov 2010), which allows the simulator to be run often enough to derive the coronal magnetic field strength profile. Gary et al. (2013) gave a successful example of this application. One can in principle solve for both B and Θ_B using this approach, as well as properties of the emitting electron energy distribution.

4.4 Plasma Emission

Plasma emission is a coherent emission mechanism, in the sense that it produces brightness temperatures far in excess of the energies of the electrons responsible for the emission. The mechanism is not direct emission of electromagnetic waves: rather, it involves several steps. The first step is the production of electrostatic Langmuir waves at the local plasma frequency f_p with large wavenumbers k. In low-density plasma, this practically occurs whenever there is free energy in the electron energy distribution, such as a positive slope in the velocity distribution parallel to the magnetic field (usually referred to as an electron beam) or even a loss-cone velocity distribution (e.g., Melrose 1985). Such free energy can produce intense levels of power in Langmuir waves. To convert the Langmuir waves to propagating electromagnetic waves that can be seen on Earth, the first step is usually the electrostatic decay of Langmuir waves into electrostatic ion sound waves and daughter Langmuir waves with wavevectors $\boldsymbol{k}$ pointing opposite to the parent Langmuir wave. Production of electromagnetic waves at the fundamental frequency $f = f_p$ then results from the combination of Langmuir waves and ion sound waves, with opposite wavevectors that nearly cancel, to produce the small k of the electromagnetic wave. Similarly, harmonic emission at $f = 2f_p$ results from the coalescence of parent and daughter Langmuir waves with nearly opposite $\boldsymbol{k}$.

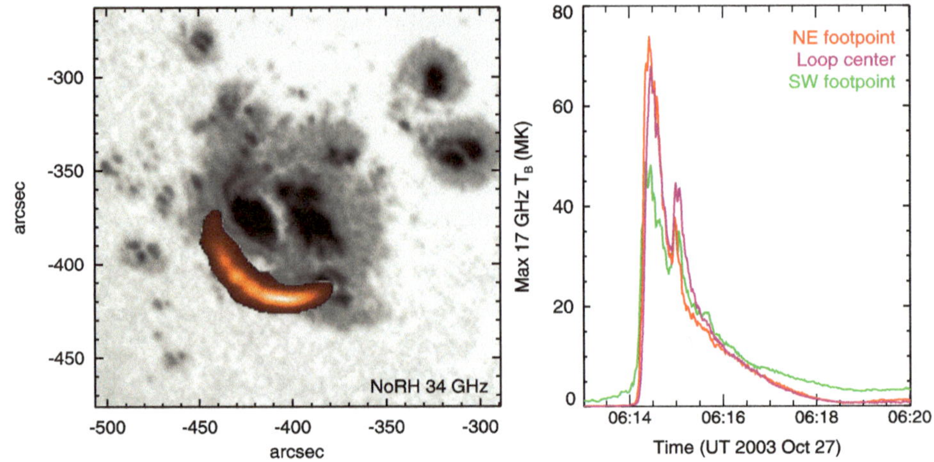

Fig. 8 Image of a microwave burst from a flare on 2003 Oct 27. The *left* panel shows emission at 34 GHz imaged with the Nobeyama Radioheliograph (red color table, resolution 8″), overlaid on a white-light image of the active region. The *right* panel shows light curves of the peak 17 GHz brightness temperature at three locations on the flaring loop: the north end (*red*), the loop top (*purple*) and the south end (*green*), showing the matching temporal behavior indicating a single electron population propagating between the ends of the loop

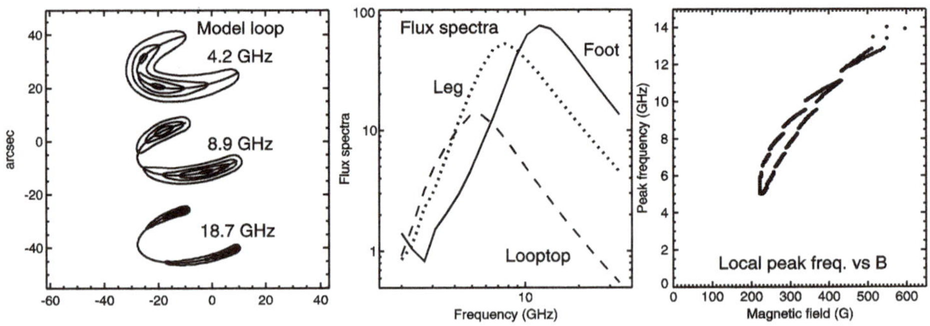

Fig. 9 Gyrosynchrotron emission from a model flare loop demonstrating the use of the spectral peaks for determining coronal magnetic field strength. The *left* panel shows how the appearance of the loop changes from low optically thick frequencies (4.2 GHz), where the loop top is brightest, to high optically thin frequencies (18.7 GHz), where the footpoints dominate because of the stronger magnetic fields there. The *middle* panel shows radio spectra at three different loop locations: at a footpoint (*solid line*), at the top of the loop (*dashed line*), and at a location half-way up the leg of the loop (*dotted line*). In the *right* panel we plot the frequency of the peak of the radio spectrum versus the actual magnetic field at the emitting location, showing that there is a nearly linear dependence of the spectral peak on the local magnetic field

Polarization of fundamental plasma emission results from the fact that only the magnetoionic *o* mode can exist below the upper hybrid frequency $f_{UH} = \sqrt{{f_p}^2 + {f_B}^2}$. The frequency of an ion acoustic wave is generally not enough to span the gap between f_p and f_{UH} ($f_{UH} - f_p \approx {f_B}^2/2f_p$ when $f_B \ll f_p$). Thus in the cold plasma limit, we might expect fundamental plasma emission to be 100% polarized in the sense of the *o* mode. In practice, this is not what is observed: fundamental Type III emission is seldom more than 60% circularly polarized and is often much less (Dulk and Suzuki 1980). This is believed to be influenced by propagation effects since there seems to be a center-to-limb variation in the observed degree of polarization, and Langmuir waves propagating through a highly inhomogeneous corona may be able to convert to fundamental emission closer to the local upper hybrid frequency. The sense of polarization is however consistent with the o mode, so fundamental emission gives us the line-of-sight direction of the magnetic field but no useful information on its strength.

Second harmonic plasma emission could in principle be more useful. As shown by Zlotnik (1981) and discussed further by Benz (1993), when the coalescing Langmuir waves have wavevectors closely aligned with the magnetic field as we expect, then harmonic emission is polarized in the sense of the o mode with degree of polarization

$$r \approx \frac{11}{48} \frac{f_B \, |\cos \Theta_{\boldsymbol{B}}|}{f_p} \tag{27}$$

If the coalescing Langmuir waves have an isotropic distribution of wavevectors relative to the magnetic field, then the polarization would be in the sense of the x mode with degree

$$r \approx -\frac{85}{48} \frac{f_B \, |\cos \Theta_{\boldsymbol{B}}|}{f_p} \tag{28}$$

In practice harmonic emission is observed to have the same sense of polarization as the fundamental, i.e., o mode, so it is likely that the Langmuir waves are closely aligned with the magnetic field and (27) applies, yielding estimates of the line-of-sight field at densities of order 10^8 cm^{-3} where Type III bursts are typically observed. A 10% degree of polarization at 180 MHz, corresponding to that density, results from a line-of-sight field of order 11 G. Since it is dependent on the (unknown) exact distribution of Langmuir wavevectors and also shows center-to-limb variation (Dulk and Suzuki 1980), this diagnostic of line-of-sight field strength will have significant uncertainties.

5 Current and Future Instrumentation for the Measurement of Coronal Magnetic Fields

The problem of determining the strength and topology of the Sun's coronal magnetic field, whether to constrain numerical models of photospheric field extrapolation, or to identify and follow the evolution of precursors of space weather events, is high in the priority list of the solar and space physics communities. Accordingly, a significant effort has been dedicated to the investigation of new diagnostic tools (see the previous Sections of this article) and to the development of coronal observing programs. In this Section, we provide a short overview of the current status of instrumentation projects that aim at the measurement of magnetic field signatures in coronal plasmas, whether the ultimate objective of these programs is the fundamental research of magneto-hydrodynamic processes in the corona, or the establishment of synoptic observations of the solar corona for the purpose of space weather monitoring.

5.1 Optical Instruments

The Coronal Solar Magnetism Observatory (CoSMO) is a suite of instruments developed by the High Altitude Observatory of the National Center for Atmospheric Research (HAO-NCAR), aimed at providing a comprehensive diagnostic set of tools for the magnetism of the solar chromosphere and corona. It comprises

- a large 1.5 m aperture refractive coronagraph, capable of a 1°-diameter FOV (i.e., $\sim 4 R_\odot$), having spectral coverage from 380 to 1450 nm that permits continuous observation of a number of visible and near-IR lines;
- the *Chromospheric and Prominence Magnetometer* (ChroMag), a full-disk chromospheric imaging spectropolarimeter fed by a 20-cm aperture refractor and working with lines in the 580–1100 nm range; and
- the *K Coronagraph* (K-Cor), a 20-cm aperture white-light coronagraph with a FOV ranging from 1.05 to 3.0 solar radii, using dual-beam polarimeters in order to measure polarized brightness in the visible spectrum between 720 and 750 nm.

The facility will be located at the NCAR Mauna Loa Solar Observatory (MLSO). The K-Cor instrument is already deployed and operational, and provides a continuous monitoring of the polarization brightness of the solar corona for density measurements, and the tracking of CMEs. The ChroMag instrument has been prototyped, and is currently being tested before deployment to the MLSO. It will provide continuous Stokes measurements of the solar atmosphere in both photospheric and chromospheric lines, with a spatial resolution of 2″. The Large Coronagraph (LC) has been designed as a high-throughput instrument to optimize its polarimetric sensitivity for Stokes V measurements of the solar corona in magnetically sensitive forbidden (M1) coronal lines (see Sect. 3.3). It will be capable of reaching a magnetic-strength measurement sensitivity of 1 G with 15 min of integration time, using the Fe XIII 1074.7 nm line.

The Daniel K. Inouye Solar Telescope (DKIST) will be the largest optical instrument dedicated to solar investigations for the next decade, and it will be managed and operated by the National Solar Observatory. One of the main science objectives of the project is to uncover the nature of the coronal magnetic field, by probing the solar corona at the highest spatial resolution compatible with the polarimetric sensitivity and temporal cadence dictated by the science.

The coronal science requirements of the DKIST have driven the conceptual design of the facility, in particular the adoption of an off-axis Gregorian optical configuration for the telescope, in order to mitigate the detrimental effects of instrumental scattered light on the measurements of the dim corona. The DKIST has one instrument specifically dedicated to performing coronal science, the Cryogenic Near Infrared Spectro-Polarimeter (Cryo-NIRSP), which is being developed by the Institute for Astronomy of the University of Hawaii (Maui, HI). This instrument will span the solar spectrum between 500 nm to 5 μm, with polarimetric capability at least up to 4 μm. It is important to remark that the Cryo-NIRSP will be able to access a region of the IR solar spectrum, above 3 μm, where the brightness of the corona is expected to be larger than the thermal background from both sky and the instrument. This instrument, as well as the Diffraction Limited Near Infrared Spectro-Polarimeter (DL-NIRSP; also a contribution of the IfA-UH), and the Visible Spectro-Polarimeter (ViSP; HAO-NCAR) will be able to perform spectro-polarimetric investigations of the solar corona as soon as DKIST operations start, which is anticipated to be in August 2019.

In addition to these ground-based instruments, developments can be expected in rocket- and satellite-borne telescopes to exploit magnetometry possibilities in the UV. Recently, the Chromospheric Lyman-Alpha Spectro-Polarimeter (CLASP) rocket experiment (Kubo et al. 2014) has demonstrated the capability to measure the linear polarization of the H I Lyα transition-region line in features on the solar disk, and addressed the technical difficulties involved in this type of measurement (see Kano et al. 2017; Ishikawa et al. 2017). Further efforts are under way to extend this capability to other UV resonance lines, such as the CLASP-2 rocket experiment aimed at detecting the polarization of the Mg II h-k doublet around 280 nm as a magnetic diagnostic of the upper chromosphere and transition region.

5.2 Radio Instruments

Both the gyroresonance and bremsstrahlung techniques for measuring coronal magnetic fields require a well-calibrated telescope capable of making high-dynamic-range, high spatial resolution images of the Sun across a broad continuous frequency range. Until recently, telescopes capable of the imaging requirements did not operate across a sufficiently broad and continuous frequency range. However, modern developments have filled in this shortcoming. In particular, the 27-element Very Large Array (VLA) radio interferometer has been upgraded to provide continuous frequency coverage from 1.0 to 50.0 GHz, and is capable of superb imaging across this entire range. However, the VLA is not solar-dedicated and its field of view at frequencies above 1.4 GHz is much smaller than the solar disk. The Frequency Agile Solar Radiotelescope (FASR) is a US-proposed solar-dedicated radio telescope with of order 80–100 elements, designed specifically to achieve the goal of routine measurement of coronal magnetic fields (e.g., Bastian 2004), but so far it has not obtained traction with the funding agencies. The Expanded Owens Valley Solar Array (EOVSA) is a solar-dedicated telescope consisting of 13 dishes operating from 2.5 to 18 GHz. EOVSA is currently nearing completion and the image quality will be sufficient to use both the gyroresonance and gyrosynchrotron techniques to measure coronal magnetic fields (e.g., Wang et al. 2015). China is also building a solar radioheliograph covering microwave frequencies (0.4–15 GHz), now called the Mingantu Ultrawide Spectral Radioheliograph (MUSR), that will have capabilities similar to those planned for FASR (Du et al. 2015) and thus will be capable of measuring coronal fields using all the methods described here.

At frequencies below 1 GHz where plasma emission dominates, there are also promising new instruments with broad frequency coverage. The LOw Frequency ARray (LOFAR) operates in the range 10–240 MHz using two types of antenna and is capable of imaging the full solar disk with excellent spatial resolution using a multiple beam-forming technique. It has been used to image Type III bursts out to a large height above the limb, and with the ability to measure circular polarization it can be used for the plasma emission diagnostic (Morosan et al. 2014, 2015). The Murchison Widefield Array (MWA, Tingay et al. 2013) has 128 elements operating in the range 80–300 MHz in a low-interference environment in outback Western Australia, and will achieve excellent dynamic range for solar observing (Oberoi et al. 2013). The Long Wavelength Array (LWA) in the US operates between the FM band at 90 MHz and the ionospheric cutoff near 10 MHz (Taylor et al. 2012), and has superb sensitivity across the entire frequency range but with no imaging ability at present. LWA can be used with effectively unlimited time and frequency resolution, and its strength will be in measuring the polarization of fast narrow-band structures such as the striae that are often seen in the fundamental band of Type III bursts (Tun Beltran et al. 2015).

Much higher frequencies than we have discussed here can be used to probe the magnetic field in the chromosphere, which aids the study of coronal magnetic fields by directly addressing the issue of how fields are affected by passing through the chromosphere. The main workhorse for this activity will be the Atacama Large Millimeter-submillimeter Array (ALMA) in Chile, which operates at frequencies in the range 84–950 GHz that probe the chromospheric layers of the solar atmosphere. ALMA will measure chromospheric fields principally via the optically-thick bremsstrahlung technique discussed above. However, the degrees of circular polarization expected at ALMA frequencies are small and these will be difficult measurements (e.g., Loukitcheva et al. 2017).

6 The Problem of Coronal Magnetic Fields

As demonstrated in the preceding discussion, optical and radio techniques are to a large extent complementary in their sensitivity to coronal magnetic fields. Radio observations utilizing the well-understood gyroresonance method to measure coronal magnetic fields operate everywhere, including on the solar disk, but are only likely to work for field strengths in excess of ~ 100 G. Optical and near-IR techniques can only be used above the solar limb due to the need to detect weak polarization signals that are washed out by the intense emission from the solar disk. These methods can measure the LOS component of $\boldsymbol{B}$ up to arbitrary field strengths using circular polarization induced by the Zeeman effect, but because of the need to use coronographic techniques, field strengths measured at the available height range are generally less than 100 G. The full vector $\boldsymbol{B}$ can be obtained using resonance polarization or the Hanle effect to address the magnetic field components in the plane of the sky, but these measurements depend on the illumination of the corona by the strong photospheric flux and consequently the analysis is somewhat model dependent.

A major reason for the difficulty in measuring coronal magnetic fields is the fact that the corona is a three-dimensional volume, and much more information is required to describe it adequately compared to a two-dimensional surface such as the photosphere. Optical techniques in the visible and IR regions of the solar spectrum assume that all coronal emission is optically thin and so must take into account an integration along the line of sight (as addressed in detail by Judge et al. 2013). The density sensitivity of these techniques makes it easier to associate them with discrete features such as loops visible in EUV and white-light images of the corona. On the other hand, gyroresonance emission has very little dependence on density, which can be an advantage for studying fields in the entire volume of interest: the gyroresonance and (at low frequencies) bremsstrahlung techniques at radio wavelengths rely on emission being optically thick, and by studying a continuous range of frequencies the radio measurements are able to measure B continuously over a wide range of heights along any given line of sight, but without knowledge of the absolute heights to which the measurements refer. Because the gyroresonance technique can measure strong fields, it can be used low down in active regions and thus is suitable to study the relationship between coronal magnetic fields and solar flares. The visible/IR techniques work better at higher altitudes and thus may be more suited for the study of the coronal magnetic field configuration leading to solar eruptions and coronal mass ejections.

The value of extrapolations of vector magnetic field measurements at the solar surface for the study of coronal fields is also clear. We argue that both optical and radio measurement techniques are essential for the development of better extrapolation techniques, since only through comparison with actual data can those techniques be validated. Extrapolations of photospheric magnetic fields have the well-known problem that they must handle the transition from a high-β plasma in the photosphere to a low-β plasma in the corona, and must somehow account for fields, often on small spatial scales, that close in the chromosphere and never reach the corona. Comparisons of extrapolations with field lines visible in coronal images provide another form of validation. All these methods are likely to be important for a better understanding of coronal fields and the activity that they produce.

7 Conclusions

The urgent need to measure the magnetic fields in the solar corona, in order to understand even the most essential observed behavior of this important region of the heliosphere, contrast with the practical difficulties in performing such measurements, which require high-sensitivity instrumentation and subtle data analysis and interpretational techniques. In this review paper, we have presented the current status of solar coronal observations, in both the optical and radio domains of the solar spectrum, and offered some new diagnostic opportunities that should be considered for the next generation of instruments dedicated to the study of the magnetism of the solar corona.

From both theoretical and observational viewpoints, which ultimately determine the optimal choice of diagnostics to pursue, there are no clear answers yet as to why the Sun is obliged to form a corona and a quasi-steady wind. Our Earth's environment is strongly modulated by the steady radiative and wind emissions from the corona, from the mesosphere, which is largely affected by the UV radiation at the coronal base (dominated by the H I Ly-α emission), to the magnetosphere, which is shaped by the solar wind through the interaction of the charged particle emission with the Earth's magnetic field.

Currently there is an enormous interest in understanding the nature and the trigger mechanisms of *perturbations* to our Earth system that occur as the geospace experiences the effects of the natural instability of the solar corona, which results in the production of CMEs on top of the quasi-steady solar wind. These perturbations, or solar transients, represent well-known risks for mankind, because of their impact on technological systems upon which our society has increasingly become dependent. The magnetized plasmas ejected from the Sun are able to induce magnetospheric reconnection, which eventually leads to large amounts of electromagnetic energy being released on various spatial and temporal scales, which can severely affect electric power grids, communication systems, GPS satellites, and pose a threat to manned as well as unmanned space missions.

Aside from these practical implications, if we could measure properties of the changing magnetic fields in the corona, then several areas of the fundamental physics of the Sun's system would become more accessible to study. Are CMEs a reflection of the need for the Sun to conserve its magnetic helicity, as argued by some MHD theorists and modelers (e.g., Low 1996)? If that is the case, then this should play an essential role in controlling the solar dynamo itself, by providing a sink for this conserved quantity, which must be generated in the solar interior. Do current sheets spontaneously form at all scales down to ionized plasma's skin depths? If so, can this explain coronal heating as conjectured by Parker (1988)? How can our investigations of the Sun's magnetic behavior inform us about the unstable behavior of laboratory plasmas, which hinders society's desire to achieve stable nuclear fusion as a possible solution to mankind's ongoing energy crisis? The methods discussed in this review will require major improvements in resolution and sensitivity before we can realistically address these important questions.

In conclusion, it is not surprising that the coronal heating problem has been with us for nearly three quarters of a century, and that the origins of CMEs and other related phenomena, such as solar flares, remain essentially unsolved. We expect they will remain so, until essential properties of the changing magnetic fields in the solar corona can be properly measured so to provide the necessary data-constrained input to the many existing modeling attempts of the Sun-Earth system.

References

J. Arnaud, Astron. Astrophys. **112**, 350 (1982)
J. Arnaud, G. Newkirk Jr., Astron. Astrophys. **178**, 263 (1987)
M.J. Aschwanden, Astrophys. J. Suppl. Ser. **224**, 25 (2016)
M.J. Aschwanden, A. Malanushenko, Sol. Phys. **287**, 345 (2013)
M.J. Aschwanden, X. Sun, Y. Liu, Astrophys. J. **785**, 34 (2014)
T.S. Bastian, in *Astrophysics and Space Science Library*. Solar and Space Weather Radiophysics, vol. 314, ed. by D.E. Gary, C.U. Keller (Kluwer Academic, Dordrecht, 2004), pp. 47–69

T.S. Bastian, A.O. Benz, D.E. Gary, Annu. Rev. Astron. Astrophys. **36**, 131 (1998)
A.O. Benz, *Plasma Astrophysics: Kinetic Processes in Solar and Stellar Coronae*. Astrophysics and Space Science Library, vol. 184 (Kluwer Academic, Dordrecht, 1993)
J.W. Brosius, S.M. White, Astrophys. J. Lett. **641**, L69 (2006)
R. Casini, Astrophys. J. **568**, 1056 (2002)
R. Casini, P.G. Judge, Astrophys. J. **522**, 524 (1999)
R. Casini, E. Landi Degl'Innocenti, in *Astrophysical Plasmas*, ed. by T. Fujimoto, A. Iwamae. Plasma Polarization Spectroscopy. Atomic, Optical, and Plasma Physics, vol. 44 (Springer, Berlin, 2008), p. 247
R. Casini, A. López Ariste, S. Tomczyk, B.W. Lites, Astrophys. J. Lett. **598**, L67 (2003)
P. Charvin, Ann. Astrophys. **28**, 877 (1965)
M.C.M. Cheung, B. De Pontieu, T.D. Tarbell et al., Astrophys. J. **801**, 83 (2015)
M.L. De Rosa, C.J. Schrijver, G. Barnes, K.D. Leka, B.W. Lites, M.J. Aschwanden, T. Amari, A. Canou, J.M. McTiernan, S. Régnier, J.K. Thalmann, G. Valori, M.S. Wheatland, T. Wiegelmann, M.C.M. Cheung, P.A. Conlon, M. Fuhrmann, B. Inhester, T. Tadesse, Astrophys. J. **696**, 1780 (2009)
J. Du, Y. Yan, W. Wang, D. Liu, Proc. Astron. Soc. Aust. **32**, 26 (2015)
G.A. Dulk, K.A. Marsh, Astrophys. J. **259**, 350 (1982)
G.A. Dulk, S. Suzuki, Astron. Astrophys. **88**, 203 (1980)
S. Fineschi, in *Magnetic Fields Across the Hertzsprung-Russell Diagram*, ed. by G. Mathys, S.K. Solanki, D.T. Wickramasinghe. Astronomical Society of the Pacific Conference Series, vol. 248 (2001), p. 597
G.H. Fisher, W.P. Abbett, D.J. Bercik et al., Space Weather **13**, 369 (2015)
G.D. Fleishman, A.A. Kuznetsov, Astrophys. J. **721**, 1127 (2010)
G.D. Fleishman, V.F. Melnikov, Astrophys. J. **587**, 823 (2003)
D.E. Gary, G.J. Hurford, Astrophys. J. **420**, 903 (1994)
D.E. Gary, C.U. Keller, *Solar and Space Weather Radiophysics—Current Status and Future Developments*. Astrophysics and Space Science Library, vol. 314 (Kluwer Academic, Dordrecht, 2004)
D.E. Gary, G.D. Fleishman, G.M. Nita, Sol. Phys. **288**, 549 (2013)
G.B. Gelfreikh, in *Solar and Space Weather Radiophysics*, ed. by D.E. Gary, C.U. Keller (Kluwer Academic, Dordrecht, 2004), pp. 115–133
S. Gibson, in *Polarimetry: From the Sun to Stars and Stellar Environments*. IAU Symposium, vol. 305 (2015), pp. 245–250
S.E. Gibson, T.A. Kucera, S.M. White, J.B. Dove, Y. Fan, B.C. Forland, L.A. Rachmeler, C. Downs, K. Reeves, Front. Astron. Space Sci. **3**, 8 (2016)
V.L. Ginzburg, S.I. Syrovatskii, Annu. Rev. Astron. Astrophys. **3**, 297 (1965)
V.L. Ginzburg, V.V. Zheleznyakov, Sov. Astron. **2**, 653 (1958)
V.L. Ginzburg, V.V. Zheleznyakov, Sov. Astron. **5**, 1 (1961)
A. Grebinskij, V. Bogod, G. Gelfreikh, S. Urpo, S. Pohjolainen, K. Shibasaki, Astron. Astrophys. Suppl. Ser. **144**, 169 (2000)
W. Hanle, Z. Phys. **30**, 93 (1924)
J.W. Harvey, PhD thesis, University of Colorado at Boulder (1969)
W. Heitler, *Quantum Theory of Radiation*. Clarendon International Series of Monographs on Physics (Clarendon, Oxford, 1954)
L.L. House, Astrophys. J. **214**, 632 (1977)
R. Ishikawa, J. Trujillo Bueno, H. Uitenbroek et al., Astrophys. J. **841**, 10 (2017)
P.G. Judge, R. Casini, in *Advanced Solar Polarimetry: Theory, Observation, and Instrumentation*, ed. by M. Sigwarth. ASP Conf. Ser., vol. 236 (2001), p. 503
P.G. Judge, S. Tomczyk, W.C. Livingston, C.U. Keller, M.J. Penn, Astrophys. J. Lett. **576**, L157 (2002)
P.G. Judge, B.C. Low, R. Casini, Astrophys. J. **651**, 1229 (2006)
P.G. Judge, S. Habbal, E. Landi, Sol. Phys. **288**, 467 (2013)
T. Kakinuma, G. Swarup, Astrophys. J. **136**, 975 (1962)
R. Kano, J. Trujillo Bueno, A. Winebarger et al., Astrophys. J. Lett. **839**, 10 (2017)
J.C. Kemp, J.H. Macek, F.W. Nehring, Astrophys. J. **278**, 863 (1984)
A. Khan, Astron. Astrophys. **545**, A52 (2012)
A. Khan, E. Landi Degl'Innocenti, Astron. Astrophys. **532**, A70 (2011)
A. Khan, E. Landi Degl'Innocenti, Astron. Astrophys. **543**, A158 (2012)
A. Khan, L. Belluzzi, E. Landi Degl'Innocenti, S. Fineschi, M. Romoli, Astron. Astrophys. **529**, A12 (2011)
J.L. Kohl et al., Astrophys. J. Lett. **501**, L127 (1998)
M. Kramar, B. Inhester, H. Lin, J. Davila, Astrophys. J. **775**, 25 (2013)
M. Kubo et al., in *Solar Polarization 7*, ed. by K.N. Nagendra, J.O. Stenflo, Q. Qu, M. Samooprna. ASP Conf. Ser., vol. 489 (2014), p. 307
J.R. Kuhn, K.S. Balasubramaniam, G. Kopp, M.J. Penn, A.J. Dombard, H. Lin, Sol. Phys. **153**, 143 (1994)
M.R. Kundu, *Solar Radio Astronomy* (Interscience, New York, 1965)

A.A. Kuznetsov, G.M. Nita, G.D. Fleishman, Astrophys. J. **742**, 87 (2011)
E. Landi Degl'Innocenti, M. Landolfi, *Polarization in Spectral Lines*. Astrophysics and Space Science Library, vol. 307 (Kluwer Academic, Dordrecht, 2004)
J. Lee, S.M. White, N. Gopalswamy, M.R. Kundu, Sol. Phys. **174**, 175 (1997)
J. Lee, A.N. McClymont, Z. Mikić, S.M. White, M.R. Kundu, Astrophys. J. **501**, 853 (1998)
J. Lee, S.M. White, M.R. Kundu, Z. Mikić, A.N. McClymont, Astrophys. J. **510**, 413 (1999)
H. Lin, J.R. Kuhn, R. Coulter, Astrophys. J. Lett. **613**, L177 (2004)
M. Loukitcheva, S.M. White, S.K. Solanki, G.D. Fleishman, M. Carlsson, Astron. Astrophys. **601**, 43 (2017)
B.C. Low, Sol. Phys. **167**, 217 (1996)
A. Malanushenko, C.J. Schrijver, M.L. DeRosa, M.S. Wheatland, Astrophys. J. **783**, 102 (2014)
M.J. Martínez González, A. Asensio Ramos, R. Manso Sainz, C. Beck, L. Belluzzi, Astrophys. J. **759**, 16 (2012)
D.J. McLean, N.R. Labrum, *Solar Radiophysics* (Cambridge University Press, Cambridge, 1985)
D.B. Melrose, *Plasma Astrophysics* (Gordon & Breach, New York, 1980)
D.B. Melrose, in *Solar Radiophysics*, ed. by D.J. McLean, N.R. Labrum (Cambridge University Press, Cambridge, 1985), p. 289
Z. Mikic, A.N. McClymont, in *Solar Active Region Evolution: Comparing Models with Observations*, ed. by K.S. Balasubramaniam, G.W. Simon. Astronomical Society of the Pacific Conference Series, vol. 68 (1994), p. 225
Y. Mok, Z. Mikić, R. Lionello, J.A. Linker, Astrophys. J. **621**, 1098 (2005)
D.E. Morosan et al., Astron. Astrophys. **568**, A67 (2014)
D.E. Morosan et al., Astron. Astrophys. **580**, A65 (2015)
V.M. Nakariakov, E. Verwichte, Living Rev. Sol. Phys. **2**, 3 (2005)
D. Oberoi et al., in *Proceedings of the International Symposium on Solar Terrestrial Physics (ISSTP)*. Astronomical Society of India Conference Series, vol. 10 (2013), pp. 131–135
E.N. Parker, Astrophys. J. **330**, 474 (1988)
J. Plowman, Astrophys. J. **792**, 23 (2014)
C.W. Querfeld, R.N. Smartt, Sol. Phys. **91**, 299 (1984)
N.-E. Raouafi, P. Lemaire, S. Sahal-Bréchot, in *8th SOHO Workshop: Plasma Dynamics and Diagnostics in the Solar Transition Region and Corona*, ed. by J.-C. Vial, B. Kaldeich-Schü. ESA Special Publication, vol. 446 (1999), p. 567
N.E. Raouafi, P. Riley, S. Gibson, S. Fineschi, S.K. Solanki, Front. Astron. Space Sci. **3**, 20 (2016)
P.A. Robinson, D.B. Melrose, Aust. J. Phys. **37**, 675 (1984)
S. Sahal-Brechot, Astron. Astrophys. **32**, 147 (1974)
S. Sahal-Brechot, Astrophys. J. **213**, 887 (1977)
S. Sahal-Brechot, V. Bommier, N. Feautrier, Astron. Astrophys. **340**, 579 (1998)
C.J. Schrijver, M.L. Derosa, T.R. Metcalf, Y. Liu, J. McTiernan, S. Régnier, G. Valori, M.S. Wheatland, T. Wiegelmann, Sol. Phys. **235**, 161 (2006)
G.B. Taylor et al., J. Astron. Instrum. **1**, 50004 (2012)
S.J. Tingay et al., J. Phys. Conf. Ser. **440**, 012033 (2013)
S. Tomczyk, in *IAU Special Session*, vol. 6 (2012), p. 214
S. Tomczyk, G.L. Card, T. Darnell, D.F. Elmore, R. Lull, P.G. Nelson, K.V. Streander, J. Burkepile, R. Casini, P.G. Judge, Sol. Phys. **247**, 411 (2008)
D. Tripathi, S.K. Solanki, R. Schwenn, V. Bothmer, M. Mierla, G. Stenborg, Astron. Astrophys. **449**, 369 (2006)
J. Trujillo Bueno, in *Magnetic Coupling Between the Interior and Atmosphere of the Sun*, ed. by S.S. Hasan, R.J. Rutten (2010), p. 118
J. Trujillo Bueno, E. Landi Degl'Innocenti, L. Belluzzi, Space Sci. Rev. (2017). doi:10.1007/s11214-016-0306-8
S.D. Tun Beltran, S. Cutchin, S. White, Sol. Phys. **290**, 2423 (2015)
Z. Wang, D.E. Gary, G.D. Fleishman, S.M. White, Astrophys. J. **805**, 93 (2015)
S.M. White, Sol. Phys. **190**, 309 (1999)
S.M. White, in *Solar and Space Weather Radiophysics*, ed. by D.E. Gary, C.U. Keller (Kluwer Academic, Dordrecht, 2004), pp. 89–113
S.M. White, M.R. Kundu, Sol. Phys. **174**, 31 (1997)
T. Wiegelmann, B. Inhester, Sol. Phys. **236**, 25 (2006)
V.V. Zheleznyakov, Sov. Astron., A.J. **6**, 3 (1962)
V.V. Zheleznyakov, *Radio Emission of the Sun and the Planets* (Pergamon, Oxford, 1970)
E.Y. Zlotnik, Sov. Astron. **12**, 245 (1968a)
E.Y. Zlotnik, Sov. Astron. **12**, 464 (1968b)
E.I. Zlotnik, Astron. Astrophys. **101**, 250 (1981)

Space Sci Rev (2017) 210:183–226
DOI 10.1007/s11214-016-0306-8

The Physics and Diagnostic Potential of Ultraviolet Spectropolarimetry

Javier Trujillo Bueno[1] · **Egidio Landi Degl'Innocenti**[2] · **Luca Belluzzi**[3,4]

Received: 11 December 2015 / Accepted: 27 October 2016 / Published online: 27 March 2017

Abstract The empirical investigation of the magnetic field in the outer solar atmosphere is a very important challenge in astrophysics. To this end, we need to identify, measure and interpret observable quantities sensitive to the magnetism of the upper chromosphere, transition region and corona. This paper provides an overview of the physics and diagnostic potential of spectropolarimetry in permitted spectral lines of the ultraviolet solar spectrum, such as the Mg II *h* and *k* lines around 2800 Å, the hydrogen Lyman-α line at 1216 Å, and the Lyman-α line of He II at 304 Å. The outer solar atmosphere is an optically pumped vapor and the linear polarization of such spectral lines is dominated by the atomic level polarization produced by the absorption and scattering of anisotropic radiation. Its modification by the action of the Hanle and Zeeman effects in the inhomogeneous and dynamic solar atmosphere needs to be carefully understood because it encodes the magnetic field information. The circular polarization induced by the Zeeman effect in some ultraviolet lines (e.g., Mg II *h* & *k*) is also of diagnostic interest, especially for probing the outer solar atmosphere in plages and more active regions. The few (pioneering) observational attempts carried out so far to measure the ultraviolet spectral line polarization produced by optically pumped atoms in the upper chromosphere, transition region and corona are also discussed. We emphasize that ultraviolet spectropolarimetry is a key gateway to the outer atmosphere of the Sun and of other stars.

Keywords Spectropolarimetry · Sun: chromosphere · Sun: transition region · Sun: corona

✉ J. Trujillo Bueno
jtb@iac.es

E. Landi Degl'Innocenti
landie@arcetri.astro.it

L. Belluzzi
belluzzi@irsol.ch

1 Instituto de Astrofísica de Canarias, 38205 La Laguna, Tenerife, Spain

2 Dipartimento di Fisica e Astronomia, Università di Firenze, 50125 Firenze, Italy

3 Istituto Ricerche Solari Locarno, 6605 Locarno Monti, Switzerland

4 Kiepenheuer-Institut für Sonnenphysik, 79104 Freiburg, Germany

1 Introduction

Since the pioneering work of Hale (Hale 1908), spectropolarimetry has shown to be a formidable diagnostic window for the investigation of the solar atmosphere, with particular emphasis on the determination of the intensity and direction of the magnetic field. This has been possible through the achievements of several generations of scientists that, on one hand, have brought the sensitivity of spectropolarimeters to unprecedented levels and, on the other hand, have developed suitable theoretical frameworks and plasma diagnostic techniques for the interpretation of the observations.

While approximately 50 % of the Sun's total energy is emitted between 4000 and 8000 Å, only about 1 % is emitted at wavelengths shorter than 3000 Å. Why then should we worry about ultraviolet spectropolarimetry? After all, some information on the magnetic fields of the solar corona can be obtained from coronal forbidden lines (in the visible and near-IR) and at radio wavelengths (see Casini et al. 2017). First of all, because of the high temperatures ($T > 10^4$ K) the primary emission of the outer solar atmosphere (upper chromosphere, transition region and corona) lies in the ultraviolet (UV; 2000–4000 Å), far ultraviolet (FUV; 1210–2000 Å) and extreme ultraviolet (EUV; 100–1210 Å) regions of the solar spectrum. Secondly, in such spectral regions there are many permitted spectral lines whose radiation can be measured when pointing either to solar disk or to off-limb positions; as reviewed in this paper, for many of these lines the linear polarization due to scattering processes is measurable and sensitive to the magnetic field strengths expected for quiet and active regions in the outer solar atmosphere.

Spectropolarimetric measurements in spectral lines with wavelengths shorter than 3000 Å can only be performed from space, and they represent an almost virgin territory. For this reason, this paper deals mainly with theoretical investigations. We start in Sect. 2 with a gentle introduction to the physical mechanisms that introduce polarization in the solar spectrum. Section 3 discusses the possibility of probing the magnetism of the (optically thin) solar corona via permitted line polarimetry, concerning both off-limb and on-disk line of sights. The optically thick case of the upper chromosphere and transition region is considered in Sect. 4, where we highlight the diagnostic potential of the solar disk radiation in several strong resonance lines that are sensitive to the Hanle and Zeeman effects. Finally, in Sect. 5 we provide information on the exposure times needed to measure polarization signals in UV, FUV and EUV lines, assuming given spatial and spectral resolutions, and we summarize the observational attempts carried out so far (e.g., the Chromospheric Lyman-Alpha SpectroPolarimeter sounding rocket experiment). The paper concludes emphasizing that ultraviolet spectropolarimetry might be our key gateway to probe the magnetism of the outer atmosphere of the Sun and of other stars.

2 A Gentle Introduction to the Physics of Polarization

There are several physical mechanisms responsible for the generation of polarized radiation in stellar atmospheres, including the atmosphere of the Sun. As far as the continuum spectrum is concerned, the polarization properties of the radiation emitted by an optically thin plasma can, in most cases, be directly derived from the laws of classical electromagnetism, or, in other words, from Maxwell's equations. Only for special situations quantum-mechanical effects have to be taken into account, which can often be done by slightly modifying the classical laws by means of suitable corrections.

The actual derivation of the polarization properties of the radiation emitted by a collection of electric charges moving arbitrarily in space was performed by Alfred-Marie Liénard and, independently, by Emil Wiechart around 1900. Starting from Maxwell's equations, working in the Lorentz gauge, and introducing the electromagnetic potentials that nowadays bear their name, Liénard and Wiechart showed that a single charge q, moving with velocity $\mathbf{u}$ and acceleration $\mathbf{a}$, produces, along the direction $\mathbf{n}$ and at a large distance R from the charge itself (in the so-called radiation zone), an electromagnetic wave whose associated electric and magnetic fields are given by the equations (for a formal derivation see, e.g., Landi Degl'Innocenti 2014)

$$\mathbf{E} = \frac{q}{c^2 \kappa^3 R} \mathbf{n} \times \left[\left(\mathbf{n} - \frac{\mathbf{u}}{c} \right) \times \mathbf{a} \right], \qquad \mathbf{B} = \mathbf{n} \times \mathbf{E}, \tag{1}$$

where

$$\kappa = 1 - \frac{\mathbf{u} \cdot \mathbf{n}}{c}, \tag{2}$$

c is the speed of light, and where all the quantities have to be evaluated at the so-called retarded time. In particular, for a non-relativistic charge ($u \ll c$, $\kappa \simeq 1$) one gets the simpler expression

$$\mathbf{E} = -\frac{q}{c^2 R} \left[\mathbf{a} - (\mathbf{a} \cdot \mathbf{n}) \mathbf{n} \right], \qquad \mathbf{B} = \mathbf{n} \times \mathbf{E}. \tag{3}$$

These equations show that the electromagnetic wave emitted by the accelerated charge is polarized. Indeed, its electric field vibrates along the component of the acceleration vector on the plane perpendicular to the propagation direction.

From Eq. (3) it is possible to find the polarization to be expected from astrophysical plasmas in a large variety of physical situations. Just specifying the geometrical nature of the physical system, it allows us to describe the polarization in Thomson scattering, Rayleigh scattering, Bremsstrahlung radiation, cyclotron radiation and, accounting for relativistic effects, in synchrotron radiation (e.g., Landi Degl'Innocenti 2014).

Particularly relevant is the case of Thomson scattering because it shows the strict connection between polarization and purely geometrical factors. Consider a free electron (charge $q = -e_0$), and suppose that it is subjected to the action of a polarized electromagnetic wave propagating along the direction $\mathbf{n}'$. Defining a pair of real unit vectors $\mathbf{e}'_1$ and $\mathbf{e}'_2$ such that they form with $\mathbf{n}'$ a right-handed triad of unit vectors, and denoting by $\mathcal{E}'_1$ and $\mathcal{E}'_2$ the complex components along the same unit vectors of the electric field carried by the wave, the electron experiences an acceleration given by

$$\mathbf{a} = -\frac{e_0}{m} \left(\mathcal{E}'_1 \mathbf{e}'_1 + \mathcal{E}'_2 \mathbf{e}'_2 \right). \tag{4}$$

The accelerated electron radiates into space. Considering the emission along the unit vector $\mathbf{n}$, defining another pair of real unit vectors $\mathbf{e}_1$ and $\mathbf{e}_2$ such that they form with $\mathbf{n}$ a right-handed triad, and denoting by $\mathcal{E}_1$ and $\mathcal{E}_2$ the complex components, along the same vectors, of the emitted electric field, by applying the Liénard and Wiechart equations the following relation is found

$$\begin{pmatrix} \mathcal{E}_1 \\ \mathcal{E}_2 \end{pmatrix} = -\frac{r_c}{R} \begin{pmatrix} \mathbf{e}_1 \cdot \mathbf{e}'_1 & \mathbf{e}_1 \cdot \mathbf{e}'_2 \\ \mathbf{e}_2 \cdot \mathbf{e}'_1 & \mathbf{e}_2 \cdot \mathbf{e}'_2 \end{pmatrix} \begin{pmatrix} \mathcal{E}'_1 \\ \mathcal{E}'_2 \end{pmatrix}, \tag{5}$$

where r_c is the classical radius of the electron ($r_c = e_0^2/(mc^2)$, with m the electron mass) and where R is the distance from the electron to the point where the emitted electric field is evaluated. This equation can be transformed in terms of the Stokes parameters. Defining such parameters for the emitted radiation with the usual convention, namely

$$I = \mathcal{K}\big(\mathcal{E}_1^*\mathcal{E}_1 + \mathcal{E}_2^*\mathcal{E}_2\big), \qquad Q = \mathcal{K}\big(\mathcal{E}_1^*\mathcal{E}_1 - \mathcal{E}_2^*\mathcal{E}_2\big)$$
$$U = \mathcal{K}\big(\mathcal{E}_1^*\mathcal{E}_2 + \mathcal{E}_2^*\mathcal{E}_1\big), \qquad V = \mathrm{i}\,\mathcal{K}\big(\mathcal{E}_1^*\mathcal{E}_2 - \mathcal{E}_2^*\mathcal{E}_1\big), \tag{6}$$

where $\mathcal{K}$ is a constant which depends on the units used to measure the Stokes parameters, and applying the same definition for the incident radiation, one finds the following result

$$\begin{pmatrix} I \\ Q \\ U \\ V \end{pmatrix} = \frac{r_\mathrm{c}^2}{2R^2} \begin{pmatrix} a^2+b^2+c^2+d^2 & a^2-b^2+c^2-d^2 & 2(ab+cd) & 0 \\ a^2+b^2-c^2-d^2 & a^2-b^2-c^2+d^2 & 2(ab-cd) & 0 \\ 2(ac+bd) & 2(ac-bd) & 2(ad+bc) & 0 \\ 0 & 0 & 0 & 2(ad-bc) \end{pmatrix}$$
$$\times \begin{pmatrix} I' \\ Q' \\ U' \\ V' \end{pmatrix}, \tag{7}$$

where

$$a = \mathbf{e}_1 \cdot \mathbf{e}'_1, \qquad b = \mathbf{e}_1 \cdot \mathbf{e}'_2, \qquad c = \mathbf{e}_2 \cdot \mathbf{e}'_1, \qquad d = \mathbf{e}_2 \cdot \mathbf{e}'_2. \tag{8}$$

This expression can be simplified by suitably choosing the unit vectors. The simplest choice is to define the unit vectors $\mathbf{e}_1$ and $\mathbf{e}'_1$ as being perpendicular to the plane defined by the unit vectors $\mathbf{n}$ and $\mathbf{n}'$ (the scattering plane), and the unit vectors $\mathbf{e}_2$ and $\mathbf{e}'_2$ as laying in the same plane. With this choice, being $a = 1$, $b = 0$, $c = 0$, $d = \mathbf{n} \cdot \mathbf{n}' = \cos\Theta$, with Θ the scattering angle, one obtains

$$\begin{pmatrix} I \\ Q \\ U \\ V \end{pmatrix} = \frac{r_\mathrm{c}^2}{2R^2} \begin{pmatrix} 1+\cos^2\Theta & \sin^2\Theta & 0 & 0 \\ \sin^2\Theta & 1+\cos^2\Theta & 0 & 0 \\ 0 & 0 & 2\cos\Theta & 0 \\ 0 & 0 & 0 & 2\cos\Theta \end{pmatrix} \begin{pmatrix} I' \\ Q' \\ U' \\ V' \end{pmatrix}. \tag{9}$$

The matrix appearing in this equation is the so-called scattering phase matrix, or Rayleigh scattering matrix, which explains a wide variety of scattering phenomena. Its basic ingredients are just geometrical factors (the scalar products $\mathbf{e}_i \cdot \mathbf{e}'_j$).

An important application of the classical formulae mentioned above concerns the intensity and polarization properties of the radiation emitted by antennas of particular geometrical shape. A linear antenna having its direction defined by the unit vector $\mathbf{k}$ radiates into the direction $\mathbf{n}$ with an intensity proportional to $\sin^2\theta$, where θ is the angle between the two unit vectors. The radiation is linearly polarized with its direction contained in the plane formed by the unit vectors $\mathbf{n}$ and $\mathbf{k}$. A circular antenna laying in the plane perpendicular to the unit vector $\mathbf{k}$ radiates into the direction $\mathbf{n}$ with an intensity proportional to $\frac{1}{2}(1+\cos^2\theta)$. The resulting polarization is, in general, elliptical. In particular, for $\mathbf{n}$ contained in the plane of the antenna, the polarization is linear, its direction being also directed along the same plane. Viceversa, for $\mathbf{n}$ perpendicular to the circular antenna plane, the polarization is purely circular with different handedness according to the fact that the radiation is directed "upwards" (along $\mathbf{k}$) or "downwards" (along $-\mathbf{k}$). It can easily be proven that three independent antennas, one linear and two circular of opposite handedness, geometrically arranged as indicated above, and of equal power, give rise to a completely isotropic and unpolarized radiation.

These classical results are particularly useful because they allow us to establish an analogy with the quantum theory of the emission (and absorption) of radiation in atomic (or molecular) spectral lines. It is well known from quantum electrodynamics that a transition between two energy levels characterized by the angular momentum J and J', respectively, is indeed the superposition of several transitions of the type $(J, M) \to (J', M')$, where M and M' are the so-called magnetic quantum numbers. Now, for an electric dipole transition, each elementary transition between the magnetic sublevels M and M' is allowed only if $\Delta M = M' - M = 0, \pm 1$, and it behaves, as far as its polarization properties are concerned, as a classical antenna of a different type. The transitions having $\Delta M = 0$ behave as linear antennas directed along the quantization axis (the axis defining M), whereas the transitions having $\Delta M = \pm 1$ behave as circular antennas of different handedness laying in the plane perpendicular to the quantization axis. Moreover, each transition has a particular relative strength given, in terms of 3-j symbols, by the expression

$$S(JM, J'M') = 3 \begin{pmatrix} J' & J & 1 \\ -M' & M & \Delta M \end{pmatrix}^2 . \tag{10}$$

This analogy of quantum transitions with antennas allows us to understand the physical basis of polarization in spectral lines. Consider for instance the case of emission of radiation by an atom. One possibility of having emission of polarized radiation is that the elementary antennas oscillate at different frequencies. This is the case of the Zeeman effect. In the single emission process the line-integrated emitted radiation is not polarized, but the magnetic field (reasoning in classical terms) shifts the frequency of one of the circular antennas by a quantity ν_{L}, the Larmor frequency, and the frequency of the other circular antenna by a quantity $-\nu_{\mathrm{L}}$, thus producing a typical polarization profile across the spectral line. The other possibility of producing polarized radiation, which works even in the absence of magnetic fields, is that the different antennas radiate with unequal power, or radiate not independently, but with well defined phase relationships between them. This results in a net amount of polarized emission. Translated in quantum mechanical terms, this last phenomenon implies that the populations of the different magnetic sublevels of the upper J-level of the transition have unequal populations, or that they have a certain amount of coherence. This physical situation is referred to in Quantum Mechanics as "Atomic Polarization".

A typical astrophysical case where atomic polarization plays a fundamental role is met in the outer layers of stellar atmospheres. Here the radiation field starts being limb darkened (brightened), the radiation propagating vertically being more (less) intense than the one propagating horizontally. This anisotropy is responsible for the creation of atomic polarization via a physical mechanism that can be understood in classical terms considering the inverse process of the one sketched above (the three antennas are differentially fed by the radiation field). This combination of differential absorption and re-emission gives rise to the so-called phenomenon of resonance polarization. In the limiting case of scattering by a two-level atom of a unidirectional beam of radiation, reasoning either classically (but taking into account the quantum mechanical result contained in Eq. (10)) or quantum-mechanically, one finds that resonance scattering in spectral lines can be described by a generalization of Eq. (9), namely

$$\begin{pmatrix} I \\ Q \\ U \\ V \end{pmatrix} = \frac{\sigma}{2R^2} (\mathbf{M}_1 + \mathbf{M}_2) \begin{pmatrix} I' \\ Q' \\ U' \\ V' \end{pmatrix} , \tag{11}$$

where σ is a suitable cross section and where

$$\mathbf{M}_1 = \frac{4}{3}(1 - W_2) \begin{pmatrix} 1 & 0 & 0 & 0 \\ 0 & 0 & 0 & 0 \\ 0 & 0 & 0 & 0 \\ 0 & 0 & 0 & 0 \end{pmatrix}, \tag{12}$$

$$\mathbf{M}_2 = \begin{pmatrix} W_2(1+\cos^2\Theta) & W_2\sin^2\Theta & 0 & 0 \\ W_2\sin^2\Theta & W_2(1+\cos^2\Theta) & 0 & 0 \\ 0 & 0 & 2W_2\cos\Theta & 0 \\ 0 & 0 & 0 & 2W_1\cos\Theta \end{pmatrix}, \tag{13}$$

with W_1 and W_2 the polarizability factors for orientation and alignment, respectively, introduced in Landi Degl'Innocenti (1984), and explicitly given in terms of 6-j symbols by the expression

$$W_K = 3(2J_u + 1) \begin{Bmatrix} 1 & 1 & K \\ J_u & J_u & J_\ell \end{Bmatrix}^2, \tag{14}$$

with J_u and J_ℓ the angular momenta of the upper and lower level, respectively. For the special case of a transition connecting a lower level with $J_\ell = 0$ and an upper level with $J_u = 1$, both W_1 and W_2 are equal to unity and the matrix appearing in Eq. (9) is recovered. Different transitions are less efficient, as far as polarization is concerned, since the polarizability factors are always less than unity and can even be zero (a typical example is the transition $J_\ell = \frac{1}{2}$, $J_u = \frac{1}{2}$ for which $W_2 = 0$). This line does not show any scattered linear polarization for whatsoever polarization of the incident radiation.

Equation (11) contains the basic physics of resonance scattering in a two-level atom. It is relatively simple to derive and has the advantage of making the physics quite transparent. However, it is of little use, except in some particular cases, to describe resonance polarization in a complicated physical environment like a stellar atmosphere. The reasons are the following.

First of all, there is a remarkable effect that can alter significantly resonance polarization. This is the so called Hanle effect (Hanle 1924), which consists in a modification of the scattering line polarization due to the presence of a magnetic field (see also Trujillo Bueno 2001). From a physical point of view, the Hanle effect can be understood returning to the antennas model developed above for describing the polarized radiation emitted by an atom. When a magnetic field is present, the circular antennas suffer a shift in frequency of $\pm\nu_\mathrm{L}$, with ν_L the Larmor frequency. As a result, the three antennas (the linear one plus the two circular ones) now oscillate at different frequencies and in the typical time t_life, the lifetime of the atom in its upper level, they result in having a phase shift $\Delta\Phi$ given, as an order of magnitude, by the expression[1]

$$\Delta\Phi \simeq 2\pi\,\nu_\mathrm{L}\,t_\mathrm{life} = \frac{e_0 B\, t_\mathrm{life}}{2mc}, \tag{15}$$

where B is the magnetic field intensity. If $\Delta\Phi \ll 1$, the dephasing of the antennas due to the magnetic field is inessential and resonance polarization is not altered. In the opposite case, when $\Delta\Phi \gg 1$, the dephasing is so large that the three antennas radiate independently, loos-

[1] Quantum mechanics is responsible for introducing in this equation an additional factor g, the level's Landé factor, so that the general equation reads $\Delta\Phi \simeq e_0 B g\, t_\mathrm{life}/2mc$.

ing any correlation that might be induced by the incident radiation field. This is the so-called regime of the "saturated Hanle effect". The intermediate regime, namely the regime where $\Delta\Phi \simeq 1$ is the most interesting from the diagnostic point of view, because the modification suffered by resonance polarization allows in principle the determination of the magnetic field (in amplitude and direction) from observations. In principle, it is possible to develop analytical formulae for the Hanle effect in order to generalize Eqs. (11)–(13). The resulting equations are, however, rather cumbersome and will not be reported here (see Eqs. (5.139) and (10.22) of Landi Degl'Innocenti and Landolfi (2004) for the simpler case where the geometry is specified by assigning the angles that the incident and the scattered radiation form with the magnetic field, and Landi Degl'Innocenti and Landi Degl'Innocenti (1988) for the more general case of arbitrary geometry).

Important modifications to the laws of resonance scattering arise from further physical facts. Indeed, even when generalized for the Hanle effect, Eq. (11) suffers from an important limitation. For this equation to hold, the ground level of the resonant line has to be unpolarized or, in other terms, it has to be naturally excited with all its sublevels sharing the same population. Moreover, it only holds for the simplest case of a two-level atom. Obviously, these are, in many cases, rather questionable assumptions that, unfortunately, cannot be released by simply working with classical analogies. To treat these more complicated cases one has to apply a more powerful theoretical framework based on Quantum Electrodynamics. This scheme involves writing the statistical equilibrium equations for the density matrix elements of the atomic system, a set of equations in which the populations of the levels and the phase interference (or coherence) between them are the unknowns, whereas the Stokes parameters of the radiation field enter the expressions of the coefficients. In the simplest instances (for example, when treating the phenomenon of resonance scattering in solar prominences), the radiation field that is illuminating the atom is known and the set of equations can be solved numerically. This allows us to find the physical state of the atom (in particular, the atomic polarization of the upper level) and to derive the ensuing emission in the four Stokes parameters. In more complex situations, like for instance in stellar atmospheres, the statistical equilibrium equations have to be coupled with the radiative transfer equations for the Stokes parameters and the combined set of equations (statistical equilibrium plus radiative transfer) has to be solved iteratively (see Trujillo Bueno 2003). This is the so-called "non-LTE problem of the 2nd kind", whose general equations are discussed in detail in the monograph by Landi Degl'Innocenti and Landolfi (2004; hereafter LL04). It is however important to note that the non-LTE problem of the 2nd kind can be solved analytically in few special cases, e.g. for treating the phenomenon of resonance scattering and the Hanle effect in a two-level atom with unpolarized lower level. The analytical solution brings to the same result as the one expressed by Eq. (11) (in the absence of magnetic fields) or by its Hanle generalization. A further remark concerns the possibility of generalizing the theoretical scheme discussed above to treat the problem of resonance scattering for complicated atomic models involving fine structure, hyperfine structure, and even a combination of the two. The important physical aspect that enters into play is that the number of unknowns entering the statistical equilibrium equations rapidly grows because it is necessary to take into account all the possible interference effects present in the generalized atomic model.

A further physical aspect which has an important impact on resonance scattering is that in the solar atmosphere the atom also experiences collisions with the particles present in the plasma. The most important collisions are those with electrons (and in some cases also with protons) and those with neutral hydrogen atoms. However, the effect of collisions is, in many cases, far from being known with sufficient accuracy. Indeed, the theory of collisions is a very difficult matter, and only few theoretical results are available. On the other hand, there is an almost complete lack of experimental results. In many applications, a distinction is made between inelastic collisions (mainly due to electrons) inducing transitions between different J-levels, well separated in energy, and elastic collisions (mainly due to neutral hydrogen atoms) inducing transitions between different M-levels of the same J-level. This distinction becomes rather vague when treating the case of multi-term atoms with hyperfine structure because the energy difference between M-levels is often comparable with the energy difference between hyperfine F-levels. Much work has still to be done on this subject, which is extremely important because collisions act on atomic systems basically with three distinct (but correlated) effects: line broadening, depolarization, and frequency redistribution in scattering events.

In this section we have tried to summarize the basic physical facts underlying the emission of polarized radiation from an astrophysical plasma, like for instance the solar atmosphere. We have seen that polarization can be produced in a spectral line either by the action of an external agent, like for instance a magnetic field, responsible for introducing an energy splitting among the different magnetic sublevels of the upper level of the transition, or, alternatively, by the presence of atomic polarization in the upper level itself. In an optically thick medium, however, there are other mechanisms at work which involve the physical state of the lower level of the transition (see Trujillo Bueno and Landi Degl'Innocenti 1997). Indeed, when this level is affected either by the presence of splittings between the magnetic sublevels, or by the presence of atomic polarization, the medium through which the radiation is propagating behaves like a dichroic and anomalously dispersive medium. These phenomena of dichroism and anomalous dispersion are suitably accounted for in the radiative transfer equation for the Stokes parameters.

Our previous considerations are general, in the sense that they apply to spectral lines irrespectively of their particular wavelength range, infrared, visible or ultraviolet. The ultraviolet lines, however, are characterized by some peculiarities that are important to stress. The first thing to remark concerns the Zeeman effect, and it is useful to recall that for not too strong fields the circular polarization signals produced by the Zeeman effect are proportional to the ratio between the Zeeman splitting and the Doppler width of the line, while the linear polarization signals are proportional to the square of such ratio. In terms of wavelength intervals, the Zeeman splitting is given by the expression

$$\Delta\lambda_{\mathrm{B}} = \frac{\lambda_0^2 e_0 B}{4\pi mc}, \tag{16}$$

where λ_0 is the wavelength of the spectral line, and B is the intensity of the magnetic field. On the other hand, the Doppler broadening of the line (neglecting the possible contribution due to microturbulent velocities) is given by

$$\Delta\lambda_{\mathrm{D}} = \frac{\lambda_0}{c}\sqrt{\frac{2k_{\mathrm{B}}T}{M}}, \tag{17}$$

where k_{B} is Boltzmann's constant, T is the temperature, and M is the mass of the element responsible for the spectral line. Considering the ratio of these two quantities one has

$$\frac{\Delta\lambda_{\mathrm{B}}}{\Delta\lambda_{\mathrm{D}}} = \frac{\lambda_0 e_0 B}{4\pi m}\sqrt{\frac{M}{k_{\mathrm{B}}T}}. \tag{18}$$

Since this ratio scales linearly with λ_0 it follows that ultraviolet lines are less sensitive to the Zeeman effect than "normal" visible lines. Moreover, if we consider a typical ultraviolet coronal line, formed in an environment where the temperature is of the order of 10^6 K, the factor $T^{1/2}$ at the denominator of the same ratio also brings to a difference of the order of 10 with respect to a visible line formed in the solar photosphere or chromosphere. Combining the two factors, we see that ultraviolet lines are in general much less sensitive to the Zeeman effect than visible lines.

The second peculiarity of ultraviolet lines is the fact that, generally speaking, the radiation field anisotropy in the ultraviolet is larger than in the visible spectrum (see, e.g., Figs. 12.4 and 12.7 in LL04). From one side this is due to the steeper behavior of the Planck function with respect to temperature, and, from the other side, it is due to the fact that ultraviolet lines generally form in the higher layers of the solar atmosphere where geometrical effects add to the center-to-limb variation (CLV) contribution to the anisotropy. Obviously, larger anisotropy implies larger scattering polarization. This fact, joined with the reduced sensitivity of ultraviolet lines to the Zeeman effect, and to the fact that the magnetic field in the formation regions of ultraviolet lines is generally weak, makes the Hanle effect in permitted ultraviolet lines a suitable diagnostic tool for investigating the magnetism of the outer solar atmosphere.

3 The Solar Corona

The solar corona is optically thin at the wavelengths of the permitted spectral lines considered in this paper. Section 3.1 studies the off-limb case of the spectral line radiation (e.g., hydrogen Ly-α) that results from the resonance scattering of solar-disk photons by residual coronal atoms (e.g., neutral hydrogen). Section 3.2 discusses a physical mechanism that introduces linear polarization in typical coronal lines (e.g., the Fe X line at 174.5 Å), produced by isotropic collisional excitation and spontaneous emission in the coronal plasma itself, which can in principle be measured even when observing coronal loops and arcades against the underlying solar disk.

3.1 Off-limb Coronal Polarimetry

A rocket experiment carried out during the solar eclipse of March 7, 1970, showed that the hydrogen Ly-α line at 1216 Å is the major contributor to the energy radiated from the corona in the height range between 0.07 and 0.7 solar radii above the visible limb, and in the wavelength interval between 850 and 2150 Å (Gabriel et al. 1971). At first sight it may appear surprising that a plasma at a temperature of the order of 10^6 K emits a line of neutral hydrogen as its strongest feature. However, one should note that H I Ly-α is the strongest emission line radiated by the underlying chromosphere, and that the corona is thus exposed to a high flux of this radiation. As pointed out by Gabriel (1971), the most important physical process responsible for the Ly-α coronal emission is in fact resonance scattering by the residual hydrogen atoms present in the solar corona. The contribution of electron impact excitation followed by radiative decay is about 10 % at low altitudes, and below 1 % at high altitudes (Gabriel 1971). Other mechanisms, such as scattering of the chromospheric radiation by electrons (Thomson scattering) or by dust particles (F-corona), are found to bring a negligible contribution. Since resonance scattering is the main physical mechanism responsible for the coronal emission of H I Ly-α, and since the chromospheric radiation that pumps the residual coronal hydrogen atoms is anisotropic (the anisotropy being simply due to geometrical effects), one can easily expect the coronal Ly-α radiation to be linearly polarized (scattering polarization). This led some authors to propose to exploit the signatures of the Hanle effect in off-limb spectropolarimetric observations of this line for investigating the elusive magnetic fields present in the solar corona (Bommier and Sahal-Bréchot 1982; Fineschi et al. 1992).

Due to its brightness, and to the fact of being almost exclusively radiatively excited, H I Ly-α is the most promising permitted line for investigating the magnetism of the solar corona through spectropolarimetric observations (Khan et al. 2011). Another FUV coronal emission line is O VI 1032 Å (Vial et al. 1980; Raymond et al. 1997). Contrary to Ly-α, the contribution of collisional excitation by electrons is not negligible in this line, which results in a decrease of the linear polarization degree of the emitted radiation (see Sect. 3.1.3). Taking the opportunity of the SOHO spacecraft rotation of March 19, 1996, Raouafi et al. (1999) pointed to a coronal hole located at about 270 arcsec above the south solar limb and observed the variation with the spacecraft rotation angle of the intensity measured in the two components of the O VI doublet at 1032 Å (polarizable line) and at 1037 Å (unpolarizable line), respectively. The ratio between such intensity signals showed a sinusoidal fluctuation with the spacecraft rotation angle, which the authors interpreted as an indication of linear polarization in the 1032 Å line. Other FUV lines of interest for the diagnostics of the magnetism of the solar corona are H I Ly-β (1025 Å) and Ly-γ (972 Å). Unfortunately, these lines are much fainter than Ly-α, and collisional excitation is comparable (Ly-β) or dominant (Ly-γ) with respect to radiative excitation. The detection of the signatures of the Hanle effect on the scattering polarization signals of these lines is thus expected to be a challenging task from the observational point of view. On the other hand, these lines are sensitive to weaker magnetic fields than Ly-α, and this property could be suitably exploited for developing multi-line methods for the diagnostics of the coronal magnetic fields (Khan et al. 2011). In this sense, another line of interest is He II 304 Å. This line can actually represent a useful reference, since its linear polarization should be immune to the Hanle effect produced by the magnetic fields expected in the solar corona (Trujillo Bueno et al. 2012).

As we shall see below, there are basically two main difficulties in determining the magnetic field vector from the signatures introduced by the Hanle effect on the linear polarization signals of the above-mentioned lines. The first one is of intrinsic character: even neglecting problems connected with the line-of-sight integration, and supposing that the magnetic field is just located in a particular region of the corona, being zero outside it, from the measurement of two Stokes parameters (namely, Q and U) it is not possible to infer the three components of the magnetic field vector. Strategies for overcoming this problem, as well as other ambiguities that are intrinsic in spectropolarimetric methods for magnetic field diagnostics (e.g., the 180° ambiguity) have been developed, mainly for the case of prominences (e.g., Bommier et al. 1981, 1994; Landi Degl'Innocenti 1982; Landi Degl'Innocenti and Bommier 1993; Trujillo Bueno et al. 2002). Most of these methods exploit the joint action of the Hanle and Zeeman effects in one or more lines. Unfortunately, in the ultraviolet and at the temperatures of the corona, the signatures of the Zeeman effect are generally negligible, and they can hardly be exploited to complement the information brought by the Hanle effect. In this sense, the estimate of the longitudinal magnetic flux performed by Lin et al. (2004) exploiting the Zeeman splitting in the forbidden infrared Fe XIII 10747 Å coronal line is a remarkable result. The other problem is that the linear polarization produced by resonance scattering is not modified only by the presence of a magnetic field through the Hanle effect, but it can also be affected in a similar manner by other competing mechanisms, such as a non-radial solar wind, or the presence of active regions (Sahal-Bréchot et al. 1986, 1998; Khan et al. 2011). For this reason, it is very important to consider, besides H I Ly-α, also other spectral lines with different sensitivities to the Hanle effect and/or to the other competing mechanisms. Also forbidden line polarimetry (see Casini et al. 2017) can represent an important complementary approach to overcome the above-mentioned difficulties.

A theoretical framework suitable for modeling scattering polarization and the Hanle effect in permitted ultraviolet coronal lines is the quantum theory of polarization described in LL04. The application of this theory to the particular case of the solar corona is discussed in some detail in the next subsections, where a series of relevant equations are derived and discussed. We recall that the theory of polarization described in LL04 is valid under the so-called flat-spectrum approximation, which requires the pumping field to be flat (i.e., independent of frequency) across intervals larger than the natural width of the levels and, when quantum interference is taken into account, larger than the frequency separation between the interfering levels. In general, the chromospheric radiation field that illuminates the atoms present in the corona shows an emission (or absorption) profile characterized by a width $\Delta\nu_p$. The error related to the flat-spectrum approximation is thus of the order of $\Delta\nu_w/\Delta\nu_p$, with $\Delta\nu_w$ the natural width of the levels or, when interference is present, the frequency separation between the interfering levels (see Sect. 13.6 of LL04). It can be verified that for H I Ly-α, as well as for all the other spectral lines mentioned above, this ratio is always very small, so that the flat-spectrum hypothesis represents a very good approximation for modeling the polarization of these lines in the solar corona.

3.1.1 The Physics of Resonance Scattering in the Solar Corona

We start considering a two-level atom with unpolarized lower level, in the absence of magnetic fields. The atom is located in the solar corona, at a given height above the solar surface, and it is illuminated by the radiation emerging from the underlying atmospheric layers. The starting equations presented in this section are equivalent to those introduced in Sect. 2, the latter being more appropriate to give an intuitive description of resonance polarization phenomena. The main difference is that the equations used in the present section are written within the theoretical framework of the spherical tensors formalism, which is particularly suitable for describing the physics of polarization (see LL04). We neglect collisional processes (as previously mentioned, this is a good approximation for Ly-α, but it is not for Ly-β, Ly-γ, and O VI 1032 Å), as well as stimulated emission (this is always a very good approximation in the solar corona). Under these assumptions, it is possible to obtain an analytical solution of the statistical equilibrium (SE) equations for the multipole moments of the density matrix of the upper level (see Eq. (10.13) of LL04). Substituting this equation into the expression of the emission coefficient for a two-level atom (see Eq. (10.15) of LL04), we obtain

$$\varepsilon_i(\nu, \mathbf{\Omega}) = k_L\, \phi(\nu_0 - \nu) \sum_{KQ} W_K(J_\ell, J_u)(-1)^Q\, \mathcal{T}_Q^K(i, \mathbf{\Omega})\, J_{-Q}^K(\nu_0), \tag{19}$$

where $\varepsilon_i(\nu, \mathbf{\Omega})$ is the emission coefficient in the four Stokes parameters (the index i can take values 0, 1, 2 and 3, standing for Stokes I, Q, U and V, respectively) at frequency ν and propagation direction $\mathbf{\Omega}$. The quantity k_L is the frequency integrated absorption coefficient, $\phi(\nu_0 - \nu)$ is the emission profile, $\mathcal{T}_Q^K$ is the polarization tensor introduced by Landi Degl'Innocenti (1984) (see also Eq. (5.157) of LL04), while W_K is the polarizability factor defined in Eq. (14). Equation (19) is valid in the atomic rest frame. The emission profile $\phi(\nu_0 - \nu)$ is thus a Lorentzian characterized by a broadening constant $\Gamma = A_{u\ell}/4\pi$, with $A_{u\ell}$ the Einstein coefficient for spontaneous emission. The tensor J_Q^K (see Eq. (5.157) of LL04 for its definition) describes the incident radiation field; note that due to the flat-spectrum approximation it is sufficient to evaluate it at the resonance frequency ν_0.

When the pumping radiation field has some spectral dependence (but can still be considered flat in the sense specified above), atoms moving with different velocities experience, because of the Doppler effect, a different radiation field. This aspect is particularly important in the corona: as one moves away from the solar surface, the solar wind accelerates and the emission peak of the chromospheric radiation, as seen from the atom, falls out of resonance (Doppler dimming effect). If collisions capable of modifying the atom velocity are negligible (as it is the case in the solar corona), the SE equations for an atom moving with velocity $\mathbf{u}$ are completely decoupled from those of atoms moving with different velocities (velocity-coherence approximation, see Sect. 13.2 of LL04). We observe that, in general, the atom velocity $\mathbf{u}$ is given by the vector sum of the thermal velocity, $\mathbf{u}_T$, and of the solar wind velocity, $\mathbf{w}$. Indicating with $f(\mathbf{u})$ the velocity distribution of the atoms, the emission coefficient in the observer's frame is given by the following equation, which generalizes Eq. (19):

$$\varepsilon_i(\nu, \boldsymbol{\Omega}) = k_L \sum_{KQ} W_K(J_\ell, J_u)(-1)^Q \, \mathcal{T}_Q^K(i, \boldsymbol{\Omega}) \times \int \mathrm{d}^3\mathbf{u}\, f(\mathbf{u})\, \delta\left(\nu_0\left(1 + \frac{\mathbf{u}\cdot\boldsymbol{\Omega}}{c}\right) - \nu\right) \left[J_{-Q}^K(\nu_0)\right]_{\mathbf{u}}, \tag{20}$$

where $[J_{-Q}^K(\nu_0)]_{\mathbf{u}}$ is the radiation field tensor in the comoving frame (i.e., as "seen" by an atom moving with velocity $\mathbf{u}$). Taking the Doppler effect into account to the first perturbative order in u/c, we have

$$\left[J_Q^K(\nu_0)\right]_{\mathbf{u}} = \int \frac{\mathrm{d}\boldsymbol{\Omega}'}{4\pi} \sum_{i=0}^{3} \mathcal{T}_Q^K(i, \boldsymbol{\Omega}')\, I_i\left(\nu_0\left(1 + \frac{\mathbf{u}\cdot\boldsymbol{\Omega}'}{c}\right), \boldsymbol{\Omega}'\right), \tag{21}$$

where I_i are the four Stokes parameters of the pumping radiation, propagating along the direction $\boldsymbol{\Omega}'$. In Eq. (20), we have made an extra approximation: taking into account that the Doppler broadening $\Delta\nu_D$ produced by any realistic velocity distribution is much larger than the natural broadening Γ of the emission profile ϕ appearing in Eq. (19), we have substituted the profile ϕ (which now, due to the Doppler effect, should be a Lorentzian centered at the frequency $\nu_0' = \nu_0(1 + \mathbf{u}\cdot\boldsymbol{\Omega}/c)$), with a Dirac-delta. This is a very good approximation for all the previously-mentioned lines.

In the solar corona, the Zeeman splittings produced by the magnetic fields are generally negligible if compared to the Doppler width of the spectral lines being considered. Under these conditions, magnetic field diagnostics is still contained, through the Hanle effect, in the frequency-integrated quantities (see Bommier and Sahal-Bréchot 1982). The frequency-integrated emission coefficient is given by

$$\tilde{\varepsilon}_i(\boldsymbol{\Omega}) \equiv \int \mathrm{d}\nu\, \varepsilon_i(\nu, \boldsymbol{\Omega}) = k_L \sum_{KQ} W_K(J_\ell, J_u)(-1)^Q \, \mathcal{T}_Q^K(i, \boldsymbol{\Omega})\, \bar{J}_{-Q}^K(\nu_0), \tag{22}$$

with

$$\bar{J}_Q^K(\nu_0) = \int \mathrm{d}^3\mathbf{u}\, f(\mathbf{u}) \left[J_Q^K(\nu_0)\right]_{\mathbf{u}}. \tag{23}$$

We observe that the previous equations are valid in an arbitrary reference frame.

We make now the following assumptions:

1. we assume that the incident radiation coming from the underlying chromosphere is unpolarized and with cylindrical symmetry along the local vertical (note that, in general, this is not the case if sunspots or active regions are present on the solar surface);
2. we assume that the velocity distribution of the atoms is either isotropic, or with cylindrical symmetry along the local vertical (note that this is not the case if the thermal velocity distribution is not isotropic, or if the solar wind is not purely radial).

Under these assumptions, exploiting the symmetry of the problem, it can be easily verified that the emitted radiation is linearly polarized either parallel or perpendicular to the nearest limb. Moreover, from Eqs. (21) and (23) it can be shown that if the quantization axis is chosen along the local vertical, then only two components of the radiation field tensor are non-zero: $\bar{J}^0_0$ and $\bar{J}^2_0$. Substituting these quantities into Eq. (22), and choosing the reference direction for positive Stokes Q the parallel to the limb, it can be easily shown that the fractional polarization is given by

$$\begin{aligned} \tilde{p}_Q &\equiv \frac{\tilde{\varepsilon}_Q}{\tilde{\varepsilon}_I} = \frac{3W_2 \sin^2\theta}{4/w + W_2(3\cos^2\theta - 1)}, \\ \tilde{p}_U &\equiv \frac{\tilde{\varepsilon}_U}{\tilde{\varepsilon}_I} = 0, \end{aligned} \tag{24}$$

where θ is the angle between the local vertical and the direction of the emitted radiation, and $w = \sqrt{2}\bar{J}^2_0/\bar{J}^0_0$ is the anisotropy factor. The maximum value of the fractional polarization is obtained for $\theta = 90°$ (i.e., for a scattering point located in the plane of the sky), and $w = 1$:

$$(\tilde{p}_Q)_{\max} = \frac{3W_2}{4 - W_2}. \tag{25}$$

We recall that the anisotropy factor is unity for a unidirectional pumping beam, a condition that in our case is reached only asymptotically by considering a scattering point located infinitely away from the solar surface.

3.1.2 The Hanle Effect

We introduce now a deterministic magnetic field, sufficiently weak so that $\nu_L \ll \Delta\nu_p$ with ν_L the Larmor frequency and $\Delta\nu_p$ the width of the emission (or absorption) line profile of the radiation illuminating the atom. Under this assumption, the flat-spectrum approximation remains verified, and the splittings between the various Zeeman components can be safely neglected (Hanle effect regime). This condition is generally well satisfied by the above-mentioned coronal lines for the typical magnetic field intensities that are expected in the solar corona (i.e., fields of a few gauss). As shown in Sects. 10.3 and 13.7 of LL04, the frequency-integrated emission coefficient for a two-level atom, accounting for the Hanle effect, is given by (see Eq. (13.46) of LL04)

$$\tilde{\varepsilon}_i(\mathbf{\Omega}) = k_L \sum_{KQ} W_K(J_\ell, J_u) \frac{1}{1 + \mathrm{i}QH_u} (-1)^Q \, \mathcal{T}^K_Q(i, \mathbf{\Omega}) \, \bar{J}^K_{-Q}(\nu_0), \tag{26}$$

where $H_u = 2\pi\nu_L g t_{\text{life}}$ is a dimensionless parameter quantifying the efficiency of the Hanle effect (H_u is the quantum analog of the phase shift $\Delta\phi$ introduced in Sect. 2, where a classical description of the Hanle effect was provided). The critical field for the onset of the

Hanle effect in a given spectral line can be defined as the field B_H for which $H_u = 1$. It can be easily seen that

$$B_H = \frac{1.137 \times 10^{-7}}{t_{\text{life}}\, g}, \tag{27}$$

where g is the Landé factor of the level under consideration (see also footnote 1), and where t_{life} is the level's lifetime expressed in seconds for calculating B_H in gauss. The value of the critical field B_H in a series of spectral lines of high diagnostic interest will be provided in Sect. 4.5. It is very important to note that Eq. (26) is only valid in a reference frame with the quantization axis directed along the magnetic field: the tensorial quantities $\mathcal{T}^K_Q(i, \mathbf{\Omega})$ and $\bar{J}^K_Q(\nu_0)$ must therefore be calculated in this reference frame.[2] Comparing Eqs. (22) and (26), we see that in this reference frame, the Hanle effect is fully described by the term $1/(1 + \mathrm{i}QH_u)$. Through the Hanle effect, the magnetic field modifies the linear polarization degree $\tilde{p} = \sqrt{\tilde{p}_Q^2 + \tilde{p}_U^2}$ with respect to the unmagnetized case (in general it produces a depolarization, but for particular geometries it can also give rise to a hyperpolarization), and rotates the plane of linear polarization. In many applications, the Hanle effect is described by means of a diagram (Hanle diagram) in which the ratio $\tilde{p}/\tilde{p}_{\mathrm{B}=0}$, and the rotation angle ϕ are plotted as a function of the intensity and orientation of the magnetic field.

3.1.3 The Impact of Collisional Excitation

The two-level atom with unpolarized lower level discussed in the previous sections is suitable for modeling the O VI line at 1032 Å, which results from a transition between an upper level with $J_u = 3/2$ and a lower level with $J_\ell = 1/2$. As previously pointed out, the contribution of collisional excitation is not negligible in this line, so that the effect of inelastic collisions with electrons has to be taken into account. As shown by Khan and Landi Degl'Innocenti (2012), when isotropic[3] inelastic collisions are considered, Eq. (26) generalizes as follows:

$$\tilde{\varepsilon}_i(\mathbf{\Omega}) = k_L \left[\sum_{KQ} W_K(J_\ell, J_u) \frac{1}{1+\varepsilon+\mathrm{i}QH_u} (-1)^Q\, \mathcal{T}^K_Q(i, \mathbf{\Omega})\, \bar{J}^K_{-Q}(\nu_0) + \frac{\varepsilon}{1+\varepsilon} B_T\, \delta_{i0} \right], \tag{28}$$

where we have introduced the parameter $\varepsilon = C_{u\ell}/(A_{u\ell} + C_{u\ell})$, with $C_{u\ell}$ the de-excitation rate of inelastic collisions with electrons. The second term in the right-hand side of Eq. (28) describes the contribution to the frequency-integrated emission coefficient brought by collisionally excited atoms (note that since collisions are assumed to be isotropic, this term only contributes to the intensity). The other effect of inelastic collisions is to modify the term describing the Hanle effect:

$$\frac{1}{1+\mathrm{i}QH_u} \rightarrow \frac{1}{1+\varepsilon+\mathrm{i}QH_u} = \frac{1}{1+\varepsilon} \frac{1}{1+\mathrm{i}QH'_u}, \tag{29}$$

with $H'_u = H_u/(1+\varepsilon)$. We thus see that, in general, the effect of inelastic collisions is to decrease the degree of linear polarization (also in the absence of magnetic fields), and to reduce the sensitivity of the line to the Hanle effect.

---

[2]Note that in this reference frame $\bar{J}^0_0$ and $\bar{J}^2_0$ are not the only non-vanishing components of the radiation field tensor, even if the incident field is characterized by cylindrical symmetry along the local vertical.

[3]The assumption of isotropic collisions is generally well justified.

3.1.4 The Impact of Fine Structure

The H I Ly-α line is due to the transition between the hydrogen levels $n = 1$ and $n = 2$, with n the principal quantum number. Taking the fine structure (FS) of hydrogen into account, the level $n = 1$ is composed of the singlet $1s\ ^2S_{1/2}$, while the upper level $n = 2$ consists of two terms: the singlet $2s\ ^2S_{1/2}$, and the doublet $2p\ ^2P_{1/2,3/2}$. Since collisional processes are negligible in the formation of this line in the solar corona, the upper level $2s\ ^2S_{1/2}$ is not populated and it does not bring any contribution to the emissivity in this line. A correct modeling of the Ly-α line can thus be obtained by considering a two-term atom with unpolarized lower term. In the following, each term will be specified by the set of quantum numbers (β, L, S), with β the electronic configuration, L the electronic orbital angular momentum, and S the electronic spin. As shown in Sect. 10.16 of LL04, the frequency-integrated emission coefficient for this atomic model is still given by Eq. (22), with the following substitution

$$W_K(J_\ell, J_u) \rightarrow \left[W_K(\beta_\ell L_\ell S \beta_u L_u)\right]_{\text{fs}}. \tag{30}$$

The polarizability factor for a two-term atom, $[W_K(\beta_\ell L_\ell S \beta_u L_u)]_{\text{fs}}$, is given by Eq. (10.133) of LL04. This factor accounts for the depolarizing effect of FS, and it can be conveniently written in the form:

$$\left[W_K(\beta_\ell L_\ell S \beta_u L_u)\right]_{\text{fs}} = W_K(L_\ell, L_u)\left[D_K(\beta_u L_u S)\right]_{\text{fs}}, \tag{31}$$

where W_K is the polarizability factor for a two-level atom defined in Eq. (14), and $[D_K(\beta_u L_u S)]_{\text{fs}}$ is the depolarizing factor due to FS, given by

$$\left[D_K(\beta_u L_u S)\right]_{\text{fs}} = \frac{1}{(2S+1)} \sum_{J_u J_u'} (2J_u + 1)(2J_u' + 1) \begin{Bmatrix} L_u & L_u & K \\ J_u & J_u' & S \end{Bmatrix}^2 \frac{1}{1 + 2\pi \mathrm{i} \nu_{J_u', J_u} / A_{u\ell}}. \tag{32}$$

The polarizability factor for a two-term atom defined in Eq. (31) accounts for the effect of interference between different FS levels of the same term. In an optically thin medium, such as the solar corona, the impact of the above-mentioned interference is however important only when the separation between the interfering levels is of the same order of magnitude as their natural width.[4] This is not the case for the upper term of Ly-α, for which

$$\frac{\Delta\nu_{\text{fs}}}{A_{u\ell}} \approx 17, \tag{33}$$

with $\Delta\nu_{\text{fs}}$ the separation between the FS levels $^2P_{1/2}$ and $^2P_{3/2}$, and $A_{u\ell}$ the Einstein coefficient for spontaneous emission from the upper to the lower term. Since this ratio is much larger than unity, we can conclude that, unless the magnetic field is sufficiently strong so to enter the incomplete Paschen-Back effect regime (as shown in Fig. 2 of Bommier and Sahal-Bréchot 1982, for Ly-α this takes place for fields larger than 1000 G), interference between different J-levels are negligible in the modeling of Ly-α in the solar corona.

In the limiting case where the splitting between the FS levels is infinitely large, and the corresponding interference is therefore completely negligible, the depolarizing factor of

[4]This is not the case for optically thick media. When modeling solar disk observations, for instance, interference between different FS levels are fundamental for explaining the polarization profiles in the wings of the components of the multiplet, even when the separation between the levels is much larger than their natural width (see Sect. 4).

Eq. (32) takes a limit value, D_K^{∞}, given by Eq. (10.139) of LL04. In this case, the polarizability factor reduces to

$$\begin{aligned}\left[W_K^{\infty}(\beta_\ell L_\ell S \beta_u L_u)\right]_{\mathrm{fs}} &= W_K(L_\ell, L_u)\left[D_K^{\infty}(LS)\right]_{\mathrm{fs}} \\ &= \frac{3(2L_u+1)}{2S+1}\begin{Bmatrix} 1 & 1 & K \\ L_u & L_u & L_\ell \end{Bmatrix}^2 \sum_{J_u}(2J_u+1)^2 \begin{Bmatrix} L_u & L_u & K \\ J_u & J_u & S \end{Bmatrix}^2. \end{aligned} \tag{34}$$

This expression neglects interference between different J-levels, but it takes into account the contribution to the emissivity brought by all the FS transitions of a given multiplet.

As shown in Sect. 10.18 of LL04, the impact of a magnetic field of arbitrary strength (from the Zeeman effect regime to the incomplete and complete Paschen-Back effect regime) on the frequency-integrated emission coefficient of a two-term atom can be taken into account by suitably generalizing the polarizability factor $[W_K(\beta_\ell L_\ell S \beta_u L_u)]_{\mathrm{fs}}$. The resulting quantity, indicated in LL04 with the symbol $[W_{KK'Q}(\beta_\ell L_\ell S \beta_u L_u;\, B)]_{\mathrm{fs}}$, with B the magnetic field intensity, is given by Eq. (10.153) of LL04. If interference between different J-levels is neglected, such equation reduces to

$$\begin{aligned}&\left[W_{KK'Q}^{\infty}(\beta_\ell L_\ell S \beta_u L_u;\, B)\right]_{\mathrm{fs}} \\ &= \frac{3(2L_u+1)}{2S+1}\sqrt{(2K+1)(2K'+1)}\begin{Bmatrix} 1 & 1 & K \\ L_u & L_u & L_\ell \end{Bmatrix}\begin{Bmatrix} 1 & 1 & K' \\ L_u & L_u & L_\ell \end{Bmatrix} \\ &\times \sum_{J_u M_u M_u'} (2J_u+1)^2 \begin{Bmatrix} L_u & L_u & K \\ J_u & J_u & S \end{Bmatrix}\begin{Bmatrix} L_u & L_u & K' \\ J_u & J_u & S \end{Bmatrix}\begin{pmatrix} J_u & J_u & K \\ -M_u & M_u' & -Q \end{pmatrix} \\ &\times \begin{pmatrix} J_u & J_u & K' \\ -M_u & M_u' & -Q \end{pmatrix}\frac{1}{1+2\pi \mathrm{i}\nu_{J_u M_u, J_u M_u'}/A_{u\ell}}. \end{aligned} \tag{35}$$

Recalling Eq. (22), the frequency-integrated emission coefficient for the Ly-α line, in the presence of a magnetic field, is thus given by

$$\tilde{\varepsilon}_i(\mathbf{\Omega}) = k_L \sum_{KQ}\left[W_K^{\infty}(\beta_\ell L_\ell S \beta_u L_u;\, B)\right]_{\mathrm{fs}}(-1)^Q\, \mathcal{T}_Q^K(i, \mathbf{\Omega})\, \bar{J}_{-Q}^K(\nu_0). \tag{36}$$

A Hanle diagram for the Ly-α line, obtained by considering a scattering point on the plane of the sky (i.e., considering a 90° scattering process), and a magnetic field, also contained in the plane of the sky and parallel to the closest limb, is shown in Fig. 5 of Bommier and Sahal-Bréchot (1982). The same authors point out that this Hanle diagram has a very little dependence on the scattering height. Indeed, although $\tilde{p}$ and $\tilde{p}_{B=0}$ vary with height, their ratio has a very small height dependence, which can be safely neglected once the measurements inaccuracies are considered (see Fig. 6 of Bommier and Sahal-Bréchot 1982), while the rotation angle ϕ is practically independent of height.

Figure 1 illustrates how the frequency-integrated fractional linear polarization of the Ly-α line is modified by the Hanle effect caused by the presence in the solar corona of random-azimuth magnetic fields inclined with respect to the solar radius vector. In agreement with the results shown in Bommier and Sahal-Bréchot (1982), Fig. 1 shows that at 1.5 solar radii above the visible solar limb, in the absence of magnetic fields, the hydrogen Ly-α line should be characterized by a linear polarization degree of more than 20 %.

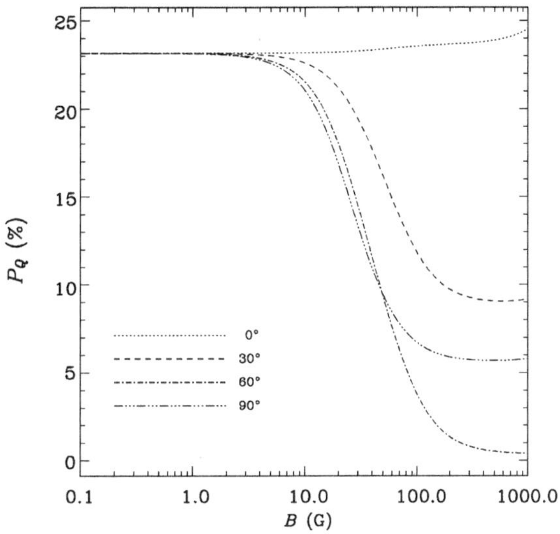

Fig. 1 Theoretical estimate of the (frequency-integrated) fractional linear polarization produced by 90° scattering processes in the hydrogen Ly-α line, taking into account the Hanle effect caused by the presence of a random-azimuth magnetic field with the inclination indicated in the figure. The model calculations correspond to a height of 1.5 solar radii above the solar visible surface. From Trujillo Bueno et al. (2005)

3.1.5 The Impact of Hyperfine Structure

Another physical ingredient that can sensibly affect scattering polarization, and that in principle should be taken into account when modeling the H I Ly-α line, is the hyperfine structure (HFS) of hydrogen. In order to estimate the impact of HFS on this line, we consider a "two-term atom with HFS". Starting from the equations derived in Casini and Manso Sainz (2005), neglecting lower term polarization and stimulated emission, after some algebra we find that in the unmagnetized case the frequency-integrated emission coefficient in the four Stokes parameters is still given by Eq. (22) with the substitution

$$W_K(J_\ell, J_u) \to \left[W_K(\beta_\ell L_\ell S \beta_u L_u, I)\right]_{\text{fs+hfs}}. \tag{37}$$

Also in this case, the polarizability factor $[W_K(\beta_\ell L_\ell S\beta_u L_u, I)]_{\text{fs+hfs}}$ can be conveniently written in the form

$$\left[W_K(\beta_\ell L_\ell S \beta_u L_u, I)\right]_{\text{fs+hfs}} = W_K(L_\ell, L_u)\left[D_K(\beta_u L_u S I)\right]_{\text{fs+hfs}}, \tag{38}$$

where the depolarizing factor due to the combined action of FS and HFS, $[D_K(\beta_u L_u S I)]_{\text{fs+hfs}}$, is given by

$$\left[D_K(\beta_u L_u S I)\right]_{\text{fs+hfs}} = \frac{1}{(2S+1)(2I+1)} \sum_{J_u J_u'} \sum_{F_u F_u'} (2J_u+1)(2J_u'+1)(2F_u+1)(2F_u'+1) \times \begin{Bmatrix} L_u & L_u & K \\ J_u & J_u' & S \end{Bmatrix}^2 \begin{Bmatrix} J_u' & J_u & K \\ F_u & F_u' & I \end{Bmatrix}^2 \frac{1}{1 + 2\pi \mathrm{i} \nu_{J_u' F_u', J_u F_u} / A_{u\ell}}, \tag{39}$$

with F the quantum number associated to the total (electronic plus nuclear) angular momentum ($\mathbf{F} = \mathbf{J} + \mathbf{I}$). It can be easily proved that if HFS is neglected (i.e., if we set $I = 0$), $[W_K(\beta_u L_u S \beta_\ell L_\ell, I)]_{\text{fs+hfs}}$ reduces to the polarizability factor of a two-term atom defined in Eq. (31).

Hyperfine structure plays an important role whenever the frequency separation among the various HFS F-levels is of the same order of magnitude as the natural width of the levels. This is not the case for the upper level ${}^2P_{3/2}$ of Ly-α, for which

$$\frac{\Delta\nu_{\rm hfs}}{A_{u\ell}} \approx 0.04 \tag{40}$$

with $\Delta\nu_{\rm hfs}$ the typical frequency separation between different HFS levels (see Sect. 3.5 of LL04). A possible way to estimate more quantitatively the impact of HFS in this line is to compare the values of $(\tilde{p}_Q)_{\rm max}$ (see Eq. (25)) obtained by taking into account and neglecting HFS (see Bommier and Sahal-Bréchot 1982). Using the values of the FS and HFS splitting of the $2p\ {}^2P$ term of hydrogen tabulated in Bashkin and Stoner (1975), we find

$$(\tilde{p}_Q)^{\rm fs}_{\rm max} = 27.3\ \%, \qquad (\tilde{p}_Q)^{\rm fs+hfs}_{\rm max} = 26.7\ \%. \tag{41}$$

Since the inaccuracies of the measurements are expected to be larger than the discrepancy between these two values, one can conclude that HFS can be safely neglected when modeling the Ly-α line, as already pointed out by Bommier and Sahal-Bréchot (1982). We recall that the quantity $(\tilde{p}_Q)_{\rm max}$ refers to the limit of a scattering point located in the plane of the sky, infinitely away from the solar surface ($w = 1$). As shown in Fig. 3 of Bommier and Sahal-Bréchot (1982), the discrepancy between the results obtained by considering and neglecting HFS decreases as one considers scattering points located closer and closer to the solar surface.

3.1.6 Integration Along the Line-of-Sight

In the previous sections, we have considered a single scattering process. Since in the spectral lines being considered the solar corona is optically thin, the Stokes parameters of the emergent radiation can be calculated by simply adding the contributions to the scattered radiation of all the points along the line-of-sight (LOS):

$$\tilde{I}_i(\mathbf{\Omega}) = \int_{LOS} \tilde{\varepsilon}_i(\mathbf{\Omega})\, {\rm d}s. \tag{42}$$

We have seen that in a single scattering process, the coronal Ly-α radiation is characterized by a measurable degree of linear polarization (on the order of 20 % at about one solar radius above the limb), and that this is appreciably modified by the Hanle effect produced by the magnetic fields expected in the solar corona. The problem is now to understand whether these results remain valid also when the "averaging" effects due to the LOS integration are taken into account. This problem has been first investigated by Fineschi et al. (1992), who considered magnetic loop structures filled by a isothermal plasma at hydrostatic equilibrium, with temperature and density higher than those of the surrounding coronal atmosphere, and calculated the Stokes parameters of the emergent Ly-α radiation performing an integration along the LOS. This work showed that the integration along the LOS does not wipe out completely the signatures of the Hanle effect, which indeed should remain measurable. The authors also pointed out that with a polarization error $\Delta\tilde{p} \approx 1\ \%$, in principle it would be possible to infer the magnetic field strength and direction with an error smaller than 30 %, and that this result does not critically depend on the model parameters and, in particular, on the relative density of the loop plasma with respect to the surrounding corona (see Figs. 3 and 4 of Fineschi et al. 1992). This investigation has been successively extended to more complex magnetic field configurations (Fineschi et al. 1993), and to H I Ly-β and Ly-γ (Fineschi et al. 1999). The above-mentioned works address the problem also from an experimental point of view, defining the relevant parameters and the required instrumental sensitivity.

Other investigations of the signatures of the Hanle effect produced by simple magnetic field configurations (e.g., dipolar fields, current sheets in the Sun's equatorial plane, photospheric field extrapolations), accounting for the integration along the LOS, have been carried out by Derouich et al. (2010) for the Ly-α line, and by Raouafi et al. (2009) for both Ly-α and Ly-β.

The integration along the LOS requires performing a series of rotations of the reference system. The expression of the emission coefficient given in Eq. (26) is actually valid in a reference frame with the quantization axis directed along the magnetic field (which of course can have a different orientation in the various points along the LOS), and the evaluation of the tensors $\mathcal{T}^K_Q$ and J^K_Q is not straightforward in this system. The powerful formalism of the polarization tensors introduced by Landi Degl'Innocenti (1984) allows us to easily perform the geometrical transformations. For the problem under consideration it is generally useful to consider three reference systems:

1. an "observer" (OBS) reference system with the origin at the center of the Sun, and defined so that the x-axis is directed along the LOS, the z-axis is in the plane of the sky, pointing, for example, to the North solar pole, and the y-axis is defined so to form a right-handed system. This is the system in which the polarization tensor $\mathcal{T}^K_Q(i, \mathbf{\Omega})$ entering the expression of the emission coefficient is more easily defined (the angles specifying $\mathbf{\Omega}$ are constant along the LOS). In this system, the position vector of the scattering point P is defined in a spherical coordinate system (r_P, θ_P, χ_P) having the polar axis directed along the z-axis.
2. a "local vertical" (LV) reference system with the origin in the scattering point, and defined so that its z'-axis is directed along the local vertical, and its x'-axis lies in the plane defined by the z and z' axes. This is the system in which the radiation field tensor describing the incident field is generally more easily defined. In this system, the magnetic field vector is specified by its modulus, B, and by the polar angles θ_B and χ_B.
3. a "local magnetic" (LB) reference system, with the origin in the scattering point and defined so that its z''-axis is directed along the magnetic field. This is the system in which the emission coefficient has its simplest form, namely that of Eq. (26).

The expression of $\mathcal{T}^K_Q$ in the LB system (needed to calculate Eq. (26)) is related to the one in the OBS system through the equation

$$\big[\mathcal{T}^K_Q(i, \mathbf{\Omega})\big]_{\rm LB} = \sum_{Q'} D^K_{Q'Q}(R)\big[\mathcal{T}^K_{Q'}(i, \mathbf{\Omega})\big]_{\rm OBS}, \tag{43}$$

where $D^K_{Q'Q}(R)$ is the rotation matrix relative to the rotation R which brings the OBS system into the LB system. The rotation R is more easily performed in two steps: first from OBS to LV (R_1) and then from LV to LB (R_2). Indeed, the Euler angles defining R_1 and R_2 are given by:

$$R_1 = (\chi_P, \theta_P, 0), \qquad R_2 = (\chi_B, \theta_B, 0), \tag{44}$$

and

$$\big[\mathcal{T}^K_Q(i, \mathbf{\Omega})\big]_{\rm LB} = \sum_{Q'Q''} D^K_{Q'Q''}(R_1) D^K_{Q''Q}(R_2)\big[\mathcal{T}^K_{Q'}(i, \mathbf{\Omega})\big]_{\rm OBS}. \tag{45}$$

Similarly, the radiation field tensor in the LB and LV systems are related by the equation

$$\left[\bar{J}^K_Q(\nu_0)\right]_{\rm LB} = \sum_{Q'} \mathcal{D}^K_{Q'Q}(R_2)\left[\bar{J}^K_{Q'}(\nu_0)\right]_{\rm LV}, \tag{46}$$

with R_2 the rotation bringing the reference system LV into LB.

3.1.7 Symmetry-Breaking Processes

As discussed in Sect. 3.1.1, if the incident radiation is unpolarized, and the problem is characterized by cylindrical symmetry around the local vertical, then the emitted radiation is linearly polarized either parallel or perpendicular to the closest limb. Under these circumstances, if the quantization axis is chosen along the local vertical, only two components of the radiation field tensor, J^0_0 and J^2_0, are non-zero. The presence of a magnetic field breaks this symmetry, thus producing a modification of the linear polarization degree, and a rotation of the plane of linear polarization with respect to the unmagnetized case (Hanle effect).

In addition to the presence of a magnetic field, there are other phenomena that are capable of breaking the symmetry of the problem, and therefore of modifying the linear polarization of the emitted radiation. The first possibility is when the velocity distribution of the atoms $f(\mathbf{u})$ is not isotropic neither cylindrically symmetric around the local vertical. This takes place either when the intrinsic velocity distribution is not isotropic or when the solar wind is not purely radial. From Eqs. (21) and (23), it can be demonstrated that in this case, besides J^0_0 and J^2_0, also the components J^2_Q, with $Q = \pm 1, \pm 2$ are non-vanishing. The inclusion of these components into Eq. (22) leads to a modification of the linear polarization of the emitted radiation, which is similar to that produced by the Hanle effect. The possibility that the velocity distribution of hydrogen and of some ions (like O VI) in the solar corona could be anisotropic was suggested by various authors in order to interpret a series of observational data provided by the Ultraviolet Coronograph Spectrometer (UVCS) instrument on board of the Solar and Heliospheric Observatory (SOHO) mission (e.g., Kohl et al. 1997). This lead Fineschi (2001), Raouafi and Solanki (2003) and, more recently, Khan and Landi Degl'Innocenti (2012) to investigate the impact of anisotropic velocity distributions on the scattering polarization profiles of the H I Ly-α line and of the O VI line at 1032 Å. The case of a non-radial solar wind was first considered by Sahal-Bréchot et al. (1998) for the O VI 1032 Å line, and more recently by Khan et al. (2011) for Ly-α, Ly-β and Ly-γ.

A second way for breaking the symmetry of the problem is through the presence of active regions, characterized by an intensity contrast with respect to the surrounding chromosphere. As shown in Sect. 12.4 of LL04, also in this case the non-zero components of the radiation field tensor are in general J^0_0, and J^2_Q, with $Q = 0, \pm 1, \pm 2$. These effects have been first investigated in the O VI line at 1032 Å by Sahal-Bréchot et al. (1986), who pointed out that by neglecting the possible presence of active regions, one can overestimate (or underestimate) the magnetic field value inferred through the Hanle effect by up to 30–40 %.

The combined effect of all these symmetry-breaking agents, namely the presence of a magnetic field (Hanle effect), active regions, and a non-radial solar wind, has been investigated by Khan et al. (2011). In this work, the authors carried out forward modeling calculations of the linear polarization signals expected in the first three lines of the Lyman series using a global coronal model, with a solar minimum-like magnetic field topology (i.e., a global dipole field directed along the Sun's polar regions, with a current-sheet-like structure in the equatorial plane, and strengths in the range of a few gauss). The authors concluded that all these symmetry-breaking agents produce an appreciable modification of the linear polarization signals of these lines, also when the integration along the LOS is performed. In the case of Ly-α, the modifications due to the various phenomena are also quantitatively similar to each other, which makes rather difficult their disentanglement. In principle, the sensitivity of Ly-β and Ly-γ to weaker fields makes these lines more suitable for magnetic field diagnostics in the solar corona. However, as previously discussed, the drawback of these lines is their faintness and their smaller degree of linear polarization, due to the larger contribution from collisional excitation. If only Ly-α can be used, in order to get reliable estimates of the coronal magnetic field through the Hanle effect, it is thus necessary to develop suitable methods for tracking active regions and for assessing the value of the solar wind velocity, so to subtract their effect on the polarization signals.

An investigation of the signatures produced by the Hanle effect on the polarization signals of the Ly-α line, in local coronal structures characterized by relatively large values of the magnetic field (e.g., coronal loops) can be found in Khan and Landi Degl'Innocenti (2011).

3.2 On-disk Coronal Polarimetry

In Sect. 3.1 we have studied the scattering polarization properties of some spectral lines (e.g., hydrogen Ly-α) originating in the solar corona from the resonance scattering of the line radiation coming from the underlying layers of the solar atmosphere. At this point, it is interesting to ask ourselves the following question: can we expect scattering polarization also in EUV coronal emission lines for which the underlying quiet solar disk is practically dark? The answer to this question may appear trivial, because it is often given for granted that atoms that are collisionally excited (mostly by electrons) are not able to emit polarized light since collisions are isotropic.

Some EUV lines share their lower level (the ground level) with that of a forbidden line at visible or near-IR wavelengths. Therefore, this lower level can be polarized due to the anisotropic illumination of the atoms at the forbidden line wavelength. Manso Sainz and Trujillo Bueno (2009) showed that this lower-level atomic polarization can be partly transferred to the upper level of the EUV line by isotropic collisional excitation. Therefore, since the upper level of the EUV line can be polarized, we may have measurable scattering polarization signals caused by the ensuing selective emission processes in the allowed EUV line. The linear polarization thus generated would be sensitive to the electronic density (N_e) and to the orientation of the magnetic field vector, although not to its strength because for permitted EUV lines $B \gg B_H(J_l)$ and $B \ll B_H(J_u)$ (with B the magnetic strength of the coronal plasma). Interestingly enough, contrary to the case of forbidden line polarimetry, the linear polarization of allowed EUV transitions would be measurable also when observing coronal structures (e.g., loops) against the underlying solar disk. This is important because it provides a way for mapping the magnetic field of the extended solar atmosphere all the way up from the photosphere to the corona.

There are several interesting EUV lines satisfying these requirements. For example, as shown in Fig. 2 (left panel), the theoretical prediction for the Fe X line at 174.5 Å is that the ensuing Q/I amplitude for a point in the plane of the sky at a height of 0.1 solar radii above the solar surface varies between about 0.1 % for $N_\mathrm{e} = 10^9$ cm^{-3} and 5 % for $N_\mathrm{e} = 10^7$ cm^{-3} in the presence of a vertical field (see left panel). The Q/I amplitude is a factor two smaller in the case of a horizontal magnetic field observed against the solar disk in forward-scattering geometry (see right panel).

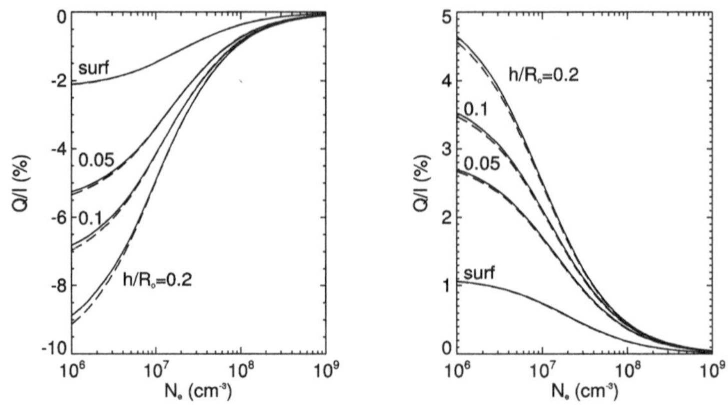

Fig. 2 Q/I signal of the permitted Fe X line at 174.5 Å for different electronic densities N_e, and two different geometrical configurations: vertical magnetic field observed perpendicularly (*left panel*); horizontal magnetic field observed over the disk center (*right panel*). The reference direction for Stokes Q is the perpendicular to the magnetic field. *Labels* indicate the height h over the solar surface in units of the solar radius $R_\odot$. *Solid curves*: calculation for an atomic model consisting of the 31 lowest levels and radiative transitions between them, and the 465 collisional transitions involving those levels. *Dashed curves*: calculation according to the simplifying assumptions in Eqs. (5)–(9) of Manso Sainz and Trujillo Bueno (2009). From Manso Sainz and Trujillo Bueno (2009)

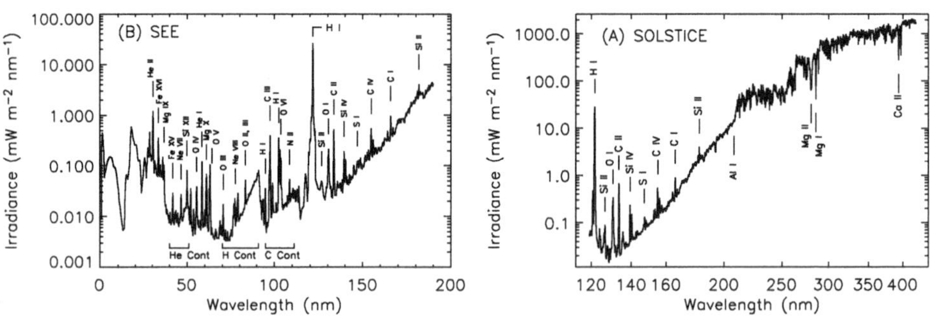

Fig. 3 The solar ultraviolet irradiance spectrum. *Left panel*: from the Thermosphere, Ionosphere, Mesosphere, Energetics, and Dynamics (TIMED) Solar Extreme ultraviolet Experiment (SEE). *Right panel*: from the Upper Atmosphere Research Satellite (UARS) Solar Stellar Irradiance Comparison Experiment (SOLSTICE). From Woods and Rottman (2002)

4 The Upper Chromosphere and Transition Region

The upper chromosphere of the Sun and the chromosphere-corona transition region are hot and rarefied plasmas with temperatures ranging from 10^4 K to 10^5 K (e.g., see Fig. 1 of Trujillo Bueno et al. 2012). The primary emission of these plasmas lies in the ultraviolet (UV; 2000–4000 Å), far ultraviolet (FUV; 1210–2000 Å) and extreme ultraviolet (EUV; 100–1210 Å) regions of the solar disk spectrum where, besides a continuum, a multitude of spectral lines is present, most of them being in emission and located at wavelengths below 2000 Å (see Fig. 3). To probe the geometrical complexity and the magnetic structure of the plasma in the upper chromosphere and transition region we need to measure and model the Stokes profiles of ultraviolet spectral lines (e.g., Stenflo 1991; Trujillo Bueno 2014).

This section deals with the polarization signals of the solar disk radiation in some strong ultraviolet resonance lines, and with their sensitivity to the presence of a magnetic field. In particular, we will focus our attention on the following spectral lines: Mg II h & k, H I Ly-α, and He II Ly-α. All these lines result from transitions between the ground term, composed of the single level $^2S_{1/2}$, and the first excited term, composed of the $^2P_{1/2}$ and $^2P_{3/2}$ levels. Since the only level of such resonance lines that can carry atomic alignment is the $^2P_{3/2}$ excited level, with angular momentum $J = 3/2$, the scattering polarization signal in the line core and its modification by the Hanle effect is solely due to the population imbalances and quantum coherence between the magnetic sublevels of this upper level. The critical field for the onset of the Hanle effect in these lines can be estimated by means of Eq. (27).

In strong resonance lines like Ly-α and Mg II h & k the upper solar chromosphere is optically thick, so that their modeling requires taking into account the effects of radiative transfer. Moreover, modeling their intensity and polarization profiles requires considering correlation effects between the frequencies of the incoming and outgoing photons in the scattering events (partial frequency redistribution, or PRD, effects). In addition, since such resonance lines pertain to multiplets in which the separation between the two upper fine-structure J-levels is relatively small, in optically thick plasmas it is generally necessary to consider the impact of quantum interference between such different J-levels (J-state interference), this latter ingredient being, however, negligible for the intensity. A correct modeling of these lines thus requires solving the full non-LTE radiative transfer problem for polarized radiation considering a two-term atom (i.e., an atomic model suitable to account for J-state interference), and including PRD effects (see Belluzzi and Trujillo Bueno 2014). The limit of complete frequency redistribution (CRD), without J-state interference, represents a suitable approximation only for estimating the scattering polarization signals at line center, but not for modeling the nearby and far wings of these lines. We finally observe that the lower level of the above-mentioned lines can be assumed to be infinitely sharp (as it is the ground level, it has a very long life time) and unpolarized (as it has $J = 1/2$ and, by definition, it cannot carry atomic alignment).

Although a fully consistent PRD theory (accounting for both collisional and Doppler redistribution) for the generation and transfer of polarized radiation in multi-term atoms has not been completed yet (e.g., Casini and Manso Sainz 2016; Bommier 2016), suitable, approximate theoretical approaches have been proposed for the particular case of a two-term atom without atomic polarization in the lower levels (see Belluzzi and Trujillo Bueno 2014, and references therein). A suitable way to describe the transfer of resonance line polarization with PRD and J-state interference is through the redistribution matrix formalism. As in the unpolarized case, the most general form of the redistribution matrix is given by a linear combination of two terms: the first one, generally referred to as R_{II} (see the terminology introduced by Hummer 1962) describes the limit of completely correlated (or coherent) scattering in the atomic rest frame; the second one, generally referred to as R_{III}, describes the limit of completely uncorrelated (CRD) scattering. According to the theoretical approach of Belluzzi and Trujillo Bueno (2014), in the absence of magnetic fields, and neglecting depolarizing collisions, the branching ratios that multiply R_{II} and R_{III} are given by α and $(1 - \alpha)$, respectively, with

$$\alpha = \frac{A_{ul} + C_{ul}}{A_{ul} + C_{ul} + Q_{\text{el.}}}, \tag{47}$$

where A_{ul} is the Einstein coefficient for spontaneous emission from the upper to the lower term (see footnote at page 314 of LL04), C_{ul} is the inelastic collisional rate for the transition from the upper to the lower term (see Eq. (38) of Belluzzi et al. 2013), and where Q_{el} is the rate of elastic collisions inducing transitions between magnetic sublevels pertaining either to the same J-level, or to different J-levels of the same term. The *coherence* fraction α is the probability that a photon (after being absorbed) is re-emitted from the same excited sublevel (before the atom suffers an elastic collision), provided that it is re-emitted at all. The fact that α is proportional to the *total* rate of transitions out of the upper level results from the definition of the redistribution matrices given by Belluzzi and Trujillo Bueno (2014), which is consistent with the above-mentioned definition of α as the conditional probability of correlated re-emission (e.g., Hubeny and Mihalas 2015). Such redistribution matrices take into account the Doppler redistribution in the laboratory frame, assuming Maxwellian velocity distribution functions.

Given that $\alpha \approx 1$ in the upper solar chromosphere (where the core of strong UV, FUV and EUV lines originates) the $R_{\rm II}$ matrix is particularly important. Its explicit expression in the laboratory frame is given by (see Belluzzi and Trujillo Bueno 2014 for a detailed derivation)

$$\begin{aligned}
&\big[R_{\rm II}(\nu', \boldsymbol{\Omega}'; \nu, \boldsymbol{\Omega})\big]_{ij} \\
&\quad = \frac{2L_u+1}{2S+1} \sum_K \sum_{J_u J_u'} \sum_{J_\ell J_\ell'} (-1)^{J_\ell - J_\ell'} 3(2J_u+1)(2J_u'+1)(2J_\ell+1)(2J_\ell'+1) \\
&\quad \times \begin{Bmatrix} L_u & L_\ell & 1 \\ J_\ell & J_u & S \end{Bmatrix} \begin{Bmatrix} L_u & L_\ell & 1 \\ J_\ell & J_u' & S \end{Bmatrix} \begin{Bmatrix} L_u & L_\ell & 1 \\ J_\ell' & J_u & S \end{Bmatrix} \\
&\quad \times \begin{Bmatrix} L_u & L_\ell & 1 \\ J_\ell' & J_u' & S \end{Bmatrix} \begin{Bmatrix} K & J_u' & J_u \\ J_\ell & 1 & 1 \end{Bmatrix} \begin{Bmatrix} K & J_u' & J_u \\ J_\ell' & 1 & 1 \end{Bmatrix} \big[P^{(K)}(\boldsymbol{\Omega}', \boldsymbol{\Omega})\big]_{ij} \\
&\quad \times \frac{1}{\pi\, \Delta\nu_D^2 \sin\Theta} \frac{1}{1+\epsilon' + 2\pi\, {\rm i}\nu_{J_u' J_u}/A_{u\ell}} \exp\left[-\frac{(\nu' - \nu + \nu_{J_\ell' J_\ell})^2}{4\,\Delta\nu_D^2 \sin^2(\Theta/2)}\right] \\
&\quad \times \frac{1}{2}\left[W\left(\frac{a}{\cos(\Theta/2)}, \frac{x_{J_u J_\ell} + x'_{J_u J_\ell'}}{2\cos(\Theta/2)}\right) + W\left(\frac{a}{\cos(\Theta/2)}, \frac{x_{J_u' J_\ell} + x'_{J_u' J_\ell'}}{2\cos(\Theta/2)}\right)^*\right].
\end{aligned} \tag{48}$$

The quantities L_u, L_ℓ, S, J_ℓ and J_u are quantum numbers that depend on the particular two-term atomic model considered. The matrix $[P^{(K)}(\boldsymbol{\Omega}', \boldsymbol{\Omega})]_{ij}$ is the so-called scattering-phase matrix, which depends on the propagation directions $\boldsymbol{\Omega}$ and $\boldsymbol{\Omega}'$, and on the angle that specifies the reference direction for positive Stokes Q. Its explicit expression can be conveniently written in terms of the polarization tensors $\mathcal{T}^K_Q(i, \boldsymbol{\Omega})$ defined in LL04:[5]

$$\big[P^{(K)}(\boldsymbol{\Omega}', \boldsymbol{\Omega})\big]_{ij} = \sum_{Q=-K}^{K} (-1)^Q\, \mathcal{T}^K_Q(i, \boldsymbol{\Omega})\, \mathcal{T}^K_{-Q}(j, \boldsymbol{\Omega}'). \tag{49}$$

The angle Θ is the angle between the directions $\boldsymbol{\Omega}$ and $\boldsymbol{\Omega}'$ (scattering angle). The Doppler width $\Delta\nu_D$ is given by

$$\Delta\nu_D = \frac{\nu_0}{c}\sqrt{\frac{2k_B T}{M}}, \tag{50}$$

[5]Note that summing over K the scattering phase matrix defined in Eq. (49) one recovers the matrix appearing in the right-hand side of Eq. (9) (multiplied by $3/4$).

where ν_0 is the frequency of the spectral line, T is the temperature of the plasma at the spatial point under consideration, and M is the mass of the atom. Note that $\varepsilon' = C_{ul}/A_{ul}$. The frequencies $\nu_{J_a J_b}$ are the Bohr frequencies corresponding to the energy interval between the levels J_a and J_b:

$$\nu_{J_a J_b} = \big(E(J_a) - E(J_b)\big)/h. \tag{51}$$

The function W is defined as

$$W(a,x) = H(a,x) + \mathrm{i}L(a,x), \tag{52}$$

where H and L are the Voigt and the Faraday-Voigt functions, respectively:

$$H(a,x) = \frac{a}{\pi}\int_{-\infty}^{+\infty} \mathrm{e}^{-y^2} \frac{1}{(x-y)^2 + a^2}\,\mathrm{d}y, \tag{53}$$

$$L(a,x) = \frac{1}{\pi}\int_{-\infty}^{+\infty} \mathrm{e}^{-y^2} \frac{x-y}{(x-y)^2 + a^2}\,\mathrm{d}y, \tag{54}$$

with a the damping parameter. Finally, the "reduced" frequencies x are defined as

$$x_{J_a J_b} = \frac{(\nu_{J_a J_b} - \nu)}{\Delta\nu_D}, \qquad x'_{J_a J_b} = \frac{(\nu_{J_a J_b} - \nu')}{\Delta\nu_D}. \tag{55}$$

The elements of the redistribution matrix are real numbers. The radiative transfer results with PRD and J-state interference shown in some of the figures below have been obtained using an approximate expression for the R_{II} matrix, obtained by averaging the frequency part of the redistribution matrix over all possible scattering angles (angle-averaged approximation, see Eq. (35) of Belluzzi and Trujillo Bueno 2014), and using the approximate form of the R_{III} matrix given by Eq. (26) of Belluzzi and Trujillo Bueno (2014).

4.1 Mg II *h* & *k*

Spectroscopic observations of the Mg II h & k lines around 2800 Å obtained with space telescopes like the Interface Region Imaging Spectrograph (IRIS; see De Pontieu et al. 2014) provide information on the thermal and dynamic complexity of the upper solar chromosphere (e.g., Pereira et al. 2014). However, with the exception of the contribution by Henze and Stenflo (1987), there are no observations of the polarization in these strong resonance lines. In spite of the fact that the polarimeter on board the Solar Maximum Mission satellite used by Henze and Stenflo (1987) was not designed to measure the Stokes profiles with high polarimetric sensitivity (at least 0.1 %) and spectral resolution (at least 0.1 Å), these authors could nevertheless measure Q/I signals in the wings of the Mg II *h* & *k* lines. Theoretical work on this topic using the approximation of coherent scattering *in the observer's frame* (see Auer et al. 1980) had predicted significant Q/I signals in the wings of the Mg II *h* & *k* lines (of the order of a few percent) and that the linear polarization should be radial (i.e., Q/I negative) in the wings between the *h* and *k* lines.[6] Although the statistical significance of the observations of Henze and Stenflo (1987) is very marginal, they could confirm the approximate theoretical predictions in the line wings on the blue side of the *k* line and in the red side of the *h* line.

[6] Note that our choice for the reference direction for linear polarization is the parallel to the nearest solar limb.

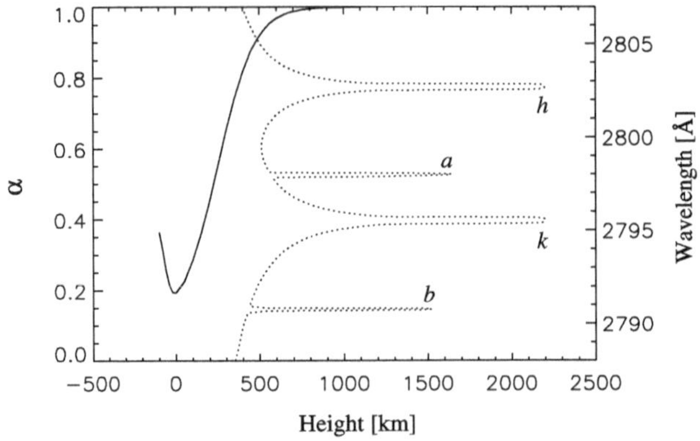

Fig. 4 The *solid curve* shows the variation with height in the solar atmosphere model C of Fontenla et al. (1993) (hereafter, FAL-C) of the coherence fraction α of Eq. (47). The *dotted curve* gives the atmospheric height where, for each wavelength, the ensuing optical depth is unity along a LOS with $\mu = 0.1$, assuming that the only blends in the wings of the h and k lines are those produced by the following Mg II lines: (a) two blending lines at 2797.930 Å and 2797.998 Å (resulting from the ${}^2P_{3/2}$–${}^2D_{3/2}$ and ${}^2P_{3/2}$–${}^2D_{5/2}$ transitions), and (b) one at 2790.777 Å (caused by the ${}^2P_{1/2}$–${}^2D_{3/2}$ transition). Note that in the FAL-C semi-empirical model the line cores of these Mg II absorption lines originate about 700 km below the Mg II h and k line cores, but still at chromospheric heights. In the FAL-C model the chromosphere-corona transition region starts around 2200 km. From Belluzzi and Trujillo Bueno (2012)

Center-to-limb observations of quiet Sun regions show a clear limb darkening in the intensity profiles of the Mg II h & k lines (Morrill and Korendyke 2008), suggesting that these lines should be linearly polarized by scattering processes (with the exception of the very center of the intrinsically unpolarizable h line, because it is a $J_l = J_u = 1/2$ transition). Recently, the physical mechanisms that control the linear and circular polarization in the Mg II h & k lines have been carefully studied by Belluzzi and Trujillo Bueno (2012); these authors took into account the joint action of PRD and J-state interference and presented detailed radiative transfer calculations in one-dimensional (1D) semi-empirical models of the solar atmosphere.

The dotted curve of Fig. 4 shows the height in a semi-empirical model of the quiet solar atmosphere where the optical depth at each of the considered wavelengths is unity along a LOS with $\mu = \cos\theta = 0.1$ (with θ the heliocentric angle). It indicates that the line-center radiation of the h and k resonance lines "originates" in the upper chromosphere of the model, just below the transition region to the 10^6 K corona. Notice that there are two other Mg II lines, in the blue and red wing of the k line, whose lower levels are the upper levels of the h and k lines. Interestingly, the line-center radiation of these other Mg II lines encodes information on the physical properties of the solar chromosphere at heights ~ 700 km below the transition region. The solid curve of Fig. 4 shows that for the Mg II h and k lines in the same semi-empirical model of the solar atmosphere the coherence fraction α (see Eq. (47)) is very close to unity at heights $h > 700$ km, where the line core and near wing wavelengths of the h and k lines originate. Therefore, the limit of completely correlated or coherent scattering *in the atomic rest frame* is a suitable approximation for modeling the intensity and polarization of the core and near wings of the Mg II h and k lines (see also Fig. 3 of Belluzzi and Trujillo Bueno 2012). The impact of $R_{\rm III}$ (which describes the limit of completely uncorrelated scattering) is more significant in the far wings of the lines, where we have weaker linear polarization signals than those obtained under the assumption of completely correlated scattering.

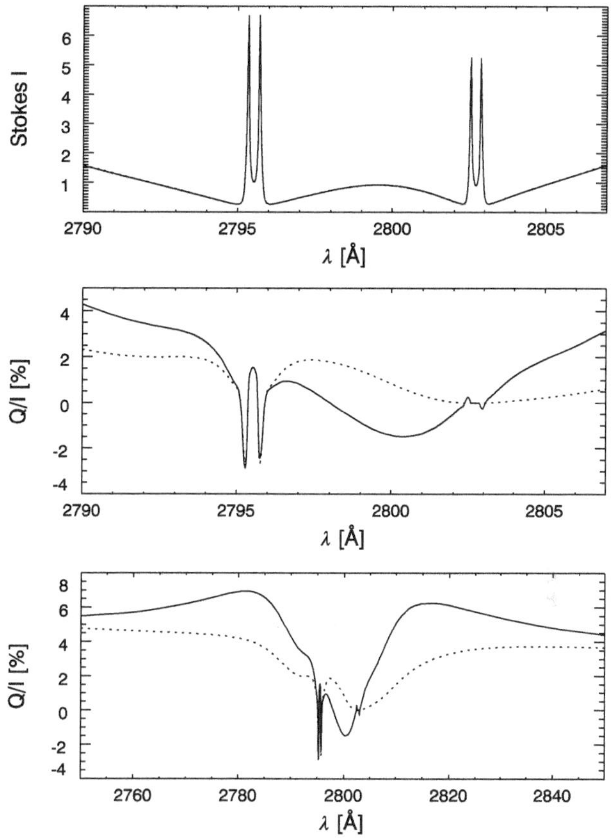

Fig. 5 Intensity (*top panel*; units of $10^{16} \times$ Number of Photons $\mathrm{cm}^{-2}\,\mathrm{s}^{-1}\,\mathrm{sr}^{-1}\,\text{Å}^{-1}$) and scattering polarization patterns (other *panels*) of the Mg II h & k lines, calculated in the FAL-C semi-empirical model of the quiet solar atmosphere for a LOS with $\mu = 0.1$ taking into account the effects of PRD with (*solid curves*) and without (*dotted curves*) J-state interference. The *top* and *middle panels* show in detail the spectral region around the core of the two lines, while the *bottom panel* shows the predicted Q/I pattern in a much broader wavelength range. The reference direction for positive Stokes Q is the parallel to the nearest limb

Figure 5 shows the Stokes-I and Q/I patterns of the Mg II h and k lines resulting from detailed radiative transfer calculations in the FAL-C semi-empirical model of the solar atmosphere, including the interaction between the line scattering processes and those responsible for the polarization of the Sun's continuous radiation (i.e., the Rayleigh and Thomson scattering processes that produce the Q/I values shown in the far wings seen in the upper panel of the figure). There is a very significant impact of PRD and J-state interference effects on the wings of the Q/I profile (see Belluzzi and Trujillo Bueno 2012). The Q/I pattern calculated in the FAL-C quiet Sun model has negative values between the h and k lines (see the solid curves). Interestingly, as shown in Fig. 6, these negative Q/I signals weaken when the radiative transfer calculations are carried out in the FAL-F and FAL-P semi-empirical models, representative of the physical conditions in network and plage regions, respectively. In order to observe such negative Q/I signals it would be necessary to develop a high-sensitivity spectropolarimeter, capable of detecting polarization signals of the order of 0.1 % or smaller (see Sect. 5.2).

It is also important to note in Fig. 5 that the joint action of PRD and J-state interference produces an appreciable Q/I signal around the line-center of the Mg II h line (compare the solid and dotted curves). Of particular interest is that at the very center of the (intrinsically unpolarizable) h line we find $Q/I = 0$ (see the middle panel), in spite of the fact that the redistribution of the spectral line radiation due to the non-coherence of the continuum scattering may in principle produce Q/I signals even in intrinsically unpolarizable lines

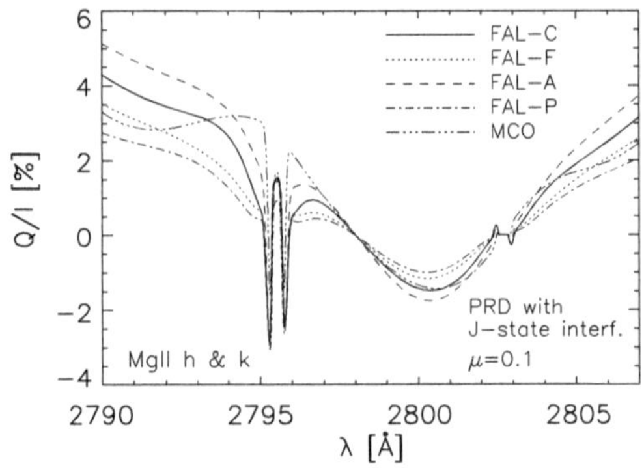

Fig. 6 The fractional scattering polarization of the Mg II h & k lines calculated in various semi-empirical models of the solar atmosphere taking into account PRD and J-state interference effects. The *labels* indicate the 1D semi-empirical models of Fontenla et al. (1993), representative of the following brightness regions on the solar surface: internetwork (A), network (F), average quiet Sun (C) and plage (P). MCO stands for the quiet Sun model of Avrett (1995), which is about 1500 K cooler than FAL-C around 1000 km above continuum optical depth unity. The reference direction for positive Stokes Q is the parallel to the nearest limb

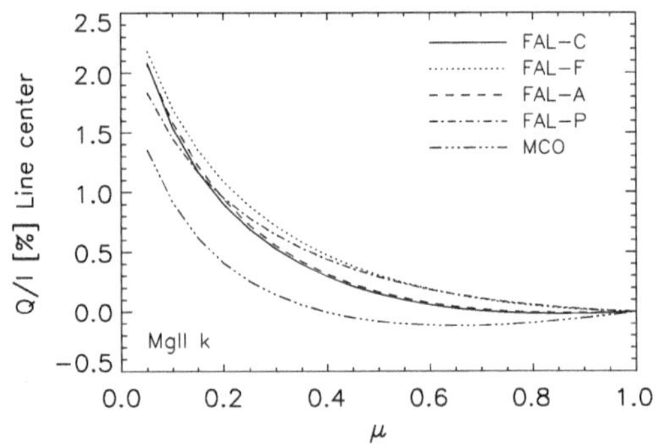

Fig. 7 The CLV of the line center scattering polarization signal of the Mg II k line calculated in various semi-empirical models of the solar atmosphere taking into account PRD and J-state interference effects. The meaning of the labels is the same as in Fig. 6. The reference direction for positive Stokes Q is the parallel to the nearest limb

(see del Pino Alemán et al. 2014). This is important because the Hanle effect operates at the center of the k line and by measuring simultaneously the scattering polarization pattern across the Mg II h & k lines one would have automatically a way to fix the zero offset of the polarization scale.

In Fig. 7 we show the CLV of the Q/I line-center amplitudes of the Mg II k line calculated in several semi-empirical models of the solar atmosphere, without including any magnetic field. It is interesting to note that the lowest polarization amplitudes are obtained in the MCO model of Avrett (1995), which is about 1500 K cooler than FAL-C around 1000 km above continuum optical depth unity.

Belluzzi and Trujillo Bueno (2012) pointed out that a two-level atom PRD approach for the Mg II k line at 2795 Å is suitable for modeling the positive line-center peak and the two negative peaks located in the nearby wings of the Q/I profile (see Fig. 8). The Mg II k line (upper level with $J = 3/2$) is the only line of the Mg II resonance doublet that

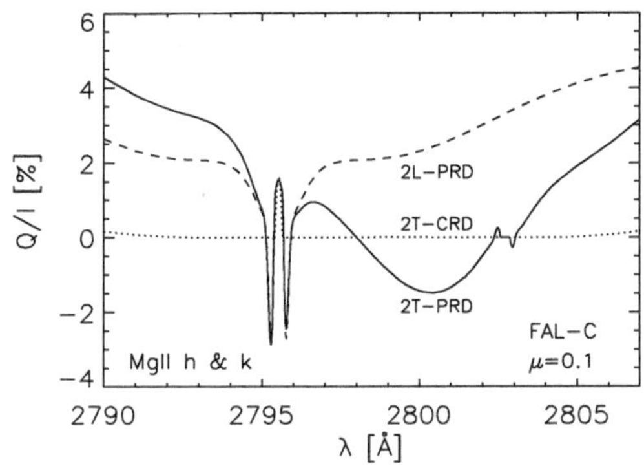

Fig. 8 The scattering polarization Q/I pattern of the Mg II h & k lines calculated in the FAL-C semi-empirical model assuming CRD (*dotted curve*) and taking into account PRD effects (*solid curve*), in both cases including the effects of J-state interference. Note that the CRD limit is suitable only for estimating the line–center linear-polarization amplitude of the k line, which is where the Hanle effect operates. The *dashed curve* indicates the two-level atom PRD solution for the Mg II k line; note that it is a reasonable approximation for modeling the positive line-center peak and the two negative peaks located in the nearby wings of the Q/I profile. The reference direction for positive Stokes Q is the parallel to the nearest limb

is sensitive to the Hanle effect. The critical magnetic field for the onset of the Hanle effect in the k line is $B_H \approx 22$ G, which explains the magnetic sensitivity seen in the core of the Q/I and U/I profiles shown Fig. 9. It is also very important to note that the Mg II k line shows a very interesting magnetic sensitivity to relatively weak fields in the *wings* of the Q/I and U/I profiles (see Fig. 9). This is due to the magneto-optical $\rho_V\, Q$ and $\rho_V\, U$ terms of the transfer equations for Stokes U and Q, respectively, which cannot be neglected because PRD and J-state interference effects produce sizable Q/I wing signals (see the solid curve of Fig. 8) and, outside the line core, ρ_V/η_I (with η_I the diagonal coefficient of the propagation matrix of the Stokes-vector transfer equation) is significant already for relatively weak fields (see Fig. 2 of Alsina Ballester et al. 2016b). The $\rho_V\, Q$ term of the transfer equation for Stokes U produces sizable U/I wing signals, which are magnetically sensitive because ρ_V scales with the Zeeman splitting. Likewise, the $\rho_V\, U$ term of the transfer equation for Stokes Q introduces magnetic sensitivity in the Q/I wings. The magnetic sensitivity in the wings of the Q/I and U/I profiles of the Mg II k line largely expands the scientific interest of this line for probing the chromosphere in quiet and active regions of the Sun; for more information see the two-level atom investigation by Alsina Ballester et al. (2016b) and the two-term atom investigation by del Pino Alemán et al. (2016). Finally, note that if the magnetic field is sufficiently strong, the familiar Zeeman effect can also produce measurable circular polarization signals (see the V/I panel of Fig. 9). Interestingly, for a longitudinal field of 50 G the Zeeman effect produces V/I signals $\sim$0.5 %, also in the two other Mg II lines located in the wings of the h & k lines.

4.2 Lyman-α of H I

The Stokes $I(\lambda)$ profile of the hydrogen Ly-α line has been measured at various positions on the solar disk with instruments on board sounding rockets (e.g., Roussel-Dupré 1982) and space probes (e.g., Curdt et al. 2008) showing its strong emission character. However, there have been only two attempts to measure the linear polarization of the Ly-α radiation of the solar disk. The first pioneering attempt, done using an instrument based on a slitless

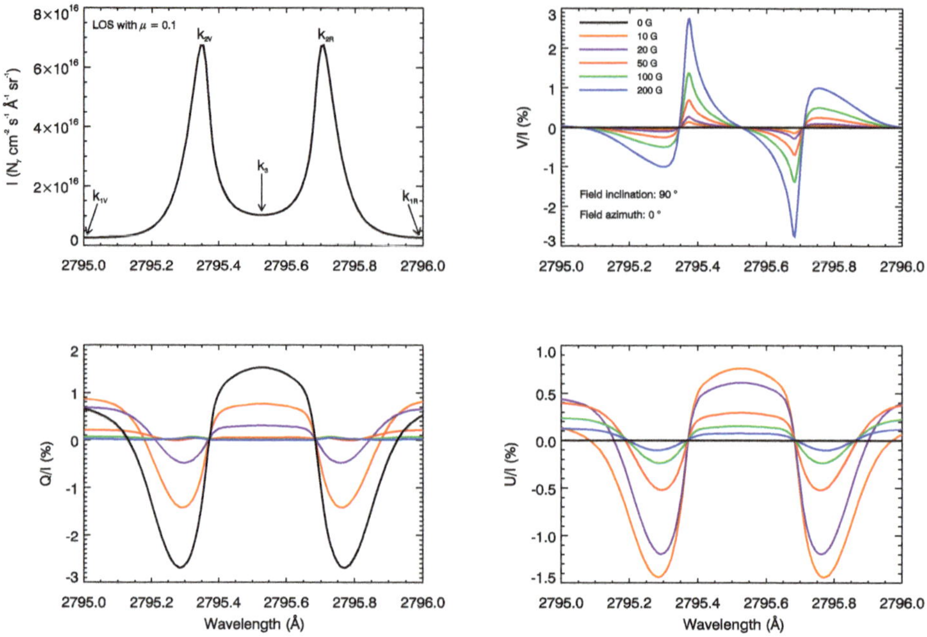

Fig. 9 The Stokes profiles of the k-line of Mg II calculated for a solar disk position close to the limb in the FAL-C semi-empirical model of the solar atmosphere, in the absence (*black curves*) and in the presence (*colored curves*) of a horizontal magnetic field with zero azimuth (i.e., on the plane defined by the local vertical and the LOS). The calculations have been carried out with a two-level atom radiative transfer code that takes into account PRD phenomena and the joint action of the Hanle and Zeeman effects for arbitrary magnetic fields (see Alsina Ballester et al. 2016a). Note the operation of the Hanle effect in the line core and of the magneto-optical effects in the wings (see text). The reference direction for positive Stokes Q is the parallel to the nearest limb. From Alsina Ballester et al. (2016b)

spectrograph, failed because the instrument efficiency degraded by about a factor of 100, presumably due to in-flight contamination (see Stenflo et al. 1980). The second attempt, which was carried out on 3 September 2015 using the Chromospheric Lyman-Alpha SpectroPolarimeter (CLASP; see Kobayashi et al. 2012), successfully measured the Q/I and U/I profiles of Ly-α (see Sect. 5.2) and confirmed the theoretical expectations described below.

In quiet regions of the Sun, the observed CLV of the spatially averaged intensity of the Ly-α line radiation is very small (e.g., Roussel-Dupré 1982; Curdt et al. 2008), which implies that the intensity of the outgoing radiation (i.e., that propagating outward) shows a very small variation with the heliocentric angle θ. However, this does not mean that the illumination of the transition region atoms is practically isotropic—which would imply negligible scattering polarization in Ly-α. As pointed out by Trujillo Bueno et al. (2011), in 1D semi-empirical and hydrodynamical models of the solar atmosphere the intensity of the *incoming* radiation (i.e., that propagating inward) shows significant CLV, so that the hydrogen atoms of the solar transition region must be illuminated by an anisotropic radiation field, even if the plasma of the upper solar chromosphere were horizontally homogeneous (see Fig. 1 of Trujillo Bueno et al. 2011).

Figure 10 shows theoretical Q/I and U/I line-core signals of the hydrogen Ly-α line, resulting from a detailed radiative transfer calculation in the FAL-C semi-empirical model

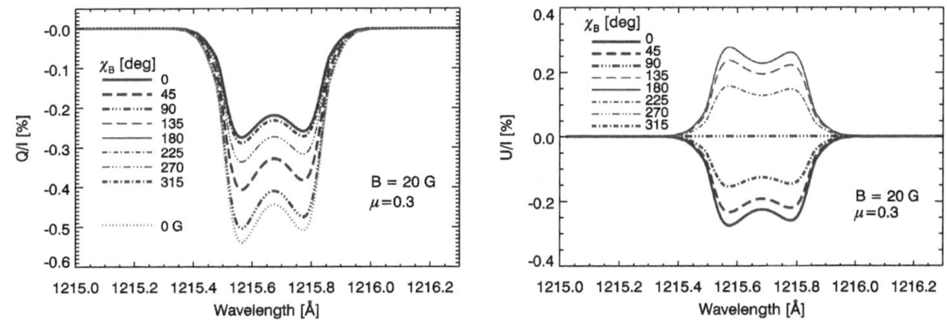

Fig. 10 The Lyα scattering polarization Q/I and U/I line core signals for a LOS with $\mu = 0.3$ assuming the presence in the FAL-C semi-empirical model atmosphere of a 20 G horizontal magnetic field with the indicated azimuth. The *dotted curve* in the Q/I panel shows the zero field reference case, for which $U/I = 0$ (in 1D models). The reference direction for positive Stokes Q is the parallel to the nearest limb. From Trujillo Bueno et al. (2011)

of the solar atmosphere assuming the presence of a 20 G horizontal magnetic field with the indicated azimuth. Note that in this close to the limb scattering geometry the Q/I amplitude calculated without (dotted curve) and with (other curves) magnetic field in the FAL-C semi-empirical model is always negative and smaller than 1 %, and that the Q/I amplitude depends on the azimuth of the horizontal magnetic field. The amplitude and sign of U/I depend on the magnetic field azimuth. Therefore, for the case of an unresolved magnetic field with uniformly distributed azimuths within the spatio-temporal resolution element of the observation we would have $U/I = 0$, but a Q/I amplitude decreasing with the magnetic strength. We point out that $B_H \approx 53$ G for the Ly-α line of hydrogen at 1216 Å.

The investigation by Trujillo Bueno et al. (2011) was based on the quantum theory of spectral line polarization described in the monograph of LL04, which neglects correlations between the frequencies of the incoming and outgoing photons in the scattering events (i.e., PRD effects). In addition, Trujillo Bueno et al. (2011) neglected quantum interference between the $^2P_{1/2}$ and $^2P_{3/2}$ upper levels of the Ly-α line. Belluzzi et al. (2012) have investigated the Ly-α scattering polarization problem taking into account the effects of PRD and J-state interference, showing that the joint action of these effects produces much more complex Q/I profiles, with large amplitudes in the wings (see Fig. 11), and that the sensitivity of the Q/I profiles to the thermal structure of the solar chromosphere is particularly important in the line wings (see left panel of Fig. 12). The wing signals of the Q/I profiles show a clear CLV, approximately following the expected $(1 - \mu^2)$ behavior (see right panel of Fig. 12).

Belluzzi et al. (2012) showed also that the approximations used by Trujillo Bueno et al. (2011) (i.e., CRD without J-state interference) are suitable for estimating the scattering polarization amplitudes at the line center, which is where the Hanle effect in Ly-α operates (see Fig. 13). This figure shows that the PRD line-center absolute values of the Q/I profile are slightly smaller than the CRD ones and that the Q/I signals are significantly weaker in the quiet-Sun model C of Fontenla et al. (1993) (with $Q/I \approx 0.35$ % at $\mu = 0.3$) than in the MCO quiet-Sun model of Avrett (1995) (with $Q/I \approx 0.85$ % at $\mu = 0.3$). Interestingly, in 1D semi-empirical models the largest Q/I line-center signals are obtained for line of sights with $0.2 \le \mu \le 0.4$ (see Fig. 14). Note also that among the semi-empirical models considered the weakest Q/I line-center signal corresponds to FAL-C and the largest to MCO.

Due to the shorter wavelength, the Zeeman effect produces in Ly-α very weak circular polarization signals, about one order of magnitude smaller than in the Mg II h & k lines.

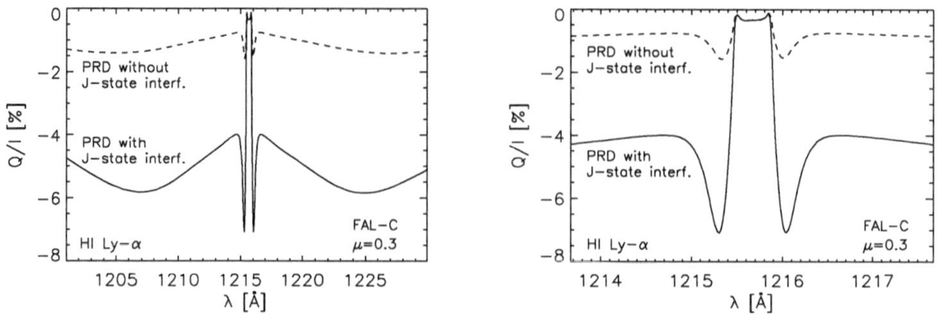

Fig. 11 The PRD Q/I profile across the hydrogen Ly-α line, calculated in the FAL-C model atmosphere for a LOS with $\mu = 0.3$, taking into account (*solid curve*) and neglecting (*dashed curve*) J-state interference. The *right panel* shows in more detail the line core region. The reference direction for Stokes Q is the parallel to the nearest limb. From Belluzzi et al. (2012)

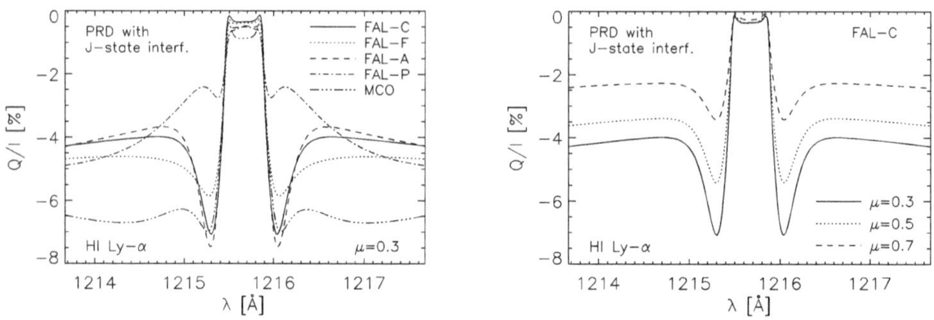

Fig. 12 Theoretical Q/I profiles of the hydrogen Ly-α line. Left panel: calculations in various solar atmospheric models for a LOS with $\mu = 0.3$, taking into account PRD and J-state interference. The 1D semi-empirical models used (see the *labels*) are those of Fontenla et al. (1993), representative of the following brightness regions on the solar disk: internetwork (A), network (F), average quiet Sun (C) and plage (P), and the MCO quiet-Sun model of Avrett (1995), which is about 1500 K cooler than FAL-C around 1000 km above continuum optical depth unity. *Right panel*: calculations for various line of sights in the FAL-C semi-empirical model. The reference direction for Stokes Q is the parallel to the nearest limb. From Belluzzi et al. (2012)

4.3 The Scattering Polarization of Lyman-α and Mg II k in 3D Models of the Solar Atmosphere

The scattering polarization profiles shown in the previous sections result from radiative transfer calculations in 1D semi-empirical models of the solar atmosphere. In 1D static models (e.g., Fontenla et al. 1993; Avrett 1995), or in 1D theoretical models with radial macroscopic velocities (e.g., Carlsson and Stein 1997), the incident radiation field at each point within the medium has axial symmetry around the local vertical direction. In other words, in such 1D models the only way to break the axial symmetry of the pumping radiation field is through the presence of an inclined magnetic field which modifies the zero-field atomic level polarization via the Hanle effect. Under such idealized circumstances (1D model atmospheres with or without radial velocities) the theoretical predictions on the scattering polarization in Ly-α and Mg II h & k are the following (see Trujillo Bueno et al. 2011; Belluzzi et al. 2012; Belluzzi and Trujillo Bueno 2012):

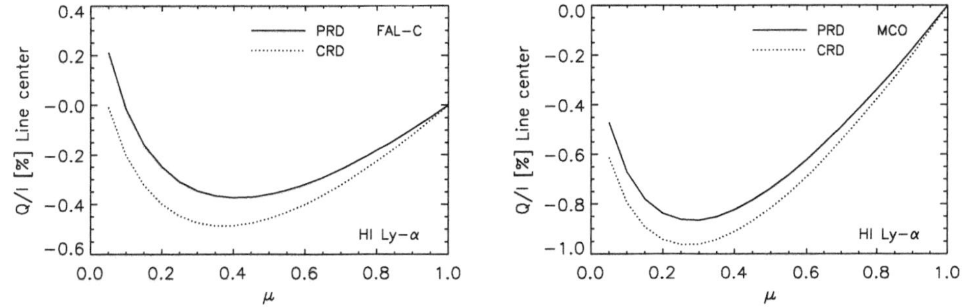

Fig. 13 *Left panel*: the CLV of the line center scattering polarization signal of the hydrogen Ly-α line, calculated in the FAL-C semi-empirical model of the solar atmosphere assuming CRD (*dotted curve*) and taking into account PRD and J-state interference effects (*solid curve*). *Right panel*: the same but for the MCO model of Avrett (1995). The reference direction for Stokes Q is the parallel to the nearest limb

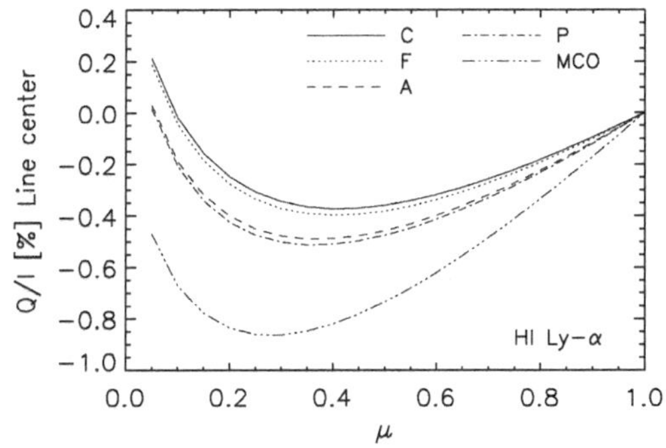

Fig. 14 The CLV of the line center scattering polarization signal of the hydrogen Ly-α line calculated in various semi-empirical models of the solar atmosphere taking into account PRD and J-state interference effects. The *labels* indicate the 1D semi-empirical models of Fontenla et al. (1993), representative of the following brightness regions on the solar surface: internetwork (A), network (F), average quiet Sun (C) and plage (P). MCO stands for the quiet Sun model of Avrett (1995). The reference direction for Stokes Q is the parallel to the nearest limb

- The Q/I signals of hydrogen Ly-α are mainly negative.[7] The line-center amplitudes are always below the 1 % level, while close to the limb the wing signals of the Q/I profiles are significantly larger than 1 %.
- The Q/I line-center signals of the Mg II k line are mainly positive, while $Q/I < 0$ at a range of wavelengths between the centers of the k and h lines.
- In forward scattering geometry (i.e., when observing the solar disk center) the only way to produce scattering polarization in a 1D axially symmetric model is via the Hanle effect of an inclined (non vertical) magnetic field. For symmetry reasons, there is no forward scattering polarization in the absence of an inclined magnetic field.

The upper solar chromosphere, where the line core radiation of the above-mentioned lines originates, is highly inhomogeneous and dynamic, and with sizable horizontal fluctuations like in the three-dimensional (3D) model of Fig. 15. In 3D models of stellar atmo-

[7] Recall that we are always choosing the parallel to the nearest solar limb as the reference direction for linear polarization.

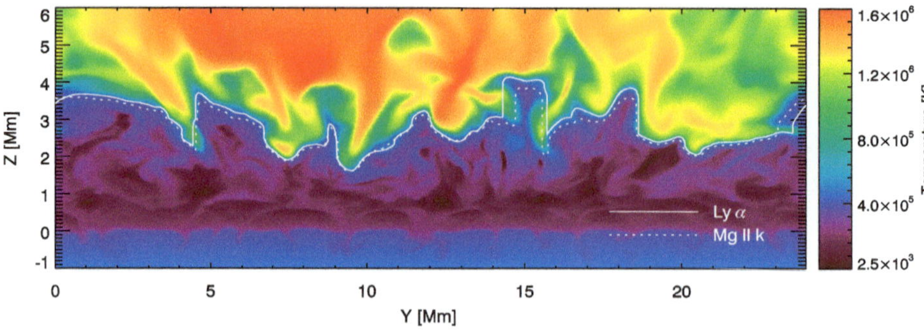

Fig. 15 The figure visualizes the temperature structure across a vertical slice through a three-dimensional snapshot model of the extended solar atmosphere, taken from a state-of-the-art magneto-hydrodynamical simulation of the chromosphere-corona transition region in an enhanced network region (see Carlsson et al. 2016). The *solid* and *dotted curves* show the heights (Z) in this model where the lines Mg II k and hydrogen Ly-α reach line-center optical depth unity, respectively

spheres the incident radiation field at each point within the medium is not axially symmetric around the local vertical direction, even in the absence of a magnetic field. The mere presence of horizontal inhomogeneities in the temperature and density of the model's atmospheric plasma breaks the axial symmetry of the pumping radiation field (Manso Sainz and Trujillo Bueno 2011, and more references therein). Non-radial macroscopic velocity fields may also break the axial symmetry of the spectral line radiation (e.g., Štěpán and Trujillo Bueno 2016). Such non-magnetic sources of symmetry breaking may have a significant impact on the Q/I and U/I scattering polarization signals, which should not be confused with the Hanle effect (the magnetic-field-induced modification of the scattering line polarization). To properly investigate this type of problems it is critical to apply a 3D radiative transfer code like PORTA (see Štěpán and Trujillo Bueno 2013), capable of performing simulations of the intensity and polarization in spectral lines taking into account the possible impact of the model's dynamical state on the anisotropic radiation pumping (e.g., Carlin et al. 2012; Štěpán and Trujillo Bueno 2016) and the symmetry breaking effects of horizontal radiative transfer.

Figure 16 shows maps of the I, Q/I and U/I line-center signals of the hydrogen Ly-α line calculated for a LOS with $\mu = 0.4$ in the 3D model of the chromosphere-corona transition region illustrated in Fig. 15, which results from a radiation magneto-hydrodynamics simulation by Carlsson et al. (2016) that can be considered representative of an enhanced network region. The model's magnetic field, which has a bipolar-like structure, has a mean field strength of about 15 G in the corrugated surface defined by the atmospheric heights where the line-center optical depth is unity; as shown in Fig. 15, this crumpled surface delineates the base of the model's transition region. It is interesting to note that Q/I shows both negative and positive values, but with a clear predominance of negative values for close to the limb line of sights. As expected, the U/I signals also fluctuate across the field of view but both signs are present with more or less the same likelihood. The standard deviation of the Q/I and U/I horizontal fluctuations is very sensitive to the assumed spatial resolution, decreasing as the spatial resolution is degraded (Štěpán et al. 2015). The spatially averaged Stokes profiles obtained at each LOS yield the CLV results shown in Fig. 17 for Q/I, while $U/I \approx 0$.

Figure 18 shows 3D results for the Mg II k line, but for the case of forward scattering geometry (LOS with $\mu = 1$). In addition to the intensity pattern shown in the top left panel,

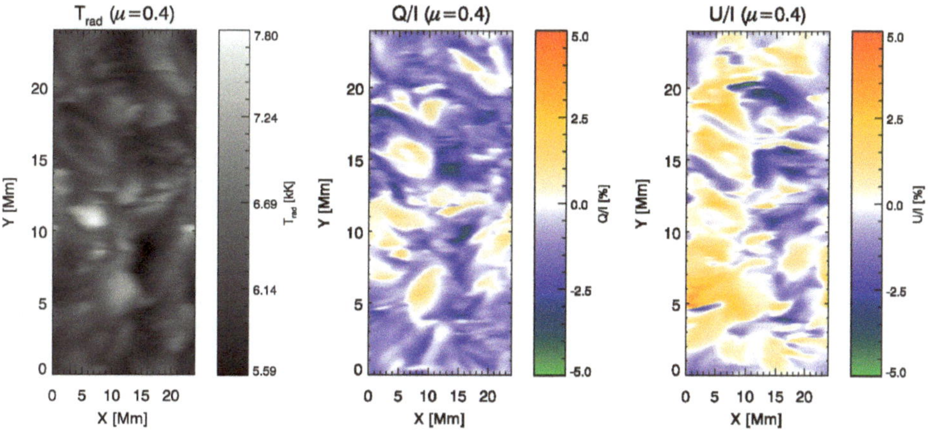

Fig. 16 The spatial variation of the intensity (*left panel*; given in terms of the radiation temperature), Q/I (*middle panel*) and U/I (*right panel*) line-center signals of the hydrogen Ly-α line calculated for a LOS with $\mu = 0.4$ in the 3D model of the chromosphere-corona transition region of Fig. 15. The reference direction for positive Stokes Q is the parallel to the nearest limb (i.e., along the Y-axis). From Štěpán et al. (2015)

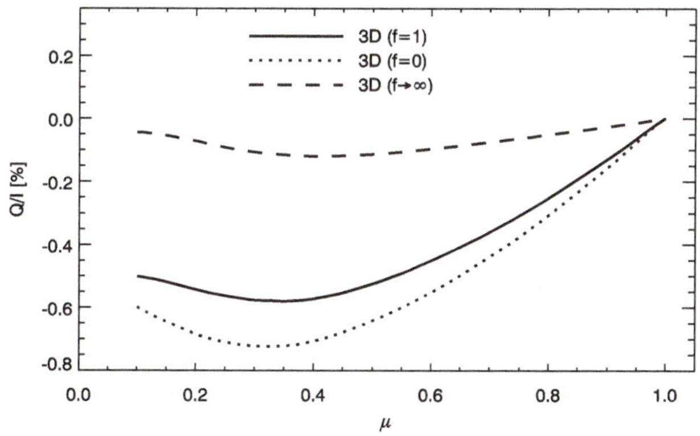

Fig. 17 The CLV of the line-center Q/I signals of the hydrogen Ly-α line after spatially averaging the Stokes profiles of the emergent spectral line radiation. The radiative transfer calculations were carried out in the 3D model of the chromosphere-corona transition region mentioned in Fig. 15, for three scaling factors (f) of the magnetic field strength at each spatial grid point (see text for the meaning of f). The reference direction for positive Stokes Q is the parallel to the nearest limb. From Štěpán et al. (2015)

the figure shows the spatial variation of the fractional total linear polarization amplitude ($P = \sqrt{Q^2 + U^2}/I$) for three different levels of magnetization. The various magnetizations of the 3D atmospheric model were obtained by scaling the magnetic strength at each grid point by a factor f, with $f = 0$ (top right panel), $f = 1$ (bottom left panel) and $f \gg 1$ (bottom right panel). As seen in the bottom left panel, the model's magnetic field is sufficient to produce, via the Hanle effect, significant modifications of the scattering polarization amplitudes with respect to the zero-field case shown in the top right panel. It is interesting to note that, with the exception of a few points in the field of view, the largest polarization signals are obtained for the zero magnetic field case and the lowest ones for the Hanle saturation regime shown in the bottom right panel. This behavior is opposite to the one corresponding to the idealized

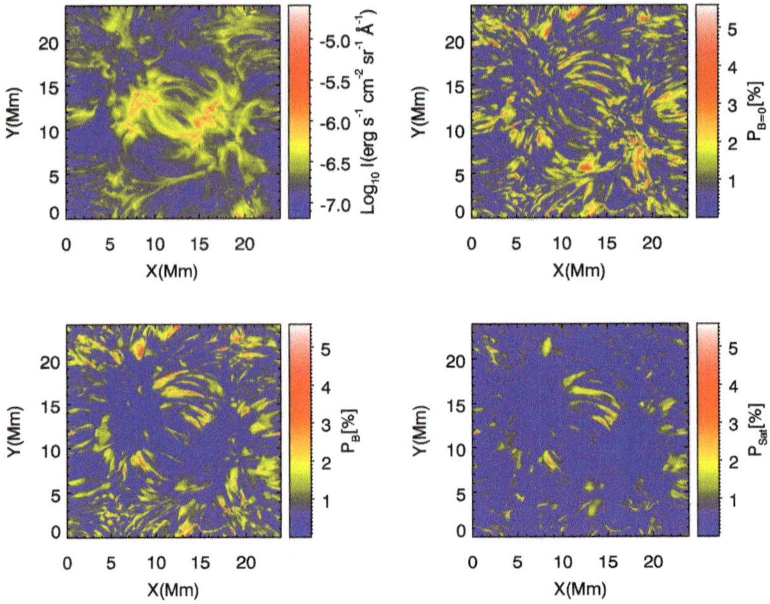

Fig. 18 The spatial variation of the intensity (*top left panel*) and of the fractional total linear polarization ($P = \sqrt{Q^2 + U^2}/I$) of the Mg II k line calculated for a LOS with $\mu = 1$ (forward scattering geometry) in the 3D model of the extended solar atmosphere of Fig. 15. At each point of the field of view the wavelength chosen is that for which P is maximum. Each of the P-panels corresponds to a different level of the model's magnetization (see the main text). From del Pino Alemán et al. (2016, in preparation)

case of a 1D model atmosphere characterized by horizontally homogeneous plane-parallel layers, where there is no forward scattering polarization unless the 1D model atmosphere is permeated by a magnetic field inclined with respect to the local vertical direction. The same happens for the hydrogen Ly-α line (see Fig. 6 of Štěpán et al. 2015). The reason is the following.

In unmagnetized 1D model atmospheres, either static or with radial velocities, the incident radiation field at each height in the model has axial symmetry around the local vertical direction. Under such circumstances we have zero scattering polarization for a LOS with $\mu = 1$ (disk center observation) and, in general, non-zero scattering polarization for any LOS with $\mu < 1$ (in particular, for close to the limb observations). In the regime of the Hanle effect, where there is no contribution of the Zeeman effect to the linear polarization of the spectral line radiation, nothing changes in the presence of a vertical magnetic field. However, in the presence of an inclined magnetic field the ensuing Hanle effect produces linear polarization in forward scattering geometry (disk center observation) and a depolarization of the zero-field scattering polarization that we have in close to the limb observations. A clear observational example where this applies can be found in the paper by Trujillo Bueno et al. (2002), which shows observations and theoretical modeling of the polarization produced by quiescent prominences and filaments in the He I 10830 Å triplet.

In general, in unmagnetized 3D model atmospheres the incident radiation field at each point within the medium does *not* have axial symmetry around the local vertical direction. A rough approximation for strong spectral lines like Ly-α, whose line-center radiation originates mainly in the corrugated surface of optical depth unity that delineates the base of the chromosphere-corona transition region (see Fig. 15), is that at each point on this corrugated

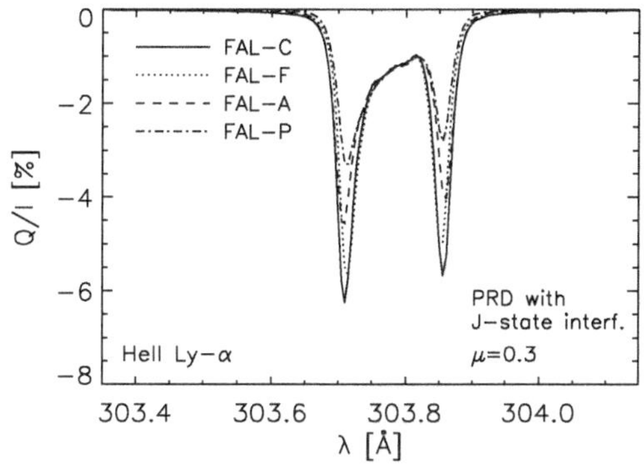

Fig. 19 The Q/I profile of the He II 304 Å line calculated in various semi-empirical models of the solar atmosphere (see the legend of Fig. 12), taking into account PRD and J-state interference effects. The reference direction for positive Stokes Q is the parallel to the nearest limb. From Belluzzi et al. (2012)

surface the unit vector $\vec{N}$ perpendicular to the local transition region defines the symmetry axis of the incident radiation at such point (cf., Štěpán et al. 2015). Therefore, in general, any given LOS will not have the same inclination with respect to the local unit vector $\vec{N}$, given that the direction of $\vec{N}$ changes from point to point on such corrugated surface (see Fig. 15). Therefore, for any LOS including that with $\mu = 1$ both Q/I and U/I may be non-zero, without the need of a magnetic field. It is thus not surprising that in forward scattering geometry (LOS with $\mu = 1$) we find many points on the field of view where the Hanle effect produced by the model's magnetic field depolarizes the zero-field scattering polarization that results from the symmetry breaking caused by the horizontal atmospheric inhomogeneities.

4.4 Lyman-α of He II

The polarization properties of the Ly-α line of He II at 304 Å have been investigated by Trujillo Bueno et al. (2012) and Belluzzi et al. (2012) in 1D semi-empirical models. Similarly to the case of the hydrogen Ly-α line, this EUV line also results from two blended transitions between the single level $^2S_{1/2}$ of the ground term and the two levels, $^2P_{1/2}$ and $^2P_{3/2}$, of the first excited term of He II. Figure 19 shows the scattering polarization pattern of the Ly-α line of He II calculated in four solar semi-empirical models for a LOS with $\mu = 0.3$, taking into account PRD and J-state interference effects. Note that the line-center amplitude of the Q/I profile is about a factor 3 larger than in the hydrogen Ly-α line.

The critical magnetic field strength for the onset of the Hanle effect in this He II line is $B_H \approx 855$ G (i.e., 16 times larger than the B_H value corresponding to the hydrogen Ly-α line). This is because for hydrogenic atoms both the fine structure (FS) splitting and the Einstein coefficient for spontaneous emission are proportional to the fourth power of the nuclear charge. As a result, the Einstein coefficients of the Ly-α transitions and the FS splitting between their $^2P_{1/2}$ and $^2P_{3/2}$ upper levels is 16 times larger in He II than in H I. The sizable difference between the B_H values of the two Ly-α lines implies that in the magnetic strength regime where the scattering polarization of the Ly-α line of hydrogen is sensitive to the Hanle effect (i.e., between 10 and 100 G) there is little Hanle effect in the Ly-α line of He II.

We point out that the whole Ly-α line of He II forms in a very thin layer of the transition region (approximately, between about 2000 and 2200 km in the FAL-C model), and that at such heights the coherence fraction α of the He II 304 Å line is essentially equal to unity; therefore, the assumption of purely coherent scattering in the atom rest frame is an excellent approximation for modeling this line (Belluzzi et al. 2012).

Finally, we point out that the circular polarization caused by the Zeeman effect in the He II 304 Å line is even weaker than in the Ly-α line of hydrogen.

4.5 Other Ultraviolet Lines

In the previous sections, we already pointed out that the critical magnetic field strengths for the onset of the Hanle effect are $B_H \approx 22$ G for Mg II k, $B_H \approx 53$ G for hydrogen Ly-α and $B_H \approx 855$ G for the He II line at 304 Å. There are additional FUV and EUV spectral lines of diagnostic interest for probing the chromosphere-corona transition region (e.g., Stenflo and Stenholm 1976). The following list gives some additional examples including their critical magnetic field strength for the onset of the Hanle effect:

- $B_H \approx 23$ G for the C IV line at 1548.20 Å.
- $B_H \approx 35$ G for the O VI line at 1032 Å.
- $B_H \approx 14$ G for the Ly-β line of hydrogen at 1025 Å.
- $B_H \approx 6$ G for the Ly-γ line of hydrogen at 972 Å.

All these lines are much fainter than the Ly-α line of hydrogen and have a significant collisional excitation contribution. Although very challenging, it would be interesting to measure their scattering polarization signals because, via the Hanle effect, such lines are sensitive to significantly weaker magnetic fields than Ly-α.

It is also interesting to note that in the blue wing of the hydrogen Ly-α line there is a Si III line at 1206 Å. This resonance line with $J_l = 0$ and $J_u = 1$ has the largest possible polarizability coefficient (i.e., $W_2 = 1$; see Eq. 14). Moreover, it has $B_H \approx 290$ G as critical field for the onset of the Hanle effect. The available spectroscopic observations (e.g., Pinfield et al. 1999) indicate that its intensity profiles show limb brightening and that the line's optical thickness in the chromosphere-corona transition region is of the order of unity. All these factors suggest that the Si III line at 1206 Å should be significantly linearly polarized due to scattering processes, with $Q/I < 0$. Interestingly, the Chromospheric Lyman-Alpha Spectro-Polarimeter (CLASP) has indeed detected negative Q/I polarization in the Si III line with a clear CLV (Ishikawa et al. 2016; in preparation). Outside active regions this line is expected to have a nearly negligible magnetic sensitivity. At present, a detailed radiative transfer investigation is being performed by del Pino Alemán et al. (2016; in preparation) to determine if this Si III line (whose $B_H \approx 290$ G) can be used as a suitable reference line for facilitating the detection of the Hanle effect in the hydrogen Ly-α line (whose $B_H \approx 53$ G).

Similarly to the hydrogen Ly-α line, the above-mentioned FUV lines have relatively little sensitivity to the Zeeman effect. In this respect, a possibly suitable line is the C IV line at 1548.20 Å whose circular polarization above a sunspot could be measured by Henze et al. (1982) using the Ultraviolet Spectrometer and Polarimeter (UVSP). Recently, Peter (2013) carried out a forward modeling study of the circular polarization produced by the Zeeman effect in this C IV line and concluded that in active regions its Stokes-V signal should be measurable with high-sensitivity spectropolarimeters capable of reaching a polarimetric sensitivity of 0.1 %.

5 Instrumental Developments for Solar Ultraviolet Spectropolarimetry

In the previous sections we have provided information on observable quantities sensitive to the physical conditions of the outer solar atmosphere (upper chromosphere, transition region and corona), with emphasis on their magnetic sensitivity via the Hanle effect. The theoretical calculations carried out to provide such information required the formulation of the relevant equations for each of the problems considered, along with the development of efficient numerical codes for computing the intensity and polarization of the emergent spectral line radiation. These codes can be applied to do forward modeling research (given a model of the atmospheric region of interest to compute the emergent Stokes profiles to be confronted with spectropolarimetric observations) and, moreover, they form the basis for further developments (e.g., Stokes inversion techniques). Obviously, an extra crucial step is to develop the telescopes and instrumentation needed to measure the observables. Given that the required observations of the ultraviolet solar spectrum have to be done above the Earth atmosphere, progress in this last aspect of the problem has been slow. For a competitive proposal to the European Space Agency see Peter et al. (2012). Here we provide information on the exposure times needed to measure fractional polarization signals in some UV, FUV and EUV lines and a brief summary of some of the pioneering spectropolarimetric observations that have been attempted so far.

5.1 Exposure Times Needed to Detect Polarization Signals in Ultraviolet Lines

Concerning the case of the solar corona observed off-limb (Sect. 3.1), useful information can be found in the paper by Khan et al. (2011). These authors point out that the exposure times for measuring the wavelength-integrated linear polarization of the hydrogen Lyman lines at 1.1 solar radii, assuming a 25 cm telescope, 20 arcsec pixels and an instrumental efficiency of 1 % are about 20 minutes for Lyman-α, 1.5 hours for Lyman-β, and 5 hours for Lyman-γ. Likewise, Manso Sainz and Trujillo Bueno (2009) provide information on the technical requirements to detect the linear polarization of the Fe X coronal line at 174.5 Å, discussed in Sect. 3.2.

With regard to the case of the upper chromosphere and transition region discussed in Sect. 4, the approximate exposure times needed for detecting in the radiation of the quiet solar disk 0.1 % fractional polarization line-center signals, assuming a 1 m telescope, an instrumental efficiency of 1 %, a spectral resolution of 0.1 Å, and 1 arcsec pixels are 8, 80 and 440 seconds for Mg II k, H I 1216 Å, and He II 304 Å, respectively. For the case of active regions, these time intervals become 2, 10 and 36 seconds, respectively.

The number of photons, N_{photons}, emitted by the Sun per square centimeter, per second, per steradian, per Angstrom can be estimated from theoretical calculations in given models of the solar atmosphere (e.g., see the top left panel of Fig. 9) and/or from observations (e.g., see Fig. 3). To estimate the exposure time T (in seconds) for other combinations of the relevant parameters, it suffices to apply the following formula:

$$\mathrm{T} = \frac{10^7}{\varepsilon\,(\mathrm{PIXEL})^2\, 2.35\times10^{-11}\, (\pi D^2/4)\, N_{\text{photons}}\, \Delta\lambda}, \tag{56}$$

where 10^7 is the number of photons we need to accumulate to detect, at the 3σ level, fractional polarization signals of 0.1 %, while ε is the efficiency of the instrument, D the diameter of the telescope in centimeters, $\Delta\lambda$ the selected wavelength interval in Å, 2.35×10^{-11} is the solid angle corresponding to 1 arcsec^2, and PIXEL $= X$ for pixels of X arcsec (e.g., $X = 1$ for 1 arcsec pixels). Therefore, the exposure time can be reduced by a factor 100 by downgrading the spatial resolution by a factor 10.

5.2 Pioneering Observations of Scattering Polarization in Ultraviolet Lines

As already mentioned in Sect. 3.1.1, taking the opportunity of the SOHO spacecraft rotation on 1996, March 19, Raouafi et al. (1999) pointed to a coronal hole located at about

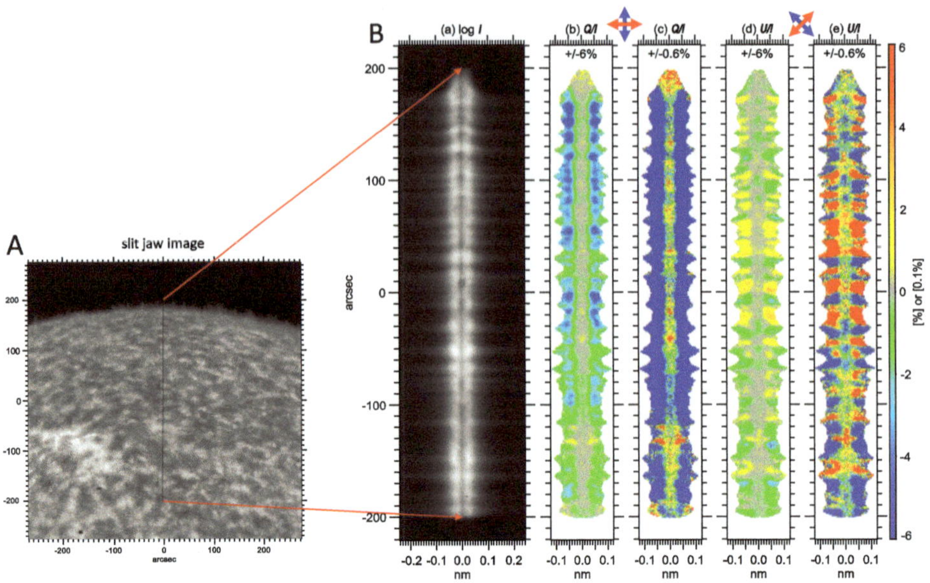

Fig. 20 Panel **A**: Ly-α image taken by the CLASP slit jaw camera. Panel **B**: I, Q/I and U/I signals measured by CLASP when observing the "quiet" Sun with the 400 arcsec spectrograph's slit oriented radially from 20 arcsec off-limb (*top* part of the *panels*) towards the solar disk center. Note that the *color scale* is ± 6 % for visualizing the wing signals in the *second* and *fourth panels* and ± 0.6 % for visualizing the line-center signals in the *third* and *fifth panels*. The reference direction for positive Stokes Q is the parallel to the limb. From Kano et al. (2017)

270 arcsec above the south solar limb and observed the variation with the rotation angle of the intensity measured in the two components of the O VI doublet at 1032 Å (polarizable line) and at 1037 Å (unpolarizable line). These authors found that the variation with the spacecraft rotation angle of the ratio of such intensity signals shows a sinusoidal fluctuation (with an amplitude of 3.8 %), which they interpreted as an indication of the presence of a certain amount of linear scattering polarization in the 1032 Å line.

The polarimeter on board the Solar Maximum Mission satellite used by Henze and Stenflo (1987) to observe the chromospheric Mg II h & k lines was not designed to measure the Stokes profiles with high polarimetric sensitivity and spectral resolution. Nevertheless, although the statistical significance of their observations is very marginal, they could detect the predicted Q/I signals in the blue wing of the k line and in the red wing of the h line.

In 1976 it took place the first attempt to measure the (spatially averaged) scattering polarization of the hydrogen Ly-α line (see Stenflo et al. 1980). The authors developed an instrument based on a slitless spectrograph (see Stenflo et al. 1976). This attempt failed because the instrument efficiency degraded by about a factor 100, presumably due to in-flight contamination.

The theoretical results summarized in Sects. 4.2 and 4.3 motivated the development of the Chromospheric Lyman-Alpha SpectroPolarimeter (CLASP), a sounding rocket experiment proposed by an international team of scientists from USA, Japan and Europe (see Kobayashi et al. 2012). Its goal has been to probe the chromosphere-corona transition region by measuring and modeling the linear polarization of the hydrogen Ly-α line of the solar disk radiation, produced by the absorption and scattering of anisotropic radiation, along with its possible modification by the Hanle effect. This first exploratory and challenging measurement, with an observing time of only 5 minutes, was carried out on 3 September 2015 using a vacuum-ultraviolet telescope and a spectropolarimeter launched by a NASA sounding rocket. The experiment was a true success and CLASP measured for the first time the complex linear polarization pattern that the absorption and scattering of anisotropic radiation by the hydrogen atoms of the upper solar chromosphere produce in the Ly-α line radiation of the solar disk (see Fig. 20). The observational results essentially confirm the theoretical fractional linear polarization profiles discussed in Sects. 4.2 and 4.3 (see Kano et al. 2017). Recently, the theoretical results reviewed in Sect. 4.1 and the success of the CLASP experiment mentioned above led to a new proposal submitted to NASA for dedicating the second flight of CLASP to observe the four Stokes parameters in the wavelength region of the Mg II *h* and *k* lines around 2800 Å. The development of a space telescope designed to carry out simultaneous spectropolarimetric observations in the Mg II *h* & *k* lines, hydrogen Ly-α and the Si III 1206 Å line would open up a novel diagnostic window into the magnetism and geometric structure of the upper solar chromosphere.

6 Concluding Comments

The empirical investigation of the magnetic field in the upper chromosphere, transition region and corona of the Sun is a very important challenge in astrophysics. To this end, we need (1) to identify observable quantities sensitive to the physical conditions of the outer solar atmosphere, (2) to build the telescopes and instruments needed for measuring such observables, and (3) to develop suitable plasma diagnostic techniques to infer the relevant physical quantities (e.g., the strength and orientation of the magnetic field) from the observations.

In this paper we have tried to provide an overview of the work done in such research frontiers, with emphasis on the polarization produced by anisotropic radiation pumping and the Hanle and Zeeman effects in permitted UV, FUV and EUV lines. Obviously, spectropolarimetric measurements in this spectral region can only be done from space, and they represent an almost virgin territory. However, we must pursue this goal because ultraviolet spectropolarimetry might be our key gateway for probing the outer atmospheres of the Sun and of other stars.

Acknowledgements This is the last paper that our friend Egidio Landi Degl'Innocenti wrote with us before he unexpectedly passed away on 12 February 2017. We are deeply grateful to Egidio for the extraordinary scientific legacy that he has left to us and to future generations of scientists. Financial support by the Spanish Ministry of Economy and Competitiveness through projects AYA2014-60476-P and AYA2014-55078-P is gratefully acknowledged, as well as the computing grants provided by the Barcelona Supercomputing Center (National Supercomputing Center, Barcelona, Spain). L.B. gratefully acknowledges the Swiss National Science Foundation through grant 200021-163405.

References

E. Alsina Ballester, L. Belluzzi, J. Trujillo Bueno, The transfer of resonance line polarization with partial frequency redistribution in the general Hanle-Zeeman regime. Astrophys. J. (2016a in press)

E. Alsina Ballester, L. Belluzzi, J. Trujillo Bueno, The magnetic sensitivity of the Mg II *k* line to the joint action of Hanle, Zeeman and magneto-optical effects. Astrophys. J. **831**, L15 (2016b)

L.H. Auer, D. Rees, J.O. Stenflo, Resonance line polarization. Line wing transfer calculations including excited state interference. Astron. Astrophys. **88**, 302 (1980)

E.H. Avrett, Two-component modeling of the solar IR CO lines, in *Infrared Tools for Solar Astrophysics. What's Next?* ed. by J.R. Kuhn, M.J. Penn. Singapore (World Scientific, Singapore, 1995), p. 303

S. Bashkin, J.O. Stoner Jr., *Atomic Energy Levels and Grotrian Diagrams* (North-Holland, Amsterdam, 1975)
L. Belluzzi, J. Trujillo Bueno, The polarization of the solar Mg II h & k lines. Astrophys. J. **750**, L11 (2012)
L. Belluzzi, J. Trujillo Bueno, The transfer of resonance line polarization. Astron. Astrophys. **564**, A16 (2014)
L. Belluzzi, J. Trujillo Bueno, J. Štěpán, The scattering polarization of the Lyα lines of H I and He II taking into account PRD and J-state interference effects. Astrophys. J. **755**, L2 (2012)
L. Belluzzi, E. Landi Degl'Innocenti, J. Trujillo Bueno, Isotropic inelastic and superelastic collisional rates in a multiterm atom. Astron. Astrophys. **551**, A84 (2013)
V. Bommier, Master equation theory applied to the redistribution of polarized radiation in the weak radiation field limit. III. Theory for the multilevel atom. Astron. Astrophys. **591**, 59 (2016)
V. Bommier, S. Sahal-Bréchot, The Hanle effect of the coronal L-alpha line of hydrogen—theoretical investigation. Sol. Phys. **78**, 157 (1982)
V. Bommier, J.L. Leroy, S. Sahal-Bréchot, Determination of the complete vector magnetic field in solar prominences, using the Hanle effect. Astron. Astrophys. **100**, 231 (1981)
V. Bommier, E. Landi Degl'Innocenti, J.L. Leroy, S. Sahal-Bréchot, Complete determination of the magnetic field vector and of the electron density in 14 prominences from linear polarization measurements in the HeI D3 and H-alpha lines. Sol. Phys. **154**, 231 (1994)
E.S. Carlin, R. Manso Sainz, A. Asensio Ramos, J. Trujillo Bueno, Scattering polarization in the Ca II infrared triplet with velocity gradients. Astrophys. J. **751**, 5 (2012)
M. Carlsson, B. Stein, Formation of solar calcium H and K bright grains. Astrophys. J. **481**, 500 (1997)
M. Carlsson, V. Hansteen, B. Gudiksen, J. Leenaarts, B. De Pontieu, A publicly available simulation of an enhanced network region of the Sun. Astron. Astrophys. **585**, A4 (2016)
R. Casini, R. Manso Sainz, Line formation theory for the multiterm atom with hyperfine structure in a magnetic field. Astrophys. J. **624**, 1025 (2005)
R. Casini, R. Manso Sainz, Frequency redistribution of polarized light in the Λ-type multi-term polarized atom. Astrophys. J. **824**, 135 (2016)
R. Casini, S.M. White, P. Judge, Magnetic diagnostics of the solar corona: Unifying optical and radio techniques. Space Sci. Rev. (2017 this issue)
W. Curdt, H. Tian, L. Teriaca, U. Schühle, P. Lemaire, The Lyman-α profile and center-to-limb variation of the quiet Sun. Astron. Astrophys. **492**, L9 (2008)
B. De Pontieu, A. Title, J.R. Lemen et al., The Interface Region Imaging Spectrograph (IRIS). Sol. Phys. **289**, 2733 (2014)
T. del Pino Alemán, R. Manso Sainz, J. Trujillo Bueno, Non-coherent continuum scattering as a line polarization mechanism. Astrophys. J. **784**, 46 (2014)
T. del Pino Alemán, R. Casini, R. Manso Sainz, Magnetic diagnostics of the solar chromosphere with the Mg II h–k lines. Astrophys. J. **830**, L24 (2016)
M. Derouich, F. Auchère, J.C. Vial, M. Zhang, Hanle signatures of the coronal magnetic field in the linear polarization of the hydrogen Ly-α line. Astron. Astrophys. **511**, 7 (2010)
S. Fineschi, Space-based instrumentation for magnetic field studies of solar and stellar atmospheres, in *Magnetic Fields Across the Hertzsprung-Russell Diagram*, ed. by G. Mathys, S.K. Solanki, D.T. Wickramasinghe. ASP Conf. Ser., vol. 248 (2001), p. 597
S. Fineschi, R.B. Hoover, A.B.C. Walker Jr., Hydrogen Lyman-alpha coronagraph/polarimeter. Proc. SPIE **1546**, 402 (1992)
S. Fineschi, R.B. Hoover, M. Zukic, K. Jongmin, A.B.C. Walker Jr., P.C. Baker, Polarimetry of the H I Lyman-α for coronal magnetic field diagnostics. Proc. SPIE **1742**, 423 (1993)
S. Fineschi, A. van Ballegoijen, J.L. Kohl, Coronal magnetic field diagnostics with UV spectropolarimetry. ESA SP **446**, 317 (1999)
J.M. Fontenla, E.H. Avrett, R. Loeser, Energy balance in the solar transition region. III. Helium emission in hydrostatic, constant-abundance models with diffusion. Astrophys. J. **406**, 319 (1993)
A.H. Gabriel, Measurements on the Lyman-α corona. Sol. Phys. **21**, 392 (1971)
A.H. Gabriel, et al., Rocket observations of the ultraviolet solar spectrum during the total eclipse of 1970 March 7. Astrophys. J. **169**, 595 (1971)
G.E. Hale, On the probable existence of a magnetic field in sunspots. Astrophys. J. **28**, 315 (1908)
W. Hanle, Über magnetische Beeinflussung der Polarisation der Resonanzfluoreszenz. Z. Phys. **30**, 93 (1924)
W. Henze, J.O. Stenflo, Polarimetry in the Mg II h and k lines. Sol. Phys. **111**, 243 (1987)
W. Henze et al., Observations of the longitudinal magnetic field in the transition region and photosphere of a sunspot. Sol. Phys. **81**, 231 (1982)
I. Hubeny, D. Mihalas, *Theory of Stellar Atmospheres* (Princeton University Press, Princeton/Oxford, 2015)
D.G. Hummer, Non-coherent scattering: I. The redistribution function with Doppler redistribution. Mon. Not. R. Astron. Soc. **125**, 21 (1962)

R. Kano, J. Trujillo Bueno, A. Winebarger et al., Discovery of scattering polarization in the hydrogen Lyman-α line of the solar disk radiation. Astrophys. J. (2017, in press)

A. Khan, E. Landi Degl'Innocenti, Solar coronal magnetic field diagnostics through polarimetric forward modelling of the Hanle effect. Astron. Astrophys. **532**, 70 (2011)

A. Khan, E. Landi Degl'Innocenti, Spectropolarimetric signatures of anisotropic velocity distributions of optically thin coronal UV lines. Astron. Astrophys. **543**, 158 (2012)

A. Khan, L. Belluzzi, E. Landi Degl'Innocenti, S. Fineschi, M. Romoli, Spectropolarimetric forward modelling of the lines of the Lyman-series using a self-consistent, global, solar coronal model. Astron. Astrophys. **529**, 12 (2011)

K. Kobayashi, R. Kano, J. Trujillo Bueno et al., The Chromospheric Lyman-Alpha SpectroPolarimeter: CLASP, in *The Fifth Hinode Science Meeting*, ed. by L. Golub, I. De Moortel, T. Shimizu. Astronomical Society of the Pacific Conference Series, vol. 456 (2012), p. 233

J.L. Kohl et al., First results from the SOHO ultraviolet coronagraph spectrometer. Sol. Phys. **175**, 613 (1997)

E. Landi Degl'Innocenti, The determination of vector magnetic fields in prominences from the observations of the Stokes profiles in the D3 line of helium. Sol. Phys. **79**, 291 (1982)

E. Landi Degl'Innocenti, Polarization in spectral lines. III. Resonance polarization in the non-magnetic, collisionless regime. Sol. Phys. **91**, 1 (1984)

E. Landi Degl'Innocenti, *Atomic Spectroscopy and Radiative Processes* (Springer, Milan, 2014)

E. Landi Degl'Innocenti, V. Bommier, A spectroscopic method for the solution of the 180° azimuth ambiguity in magnetograms. Astrophys. J. **411**, L49 (1993)

M. Landi Degl'Innocenti, E. Landi Degl'Innocenti, An analytical expression for the Hanle-effect scattering phase matrix. Astron. Astrophys. **192**, 374 (1988)

E. Landi Degl'Innocenti, M. Landolfi, *Polarization in Spectral Lines* (Kluwer, Dordrecht, 2004)

H. Lin, J.R. Kuhn, R. Coulter, Coronal magnetic field measurements. Astrophys. J. **613**, 177 (2004)

R. Manso Sainz, J. Trujillo Bueno, A possible polarization mechanism of EUV coronal lines, in *Solar Polarization 5*, ed. by S.V. Berdyugina, K.N. Nagendra, R. Ramelli. Astronomical Society of the Pacific Conference Series, vol. 405 (2009), p. 423

R. Manso Sainz, J. Trujillo Bueno, Scattering polarization and Hanle effect in stellar atmospheres with horizontal inhomogeneities. Astrophys. J. **743**, 12 (2011)

J.S. Morrill, C.M. Korendyke, High-resolution center-to-limb variation of the quiet solar spectrum near Mg II. Astrophys. J. **687**, 646 (2008)

T. Pereira, B. De Pontieu, M. Carlsson et al., An Interface Region Imaging Spectrograph first view on spicules. Astrophys. J. **792**, L15 (2014)

H. Peter, Magnetic field diagnostics and spatio-temporal variability of the solar transition region. Sol. Phys. **288**, 531 (2013)

H. Peter et al., Solar magnetism eXplorer (SolmeX). Exploring the magnetic field in the upper atmosphere of our closest star. Exp. Astron. **33**, 271 (2012)

D.J. Pinfield, F.P. Keenan, M. Mathioudakis, K.J.H. Phillips, W. Curdt, K. Wilhelm, Evidence for non-Maxwellian electron energy distributions in the solar transition region: Si III line ratios from SUMER. Astrophys. J. **527**, 1000 (1999)

N.E. Raouafi, S.K. Solanki, Effect of anisotropic velocity distribution on the linear polarization of coronal lines. Does the ion cyclotron exist in the inner corona? Astron. Astrophys. **412**, 271 (2003)

N.E. Raouafi, P. Lemaire, S. Sahal-Bréchot, Detection of the O VI 103.2 nm line polarization by the SUMER spectrometer on the SOHO spacecraft. Astron. Astrophys. **345**, 999 (1999)

N.E. Raouafi, S.K. Solanki, T. Wiegelmann, Hanle effect diagnostics of the coronal magnetic field: A test using realistic magnetic field configurations, in *Solar Polarization 5*, ed. by S.V. Berdyugina, K.N. Nagendra, R. Ramelli. ASP Conf. Ser., vol. 405 (2009), p. 429

J.C. Raymond et al., Composition of coronal streamers from the SOHO ultraviolet coronagraph spectrometer. Sol. Phys. **175**, 645 (1997)

D. Roussel-Dupré, Skylab observations of H I Lyman-α. Astrophys. J. **256**, 284 (1982)

S. Sahal-Bréchot, M. Malinovski, V. Bommier, The polarization of the O VI 1032 Å line as a probe for measuring the coronal vector magnetic field via the Hanle effect. Astron. Astrophys. **168**, 284 (1986)

S. Sahal-Bréchot, V. Bommier, N. Feautrier, Doppler redistribution of anisotropic radiation and resonance polarization in moving scattering media. I. Theory revisited in the density matrix formalism. Astron. Astrophys. **340**, 579 (1998)

J.O. Stenflo, Applications of the Hanle effect in solar physics, in *The Hanle Effect and Level-Crossing Spectroscopy*, ed. by G. Moruzzi, F. Strumia (Plenum, New York, 1991), p. 237

J.O. Stenflo, L. Stenholm, Resonance line polarization. II Calculations of linear polarization in solar UV emission lines. Astron. Astrophys. **46**, 69 (1976)

J.O. Stenflo, H. Biverot, L. Stenmark, Ultraviolet polarimeter to record resonance-line polarization in the solar spectrum around 130—150 nm. Appl. Opt. **15**, 1188 (1976)

J.O. Stenflo, D. Dravins, N. Wihlborg et al., Search for spectral line polarization in the solar vacuum ultraviolet. Sol. Phys. **66**, 13 (1980)

J. Štěpán, J. Trujillo Bueno, PORTA: a three-dimensional multilevel radiative transfer code for modeling the intensity and polarization of spectral lines with massively parallel computers. Astron. Astrophys. **557**, A143 (2013)

J. Štěpán, J. Trujillo Bueno, The Hanle and Zeeman polarization signals of the solar Ca ii 8542 Å line. Astrophys. J. **826**, L10 (2016)

J. Štěpán, J. Trujillo Bueno, J. Leenaarts, M. Carlsson, Three-dimensional radiative transfer simulations of the scattering polarization of the hydrogen Ly-α line in a magnetohydrodynamic model of the chromosphere-corona transition region. Astrophys. J. **803**, 65 (2015)

J. Trujillo Bueno, Atomic polarization and the Hanle effect, in *Advanced Solar Polarimetry. Theory, Observation and Instrumentation*, ed. by M. Sigwarth. Astronomical Society of the Pacific Conference Series, vol. 236 (2001), p. 161

J. Trujillo Bueno, The generation and transfer of polarized radiation in stellar atmospheres, in *Stellar Atmosphere Modeling*, ed. by I. Hubeny, D. Mihalas, K. Werner. Astronomical Society of the Pacific Conference Series, vol. 288 (2003), p. 551

J. Trujillo Bueno, Polarized radiation observables for probing the magnetism of the outer solar atmosphere, in *Solar Polarization 7*, ed. by K.N. Nagendra, J.O. Stenflo, Z.Q. Qu, M. Sampoorna. Astronomical Society of the Pacific Conference Series, vol. 489 (2014), p. 137

J. Trujillo Bueno, E. Landi Degl'Innocenti, Linear polarization due to lower-level depopulation pumping in stellar atmospheres. Astrophys. J. **482**, L183 (1997)

J. Trujillo Bueno, E. Landi Degl'Innocenti, M. Collados, L. Merenda, R. Manso Sainz, Selective absorption processes as the origin of puzzling spectral line polarization from the Sun. Nature **415**, 403 (2002)

J. Trujillo Bueno, E. Landi Degl'Innocenti, R. Casini, V. Martí nez Pillet, The scientific case for spectropolarimetry from space: A novel diagnostic window on cosmic magnetic fields, in *39th ESLAB Symposium on Trends in Space Science and Cosmic Vision 2020*, ed. by F. Favata, J. Sanz-Forcada, A. Giménez, B. Battrick. ESA Publications Division (ESA SP-588) (2005), p. 203

J. Trujillo Bueno, J. Štěpán, R. Casini, The Hanle effect of the hydrogen Ly-α line for probing the magnetism of the solar transition region. Astrophys. J. **738**, L11 (2011)

J. Trujillo Bueno, J. Štěpán, L. Belluzzi, The Ly-α lines of H i and He ii: a differential Hanle effect for exploring the magnetism of the solar transition region. Astrophys. J. **746**, L9 (2012)

J.C. Vial, P. Lemaire, G. Artzner, P. Gouttebroze, O vi ($\lambda = 1032$ Å) profiles in and above an active region prominence, compared to quiet Sun center and limb profiles. Sol. Phys. **68**, 187 (1980)

T.N. Woods, G.J. Rottman, Solar ultraviolet variability over time periods of aeronomic interest, in *Comparative Aeronomy in the Solar System*, ed. by M. Mendillo, A. Nagy, J. Hunter Waite Jr. Geophys. Monograph Series, vol. 130 (American Geophysical, Union, Washington DC, 2002), p. 221

Space Sci Rev (2017) 210:227–247
DOI 10.1007/s11214-015-0224-1

Minimal Magnetic States of the Sun and the Solar Wind: Implications for the Origin of the Slow Solar Wind

E.W. Cliver[1] · R. von Steiger[2,3]

Received: 1 September 2015 / Accepted: 17 November 2015 / Published online: 18 December 2015

Abstract During the last decade it has been proposed that both the Sun and the solar wind have minimum magnetic states, lowest order levels of magnetism that underlie the 11-yr cycle as well as longer-term variability. Here we review the literature on basal magnetic states at the Sun and in the heliosphere and draw a connection between the two based on the recent deep 2008–2009 minimum between cycles 23 and 24. In particular, we consider the implications of the low solar activity during the recent minimum for the origin of the slow solar wind.

Keywords Sun · Solar wind · Magnetic field

1 Introduction

The Sun has an 11-yr cycle of activity (Harvey 1992; Balogh et al. 2015; Hathaway 2015) characterized by the out-of-phase waxing and waning of sunspots and polar magnetic fields. The apparent regularity of sunspot activity over the last three centuries was preceded by the Maunder Minimum (1645–1715; Spörer 1887, 1889; Maunder 1890, 1894, 1922; Eddy 1976; Ribes and Nesme-Ribes 1993; Usoskin 2013; Riley et al. 2015) during which the envelope of solar activity dipped significantly below recent levels (Usoskin et al. 2015; Svalgaard and Schatten 2015; cf., Zolotova and Ponyavin 2015). Long-term solar variability naturally begs the question about its extremes, both high and low. In this paper we consider the latter, specifically the question whether there is a minimal magnetic state at the Sun and in the heliosphere.

The deep solar minimum in 2007–2009 (Mewaldt et al. 2010; Russell et al. 2010; Cranmer et al. 2010; Mandrini and Webb 2012) provided a unique, well-observed, opportunity to examine this point. In Sects. 2 and 3, we review recent suggestions and evidence for

✉ E.W. Cliver
ecliver@nso.edu

[1] National Solar Observatory, Boulder, CO 80303, USA

[2] International Space Science Institute, Hallerstrasse 6, 3012 Bern, Switzerland

[3] Physikalisches Institut, University of Bern, Sidlerstrasse 5, 3012 Bern, Switzerland

minimal magnetic states on the Sun (Schrijver et al. 2011) and in the solar wind (Svalgaard and Cliver 2007), respectively. In Sect. 4 we consider the possibility that the slow solar wind is a by-product of the Sun's basal, non-cycling, magnetic state. In Sect. 5 we summarize our views on minimal magnetic states on the Sun and in the solar wind and the implications of such basal states for the origin of the slow solar wind. We then list the various questions addressed in this review and note the mixed evidence attending each of them.

2 The Sun's Minimal Magnetic State

Schrijver et al. (2011) argued that the Sun approached its minimum magnetic state globally in 2008–2009, during a period when significant sunspot activity was absent from the disk for an extended interval. From a long-running series of Ca II K line intensity observations (Livingston et al. 2007), Schrijver et al. argued that the quietest parts of the Sun exhibited similar magnetic quiescence during the preceding ~ 35 years, following intervals of extended sunspot inactivity. In other words we had our best look in the modern era of Maunder Minimum conditions in 2008–2009 and could glimpse Maunder-Minimum-like conditions on localized regions of the Sun during previous solar minima. The identification of the Maunder Minimum with the 2009 Sun applies mainly to the low- and mid-latitudes because the solar polar fields in 2009 are inferred to have been considerably stronger—and consequently the polar coronal holes larger—than those during the Maunder Minimum. This reasoning is based on the precursor relationship used to predict the peak 11-yr sunspot number that is based on the polar magnetic field strength (or a proxy) at cycle onset (Schatten et al. 1978; Svalgaard et al. 2005; Wang and Sheeley 2009; Muñoz-Jaramillo et al. 2013). Nonetheless, the possibility that in 2008–2009 we may have seen large portions of the Sun's photosphere at or near its quietest state (Fig. 1) is remarkable.

Schrijver et al. (2011) noted that, "In 2008 and 2009... thousands of small magnetic bipolar regions (the "ephemeral regions" with unsigned fluxes from $\sim 10^{18}$ Mx to a few times 10^{19} Mx) emerged every day (Hagenaar et al. 2003), and many more even smaller flux bundles (Harvey et al. 2007)." They used a surface flux transport model (Schrijver et al.

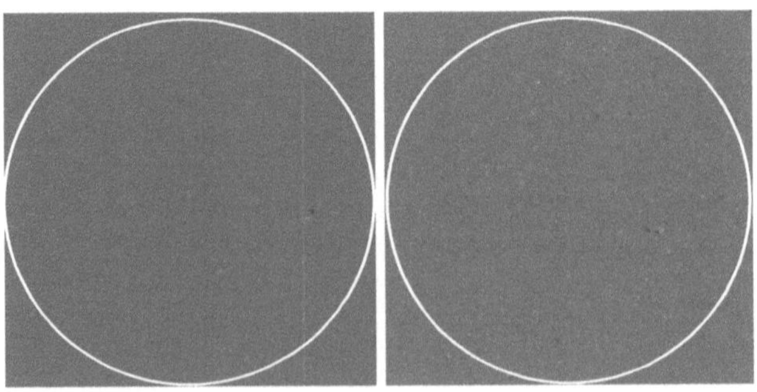

Fig. 1 (*Left*) Solar and Heliospheric Observatory (SOHO; Domingo et al. 1995) Michelson Doppler Imager (MDI; Scherrer et al. 1995) magnetogram of the Sun's disk center on 15 September 2009, at which time its magnetic state may have been similar to that during the Maunder Minimum. From 1 April 2008 to 20 September 2009, there was only one day with as many as three spot groups on the Sun and for 79 % of these days no spots were observed. (*Right*) Comparison MDI magnetogram for 1 November 1996 at the preceding solar minimum. Note the relative absence of "salt and pepper" in 2009. The *white circles* encompass a longitude range from $-45°$ to $+45°$ and a latitude range from $\sim -40°$ to $\sim +50°$

2002; Schrijver and De Rosa 2003) with the sunspot number as driving input prior to July 1996 and assimilated magnetic maps thereafter to infer that the entire Sun for much of 2008–2009 had a minimal magnetic state characterized by an unsigned photospheric flux of $\sim 1.5 \times 10^{23}$ Mx. This value is similar to the $\sim 1.8 \times 10^{23}$ Mx (corresponding to an average surface field of ~ 3 G) deduced by Worden and Harvey (2000) for quiet solar intervals. Stenflo (2012) obtained a basal magnetic field of 3.5 G (2.2×10^{23} Mx) from *Hinode* observations. The years 2007, 2008, and 2009 had 163, 266, and 264 days without sunspots, respectively (ranking 3rd, 1st, and 2nd during the space age).

Thibault et al. (2014) challenged the notion that photospheric conditions in 2009 were similar to those at the depths of the Maunder Minimum. They used a numerical simulation of the photospheric network over a solar cycle to obtain a network relaxation time of 2.9 years indicating that the Sun would not have been devoid of the flux injected during cycle 23 until October 2011. Recent related work by Muñoz-Jaramillo et al. (2015a, 2015b) initially indicated that the Sun reached its minimum magnetic state in both hemispheres during the recent minimum, but out of phase. Upon further consideration (remarked in Muñoz-Jaramillo et al. 2015b), however, the authors concluded that the baseline state they inferred could not be taken as a hard lower limit for the Sun.

2.1 The Sun's Magnetic Carpet

Magnetic bipoles on the Sun range from large sunspot-containing active regions to internetwork elements. Sunspots follow well-established rules (Pevtsov et al. 2014; Hathaway 2015) in terms of occurrence frequency (11-yr cycle), latitude range (butterfly diagram), and magnetic orientation (Hale's law and Joy's law). Small-scale magnetic features (Martin 1988) in the "quiet Sun" include ephemeral regions (Harvey and Martin 1973; total (unsigned) flux range from $\sim 3 \times 10^{18}$ Mx to $\sim 4 \times 10^{20}$ Mx (Close et al. 2004) with spatial scale $\sim 3 \times 10^{4}$ km), internetwork fields (IN; Livingston and Harvey 1971, 1975; Smithson 1975; $\sim 10^{16}$ to 10^{18} Mx, $\sim 2 \times 10^{3}$ km) in the interior of supergranules, and network features at the boundaries of supergranules (Simon and Leighton 1964; $\sim 10^{18}$ to 10^{19} Mx). These features (Schrijver et al. 1997) are collectively referred to as the Sun's "magnetic carpet" (Title and Schrijver 1998; Schrijver and Title 2002).

Recently, as a result of high-resolution observations of magnetic features on the Sun both from Earth and space, there has been a surge in understanding of small-scale magnetic fields on the Sun. The orientation (vertical or horizontal) of the photospheric fields in internetwork field has been an area of intense research. After decades in which solar magnetic fields in quiet regions were assumed to be radial at the surface, in part for ease of calculation, ground-based, balloon-based, and space-based observations by the Solar Optical Telescope (SOT; Tsuneta et al. 2008a; Suematsu et al. 2008; Shimizu et al. 2008; Ichimoto et al. 2008) on *Hinode* (Kosugi et al. 2007) have identified ubiquitous internetwork transverse fields on the Sun (e.g., Lites et al. 1996; Harvey et al. 2007; Centeno et al. 2007; Orozco Suárez et al. 2007a, 2007b; Lites et al. 2008; Ishikawa et al. 2008; Martínez González and Bellot Rubio 2009; Ishikawa et al. 2010; Orozco Suárez and Bellot Rubio 2012). Determining the strength of these fields has been challenging (Solanki 2009; Danilovic et al. 2010; Shchukina and Trujillo Bueno 2011; Asensio Ramos and Martínez González 2014). Although agreement is not universal, e.g., Stenflo (2013), the current synopsis (Martínez Pillet 2013) is that: "The comparison between the internetwork fields observed by *Hinode* and the magnetism inferred from Hanle measurements are converging towards a similar description. They are both described as randomly oriented, largely transverse fields in the several hecto-Gauss range." These fields are generally described as ubiquitous and identical in active regions, polar holes, and quiet Sun (e.g., Tsuneta et al. 2008b; Ito et al. 2010;

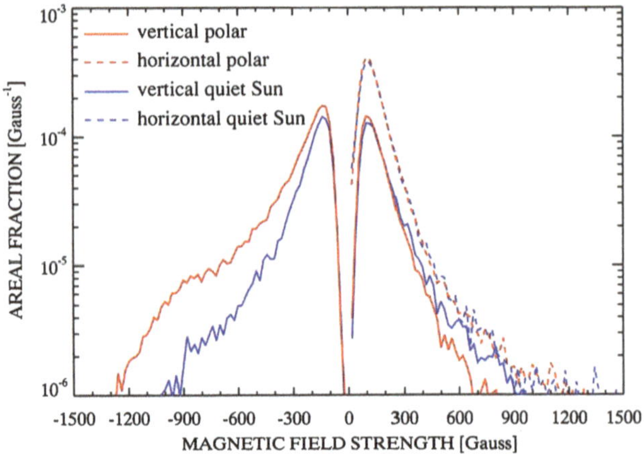

Fig. 2 Areal fraction of magnetic field strength for a quiet Sun region (*blue*) and the north polar region (*red*). The horizontal magnetic flux is unsigned. The vertical axis is the number of pixels with field of a given strength divided by the total number of pixels in the field of view (from Ito et al. 2010)

Ishikawa and Tsuneta 2009). As one moves from sunspots to the small-scale elements of magnetic field the well-defined temporal, spatial, and magnetic orientation properties that characterize sunspots gradually disappear or weaken (Hagenaar et al. 2003; Jin et al. 2011; Jin and Wang 2015; Yang and Zhang 2014).

It is clear that the non-active region fields ($\sim 10^{16}$ to $\sim 10^{20}$ Mx) represent a copious amount of emerging flux, $\sim 10^{25}$ Mx day^{-1} (e.g., Jin et al. 2011; Thornton and Parnell 2011), comparable to the amount that emerges in active regions (with average fluxes $\sim 10^{22}$ Mx) over the course of a cycle (Mosher 1977; Schrijver and Harvey 1994). Thus the small flux concentrations must be taken into account in the magnetic flux budget of the photosphere. Their high emergence rate requires equally high loss rates to prevent flux build-up that would inhibit or halt convection. Estimates for the replacement time for the magnetic carpet range from 1.5–3 days (Schrijver et al. 1997) to ~ 14 hours (Hagenaar 2001) (or 8–19 hours; Hagenaar et al. 2003), with the time decreasing as smaller magnetic elements are considered. The sub arcsec magnetic features observed by IMaX (Martínez Pillet et al. 2011) on SUNRISE have typical lifetimes on the order of a few minutes (Solanki et al. 2010).

The origin of the magnetic carpet remains an open question: The leading candidates (de Wijn et al. 2009) are: (1) the global dynamo (reviewed by Charbonneau 2010); (2) recycled flux of decay products of active regions and ephemeral regions produced by the global dynamo (Nordlund et al. 1992); and (3) a local turbulent dynamo (also referred to as a shallow, or surface, dynamo) that operates independently of the global dynamo (Petrovay and Szakaly 1993; Cattaneo 1999; Vögler and Schüssler 2007). The principal observational arguments for a local dynamo involve the perceived time constancy of the small magnetic elements in relation to the sunspot (or longer) cycles (e.g., Trujillo Bueno et al. 2004; Livingston et al. 2007; Ishikawa and Tsuneta 2009; Schrijver et al. 2011; Buehler et al. 2013), their random magnetic field orientations (e.g., Hagenaar et al. 2003; Yang and Zhang 2014), and their near uniform distribution over the Sun (e.g., Lites et al. 2008; Tsuneta et al. 2008b; Ito et al. 2010; Ishikawa and Tsuneta 2009; cf., Muñoz-Jaramillo et al. 2015b).

The comparison by Ito et al. (2010) of *Hinode* SOT Spectro-Polarimeter (SP; Lites et al. 2001) observations in late 2007 of the north polar region and a quiet region on the east limb of the Sun (Fig. 2) are particularly informative in that both horizontal and vertical fields were considered. The histograms show that the distributions of the horizontal unsigned flux (dashed lines) were identical in both the north polar region and in the quiet Sun. For the vertical flux, the distributions in the two regions were identical for the positive polarity flux

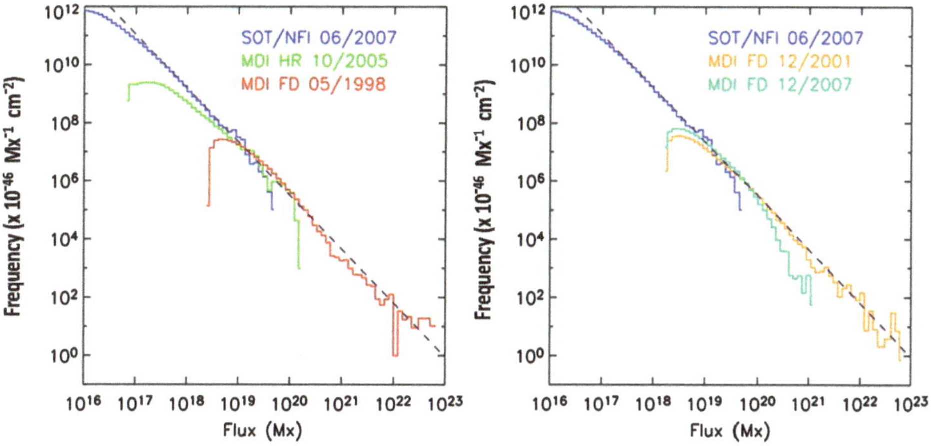

Fig. 3 Histograms of flux concentrations observed by *Hinode* SOT and SOHO MDI (in high-resolution (HR) and full-disk (FD) magnetograms) for different time intervals (from Parnell et al. 2009)

but in the negative polarity northern coronal hole there was an excess of negative polarity flux, particularly for the stronger fields. Evidence that the uniformity of the background bipole distribution seen in Fig. 2 applies at all phases of the solar cycle has been provided by the long-term study of Livingston et al. (2007) of Ca II K and C I 5380 Å intensities and the more recent work of Buehler et al. (2013) based on *Hinode* SOT/SP images. Thus there appears to be a population of small-scale magnetic bipoles on the Sun that is independent of the solar cycle and, given the depth of the 2007–2009 11-yr minimum, perhaps longer-term variation as well.

2.2 One or Two Dynamos on the Sun?

The evidence for two distinct populations of magnetic elements on the Sun, one behaving in concert with the solar cycle and one independent of the cycle, can be interpreted (e.g., Ito et al. 2010; Buehler et al. 2013; Jin and Wang 2015) in terms of two dynamos operating on the Sun: a local dynamo that gives essentially identical distributions of small-scale horizontal and vertical flux everywhere that serves as background for the flux generated by the global solar dynamo. Thus a local dynamo is an attractive candidate to produce the minimal magnetic state on the Sun deduced by Schrijver et al. (2011) to have applied to large portions of the photosphere in 2009.

The main counter-argument to the notion of two separate dynamos is the fact that the size distribution of flux concentrations on the Sun extends with a smooth power law of slope -1.85 ± 0.14 over nearly six orders of magnitude, from 2×10^{17} Mx to 10^{23} Mx as shown in Fig. 3 (Parnell et al. 2009). Parnell et al. suggested two possibilities to explain this remarkable result: "(1) all magnetic features are created by a solar dynamo (extending throughout the convection zone (termed a "spatially nonlocal small-scale dynamo" by Lamb et al. 2014)) that acts in the same way on all scales, or (2) after their emergence into the solar atmosphere all magnetic features are dominated by surface processes ... which somehow (e.g., see Thibault et al. 2012) creates a single distribution of feature fluxes". Thornton and Parnell (2011) addressed (2) by constructing a histogram of emerging flux which, like the flux distributions in Fig. 3, is a power law extending over six orders of magnitude, with an index of ~ -2.7. As Stein (2012) points out, a global dynamo of the type envisioned by

Parnell et al. (2009) and Lamb et al. (2014), "has no preferred scale ... it acts throughout the convection zone with each scale of convective motions generating new flux on that scale (Parnell et al. 2009; Thornton and Parnell 2011). That is, all the surface magnetic features are produced by a common process (which can not be all dominated by surface effects since the sunspots and active regions clearly are not)." Stenflo and Holzreuter (2003) and Stenflo (2010) have similarly called attention to the scale-invariant fractal-like appearance of the Sun's magnetic field. In regard to the distributed scale-free dynamo, Parnell et al. (2009) suggest that the larger flux concentrations ($> 10^{20}$ Mx) related to sunspots produced at the tachocline would "be choked off" at solar minimum but that smaller features would be continuously produced throughout the convection zone during the entire cycle. Although no elaboration is given on the choking off of tachocline-generated flux at solar minimum, this time variability has the same effect at the surface as two separate dynamos, one for sunspots and one for small-flux features. In this picture, the global dynamo would not disappear (or nearly disappear) during Grand Minima. Only the part involving the tachocline that causes the 11-yr cycle would vanish or diminish and the convection zone above the tachocline would produce a minimal (or near minimal) magnetic state consisting of small bipoles. It has also been suggested that the continuous distribution of flux concentrations such as seen in Fig. 3 is evidence of a weak coupling between local and global dynamos (Hagenaar et al. 2003).

3 Floor in the Solar Wind Magnetic Field

The notion that there might be a floor in the solar wind magnetic field was first explicitly proposed by Svalgaard and Cliver (2007). They used a long-term (~ 130 year) reconstruction of solar wind magnetic field strength (B) at Earth based on the IDV index of geomagnetic activity (Svalgaard and Cliver 2005) and a correlation between annual means of B and the square root of the sunspot number to obtain a floor value of ~ 4.6 nT for a yearly sunspot number of zero. The concept of a floor is implicit earlier in the work of Owens and Crooker (2006) who simulated the heliospheric open flux in terms of a constant background and a time-varying component due to coronal mass ejection (CME) activity (cf., Wang and Sheeley 2015). Owens and Crooker obtained a constant background of ~ 4.8 nT vs. the floor of ~ 4.6 nT proposed by Svalgaard and Cliver. Owens et al. (2008) subsequently proposed a "no-CME" floor of $\sim 4.0 \pm 0.3$ nT. These various values were approached/broached during the recent extended minimum for which B values at Earth orbit of 4.5 nT, 4.2 nT, and 3.9 nT were recorded from 2007–2009, respectively. As a result, Crooker and Owens (2010) dropped their estimate of the floor to ~ 3.7 nT to accommodate the 2008–2009 minimum while Cliver and Ling (2011) lowered theirs to ~ 2.8 nT to accommodate conditions of zero peak 11-yr sunspot maximum (i.e., disappearance of the solar cycle) and zero polar field strength at 11-yr minima, close to those thought to apply during the Maunder Minimum (Cliver 2012).

In retrospect, the Svalgaard and Cliver (2007) floor of ~ 4.6 nT was too high at the outset because it did not allow for the drop in B values required by precursor relations (Schatten et al. 1978; Svalgaard et al. 2005; Wang and Sheeley 2009; Cliver and Ling 2011) to predict the peak 11-yr sunspot numbers inferred/observed for the Maunder Minimum. Both ^{10}Be data (Beer et al. 1998; Berggren et al. 2009; Poluianov et al. 2014) and sunspot data indicate that the solar cycle continued during the Maunder Minimum, although at its depths, from ~ 1660–1700, the weakest solar maxima were very low with peak annual sunspot numbers (SSNs) ranging from ≤ 1 (Hoyt and Schatten 1998) to ~ 10 (Svalgaard and Schatten 2015).

This implies that the polar fields during 11-yr minima of the Maunder Minimum, and the corresponding values of B at Earth, must have been significantly lower than for minima during the 1872–2004 period considered by Svalgaard and Cliver for which the smallest cycle (No. 14; 1901–1913) had an amplitude of ~ 65.

3.1 Additional Support for a Floor in the Solar Wind Magnetic Field

3.1.1 Consideration of New Long-Term Sunspot Number and B Time Series

Cliver and Ling (2011) based their revised estimate for the floor of ~ 2.8 nT on a precursor relationship (linearly extrapolated to zero peak SSN) that used Schwabe cycle minimum values of B for cycles 14–23 as a proxy for the solar polar field strength and a correlation between solar polar field strength and solar wind B at the last four minima (also extrapolated to zero SSN).

Recently both the long-term sunspot number (Cliver et al. 2013, 2015; Clette et al. 2014; Svalgaard and Schatten 2015; Clette and Lefèvre 2015) and solar wind magnetic field time series (Owens et al. 2015) have undergone major revisions. The group and international sunspot numbers, previously divergent for years prior to ~ 1885 have now been reconciled and the consensus reconstruction of solar wind B back to ~ 1900 (Svalgaard and Cliver 2010; Lockwood and Owens 2011) has now been extended back to 1845.

We used these new data series to check the precursor relation used by Cliver and Ling (2011) to infer the floor level of B in the solar wind. We obtained a B value of ~ 3.1 nT, close enough to the ~ 2.9 nT obtained by Cliver and Ling (2011) that given the uncertainty, we do not change our working value for the floor of 2.8 nT obtained from the two separate correlations noted above.

The new sunspot number and B time series can also be used to re-examine the relationship between the sunspot number and solar wind B (from Owens et al. 2015) over the 1845–2013 interval. From theoretical considerations (Wang and Sheeley 2003b; Wang et al. 2005), solar wind B should vary as the square root of the sunspot number (cf., Owens et al. 2015). In Fig. 4(a) we see that the power law fit to a scatter plot of B vs. the new international sunspot number (S_N; Clette and Lefèvre 2015) has an exponent of 0.15. Figure 4(b) gives a plot of B with 2.8 nT subtracted (floor removed) for which the exponent of the power law fit is 0.29, in better agreement with, but still below, the theoretically expected value of 0.5. For an exponent of ~ 0.5, it is necessary to subtract a floor of 3.8 nT, just below the lowest observed/inferred value in the annual B-time series of 3.93 nT in 2009. While the square root model may be incorrect, it seems equally plausible that the marked discrepancy between theory and observation in Fig. 4(a) is due, at least in part, to a floor in solar wind B.

3.1.2 Comparison with the ^{10}Be-Based Time Series for B

Steinhilber et al. (2010) noted that their ^{10}Be-based reconstruction of 40-yr averages of solar wind B over the last ~ 9000 years repeatedly (~ 40 times or more) violated, i.e., fell below, the floor of ~ 4.6 nT proposed by Svalgaard and Cliver (2007). The revised floor of 2.8 nT proposed by Cliver and Ling (2011) still lay above the B values inferred by Steinhilber et al. for at least 15 intervals, nine of which had average values of B of approximately zero nT. If any one of these nine deep minima holds up, the question of a floor will be settled in the negative. Recently, McCracken and Beer (2014), using a substantially larger database (derived from seven ice-cores (^{10}Be) and the ^{14}C record from tree-rings)

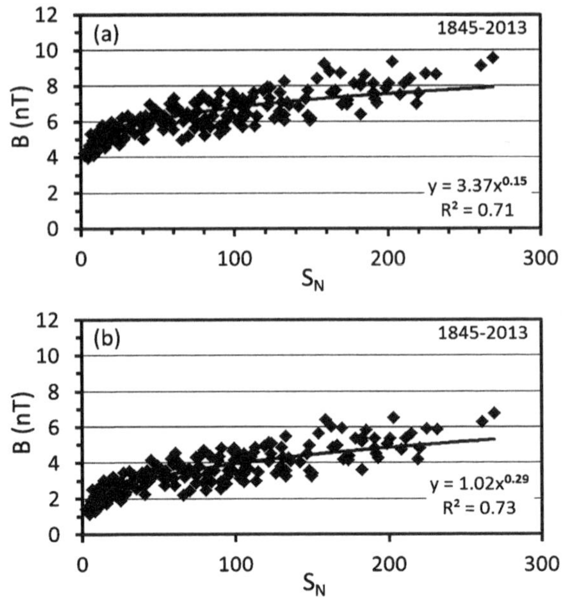

Fig. 4 (**a**) Scatter plot of annual averages of solar wind B vs. the corresponding annual S_N values for 1845–2013, with a power law fit. (**b**) Same as (**a**) except here the floor value of 2.8 nT has been subtracted from yearly averages of B. B values for 1845–1962 are from Owens et al. (2015) after which observed values are used

than that of Steinhilber et al. (2010), estimated that the lowest annual field strength during the Spörer Minimum (1420–1540 AD; one of the nine cases noted above) was 2.4 nT, closer to the 2.8 nT value inferred by Cliver and Ling (2011). The full 9300 year time series is presently being analyzed, and will establish the minimum B field strength during the other eight intervals of near zero field reported by Steinhilber et al. (2010) (McCracken and Beer 2015).

3.1.3 Solar Wind B and the Sun's 10.7-cm Radio Emission

The sunspot number is highly correlated with the Sun's 10.7-cm (2800 MHz) radio emission (e.g., Kundu 1965). Unlike the sunspot number, however, the 10.7-cm flux density does not approach zero at sunspot minima but rather settles at a finite annually-averaged value of ~ 60 solar flux units (1 solar flux unit (sfu) $= 10^{-22}\ \mathrm{W\,m^{-2}\,Hz^{-1}}$). The Sun's background 2800 MHz emission has approached this value at every minimum after 1947 when systematic recording began (http://www.spaceweather.gc.ca/solarflux/sx-en.php).

Following Fig. 2 from Feynman (1982), in Fig. 5 we plot solar wind magnetic field strength vs. the 10.7-cm radio emission for 1947–2014, indicating solar maximum and minimum years during this interval. A power law regression line through the data points for years of sunspot maxima intersects the linear fit for years of minima at a B-value of ~ 2.7 nT. The caution, here as elsewhere, is that this value is based on extrapolations. That said, the intersection of the two lines near the inferred ~ 2.8 nT level of the floor is suggestive because the point at which 10.7-cm flux density at solar minimum equals that at solar maximum indicates the disappearance of the solar cycle.

Recently, Svalgaard (2015) has reported that the annually-averaged value of the regular variation of the eastward component (rY) of the Earth's magnetic field also has an apparent floor which has been reached/approached at each solar minimum since 1840. This variation of Earth's field is attributed to E-region ionization by solar EUV emission. Annual rY and 10.7-cm flux averages are highly correlated. A scatterplot of B vs. rY for 1845–2014 shows

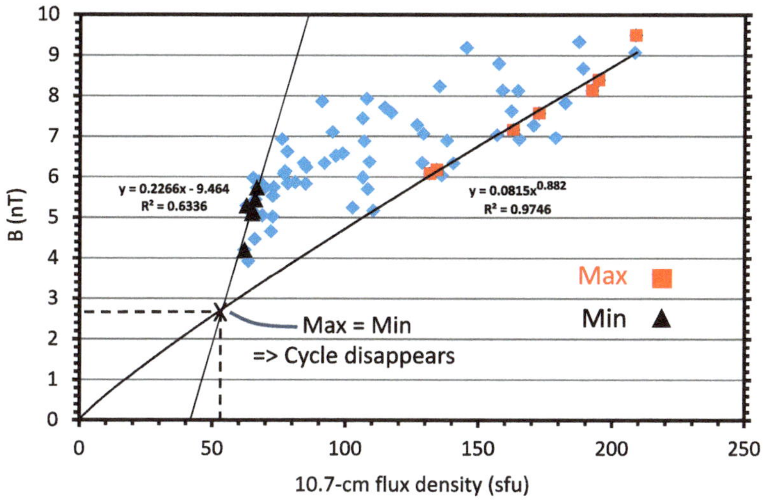

Fig. 5 Scatter plot of annual averages of solar wind B vs. yearly averages of the Sun's 10.7-cm radio emission for 1947–2014. Solar wind B data are from Owens et al. (2015) for 1947–1964 and from the OMNI data base thereafter

Table 1 Parameters of the slow solar wind for recent 11-yr minima and for an extrapolation to the inferred floor value of B (∼ 2.8 nT) (from Cliver and Ling 2011)

Min. year(s)	Total B_{min}	Slow solar wind		
		(% time)	⟨B⟩	⟨V⟩
(1976, 1986, 1996)	5.43 nT	43.3 %	4.85 nT	372 km/s
2009	3.93 nT	71.3 %	3.65 nT	336 km/s
Floor	∼ 2.8 nT	(92 %)	(2.75 nT)	(309 km/s)

more scatter for years of maximum than in Fig. 5 and has a B value of ∼ 4 nT at the intersection of the lines fitted though the maximum and minimum data points.

3.2 The Slow Solar Wind in 2009 and During the Maunder Minimum

Richardson et al. (2000, 2002) decomposed the solar wind from 1972–2000 into slow solar wind (SSW), high-speed streams from coronal holes, and CMEs. This decomposition was updated through 2009 in Cliver and Ling (2011). Characteristics of the slow solar wind used by Richardson in these analyses included: speed $<\sim 400\ \mathrm{km\,s^{-1}}$, generally high density, and wide variations in the plasma parameters. Cliver and Ling (2011) noted that Earth was embedded in slow solar wind ∼ 70 % of the time during 2009. This was an increase from values of ∼ 45 % in 2007 and 2008 when there were large low-latitude coronal holes on the Sun. By 2009 these holes had largely disappeared and were replaced by small mid-latitude coronal holes (de Toma 2011). During the three preceding 11-yr minima in 1976, 1986, and 1996, Earth was embedded in slow solar wind ∼ 43 % of the time on average (Table 1, from Cliver and Ling 2011). The significant increase in the SSW at Earth in 2009, relative to these earlier minima, was a direct reflection of the weaker solar polar field strength at the end of cycle 23 which was ∼ 40 % lower than that for the 1996 minimum (Svalgaard et al. 2005; Wang et al. 2009b; Cliver and Ling 2011). The weaker polar fields strengths (and smaller polar coronal hole areas; de Toma 2011; Wang et al. 2009b) were accompanied by a reduction of high-speed streams at Earth; the percentage of time the Earth spent in high-speed

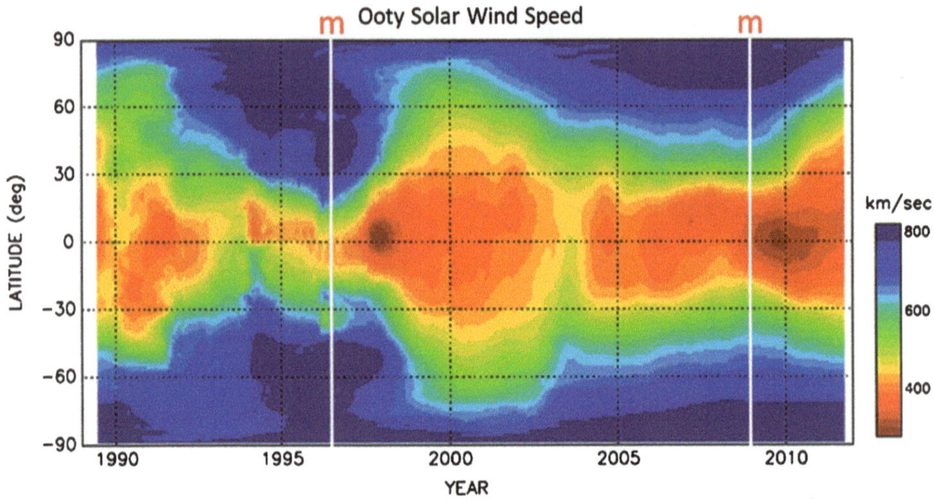

Fig. 6 Latitudinal distribution of the solar wind, 1989–2011 (from Manoharan 2012)

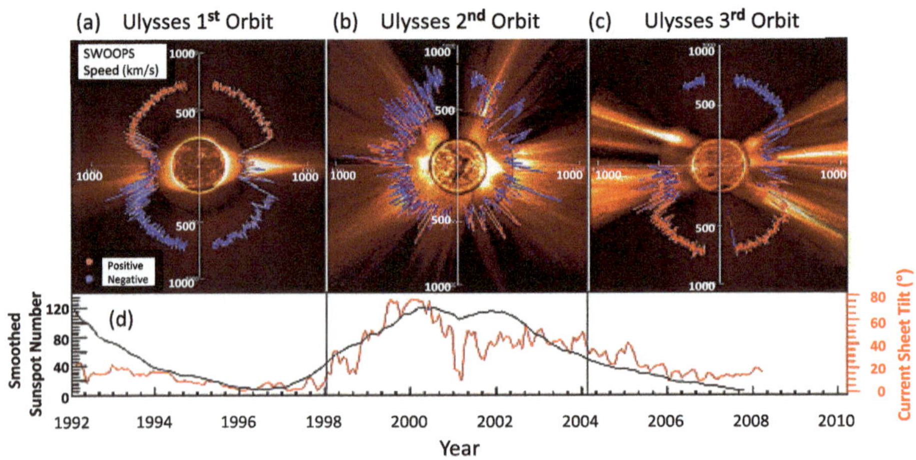

Fig. 7 (**a**)–(**c**) Polar plots of the solar wind speed, as measured by the Solar Wind Observations Over the Poles of the Sun (SWOOPS) plasma experiment (Bame et al. 1992) on Ulysses, 1992–2008, with magnetic polarity indicated, for the three Ulysses polar orbits. The polar plots are superimposed on representative images for each orbit from the SOHO Large Angle Spectroscopic Coronagraph (LASCO; Brueckner et al. 1995), the Mauna Loa K coronameter (Fisher et al. 1981), and the Extreme-Ultraviolet Imaging Telescope (EIT; Delaboudinière et al. 1995). (**d**) Values for the smoothed sunspot number and heliospheric current sheet tilt over this interval (adapted from McComas et al. 2008)

streams decreased from ~ 50 % in 1976/1986/1996 to ~ 25 % for 2009 (Cliver and Ling 2011). The reduction in the latitudinal extent of the fastest high-speed streams from polar holes can be seen in Fig. 6 from Manoharan (2012) based on Ooty interplanetary scintillation measurements (IPS) at 327 MHz. In 1996 the inferred 800 $\mathrm{km\,s^{-1}}$ flow from the poles extended on average down to latitudes of ~ 35° (encompassing 40 % of the heliosphere) vs. ~ 65° (10 %) in 2009.

The direct observations from Ulysses (Wenzel et al. 1992) during its third polar orbit from March 2004 to February 2008 (Fig. 7(c)) showed that the spacecraft did not become fully immersed in the high-speed wind of the southern polar hole until it reached nearly 40° (McComas et al. 2006, 2008; von Steiger 2008). During the first polar orbit of Ulysses (Fig. 7(a)), spanning the solar minimum between cycles 22 and 23, it left the low-speed wind of the streamer belt at or beyond latitudes $\sim 30°$. The broader width of the streamer belt during the third polar orbit is attributed to: (1) the weaker solar polar fields during the recent solar minimum, which were also observed *in situ* in the high-latitude heliosphere (Smith and Balogh 2008), and (2) large low- to-mid-latitude coronal holes (see Fig. 4 in de Toma 2011) that kept the tilt angle (Smith and Thomas 1986) of the heliospheric current sheet (Fig. 7(d)) at relatively high values through 2008 (von Steiger 2008).

A comparison in Table 1 of average B in the slow solar wind for the combined solar minimum years of 1976/1986/1996 with that of 2009, the year of lowest B during the recent unusual solar minimum, indicated that if one extrapolated further to the inferred floor value of ~ 2.8 nT, Earth would be immersed in SSW, with an average speed $\sim 300\ \mathrm{km\,s^{-1}}$, $\sim 90\ \%$ of the time with an SSW B value essentially equal to that of the floor. This raised the possibility that slow solar wind and the floor were one and the same, and that should the solar cycle ever disappear and the Sun reach its minimal magnetic state, one would be left with only slow solar wind.

4 The Sun's Minimal Magnetic State and the Origin of the Slow Solar Wind

The origin of the slow solar wind is recognized as an important long-standing problem in solar physics (Antiochos et al. 2011) and has received much attention lately. The two principal hypotheses for generation of slow solar wind can be grouped as wave-turbulence driven (WTD) models and reconnection/loop-opening (RLO) models (see review by Cranmer 2012). WTD models posit that both slow and fast solar wind originates in coronal holes, with the slow wind hypothesized to emanate near coronal hole boundaries in flux tubes with large expansion factors (Wang and Sheeley 1990, 1992, 2003a; Arge and Pizzo 2000; Cranmer et al. 2007; Wang et al. 2009a; Cranmer 2010). In RLO models (e.g., Wang et al. 1998; Fisk et al. 1999a, 1999b; Fisk and Schwadron 2001; Fisk 2005; Zhao et al. 2009, 2014; Moore et al. 2011; Antiochos et al. 2011, 2012; Antiochos 2013; Riley and Luhmann 2012; Crooker et al. 2012; Rappazzo et al. 2012; Edmondson 2012; Viall and Vourlidas 2015; D'Amicis and Bruno 2015; Brooks et al. 2015), solar plasma is ejected into the corona when closed magnetic loops interchange reconnect with adjacent open field lines. In some of the RLO models, the closed loops arise from the magnetic carpet of the quiet Sun, on supergranular scales (e.g., Fisk et al. 1999b; Fisk 2003).

4.1 Compositional Signatures of the Slow Solar Wind in 2009

The signatures used by Richardson et al. (2000, 2002) to identify intervals of slow solar wind at Earth (speed, plasma parameter variability) did not include the elemental abundance and charge state signatures of such wind (e.g., Geiss et al. 1995; von Steiger et al. 2000; Zurbuchen et al. 2002; von Steiger 2008; von Steiger et al. 2010) that were developed using measurements from the Solar Wind Ion Composition Spectrometer (SWICS, Gloeckler et al. 1998) on the Advanced Composition Explorer (ACE, Stone et al. 1998) and from SWICS (Gloeckler et al. 1992) on Ulysses. These new discriminators between fast and slow solar wind include the ratio of O^{7+}/O^{6+} (Bürgi and Geiss 1986; von Steiger et al. 1997; Zurbuchen et al. 2000; Gloeckler et al. 2003; Fisk 2003) (or C^{6+}/C^{5+}; von Steiger 2008) as a proxy for the low-coronal temperature, and the strength of the first ionization potential (FIP) effect (Meyer 1985; Geiss et al. 1995). Slow solar wind has been defined by a O^{7+}/O^{6+} ratio >0.145 (Zhao et al. 2009) and a pattern of enhanced (by factors of $\sim$ 3–4) abundances of elements with a low (< 10 eV) first ionization potential (FIP) relative to the photospheric abundances of these elements (Feldman and Laming 2000; Asplund et al. 2009; Caffau et al. 2011; Schmelz et al. 2012). This observed FIP pattern is similar to that deduced spectroscopically in large-scale coronal loops above quiet regions on the Sun (Feldman et al. 1999, 2005).

How did the above compositional criteria for defining intervals of slow solar wind hold up in 2009? Using the fixed threshold value of 0.145 for the O^{7+}/O^{6+} ratio for 2009, in part to maintain consistency with the procedure in Zhao et al. (2009), Zhao et al. (2014) determined that $<\sim 20$ % of the solar wind in this year was of the slow variety, in marked contrast to the ~ 70 % value given in Table 1. As noted by Lepri et al. (2013), however, the use of a fixed ratio over the solar cycle is inadvisable. Quoting from their paper, "While heavy ion charge state ratios and average charge states are excellent local identifiers of changes in solar wind source within a Carrington rotation, or even across a given year, they cannot be used as an absolute discriminator of solar wind source across the entire solar cycle as the values evolve considerably across the solar cycle." The low fraction of slow solar wind deduced using a fixed threshold for O^{7+}/O^{6+} is inconsistent with the low average wind speed of the solar wind in 2009 as well as the disappearance of the strong low-to-mid-latitude coronal holes that contributed to average solar wind speeds of 440 km s^{-1} and 450 km s^{-1} in 2007 and 2008, respectively (Cliver and Ling 2011). Moreover, the solar wind density increased from 4.9 cm^{-3} in 2008 to 6.0 cm^{-3} in 2009, consistent with the increased time Earth spent in the slow solar wind. In regard to the FIP effect, Lepri et al. (2013) noted that while the FIP effect for the fast solar wind varied from its normal photospheric pattern in 2009, the FIP bias in the slow solar wind was not appreciably different from normal. From this result, Lepri et al. (2014) inferred the long-term stability of the mechanism that produces the FIP effect in the slow solar wind. We note that such stability is also consistent with the slow solar wind originating in a baseline magnetic state of the Sun. In addition, the apparent drop in the O^{7+}/O^{6+} ratio for such wind in 2009 suggests the involvement of smaller magnetic loops than is generally thought to be the case for the generation of slow solar wind (e.g., Feldman et al. 2005).

4.2 A Lower Limit of ~ 250 km s^{-1} for the Solar Wind Speed?

Whereas the focus of this paper is on solar and solar wind minimal magnetic fields, there is evidence that solar wind speed may have a minimal state as well. Early in the space age, it was noted that the solar wind rarely reached values as low as $\sim$ 250 km s^{-1} (Hundhausen 1972). More recently, Kasper et al. (2007, 2012) called renewed attention to this apparent lower limit of $\sim$ 250 km s^{-1} for the solar wind speed as shown in Fig. 8. Kasper et al. (2012) noted that the abundance of helium in the solar wind approached zero as solar wind speed neared this apparent asymptotic value and speculated that coronal helium might play a role in establishing the apparent minimum wind speed.

Figure 9 gives histograms of wind speed for 1997 and 2009, the only two solar minimum or near minimum years with low (< 400 km s^{-1}) wind speeds during the space age. Average solar wind B (speed) in 1997 and 2009 was 5.54 nT (381 km s^{-1}) and 3.93 nT (364 km s^{-1}),

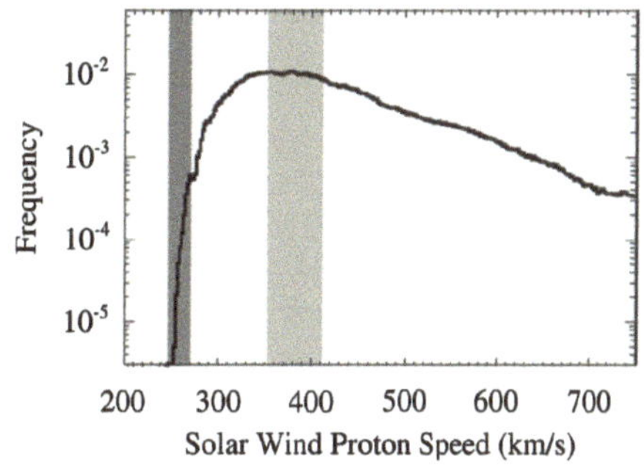

Fig. 8 Occurrence frequency of hourly solar wind speeds as measured by the Wind spacecraft from 1995–2005 (from Kasper et al. 2007)

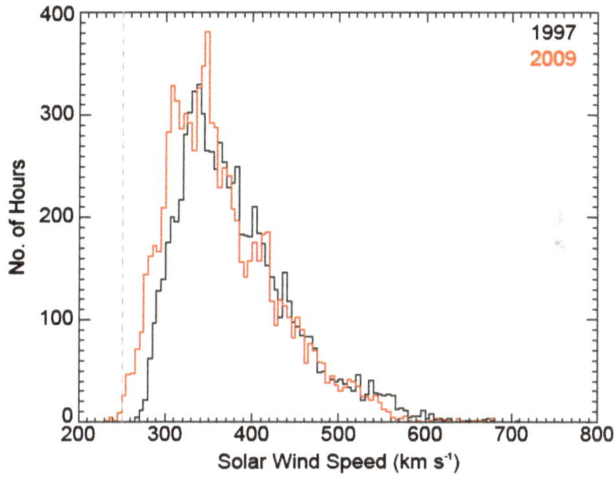

Fig. 9 Histograms of hourly solar wind speed for the only two years during the space age, both near solar minimum, for which the solar wind had average wind speeds < 400 km s^{-1}. The *vertical dashed gray line* is drawn at ~ 250 km s^{-1}

respectively. The drop in B by ~ 1.5 nT between these two minima was accompanied by a shift of the 2009 speed distribution to the left by ~ 25 km s^{-1}, where it rises sharply near the dashed gray line at ~ 250 km s^{-1}. A further reduction of ~ 1 nT to the putative floor of 2.8 nT in the solar wind magnetic field will test the conjecture that the ~ 250 km s^{-1} level has a physical basis related to helium abundance in the corona.

4.3 Modelling the Solar Wind During the Maunder Minimum

Cliver (2012) pointed out that the existence of a minimal magnetic state on the Sun (Schrijver et al. 2011) is a necessary condition for a floor in the solar wind magnetic field. Is it a sufficient condition?

Various modeling studies indicate that it is not. The small-scale fields of the minimal magnetic state of the Sun appear to be incapable of producing the solar wind obtained by extrapolating the solar wind parameters observed in 2009 to Maunder Minimum conditions (Table 1). Cranmer and van Ballegooijen (2010) specifically asked the question "Can the solar wind be driven by magnetic reconnection in the Sun's magnetic carpet?" Using Monte Carlo simulations of the magnetic carpet, they found that in quiet regions and mixed-polarity coronal holes, the energy fluxes released in reconnection events were found to be too low to accelerate the solar wind. In a similar vein, Wang and Sheeley (2013) determined that unless the tilt angles of the ephemeral regions were all aligned, rather than randomized (e.g., Hagenaar et al. 2003), they would not produce a solar wind. Wang and Sheeley noted that if the tilt angles were all aligned they would produce an axial dipole component of the solar magnetic field approximately 7500 times larger than that for the random case and a solar wind with an "extremely strong" field strength, but did not explore the case of partial or localized alignment. In addition, Ito et al. (2010) used a potential source surface model based on *Hinode* vertical field observations to show that almost all of the field lines above the quiet sun region in November 2007 that they examined were closed, reflecting the flux balance in the quiet Sun. Finally, Klimchuk (2012) has questioned the energetics of the Moore et al. (2011) model for the solar wind that is based on interchange reconnection between granule size emerging bipoles and open field lines.

Recently, however, other studies of the solar wind during the Maunder Minimum—when the Sun was presumably very close to its minimum magnetic state—have implied more substantial solar winds than those indicated by the above analyses. Using a continuity equation for the open solar flux with a source term proportional to the sunspot number and a loss term related to the inclination of the heliospheric current sheet (Owens and Lockwood 2012), Lockwood and Owens (2014) deduced a solar wind B of ~ 1 nT (with speed ~ 250–$275\ \mathrm{km\,s^{-1}}$) for 11-yr minima during the Maunder Minimum. From the ^{10}Be record in polar ice cores, McCracken and Beer (2014) obtained a minimum B value of $\lesssim 2.0$ nT in 1697. Riley et al. (2015) used a global thermodynamic MHD simulation that assumed a photosphere composed of only parasitic polarity of average strength ± 3.3 G, comparable to the average B value inferred for the minimal magnetic state of Schrijver et al. (2011), and no polar fields (no cycle) to obtain a B_R value of 0.08 nT which translates to an open flux of $\sim 10^{21}$ Mx. From the relationship between the open flux and B in Fig. 7 of Lockwood and Owens (2011), this implies a solar wind B of ~ 0.5 nT. While all of these values of solar wind B (~ 0.5 nT, ~ 1 nT, $\lesssim 2.0$ nT) fall below the ~ 2.8 nT floor value proposed by Cliver and Ling (2011), each of them—obtained by completely different methods—is considerably above the orders of magnitude shortfalls for B deduced by Cranmer and van Ballegooijen (2010) and Wang and Sheeley (2013) for a basal Sun. The estimates of Lockwood and Owens (2014) and Riley et al. (2015), as well as that of Cliver and Ling (2011), all fall within factors of 2–4 of the ^{10}Be-based value of McCracken and Beer (2014), the closest thing we have to an actual measurement.

As noted in Sect. 2, Schrijver et al. (2011) estimated the minimum total unsigned photospheric flux of a basal Sun to be $\sim 1.5 \times 10^{23}$ Mx based on SOHO MDI observations during 2009. For floor values in the solar wind of $B = 2.8$ nT and $V = 300\ \mathrm{km\,s^{-1}}$ Cliver and Ling (2011) deduced a minimum solar wind open flux of $\sim 8 \times 10^{21}$ Mx, vs. the $\sim 10^{21}$ Mx from the B_R estimate of Riley et al. (2015). These values correspond to ~ 10 % and ~ 1 %, respectively, of the estimated signed solar surface flux.

Brooks et al. (2015) recently used full-disk *Hinode* EUV Imaging Spectrometer (Culhane et al. 2007) images and the FIP effect to image the sources of the slow solar wind for a 48-hr interval on 16–18 January 2013, during the maximum phase of the current weak cycle. They concluded that the dominant slow wind sources were in open field lines at the edges of active regions (Sakao et al. 2007; Harra et al. 2008). For a similar circumstance in February 2007, Sakao et al. (2007) noted that the source of the outflows corresponded to "an ensemble of small sunspots (or pores)". Similarly, Doschek et al. (2008) reported that "the large outflow regions (in active regions) are more representative of quiet-Sun regions than active regions." Such statements, combined with the Brooks et al. result, support our working hypothesis, based on extrapolation to a non-cycling Sun, that open field lines emanating from a carpet of small-scale bipoles are a source of the slow solar wind under normal conditions.

5 Conclusion

5.1 Summary

It has recently been suggested that both the Sun and the solar wind have minimal states of magnetism. A minimal magnetic state on the Sun is a prerequisite for a floor in the magnetism of the solar wind. A basal Sun would lack a solar cycle and be devoid of the large-scale magnetic features (coronal holes, active regions, CMEs) associated with the global dynamo. Extrapolating from the increase in slow solar wind observed in 2009 (Table 1 and Fig. 6) relative to previous 11-yr minima, the corresponding heliosphere would be completely filled with slow solar wind. Under these premises, it is a small step to infer that the magnetic carpet generated by a local dynamo is the source of the floor in the solar wind and, moreover, that these small-scale fields contribute to the slow solar wind when the global dynamo is operating. This picture requires a base-level amount of open flux (estimated to be $\sim 10^{22}$ Mx) because in the RLO process, which is generally preferred for the slow solar wind (e.g., Schwadron et al. 1999; Zurbuchen et al. 2002), interchange reconnection does not create or destroy field lines.

5.2 Open Questions

The path from a minimal magnetic state on the Sun to the origin of the slow solar wind involves a series of open questions: (1) Does the Sun have a minimal (non-cycling) magnetic state and was it reached/approached in 2009 or during the Maunder Minimum? (2) Are there two dynamos on the Sun, a global dynamo responsible for the 11-yr cycle and an independent shallow dynamo responsible for the minimal magnetic state? (3) Does the solar wind have a floor in its magnetic field strength and speed? (4) Can a ground-level solar magnetic state consisting only of small-scale dipoles produce a solar wind? (5) Would a baseline solar wind consist entirely of what we currently identify as slow solar wind?

For each of these questions, there is evidence in the literature and the observations we have reviewed for either a positive or a negative answer. Thus, for question (1), Thibault et al. (2014) reckon that the magnetic flux injected during cycle 23 would not have completely disappeared from the Sun until late 2011, contrary to the report of Schrijver et al. (2011) that the Sun approached its minimal magnetic state globally during the deep minimum of 2008–2009. Similar conflicting evidence lies on either side of the question of one or two dynamos. Ito et al. (2010), Buehler et al. (2013), and Jin and Wang (2015) find support for an independent local or surface dynamo in *Hinode* observations of the spatial/temporal constancy of quiet Sun magnetic fields. Conversely, the power-law size distribution of magnetic flux concentrations on the Sun constructed by Parnell et al. (2009) that extends smoothly over nearly six orders of magnitude argues for a single global dynamo. In regard to question (3), proposed lower limits of $\sim 250\ \mathrm{km\,s^{-1}}$ in solar wind speed (Kasper et al. 2007) and ~ 4.6 nT in magnetic field strength (Svalgaard and Cliver 2007) were challenged and negated, respectively, by direct measurements during the deep 2007–2009 solar minimum, necessitating a downward revision of the floor in B (Cliver and Ling 2011). For question (4), Cranmer and van Ballegooijen (2010) simulated the Sun's magnetic carpet and found it incapable of producing a solar wind, while Riley et al. (2015) used an alternative simulation that also started from the solar magnetic ground state to generate a magnetically weak wind. In response to question (5), Cliver and Ling (2011) and Cliver (2012), extrapolating from solar and solar wind conditions during 2009, have argued that a baseline solar wind would be a weaker version of the omnipresent slow solar wind, originating in the Sun's quiet magnetic field in the absence (or substantial reduction) of active regions and the solar polar fields. Conversely, Wang and Sheeley (2013), present observational evidence (the persistence of the 11-yr cycle during the Maunder Minimum; randomness of tilt angles in ephemeral regions) and physical arguments to conclude that the solar wind during the Maunder Minimum did not emanate from solar quiet regions but from the equatorward edges of polar coronal holes.

The breadth of the above questions, the mixed evidence, and the spate of recent observations and suggestions bearing on the origin of the slow solar wind (e.g., persistent slow outflow at active region edges, Harra et al. 2008; web of separatrices, Antiochos et al. 2011; Type II spicules, Moore et al. 2011; pseudo-streamers, Riley and Luhmann 2012; helicity condensation, Antiochos 2013; new class of coronal features observed during eclipses, Druckmüller et al. 2014; prominence-corona interface, Habbal et al. 2014) mark a particularly exciting time in solar and solar wind studies. Much hinges on the exceptionally deep solar minimum between cycles 23 and 24 which provides the most suggestive evidence thus far for: (1) a minimal magnetic state on the Sun and in the solar wind, (2) the existence of a local or surface solar dynamo, and (3) a quiet Sun origin for the slow solar wind. There are hints in the recent weakened "rush-to-the-poles" phenomenon of coronal green line emission (Altrock 2014) and its associated stunted high-latitude zonal flow (Komm et al. 2014) that the approaching solar minimum may reach a greater depth than in 2009 (Cliver 2014). If so, that would provide an important test of the validity of these various concepts.

Acknowledgements We thank André Balogh for the opportunity to participate in this informative and stimulating workshop. We acknowledge helpful comments from an anonymous referee. E.W. Cliver is grateful for support from the ISSI Visitor program.

References

R.C. Altrock, Sol. Phys. **289**, 623 (2014). doi:10.1007/s11207-012-0216-1
S.K. Antiochos, Astrophys. J. **772**, 72 (2013). doi:10.1088/0004-637X/772/1/72
S.K. Antiochos, Z. Mikić, V.S. Titov, R. Lionello, J.A. Linker, Astrophys. J. **731**, 112 (2011). doi:10.1088/0004-637X/731/2/112
S.K. Antiochos, J.A. Linker, R. Lionello, Z. Mikić, V. Titov, T.H. Zurbuchen, Space Sci. Rev. **172**, 169 (2012). doi:10.1007/s11214-011-9795-7
C.N. Arge, V.J. Pizzo, J. Geophys. Res. **105**, 10465 (2000). doi:10.1029/1999JA000262
A. Asensio Ramos, M.J. Martínez González, Astron. Astrophys. **572**, A98 (2014). doi:10.1051/0004-6361/201423860
M. Asplund, N. Grevesse, A.J. Sauval, P. Scott, Annu. Rev. Astron. Astrophys. **47**, 481 (2009). doi:10.1146/annurev.astro.46.060407.145222
A. Balogh, H. Hudson, K. Petrovay, R. von Steiger (eds.), in *The Solar Activity Cycle*. Space Sciences Series of ISSI, vol. 53 (Springer, New York, 2015). doi:10.1007/978-1-4939-2584-1
S.J. Bame, D.J. McComas, B.L. Barraclough, J.L. Phillips, K.J. Sofaly, J.C. Chavez, B.E. Goldstein et al., Astron. Astrophys. Suppl. Ser. **92**, 237 (1992)
J. Beer, S. Tobias, N. Weiss, Sol. Phys. **181**, 237 (1998)
A.-M. Berggren, J. Beer, G. Possnert, A. Aldahan, P. Kubik, M. Christl, S. Johnsen et al., Geophys. Res. Lett. **36**, L11801 (2009). doi:10.1029/2009GL038004
D.H. Brooks, I. Ugarte-Urra, H.P. Warren, Nat. Commun. **6**, 5947 (2015). doi:10.1038/ncomms6947
G.E. Brueckner, R.A. Howard, M.J. Koomen, C.M. Korendyke, D.J. Michels, J.D. Moses, D.G. Socker et al., Sol. Phys. **162**, 357 (1995). doi:10.1007/BF00733434
D. Buehler, A. Lagg, S.K. Solanki, Astron. Astrophys. **555**, A33 (2013). doi:10.1051/0004-6361/201321152
A. Bürgi, J. Geiss, Sol. Phys. **103**, 347 (1986)
E. Caffau, H.-G. Ludwig, M. Steffen, B. Freytag, P. Bonifacio, Sol. Phys. **268**, 255 (2011). doi:10.1007/s11207-010-9541-4
F. Cattaneo, Astrophys. J. Lett. **515**, L39 (1999). doi:10.1086/311962
R. Centeno, H. Socas-Navarro, B. Lites, M. Kubo, Z. Frank, R. Shine, T. Tarbell et al., Astrophys. J. Lett. **666**, L137 (2007). doi:10.1086/521726

P. Charbonneau, Living Rev. Sol. Phys. **7**, 3 (2010). doi:10.12942/lrsp-2010-3

F. Clette, L. Lefèvre, Sol. Phys. (2015, submitted)

F. Clette, L. Svalgaard, J.M. Vaquero, E.W. Cliver, Space Sci. Rev. **186**, 35 (2014). doi:10.1007/s11214-014-0074-2

E.W. Cliver, in *Comparative Magnetic Minima: Characterizing Quiet Times in the Sun and Stars*, ed. by C.H. Mandrini, D.F. Webb. IAU Symposium, vol. 286 (Cambridge University Press, Cambridge, 2012), p. 179. doi:10.1017/S1743921312004814

E.W. Cliver, Space Sci. Rev. **186**, 169 (2014). doi:10.1007/s11214-014-0093-z

E.W. Cliver, A.G. Ling, Sol. Phys. **274**, 285 (2011). doi:10.1007/s11207-010-9657-6

E.W. Cliver, F. Clette, L. Svalgaard, Cent. Eur. Astrophys. Bull. **37**, 401 (2013)

E.W. Cliver, F. Clette, L. Svalgaard, J.M. Vaquero, Cent. Eur. Astrophys. Bull. **39**, 1 (2015)

R.M. Close, C.E. Parnell, D.W. Longcope, E.R. Priest, Astrophys. J. Lett. **612**, L81 (2004). doi:10.1086/424659

S.R. Cranmer, Astrophys. J. **710**, 676 (2010). doi:10.1088/0004-637X/710/1/676

S.R. Cranmer, Space Sci. Rev. **172**, 145 (2012). doi:10.1007/s11214-010-9674-7

S.R. Cranmer, A.A. van Ballegooijen, Astrophys. J. **720**, 824 (2010). doi:10.1088/0004-637X/720/1/824

S.R. Cranmer, A.A. van Ballegooijen, R.J. Edgar, Astrophys. J. Suppl. **171**, 520 (2007). doi:10.1086/518001

S.R. Cranmer, J.T. Hoeksema, J.L. Kohl (eds.), *SOHO-23: Understanding a Peculiar Solar Minimum*. ASP Conference Series, vol. 428 (Astronomical Society of the Pacific, San Francisco, 2010)

N.U. Crooker, M.J. Owens, in *Proc. of SOHO 23: Understanding a Peculiar Solar Minimum*, ed. by S.R. Cranmer, J. Todd Hoeksema, J.L. Kohl. ASP Conference Series, vol. 428 (Astronomical Society of the Pacific, San Francisco, 2010), p. 279

N.U. Crooker, S.K. Antiochos, X. Zhao, M. Neugebauer, J. Geophys. Res. **117**, A04104 (2012). doi:10.1029/2011JA017236

J.L. Culhane, L.K. Harra, A.M. James, K. Al-Janabi, L.J. Bradley, R.A. Chaudry, K. Rees et al., Sol. Phys. **243**, 19 (2007). doi:10.1007/s01007-007-0293-1

R. D'Amicis, R. Bruno, Astrophys. J. **805**, 84 (2015). doi:10.1088/0004-637X/805/1/84

S. Danilovic, M. Schüssler, S.K. Solanki, Astron. Astrophys. **513**, A1 (2010). doi:10.1051/0004-6361/200913379

G. de Toma, Sol. Phys. **274**, 195 (2011). doi:10.1007/s11207-010-9677-2

A.G. de Wijn, J.O. Stenflo, S.K. Solanki, S. Tsuneta, Space Sci. Rev. **144**, 275 (2009). doi:10.1007/s11214-008-9473-6

J.-P. Delaboudinière, G.E. Artzner, J. Brunaud, A.H. Gabriel, J.F. Hochedez, F. Millier, X.Y. Song et al., Sol. Phys. **162**, 291 (1995). doi:10.1007/BF00733432

V. Domingo, B. Fleck, A.I. Poland, Sol. Phys. **162**, 1 (1995). doi:10.1007/BF00733425

G.A. Doschek, H.P. Warren, J.T. Mariska, K. Muglach, J.L. Culhane, H. Hara, T. Watanabe, Astrophys. J. **686**, 1362 (2008). doi:10.1086/591724

M. Druckmüller, S.R. Habbal, H. Morgan, Astrophys. J. **785**, 14 (2014). doi:10.1088/0004-637X/785/1/14

J.A. Eddy, Science **192**, 1189 (1976)

J.K. Edmondson, Space Sci. Rev. **172**, 209 (2012). doi:10.1007/s11214-011-9767-y

U. Feldman, J.M. Laming, Phys. Scr. **61**, 222 (2000). doi:10.1238/Physica.Regular.061a00222

U. Feldman, G.A. Doschek, U. Schühle, K. Wilhelm, Astrophys. J. **518**, 500 (1999). doi:10.1086/307252

U. Feldman, E. Landi, N.A. Schwadron, J. Geophys. Res. **110**, A07109 (2005). doi:10.1029/2004JA010918

J. Feynman, J. Geophys. Res. **87**, 6153 (1982). doi:10.1029/JA087iA08p06153

R.R. Fisher, R.H. Lee, R.M. MacQueen, A.I. Poland, Appl. Opt. **20**, 1094 (1981). doi:10.1364/AO.20.001094

L.A. Fisk, J. Geophys. Res. **108**, 1157 (2003). doi:10.1029/2002JA009284

L.A. Fisk, Astrophys. J. **626**, 563 (2005). doi:10.1086/429957

L.A. Fisk, N.A. Schwadron, Astrophys. J. **560**, 425 (2001). doi:10.1086/322503

L.A. Fisk, T.H. Zurbuchen, N.A. Schwadron, Astrophys. J. **521**, 868 (1999a). doi:10.1086/307556

L.A. Fisk, N.A. Schwadron, T.H. Zurbuchen, J. Geophys. Res. **104**, 19765 (1999b). doi:10.1029/1999JA900256

J. Geiss, G. Gloeckler, R. von Steiger, Space Sci. Rev. **72**, 49 (1995). doi:10.1007/BF00768753

G. Gloeckler, J. Geiss, H. Balsiger, P. Bedini, J.C. Cain, J. Fischer, L.A. Fisk, Astron. Astrophys. Suppl. Ser. **92**, 267 (1992)

G. Gloeckler, J. Cain, F.M. Ipavich, E.O. Tums, P. Bedini, L.A. Fisk, T.H. Zurbuchen et al., Space Sci. Rev. **86**, 497 (1998). doi:10.1023/A:1005036131689

G. Gloeckler, T.H. Zurbuchen, J. Geiss, J. Geophys. Res. **108**, 1158 (2003). doi:10.1029/2002JA009286

S.R. Habbal, H. Morgan, M. Druckmüller, Astrophys. J. **793**, 119 (2014). doi:10.1088/004-637X/793/2/119

H.J. Hagenaar, Astrophys. J. **555**, 448 (2001). doi:10.1086/321448

H.J. Hagenaar, C.J. Schrijver, A.M. Title, Astrophys. J. **584**, 1107 (2003). doi:10.1086/345792

L.K. Harra, T. Sakao, C.H. Mandrini, H. Hara, S. Imada, P.R. Young, L. van Driel-Gesztelyi, D. Baker, Astrophys. J. Lett. **676**, L147 (2008). doi:10.1086/587485
K.L. Harvey (ed.), *The Solar Cycle*. ASP Conference Series, vol. 27 (Astronomical Society of the Pacific, San Francisco, 1992)
K.L. Harvey, S.F. Martin, Sol. Phys. **32**, 389 (1973). doi:10.1007/BF00154951
J.W. Harvey, D. Branston, C.J. Henney, C.U. Keller (SOLIS and GONG Teams), Astrophys. J. **659**, L177 (2007). doi:10.1086/518036
D.H. Hathaway, Living Rev. Sol. Phys. **12**, 4 (2015). doi:10.1007/lrsp-2015-4
D.V. Hoyt, K.H. Schatten, Sol. Phys. **179**, 189 (1998). doi:10.1023/A:1005007527816
A.J. Hundhausen, *Solar Wind and Coronal Expansion* (Springer, New York, 1972), p. 38
K. Ichimoto, B. Lites, D. Elmore, Y. Suematsu, S. Tsuneta, Y. Katsukawa, T. Shimizu et al., Sol. Phys. **249**, 233 (2008). doi:10.1007/s11207-008-9169-9
R. Ishikawa, S. Tsuneta, Astron. Astrophys. **495**, 607 (2009). doi:10.1051/0004-6361:200810636
R. Ishikawa, S. Tsuneta, K. Ichimoto, H. Isobe, Y. Katsukawa, B.W. Lites, S. Nagata et al., Astron. Astrophys. **481**, L25 (2008). doi:10.1051/0004-6361:20079022
R. Ishikawa, S. Tsuneta, J. Jurčák, Astrophys. J. **713**, 1310 (2010). doi:10.1088/0004-637X/713/2/1310
H. Ito, S. Tsuneta, D. Shiota, M. Tokumaru, K. Fujiki, Astrophys. J. **719**, 131 (2010). doi:10.1088/0004-637X/719/1/131
C. Jin, J. Wang, Astrophys. J. **806**, 174 (2015). doi:10.1088/0004-637X/806/2/174
C.L. Jin, J.X. Wang, Q. Song, H. Zhao, Astrophys. J. **731**, 37 (2011). doi:10.1088/0004-637X/731/1/37
J.C. Kasper, M.L. Stevens, A.J. Lazarus, J.T. Steinberg, K.W. Ogilvie, Astrophys. J. **660**, 901 (2007). doi:10.1086/510842
J.C. Kasper, M.L. Stevens, K.E. Korreck, B.A. Maruca, K.K. Kiefer, N.A. Schwadron, S.T. Lepri, Astrophys. J. **745**, 162 (2012). doi:10.1088/0004-637X/745/2/162
J.A. Klimchuk, J. Geophys. Res. **117**, A12102 (2012). doi:10.1029/2012JA018170
R. Komm, R. Howe, I. González Hernández, F. Hill, Sol. Phys. **289**, 343 (2014). doi:10.1007/s11207-014-0490-1
T. Kosugi, K. Matsuzaki, T. Sakao, T. Shimizu, Y. Sone, S. Tachikawa, T. Hashimoto et al., Sol. Phys. **243**, 3 (2007). doi:10.1007/s11207-007-9014-6
M.R. Kundu, *Solar Radio Astronomy* (Interscience, New York, 1965)
D.A. Lamb, T.A. Howard, C.E. DeForest, Astrophys. J. **788**, 7 (2014). doi:10.1088/0004-637X/788/1/7
S.T. Lepri, E. Landi, T.H. Zurbuchen, Astrophys. J. **768**, 94 (2013). doi:10.1088/0004-637X/768/1/94
S.T. Lepri, E. Landi, T.H. Zurbuchen, in *Outstanding Problems in Heliophysics: From Coronal Heating to the Edge of the Heliosphere*, ed. by Q. Hu, G.P. Zank. ASP Conference Series, vol. 484 (Astronomical Society of the Pacific, San Francisco, 2014), p. 104
B.W. Lites, K.D. Leka, A. Skumanich, V. Martinez Pillet, T. Shimizu, Astrophys. J. **460**, 1019 (1996). doi:10.1086/177028
B.W. Lites, D.F. Elmore, K.V. Streander, in *Advanced Solar Polarimetry-Theory, Observation, and Instrumentation*, ed. by M. Sigwarth. ASP Conference Series, vol. 236 (Astronomical Society of the Pacific, San Francisco, 2001), p. 33
B.W. Lites, M. Kubo, H. Socas-Navarro, T. Berger, Z. Frank, R. Shine, T. Tarbell, Astrophys. J. **672**, 1237 (2008). doi:10.1086/522922
W. Livingston, J. Harvey, in *Solar Magnetic Fields*, ed. by R. Howard. IAU Symposium, vol. 43 (Reidel, Dordrecht, 1971), p. 51
W. Livingston, J. Harvey, Bull. Am. Astron. Soc. **7**, 346 (1975)
W. Livingston, L. Wallace, O.R. White, M.S. Giampapa, Astrophys. J. **657**, 1137 (2007). doi:10.1086/511127
M. Lockwood, M.J. Owens, J. Geophys. Res. **116**, A04109 (2011). doi:10.1029/2010JA016220
M. Lockwood, M.J. Owens, Astrophys. J. Lett. **781**, L7 (2014). doi:10.1088/2041-8205/781/1/L7
C.H. Mandrini, D.F. Webb (eds.), *Comparative Magnetic Minima: Characterizing Quiet Times in the Sun and Stars*. Proceedings of IAU Symposium, vol. 286 (Cambridge University Press, Cambridge, UK, 2012)
P.K. Manoharan, Astrophys. J. **751**, 128 (2012). doi:10.1088/0004-637X/751/2/128
S.F. Martin, Sol. Phys. **117**, 243 (1988). doi:10.1007/BF00147246
M.J. Martínez González, L.R. Bellot Rubio, Astrophys. J. **700**, 1391 (2009). doi:10.1088/0004-637X/700/2/1391
V. Martínez Pillet, Space Sci. Rev. **178**, 141 (2013). doi:10.1007/s11214-013-9967-8
V. Martínez Pillet, J.C. del Toro Iniesta, A. Álvarez-Herrero, V. Domingo, J.A. Bonet, L. González Fernández, A. López Jiménez et al., Sol. Phys. **268**, 57 (2011). doi:10.1007/s11207-010-9644-y
E.W. Maunder, Mon. Not. R. Astron. Soc. **50**, 251 (1890)
E.W. Maunder, Knowledge **17**, 173 (1894)
E.W. Maunder, J. Br. Astron. Assoc. **32**, 140 (1922)

D.J. McComas, H.A. Elliott, J.T. Gosling, R.M. Skoug, Geophys. Res. Lett. **33**, L09102 (2006). doi:10.1029/2006GL025915
D.J. McComas, R.W. Ebert, H.A. Elliott, B.E. Goldstein, J.T. Gosling, N.A. Schwadron, R.M. Skoug, Geophys. Res. Lett. **35**, L18103 (2008). doi:10.1029/2008GL034896
K.G. McCracken, J. Beer, J. Geophys. Res. **119**, 2379 (2014). doi:10.1002/2013JA019504
K.G. McCracken, J. Beer, Personal communication (2015)
R.A. Mewaldt, A.J. Davis, K.A. Lave, R.A. Leske, E.C. Stone, M.E. Wiedenbeck, W.R. Binns et al., Astrophys. J. Lett. **723**, L1 (2010). doi:10.1088/2041-8205/723/1/L1
J.-P. Meyer, Astrophys. J. Suppl. Ser. **57**, 173 (1985). doi:10.1086/191001
R.L. Moore, A.C. Sterling, J.W. Cirtain, D.A. Falconer, Astrophys. J. Lett. **731**, L18 (2011). doi:10.1088/2041-8205/731/1/L18
J.M. Mosher, The Magnetic History of Active Regions. Ph.D. Thesis, California Inst. of Tech., Pasadena (1977)
A. Muñoz-Jaramillo, M. Dasi-Espuig, L.A. Balmaceda, E.E. DeLuca, Astrophys. J. Lett. **767**, L25 (2013). doi:10.1088/2041-8205/767/2/L25
A. Muñoz-Jaramillo, R.R. Senkpeil, J.C. Windmueller, E.C. Amouzou, D.W. Longcope, A.G. Tlatov, Y.A. Nagovitsyn et al., Astrophys. J. **800**, 48 (2015a). doi:10.1088/0004-637X/800/1/48
A. Muñoz-Jaramillo, R.R. Senkpeil, D.W. Longcope, A.G. Tlatov, A.A. Pevtsov, L.A. Balmaceda, E.E. DeLuca et al., Astrophys. J. **804**, 68 (2015b). doi:10.1088/0004-637X/804/1/68
A. Nordlund, A. Brandenburg, R.L. Jennings, M. Rieutord, J. Ruokolainen, R.F. Stein, I. Tuominen, Astrophys. J. **392**, 647 (1992). doi:10.1086/171465
D. Orozco Suárez, L.R. Bellot Rubio, Astrophys. J. **751**, 2 (2012). doi:10.1088/0004-637X/751/1/2
D. Orozco Suárez, L.R. Bellot Rubio, J.C. del Toro Iniesta, S. Tsuneta, B.W. Lites, K. Ichimoto, Y. Katsukawa et al., Astrophys. J. Lett. **670**, L61 (2007a). doi:10.1086/524139
D. Orozco Suárez, L.R. Bellot Rubio, J.C. del Toro Iniesta, S. Tsuneta, B. Lites, K. Ichimoto, Y. Katsukawa et al., Publ. Astron. Soc. Jpn. **59**, S837 (2007b). doi:10.1093/pasj/59.sp3.S837
M.J. Owens, N.U. Crooker, J. Geophys. Res. **111**, A10104 (2006). doi:10.1029/2006JA011641
M.J. Owens, M. Lockwood, J. Geophys. Res. **117**, A04102 (2012). doi:10.1029/2011JA017193
M.J. Owens, N.U. Crooker, N.A. Schwadron, T.S. Horbury, S. Yashiro, H. Xie, O.C. St. Cyr et al., Geophys. Res. Lett. **35**, L20108 (2008). doi:10.1029/2008GL035813
M.J. Owens, E.W. Cliver, K.G. McCracken, M. Lockwood, J. Beer, A. Balogh, P. Charbonneau et al., J. Geophys. Res. (2015, to be submitted)
C.E. Parnell, C.E. DeForest, H.J. Hagenaar, B.A. Johnston, D.A. Lamb, B.T. Welsch, Astrophys. J. **698**, 75 (2009). doi:10.1088/0004-637X/698/1/75
K. Petrovay, G. Szakaly, Astron. Astrophys. **274**, 543 (1993)
A.A. Pevtsov, M.A. Berger, A. Nindos, A.A. Norton, L. van Driel-Gesztelyi, Space Sci. Rev. **186**, 285 (2014). doi:10.1007/s11214-014-0082-2
S.V. Poluianov, I.G. Usoskin, G.A. Kovaltsov, Sol. Phys. **289**, 4701 (2014). doi:10.1007/s11207-014-0587-6
A.F. Rappazzo, W.H. Matthaeus, D. Ruffolo, S. Servidio, M. Velli, Astrophys. J. Lett. **758**, L14 (2012). doi:10.1088/2041-8205/758/1/L14
J.C. Ribes, E. Nesme-Ribes, Astron. Astrophys. **276**, 549 (1993)
I.G. Richardson, E.W. Cliver, H.V. Cane, J. Geophys. Res. **105**, 18203 (2000). doi:10.1029/1999JA000400
I.G. Richardson, H.V. Cane, E.W. Cliver, J. Geophys. Res. **107**, 1187 (2002). doi:10.1029/2001JA000504
P. Riley, J.G. Luhmann, Sol. Phys. **277**, 355 (2012). doi:10.1007/s11207-011-9909-0
P. Riley, R. Lionello, J.A. Linker, E. Cliver, A. Balogh, J. Beer, P. Charbonneau et al., Astrophys. J. **802**, 105 (2015). doi:10.1088/0004-637X/802/2/105
C.T. Russell, J.G. Luhmann, L.K. Jian, Rev. Geophys. **48**, RG2004 (2010). doi:10.1029/2009RG000316
T. Sakao, R. Kano, N. Narukage, J. Kotoku, T. Bando, E.E. DeLuca, L.L. Lundquist et al., Science **318**, 1585 (2007). doi:10.1126/science.1147292
K.H. Schatten, P.H. Scherrer, L. Svalgaard, J.M. Wilcox, Geophys. Res. Lett. **5**, 411 (1978)
P.H. Scherrer, R.S. Bogart, R.I. Bush, J.T. Hoeksema, A.G. Kosovichev, J. Schou, W. Rosenberg et al., Sol. Phys. **162**, 129 (1995). doi:10.1007/BF00733429
J.T. Schmelz, D.V. Reames, R. von Steiger, S. Basu, Astrophys. J. **755**, 33 (2012). doi:10.1088/0004-637X/755/1/33
C.J. Schrijver, M.L. De Rosa, Sol. Phys. **212**, 165–200 (2003). doi:10.1023/A:1022908504100
C.J. Schrijver, K.L. Harvey, Sol. Phys. **150**, 1 (1994). doi:10.1007/BF00712873
C.J. Schrijver, A.M. Title, Sol. Phys. **207**, 223 (2002). doi:10.1023/A:1016295516408
C.J. Schrijver, A.M. Title, A.A. van Ballegooijen, H.J. Hagenaar, R.A. Shine, Astrophys. J. **487**, 424 (1997). doi:10.1023/A:1004901916229
C.J. Schrijver, M.L. De Rosa, A.M. Title, Astrophys. J. **577**, 1006 (2002). doi:10.1086/342247

C.J. Schrijver, W.C. Livingston, T.N. Woods, R.A. Mewaldt, Geophys. Res. Lett. **38**, L06701 (2011). doi:10.1029/2011GL046658
N.A. Schwadron, L.A. Fisk, T.H. Zurbuchen, Astrophys. J. **521**, 859 (1999). doi:10.1086/307575
N. Shchukina, J. Trujillo Bueno, Astrophys. J. Lett. **731**, L21 (2011). doi:10.1088/2041-8205/731/1/L21
T. Shimizu, S. Nagata, S. Tsuneta, T. Tarbell, C. Edwards, R. Shine, C. Hoffmann et al., Sol. Phys. **249**, 221 (2008). doi:10.1007/s11207-007-9053-z
G.W. Simon, R.B. Leighton, Astrophys. J. **140**, 1120 (1964). doi:10.1086/148010
E.J. Smith, A. Balogh, Geophys. Res. Lett. **35**, L22103 (2008). doi:10.1029/2008GL035345
E.J. Smith, B.T. Thomas, J. Geophys. Res. **91**, 2933 (1986). doi:10.1029/JA091iA03p02933
R.C. Smithson, Bull. Am. Astron. Soc. **7**, 346 (1975)
S.K. Solanki, in *Solar Polarization 5: in Honor of Jan Stenflo*, ed. by S.V. Berdyugina, K.N. Nagendra, R. Ramelli. ASP Conference Series, vol. 405 (Astronomical Society of the Pacific, San Francisco, 2009), p. 135
S.K. Solanki, P. Barthol, S. Danilovic, A. Feller, A. Gandorfer, J. Hirzberger, T.L. Riethmüller et al., Astrophys. J. Lett. **723**, L127 (2010). doi:10.1088/2041-8205/723/2/L127
F.W.G. Spörer, Vierteljahrsschr. Astron. Ges. (Leipzig) **22**, 323 (1887)
F.W.G. Spörer, Bull. Astron. **6**, 60 (1889)
R.F. Stein, Living Rev. Sol. Phys. **9**, 4 (2012). doi:10.12942/lrsp-2012-4
F. Steinhilber, J.A. Abreu, J. Beer, K.G. McCracken, J. Geophys. Res. **115**, A01104 (2010). doi:10.1029/2009JA014193
J.O. Stenflo, Astron. Astrophys. **517**, A37 (2010). doi:10.1051/0004-6361/200913972
J.O. Stenflo, Astron. Astrophys. **547**, A93 (2012). doi:10.1051/0004-6361/201219833
J.O. Stenflo, Astron. Astrophys. **555**, A132 (2013). doi:10.1051/00046361/201321608
J.O. Stenflo, R. Holzreuter, in *Current Theoretical Models and Future High Resolution Solar Observations: Preparing for ATST*, ed. by A.A. Pevtsov, H. Uitenbroek. ASP Conf. Ser., vol. 286A (Astronomical Society of the Pacific, San Francisco, 2003), p. 169
E.C. Stone, A.M. Frandsen, R.A. Mewaldt, E.R. Christian, D. Margolies, J.F. Ormes, F. Snow, Space Sci. Rev. **86**, 1 (1998). doi:10.1023/A:1005082526237
Y. Suematsu, S. Tsuneta, K. Ichimoto, T. Shimizu, M. Otsubo, Y. Katsukawa, M. Nakagiri et al., Sol. Phys. **249**, 197 (2008). doi:10.1007/s11207-008-9129-4
L. Svalgaard, Sol. Phys. (2015, submitted)
L. Svalgaard, E.W. Cliver, J. Geophys. Res. **110**, A12103 (2005). doi:10.1029/2005JA011203
L. Svalgaard, E.W. Cliver, Astrophys. J. Lett. **661**, L203 (2007). doi:10.1086/518786
L. Svalgaard, E.W. Cliver, J. Geophys. Res. **115**, A09111 (2010). doi:10.1029/2009JA015069
L. Svalgaard, K.H. Schatten, Sol. Phys. (2015, in press)
L. Svalgaard, E.W. Cliver, Y. Kamide, Geophys. Res. Lett. **32**, L01104 (2005). doi:10.1029/2004GL021664
K. Thibault, P. Charbonneau, A.D. Crouch, Astrophys. J. **757**, 187 (2012). doi:10.1088/0004-637X/757/2/187
K. Thibault, P. Charbonneau, M. Béland, Astrophys. J. **796**, 19 (2014). doi:10.1088/0004-637X/796/1/19
L.M. Thornton, C.E. Parnell, Sol. Phys. **269**, 13 (2011). doi:10.1007/s11207-010-9656-7
A.M. Title, C.J. Schrijver, in *The Tenth Cambridge Workshop on Cool Stars, Stellar Systems and the Sun*, ed. by R.A. Donahue, J.A. Bookbinder. ASP Conf. Ser., vol. 154 (Astronomical Society of the Pacific, San Francisco, 1998), p. 345
J. Trujillo Bueno, N. Shchukina, A. Asensio Ramos, Nature **430**, 326 (2004). doi:10.1038/nature02669
S. Tsuneta, K. Ichimoto, Y. Katsukawa, S. Nagata, M. Otsubo, T. Shimizu, Y. Suematsu et al., Sol. Phys. **249**, 167 (2008a). doi:10.1007/s11207-008-9174-z
S. Tsuneta, K. Ichimoto, Y. Katsukawa, B.W. Lites, K. Matsuzaki, S. Nagata, D. Orozco Suárez et al., Astrophys. J. **688**, 1374 (2008b). doi:10.1086/592226
I.G. Usoskin, Living Rev. Sol. Phys. **10**, 1 (2013). 1. doi:10.12942/lrsp-2013
I.G. Usoskin, R. Arlt, E. Asvestari, E. Hawkins, M. Käpylä, G.A. Kovaltsov, N. Krivova et al., Astron. Astrophys. **581**, A95 (2015). doi:10.1051/0004-6361/201526652
N.M. Viall, A. Vourlidas, Astrophys. J. **807**, 176 (2015). doi:10.1088/0004-637X/807/2/176
A. Vögler, M. Schüssler, Astron. Astrophys. **465**, L43 (2007). doi:10.1051/0004-6361:20077253
R. von Steiger, in *The Heliosphere Through the Solar Activity Cycle*, ed. by A. Balogh, L.J. Lanzerotti, S.T. Suess (Springer-Praxis, New York, 2008), p. 41. doi:10.1007/978-3-540-74302-6_3
R. von Steiger, J. Geiss, G. Gloeckler, in *Cosmic Winds and the Heliosphere*, ed. by J.R. Jokipii, C.P. Sonett, M.S. Giampapa (University of Arizona Press, Tucson, 1997), p. 581
R. von Steiger, N.A. Schwadron, L.A. Fisk, J. Geiss, G. Gloeckler, S. Hefti, B. Wilken, R.F. Wimmer-Schweingruber et al., J. Geophys. Res. **105**, 27217 (2000). doi:10.1029/1999JA000358
R. von Steiger, T.H. Zurbuchen, D.J. McComas, Geophys. Res. Lett. **37**, L22101 (2010). doi:10.1029/2010GL045389

Y.-M. Wang, N.R. Sheeley Jr., Astrophys. J. **355**, 726 (1990). doi:10.1086/168805
Y.-M. Wang, N.R. Sheeley Jr., Astrophys. J. **392**, 310 (1992). doi:10.1086/171430
Y.-M. Wang, N.R. Sheeley Jr., Astrophys. J. **587**, 818 (2003a). doi:10.1086/368302
Y.-M. Wang, N.R. Sheeley Jr., Astrophys. J. **590**, 1111 (2003b). doi:10.1086/375026
Y.-M. Wang, N.R. Sheeley Jr., Astrophys. J. Lett. **694**, L11–L15 (2009). doi:10.1088/0004-637X/694/1/L11
Y.-M. Wang, N.R. Sheeley Jr., Astrophys. J. **764**, 90 (2013). doi:10.1088/0004-637X/764/1/90
Y.-M. Wang, N.R. Sheeley Jr., Astrophys. J. Lett. **809**, L24 (2015). doi:10.1088/2041-8205/809/2/L24
Y.-M. Wang, N.R. Sheeley Jr., J.H. Walters, G.E. Brueckner, R.A. Howard, D.J. Michels, P.L. Lamy et al., Astrophys. J. **498**, L165–L168 (1998). doi:10.1086/311321
Y.-M. Wang, J.L. Lean, N.R. Sheeley Jr., Astrophys. J. **625**, 522 (2005). doi:10.1086/429689
Y. Wang, Y. Ko, R. Grappin, Astrophys. J. **691**, 760 (2009a). doi:10.1088/0004-637X/691/1/760
Y.-M. Wang, E. Robbrecht, N.R. Sheeley Jr., Astrophys. J. **707**, 1372 (2009b). doi:10.1088/0004-637X/707/2/1372
K.P. Wenzel, R.G. Marsden, D.E. Page, E.J. Smith, Astron. Astrophys. Suppl. Ser. **92**, 207 (1992)
J. Worden, J. Harvey, Sol. Phys. **195**, 247 (2000). doi:10.1023/A:1005272502885
S. Yang, J. Zhang, Astrophys. J. **781**, 7 (2014). doi:10.1088/0004-637X/781/1/7
L. Zhao, T.H. Zurbuchen, L.A. Fisk, Geophys. Res. Lett. **36**, L14104 (2009). doi:10.1029/2009GL039181
L. Zhao, E. Landi, T.H. Zurbuchen, L.A. Fisk, S.T. Lepri, Astrophys. J. **793**, 44 (2014). doi:10.1088/0004-637X/793/1/44
N.V. Zolotova, D.I. Ponyavin, Astrophys. J. **800**, 42 (2015). doi:10.1088/0004-637X/800/1/42
T.H. Zurbuchen, S. Hefti, L.A. Fisk, G. Gloeckler, N.A. Schwadron, J. Geophys. Res. **105**, 18327 (2000). doi:10.1029/1999JA000427
T.H. Zurbuchen, L.A. Fisk, G. Gloeckler, R. von Steiger, Geophys. Res. Lett. **29**, 1352 (2002). doi:10.1029/2001GL01394

Space Sci Rev (2017) 210:249–274
DOI 10.1007/s11214-015-0178-3

Coronal Magnetic Field Models

Thomas Wiegelmann[1] · **Gordon J.D. Petrie**[2] · **Pete Riley**[3]

Received: 29 May 2015 / Accepted: 22 June 2015 / Published online: 31 July 2015

Abstract Coronal magnetic field models use photospheric field measurements as boundary condition to model the solar corona. We review in this paper the most common model assumptions, starting from MHD-models, magnetohydrostatics, force-free and finally potential field models. Each model in this list is somewhat less complex than the previous one and makes more restrictive assumptions by neglecting physical effects. The magnetohydrostatic approach neglects time-dependent phenomena and plasma flows, the force-free approach neglects additionally the gradient of the plasma pressure and the gravity force. This leads to the assumption of a vanishing Lorentz force and electric currents are parallel (or anti-parallel) to the magnetic field lines. Finally, the potential field approach neglects also these currents. We outline the main assumptions, benefits and limitations of these models both from a theoretical (how realistic are the models?) and a practical viewpoint (which computer resources to we need?). Finally we address the important problem of noisy and inconsistent photospheric boundary conditions and the possibility of using chromospheric and coronal observations to improve the models.

Keywords Sun · Corona · Magnetic Field · Active Region · MHD

1 Why Do We Need to Model the Coronal Field?

Because the most spectacular and dangerous manifestations of solar activity, flares and coronal mass ejections (CMEs), occur in the corona, and are almost certainly powered by free magnetic energy stored in the coronal magnetic field, modeling and measurement of the coronal field have been urgent goals of solar physics for many years. Only in the last decade

✉ T. Wiegelmann
wiegelmann@mps.mpg.de

1 Max-Planck-Institut für Sonnensystemforschung, Justus-von-Liebig-Weg 3, 37077 Göttingen, Germany

2 National Solar Observatory, 950 N. Cherry Ave., Tucson, AZ, 85719, USA

3 Predictive Science, 9990 Mesa Rim Road, Suite 170, San Diego, CA, USA

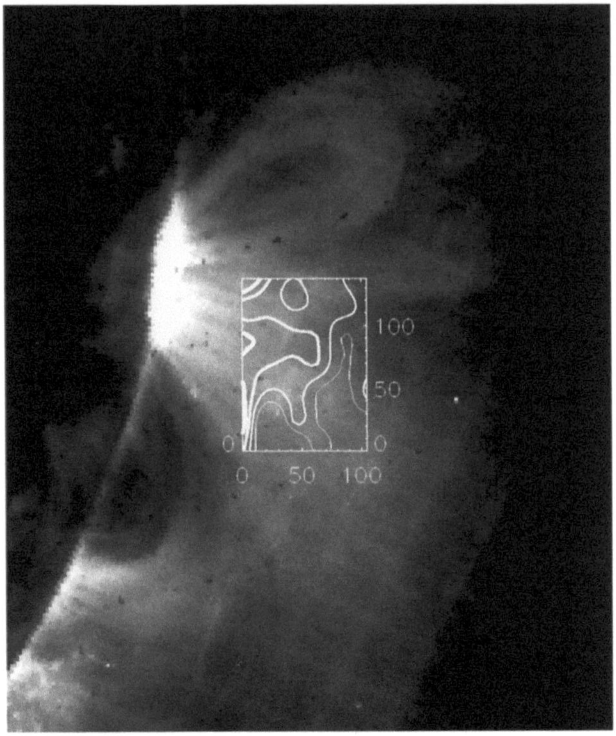

Fig. 1 Contour map of the measured coronal magnetic field strength using Fe XIII overplotted on the EIT Fe XV image. The thick contour corresponds to 4 G, with additional contours corresponding to flux densities of 2, 0, and −2 G. Here the line-of-sight magnetic flux density and transverse field orientation were mapped in two dimensions with a sensitivity of about 1 G and 20″ spatial resolution after 70 minutes of integration. These measurements reveal the line-of-sight coronal magnetic field 100 Mm above an active region to have a flux density of about 4 G. From Lin et al. (2004), Fig. 3. © AAS. Reproduced with permission

or so has Zeeman splitting of magnetically sensitive lines been successfully used to deduce the coronal magnetic flux density. Lin et al. (2004) mapped the line-of-sight flux density and transverse field orientation in two dimensions after a 70-minute integration with 1 G sensitivity and 20″ spatial resolution. Their coronal magnetogram is shown in Fig. 1, over-plotted on an SoHO/EIT EUV intensity image showing magnetic loop structure. The magnetogram indicates that the line-of-sight flux density of the loop system is decreasing with height, as is to be expected in a stratified atmosphere.

On the face of it, coronal magnetic field measurements might be expected to be quite straightforward. The structures are large and in plain sight. On the other hand, they are optically thin, making their geometrical interpretation difficult. Also the faintness of the signal in coronal spectro-polarimetric measurements makes it challenging to determine the coronal magnetic field. Also there are ambiguities in the transverse field measurements, specifically the 180° ambiguity and the Van Vleck ambiguity (Rachmeler et al. 2013). The photospheric and chromospheric signals are much stronger and the field measurements easier to interpret. Regular measurements have been available from the photosphere since the 1960s and 1970s (Howard 1967; Livingston et al. 1976; Svalgaard et al. 1978), and from the chromosphere since the 1980s (Jones 1985). Quantitative information for the coronal magnetic field has mainly been provided by models extrapolated from more accessible measurements deriving from lower atmospheric layers, usually photospheric but sometimes chromospheric. These models are based on simplifying assumptions regarding both their governing equations and the treatment of the lower-boundary data. The basic assumptions are that surface magnetograms can be used to represent accurately the flux magnetic field distribution across the base of the corona, and the magnetic structure of the corona can then be effectively approximated by a simplified model extrapolated from these lower-boundary data.

Some comparisons between extrapolated coronal field models and direct coronal field measurements have been carried out (Liu and Lin 2008). Usually, however, the structure of the coronal magnetic field is interpreted from, e.g., EUV intensity images as in Fig. 1. For example, the fibril loop structure seen in coronal images is often assumed to be aligned with the magnetic vector field. This assumption is based on the facts that the plasma is electrically highly conducting so that there is no plasma flow across field lines, the plasma is hot enough for the gradients along the loop to be much smaller than those across, and the thermal conduction is much more efficient along fields than across, keeping cross-field temperature gradients much higher than field-aligned temperature gradients.

In the following sections we will focus on the four main types of coronal model, the magnetohydrodynamics (MHD) model, the magnetohydrostatic (MHS) model, the nonlinear force-free field (NLFFF) model and the potential-field source-surface (PFSS) model, in decreasing order of sophistication. While these four models differ enormously in complexity, they all represent simplifying approximations of the real corona. Within this review we go from complex to simple models, compared to the usual direction of such reviews because we discuss more and more physical assumptions (instead of physical ingredients) as we proceed. The intention is to emphasize each new assumption's effects (strengths and limitations) at each step. We will describe the physical assumptions underlying these approximations, and we will illustrate the approximations with examples. We will proceed in descending order of complexity, beginning with MHD and ending with potential fields, describing at each step the additional assumptions behind each successive approximation.

2 Magnetohydrodynamics in the Solar Corona

MHD describes the interaction between a magnetized fluid, i.e., a plasma, and a magnetic field. Many descriptions of such interactions exist, offering different compromises between physical complexity and practical simplicity, from a full N-body problem to single-fluid MHD. Here we confine ourselves to single-fluid MHD, since this is the most widely used plasma physics approximation in coronal modeling, and the only self-consistent approach for modeling the interaction between the global coronal magnetic field and plasma that is viable with present-day computational resources.

The validity of the MHD approximation relies on the plasma behaving like a fluid. The usual condition for a fluid description is that the mean free path λ_c is much smaller than the characteristic length scales of the system, $\lambda_c \ll \nabla f/f$ for all quantities f. But $\lambda_c \propto T^2$ where T is the temperature, so this fluid condition may be violated for hot plasmas, e.g., the several-MK plasmas in the solar corona.

Fortunately, as far as the corona is concerned, this argument is misleading. At coronal temperatures, the particles are almost all ionized, and electrostatic interaction prevents large-scale separation of opposite charges, so the large-scale dynamics are controlled by the magnetic field **B**.

In a uniform field **B**, a charged particle moves helically along a field line with Larmor radius $r_L = vm/ZeB$ and with gyrofrequency $\Omega = ZeB/m$, where $B = |\mathbf{B}|$, v is the speed of the particle in the plane of the circle associated with r_L, m and Ze are its mass and charge, and e is the charge of an electron. This determines the length scale of the particle motion in the plane perpendicular to the field.

Along the field, the particle drifts freely until it collides with another particle. Coulomb collisions, near-encounters with other particles that scatter the particle from its helical path, occur at a frequency $\nu_c = 1.3 \times 10^{-20} n/(m^{1/2} T^{3/2})$ where n is the particle number density. The drift distance between collisions is the mean free path $\lambda_c = v_{\mathrm{th}}/\nu_c$ where $v_{\mathrm{th}} = (kBT/m)^{1/2}$ is the typical thermal speed of the particles.

In the corona, the Larmor radius $r_L \approx 1$ m and 20 mm, and the gyrofrequency $\Omega = 10^5$ rad s^{-1} and 10^7 rad s^{-1} for ions and electrons, respectively. Collision frequencies are $0.1\ \mathrm{s}^{-1}$ and $5\ \mathrm{s}^{-1}$, giving a mean free path of order 1000 km along the field.

In cool loops and prominences the collision frequency is much higher, and the mean free path much smaller, because of the lower temperature and higher density. These length scales are much smaller than the characteristic sizes of structures featuring in the coronal models under discussion, and the timescales are much shorter than those of coronal phenomena except transient flare phenomena. Furthermore, the plasma length scales are strongly anisotropic because of the effects of the magnetic field. Gradients parallel to the field, in which direction the mean free path is long, tend to be much weaker than in the perpendicular direction. Under typical coronal conditions, the MHD approximation is therefore appropriate. For a more detailed discussion on MHD in general we refer to Braginskii (1965) and for applying MHD to the coronal plasmas to Petrie (2000) and Marsch (2006).

Figures 2 and 3 show examples of global coronal MHD models compared with observations of coronal magnetic structures. The MHD model includes a self-consistent description of the structure of both the magnetic field and the plasma. Synthetic images can therefore be derived from the model plasma state parameters, and these can be directly compared with observations. An example is shown in Fig. 2, where a synthetic image for the coronal polarization brightness, based on a steady-state MHD model, is compared with an eclipse photograph. Also shown are selected field lines from the model. It is clear that bright features in the white-light corona correspond to regions with closed magnetic field in the model.

Synthetic EUV images can also be derived from MHD models, for comparison with EUV observations of the corona from, e.g., SoHO/EIT, STEREO/SECCHI/EUVI or SDO/AIA. Figure 3 shows a comparison of observations from SDO/AIA in four different wavelength (171 Å, 193 Å, 211 Å, 335 Å) with synthetic images created from a thermodynamic solution of an MHD-model.

This exercise illustrates the degree to which the MHD model plasma state variables are under-constrained by the observations. On the other hand, insight can be gained regarding the heating distribution in the corona: it can be argued that the heating model producing the coronal hole distribution in best agreement with observations is closest to the real heating distribution of the corona.

Solving the full MHD equations on a sizable three-dimensional grid requires computational resources beyond a present-day personal workstation. Also, MHD models are difficult to constrain observationally, since there are many free parameters representing quantities that are difficult to measure. Fortunately, MHD is not necessary for all coronal applications, and there are further useful approximations where models can be produced more easily and cheaply, and constrained only by surface magnetic field measurements.

The first such approximation to consider is the steady-state approximation, where temporal variations are ignored. Because the characteristic speed of the corona is high, the corona tends to evolve in a quasi-static manner, punctuated by episodes of abrupt dynamical change. It is the equilibrium states that we are interested in here. The Alfvén speed in the corona is estimated to be around 1000 km s^{-1} or higher (Aschwanden 2005). For an active region of size 200 or 300 Mm the Alfvén transit time is a few to five minutes. Therefore, if a region's magnetic structure remains essentially unchanged over timescales significantly longer than five minutes, e.g., on timescales of order 1 hr, then the magnetohydrostatic approximation becomes relevant.

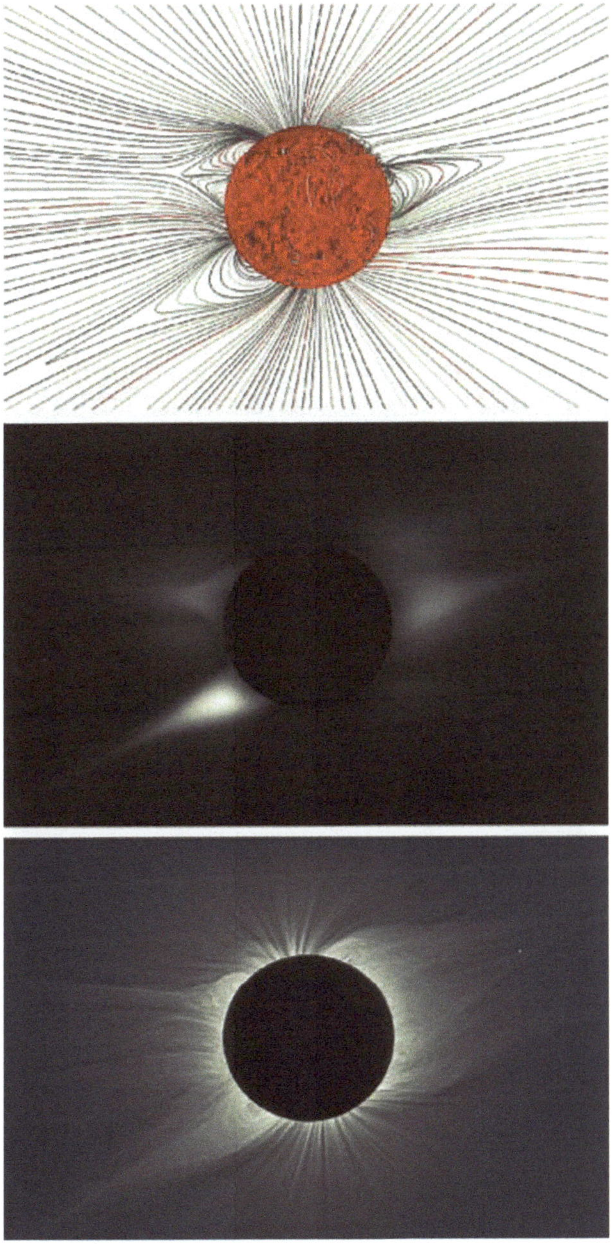

Fig. 2 Comparison between the predicted solar magnetic field lines from an MHD model (*top panel*) and corona polarization brightness (*middle panel*) for the 2008 August 1 total solar eclipse from an MHD model and the sharpened white-light eclipse image (*bottom panel*) from Bor Udzuur, Mongolia. The images are oriented with solar north 12°.1 counterclockwise from the vertical direction. It is evident that the bright features in the white-light corona correspond to regions with closed magnetic field lines. From Rušin et al. (2010), Fig. 5. Reproduced with permission of authors

To quantify the relative influence of the time-dependent versus the time-independent terms, and the terms associated with the magnetic field versus those associated with the plasma, it is helpful to define some dimensionless constants as follows. The Alfvén Mach number $M = v/v_A$ gives the flow speed in terms of the Alfvén speed $v_A = B_0/(4\pi\rho_0)^{1/2}$ for typical field strength B_0 and density ρ_0. The plasma beta $\beta = 8\pi p_0/B_0^2$, the ratio of the plasma pressure p_0 and magnetic pressure $B_0^2/8\pi$, measures the dominance of the mag-

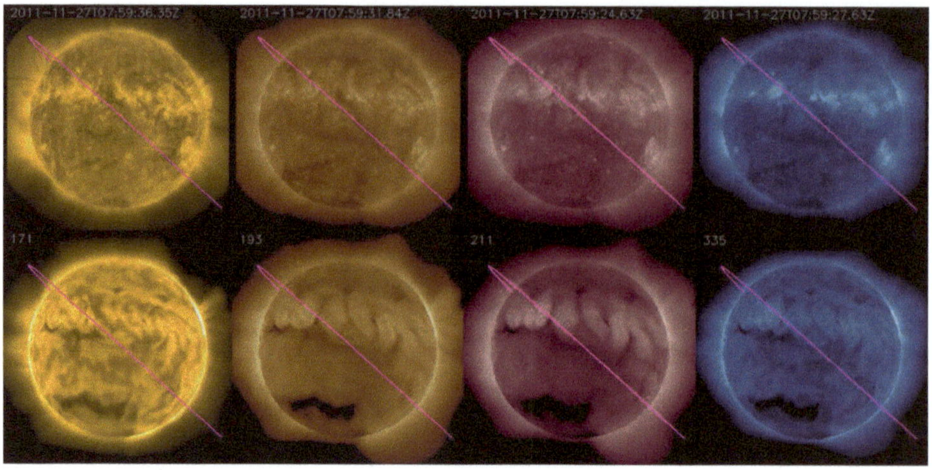

Fig. 3 A comparison of observations from SDO/AIA in different wavelength taken on November 27th 2011 (*top panels*) with synthetic images created from an MHD-model. From Downs et al. (2013), Supplementary material S3. The images where created as part of a study to investigate a Sun-Grazing Comet. A discussion of this topic is well outside the scope of this review, however. Reproduced with permission of authors

netic field. We define a similar ratio between the gravitational energy density and magnetic pressure, $\beta_g = 8\pi\rho_0\psi_0/B_0^2$, for $g = \nabla\psi$, where ψ is the gravitational potential.

For length scale L and Alfvén time $t_A = L/v_A$, the MHD equations normalize as,

$$\tilde{\rho}\left(\frac{t_A}{t_0}\frac{t_0}{t_A}\frac{\partial\tilde{\mathbf{v}}}{\partial\tilde{t}} + \frac{v_0^2}{v_A^2}\tilde{\mathbf{v}}\cdot\tilde{\nabla}\tilde{\mathbf{v}}\right) = \tilde{\mathbf{j}}\times\tilde{\mathbf{B}} - \frac{\beta}{2}\tilde{\nabla}\tilde{p} - \frac{\beta_g}{2}\tilde{\rho}\tilde{\nabla}\tilde{\psi}, \tag{1}$$

$$\frac{L}{t_0 v_0}\frac{\partial\tilde{\mathbf{B}}}{\partial\tilde{t}} = \tilde{\nabla}\times(\tilde{\mathbf{v}}\times\tilde{\mathbf{B}}), \tag{2}$$

$$\frac{L}{t_0 v_0}\frac{\partial\tilde{\rho}}{\partial\tilde{t}} + \tilde{\nabla}\cdot(\tilde{\rho}\tilde{\mathbf{v}}) = 0, \tag{3}$$

$$\tilde{\nabla}\cdot\tilde{\mathbf{B}} = 0. \tag{4}$$

In Eqs. (1)–(4), tildes indicate normalized variables.

3 Magnetohydrostatic Models

If flows are sub-Alfvénic, i.e.,

$$\frac{t_A}{t_0} = \frac{v_0}{v_A} = \varepsilon \ll 0, \tag{5}$$

then Eq. (1) becomes,

$$\mathbf{0} = \tilde{\mathbf{j}}\times\tilde{\mathbf{B}} - \frac{\beta}{2}\tilde{\nabla}\tilde{p} - \frac{\beta_g}{2}\tilde{\rho}\tilde{\nabla}\tilde{\psi}, \tag{6}$$

and Eqs. (2), (3) become trivial. Equation (6) is the magnetohydrostatic equation, describing the balance between the Lorentz force and the plasma pressure gradient and gravitational forces.

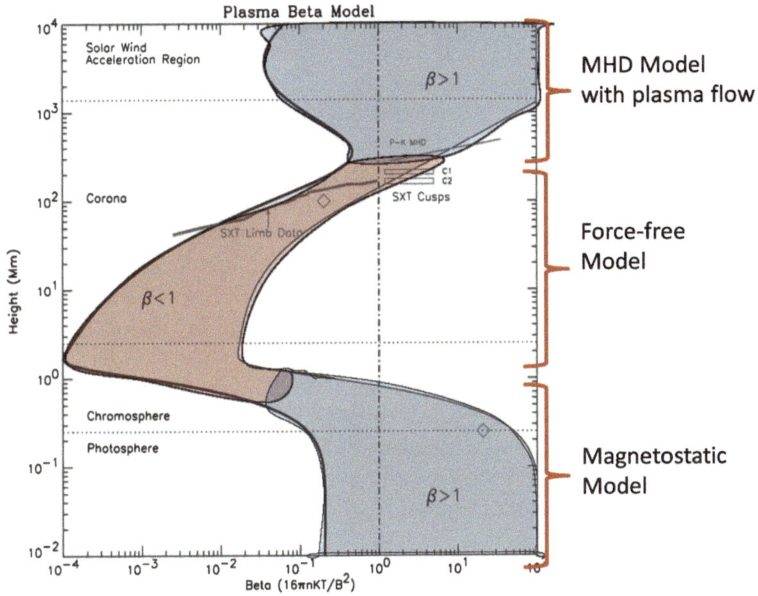

Fig. 4 Plasma β in the solar atmosphere. Modified version from Gary (2001), Fig. 3. Reprinted with kind permission of Springer Science and Business

While the corona contains a low plasma β, this is not the case in the lowest layers of the solar atmosphere (photosphere and lower chromosphere) and also not in the area beyond the source surface at about 2.5 Rs, see also Sect. 4 and Fig. 4. Consequently non-magnetic forces like pressure gradients, gravity and the ram pressure of the solar wind have to be taken into account for this regions. Here we briefly discuss magneto-hydro-static approach (no plasma flows), which is a reasonable assumption for flow velocities below the Alfvén and sound speed.

In the generic, non-linear case, it is in principle possible to solve the MHS-equation (6) by minimizing a functional

$$L(\mathbf{B}, p, \rho) = \int \left[\frac{|(\nabla \times \mathbf{B}) \times \mathbf{B} - \nabla p - \rho \nabla \Psi|^2}{B^2} + |\nabla \cdot \mathbf{B}|^2 \right] dV \tag{7}$$

and derive iterative equations for the magnetic field **B**, plasma pressure p, and plasma density ρ (see Wiegelmann et al. 2007, for details). The MHS-optimization principle generalizes (and has it heritage) from optimization principles for nonlinear force-free fields, see Sect. 4. For complete and consistent boundary conditions this optimization approach works well and reconstructs analytic test solutions with high accuracy, but significant slower convergence time compared with force-free approaches. In particular mixed plasma β regions lead to long computing times, because **B**, p, and ρ have to be iterated simultaneously and low plasma β regions with small (but non-vanishing) plasma forces require strongly reduced iteration time steps.

Somewhat more popular (but less realistic) are MHS-approaches with further restrictions, e.g. on the electric current density in the form:

$$\nabla \times \mathbf{B} = \alpha_0 \mathbf{B} + f(r)\big(\nabla(\mathbf{r} \cdot \mathbf{B})\big) \times \mathbf{r}. \tag{8}$$

In this approach the electric current contains a linear force-free part $\alpha_0\mathbf{B}$ and a current on spherical shells strictly perpendicular to gravity. α_0 is the global constant linear force-free parameter and $f(r)$ a free function. An advantage of representing the current in the form of Eq. (8) is that it linearizes the mathematical problem and separable solutions have been found in different geometries. Naturally (global) solar applications have some preference for spherical geometry (see Bogdan and Low 1986, for solutions without the linear force-free part) and (Neukirch 1995, with both parts of the current). A principal limitation of this approach is that splitting the electric current in this form is motivated by mathematical simplicity (one aims to find separable solutions) and not by physical reasoning. The force-free parameter α_0 has to be globally constant, which is not a very realistic assumption for modelling the entire corona (but might be reasonable for localized regions) and also the free function $f(r)$ has to be globally fixed, which causes additional limitations. Linear MHS models are moderately popular (see, e.g., Bagenal and Gibson 1991; Gibson and Bagenal 1995; Gibson et al. 1996; Zhao and Hoeksema 1994; Zhao et al. 2000; Ruan et al. 2008, for solar applications), but the amount of studies based on this model is much lower compared to the simpler PFSS-models described in Sect. 5. Separable solution in cartesian geometry have been computed by Fourier transform (Low 1991) and Greens function methods (Petrie and Neukirch 2000).

4 Force-Free Models

It is well known that the magnetic field dominates the plasma in the corona, unlike in the photosphere where the dense plasma dominates the dynamics (except arguably in sunspots), or in the heliosphere where the solar wind is dominant. Gary (2001) constructed a simple one-dimensional model for the magnetic stratification of the solar atmosphere and constrained the magnetic and plasma pressures of this model using numerous observations at various heights, see Fig. 4. He argued that there is a range of heights in the solar atmosphere, between the chromosphere and about 100 Mm, where the plasma $\beta \ll 1$. Within this height range, the magnetic field is so dominant in the force balance that the effect on the field structure of the interaction between the Lorentz force and the plasma forces is negligible, i.e., the force-free approximation can apply.

Thus in the magnetically dominated corona, the plasma pressure and gravity forces can be neglected, so that Eq. (6) becomes $\tilde{\mathbf{j}} \times \tilde{\mathbf{B}} = \mathbf{0}$, resulting in the force-free equation,

$$\nabla \times \mathbf{B} = \alpha(\mathbf{x})\mathbf{B}, \tag{9}$$

where the spatially-dependent scalar function $\alpha(\mathbf{x})$ scales as $1/L$. The function $\alpha(\mathbf{x})$ may be interpreted as the magnetic twist per unit length. Since, from Eq. (4) and the divergence of Eq. (9), $\mathbf{B} \cdot \nabla\alpha = 0$, α is constant along magnetic field lines. The force-free model is referred to as the linear force-free field (LFFF) model if α is globally constant, and as the nonlinear force-free field model (NLFFF) if α varies from field line to field line.

Since $\alpha(\mathbf{x})$ scales as $1/L$, this parameter is unimportant for large structures, and significant α values are confined to closed fields within active regions, such as the filament labeled in Fig. 5. In force-free models, non-zero α represents Maxwell stresses trapped within the field in the form of twist or shear. Such a field is not in its lowest energy state: in principle, the Maxwell stress can be released in an eruption such as a flare or CME, fueled by the free magnetic energy contained in the stressed field. The force-free model is the minimal requirement for reconstructing the coronal field of active regions with significant free magnetic energy.

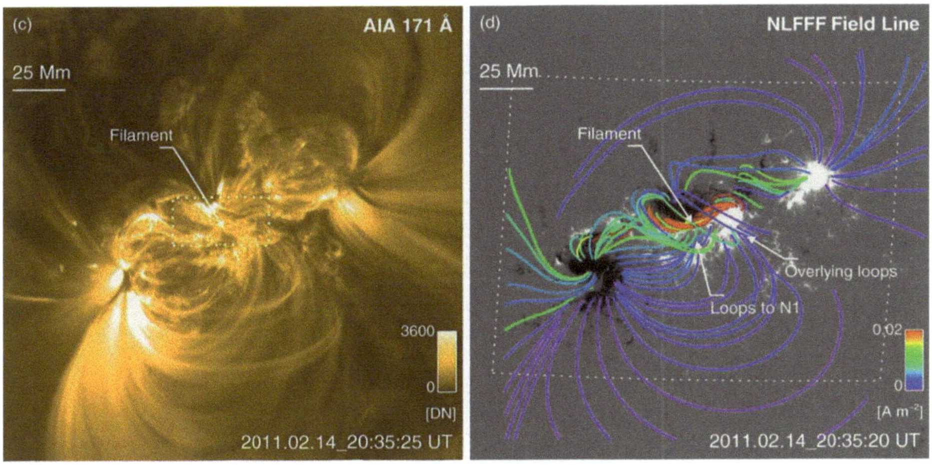

Fig. 5 Observations and modeling results for AR 11158 on 2011 February 14 20:35 UT, about 5 hr before the X2.2 flare. *Left*: Image from AIA 171-band showing the corona magnetic structures. *Right*: Selected field lines from the NLFFF extrapolation plotted over a cutout from the vertical field map. The lines are color-coded by the vertical current density at their footpoints (see the color bar); *red field lines* correspond to strong current density. [From Sun et al. (2012), Fig. 1. © AAS. Reproduced with permission]

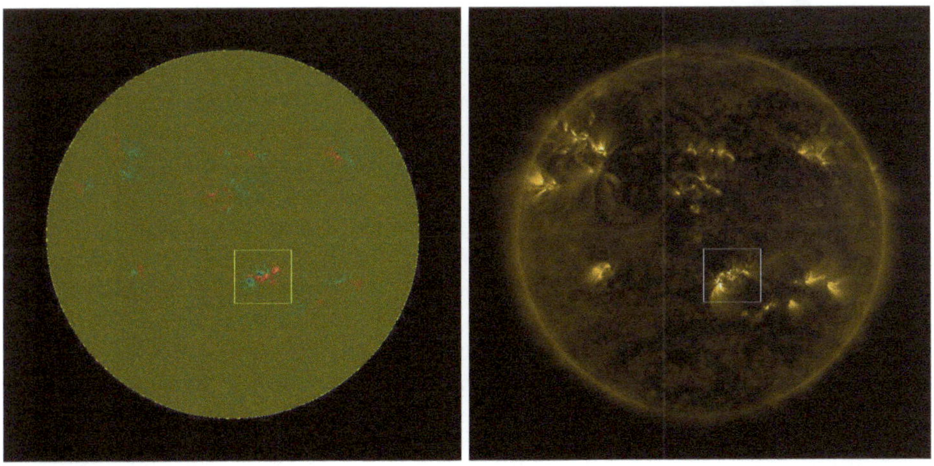

Fig. 6 The *left panel* shows a magnetogram from SDO/HMI and the *right panel* a coronal image from SDO/AIA. The *marked area* corresponds to an active region which was reconstructed with a cartesian force-free code. It takes typical about 5–10 h to compute a NLFFF-model for an active region on a desktop PC. [Source: Wiegelmann et al. (2012), Fig. 1. Reprinted with kind permission of Springer Science and Business]

The force-free approximation allows us to extrapolate a solution from measurements in the lower atmosphere independently of the state of the plasma. The force-free model has no model for the plasma structure, and so it is usually validated by comparing chosen model field lines with observed magnetic structures in, e.g., EUV images such as the examples shown in Figs. 5 and 6. Since full disk vector magnetograms are routinely available from

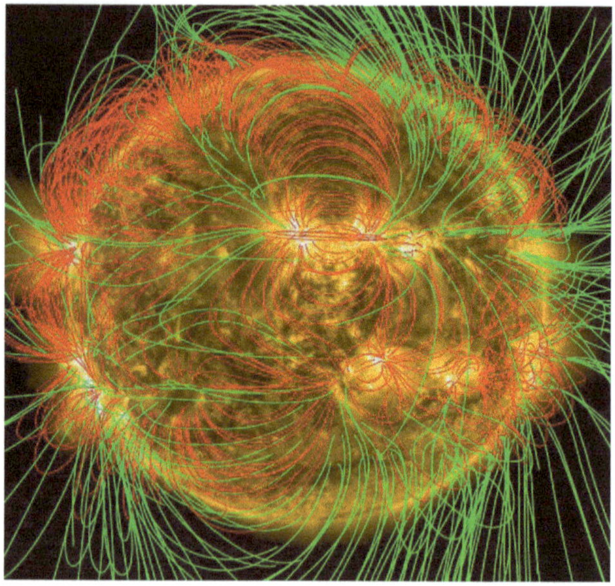

Fig. 7 Global nonlinear force-free coronal magnetic field model based on a synoptic vector magnetogram from SOLIS (June 2012). *Green* and *red lines* correspond to open and closed field lines. The background image shows the coronal plasma observed with SDO/AIA. [Source: Fig. 2 in Tadesse et al. (2014a). Reprinted with kind permission of Springer Science and Business Media]

SDO/HMI, NLFFF models are available also on global scales, see Fig. 7. While the numerical more expensive NLFFF models resemble the observed structure (e.g. as visible in SDO/AIA) somewhat better than simpler potential field models (see Tadesse et al. 2014c, for a comparison of both models). The study by Tadesse et al. (2014c) revealed that active regions are magnetically connected, but hardly share a significant amount of electric currents. A study by Tadesse et al. (2014b) based on synoptic vector magnetograms from SOLIS, revealed that the total magnetic energy for the global NLFFF model was about 10 % than in a PFSS model and the free magnetic energy was located mainly within active regions. Consequently NLFFF models for the global coronal field tend to be close to potential fields, except within active regions. Figure 8 shows a global NLFFF model by Amari et al. (2014), where the global field is untwisted and unsheared, even though there is a highly twisted flux rope present in the model. Because the force free function α is inversely dependent on length scale, large scale (or global) structures are hardly influenced by moderate changes of α. This illustrates why potential field models have been successful in modeling the global coronal magnetic field.

4.1 Methods for Computing Nonlinear Force-Free Fields

Here we briefly describe different methods how non-linear force-free coronal magnetic fields can be computed. For earlier reviews on force-free fields see Amari et al. (1997); Wiegelmann (2008); Wiegelmann and Sakurai (2012) and an overview of magnetic fields in the upper solar atmosphere has been given recently in Wiegelmann et al. (2014).

The force-free equations are defined by a vanishing Lorentz-force leading to

$$(\nabla \times \mathbf{B}) \times \mathbf{B} = \mathbf{0}, \tag{10}$$

$$\nabla \cdot \mathbf{B} = 0, \tag{11}$$

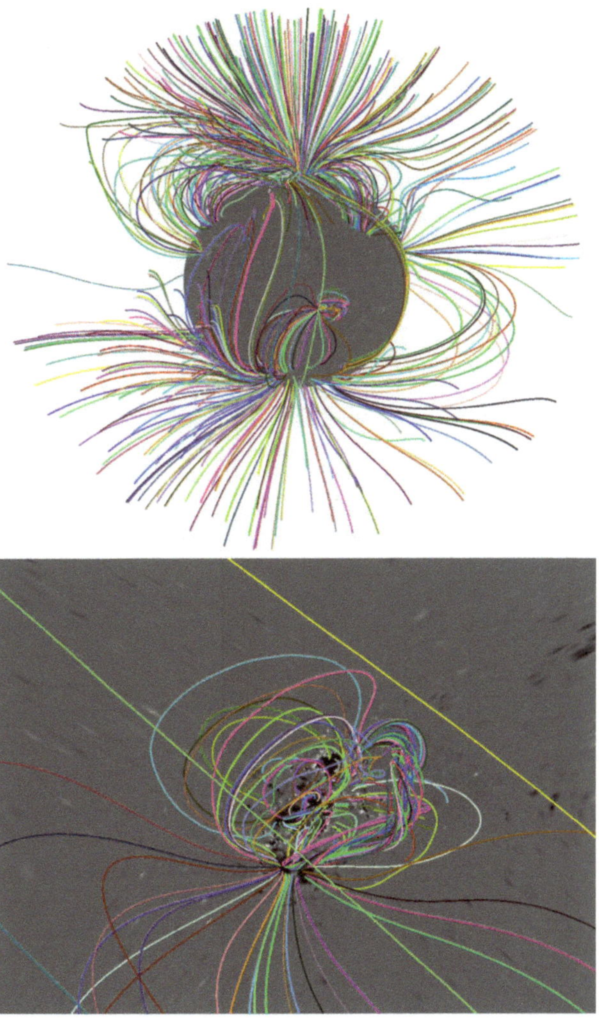

Fig. 8 *Top*: Set of field lines of the reconstructed global solar force-free magnetic field. The calculation is performed with resolution 171 × 251 × 363 and use active-region and full-disk magnetograms from SDO/HMI, and synoptic map from SOLIS. *Bottom*: The structure at the active region shown in the plot of the global solution in the left picture. A twisted flux rope in equilibrium is clearly seen to be present. [From Amari et al. (2014), Figs. 2 and 3. Reproduced with permission of authors]

where **B** is the magnetic field. This set of equations is equivalent to

$$\nabla \times \mathbf{B} = \alpha \mathbf{B}, \tag{12}$$

$$\mathbf{B} \cdot \nabla \alpha = 0, \tag{13}$$

where the force-free function α is constant along magnetic field lines. These sets of equations, either (10) and (11) or (12) and (13) are solved numerically with the measured magnetic field vector in the photosphere as boundary condition. As a first step for a force-free model a current-free potential field is computed, which requires only the vertical magnetic field component as boundary condition. This potential field model is then used as an initial equilibrium before numerical schemes include electric currents. As boundary condition one uses either the photospheric magnetic field vector more or less directly (usually after some preprocessing) and solves Eqs. (10) and (11) by MHD-relaxation or optimization. Alternatively, one uses only the vertical photospheric field and the vertical current density (the ratio of these quantities corresponds to the distribution of α on the photosphere) as boundary conditions and solves (12) and (13) with Grad-Rubin codes. Other approaches, like direct upward integration and the boundary element methods use both the photospheric field vector and the distribution of α. Current joint approaches of the nonlinear force-free consortium (a group of scientists comparing, testing and evaluating NLFFF models since 2004) concentrate on Grad-Rubin, MHD-relaxation and optimization, however and so do we in this review. See Wiegelmann (2008) and references therein for the upward integration and the boundary element methods.

4.2 Grad-Rubin Codes

The Grad-Rubin approach, originally proposed by Grad and Rubin (1958) for fusion plasmas, uses the vertical magnetic field and the α distribution for one polarity (positive or negative) as boundary condition. As initial condition a potential field is used and the electric current (or α) is injected into the computational domain by solving Eq. (13). This naturally changes the magnetic field structure as re-computed from Eq. (13). The iterative process continues until a stationary state is reached. Mathematicians like the Grad-Rubin very much, because it is based on a well posed mathematical problem (see, e.g., Amari et al. 2006) and at least for moderate values and gradients of α it has been shown that a solution exists and is unique (Bineau 1972). For α derived from measurements, which can have high values and strong gradients due to localized current concentrations convergence of the Grad-Rubin method is, however, not ensured. An additional pitfall is that one has two distinct well posed problems: (1) vertical magnetic field and vertical current computed from the positive polarity and (2) vertical magnetic field and vertical current computed from the negative polarity. And the solutions of these two problems usually differ significantly for measured data (see, e.g., Schrijver et al. 2008). In a subsequent study Wheatland and Régnier (2009) showed that it is possible to extend the Grad-Rubin code by an outer iteration circle, which computes the photospheric α distribution from both polarities and includes the measurement errors of the horizontal magnetic field to weight between α-values computed from both polarities.

4.3 MHD-Relaxation and Optimization

These code have many similarities. Starting from a potential field, they solve the force-free equations by iterating

$$\frac{\partial \mathbf{B}}{\partial t} = \mu \mathbf{F} \tag{14}$$

where the pseudo force $\mathbf{F}$ is derived from a reformulation of the MHD-equations (relaxation codes, see, e.g., Valori et al. 2005) or from minimizing a functional similar as in the MHS case (7), but here the plasma terms vanish (optimization codes, see, e.g., Wheatland et al. 2000).

As initial condition the computational box contains a potential field, which is then disturbed by an injection of the measured photospheric magnetic field vector. Originally both approaches fixed the magnetic field vector on the lower boundary of the computational box, but newer codes allow now a more gradual introduction of these boundary conditions and allow to change the horizontal field on the boundary in accordance with measurement errors. The iteration of Eq. (14) continues until a stationary state is reached and the pseudo force $\mathbf{F}$ vanishes. A principal problem with the original approach (magnetic field vector fixed on the lower boundary) is that for measured data, which include noise and inconsistencies, a solution of (10) and (11) cannot be guaranteed. Preprocessing of the measurements reduces but

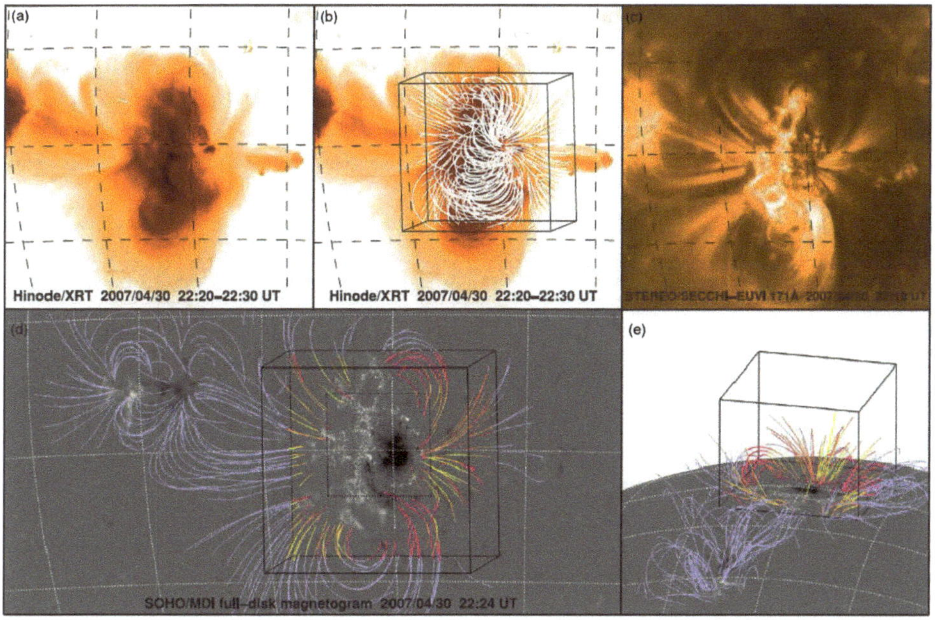

Fig. 9 Comparison of NLFFF-models and stereo-loops. Panel **a** shows a Hinode soft X-ray image of AR10953, in panel **b** overlaid with NLFFF-fieldlines. Panel **c** depicts one of the STEREO-SECCHI images, which has been used for stereoscopy. Finally in panels **d** and **e** the stereoscopic loops *(in blue)* are compared with NLFFF loops. The *dotted area* depicts the region where photospheric vector field measurements from Hinode have been available, the *solid black lines* the NLFFF computational domain (outside the Hinode-FOV only LOS-measurements from SOHO/MDI have been available). [Source: DeRosa et al. (2009), Fig. 1. © AAS. Reproduced with permission]

does not entirely solve this problem. Optimization codes have been developed in cartesian and spherical geometry, see Figs. 7 and 8 for global spherical NLFFF-models based on an optimization code and a Grad-Rubin method, respectively.

4.4 Comparison and Evaluation of NLFFF-Codes

In a series of workshops/publications a group of scientists have compared and evaluated NLFFF-codes since about a decade—henceforth referred to as the *NLFFF consortium*. The comparison with synthetic data (for which the exact solutions where known, see Schrijver et al. 2006; Metcalf et al. 2008) revealed that the codes are very reliable and accurate if fed with complete and consistent boundary conditions. Complications occur, when the boundary conditions are neither complete, nor force-free consistent, which is unfortunately the case for a number of photospheric measurements (see also Sect. 7). For such inconsistent boundary conditions (a relative small FOV measured by Hinode) the model fields differed in geometry, magnetic energy content, and degree of force-freeness (see Schrijver et al. 2008). In a subsequent study (DeRosa et al. 2009) coronal magnetic field lines obtained from NLFFF-models have been compared with stereoscopically reconstructed coronal loops, see Fig. 9. The comparison was not satisfactory. NLFFF models did not perform better than potential fields in this study. Possible reasons are the small FOV of the vector data from Hinode, which only covered about 10 % of the area spanned by the stereoscopic reconstructed loops. Also the separation angle of the two spacecraft was very small (7°) which leads to a large stereoscopic reconstruction error. DeRosa et al. (2009) concluded that a successful application of NLFFF models requires:

1. Large model volumes at high resolution that accommodate most of the connectivity within a region and to its surroundings;
2. Accommodation of measurement uncertainties (in particular in the transverse field component) in boundary condition;
3. *Preprocessing* of the lower-boundary vector field for a realistic approximation of the high-chromospheric, near force-free field;
4. Force-free models should be compared (or even improved) with coronal observations

Boundary conditions for large model volumes are routinely available since the launch of SDO in 2010 and recent updates of NLFFF-codes are suitable to incorporate measurement errors (see Wheatland and Leka 2011, for Grad-Rubin codes and Wiegelmann and Inhester 2010, for optimization). We will discuss preprocessing in Sect. 7, which is dedicated to boundary conditions (for all methods, not limited to NLFFF). Let us note that some datasets of SDO/HMI are remarkably force-free consistent in the photosphere and do not require preprocessing (Wiegelmann et al. 2012). A comparison of extrapolation results with coronal images (like SDO/AIA) is now routinely performed, see Fig. 7. We briefly address recent methods to improve NLFFF models with coronal observations in Sect. 7.

5 Potential-Field Models for the Global Coronal Field

When we neglect electric currents from the coronal model, we can represent the field using a scalar potential, $\mathbf{B} = \nabla\phi$, so that Eq. (4) leads to a Laplace equation

$$\nabla^2\phi = 0 \tag{15}$$

with the line-of-sight or normal derivative of ϕ determined on the lower boundary by the observational line-of-sight or radial field component (Altschuler and Newkirk 1969; Schatten et al. 1969). Force-free electric currents can be significant in active regions, as we have seen in Figs. 5 and 8. The other major location where significant electric currents are likely to flow is the volume above the low-β region in the corona, corresponding to the volume above heights of order 100 Mm in the model of Gary (2001). At these heights the thermal and dynamical pressures of the expanding solar wind can exceed the coronal magnetic pressure, under which conditions the fields cannot return to the Sun. Instead they are dragged out into the heliosphere. In the PFSS model (Altschuler and Newkirk 1969; Schatten et al. 1969), this effect of the solar wind on the coronal field is modeled by setting the scalar potential to a constant value on an outer boundary surface, called the source surface, forcing the model field to be radially directed there. The radius of the source surface is usually set to between 2 and 3 solar radii (Hoeksema 1984). Within the source surface the field is dominant and current-free. Outside the source surface the field is dominated by the solar wind and follows the spiral trajectory of this wind.

Equation (15) can be solved analytically as an expansion of spherical harmonics (Schatten et al. 1969; Altschuler and Newkirk 1969) or numerically using a finite difference approach (e.g. Riley et al. 2006). In a recent study Tóth et al. (2011) compared the spherical harmonic decomposition method with a finite difference approach and the authors found significant differences. A spherical harmonic decomposition is prone to artifacts like rings around strongly localized structures. Another problem is that polar field measurements are

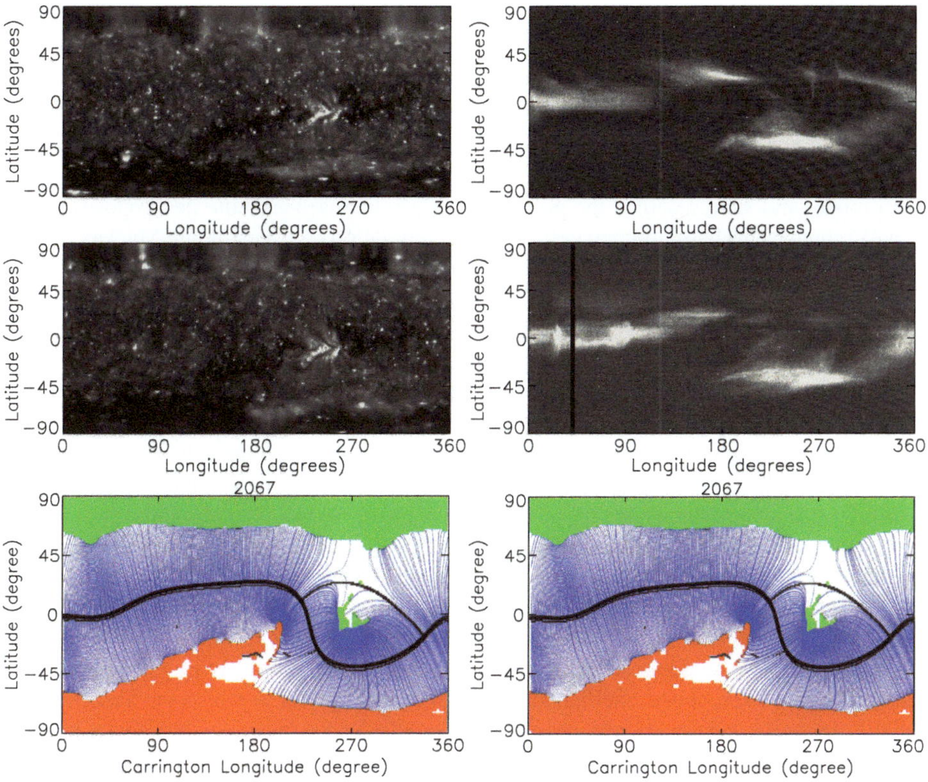

Fig. 10 STEREO/SECCHI/EUVI 171 Å synoptic maps for Carrington rotation 206, from the ahead (*top left*) plot and behind (*middle left plot*) spacecraft. STEREO/SECCHI/COR1 synoptic maps for Carrington rotation 2067 from the ahead (*top right plot*) and behind (*middle right plot*) spacecraft. These maps show the streamer brightness at 2.6 solar radii off the east limb. PFSS models extrapolated from the GONG synoptic magnetograms Carrington rotations 2067 (*bottom pictures*, which are identical). Positive and negative model coronal holes are colored *red* and *green*. The streamer-belt neutral lines are represented by *thick black lines* and pseudo-streamer locations by *thin black lines*. Streamer-belt fields are plotted in *blue*. [From Petrie et al. (2011). Figs. 2, 3 and 4. Reprinted with kind permission of Springer Science and Business Media]

difficult because of, among other things, the very large projection angle associated with the small angle between the solar rotation axis and the ecliptic plane, the fact that each pole is visible for <6 months each year, the intrinsic weakness of the polar fields, and limb noise. These difficulties are discussed by Petrie et al. (2014). For both numeric and analytic PFSS methods, the radial magnetic field at the inner boundary is derived from observed photospheric magnetic field measurements (typically by using precomputed synoptic maps). At the outer radial boundary (typically 1.5–3.5 R_S), the field is forced to become radial. This outer boundary is, in effect, a tunable parameter in the model, allowing the modeler to better match observed features in the corona (e.g., the location of coronal hole boundaries) or the solar wind (the amount of open flux inferred from interplanetary spacecraft).

Though the PFSS model is simple, its phenomenology is rich. Selected classes of field lines correspond to well-known large-scale coronal features. Open field lines represent coronal holes, and the set of footpoints of open fields can be compared to coronal hole observations in EUV or He I 10830 Å. The set of tallest closed field lines can be compared to the main streamer belt and the set of pseudostreamers seen in coronagraph images. Comparisons of both coronal hole and streamer locations are shown in Fig. 10.

Here we have only mentioned the most popular methods for reconstructing the coronal magnetic field structure from photospheric or chromospheric field measurements. Other methods include the extension of the PFSS model to include a heliospheric current sheet by Schatten (1971), and the linear magnetohydrostatic approach (8) is a generalization of the PFSS model and shares with it the possibility of a spherical harmonic decomposition. For global computations based on magnetofrictional simulations see Yeates and Mackay (2012).

6 Comparison Between PFSS and MHD Models

As we have noted, the large-scale, quasi-static magnetic field within the corona is computed using PFSS, NLFFF, MHS or MHD models. All techniques rely on measurements of the photospheric magnetic field (line-of-sight field or vector field). In the following we want to compare the simplest (PFSS) and most sophisticated approach (MHD) in some detail.

PFSS models have the advantage of being simple to develop and apply, and do not require significant computational resources (Riley et al. 2006). Additionally, they are, in principle, capable of resolving spatial scales beyond those currently seen with MHD solutions. However, they are limited because their basic assumptions are rarely, if ever met in reality. In particular: (1) magnetic fields in the low corona often depart significantly from a potential configuration; (2) time-dependent effects (e.g., reconnection) are likely important processes that reconfigure the magnetic field; and (3) the reference surface at which the field becomes radial is generally not spherically-shaped. In the following subsections, we summarize the PFSS and MHD techniques, and explore each of these limitations in more detail.

MHD models can be thought of as a refinement (albeit substantial) to the PFSS approach. They solve the full set of resistive MHD equations, typically in spherical coordinates, and can incorporate a wide range of physical processes. Most recently, self-consistent descriptions of waves/turbulence have been incorporated with the aim of prescribing both the heating of the corona and the acceleration of the solar wind in a self-consistent fashion (e.g. Lionello et al. 2014). For the purposes of comparing with PFSS results, however, we limit ourselves to the simpler polytropic version of the model, in which the energy equation is simplified by an adiabatic equation, where the polytropic index, γ, is set to 1.05, to mimic the near isothermal properties of the corona. Although this results in plasma values and variations that do not match well with observations, the magnetic structure of the corona appears to be well captured.

The MHD equations have been described in Sect. 2 and the methods used to compute the MHD solutions can be found in Mikić and Linker (1994), Lionello et al. (1998), Riley et al. (2006). Here, we note only that: (1) the coronal model computes solutions between 1 R_S and 20–30 R_S; (2) we use the same radial photospheric magnetic field to drive the MHD model; and (3) although the solutions are time-dependent, we advance the solution forward in time until a steady-state equilibrium is reached.

Figure 11 compares the MHD and PFSS solutions for two Carrington rotations representative of solar minimum (CR1910) and solar maximum (CR1969) conditions. In each panel, the same starting points were used to trace the field lines; thus, differences in the location and shape of the field lines represent differences in the model solutions. We infer that the two approaches appear match reasonably well, at least qualitatively, at both solar minimum and maximum. However, there are several noteworthy differences. For example, the PFSS model does not reproduce the cusp-like features of the streamer belt, which merge into the

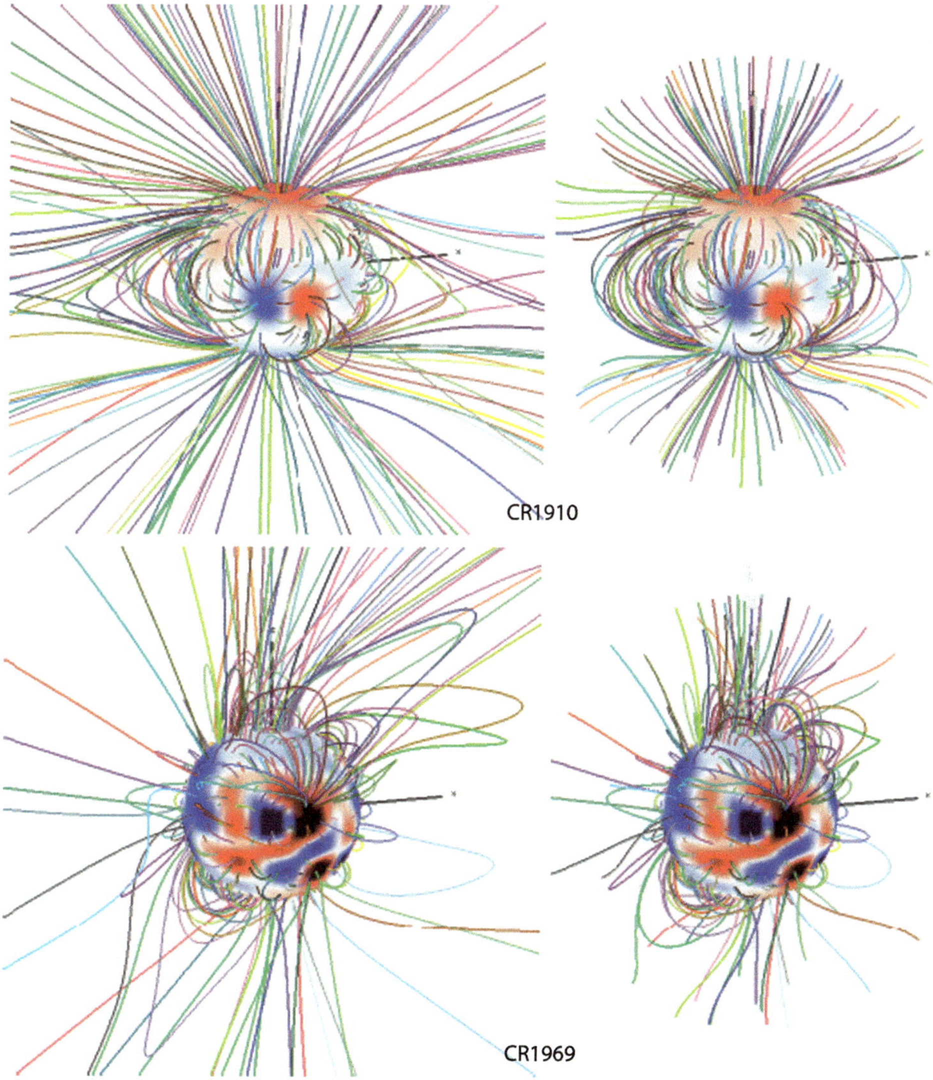

Fig. 11 Comparison between MHD solution (*left*) and PFSS model (*right*) for two solar rotations illustrative of solar minimum (CR1910) and solar maximum (CR1969) conditions. The solar surface is colored according to the radial component of the magnetic field at the photosphere. Field lines have been colored arbitrarily for clarity. [Adapted from Riley et al. (2006), Fig. 1. © AAS. Reproduced with permission]

heliospheric current sheet in the MHD solution. Additionally, the PFSS model appears to underestimate the amount of open solar flux, as inferred from the total number of field lines that open up into the heliosphere. Finally, in general, field lines in the PFSS solutions are shorter than their MHD counterparts.

6.1 Is the Source Surface Really Spherical?

One of the basic assumptions of the PFSS model is that, at the outer radial boundary, the field lines are radial, i.e., this surface is a spherical equipotential. Figure 12 tests this assumption.

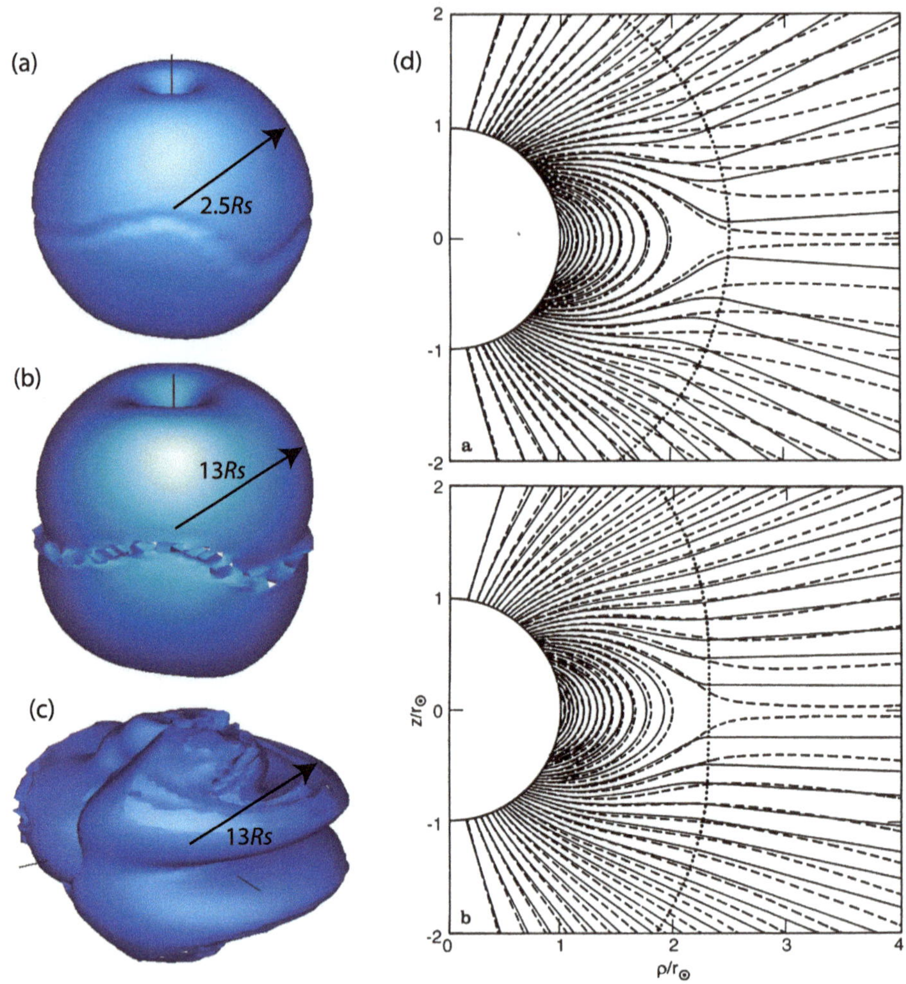

Fig. 12 Isosurfaces of $|B_r|/|B| = 0.97$ for: (**a**) PFSS solution for CR1910; (**b**) MHD solution for CR1910; and (**c**) MHD solution for CR1969. The Sun's rotation axis is indicated by the *vertical black line*. (**d**) PFSS solutions using a spherical source surface (*top*) and "dipole-isogauss" source surface (*bottom*). The *dashed lines* in both panels of (**d**) correspond to MHD model solutions. (**a**)–(**c**) were adapted from Riley et al. (2006), Figs. 4 and 5. (**d**) [from Schulz et al. (1978), Figs. 1 and 2. Panels a–c: © AAS. Reproduced with permission; Panel d: Reprinted with kind permission of Springer Science and Business Media]

In panel (a), we have plotted the isosurface of $|B_r|/|B| = 0.97$ for the PFSS solution of CR1910. Not surprisingly, the field lines have become radial by 2.5 R_S. The sinusoidal variation about the equator marks the boundary between outwardly and inwardly directed fields. Although there are no current sheets in the PFSS model, this corresponds to where the neutral line would lie. The polar dimples illustrate the fact that while the model forces the field lines to become radial by 2.5 R_S, there is no requirement that it cannot occur closer to the Sun, as it does when there are large unipolar regions. Panels (b) and (c) show the same isosurface for the MHD solutions at solar minimum and maximum. We note that the field lines do not become substantially radial until beyond 10 R_S. Moreover, at solar minimum, the shape of the source surface is prolate spheroid, again with dimples at the poles. At solar

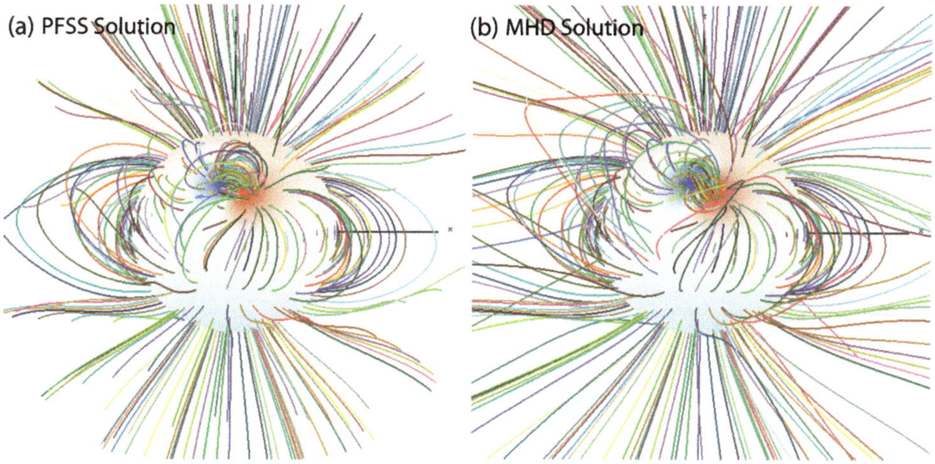

Fig. 13 (**a**) PFSS and (**b**) MHD solutions for an idealized large-scale dipole configuration with an embedded active region. See text for more details. [Adapted from Riley et al. (2006), Figs. 8 and 9. © AAS. Reproduced with permission]

maximum, the isosurface displays a more complex structure that changes from one rotation to the next. On average, however, it could be argued that it tends to be more spherical. The main point, however, is that the MHD model implies that the Sun's real 'source surface' is located further away than the canonical 2.5 R_S used in PFSS models.

Figure 12(d) summarizes two analytic PFSS solutions relying on (top) a spherical source surface and (bottom) a "dipole-isogauss" (or prolate spheroid) isosurface. The first of these matches the shape in panel (a) while the latter more closely resembles (b). Thus, we conclude that even during solar minimum conditions an important, yet simple refinement to PFSS models is to employ a "dipole-isogauss" surface, as suggested by Schulz et al. (1978) for axisymmetric cases.

6.2 Is the Field Really Potential?

To explore the possible effects of non-potentiality on the solutions, we extracted a snapshot from an idealized time-dependent MHD solution, in which an active region was being emerged. We computed a PFSS solution using the radial magnetic field at the base of the MHD solution and compared the results with the MHD model (Fig. 13). The comparison is somewhat artificial in the sense that had we re-run the MHD model with this radial boundary condition, allowing it to reach a steady-state equilibrium, the comparison between the PFSS and MHD solutions would have been more similar to that in Fig. 11. However, our point is to better understand the differences that non-potentiality can introduce into the solutions. We remark that the overall, large-scale features are qualitatively similar. The MHD fields tend to be more inflated, and, hence, more opened. The largest closed fields in the MHD solution also display a cusp-like geometry, whereas the PFSS solution does not. The most striking differences, however, are related to the active region, where the PFSS solution has not reproduced any of the sheared magnetic field lines that are aligned with the neutral line, which separates the two polarities of the bipole.

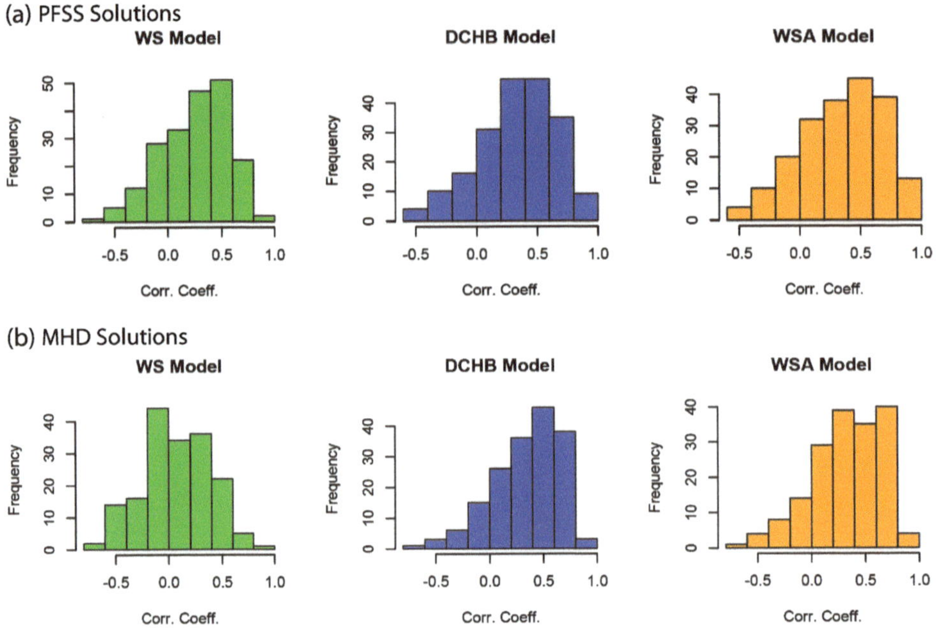

Fig. 14 Histograms of Pearson correlation coefficients of modeled and observed solar wind speed at 1 AU in the ecliptic plane for an interval running from 1996 through 2011. (**a**) Results when the model is driven by a PFSS solution, using (*left*) the WS model, (*middle*) the DCHB model, and (*right*) the WSA model. (**b**) The same empirical models are compared when driven by MHD solutions. See text for more details. [Adapted from Riley et al. (2015). Reproduced with permission of authors]

6.3 The PFSS and MHD Approaches in Action: Predicting Solar Wind Speed

Riley et al. (2015) developed a set of empirically-based models to better understand the relationship between solar wind speed models. In particular, they compared the predicted speeds of solar wind plasma at 1 AU in the ecliptic plane over a 15-year period. They found that, in general, a model relying exclusively on the expansion factor of the coronal magnetic fields to estimate the speed of the plasma (Wang and Sheeley 1990), did not perform as well as a 'boundary layer' model, where the speed of the solar wind was everywhere fast, except in a band at the edge between open and closed field lines (Riley et al. 2001). Of relevance here, they also found that the expansion factor model performed significantly better when it was driven by a PFSS model as oppose to an MHD solution. Since, based on the fact that the MHD model has generally been found to produce a better match with a wide range of remote solar observations, this suggests that the expansion factor and PFSS techniques are somehow conspiring together to produce better results.

Figure 14 compares Pearson correlation coefficients estimated by comparing modeled and observed solar wind speed, rotation by rotation, over a period of 15 years. In each row, the coefficients are compared for the WS (expansion factor), DCHB (boundary layer), and WSA (hybrid approach) techniques. When the PFSS model is used in all three cases, the correlations are comparable. When the MHD solutions are instead used, the DCHB and WSA correlations improve modestly, but the PFSS correlations drop dramatically. This can be understood in terms of the assumptions of the PFSS model. Specifically, by requiring that the field become radial by 2.5 R_S, the modeled flux tubes do not have the correct properties, particularly near the boundary between open and closed field lines, where the expansion factors are larger over a wider distance from the open/closed boundary.

In conclusion, the PFSS model continues to offer a simple method for computing large-scale coronal magnetic fields under certain conditions, and subject to some limitations. However, as computers become more capable, and global MHD algorithms become generally more available to the scientific community, their advantages may be outweighed by their limitations.

6.4 Possible Improvements of MHD Models

Global MHD models of the solar corona are continually being refined and improved. These developments can be categorized in terms of *numerical* improvements or *scientific* improvements. The former, include the ability to run larger runs, using more processors, allowing us to explore finer spatial scales at increased temporal resolution. Additionally, as new computational hardware becomes available, such as GPUs, additional leaps into new regimes may be possible. While these developments can and should lead to a better understanding of the physical processes at work, it is through the incorporation of better physics that we expect to learn the most. Currently, coronal models are run in either the so-called *polytropic* or *thermodynamic* approximations: The latter allows a more realistic treatment of energy transport processes, but still relies on ad hoc prescriptions for the heating of the corona and acceleration of the solar wind (Riley et al. 2012a, 2012b; Riley and Luhmann 2012). Most recently, models incorporating self-consistent descriptions of wave-turbulence (Cranmer et al. 2007, e.g.) have been developed (Lionello et al. 2014). These promise to shed new insight to the underlying mechanisms that heat the corona and accelerate the solar wind. Additionally, models now can address multiple species, with different gyrotropic temperatures (van der Holst et al. 2014, e.g.).

7 A Note on Boundary Conditions

One inescapable fact of extrapolating coronal field models from photospheric measurements is that the physical conditions of the high-β photosphere do not match those of the low-β corona. This means that direct application of observed photospheric data as coronal model boundary data is not a physically consistent approach. The photospheric field is frozen into a high-β plasma that dominates its dynamics, tending to keep it far from a force-free state, whereas the coronal plasma is low-β and the coronal field nearly force-free. Though many photospheric fields, e.g., sunspot penumbrae or flaring neutral-line fields, are nearly horizontal, most photospheric fields have been found to be approximately vertical to within about 10° (Svalgaard et al. 1978; Wang and Sheeley 1992; Petrie and Patrikeeva 2009; Gosain and Pevtsov 2013). For this reason, computing the radial field component B_r from the line-of-sight component B_{LOS} as $B_r = B_{\mathrm{LOS}}/\cos(\rho_{\mathrm{h}})$, where ρ_{h} is the heliospheric angle, gives a reasonable estimate for the radial flux distribution over the photospheric surface. This gives a continuous B_r component, but discontinuous tangential components $\mathbf{B}_{\mathrm{h}}$, across the photospheric surface. The discontinuity is interpreted as a transition from the nearly radial, non-potential photospheric field to the nearly potential, non-radial coronal field: the transition region is mathematically idealized as a current layer (Wang and Sheeley 1992). This radial-field approximation has become an essential part of the standard method for applying line-of-sight photospheric data as boundary data for PFSS and MHD models, including those in Figs. 2, 3 and 10. It not only makes more coherent physical sense than applying

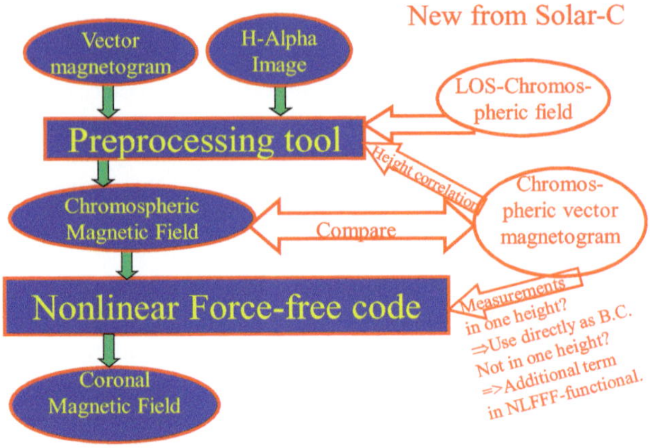

Fig. 15 The original concept of NLFFF-modeling from photospheric measurements (*blue boxes*) and future plans to incorporate chromospheric measurements from Solar-C (*red*)

line-of-sight measurements directly as boundary data for the line-of-sight field component of the model, but also results in more successful models (Wang and Sheeley 1992).

For the NLFFF model, the problem of applying photospheric boundary data as lower-coronal boundary conditions is more complex. The NLFFF model involves solving the vector equation (9) (and $\nabla \cdot \mathbf{B} = 0$), and so the solution is over-constrained if we impose the three observed photospheric vector field components as boundary conditions. In this sense the problem is mathematically ill-posed. There also remains the problem that the measurements derive from the high-β photosphere, and are therefore associated with significant Lorentz forces in general, whereas the coronal model is force-free by assumption. In practice, the boundary data are modified to be approximately force-free (Wiegelmann et al. 2006, 2008) or are reduced to boundary data for B_z and $(\nabla \times \mathbf{B})_z/B_z$ to form a mathematically well-posed problem with boundary conditions consistent with a force-fee field. Recently, methods have been developed that ignore the horizontal photospheric field information and instead use intensity patterns in chromospheric or coronal images to guide the modeling (Aschwanden 2013; Malanushenko et al. 2014). In a recent approach (Chifu et al. 2015) demonstrated how photospheric vector magnetograms and stereoscopic coronal observations can be combined within the NLFFF approach. This new approach does not ignore the horizontal photospheric field, but stabilizes the NLFFF computation for cases where the field measurements contain large measurement uncertainties.

7.1 Use Chromospheric Measurements, e.g., from Solar-C

As seen in Fig. 4 the plasma-β in the photosphere is about unity or more and consequently the photospheric magnetic field is not necessarily force-free. For force-free coronal magnetic field extrapolations it would be preferable to measure the field at the base of the corona or in the upper chromosphere, where β is small. Currently we estimate the field in the upper solar chromosphere by a preprocessing of photospheric measurements, which allows also the incorporation of direct chromospheric observations like H-α images (Wiegelmann et al. 2008) and the use of line-of-sight chromospheric fields as measured with Solis (Jing et al. 2010). It would be preferable, however, to measure the chromospheric magnetic field vector directly, as planned for the Solar-C mission. A concept how these measurements can be incorporated into NLFFF-modeling is shown in Fig. 15. A problem is that the exact height above the photosphere of the chromospheric measurements is a priori unknown and one needs to incorporate a correlation analysis with extrapolations from the photosphere to specify this height.

7.2 Incorporate Coronal Measurements

Direct coronal measurements have carried out for magnetic sensitive coronal lines. Several presentations have been given on the ISSI-2015 meeting and corresponding publications are planned in the same special issues as this paper, e.g. by Phil Judge, Stephen White, Roberto Casini on *Coronal measurements* and by Javier Trujillo Bueno, Egidio Landi degl-Innocenti on *UV Polarized Radiation Observables for Probing the Upper Chromosphere and Corona.* Due to the optical thin coronal plasma these measurements have a line-of-sight integrated character and require vector tomography methods (see Kramar et al. 2006, 2013) to derive the 3D coronal magnetic field structure. A prosperous approach is to combine extrapolation methods from photospheric measurements with a vector tomographic inversion of coronal measurements, because algorithms for both can be derived from optimization principles.

Acknowledgements We acknowledge warm hospitality and fruitful discussions during a workshop on *Solar Magnetic Fields: From Measurements Towards Understanding* at ISSI in Bern in January 2015.

References

M.D. Altschuler, G. Newkirk, Magnetic fields and the structure of the solar corona. I: methods of calculating coronal fields. Sol. Phys. **9**, 131–149 (1969). doi:10.1007/BF00145734

T. Amari, J.J. Aly, J.F. Luciani, T.Z. Boulmezaoud, Z. Mikic, Reconstructing the solar coronal magnetic field as a force-free magnetic field. Sol. Phys. **174**, 129–149 (1997). doi:10.1023/A:1004966830232

T. Amari, T.Z. Boulmezaoud, J.J. Aly, Well posed reconstruction of the solar coronal magnetic field. Astron. Astrophys. **446**, 691–705 (2006). doi:10.1051/0004-6361:20054076

T. Amari, J.J. Aly, P. Chopin, A. Canou, Z. Mikic, Large scale reconstruction of the solar coronal magnetic field. J. Phys. Conf. Ser. **544**(1), 012012 (2014). doi:10.1088/1742-6596/544/1/012012

M.J. Aschwanden, *Physics of the Solar Corona. An Introduction with Problems and Solutions*, 2nd edn. (2005)

M.J. Aschwanden, Nonlinear force-free magnetic field fitting to coronal loops with and without stereoscopy. Astrophys. J. **763**, 115 (2013). doi:10.1088/0004-637X/763/2/115, arXiv:1212.2996

F. Bagenal, S. Gibson, Modeling the large-scale structure of the solar corona. J. Geophys. Res. **96**, 17,663 (1991). doi:10.1029/90JA02625

M. Bineau, Existence of force-free magnetic-fields. Commun. Pure Appl. Math. **25**, 77 (1972)

T.J. Bogdan, B.C. Low, The three-dimensional structure of magnetostatic atmospheres. II—Modeling the large-scale corona. Astrophys. J. **306**, 271–283 (1986). doi:10.1086/164341

S.I. Braginskii, Transport processes in a plasma. Rev. Plasma Phys. **1**, 205 (1965)

I. Chifu, B. Inhester, T. Wiegelmann, Coronal magnetic field modeling using stereoscopy constraints. Astron. Astrophys. **577**, A123 (2015). doi:10.1051/0004-6361/201322548

S.R. Cranmer, A.A. van Ballegooijen, R.J. Edgar, Self-consistent coronal heating and solar wind acceleration from anisotropic magnetohydrodynamic turbulence. Astrophys. J. Suppl. Ser. **171**, 520–551 (2007). doi:10.1086/518001, arXiv:astro-ph/0703333

M.L. DeRosa, C.J. Schrijver, G. Barnes, K.D. Leka, B.W. Lites, M.J. Aschwanden, T. Amari, A. Canou, J.M. McTiernan, S. Régnier, J.K. Thalmann, G. Valori, M.S. Wheatland, T. Wiegelmann, M.C.M. Cheung, P.A. Conlon, M. Fuhrmann, B. Inhester, T. Tadesse, A critical assessment of nonlinear force-free field modeling of the solar corona for active region 10953. Astrophys. J. **696**, 1780–1791 (2009). doi:10.1088/0004-637X/696/2/1780

C. Downs, J.A. Linker, Z. Mikić, P. Riley, C.J. Schrijver, P. Saint-Hilaire, Probing the solar magnetic field with a sun-grazing comet. Science **340**, 1196–1199 (2013). doi:10.1126/science.1236550

G.A. Gary, Plasma beta above a solar active region: rethinking the paradigm. Sol. Phys. **203**, 71–86 (2001). doi:10.1023/A:1012722021820

S.E. Gibson, F. Bagenal, Large-scale magnetic field and density distribution in the solar minimum corona. J. Geophys. Res. **100**, 19,865–19,880 (1995). doi:10.1029/95JA01905
S.E. Gibson, F. Bagenal, B.C. Low, Current sheets in the solar minimum corona. J. Geophys. Res. **101**, 4813–4824 (1996). doi:10.1029/95JA03477
S. Gosain, A.A. Pevtsov, Resolving azimuth ambiguity using vertical nature of solar quiet-sun magnetic fields. Sol. Phys. **283**, 195–205 (2013). doi:10.1007/s11207-012-0135-1, arXiv:1210.6691
H. Grad, H. Rubin, in *Proc. 2nd Intern. Conf. on Peaceful Uses of Atomic Energy*, vol. 31, pp. 190–+ (1958)
J.T. Hoeksema, Structure and evolution of the large scale solar and heliospheric magnetic fields. Ph.D. thesis, Stanford Univ., CA (1984)
R. Howard, Magnetic field of the Sun (observational). Annu. Rev. Astron. Astrophys. **5**, 1 (1967). doi:10.1146/annurev.aa.05.090167.000245
J. Jing, C. Tan, Y. Yuan, B. Wang, T. Wiegelmann, Y. Xu, H. Wang, Free magnetic energy and flare productivity of active regions. Astrophys. J. **713**, 440–449 (2010). doi:10.1088/0004-637X/713/1/440
H.P. Jones, Recent studies of magnetic canopies. Aust. J. Phys. **38**, 919–928 (1985)
M. Kramar, B. Inhester, S.K. Solanki, Vector tomography for the coronal magnetic field. I. Longitudinal Zeeman effect measurements. Astron. Astrophys. **456**, 665–673 (2006). doi:10.1051/0004-6361:20064865
M. Kramar, B. Inhester, H. Lin, J. Davila, Vector tomography for the coronal magnetic field. II. Hanle effect measurements. Astrophys. J. **775**, 25 (2013). doi:10.1088/0004-637X/775/1/25
H. Lin, J.R. Kuhn, R. Coulter, Coronal magnetic field measurements. Astrophys. J. Lett. **613**, L177–L180 (2004). doi:10.1086/425217
R. Lionello, Z. Mikić, J.A. Linker, Magnetohydrodynamics of solar coronal plasmas in cylindrical geometry. J. Comput. Phys. **140**, 172 (1998)
R. Lionello, M. Velli, C. Downs, J.A. Linker, Z. Mikić, A. Verdini, Validating a time-dependent turbulence-driven model of the solar wind. Astrophys. J. **784**, 120 (2014). doi:10.1088/0004-637X/784/2/120, arXiv:1402.4188
Y. Liu, H. Lin, Observational test of coronal magnetic field models. I. Comparison with potential field model. Astrophys. J. **680**, 1496–1507 (2008). doi:10.1086/588645, arXiv:0710.3223
W.C. Livingston, J. Harvey, A.K. Pierce, D. Schrage, B. Gillespie, J. Simmons, C. Slaughter, Kitt peak 60-cm vacuum telescope. Appl. Opt. **15**, 33–39 (1976). doi:10.1364/AO.15.000033
B.C. Low, Three-dimensional structures of magnetostatic atmospheres. III—A general formulation. Astrophys. J. **370**, 427–434 (1991). doi:10.1086/169829
A. Malanushenko, C.J. Schrijver, M.L. DeRosa, M.S. Wheatland, Using coronal loops to reconstruct the magnetic field of an active region before and after a major flare. Astrophys. J. **783**, 102 (2014). doi:10.1088/0004-637X/783/2/102, arXiv:1312.5389
E. Marsch, Kinetic physics of the solar corona and solar wind. Living Rev. Sol. Phys. **3**, 1 (2006). doi:10.12942/lrsp-2006-1
T.R. Metcalf, M.L. De Rosa, C.J. Schrijver, G. Barnes, A.A. van Ballegooijen, T. Wiegelmann, M.S. Wheatland, G. Valori, J.M. McTtiernan, Nonlinear force-free modeling of coronal magnetic fields. II. Modeling a filament arcade and simulated chromospheric and photospheric vector fields. Sol. Phys. **247**, 269–299 (2008). doi:10.1007/s11207-007-9110-7
Z. Mikić, J.A. Linker, Disruption of coronal magnetic field arcades. Astrophys. J. **430**, 898 (1994)
T. Neukirch, On self-consistent three-dimensional analytic solutions of the magnetohydrostatic equations. Astron. Astrophys. **301**, 628 (1995)
G.J.D. Petrie, Three-dimensional Equilibrium Solutions to the Magnetohydrodynamic Equations and their Application to Solar Coronal Structures. Ph.D. thesis, Department of Mathematics and Statistics, University of St. Andrews, North Haugh, St Andrews KY16 9SS (2000)
G.J.D. Petrie, T. Neukirch, The Green's function method for a special class of linear three-dimensional magnetohydrostatic equilibria. Astron. Astrophys. **356**, 735–746 (2000)
G.J.D. Petrie, I. Patrikeeva, A comparative study of magnetic fields in the solar photosphere and chromosphere at equatorial and polar latitudes. Astrophys. J. **699**, 871–884 (2009). doi:10.1088/0004-637X/699/1/871, arXiv:1010.6041
G.J.D. Petrie, A. Canou, T. Amari, Nonlinear force-free and potential-field models of active-region and global coronal fields during the whole heliosphere interval. Sol. Phys. **274**, 163–194 (2011). doi:10.1007/s11207-010-9687-0, arXiv:1010.6283
G.J.D. Petrie, K. Petrovay, K. Schatten, Solar polar fields and the 22-year activity cycle: observations and models. Space Sci. Rev. **186**, 325–357 (2014). doi:10.1007/s11214-014-0064-4
L.A. Rachmeler, S.E. Gibson, J.B. Dove, C.R. DeVore, Y. Fan, Polarimetric properties of flux ropes and sheared arcades in coronal prominence cavities. Sol. Phys. **288**, 617–636 (2013). doi:10.1007/s11207-013-0325-5, arXiv:1304.7594
P. Riley, J.G. Luhmann, Interplanetary signatures of unipolar streamers and the origin of the slow solar wind. Sol. Phys. **277**, 355–373 (2012). doi:10.1007/s11207-011-9909-0

P. Riley, J.A. Linker, Z. Mikić, An empirically-driven global MHD model of the corona and inner heliosphere. J. Geophys. Res. **106**, 15,889 (2001). doi:10.1029/2000JA000121

P. Riley, J.A. Linker, Z. Mikić, R. Lionello, S.A. Ledvina, J.G. Luhmann, A comparison between global solar magnetohydrodynamic and potential field source surface model results. Astrophys. J. **653**, 1510 (2006). doi:10.1086/508565

P. Riley, J.A. Linker, R. Lionello, Z. Mikic, Corotating interaction regions during the recent solar minimum: The power and limitations of global MHD modeling. J. Atmos. Sol.-Terr. Phys. **83**, 1–10 (2012a). doi:10.1016/j.jastp.2011.12.013

P. Riley, R. Lionello, J.A. Linker, Z. Mikic, J. Luhmann, J. Wijaya, Global MHD modeling of the solar corona and inner heliosphere for the whole heliosphere interval. Sol. Phys. **274**, 361–3775 (2012b). doi:10.1007/s11207-010-9698-x

P. Riley, J.A. Linker, C.N. Arge, On the role played by magnetic expansion factor in the prediction of solar wind speed. Accepted for publication in Space Weather (2015)

P. Ruan, T. Wiegelmann, B. Inhester, T. Neukirch, S.K. Solanki, L. Feng, A first step in reconstructing the solar corona self-consistently with a magnetohydrostatic model during solar activity minimum. Astron. Astrophys. **481**, 827–834 (2008). doi:10.1051/0004-6361:20078834

V. Rušin, M. Druckmüller, P. Aniol, M. Minarovjech, M. Saniga, Z. Mikić, J.A. Linker, R. Lionello, P. Riley, V.S. Titov, Comparing eclipse observations of the 2008 August 1 solar corona with an MHD model prediction. Astron. Astrophys. **513**, A45 (2010). doi:10.1051/0004-6361/200912778

K.H. Schatten, Current sheet magnetic model for the solar corona. Cosm. Electrodyn. **2**, 232–245 (1971)

K.H. Schatten, J.M. Wilcox, N.F. Ness, A model of interplanetary and coronal magnetic fields. Sol. Phys. **6**, 442–455 (1969). doi:10.1007/BF00146478

C.J. Schrijver, M.L. De Rosa, T.R. Metcalf, Y. Liu, J. McTiernan, S. Régnier, G. Valori, M.S. Wheatland, T. Wiegelmann, Nonlinear force-free modeling of coronal magnetic fields part I: a quantitative comparison of methods. Sol. Phys. **235**, 161–190 (2006). doi:10.1007/s11207-006-0068-7

C.J. Schrijver, M.L. De Rosa, T. Metcalf, G. Barnes, B. Lites, T. Tarbell, J. McTiernan, G. Valori, T. Wiegelmann, M.S. Wheatland, T. Amari, G. Aulanier, P. Démoulin, M. Fuhrmann, K. Kusano, S. Régnier, J.K. Thalmann, Nonlinear force-free field modeling of a solar active region around the time of a major flare and coronal mass ejection. Astrophys. J. **675**, 1637–1644 (2008). doi:10.1086/527413. arXiv:0712.0023

M. Schulz, E.N. Frazier, D.J. Boucher, Coronal magnetic-field model with non-spherical source surface. Sol. Phys. **60**, 83–104 (1978)

X. Sun, J.T. Hoeksema, Y. Liu, T. Wiegelmann, K. Hayashi, Q. Chen, J. Thalmann, Evolution of magnetic field and energy in a major eruptive active region based on SDO/HMI observation. Astrophys. J. **748**, 77 (2012). doi:10.1088/0004-637X/748/2/77, arXiv:1201.3404

L. Svalgaard, T.L.Jr. Duvall, P.H. Scherrer, The strength of the sun's polar fields. Sol. Phys. **58**, 225–239 (1978). doi:10.1007/BF00157268

T. Tadesse, A.A. Pevtsov, T. Wiegelmann, P.J. MacNeice, S. Gosain, Global solar free magnetic energy and electric current density distribution of Carrington rotation 2124. Sol. Phys. **289**, 4031–4045 (2014a). doi:10.1007/s11207-014-0581-z

T. Tadesse, T. Wiegelmann, S. Gosain, P. MacNeice, A.A. Pevtsov, First use of synoptic vector magnetograms for global nonlinear, force-free coronal magnetic field models. Astron. Astrophys. **562**, A105 (2014b). doi:10.1051/0004-6361/201322418, arXiv:1309.5853

T. Tadesse, T. Wiegelmann, P.J. MacNeice, B. Inhester, K. Olson, A. Pevtsov, A comparison between nonlinear force-free field and potential field models using full-disk SDO/HMI magnetogram. Sol. Phys. **289**, 831–845 (2014c). doi:10.1007/s11207-013-0364-y, arXiv:1212.5639

G. Tóth, B. van der Holst, Z. Huang, Obtaining potential field solutions with spherical harmonics and finite differences. Astrophys. J. **732**, 102 (2011). doi:10.1088/0004-637X/732/2/102, arXiv:1104.5672

G. Valori, B. Kliem, R. Keppens, Extrapolation of a nonlinear force-free field containing a highly twisted magnetic loop. Astron. Astrophys. **433**, 335–347 (2005). doi:10.1051/0004-6361:20042008

B. van der Holst, I.V. Sokolov, X. Meng, M. Jin, W.B. Manchester IV, G. Tóth, T.I. Gombosi, Alfvén Wave Solar Model (AWSoM): coronal heating. Astrophys. J. **782**, 81 (2014). doi:10.1088/0004-637X/782/2/81, arXiv:1311.4093

Y.M. Wang, N.R. Sheeley, Solar wind speed and coronal flux-tube expansion. Astrophys. J. **355**, 726 (1990). doi:10.1086/168805

Y.M. Wang, N.R.Jr. Sheeley, On potential field models of the solar corona. Astrophys. J. **392**, 310–319 (1992). doi:10.1086/171430

M.S. Wheatland, K.D. Leka, Achieving self-consistent nonlinear force-free modeling of solar active regions. Astrophys. J. **728**, 112 (2011). doi:10.1088/0004-637X/728/2/112, arXiv:1012.3503

M.S. Wheatland, S. Régnier, A self-consistent nonlinear force-free solution for a solar active region magnetic field. Astrophys. J. Lett. **700**, L88–L91 (2009). doi:10.1088/0004-637X/700/2/L88, arXiv:0906.4414

M.S. Wheatland, P.A. Sturrock, G. Roumeliotis, An optimization approach to reconstructing force-free fields. Astrophys. J. **540**, 1150–1155 (2000). doi:10.1086/309355

T. Wiegelmann, Nonlinear force-free modeling of the solar coronal magnetic field. J. Geophys. Res. Space Phys. **113**, A03S02 (2008). doi:10.1029/2007JA012432

T. Wiegelmann, B. Inhester, How to deal with measurement errors and lacking data in nonlinear force-free coronal magnetic field modelling? Astron. Astrophys. **516**, A107 (2010). doi:10.1051/0004-6361/201014391

T. Wiegelmann, T. Sakurai, Solar force-free magnetic fields. Living Rev. Sol. Phys. **9**, 5 (2012). doi:10.12942/lrsp-2012-5, arXiv:1208.4693

T. Wiegelmann, B. Inhester, T. Sakurai, Preprocessing of vector magnetograph data for a nonlinear force-free magnetic field reconstruction. Sol. Phys. **233**, 215–232 (2006). doi:10.1007/s11207-006-2092-z, arXiv:astro-ph/0612641

T. Wiegelmann, T. Neukirch, P. Ruan, B. Inhester, Optimization approach for the computation of magnetohydrostatic coronal equilibria in spherical geometry. Astron. Astrophys. **475**, 701–706 (2007). doi:10.1051/0004-6361:20078244, arXiv:0801.2916

T. Wiegelmann, J.K. Thalmann, C.J. Schrijver, M.L. De Rosa, T.R. Metcalf, Can we improve the preprocessing of photospheric vector magnetograms by the inclusion of chromospheric observations? Sol. Phys. **247**, 249–267 (2008). doi:10.1007/s11207-008-9130-y, arXiv:0801.2707

T. Wiegelmann, J.K. Thalmann, B. Inhester, T. Tadesse, X. Sun, J.T. Hoeksema, How should one optimize nonlinear force-free coronal magnetic field extrapolations from SDO/HMI vector magnetograms? Sol. Phys. **281**, 37–51 (2012). doi:10.1007/s11207-012-9966-z, arXiv:1202.3601

T. Wiegelmann, J.K. Thalmann, S.K. Solanki, The magnetic field in the solar atmosphere. Astron. Astrophys. Rev. **22**, 78 (2014). doi:10.1007/s00159-014-0078-7, arXiv:1410.4214

A.R. Yeates, D.H. Mackay, Chirality of high-latitude filaments over Solar Cycle 23. Astrophys. J. Lett. **753**, L34 (2012). doi:10.1088/2041-8205/753/2/L34, arXiv:1206.2327

X. Zhao, J.T. Hoeksema, A coronal magnetic field model with horizontal volume and sheet currents. Sol. Phys. **151**, 91–105 (1994). doi:10.1007/BF00654084

X.P. Zhao, J.T. Hoeksema, P.H. Scherrer, Modeling the 1994 April 14 polar crown SXR arcade using three-dimensional magnetohydrostatic equilibrium solutions. Astrophys. J. **538**, 932–939 (2000). doi:10.1086/309166

Space Sci Rev (2017) 210:275–316
DOI 10.1007/s11214-015-0204-5

Solar Magnetoconvection and Small-Scale Dynamo

Recent Developments in Observation and Simulation

**J.M. Borrero[1] · S. Jafarzadeh[2] · M. Schüssler[3] ·
S.K. Solanki[3,4]**

Received: 14 July 2015 / Accepted: 22 September 2015 / Published online: 17 December 2015

Abstract A number of observational and theoretical aspects of solar magnetoconvection are considered in this review. We discuss recent developments in our understanding of the small-scale structure of the magnetic field on the solar surface and its interaction with convective flows, which is at the centre of current research. Topics range from plage areas in active regions over the magnetic network shaped by supergranulation to the ubiquitous 'turbulent' internetwork fields. On the theoretical side, we focus upon magnetic field generation by small-scale dynamo action.

Keywords Sun · Convection · Magnetic field

1 Introduction

New observational facilities and instruments (on the ground and in space), innovative methods for data analysis, and the rapid growth of computing power have led to considerable progress in our understanding of solar magnetoconvective processes during the last decade. This ranges all the way from the smallest observable magnetic structures in 'quiet' areas to the dynamical fine structure of sunspot umbrae and penumbrae. The interplay of observations and radiative 3D-MHD simulations is crucial for this progress: simulations permit us to identify the physical processes leading to the observed phenomena once their consistency

✉ M. Schüssler
schuessler@mps.mpg.de

1 Kiepenheuer-Institut für Sonnenphysik, Schöneckstr 6, 79104 Freiburg, Germany

2 Institute of Theoretical Astrophysics, University of Oslo, P.O. Box 1029 Blindern, 0315 Oslo, Norway

3 Max-Planck-Institut für Sonnensystemforschung, Justus-von-Liebig-Weg 3, 37077 Göttingen, Germany

4 School of Space Research, Kyung Hee University, Yongin, Gyeonggi-Do, 446-701, Republic of Korea

with the observations is shown by comparing 'synthetic' with real observations. The simulations then can sharpen the questions addressed by the observations by predicting further properties of the magnetic, thermodynamic, and flow structure.

In this review, we consider a subset of the solar magnetoconvective phenomena, also focussing on more recent developments. Plage and network fields are of considerable interest, not only because of their intrinsic importance for understanding the interaction of magnetic field and convective flows at various spatial scales, but also because of their crucial role for solar irradiance variations with potential impact on terrestrial climate variations. Internetwork fields in the 'quiet' Sun probably represent a huge reservoir of restless 'turbulent' magnetic flux, possibly heralding a fully magnetized state (in the sense of equipartition between magnetic and kinetic energy of the convective motions) of the whole convection zone with important implications for the solar dynamo and the generation of differential rotation. A plausible mechanism for the generation of the internetwork field is small-scale dynamo action by a flow with chaotic streamlines, a truly 'turbulent' dynamo. Numerical simulations suggest that such a dynamo could be active throughout the convection zone, but these simulations can only be run for effective Reynolds numbers and magnetic Prandtl numbers far away from solar values. Nevertheless, state-of-the-art radiative 3D-MHD simulations provide results which are consistent with observations. In what follows, we give an overview of the state of research in the three areas mentioned above: plage and network fields (Sect. 2), internetwork fields (Sect. 3), and small-scale dynamo (Sect. 4).

2 Plage and Network Fields

2.1 General Properties of Plage and Network Fields

At photospheric layers active regions are mainly composed of large, cool magnetic features such as sunspots and pores and smaller, bright magnetic elements forming the plage regions, named after the bright appearance of such regions at chromospheric heights. They are sometimes also called facular fields after the bright faculae visible mainly near the limb in white light (or in other photospheric continuum radiation). Similarly, network features are named after the network of brightenings seen in the cores of chromospheric lines and covering the quiet Sun.

Plage and network magnetic fields are intermediate between the large and dark sunspots and pores on the one hand, and the small weak internetwork magnetic features on the other hand. They are distinguished from the pores by the fact that they are considerably brighter in white light, being on average roughly the same brightness as the average quiet Sun (around solar disc centre), or brighter (near the limb). Their larger magnetic fluxes and their arrangement inside active regions or at the boundaries of supergranule cells distinguishes them from the internetwork elements.

Earlier reviews covering the physics of photospheric flux tubes and in particular plage and network magnetic fields have been written by Solanki (1993); Stenflo (1994); Solanki et al. (2006); de Wijn et al. (2009); see also Stein (2012) for a review of magnetoconvection, which is central to the physics of faculae and the network, and Wiegelmann et al. (2014) for a more recent overview of magnetic fields in the solar atmosphere, but focusing more on the corona.

Plage and network magnetic fields are distinguished from each other by the fact that the density of magnetic flux typically present inside plage is higher, and by their location, with the former being found inside active regions, while network fields are present all over the Sun (including among the decay products of active regions, which form the so-called enhanced network), but concentrated mainly at the edges of supergranules with a length scale of 15–30 Mm. This implies that the plage fields are restricted to the activity belts, i.e. to latitudes lower than roughly $\pm 30^{\circ}$ (see Hale and Nicholson 1925), while the network is seen at all latitudes (Muller and Roudier 1994). In addition, plage and network show a markedly different behaviour over the solar activity cycle, with the solar surface area covered by plage regions waxing and waning strongly in phase with the number of sunspots, while the network fields display a much weaker variation with time. Indeed, there is some controversy about the extent to which network fields vary in phase or in antiphase with the solar cycle (Harvey 1993; Harvey and Zwaan 1993; Hagenaar et al. 2003). Jin et al. (2011) have proposed that the phase vs. anti-phase behaviour depends on the size and flux of the network magnetic features, with the larger features varying in phase, while the smaller features vary in antiphase with the sunspot cycle.

Also, plage and network fields have partly different origins. A large fraction of the plage fields emerge in place, but some are the decay products of the sunspots in their active regions (Petrovay et al. 1999). Therefore, plage fields are generally produced by the global dynamo. The network in turn is partly the result of the decay of active region plage. To a larger extent, however, the network is fed by ephemeral regions, also called ephemeral active regions, small bipolar regions without sunspots with magnetic fluxes of roughly 10^{18}–10^{20} Mx (Dodson 1953; Harvey and Harvey 1973; Harvey and Martin 1973).

The constant emergence of new magnetic flux in the form of ephemeral regions over the solar surface, at a rate of 3×10^{21} Mx per hour (assuming a homogeneous emergence rate over the entire Sun; Schrijver et al. 1997), implies a constant removal of magnetic flux at the same rate. This has led to estimates of the flux turnover time in the magnetic network of 40 h by Schrijver et al. (1998) and 14 h by Hagenaar (2001).

Recently, the importance of the contribution to the network by magnetic flux emerging in the intranetwork has been pointed out (Gošić et al. 2014). Although both, merging and cancellation processes take place (Iida et al. 2012), the former was found to strongly dominate, resulting in a rate of net flux transfer of 1.5×10^{24} Mx per day from the internetwork to the network. Gošić et al. (2014) argue that this implies internetwork fields replace the entire flux in the network within 18–24 hours, on the same order as the replacement time deduced by Hagenaar (2001) based on ephemeral regions. The authors do not discuss the pressing question of why, if the internetwork flux in the two polarities is well balanced while the network flux in their observations clearly is not, merging dominates so strongly over cancellation. Also unclear is what happens with the internetwork flux of the opposite polarity, which is not merged with the network. Also unclear is why, if the flux contributed by the internetwork to the network is an order of magnitude larger than that contributed by ephemeral regions, the turnover times for the network flux found by Hagenaar (2001) (based only on ephemeral regions) is similar to that obtained by Gošić et al. (2014) (including also internetwork features). This could imply that the network in the region considered by Gošić et al. (2014) carried a particularly large amount of flux. Furthermore, the decay or fragmentation of network elements provides an additional source of internetwork flux, so that the net contribution of the internetwork to the network field could actually vanish.

2.2 Observations of Magnetic Concentrations Making Up Plage and the Network

In photospheric layers, where most of the measurements of solar magnetic fields have been made, both, plage regions and the network are composed of groups of more or less discrete magnetic flux concentrations. These concentrations display a range of sizes (e.g. Utz et al. 2009; Feng et al. 2013) and magnetic fluxes, whose probability distribution function follows a power law with an exponent of -1.85 according to Parnell et al. (2009), or -2 according to Harvey and Zwaan (1993). These values indicate that large and small magnetic concentrations contribute roughly the same amount to the total unsigned magnetic flux present at any given instant on the solar surface (with the larger features dominating if the value of Parnell et al. 2009 is used). Owing to the much shorter lifetime of ephemeral regions, this approximate equality implies they bring magnetic flux to the solar surface at a much higher rate than the larger active regions. Zirin (1987) has estimated that there is a factor of 100 difference in the fluxes emerging over a solar cycle in these two types of bipolar features (cf. Thornton and Parnell 2011; Zhou et al. 2013).

Magnetic fields in active regions and the network are mainly strong, i.e. on the order of 1–2 kG (Stenflo 1973; Wiehr 1978; Dara-Papamargaritis and Koutchmy 1983; Rabin 1992a; Rabin 1992b; Rüedi et al. 1992b; Grossmann-Doerth et al. 1996; Martínez Pillet et al. 1997; Sánchez Almeida and Lites 2000; Viticchié et al. 2011; Bühler et al. 2015), also see Blanco Rodríguez and Kneer (2010) and Kaithakkal et al. (2013) for polar faculae. These fields are close to vertical (e.g., Sanchez Almeida and Martinez Pillet 1994; Bernasconi et al. 1995; Martínez Pillet et al. 1997; Bühler et al. 2015); cf. Leighton (1959), albeit with limitations of their instrument at the time. Note that the field strength is a strong function of height, dropping from high values of around 1500–2000 G in the deepest observable layers (Rabin 1992a; Rüedi et al. 1992b; Bühler et al. 2015) to a field strength of 250–500 G in the upper photosphere (Zirin and Popp 1989; Bruls and Solanki 1995; Bühler et al. 2015); see also Moran et al. (2000). In addition to the kG flux concentrations, small patches of transient horizontal field are also found in active regions (Ishikawa et al. 2008; Ishikawa and Tsuneta 2009), with properties that are similar to corresponding patches in the quiet Sun. In particular, the smaller ones are isotropically oriented, although the larger ones appear to have a slight preference for the orientation of the active region.

According to Abramenko and Longcope (2005) magnetic flux concentrations within active regions possess fluxes of 10^{18}–10^{20} Mx, whereby features with fluxes larger than 10^{19} Mx are likely darker than the surroundings (e.g., Grossmann-Doerth et al. 1994). The network is composed of smaller magnetic features hosting less magnetic flux, with kG flux tubes carrying a flux of around 10^{17} Mx being the smallest such entities that have been isolated so far. Even smaller magnetic features are regularly seen in radiation MHD simulations and are also likely to be present on the Sun. Solanki et al. (1999) pointed out that magnetic features with magnetic fluxes differing by up to 6 orders of magnitude have roughly the same field strength of 1000–2000 G when averaged over their cross-sections. Such magnetic features range from the smallest kG flux tubes (magnetic elements) to large sunspots. Note that only the field strength averaged over the whole cross-section remains roughly independent of magnetic flux, the field strength at the core of the features changes by a considerable amount. Magnetic features with flux below a few times 10^{16} Mx have weaker fields (e.g. Solanki et al. 1996; Khomenko et al. 2003; Orozco Suárez et al. 2008; Ishikawa and Tsuneta 2009; Stenflo 2011; Utz et al. 2013). Note that the field strength of individual magnetic structures may also be a function of time, e.g. emerging magnetic flux is usually associated with a weaker field (e.g. Zwaan 1985; Zhang and Song 1992; Lites et al. 1998; Centeno et al. 2007; Martínez González and Bellot Rubio 2009; Cheung and Isobe 2014), but in small features the field strength may also vary significantly after that (e.g. Martínez González et al. 2011a; Requerey et al. 2014).

The strong magnetic fields in plage decrease the convective blueshift of spectral lines (Livingston 1982; Cavallini et al. 1985; Brandt and Solanki 1990) in agreement with the smaller upflow velocities in granules along with the faster downflows in intergranular lanes seen by Kostik and Khomenko (2012). These authors also find that the convective flows reach greater heights in facular areas. The contrast of the granulation and the sizes of individual granules become smaller in plage while their lifetimes increase, resulting in what is called abnormal granulation (Title et al. 1992; Berger et al. 1998; Narayan and Scharmer 2010).

Martínez Pillet et al. (1997) noted from Stokes vector observations with a spatial resolution of around 1 arc sec obtained with the Advanced Stokes Polarimeter (ASP) that the magnetic concentrations in plage displayed no significant downflows beyond about 250 m/s, confirming a result first found by Solanki (1986). The absence of downflows was later challenged by Bellot Rubio et al. (2000), who argued that the tiny wavelength shifts of the many lines in a Fourier-Transform-Spectrometer (FTS) Stokes *V* spectrum could only be interpreted in terms of a strong downflow inside the magnetic features. This was in turn criticised by Frutiger and Solanki (2001), who showed that this signature could also be produced by velocities not requiring a net flow. Consequently, spatially unresolved data have not been able to provide a definitive answer to the question of flows within magnetic elements. Upflows of around 1–2 km/s inside plage magnetic elements, surrounded by 1.5–3.3 km/s downflows, were reported by Langangen et al. (2007). Recently, Bühler et al. (2015) have confirmed the location of the magnetic concentrations inside the downflow lanes separating granules, but without harbouring strong downflows themselves. At individual locations up- and downflows within the magnetic features are present, but in an average sense they are close to being at rest. These authors also discovered that the downflows immediately surrounding strong flux concentrations can reach supersonic values at the solar surface, unlike in normal intergranular lanes, which display only subsonic downflows in similar data. This provides observational confirmation that the additional cooling of the gas in the immediate surroundings of magnetic concentrations causes the surrounding gas to sink faster, as initially pointed out by Deinzer et al. (1984).

An early discovery of the observation of full Stokes *V* profiles was their asymmetry. The blue and red wings of the profiles having different amplitudes and areas (Stenflo et al. 1984; Solanki and Stenflo 1984; Solanki and Stenflo 1985), with the asymmetry being in general larger in network features than in plage. Stenflo et al. (1987) found the sign of the area asymmetry to change near the solar limb. These results were confirmed by Martínez Pillet et al. (1997) for plage regions. The Stokes *V* area asymmetry has been interpreted in terms of (nearly) static gas inside flux tubes whose field expands with height thus forming a canopy overlying the downflowing lanes of the surrounding convective cells (Grossmann-Doerth et al. 1988; Solanki 1989; Steiner et al. 1999; Shelyag et al. 2007). This geometry is now well-established and easily visible in high-resolution images (Berger et al. 2004; Pietarila et al. 2010), see also the previous paragraph. This picture could also explain the change in sign of the asymmetry close to the limb (Bünte et al. 1993) and high resolution observations that show Stokes *V* to be less asymmetric in the interiors of magnetic concentrations, but strongly asymmetric near the edges, i.e. where the canopy is expected to be located (Rezaei et al. 2007; Martínez González et al. 2012b). Rezaei et al. (2007) even find an opposite sign of the asymmetry inside magnetic features in the examples considered by them. Such observations are a sign that magnetic flux concentrations in plages and the network are becoming spatially resolved and their internal structure probed. Some Stokes *V* profiles are anomalous in shape, i.e. having only a single lobe, or three or four lobes. Such profiles are most common in the internetwork of the quiet Sun, but are to some extent also found in active-region plage. They can have different origins, with Sigwarth (2001) proposing network and plage fields as either a mixture of kG flux tubes and a sub-kG field.

This interpretation is supported by high resolution observations that indicate weak opposite polarity fields surrounding kG magnetic features.

Martínez Pillet et al. (1997) also found that plage fields contain many so-called azimuth centers, i.e. regions where, starting from the center of the feature, the magnetic azimuth points roughly isotropically in all horizontal directions. They are thought to be signatures of intermediate-sized magnetic features described by flux tubes expanding with height. Inversions of data obtained by Hinode by Bühler et al. (2015) have confirmed the presence of numerous azimuth centres in plage areas. These authors also showed that the larger of these are associated with micropores. Also visible are weak opposite polarity fields located just outside the magnetic features in the deepest observable layers, underlying the canopy of expanding fields in the upper photosphere (Zayer et al. 1989; Narayan 2011; Scharmer et al. 2013; Bühler et al. 2015).

2.3 Brightness of Magnetic Features in Plage and the Network

The brightness of magnetic concentrations in plage and the network has been a subject of intense study. The brightness gives insight into the transport of energy within magnetic features, e.g., by the influx of radiation from the surrounding granules (e.g. Spruit 1976; Stenholm and Stenflo 1977; Vögler et al. 2005; Holzreuter and Solanki 2012). Another strong motivation comes from the contribution of facular and plage brightness to changes in the total and spectral solar irradiance, the solar drivers of the Earth's climate (e.g. Foukal et al. 2004; Domingo et al. 2009; Gray et al. 2010; Ermolli et al. 2013; Solanki et al. 2013) along with other natural and man-made causes. Until recently the highest spatial resolution was reached only, or at least the easiest, in broad-band images taken in rapid bursts.

The variation over the solar cycle of solar surface area coverage by plage also changes the intensity close to the solar limb (due to the enhanced brightness of plage there) and hence produces variations in the apparent solar radius (Bruls and Solanki 2004, cf. Ulrich and Bertello 1995).

The contrast of magnetic elements with respect to the quiet Sun increases rapidly with height in the photosphere (e.g. see Riethmüller et al. 2010). At the height of formation of the continuum in visible light the contrast is low and can be of either sign. It is generally much stronger at higher layers, sampled by, e.g., line cores (e.g. Title et al. 1992; Stangl and Hirzberger 2005; Yeo et al. 2013). The continuum and line core contrasts also display a totally different centre-to-limb behaviour. In the continuum the contrast is low at disk centre. According to Topka et al. (1992, 1997) it is even negative for all types of magnetic features. Negative contrasts of some types of small-scale magnetic features have also been seen by Ortiz et al. (2002); Berger et al. (2007); Yeo et al. (2013). The continuum contrast increases toward the solar limb (Domingo et al. 2005; Kobel et al. 2009). Such a behaviour is also observed for polar faculae (Blanco Rodríguez et al. 2007). In many observations, the continuum contrast increases from disk centre only up to a given heliocentric angle (θ), and then decreases towards the limb. Thus Auffret and Muller (1991) find that it peaks at around $\mu = \cos\theta \approx 0.35$, while Ortiz et al. (2002) and Yeo et al. (2013) obtain that this angle depends on the average flux density, i.e. probably the size of the magnetic feature, which is in good agreement with results from flux-tube models.

In line cores, however, the largest contrasts are found near solar disk centre, with the contrast decreasing towards the limb (Yeo et al. 2013). Hirzberger and Wiehr (2005) found no significant variation of the contrast of faculae with heliocentric angle, from observations in G-band and 587.5 ± 1.5 nm continuum. In the G-band this can be explained by the mixture of lines and continuum, at 587.5 nm, it is more surprising.

The continuum contrast of magnetic elements is, on average, larger in the network than in the plage (Ortiz et al. 2006; Kobel et al. 2011; Romano et al. 2012). This is mainly caused by the larger average size of magnetic elements in these regions, with magnetic features in plages being larger on average (Grossmann-Doerth et al. 1994). Larger features are generally less bright around disk centre.

Correspondence of localised brightness enhancements to the magnetic flux concentrations have been shown in MHD simulations (Vögler et al. 2005; Shelyag et al. 2007) as well as in the observations. Observations at solar disc centre display a decrease in the contrast of magnetic features with field strength, either starting right from small field-strength values (Topka et al. 1992, 1997; Title et al. 1992), or after an initial increase, i.e. displaying a "knee" in brightness (Lawrence et al. 1993; Stangl and Hirzberger 2005; Narayan and Scharmer 2010; Schnerr and Spruit 2011; Kobel et al. 2011). Note, however, that observations made by Schnerr and Spruit (2011) with the very high resolution SST (Swedish Solar Telescope; Scharmer et al. 2003) display an increasing brightness right up to the largest field strengths that they plot. This suggests that the spatial resolution does play a role in the relationship between brightness and magnetic flux. Danilovic et al. (2013) studied the effect of degrading simulated images to the spatial resolution of Hinode/SP images (i.e. of limiting the spatial resolution; see also Röhrbein et al. 2011). With the help of the appropriate point-spread-function Danilovic et al. (2013) could reconcile the different rms contrasts obtained from MHD simulations and observations. The apparent difference between the contrast of magnetic features in observations and MHD simulations was pointed out and explained by Röhrbein et al. (2011). They showed that the application of an appropriate point-spread-function turned the monotonic relation between brightness and field strength found at the original resolution of their MHD simulations into a relation with a maximum at intermediate field strength.

A very high contrast is achieved by imaging within molecular bands, which have the advantage that owing to the large density of lines, relatively broad filters can be used, allowing a good signal-to-noise ratio to be reached within a short exposure time. The most commonly used molecular band has been the Fraunhofer G-band of CH molecular absorption (Muller and Roudier 1984; Muller and Mena 1987; Berger et al. 1995; van Ballegooijen et al. 1998; Berger and Title 2001; Nisenson et al. 2003; Langhans et al. 2004; Rouppe van der Voort et al. 2005; Viticchié et al. 2010; Bodnárová et al. 2014). The bright points seen in this wavelength range are associated with magnetic fields (Berger and Title 2001; Bharti et al. 2006; Beck et al. 2007), though not all areas with a large magnetic flux can be necessarily associated with G-band bright points (Ishikawa et al. 2007). Although the CN-bandhead at 388 nm offers a larger density of lines and hence produces stronger contrasts (e.g. Rutten et al. 2001; Berdyugina et al. 2003; Zakharov et al. 2005; Uitenbroek and Tritschler 2007; but see Uitenbroek and Tritschler 2006 for a deviating view), the G-band profits from the often higher camera sensitivity and the larger number of solar photons. The high contrast of the molecular bands comes from the dissociation of the CH (respectively the CN) molecule in the higher temperature of magnetic elements (Steiner et al. 2001; Sánchez Almeida et al. 2001), which is well reproduced by radiation-MHD simulations (Schüssler et al. 2003; Shelyag et al. 2004).

A small portion of an active region is imaged in the photospheric G-band and the low chromospheric passband centred on the Ca II H line core in Figs. 1A and 1B, respectively (from hight-resolution observations with SST; from Berger et al. 2004). They clearly demonstrate the increasing contrast with height. The Fe I 630.25 nm magnetogram is displayed in Fig. 1C, for comparison with the location of the bright points. Also, Fig. 1D presents the corresponding Dopplergram linearly scaled between -1.5 and 0.8 km/s.

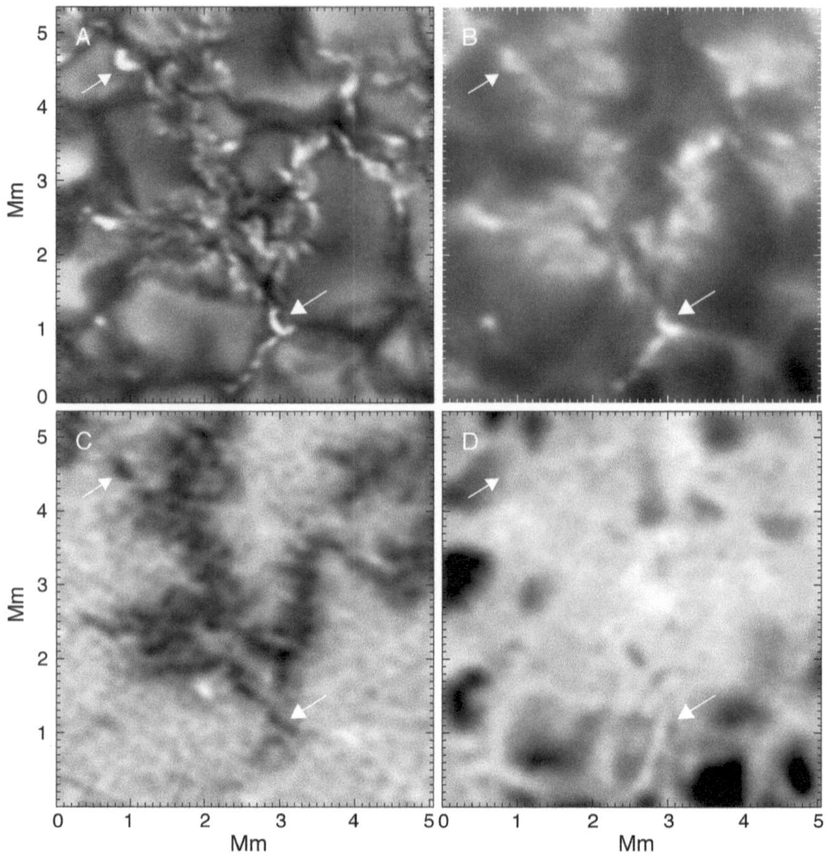

Fig. 1 Brightness contrast in G-band (**A**) and Ca II H (**B**) passbands, from high-resolution observations with SST, as well as maps of magnetic flux density (**C**) and of line-of-sight velocity (**D**). The magnetic flux density and the line-of-sight velocity are linearly scaled, ranging from -1090 to 330 $\mathrm{Mx/cm^2}$ and from -1.5 to 0.8 km/s, respectively. The *arrows* indicate loop-like emission structures over downflowing lanes studied by Berger et al. (2004) (see Sect. 2.2). Reproduced with permission from Berger et al. (2004), Astronomy & Astrophysics, © ESO

The main difference between plage and network magnetic features is their size (e.g. Kobel et al. 2014), i.e. the amount of magnetic flux that they carry, as well as their motions. The brightest magnetic elements may have, however, similar sizes in plage and the network (Kobel et al. 2011). Some of the other differences are probably related to the size difference. An example is their lifetimes, which are likely to be different in the network and in plage, since larger magnetic features tend to live longer (Meunier and Zhao 2009).

A number of studies have found a correlation between the intrinsic field strength of magnetic elements in network and plage and the magnetic filling factor, which is a proxy of the amount of magnetic flux in a given spatial resolution element (Stenflo and Harvey 1985; Schüssler and Solanki 1988; Zayer et al. 1990; Keller et al. 1990; Rüedi et al. 1992b; Grossmann-Doerth et al. 1996; Martínez Pillet et al. 1997; Sigwarth et al. 1999). In most of these studies the resolution element was considerably larger than the individual magnetic features. The dependence may, again, be related to the larger average size of magnetic features in regions with higher magnetic flux density.

Fig. 2 Sketch of the vertical cross-section of a thin flux tube (*shaded, gray*). The Wilson depression, ΔZ, is the geometrical difference between the same optical depth ($\tau_{5000} = 1$) inside and outside of the flux tube (see main text). The radiating hot walls are indicated by *yellow blocks*. The *green arrows* illustrate the convection. The cartoon is not to scale. Inspired by a similar sketch by Schrijver and Zwaan (2000)

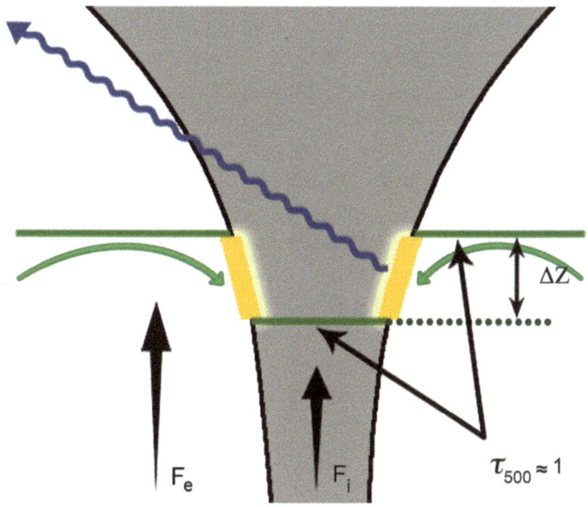

2.4 Theoretical Description of Plage and Network Magnetic Concentrations

Solar magnetic flux concentrations have generally been described by a flux-tube model (see, e.g., Spruit 1976, 1977, 1981; Deinzer et al. 1984; see Solanki 1993, for a review of early work). It assumes that the magnetic features can be described by a bundle of ordered field lines that are enclosed by a topologically simple surface. In addition, the features are assumed to be stationary and in force and energy balance with their surroundings. Although reality is expected to be more complex than this, for many purposes further simplifications are introduced. For example, the shape of the cross-section of (vertical) flux tubes is often considered to be round (i.e. having axial symmetry). This geometry also builds on the near axial symmetry of the most regular sunspots. Alternatively, sometimes flux sheets, with translational and mirror symmetry, are considered (e.g., Holzreuter and Solanki 2012). This geometry describes better magnetic structures in regions with a larger concentration of magnetic flux as in active-region plage. There the flux has a tendency to fill the intergranular lanes, i.e. to concentrate into "ribbon-like" structures (Bushby and Houghton 2005). In general, it is assumed that a flux tube (used as a generic term also to describe features with a non-circular cross-section) are surrounded by an electric current layer where the magnetic field drops rapidly. The width of this layer is estimated to be a few km (Schüssler 1986).

A further and very commonly used simplification is the thin-tube approximation (Parker 1955; Defouw 1976; Pneuman et al. 1986; Ferriz-Mas et al. 1989). In its simplest form it reduces force balance to a simple horizontal and vertical balance of the total pressure (composed of gas and magnetic pressure, with sometimes the inclusion of turbulent pressure as well, which, however, is small compared to the other two in the solar photosphere). The assumption is strictly valid only for flux tubes whose radii are much smaller than the pressure scale height (which in the photosphere is on the order to 100 km). Due to the presence of magnetic pressure inside the tube, the gas pressure and hence also the density is lower there, so that we see deeper layers (an effect known as the Wilson depression in sunspots, where it can be directly observed; e.g. see Solanki 2003 for a review; also Lites et al. 2004 for observations at high spatial resolution).

A sketch outlining the main points of a photospheric flux tube is given in Fig. 2. It illustrates a typical, small-scale magnetic flux concentration that appears bright. The evacuation inside the flux tube lowers the optical depth inside the tube (Wilson depression) allowing the radiation from the hotter gas inside the tube (compared to the external gas at equal optical depth although not necessarily at equal geometric height) to escape and hence be observed. The gas inside the tube is heated by radiation from the hot walls of the tube (yellow in the sketch) through which the hot sub-surface gas in the surroundings radiates excess energy, compared to the surroundings intergranular areas (Spruit 1976).

Convective energy transport is strongly reduced or completely quenched by the strong magnetic field inside magnetic flux tubes with kG fields (Gough and Tayler 1966; Ferriz-Mas and Schüssler 1994). Since the vertical energy transport by radiation is comparatively inefficient in the solar convection zone, for $\tau \gg 1$ the vertical energy flux density inside the flux tube, F_i, is much smaller than that in the surroundings, F_e. Consequently, most of the energy influx into the interior of a flux tube comes through radiation flowing in from its walls (between the $\tau = 1$ levels inside and outside the flux tube), as originally proposed by Spruit (1976) and later demonstrated with the help of numerical simulations by, e.g., Deinzer et al. (1984); Knoelker and Schüssler (1988); Vögler et al. (2005), cf. Schrijver and Zwaan (2000).

As the gas pressure drops with height, so must the magnetic field in order to maintain pressure balance. Magnetic flux conservation then causes the magnetic field to expand with height, creating a canopy of field overlying the convecting gas. Above a certain height the field of neighbouring magnetic features reaches each other and if they are of the same polarity (which is common in active region plage), they merge (Spruit 1983). The rate of expansion of the field is given mainly by the relative temperature inside and outside the flux tube (Solanki and Steiner 1990). We note that the relative rates of expansion with height of large and small flux tubes are similar (Solanki et al. 1999).

The MHD simulations clearly show that the magnetic flux concentrations are neither mainly circular, nor always sheet-like (although examples of features similar to both types can be found, see Figs. 3 and 4). Rather they constantly evolve, changing in size and shape, and interact with each other. Although the shape of the cross-section does not agree with that underlying the thin-tube approximation, the main assumption, namely of horizontal pressure balance, turns out to be satisfied if the second order expansion of the thin-tube approximation, as derived by Pneuman et al. (1986) and Ferriz-Mas et al. (1989), is used (see Yelles Chaouche et al. 2009). Figure 3 presents results of the MHD simulations by Yelles Chaouche et al. (2009), where the vertical component of the magnetic fields, gas pressure, and total pressure are plotted for three different photospheric layers, with $z = 0$ corresponding to the spatially averaged optical depth unity at 500 nm.

2.5 Comparison of Theory and Observations

How does the model of the thin flux tube (or flux sheet) live up to observations? One great success of this model has been that it could reproduce the centre-to-limb variation of the brightness of magnetic features (or equivalently of their contrast relative to the quiet Sun).

Now, thanks to the high-resolution data available from sources such as Hinode (Tsuneta et al. 2008), the Swedish Solar Telescope (Scharmer et al. 2003) and the Sunrise balloon-borne observatory (Solanki et al. 2010; Barthol et al. 2011; Berkefeld et al. 2011; Gandorfer et al. 2011; Martínez Pillet et al. 2011) we can test with much higher fidelity to what extent such descriptions live up to reality.

Thus, the presence of weak opposite-polarity fields surrounding magnetic flux concentrations has been established by MHD simulations, as can be seen in Fig. 4 taken from Vögler et al. (2005), in good agreement with the observations of such fields (Zayer et al. 1989; Narayan 2011; Scharmer et al. 2013; Bühler et al. 2015). The simulations reveal the geometry of the opposite polarity fields (which is not accessible to the observations). These fields

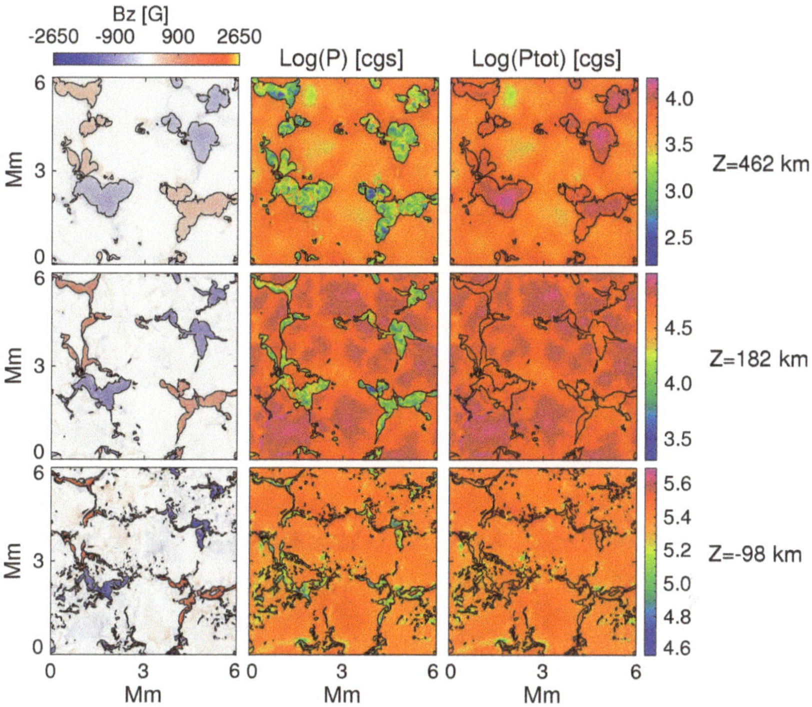

Fig. 3 Stratification of the longitudinal component of the magnetic field (B_Z; *left column*), gas pressure (P; *middle column*), and total pressure (*Ptot*; *right column*), with respect to the optical depth unity at 500 nm. The *black contours* include areas where $|B_Z| > 500$ G at -98 km, $|B_Z| > 400$ G at 182 km, and $|B_Z| > 300$ G at 462 km. Reproduced with permission from Yelles Chaouche et al. (2009), Astronomy & Astrophysics, © ESO

are due to field lines anchored within the magnetic concentration that are turned over and pulled down by the surrounding downflows. In general, they do not extend higher up in the atmosphere, but rather most of these turn back at relatively low heights in the photosphere, just as in the observations. Illustrated in Fig. 5 is a side-view of such structures from MHD simulations (Vögler et al. 2005), where concentrations of the field lines represent along-side-view of a thin magnetic sheet, surrounded by the low-lying field lines.

A remarkable success of MHD simulations has been the agreement of the brightness (or brightness contrast) of simulated magnetic features with that of observed ones. The centre-to-limb variation of the contrast at various wavelengths has been modelled by Carlsson et al. (2004), Keller et al. (2004), and Penza et al. (2004), cf. Steiner (2005). The distribution of the brightness of magnetic bright points in the G- and CN-bands has been successfully compared with the high-resolution Sunrise observations by Riethmüller et al. (2010). The RMS contrast in the same images is dominated by granulation and was also reproduced by the same simulations (Hirzberger et al. 2010). Riethmüller et al. (2014) went a full step further and compared not just the brightness in broad wavelength bands, but also a host of further parameters, such as the Stokes V amplitude, the Doppler shift, the line width, etc. and found a good agreement (with some slight disagreement only in the width of the distribution of Doppler shift values) between the values found in the quiet Sun by Sunrise and those from an MHD simulation with an average vertical field of 30 G. The evolution of the fine-structure of faculae has been investigated by, e.g., De Pontieu et al. (2006), who

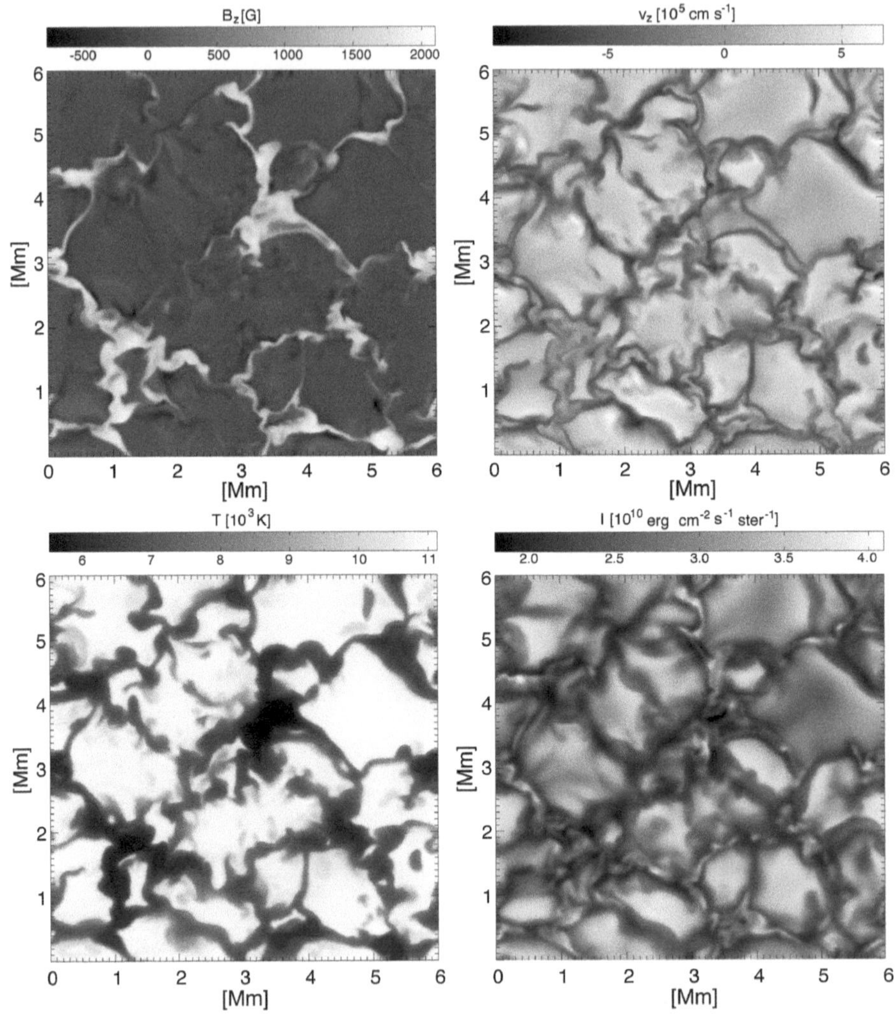

Fig. 4 Maps of longitudinal magnetic field (*upper left*), line-of-sight velocity (*upper right*), temperature (*bottom left*), and intensity (*bottom right*) at the average geometrical height corresponding to optical depth unity (at 500 nm), from MHD simulations made with the MuRAM code. The strong magnetic field concentrations, in downflow areas, are surrounded by narrow lanes of weak opposite-polarity magnetic flux. Reproduced with permission from Vögler et al. (2005), Astronomy & Astrophysics, © ESO

found, using both high-resolution observations and MHD simulations, that faculae change on a time-scale of minutes and that this change is mainly produced by the evolution of the granules lying just behind them, whose hot sides are better seen due to the evacuation of the faculae.

Disagreements in the observed and simulated rms contrasts found by, e.g., Uitenbroek et al. (2007), are very likely due to the effect of scattered light that is pervasive in ground-based data and even affects space-based observations (Danilovic et al. 2008). The best observations with low scattered light, e.g., Hirzberger et al. (2010), display a good agreement with simulated rms contrasts. These results indicate the importance of the scattered light in decreasing the contrast of high-resolution observations.

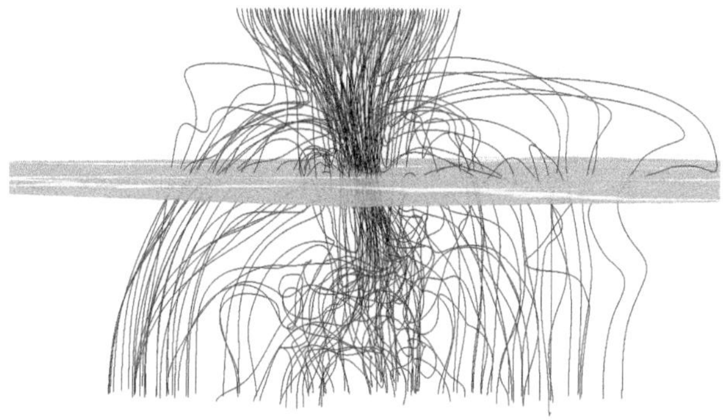

Fig. 5 A side-view from MHD simulations, illustrating geometry of the field lines at magnetic flux concentrations (along a thin magnetic sheet) surrounded by weaker low-lying magnetic loops. The *grey plane* indicates the average geometrical height corresponding to $\tau_{5000} = 1$. Reproduced with permission from Vögler et al. (2005), Astronomy & Astrophysics, © ESO

2.6 Concentration of Magnetic Flux: Convective Collapse

The formation of strong-field magnetic features was for a long time studied mainly with theoretical methods. Two main mechanisms for the concentration of magnetic flux were proposed. The first, flux expulsion, represents a self-organized separation of magnetic field from convection. Any vertically oriented magnetic field at the solar surface is quickly transported to the intergranular lanes by the horizontal flows within granules. This allows granular convection to continue operating even in the presence of a significant amount of magnetic flux (although the convection cells do change, becoming smaller, less turbulent with smaller horizontal velocities, as well as more ordered with a longer lifetime, e.g. Title et al. 1989, 1992; Narayan and Scharmer 2010). The fields produced by this mechanism have roughly the same energy density as the convective flows (i.e. the field is in equipartition with the kinetic energy of the flow), which corresponds to roughly 200–400 G at the solar surface.

The convective instability, acting on regions with such equipartition fields can then intensify them to kG levels (Parker 1978; Webb and Roberts 1978; Spruit 1979; Spruit and Zweibel 1979; Schüssler 1990; Takeuchi 1997; Grossmann-Doerth et al. 1998; Steiner 2003; Danilovic et al. 2010b; Hewitt et al. 2014). The process can be idealized by the adiabatic instability of a static thin magnetic flux tube with an equipartition magnetic field embedded in a superadiabatically stratified medium. The instability leads to accelerated downflow along the tube, which entails a reduction of the internal gas pressure. The excess pressure in the surroundings then compresses and enhances the field until the magnetic pressure becomes sufficiently large to compensate for the loss of gas pressure and a new equilibrium is reached (see Spruit and Zweibel 1979 for a discussion of the non-linear development of the instability). The final state needs not to be a stationary one, with Hasan (1985) finding an oscillatory end state in his 1D model.

In reality, a static equilibrium with equipartition field will most probably never be realized. Instead, the magnetic flux advected by the horizontal granular flow accumulates in intergranular downflow regions. Once the horizontal flow is suppressed by the growing field strength, the downflow is no longer supplied by these flows, yet continues owing to the superadiabatic stratification and the ongoing surface cooling by radiation. This leads to partial evacuation and growth of the field strength (Schüssler 1990).

The 3D radiation MHD simulations of Danilovic et al. (2010b) and Hewitt et al. (2014) followed all steps of the process from the gathering of flux at the intergranular lanes, followed by an evacuation produced by a downflow and the associated strengthening of the field that then quenches the downflow. The field can get weakened again, in some cases possibly due to the following upflows (Grossmann-Doerth et al. 1998;

Danilovic et al. 2010b), but at least in some cases without the need of an upflow (Hewitt et al. 2014).

If magnetic features are below a certain size (or below a certain amount of magnetic flux) then the radiative exchange of energy with the surroundings becomes strong enough to reduce or to completely compensate for the cooling of the downflowing material, so that the collapse does not occur. Hence according to theory the features with the smallest magnetic flux should possess weak fields (Venkatakrishnan 1986; Solanki et al. 1996; Grossmann-Doerth et al. 1998).

Observations have in general been lagging behind the theory, with the first suggestion of a correlation between the magnetic flux of a feature and its intrinsic field strength being found by Solanki et al. (1996). The observed strengthening of the field was first interpreted as a convective collapse by Bellot Rubio et al. (2001). The increase in the field strength in this case was from 400 G to only 600 G. Larger increases have been obtained by the more recent investigations by Shimizu et al. (2008); Nagata et al. (2008); Fischer et al. (2009); Narayan (2011); Requerey et al. (2014); Utz et al. (2014). The final two papers study the evolution of small magnetic elements in the quiet Sun. They find that, at least for these features with relatively small amounts of flux, the strong-field state does not last very long, with the field strength fluctuating in an almost periodic fashion (cf. Martínez González et al. 2011a). These studies suggest that the evolution of the field strength of the larger and stronger magnetic concentrations making up the bulk of active region plage would be a promising topic of study.

3 Recent Discoveries on Internetwork Magnetic Fields

Despite being much more difficult to detect and measure than magnetic fields in sunspots, plage and network regions, internetwork (IN in what follows) magnetic fields are of great importance for several reasons. Firstly, there is the possibility that a considerable amount of magnetic flux in the Sun's quietest regions (i.e., the internetwork) is provided by a turbulent small-scale dynamo. Since numerical simulations cannot be carried out for the extreme solar values of Reynolds numbers and magnetic Prandtl number (see Sect. 4), observations are indispensable to test the validity of the computational results and possibly prove the existence of small-scale dynamo action. Secondly, the relatively weak internetwork magnetic fields (see Fig. 6) interact in a highly dynamical fashion with convective motions in the solar surface layers. This gives rise to highly interesting small-scale phenomena and convoluted magnetic topologies (Amari et al. 2015). Thirdly, if IN magnetic fields are produced by a small-scale dynamo and considering that at least 90 % of the solar surface is covered by the IN,[1] this would mean that most of the unsigned photospheric magnetic flux does not vary over the 11-year magnetic cycle. This could have important consequences for our understanding of the solar irradiance variations, which contribute to terrestrial climate variations (Solanki et al. 2013; Yeo et al. 2014).

The main question is whether IN magnetic fields do actually arise from a turbulent small-scale dynamo or are leftovers of the magnetic fields from decaying active regions, being spread out over the solar surface by advection due to supergranulation, meridional flow, and differential rotation. Therefore, characterizing the internetwork magnetic fields can be considered as one of the most pressing issues in solar physics.

[1] This number should be considered as a lower threshold as sometimes it is not easy to distinguish between strong isolated IN fields and the network.

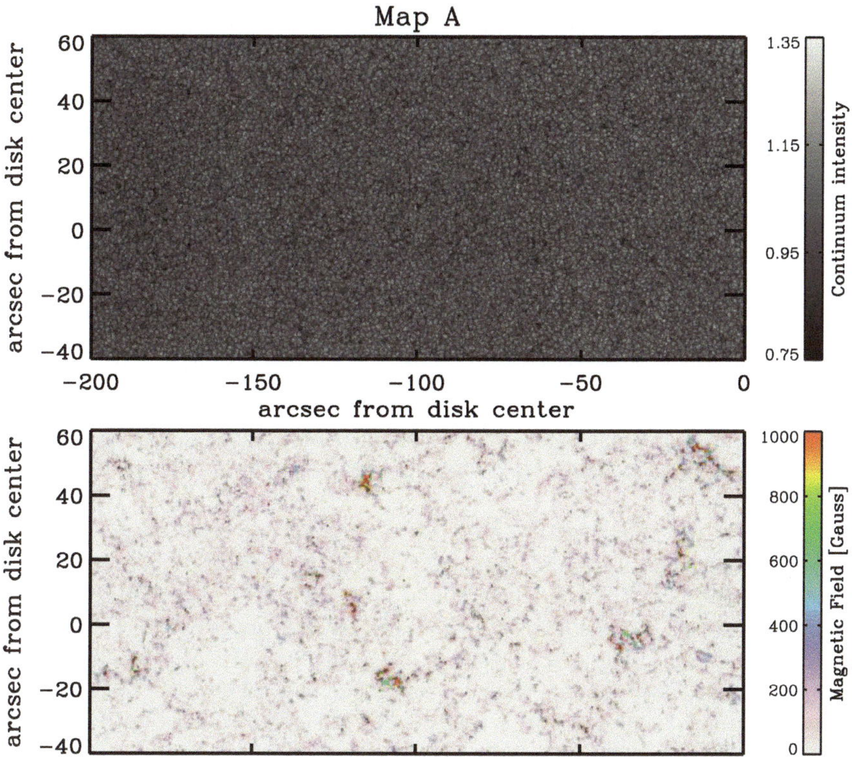

Fig. 6 Continuum intensity normalized to its average value over the entire field of view (*top*) and magnetic field strength (bottom) inferred from Hinode observations of the quiet Sun very close to disk center (Borrero and Kobel 2011). The network on the lower map can been identified in *green/orange/red* colors, whereas the internetwork can be seen as the regions in *pink/white* colors. The latter covers at least 90 % of the total observed area. This particular map has been employed by many authors to study the angular distribution of the magnetic field in the internetwork (see Sect. 3.1). Reproduced with permission from Astronomy & Astrophysics, ©ESO

3.1 Angular Distribution of Internetwork Fields

A small-scale dynamo operating in a stratified medium provides an isotropic magnetic field at scales smaller than the pressure scale height. Numerical simulations confirm this expectation in the sub-surface layers, but predict a strong anisotropy in favor of horizontal fields in the photosphere (i.e., in convectively stable layers), both for the small-scale dynamo (Schüssler and Vögler 2008; Rempel 2014) and for magnetoconvection with a mean vertical field (Steiner et al. 2008; Steiner and Rezaei 2012). Consequently, observationally confirming or ruling out such a distribution has been a major field of research over the past decade. Unfortunately, inferring the magnetic field in the internetwork is extremely challenging owing to their weak observational signatures (i.e. polarization signals barely above the noise). Not surprisingly, these difficulties have led to seemingly contradictory results. The situation is further complicated by the varied analysis techniques employed and the different kinds of data analyzed. In the following we present a summary of observational results organized in terms of the inferred angular distribution for IN magnetic fields.

3.1.1 *Preference for Horizontal Fields*

Orozco Suárez et al. (2007) and Lites et al. (2008) analyzed low-noise ($\sigma \approx 3 \times 10^{-4}$)[2] data from the Fe I line pair at 630 nm recorded at disk center with the spectropolarimeter on-board the Hinode satellite. Employing inversion techniques and different calibration curves they concluded that magnetic fields in the internetwork have a clear preference for the horizontal (i.e. parallel to the solar surface) direction, with the horizontal magnetic field being much stronger (in a spatially average sense) than the vertical one. However, Stenflo (2010) and Borrero and Kobel (2011, 2012) later argued that this result is likely due to photon noise. Indeed, if we consider a weak magnetic field whose projection along the observer's line of sight is the same as the projection on the plane perpendicular to the line of sight, $B_{\parallel} = B_{\perp}$, the 630 nm lines provide much smaller (about 10 times weaker) signal in linear polarization (Stokes Q and U) than in circular polarization (Stokes V). Consequently, the same amount of photon noise in the observed circular and linear polarizations is interpreted as a magnetic field with $B_{\perp} \gg B_{\parallel}$, i.e., a preference for horizontal orientation.

To circumvent this issue, Bellot Rubio and Orozco Suárez (2012) analyzed data from the same instrument, recorded also at disk center, but averaged in time over 24 minutes. This allowed them to decrease the photon noise to $\sigma \approx 7 \times 10^{-5}$ such that almost 70 % of the internetwork displayed linear polarization signals clearly above the noise. The magnetic fields inferred from the analysis of these regions again showed a clear preference for the horizontal direction. However, the long integration times needed to decrease the noise might introduce additional effects (such as the cancellation of signals from opposite polarities and averaging over different structures) that need to be carefully addressed.

3.1.2 *Preference for Vertical Fields*

Ishikawa and Tsuneta (2011) studied, using also calibration curves, the very same data set as Orozco Suárez et al. (2007) and Lites et al. (2008). However, they concluded that the vertical magnetic field in the internetwork is almost as strong as the horizontal field. This would indicate a larger contribution from vertical (perpendicular to the solar surface) magnetic fields than in the aforementioned works. However, we note that the low value of the horizontal magnetic flux in Ishikawa and Tsuneta (2011) follows from their procedure to ascribe a zero value to the perpendicular component of the magnetic field whenever the linear polarization was deemed to be at, or below, the noise level ($B_{\perp} = 0$ if the line-integrated unsigned linear polarization $L_{\mathrm{tot}} \leq 1.5 \times 10^{-4}$). This approach has the problem of potentially deriving vertical magnetic fields from signals that are actually produced by strongly inclined fields. This happens because the most that can be said about signals with $L_{\mathrm{tot}} \leq 1.5 \times 10^{-4}$ is that $B_{\perp}$ is below some threshold value $B_{\mathrm{thr},\perp}$. The problem is that if the value of this threshold is such that $B_{\mathrm{thr}} \gg B_{\parallel}$ then this approach will heavily underestimate the contribution of $B_{\perp}$ to the total magnetic field.

Mostly vertical fields have also been reported by Jafarzadeh et al. (2014a) using geometrical considerations in internetwork magnetic bright points seen at different photospheric layers. Unfortunately this method is not applicable to the majority of the internetwork (i.e., outside magnetic bright points).

[2]The noise level σ is given in units of the continuum intensity spatially averaged over the quiet Sun.

3.1.3 Isotropic and Quasi-isotropic Distribution

Asensio Ramos (2009) also analyzed spectropolarimetric data at disk center from Hinode (see Fig. 6), but applied a Bayesian technique to observations with a larger photon noise ($\sigma \approx 10^{-3}$) than all previously cited works. This resulted in an inferred angular distribution of the magnetic field in the internetwork that is close to isotropic. Asensio Ramos and Martínez González (2014) reached the same conclusion, although they could not exclude the existence of a non-negligible contribution from highly inclined magnetic fields.

An isotropic distribution was also favored by Martínez González et al. (2008). This work employed two spectral lines in the infrared (Fe I at 1565 nm) that are less affected by photon noise in the linear polarization, thus leading to more reliable inferences of $B_\perp$. Unlike most of the previous works, spectropolarimetric observations of the internetwork at various heliocentric viewing angles were considered. The resulting distribution of observed polarization signals (not of inferred magnetic fields) was found to be nearly independent of the viewing angle, suggesting an isotropic distribution of the IN fields. However, the authors emphasized that the observed distributions of polarimetric signals lacked a sufficiently large number of occurrences for signals above 4×10^{-3} (in units of the continuum intensity) to be considered statistically significant.

Borrero and Kobel (2013) also studied the distributions of polarimetric signals and tried to obtain more reliable statistics of spectropolarimetric observations at different heliocentric angles. To that end, they analysed a large data set from the Hinode satellite with a noise level of $\sigma \approx 3 \times 10^{-4}$. They found that the distribution of signals does in fact depend on the heliocentric angle, indicating a non-isotropic distribution of magnetic fields in the IN.

3.1.4 Bi-modal Distribution

The same data (see Fig. 6) as Asensio Ramos (2009) and Asensio Ramos and Martínez González (2014) was analyzed by Stenflo (2010), who employed a line-ratio technique aided by several calibration curves in order to disentangle thermodynamic and magnetic effects in the observed signals. He found evidence of two distinct populations of polarimetric signals in the internetwork. A population with a slope of the line-ratio[3] of $s \approx 1.15$ was attributed to strong (i.e. kilogauss) and mostly vertical magnetic fields, which remained unresolved at the spatial resolution of Hinode observations ($\approx 0.3''$). The second population of signals with a slope of $s \approx 1.66$ was ascribed to weak and isotropically distributed magnetic fields. These findings were further investigated by Steiner and Rezaei (2012), who analyzed polarimetric signals synthesized by solving the radiative transfer equation with input from three-dimensional MHD simulations. They also found the two populations with the same slopes as reported by Stenflo (2010). However, the two populations resulted from magnetic fields that, while all being weak, increase or decrease with height in the photosphere. Regardless of the physical realism of these simulations, the results by Steiner and Rezaei (2012) clearly point towards an uniqueness problem in the interpretation of the observed signals via the line-ratio technique.

3.1.5 Further Results

It has now become clear that data recorded at disk center are not sufficient to uniquely determine the angular distribution of the magnetic field in the internetwork. Indeed, the

[3]The line-ratio technique measures the ratio s of the amplitudes in the circular polarization signals in two different spectral lines. Under the assumption that both lines possess the same opacity and provided that the magnetic field is weak, then s is equal to the quotient of their Landé factors.

most recent attempts have focused on observations taken at several heliocentric angles, but a consensus has still not been reached. Using spectropolarimetric data from Hinode, Orozco Suárez and Katsukawa (2012) concluded that the distribution of the inclination of the magnetic field has a preference for the perpendicular direction (with respect to the line of sight) at all heliocentric angles studied ($\mu = 1.0, 0.5, 0.23, 0.1$). This result can be explained by magnetic fields that, being mostly horizontal at disk center, become gradually perpendicular to the solar surface for smaller μ values. We emphasize, however, that these results are possibly affected by the same bias as those of Orozco Suárez et al. (2007) (see Sect. 3.1.2). Borrero and Kobel (2013) also interpreted Hinode data in terms of IN magnetic fields that become gradually more horizontal for decreasing μ. This should not be considered a general trend because these authors employed only data at $\mu = 1.0$ and $\mu = 0.7$. In contrast, Stenflo (2013) concluded from spectropolarimetric observations of the Fe I lines at 525 nm taken with the ZIMPOL instrument that magnetic fields at $\mu = 0.5$ are more perpendicular to the solar surface (i.e., more vertical) than at $\mu = 0.1$.

Several effects can be invoked to explain why the angular distribution of IN magnetic fields changes with the heliocentric angle μ. For instance we could take into account that the height of formation of the spectral lines changes with μ because the continuum optical depth varies with the viewing angle: $\tau_c = f(\mu)$. This would readily introduce a μ-dependence in the angular distribution of IN magnetic fields if, as suggested by numerical simulations (Schüssler and Vögler 2008; Steiner et al. 2008; Steiner and Rezaei 2012; Rempel 2014), the angular distribution of the magnetic field changes with height in the photosphere. Alternatively, if τ_c is a slowly varying function of μ or if the angular distribution of the magnetic field does not vary with height on the photosphere, one could ascribe the μ-variations of the angular distribution to, for instance, an intrinsic difference between IN polar fields and those at the equator. In addition, it should be considered that the spatial resolution of the observations depends on μ, because, due to foreshortening, and given angular resolution element covers a larger piece of the solar surface near the limb than at disk centre. While in actuality all these effects are likely to play a role, it is not clear how to model them when interpreting the observations. A way to study which of the first two aforementioned effects is more important would be by means of high signal-to-noise spectropolarimetric observations along the equator and along the central meridian.

3.1.6 Future Prospects

In spite of many efforts, we are currently no closer to accurately determining the angular distribution of the magnetic field in the internetwork than we were a decade ago. This results from a combination of the effects of photon noise and signal selection, as well as from the different techniques employed to infer the magnetic field from spectropolarimetric signals produced by the Zeeman effect. Advances on this front could be achieved by means of spectropolarimetric observations in the infrared. In particular, the Fe I line pair at 1565 nm is much less affected by noise in the linear polarization (see Sect. 3.1.3). Moreover, the continuum of these spectral lines is formed deeper (about 70–100 km deeper depending on the opacity sources included in the calculation of the absorption coefficient) than in the commonly used Fe I line pair at 630 nm. Therefore, these spectral lines are best suited to study the angular distribution of the magnetic field in the lower layers of the photosphere (see Sect. 3.1). The only drawback comes from the lower spatial resolution in the infrared, but this will be alleviated in the near future with observations using large-aperture telescopes such as GREGOR and DKIST, thanks to their large primary mirrors and hence smaller diffraction limits. The former one, with a 1.5-meter aperture, is already obtaining spectropolarimetric observations in these spectral lines with a spatial resolution comparable to that of the Hinode telescope in the visible (Collados et al. 2012).

Spectral lines whose spectropolarimetric signals are produced by the Hanle effect are also worth considering for future observations and analysis. Indeed, in the saturated Hanle regime the linear polarization signals depend only on the orientation of the magnetic field, but not on its magnitude (Carlin and Asensio Ramos 2015). This removes an important source of uncertainty. In addition, Hanle signals are less affected by cancellation effects if the magnetic field is unresolved. This feature of the Hanle effect is particularly relevant for investigations of the magnetic field produced by a small-scale turbulent dynamo. It remains to be seen, however, how accurately our theoretical understanding of the Hanle effect and its intricate interactions with the thermodynamics and kinematics of the plasma (Carlin et al. 2012) allow us to determine the properties of IN magnetic fields.

3.2 Solar Cycle Variations

The presence of a small-scale turbulent dynamo that operates independently of the large-scale dynamo responsible for the 11-year activity cycle would provide a time-independent background level of unsigned magnetic flux in the quiet Sun.

From ground-based observations Sánchez Almeida (2003) concluded that there is no correlation between the strength of the magnetic field in the internetwork and the solar cycle. However, to study this particular aspect of the magnetic fields in the internetwork, it is necessary to employ high-quality long-term observations of the polarization signals in spectral lines. This is the case for the synoptic observations carried out by Hinode's spectropolarimeter. Bühler et al. (2013); Jin and Wang (2015a, 2015b) analysed these data and concluded that during the period between 2006–2015 the magnetic field in IN regions at disk center did not vary significantly. This result was later extended by Lites et al. (2014) (see Fig. 7) to cover both low and mid latitudes ($\mu \geq 0.5$). While these studies were based on the Zeeman effect in Fe I lines at 630 nm, Kleint et al. (2010) and Bianda et al. (2014) analysed spectropolarimetric data from the ZIMPOL instrument in the Sr I line at 460 nm and in molecular C_2 lines at 514 nm. The polarization signals of the recorded spectral lines are produced by the Hanle effect, which is much less affected by cancellation than the Zeeman effect for magnetic fields of opposite polarities. These authors concluded that for the period between 2007 to 2009 (solar minimum) the magnetic field strength in the IN remained the same. However, they report a possible variation with respect to earlier measurements obtained in 2000 (solar maximum).

3.3 Small-Scale Events

While most polarimetric signals in the internetwork are at or below the noise level of the observations (Sect. 3.1), some features show particularly strong signals. Detailed analysis of these revealed a variety of dynamic magnetic phenomena taking place in the internetwork. This is not surprising because the magnetic field in the internetwork is not strong enough to dominate over convective motions. Instead, these motions twist and bend the field lines into complicated patterns. This leads to large variations of the magnetic, kinematic and thermodynamic properties on small scales, which manifest themselves in complex spectropolarimetric signals featuring large asymmetries and additional components or lobes.

Since the observational signals associated with such small-scale phenomena are very dynamic and short-lived, high-cadence instruments attached to seeing-free telescopes (Hinode, Sunrise) or large-aperture telescopes at the ground (SST, GREGOR), are required to characterize them. In the following we summarize some of the recent results.

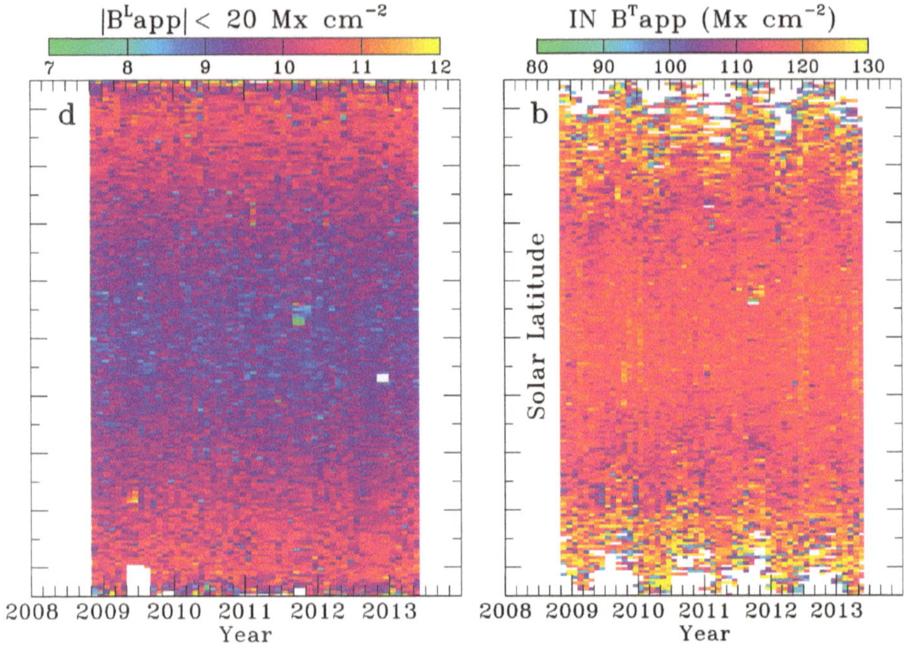

Fig. 7 Unsigned longitudinal component (*left panel*) and transversal component (*right panel*) of the magnetic field in the internetwork as a function of time (*abcissa*) and solar latitude (*ordinate*, from $-90°$ to $+90°$ with tickmarks at intervals of $10°$). The time coverage spans from 2009 (close to solar minimum) until 2013 (close to solar maximum). Figure adapted from Lites et al. (2014), reproduced with permissions from the Oxford University Press on behalf of the Astronomical Society of Japan

3.3.1 Emergence of Magnetic Loops

The emergence of magnetic Ω-loops has been studied in detail, by Centeno et al. (2007) and Martínez González and Bellot Rubio (2009) using Hinode data. Such loops are observed as single patches of linear polarization (indicating horizontal magnetic fields) that are seen prior to the appearance of two patches of opposite circular polarization (indicating magnetic fields of opposite polarities) at the endpoints of the elongated horizontal field patch. As it evolves in time, the horizontal magnetic field disappears and the footpoints of opposite polarities separate. While these observations are, in principle, compatible with both an emerging Ω-loop and a submerging inverted Ω-loop, the former interpretation is preferred whenever the initial horizontal field usually is found in upflow regions (i.e., granules). Those appearing in downflow regions are therefore associated to inverted Ω-loops (Pietarila et al. 2011).

The lifetime of a few minutes and size of 1–2″ of the features are very similar to corresponding quantities of granules. From observations of the Mg I 517 nm and Ca II 396 nm spectral lines (formed in the mid and lower chromosphere, respectively) (Martínez González et al. 2010) concluded that many of the loops reach the lower chromosphere, carrying magnetic energy at a rate of $\approx 2 \times 10^6$ erg cm^{-2} s^{-1}.

However, not all horizontal fields observed in the internetwork are part of an emergence process. Danilovic et al. (2010a) analysed spectropolarimetric observations of the Fe I line at 525 nm with the IMaX instrument on-board Sunrise. They studied a large number of internetwork horizontal field patches and concluded that, although they tend to appear at

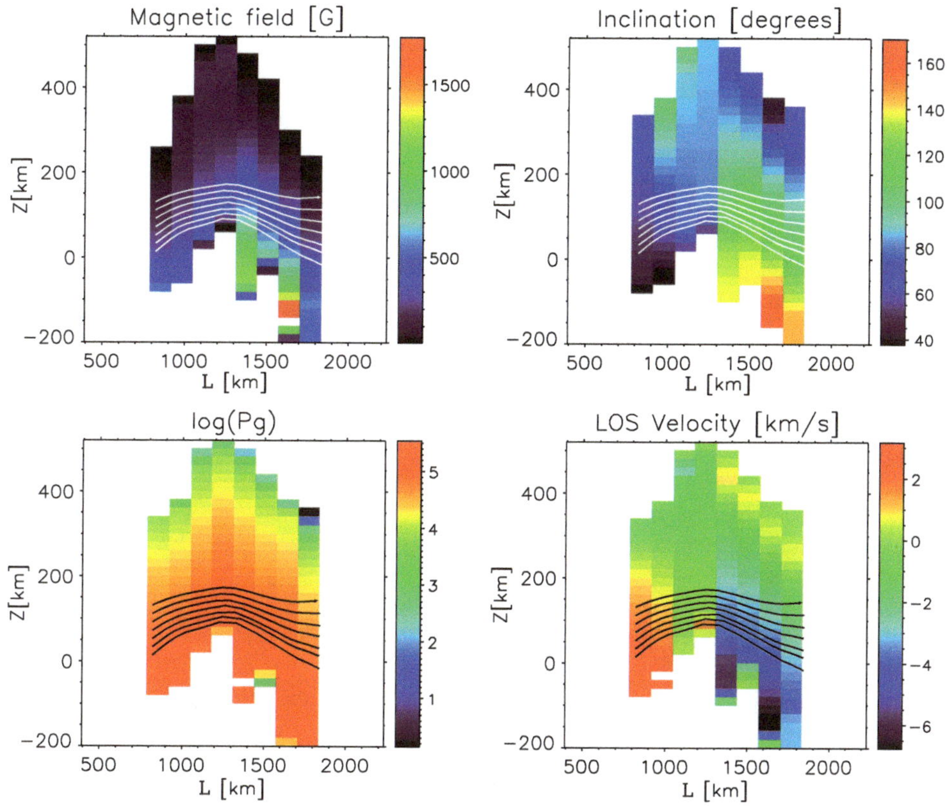

Fig. 8 Side-view of a magnetic Ω-loop seen in Hinode/SP data (Quintero Noda et al. 2014): magnetic field strength (*upper-left*), magnetic field inclination with respect to the vertical direction (*upper-right*), gas pressure (*lower-left*), line-of-sight velocity (*lower-right*). The magnetic field at the *left* and *right* footpoints of the loop points upwards and downwards, respectively. On the other hand, the velocity points *downwards* and *upwards* at the *left* and *right* footpoints, respectively, with the latter reaching values up to $-6\ \mathrm{km\,s^{-1}}$ (supersonic). Reproduced with permission from Astronomy & Astrophysics, ©ESO

the boundary of granules, most of them are caught at some point in their evolution in an intergranular lane. This makes it difficult to uniquely ascribe them to a emergence of submergence event. These authors also found that horizontal IN fields do not possess a typical life-time or size.

3.3.2 Siphon-Flows Along Magnetic Loops

Highly asymmetric, single-lobed Stokes V profiles are often observed at the footpoints of the magnetic loops discussed in the previous subsection (Sainz Dalda et al. 2012). Viticchié (2012) interpreted them as being produced by a siphon flow along the magnetic loop (Meyer and Schmidt 1968a, 1968b). This conclusion is supported by Quintero Noda et al. (2014), who also found that the flow has the opposite direction at the two footpoints and can become supersonic in between (see Fig. 8). Supersonic siphon flows had been theoretically predicted by Thomas and Montesinos (1990) and already observationally detected by Rüedi et al. (1992a) and Degenhardt et al. (1993).

3.3.3 Supersonic Upflows

Borrero et al. (2010) found highly blue-shifted Stokes V signals in Sunrise/IMaX data, appearing close to, but not exactly at, locations with large linear polarization signals (i.e., horizontal magnetic fields). They ascribed these signals to supersonic magnetic upflows occurring at the centers or edges of granules. According to Quintero Noda et al. (2013) the detection of the linear polarization signal precedes the neighbouring supersonic upflow (highly blue-shifted circular polarization). Rubio da Costa et al. (2015) studied many such events at different positions on the solar disk and found that their properties do not vary significantly with the heliocentric angle.

Borrero et al. (2013) suggested that the supersonic upflows were caused by the reconnection of an emerging Ω-loop with an ambient field of opposite polarity (Parker 1972, 1973). However, this particular interpretation has been questioned by Danilovic et al. (2015) who, analyzing synthetic spectropolarimetric data obtained from MHD simulations, found that the same observational features can also be reproduced in the absence of reconnection. Instead they propose a magnetic topology that closely resembles that of a plasma bubble filled with a strong unipolar magnetic field that rises through the photosphere. Structures with very similar magnetic topologies but moving downwards had previously been reported by Quintero Noda et al. (2014).

3.3.4 Magnetic Elements in the Internetwork

Magnetic flux concentrations sufficiently large to appear in the form of bright points are also present in the internetwork but, unlike in the network, they appear isolated and are less abundant: 2.2 bright points per Mm^2 in the network versus 0.85 in the internetwork (Sánchez Almeida et al. 2010). Note that, in addition, smaller kG flux concentrations may exist that are not resolved by the observations.

The properties of the internetwork elements are rather diverse. Some of them are observed as mostly horizontal fields with a lifetime of 1–10 minutes and typical sizes of several arcseconds. These *transient horizontal fields* (THFs) were previously detected in plage regions and extensively studied by Ishikawa et al. (2008) using Hinode data. Later, Ishikawa and Tsuneta (2009) found that THFs in in plages feature very similar distributions for the vertical and horizontal component of the magnetic field (after removing the bias caused by persistent vertical fields in the plage) than those in the internetwork. Additionally they found that, whereas THFs in plage regions are preferentially oriented along the plage's direction, internetwork THFs possess no preferred orientation. This indicates that the topology of THFs magnetic field in plages depends on the large scale structure of the surrounding field and their formation might be associated with the emergence of magnetic flux in the Photosphere. A detailed investigation of the magnetic topology in THFs was later carried out by Ishikawa et al. (2010), who found that they correspond to emerging Ω-loops similar to those described in Sect. 3.3.1.

Other isolated internetwork magnetic elements are seen mostly in the circular polarization, thus indicating the presence of mostly vertical and strong (kGs) magnetic fields. Using Sunrise/IMaX data (Martínez González et al. 2011b) discovered damped oscillations in the magnetic flux of some of these elements.

Internetwork magnetic elements have been found to migrate towards the surrounding network boundaries (de Wijn et al. 2008), being advected by supergranular flows (Orozco Suárez et al. 2012). Using Hinode/SP data Utz et al. (2010) and Manso Sainz et al. (2011) discovered that, during this migration process, the magnetic elements behave as random walkers. High spatial-resolution observations from SST, BBSO, and Sunrise revealed, however, super-diffusive trajectories for the internetwork magnetic bright points (Abramenko et al. 2011; Chitta et al. 2012; Jafarzadeh et al. 2014b). The latter study showed that their horizontal motion could be described as a random walker (due to turbulence in intergranular lanes and granular evolution) superposed on a systematic velocity (caused by granular, meso-, and super-granular flows). That analysis also revealed that the diffusion coefficients of the motion of the internetwork magnetic bright points lie within the range obtained from the decay rates of the magnetic field on the solar surface in 3D radiative MHD simulations of Cameron et al. (2011).

4 Small-Scale Dynamo

As discussed in detail in Sect. 3, observations indicate the existence of an ubiquitous small-scale "turbulent" magnetic field, which apparently does not vary in the course of the solar cycle. Petrovay and Szakaly (1993) were the first to suggest that such a field would be generated by a small-scale dynamo (SSD), which is independent of the large-scale dynamo (LSD) responsible for the global (system-scale) magnetic field of the solar cycle. For the LSD (for a review, see Charbonneau 2010), differential rotation and the Coriolis force are essential ingredients, leading to large-scale poloidal and toroidal magnetic fields. In contrast, the SSD generates a magnetic field also in the absence of rotation through stretching, twisting and folding of field lines by a flow with chaotic streamlines (such as turbulence), if the magnetic Reynolds number of the flow is high enough. The generated field is of mixed polarity on spatial scales small compared to the integral scale of the flow and its large-scale average vanishes. Therefore, the SSD is sometimes also denominated as "fluctuation dynamo". In the solar literature, the term "local dynamo" has also been used occasionally. This is potentially misleading if understood as a locality in space, e.g., in the granulation layer: SSD action probably takes place throughout the whole convection zone (Hotta et al. 2015), modified by rotational shear and Coriolis force in the deeper layers. The SSD is also not local in wavelength space since the large-scale convective patterns imprint themselves on the distribution of the SSD-generated field by flux transport and expulsion, leading to a flat spectrum at large scales for the saturated dynamo state (Rempel 2014).

First suggestions of a turbulent SSD dynamo go back to Batchelor (1950), who drew on the analogy between the hydrodynamic vorticity equation and the magnetic induction equation in the kinematic limit. For an idealized case, SSD action was first demonstrated by Kazantsev (1968). Starting with Meneguzzi et al. (1981), the operation of SSDs was later found in many simulations of forced turbulence as well as for incompressible, Boussinesq, and anelastic thermal convection (for reviews, see Brandenburg and Subramanian 2005; Brandenburg et al. 2012; Tobias et al. 2013). Most of these simulations were carried out for fluids with magnetic Prandtl number $P_M = \nu/\eta$ (ν: kinematic viscosity, η: magnetic diffusivity) of order unity or bigger. The regime $P_M \ll 1$, which is relevant for the Sun ($P_M \approx 10^{-5}$), is much more demanding computationally and thus not well covered by simulations. Heuristic arguments suggest that a SSD is more difficult to excite (i.e., requires higher magnetic Reynolds number) for low Prandtl number and it was debated for some time whether SSD action is possible at all in this regime. Several studies showed, however, that the critical magnetic Reynolds number increases only moderately for $P_M \ll 1$ (Brandenburg 2011; Buchlin 2011; Tobias et al. 2013). Therefore, SSD action is expected for solar conditions, owing to the high magnetic Reynolds numbers of solar convection.

Direct simulations of SSD action under solar conditions is computationally unfeasible owing to the extreme values of the (kinetic and magnetic) Reynolds numbers and of P_M in the convection zone. For comprehensive simulations representing the relevant physical processes in the upper solar convection zone and photosphere (compressible convection, partial ionization, proper radiative energy transport), one has therefore to resort to large-eddy simulations: artificial diffusivities are introduced such that they minimize diffusion on the numerically resolved scales and provide an efficient diffusive cutoff at the grid scale (cf. Rempel et al. 2009; Rempel 2014; Miesch et al. 2015). It can be argued that these procedures lead to effective magnetic Prandtl numbers of order unity. It is not clear, therefore, that such simulations correctly represent the putative SSD action on the Sun. On the other hand, simulations of convection, of magneto-convection with an imposed background field, as well as of sunspots are consistent with observations on all spatial scales reached by the latter (e.g., Nordlund et al. 2009; Stein 2012; Rempel and Schlichenmaier 2011). Furthermore, it is well conceivable that saturated SSD action reaches a saturated state for which all flow scales smaller than the resistive cutoff are suppressed by the Lorentz force, leading to an effective magnetic Prandtl number of order unity. In any case, comprehensive simulations provide self-consistent data for the forward modelling of observations and test cases for data analysis methods.

Comprehensive simulations carried out so far suggest that SSD action is indeed pervasive in the convection zone (Vögler and Schüssler 2007; Pietarila Graham et al. 2010; Rempel 2014; Hotta et al. 2015), at least for effective magnetic Prandtl numbers of order unity (Thaler and Spruit 2015). The structure of the generated magnetic field is in many regards similar to the turbulent internetwork field inferred from observations (Schüssler and Vögler 2008; Pietarila Graham et al. 2009; Danilovic et al. 2010b; Schüssler 2013). The amplitude of the simulated field, however, appears to fall short of the observational inferences (Danilovic et al. 2010a; Shchukina and Trujillo Bueno 2011), unless the lower boundary condition is modified (Rempel 2014, discussed further below).

Figure 9 shows a snapshot from a near-surface SSD simulation with the MURaM code. The computational box had a size of 18×18 Mm^2 horizontally and 7 Mm vertically, reaching about 1 Mm above the optical surface (average height of bolometric optical depth unity). The run was started from a relaxed hydrodynamic run with a 4×4 checkerboard sinusoidal seed field of 0.1 G amplitude. The run was then continued with stepwise increases of the grid resolution, for 10.4 hours of solar time, thus achieving a statistically stationary state. Within the full box, the saturated rms field strength was 488 G and the average unsigned vertical field at optical depth unity reached 19.4 G. The magnetic field exhibits the typical 'salt-and-pepper' structure of SSD action with mixed polarity at small scales, resulting from turbulent eddies providing the 'stretch-twist-fold' mechanism for flux generation (Zeldovich et al. 1983). This is also reflected in the fact that the magnetic flux is found predominantly at those locations where the velocity appears 'rough', indicating enhanced turbulence. Most of the complexity of the SSD-generated field is lost, however, when observed with limited spatial resolution. The lower left panel of Fig. 9 shows the result of convolving the upper-right field map with an Airy function corresponding to a 50-cm telescope (such as the one onboard Hinode) at a wavelength of 630 nm. When watching a time evolution of such maps, one detects many instances of the apparent emergence or cancellation of bipoles, but these events mostly result from the smoothing over much a more complex magnetic structure.

Spatial smoothing and the resulting cancellation of opposite-polarity fields has also drastic effects on the unsigned flux (average unsigned vertical field component, $\langle|B_z|\rangle$) detectable at a given spatial resolution. At the original grid resolution of the simulation discussed here, we have $\langle|B_z|\rangle = 19.4$ G at optical depth unity. Smoothing with Airy functions

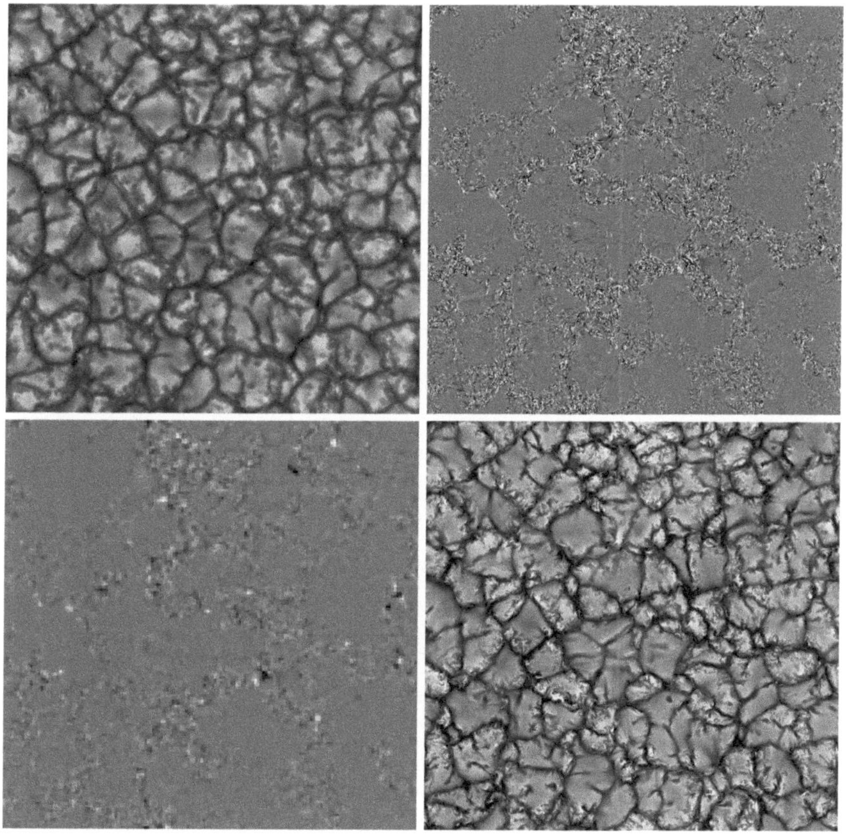

Fig. 9 Snapshot from a SSD simulation with the MURaM code. Shown are maps of the bolometric intensity (*upper left*), the vertical magnetic field component at optical depth unity at simulation resolution (*upper right*) and smoothed with an Airy function corresponding to a 50-cm telescope at 630 nm (*lower left*), both saturated at ± 100 G, and the vertical velocity component at optical depth unity (*lower right*); upflows are represented by *brighter*, downflows by *darker shades*, saturated at ± 5 km s^{-1}. The size of the computational box is 18×18 Mm^2 horizontally and 7 Mm in depth, with a grid cell size of 10 km

corresponding to spatial resolutions of $0.1''$, $0.2''$, and $1''$ leads to values of 9 G, 5.3 G, and 1.1 G, respectively. These are actually upper limits since the effects of a realistic MTF and noise would significantly decrease the values of $\langle |B_z| \rangle$ even further. On the other hand, this simulation probably underestimates the unsigned flux by about a factor 3 (see discussion further below).

The overall spatial distribution of the magnetic field reflects the scales of the convective flows. The magnetic structures are transported towards the downflow regions, where they accumulate and outline the convective flow patterns from the dominant granulation scale up to the largest flow scale in this simulation of about 6 Mm, which is determined by the depth of the computational box. As a result, the flux distribution is spatially inhomogeneous at these scales and flux-deficient 'voids' appear in the distribution of the magnetic field (Martínez González et al. 2012a). In the simulation, their size is limited by the depth of the box, but in the real Sun the maximum size of the voids is reached when the time scale of the flow pattern becomes about equal to the growth time of the SSD in the near surface layers. As the simulations show, the formation of voids is not in contradiction to the generation of mag-

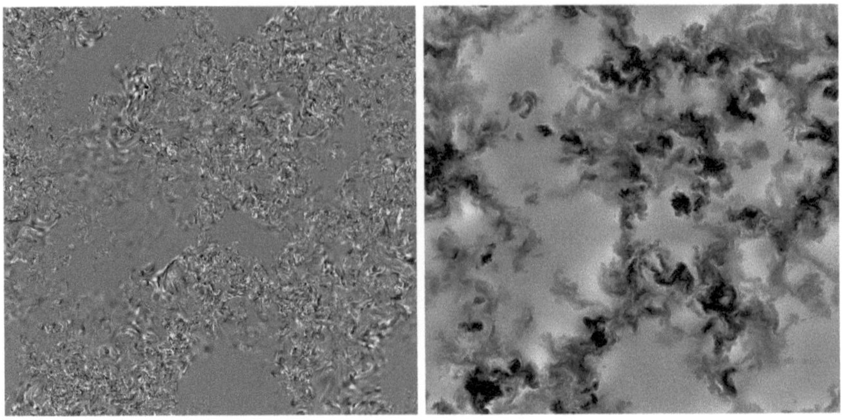

Fig. 10 Vertical components of magnetic field (*left*) and velocity (*right*) in a depth of 5 Mm below the optical surface for the same snapshot as shown in Fig. 9. The greyscales are saturated for ± 1800 G and $\pm 1.2\ \mathrm{km\,s^{-1}}$, respectively. Upflows are represented by *brighter*, downflows by *darker shades*

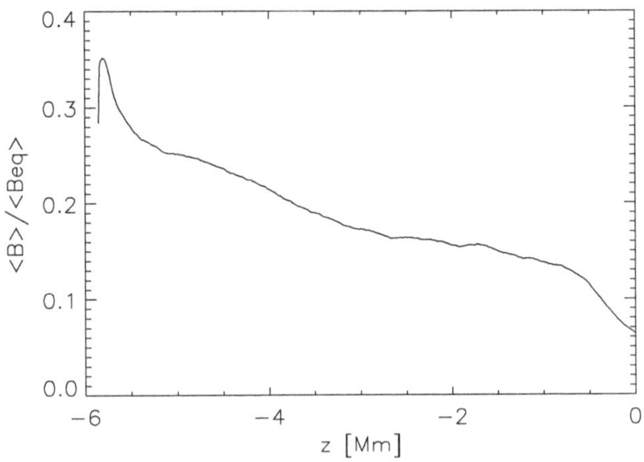

Fig. 11 Ratio of actual field strength and equipartition field strength with respect to the kinetic energy density of the convective flows (both horizontally averaged) as a function of height for the same snapshot as shown in Figs. 9 and 10. $z = 0$ refers to the optical surface

netic flux mainly in the turbulent intergranular lanes (cf. Martínez Pillet 2013): the lanes are advected by the larger-scale flows while the field is amplified by the SSD. The voids would be there even if the operation of the SSD were limited to the near-surface granulation layer in the sense of a truly 'local dynamo'. However, this is not the case: the simulations show that SSD action involves the whole convecting volume and in global simulations pervades the whole convection zone (Hotta et al. 2015).

Figure 10 shows maps of the vertical field and vertical velocity in the deeper layers of the box, at a depth of 5 Mm. The salt-and-pepper structure of the magnetic field and its association with the downflow patterns (dark in the greyscale representation) is similar to the near-surface maps shown in Fig. 9. Also the 'voids', which appear at roughly the same positions as near the surface, are clearly associated with the upflow regions of the larger-scale convective pattern at that depth.

The field strength divided by the equipartition field strength (with respect to the kinetic energy density of the convective flows) given in Fig. 11 increases with depth and shows that the SSD is rather efficient, even though the layers below the simulation box are (unrealis-

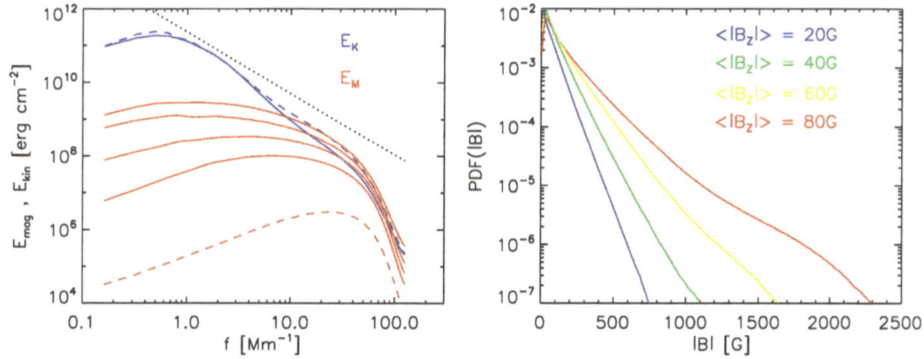

Fig. 12 *Left panel*: kinetic energy spectra (*blue curves*) and magnetic energy spectra (*red curves*, at different times during the growth phase of the dynamo, corresponding to the levels of the mean magnetic field strength indicated in the *right panel*) at the optical surface for a SSD run. *Dashed lines* indicate the kinematic phase (scaled up for the magnetic energy in order to show it on the same plot). The *full blue curve* and the *uppermost red curve* correspond to the saturated state of the dynamo. The *dotted line* indicates a slope of $-5/3$. *Right panel*: field strength distribution (PDFs) at the optical surface for various levels of the mean vertical field during the growth phase of the dynamo (from Rempel 2014, Fig. 2; ©AAS, reproduced with permission)

tically) assumed to be field-free. In total, the magnetic energy amounts to about 7.4 % of the kinetic energy in the box. In the global simulations of Hotta et al. (2015), the magnetic energy density reaches 95 % of the equipartition value near the bottom of the convection zone, with a significant suppression of the convective flow velocities due to the action of the Lorentz force. This shows that SSD action is much more efficient than the large-scale dynamo in converting kinetic to magnetic energy.

When synthetic Stokes profiles are calculated from simulations and compared with observations of the internetwork or turbulent fields (e.g., Danilovic et al. 2010a; Shchukina and Trujillo Bueno 2011), SSD simulations in shallow boxes without advection of magnetic flux from below (such as the example presented above) appear to provide too little unsigned flux. Typical values for $\langle |B_z| \rangle$ at the optical surface are 20–30 G while the observations require at least a factor 3 more magnetic flux. This may partly be due to too low values of the magnetic Reynolds number, although this may affect the saturated state much less than the kinematic growth rate (see below).

Another reason for too low surface flux from the SSD results from the lower boundary condition. In order to demonstrate proper SSD action in the simulation box, the simulations of Vögler and Schüssler (2007) and Pietarila Graham et al. (2010) did not allow the advection of magnetic flux by upflows (inflows) at the lower boundary; only flux transport out of the box was permitted. Therefore, the deeper layers of the convection zone were implicitly assumed to be field free and SSD action was artificially restricted to the volume of the box. This is not expected to be the case in the real Sun, where flux generated by SSD action in the deeper layers probably contributes to the surface field as well. In this sense, a more realistic boundary condition is symmetric for magnetic field, i.e., assuming that the internal magnetic structure continues smoothly to the deeper levels. This implies that inflows now can transport magnetic flux into the box, so that we no longer have dynamo action in a strict sense (i.e., no external sources) in such simulations, which nevertheless are probably more realistic.

Simulations with a symmetric lower boundary condition for the magnetic field have been carried out recently by Rempel (2014, hereafter referred to as R14). He tested the effect of the boundary condition by comparing two simulations with different depths of the bottom

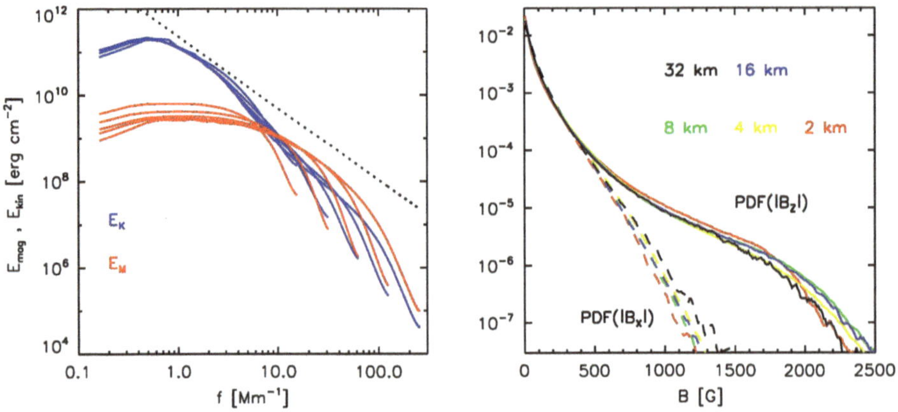

Fig. 13 Dependence on grid resolution. *Left panel*: spectra of magnetic (*red*) and kinetic (*blue*) energy. The *dotted line* indicates the Kolmogorov scaling (slope $-5/3$). *Right panel*: PDFs of the vertical field (B_z) and of one component of the horizontal field (B_x). The result for the other horizontal component is almost identical. Grid resolution varies from 32 km to 2 km (adapted from Rempel 2014, Fig. 3; ©AAS, reproduced with permission)

(2.2 Mm and 7 Mm, respectively). It turned out that the average magnetic quantities in the shallow box agreed quite well with those in the corresponding part of the deep box. This means that the symmetric boundary condition does a fairly good job in representing SSD action below the box. A simulation with a grid spacing of 4 km in a box of 6.1×6.1 Mm^2 horizontal size and 3.1 Mm depth provides $\langle|B_z|\rangle = 86$ G at the optical surface for the saturated state, an amplitude that is compatible with spectro-polarimetric observations. The magnetic structure is still of mixed polarity on small scales, with more sheet-like structures in the intergranular lanes (see Fig. 1 of R14). Figure 12 shows the time evolution of the magnetic and kinetic energy spectra and of the field strength distribution (probability density function, PDF) at the optical surface for this run, which started with a random seed field of 10^{-3} G inserted in a relaxed hydrodynamic simulation. Dashed lines indicate the early kinematic state, for which the magnetic energy peaks near the diffusive cutoff. As time progresses, the maximum moves towards larger scales and the spectrum becomes nearly flat for low wave numbers in the saturated state. This should not be misinterpreted as indicating that a large-scale component (in the sense of a mean field) develops: it just represents the organization (by flux expulsion) of the small-scale field in the patterns of convective flows, which cover all scales from granulation up to the depth of the box. In the saturated state, the Lorentz force feedback leads to a suppression of the kinetic energy on scales below 100 km. The PDFs (right panel of Fig. 12) show a growing fraction of kG fields; they correspond to intergranular flux concentrations covering about 0.5 % of the area at the optical surface, often appearing as bright points in continuum images (cf. Fig. 15). The density of such bright points in the quiet Sun can be taken as an observational measure of the amplitude of SSD action, i.e. the mean unsigned vertical field (Sánchez Almeida et al. 2010).

The dependence of the SSD results on grid resolution for the same simulation setup is shown in Fig. 13. The quasi-stationary saturated regime was reached in all cases. For increasing grid resolution, the energy spectra converge at larger scales while the small-scale part becomes more extended. As far as the large scales are concerned, the simulation has converged for a grid resolution of 8 km. The magnetic energy exceeds the kinetic energy at small scales by about a factor 2, a result also found in direct SSD simulations, i.e., simulations with physical diffusion terms (Brandenburg and Subramanian 2005). The PDFs (right

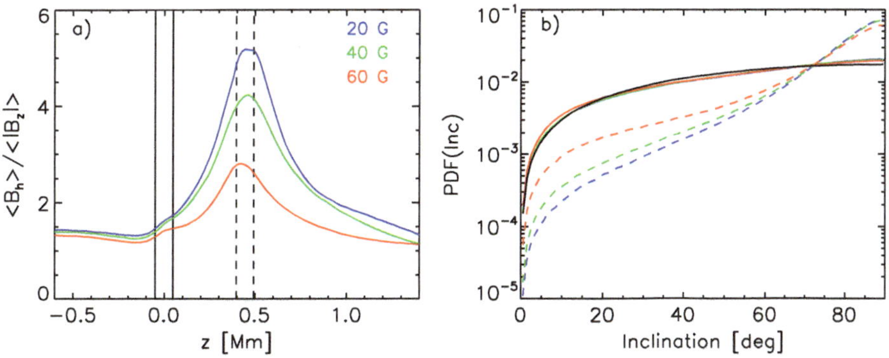

Fig. 14 Angular distribution of the field vector. Panel (**a**) gives the ratio of horizontal to vertical field strength, both horizontally averaged, as a function of height ($z = 0$ refers to the optical surface). The *line colors* correspond to simulations with different mean flux density of the vertical field at $z = 0$. (**b**) Probability distribution functions of the field inclination (with respect to the vertical direction) for two different layers indicated by *vertical lines* in panel (**a**). *Full lines* refer to a layer near the visible surface, *dashed lines* to a height around 450 km. The *black solid line* indicates a perfectly isotropic distribution (from Rempel 2014, Fig. 14; ©AAS, reproduced with permission)

panel of Fig. 13) do not change significantly with increasing grid resolution. These results indicate that, at least in the framework of the numerical approach taken, the simulation results reach an asymptotic limit on scales above about 50 km. Increasing resolution only smoothly extends the spectra to smaller scales.

In an unstratified medium without a preferred direction, the orientations of the magnetic field generated by a SSD are isotropically distributed. This is also to be expected in a stratified medium as long as scales smaller than the pressure scale height are considered. SSD simulations under solar conditions in fact exhibit isotropic field distributions in the subsurface layers, but show a strong preference for horizontal fields in the middle photosphere (Schüssler and Vögler 2008; Rempel 2014). This would be in accordance with some observational inferences (e.g., Harvey et al. 2007; Lites et al. 2008; Orozco Suárez and Katsukawa 2012, see the detailed discussion in Sect. 3.1). Figure 14 illustrates this result on the basis of the SSD simulations of Rempel (2014). The ratio of the horizontally averaged horizontal and vertical field components (left panel) changes from values consistent with an isotropic distribution at the optical surface and below to a strong dominance of horizontal fields in a layer centered at about 450 km height in the photosphere. The preference for horizontal field is less pronounced for higher mean vertical flux densities. The right panel of Fig. 14 gives probability distribution functions of the field orientation for two layers: around the optical surface (full lines) and between 450 km and 500 km (dashed lines). The distributions are nearly isotropic in the lower layer, but dominated by inclined fields in the upper layer. Two physical mechanisms have been suggested to contribute to the dominance of horizontal fields in the middle photosphere: (1) since the SSD-generated field is of mixed polarity on small scales, the magnetic loops connecting nearby opposite-polarity patches reach maximal heights comparable to their footpoint separations (Schüssler and Vögler 2008), and (2) by flux expulsion, the overturning flows of granular convection concentrate the horizontal field in the middle photosphere (Steiner et al. 2008; Steiner and Rezaei 2012).

Figure 15 shows a comparison of SSD simulations in a small, shallow box ($6.1 \times 6.1 \times 3.1\ \mathrm{Mm}^3$) and in a big, deep box ($24.6 \times 24.6 \times 7.7\ \mathrm{Mm}^3$), both at a grid resolution of 16 km. The figure shows maps of the bolometric brightness (left panel) and the vertical magnetic

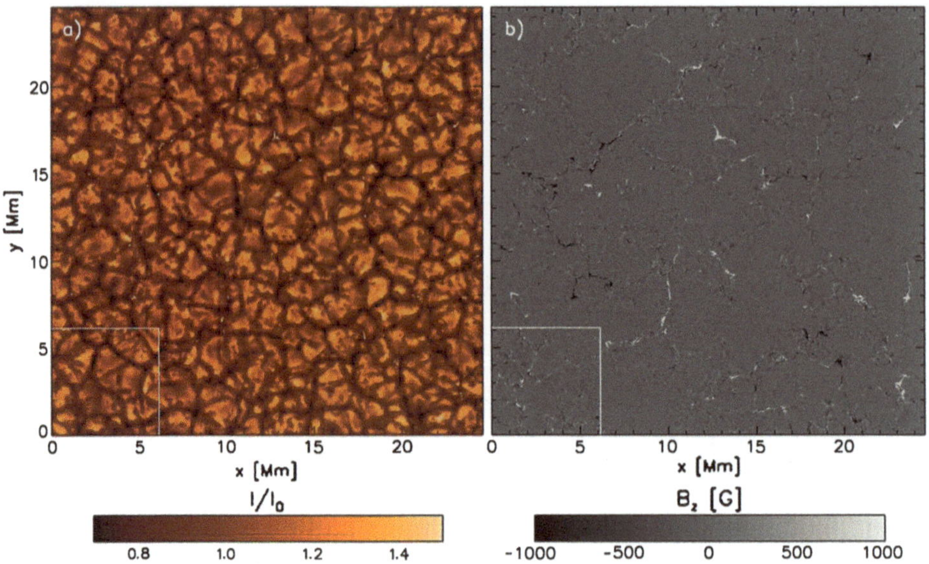

Fig. 15 Maps of bolometric intensity (*left*) and vertical magnetic field at the optical surface (*right*) for a SSD simulation in a big and 7.7 Mm deep box, compared to the result of a simulation in a smaller box of 3.1 Mm depth (inserted at the *lower left corner* of the maps). The larger-scale convective flow patterns in the big box lead to a network-like structure of the field with bigger patches of positive and negative flux appearing as bright structures (from Rempel 2014, Fig. 14; ©AAS, reproduced with permission)

field component at the optical surface (right panel). A snapshot from the simulation in the small, shallow box is inserted at the lower left corner of both maps. In the big box, a network structure of the magnetic flux on a scale of 5–10 Mm has developed, which is a result of flux advection by convective flows on the corresponding spatial scales. Larger patches of concentrated magnetic flux appear as bright structures in the intensity image. These are absent in the simulation in the small box, where the responsible longer-lived, larger-scale flows cannot develop. The energy spectra of both simulations differ only in their extension towards larger scales. The PDFs of the vertical magnetic field shows a pronounced strong-field bump in the case of the big box, while those of the horizontal field do not significantly differ. The two simulations are also consistent with each other in terms of the depth-dependence of horizontally averaged RMS field strength, indicating that the symmetric magnetic boundary condition provides a reasonable representation of the SSD field below the simulation box (Figs. 9 and 10 in Rempel 2014, not shown here).

5 Concluding Remarks and Outlook

Although the heliographic distribution of plage and network magnetic fields is rather different, as is the source of these fields, the properties of the individual magnetic features therein are relatively similar, suggesting that their properties are determined mainly by local effects (such as the efficiency of the flux expulsion and convective collapse processes) rather than global parameters. The largest advances in recent years in attaining knowledge and an understanding of the structure of and the processes acting within plage and network magnetic elements have been driven by high resolution observations and by 3D MHD simulations.

Thus the combination of high-resolution observations and MHD simulations suggests that the simple classical description of these flux concentrations in terms of thin flux tubes still holds for many purposes, although additional features are being uncovered, such as return-flux in the surroundings of the flux concentrations or the fact that the kG fields within the magnetic features are not in all cases long-lived. As the spatial resolution of observations continues to increase and simulations become ever more sophisticated, we expect to obtain further insights into these magnetic features. In particular, the role played by a small-scale dynamo in feeding the plage and network fields is open and promises to become an exciting field of research.

The lack of sufficient spatial resolution and the insufficient signal-to-noise ratio of spectropolarimetric observations still limit our capability to correctly determine the properties of internetwork magnetic fields. This has been aggravated by the use of analysis methods that are biased towards certain sorts of distributions. While better observations will certainly help, attention must also be paid to data recorded at different heliocentric positions and in spectral lines formed at different heights, as well as in lines whose polarization signals are produced by mechanisms other than the Zeeman effect. Last but not least, we must continue to make use of forward analysis employing known distributions of magnetic fields, be it by means of simple analytical distributions or by means of more complex (i.e., realistic) distributions from MHD simulations, in order to understand the various biases and uncertainties introduced by noise, analysis technique, center-to-limb effects, and so forth. Although some of these needs will be met in the near future with the arrival of new space missions and large-aperture telescopes, the remaining ones still require further and detailed investigations.

Small-scale dynamo action is a plausible candidate for the source of the observed internetwork and 'turbulent' fields. It may contribute to the flux supply of network fields and it may affect the operation of the large-scale dynamo (Cattaneo and Tobias 2014; Squire and Bhattacharjee 2015). Small-scale dynamo action is found in near-surface local-box simulations and in global spherical simulations. Although their results are consistent with a number of observed properties of the internetwork field, the simulations are carried for (effective) Reynolds numbers and magnetic Prandtl numbers that are far from the regime in which the Sun operates. Nonlinear effects, i.e., the suppression of small-scale motions by the generated magnetic field might alleviate this problem. This point needs to be clarified also by direct numerical simulations (using explicit diffusivities) with a sufficiently high resolution so that magnetic Prandtl numbers significantly below unity can be reached. As observations of the internetwork field grow more sensitive and better resolved with the new telescopes and innovative polarimetric instrumentation that are now becoming available, a much more detailed comparison between observation and simulation will be possible. This will certainly put the realism of the solar small-scale dynamo simulations to a critical test. This is the more important since the small-scale dynamo is a rather fundamental hydromagnetic process, which has potential implications for a considerable number of astrophysical systems. The combination of solar observations and simulations so far provide the only possibility to study this process in a natural environment.

Author contributions S. Jafarzadeh and S.K. Solanki: Chap. 2, J.M. Borrero: Chap. 3, M. Schüssler: Chap. 4.

Acknowledgements This work was partially supported by the BK21 plus program through the National Research Foundation (NRF) funded by the Ministry of Education of Korea.

References

V.I. Abramenko, D.W. Longcope, Distribution of the magnetic flux in elements of the magnetic field in active regions. Astrophys. J. **619**, 1160–1166 (2005)

V.I. Abramenko, V. Carbone, V. Yurchyshyn, P.R. Goode, R.F. Stein, F. Lepreti, V. Capparelli, A. Vecchio, Turbulent diffusion in the photosphere as derived from photospheric bright point motion. Astrophys. J. **743**, 133 (2011)

T. Amari, J.-F. Luciani, J.-J. Aly, Small-scale dynamo magnetism as the driver for heating the solar atmosphere. Nature **522**, 188–191 (2015). doi:10.1038/nature14478

A. Asensio Ramos, Evidence for quasi-isotropic magnetic fields from Hinode quiet-Sun observations. Astrophys. J. **701**, 1032–1043 (2009). doi:10.1088/0004-637X/701/2/1032

A. Asensio Ramos, M.J. Martínez González, Hierarchical analysis of the quiet-Sun magnetism. Astron. Astrophys. **572**, 98 (2014). doi:10.1051/0004-6361/201423860

H. Auffret, R. Muller, Center-to-limb variation of the network bright points in the solar photosphere. Astron. Astrophys. **246**, 264–279 (1991)

P. Barthol, A. Gandorfer, S.K. Solanki, M. Schüssler, B. Chares, W. Curdt, W. Deutsch, A. Feller, D. Germerott, B. Grauf, K. Heerlein, J. Hirzberger, M. Kolleck, R. Meller, R. Müller, T.L. Riethmüller, G. Tomasch, M. Knölker, B.W. Lites, G. Card, D. Elmore, J. Fox, A. Lecinski, P. Nelson, R. Summers, A. Watt, V. Martínez Pillet, J.A. Bonet, W. Schmidt, T. Berkefeld, A.M. Title, V. Domingo, J.L. Gasent Blesa, J.C. del Toro Iniesta, A. López Jiménez, A. Álvarez-Herrero, L. Sabau-Graziati, C. Widani, P. Haberler, K. Härtel, D. Kampf, T. Levin, I. Pérez Grande, A. Sanz-Andrés, E. Schmidt, The Sunrise mission. Sol. Phys. **268**, 1 (2011)

G.K. Batchelor, On the spontaneous magnetic field in a conducting liquid in turbulent motion. Proc. R. Soc. Lond., Ser. A **201**, 405–416 (1950). doi:10.1098/rspa.1950.0069

C. Beck, L.R. Bellot Rubio, R. Schlichenmaier, P. Sütterlin, Magnetic properties of G-band bright points in a sunspot moat. Astron. Astrophys. **472**, 607–622 (2007)

L.R. Bellot Rubio, D. Orozco Suárez, Pervasive linear polarization signals in the quiet Sun. Astrophys. J. **757**, 19 (2012). doi:10.1088/0004-637X/757/1/19

L.R. Bellot Rubio, B. Ruiz Cobo, M. Collados, Structure of plage flux tubes from the inversion of Stokes spectra. I. Spatially averaged Stokes I and V profiles. Astrophys. J. **535**, 489–500 (2000)

L.R. Bellot Rubio, I. Rodríguez Hidalgo, M. Collados, E. Khomenko, B. Ruiz Cobo, Observation of convective collapse and upward-moving shocks in the quiet Sun. Astrophys. J. **560**, 1010–1019 (2001)

S.V. Berdyugina, S.K. Solanki, C. Frutiger, The molecular Zeeman effect and diagnostics of solar and stellar magnetic fields. II. Synthetic Stokes profiles in the Zeeman regime. Astron. Astrophys. **412**, 513 (2003)

T.E. Berger, A.M. Title, On the relation of G-band bright points to the photospheric magnetic field. Astrophys. J. **553**, 449 (2001)

T.E. Berger, C.J. Schrijver, R.A. Shine, T.D. Tarbell, A.M. Title, G. Scharmer, New observations of subarcsecond photospheric bright points. Astrophys. J. **454**, 531 (1995)

T.E. Berger, M.G. Löfdahl, R.S. Shine, A.M. Title, Measurements of solar magnetic element motion from high-resolution filtergrams. Astrophys. J. **495**, 973–983 (1998)

T.E. Berger, L.H.M. Rouppe van der Voort, M.G. Löfdahl, M. Carlsson, A. Fossum, V.H. Hansteen, E. Marthinussen, A. Title, G. Scharmer, Solar magnetic elements at 0.1 arcsec resolution. General appearance and magnetic structure. Astron. Astrophys. **428**, 613–628 (2004)

T.E. Berger, L. Rouppe van der Voort, M. Löfdahl, Contrast analysis of solar faculae and magnetic bright points. Astrophys. J. **661**, 1272–1288 (2007)

T. Berkefeld, W. Schmidt, D. Soltau, A. Bell, H.P. Doerr, B. Feger, R. Friedlein, K. Gerber, F. Heidecke, T. Kentischer, O. von der Lühe, M. Sigwarth, E. Wälde, P. Barthol, W. Deutsch, A. Gandorfer, D. Germerott, B. Grauf, R. Meller, A. Álvarez-Herrero, M. Knölker, V. Martínez Pillet, S.K. Solanki, A.M. Title, The wave-front correction system for the Sunrise balloon-borne solar observatory. Sol. Phys. **268**, 103 (2011)

P.N. Bernasconi, C.U. Keller, H.P. Povel, J.O. Stenflo, Direct measurements of flux tube inclinations in solar plages. Astron. Astrophys. **302**, 533 (1995)

L. Bharti, R. Jain, C. Joshi, S.N.A. Jaaffrey, G-band bright points and photospheric magnetic fields, in *Astronomical Society of the Pacific Conference Series*, ed. by R. Casini, B.W. Lites. Astronomical Society of the Pacific Conference Series, vol. 358 (2006), p. 61

M. Bianda, R. Ramelli, D. Gisler, J.O. Stenflo, Solar cycle variations of the second solar spectrum, in *Solar Polarization 7*, ed. by K.N. Nagendra, J.O. Stenflo, Q. Qu, M. Samooprna. Astronomical Society of the Pacific Conference Series, vol. 489 (2014), p. 167

J. Blanco Rodríguez, F. Kneer, Faculae at the poles of the Sun revisited: Infrared observations. Astron. Astrophys. **509**, 92 (2010)

J. Blanco Rodríguez, O.V. Okunev, K.G. Puschmann, F. Kneer, B. Sánchez-Andrade Nuño, On the properties of faculae at the poles of the Sun. Astron. Astrophys. **474**, 251–259 (2007)

M. Bodnárová, D. Utz, J. Rybák, On dynamics of G-band bright points. Sol. Phys. **289**, 1543–1556 (2014)

J.M. Borrero, P. Kobel, Inferring the magnetic field vector in the quiet Sun. I. Photon noise and selection criteria. Astron. Astrophys. **527**, 29 (2011). doi:10.1051/0004-6361/201015634

J.M. Borrero, P. Kobel, Inferring the magnetic field vector in the quiet Sun. II. Interpreting results from the inversion of Stokes profiles. Astron. Astrophys. **547**, 89 (2012). doi:10.1051/0004-6361/201118238

J.M. Borrero, P. Kobel, Inferring the magnetic field vector in the quiet Sun. III. Disk variation of the Stokes profiles and isotropism of the magnetic field. Astron. Astrophys. **550**, 98 (2013). doi:10.1051/0004-6361/201118239

J.M. Borrero, V. Martínez-Pillet, R. Schlichenmaier, S.K. Solanki, J.A. Bonet, J.C. del Toro Iniesta, W. Schmidt, P. Barthol, A. Gandorfer, V. Domingo, M. Knölker, Supersonic magnetic upflows in granular cells observed with SUNRISE/IMAX. Astrophys. J. **723**, 144–148 (2010). doi:10.1088/2041-8205/723/2/L144

J.M. Borrero, V. Martínez Pillet, W. Schmidt, C. Quintero Noda, J.A. Bonet, J.C. del Toro Iniesta, L.R. Bellot Rubio, Is magnetic reconnection the cause of supersonic upflows in granular cells? Astrophys. J. **768**, 69 (2013). doi:10.1088/0004-637X/768/1/69

A. Brandenburg, Nonlinear small-scale dynamos at low magnetic Prandtl numbers. Astrophys. J. **741**, 92 (2011). doi:10.1088/0004-637X/741/2/92

A. Brandenburg, K. Subramanian, Astrophysical magnetic fields and nonlinear dynamo theory. Phys. Rep. **417**, 1–209 (2005). doi:10.1016/j.physrep.2005.06.005

A. Brandenburg, D. Sokoloff, K. Subramanian, Current status of turbulent dynamo theory. From large-scale to small-scale dynamos. Space Sci. Rev. **169**, 123–157 (2012). doi:10.1007/s11214-012-9909-x

P.N. Brandt, S.K. Solanki, Solar line asymmetries and the magnetic filling factor. Astron. Astrophys. **231**, 221–234 (1990)

J.H.M.J. Bruls, S.K. Solanki, Infrared lines as probes of solar magnetic features. IX. MgI 12 μm diagnostics of solar plage. Astron. Astrophys. **293**, 240–251 (1995)

J.H.M.J. Bruls, S.K. Solanki, Apparent solar radius variations. The influence of magnetic network and plage. Astron. Astrophys. **427**, 735–743 (2004)

E. Buchlin, Intermittent turbulent dynamo at very low and high magnetic Prandtl numbers. Astron. Astrophys. **534**, 9 (2011). doi:10.1051/0004-6361/201117890

D. Bühler, A. Lagg, S.K. Solanki, Quiet Sun magnetic fields observed by Hinode: Support for a local dynamo. Astron. Astrophys. **555**, 33 (2013). doi:10.1051/0004-6361/201321152

D. Bühler, A. Lagg, S.K. Solanki, M. van Noort, Properties of solar plage from a spatially coupled inversion of Hinode SP data. Astron. Astrophys. **576**, 27 (2015)

M. Bünte, S.K. Solanki, O. Steiner, Centre-to-limb variation of the Stokes V asymmetry in solar magnetic flux tubes. Astron. Astrophys. **268**, 736–748 (1993)

P.J. Bushby, S.M. Houghton, Spatially intermittent fields in photospheric magnetoconvection. Mon. Not. R. Astron. Soc. **362**, 313–320 (2005)

R. Cameron, A. Vögler, M. Schüssler, Decay of a simulated mixed-polarity magnetic field in the solar surface layers. Astron. Astrophys. **533**, 86 (2011)

E.S. Carlin, A. Asensio Ramos, Chromospheric diagnosis with Ca II lines: Forward modeling in forward scattering. I. Astrophys. J. **801**, 16 (2015). doi:10.1088/0004-637X/801/1/16

E.S. Carlin, R. Manso Sainz, A. Asensio Ramos, J. Trujillo Bueno, Scattering polarization in the Ca II infrared triplet with velocity gradients. Astrophys. J. **751**, 5 (2012). doi:10.1088/0004-637X/751/1/5

M. Carlsson, R.F. Stein, Å. Nordlund, G.B. Scharmer, Observational manifestations of solar magnetoconvection: Center-to-limb variation. Astrophys. J. Lett. **610**, 137–140 (2004)

F. Cattaneo, S.M. Tobias, On large-scale dynamo action at high magnetic Reynolds number. Astrophys. J. **789**, 70 (2014). doi:10.1088/0004-637X/789/1/70

F. Cavallini, G. Ceppatelli, A. Righini, Asymmetry and shift of three Fe I photospheric lines in solar active regions. Astron. Astrophys. **143**, 116 (1985)

R. Centeno, H. Socas-Navarro, B. Lites, M. Kubo, Z. Frank, R. Shine, T. Tarbell, A. Title, K. Ichimoto, S. Tsuneta, Y. Katsukawa, Y. Suematsu, T. Shimizu, S. Nagata, Emergence of small-scale magnetic loops in the quiet-Sun internetwork. Astrophys. J. **666**, 137–140 (2007). doi:10.1086/521726

P. Charbonneau, Dynamo models of the solar cycle. Living Rev. Sol. Phys. **7**, 3 (2010). doi:10.12942/lrsp-2010-3, http://www.livingreviews.org/lrsp-2010-3

M.C.M. Cheung, H. Isobe, Flux emergence (theory). Living Rev. Sol. Phys. **11**, 3 (2014)

L.P. Chitta, A.A. van Ballegooijen, L. Rouppe van der Voort, E.E. DeLuca, R. Kariyappa, Dynamics of the solar magnetic bright points derived from their horizontal motions. Astrophys. J. **752**, 48 (2012)

M. Collados, R. López, E. Páez, E. Hernández, M. Reyes, A. Calcines, E. Ballesteros, J.J. Díaz, C. Denker, A. Lagg, R. Schlichenmaier, W. Schmidt, S.K. Solanki, K.G. Strassmeier, O. von der Lühe, R. Volkmer, GRIS: The GREGOR Infrared Spectrograph. Astron. Nachr. **333**, 872 (2012). doi:10.1002/asna.201211738

S. Danilovic, A. Gandorfer, A. Lagg, M. Schüssler, S.K. Solanki, A. Vögler, Y. Katsukawa, S. Tsuneta, The intensity contrast of solar granulation: Comparing Hinode SP results with MHD simulations. Astron. Astrophys. **484**, 17 (2008)

S. Danilovic, B. Beeck, A. Pietarila, M. Schüssler, S.K. Solanki, V. Martínez Pillet, J.A. Bonet, J.C. del Toro Iniesta, V. Domingo, P. Barthol, T. Berkefeld, A. Gandorfer, M. Knölker, W. Schmidt, A.M. Title, Transverse component of the magnetic field in the solar photosphere observed by SUNRISE. Astrophys. J. Lett. **723**, 149–153 (2010a). doi:10.1088/2041-8205/723/2/L149

S. Danilovic, M. Schüssler, S.K. Solanki, Probing quiet Sun magnetism using MURaM simulations and Hinode/SP results: Support for a local dynamo. Astron. Astrophys. **513**, 1 (2010b). doi:10.1051/0004-6361/200913379

S. Danilovic, D. Röhrbein, R.H. Cameron, M. Schüssler, On the relation between continuum brightness and magnetic field in solar active regions. Astron. Astrophys. **550**, 118 (2013)

S. Danilovic, R.H. Cameron, S.K. Solanki, Simulated magnetic flows in the solar photosphere. Astron. Astrophys. **574**, 28 (2015). doi:10.1051/0004-6361/201423779

H. Dara-Papamargaritis, S. Koutchmy, Photospheric faculae-III-intensity, and magnetic field mapping of a typical element of the photospheric network. Astron. Astrophys. **125**, 280–286 (1983)

B. De Pontieu, M. Carlsson, R. Stein, L. Rouppe van der Voort, M. Löfdahl, M. van Noort, Å. Nordlund, G. Scharmer, Rapid temporal variability of faculae: High-resolution observations and modeling. Astrophys. J. **646**, 1405–1420 (2006)

A.G. de Wijn, B.W. Lites, T.E. Berger, Z.A. Frank, T.D. Tarbell, R. Ishikawa, Hinode observations of magnetic elements in internetwork areas. Astrophys. J. **684**, 1469–1476 (2008). doi:10.1086/590237

A.G. de Wijn, J.O. Stenflo, S.K. Solanki, S. Tsuneta, Small-scale solar magnetic fields. Space Sci. Rev. **144**, 275 (2009)

R.J. Defouw, Wave propagation along a magnetic tube. Astrophys. J. **209**, 266–269 (1976)

D. Degenhardt, S.K. Solanki, B. Montesinos, J.H. Thomas, Evidence for siphon flows with shocks in solar magnetic flux tubes. Astron. Astrophys. **279**, 29–32 (1993)

W. Deinzer, G. Hensler, M. Schüssler, E. Weisshaar, Model calculations of magnetic flux tubes. II. Stationary results for solar magnetic elements. Astron. Astrophys. **139**, 435 (1984)

H.W. Dodson, High latitude sunspot, August 13, 1953. Publ. Astron. Soc. Pac. **65**, 256 (1953)

V. Domingo, A. Ortiz, B. Sanahuja, I. Cabello, Centre-to-limb variation of photospheric facular radiance and image resolution. Adv. Space Res. **35**, 345–349 (2005)

V. Domingo, I. Ermolli, P. Fox, C. Fröhlich, M. Haberreiter, N. Krivova, G. Kopp, W. Schmutz, S.K. Solanki, H.C. Spruit, Y. Unruh, A. Vögler, Solar surface magnetism and irradiance on time scales from days to the 11-year cycle. Space Sci. Rev. **145**, 337–380 (2009)

I. Ermolli, K. Matthes, T. Dudok de Wit, N.A. Krivova, K. Tourpali, M. Weber, Y.C. Unruh, L. Gray, U. Langematz, P. Pilewskie, E. Rozanov, W. Schmutz, A. Shapiro, S.K. Solanki, T.N. Woods, Recent variability of the solar spectral irradiance and its impact on climate modelling. Atmos. Chem. Phys. **13**, 3945–3977 (2013)

S. Feng, L. Deng, Y. Yang, K. Ji, Statistical study of photospheric bright points in an active region and quiet Sun. Astrophys. Space Sci. **348**, 17–24 (2013)

A. Ferriz-Mas, M. Schüssler, Waves and instabilities of a toroidal magnetic flux tube in a rotating star. Astrophys. J. **433**, 852–866 (1994)

A. Ferriz-Mas, M. Schüssler, V. Anton, Dynamics of magnetic flux concentrations—the second-order thin flux tube approximation. Astron. Astrophys. **210**, 425–432 (1989)

C.E. Fischer, A.G. de Wijn, R. Centeno, B.W. Lites, C.U. Keller, Statistics of convective collapse events in the photosphere and chromosphere observed with the Hinode SOT. Astron. Astrophys. **504**, 583 (2009)

P. Foukal, P. Bernasconi, H. Eaton, D. Rust, Broadband measurements of facular photometric contrast using the solar bolometric imager. Astrophys. J. Lett. **611**, 57–60 (2004)

C. Frutiger, S.K. Solanki, Empirical models of solar magnetic flux-tubes and their non-magnetic surroundings. Astron. Astrophys. **369**, 646–659 (2001)

A. Gandorfer, B. Grauf, P. Barthol, T.L. Riethmüller, S.K. Solanki, B. Chares, W. Deutsch, S. Ebert, A. Feller, D. Germerott, K. Heerlein, J. Heinrichs, D. Hirche, J. Hirzberger, M. Kolleck, R. Meller, R. Müller, R. Schäfer, G. Tomasch, M. Knölker, V. Martínez Pillet, J.A. Bonet, W. Schmidt, T. Berkefeld, B. Feger, F. Heidecke, D. Soltau, A. Tischenberg, A. Fischer, A. Title, H. Anwand, E. Schmidt, The filter imager SuFI and the Image Stabilization and Light Distribution system ISLiD of the Sunrise balloon-borne observatory: Instrument description. Sol. Phys. **268**, 35 (2011)

M. Gošić, L.R. Bellot Rubio, D. Orozco Suárez, Y. Katsukawa, J.C. del Toro Iniesta, The solar internetwork. I. Contribution to the network magnetic flux. Astrophys. J. **797**, 49 (2014)
D.O. Gough, R.J. Tayler, The influence of a magnetic field on Schwarzschild's criterion for convective instability in an ideally conducting fluid. Mon. Not. R. Astron. Soc. **133**, 85 (1966)
L.J. Gray, J. Beer, M. Geller, J.D. Haigh, M. Lockwood, K. Matthes, U. Cubasch, D. Fleitmann, G. Harrison, L. Hood, J. Luterbacher, G.A. Meehl, D. Shindell, B. van Geel, W. White, Solar influences on climate. Rev. Geophys. **48**, 4001 (2010)
U. Grossmann-Doerth, M. Schüssler, S.K. Solanki, Unshifted, asymmetric Stokes V-profiles—possible solution of a riddle. Astron. Astrophys. **206**, 37–39 (1988)
U. Grossmann-Doerth, M. Knoelker, M. Schüssler, S.K. Solanki, The deep layers of solar magnetic elements. Astron. Astrophys. **285**, 648–654 (1994)
U. Grossmann-Doerth, C.U. Keller, M. Schüssler, Observations of the quiet Sun's magnetic field. Astron. Astrophys. **315**, 610–617 (1996)
U. Grossmann-Doerth, M. Schüssler, O. Steiner, Convective intensification of solar surface magnetic fields: Results of numerical experiments. Astron. Astrophys. **337**, 928–939 (1998)
H.J. Hagenaar, Ephemeral regions on a sequence of full-disk Michelson Doppler imager magnetograms. Astrophys. J. **555**, 448–461 (2001)
H.J. Hagenaar, C.J. Schrijver, A.M. Title, The properties of small magnetic regions on the solar surface and the implications for the solar dynamo(s). Astrophys. J. **584**, 1107–1119 (2003)
G.E. Hale, S.B. Nicholson, The Law of Sun-spot polarity. Astrophys. J. **62**, 270 (1925)
K.L. Harvey, Magnetic bipoles on the Sun, PhD thesis, Univ. Utrecht (1993)
K. Harvey, J. Harvey, Observations of moving magnetic features near sunspots. Sol. Phys. **28**, 61–71 (1973)
K.L. Harvey, S.F. Martin, Ephemeral active regions. Sol. Phys. **32**, 389–402 (1973)
K.L. Harvey, C. Zwaan, Properties and emergence of bipolar active regions. Sol. Phys. **148**, 85–118 (1993)
J.W. Harvey, D. Branston, C.J. Henney, C.U. Keller (SOLIS and GONG Teams), Seething horizontal magnetic fields in the quiet solar photosphere. Astrophys. J. Lett. **659**, 177–180 (2007). doi:10.1086/518036
S.S. Hasan, Convective instability in a solar flux tube. II—Nonlinear calculations with horizontal radiative heat transport and finite viscosity. Astron. Astrophys. **143**, 39–45 (1985)
R.L. Hewitt, S. Shelyag, M. Mathioudakis, F.P. Keenan, Plasma properties and Stokes profiles during the lifetime of a photospheric magnetic bright point. Astron. Astrophys. **565**, 84 (2014)
J. Hirzberger, E. Wiehr, Solar limb faculae. Astron. Astrophys. **438**, 1059–1065 (2005)
J. Hirzberger, A. Feller, T.L. Riethmüller, M. Schüssler, J.M. Borrero, N. Afram, Y.C. Unruh, S.V. Berdyugina, A. Gandorfer, S.K. Solanki, P. Barthol, J.A. Bonet, V. Martínez Pillet, T. Berkefeld, M. Knölker, W. Schmidt, A.M. Title, Quiet-Sun intensity contrasts in the near-ultraviolet as measured from Sunrise. Astrophys. J. Lett. **723**, 154 (2010)
R. Holzreuter, S.K. Solanki, Three-dimensional non-LTE radiative transfer effects in Fe I lines. I. Flux sheet and flux tube geometries. Astron. Astrophys. **547**, 46 (2012)
H. Hotta, M. Rempel, T. Yokoyama, Efficient small-scale dynamo in the solar convection zone. Astrophys. J. **803**, 42 (2015). doi:10.1088/0004-637X/803/1/42
Y. Iida, H.J. Hagenaar, T. Yokoyama, Detection of flux emergence, splitting, merging, and cancellation of network field. I. Splitting and merging. Astrophys. J. **752**, 149 (2012)
R. Ishikawa, S. Tsuneta, Comparison of transient horizontal magnetic fields in a plage region and in the quiet Sun. Astron. Astrophys. **495**, 607–612 (2009). doi:10.1051/0004-6361:200810636
R. Ishikawa, S. Tsuneta, The relationship between vertical and horizontal magnetic fields in the quiet Sun. Astrophys. J. **735**, 74 (2011). doi:10.1088/0004-637X/735/2/74
R. Ishikawa, S. Tsuneta, Y. Kitakoshi, Y. Katsukawa, J.A. Bonet, S. Vargas Domínguez, L.H.M. Rouppe van der Voort, Y. Sakamoto, T. Ebisuzaki, Relationships between magnetic foot points and G-band bright structures. Astron. Astrophys. **472**, 911–918 (2007)
R. Ishikawa, S. Tsuneta, K. Ichimoto, H. Isobe, Y. Katsukawa, B.W. Lites, S. Nagata, T. Shimizu, R.A. Shine, Y. Suematsu, T.D. Tarbell, A.M. Title, Transient horizontal magnetic fields in solar plage regions. Astron. Astrophys. **481**, 25–28 (2008). doi:10.1051/0004-6361:20079022
R. Ishikawa, S. Tsuneta, J. Jurčák, Three-dimensional view of transient horizontal magnetic fields in the photosphere. Astrophys. J. **713**, 1310–1321 (2010). doi:10.1088/0004-637X/713/2/1310
S. Jafarzadeh, S.K. Solanki, A. Lagg, L.R. Bellot Rubio, M. van Noort, A. Feller, S. Danilovic, Inclinations of small quiet-Sun magnetic features based on a new geometric approach. Astron. Astrophys. **569**, 105 (2014a). doi:10.1051/0004-6361/201423414
S. Jafarzadeh, R.H. Cameron, S.K. Solanki, A. Pietarila, A. Feller, A. Lagg, A. Gandorfer, Migration of Ca II H bright points in the internetwork. Astron. Astrophys. **563**, 101 (2014b)
C. Jin, J. Wang, Does the variation of solar inter-network horizontal field follow sunspot cycle? ArXiv e-prints (2015a)

C. Jin, J. Wang, Solar cycle variation of the inter-network magnetic field. Astrophys. J. **806**, 174 (2015b). doi:10.1088/0004-637X/806/2/174

C.L. Jin, J.X. Wang, Q. Song, H. Zhao, The Sun's small-scale magnetic elements in Solar Cycle 23. Astrophys. J. **731**, 37 (2011)

A.J. Kaithakkal, Y. Suematsu, M. Kubo, D. Shiota, S. Tsuneta, The association of polar faculae with polar magnetic patches examined with Hinode observations. Astrophys. J. **776**, 122 (2013)

A.P. Kazantsev, Enhancement of a magnetic field by a conducting fluid. Sov. Phys. JETP **26**, 1031 (1968)

C.U. Keller, J.O. Stenflo, S.K. Solanki, T.D. Tarbell, A.M. Title, Solar magnetic field strength determinations from high spatial resolution filtergrams. Astron. Astrophys. **236**, 250–255 (1990)

C.U. Keller, M. Schüssler, A. Vögler, V. Zakharov, On the origin of solar faculae. Astrophys. J. Lett. **607**, 59–62 (2004)

E.V. Khomenko, M. Collados, S.K. Solanki, A. Lagg, J. Trujillo Bueno, Quiet-Sun inter-network magnetic fields observed in the infrared. Astron. Astrophys. **408**, 1115–1135 (2003)

L. Kleint, S.V. Berdyugina, A.I. Shapiro, M. Bianda, Solar turbulent magnetic fields: Surprisingly homogeneous distribution during the solar minimum. Astron. Astrophys. **524**, 37 (2010). doi:10.1051/0004-6361/201015285

M. Knoelker, M. Schüssler, Model calculations of magnetic flux tubes. IV—Convective energy transport and the nature of intermediate size flux concentrations. Astron. Astrophys. **202**, 275–283 (1988)

P. Kobel, J. Hirzberger, S.K. Solanki, A. Gandorfer, V. Zakharov, Discriminant analysis of solar bright points and faculae. I. Classification method and center-to-limb distribution. Astron. Astrophys. **502**, 303–314 (2009)

P. Kobel, S.K. Solanki, J.M. Borrero, The continuum contrast of magnetic elements as a function of magnetic field (disk center): Early studies and Hinode/sp results, in *Solar Polarization 6*, ed. by J.R. Kuhn, D.M. Harrington, H. Lin, S.V. Berdyugina, J. Trujillo-Bueno, S.L. Keil, T. Rimmele. Astronomical Society of the Pacific Conference Series, vol. 437 (2011), p. 297

P. Kobel, J. Hirzberger, S.K. Solanki, Discriminant analysis of solar bright points and faculae II. Contrast and morphology analysis. ArXiv e-prints (2014)

R. Kostik, E.V. Khomenko, Properties of convective motions in facular regions. Astron. Astrophys. **545**, 22 (2012)

Ø. Langangen, M. Carlsson, L. Rouppe van der Voort, R.F. Stein, Velocities measured in small-scale solar magnetic elements. Astrophys. J. **655**, 615–623 (2007)

K. Langhans, W. Schmidt, T. Rimmele, Diagnostic spectroscopy of G-band brightenings in the photosphere of the sun. Astron. Astrophys. **423**, 1147–1157 (2004)

J.K. Lawrence, K.P. Topka, H.P. Jones, Contrast of faculae near the disk center and solar variability. J. Geophys. Res. **98**, 18911 (1993)

R.B. Leighton, Observations of solar magnetic fields in plage regions. Astrophys. J. **130**, 366 (1959)

B.W. Lites, A. Skumanich, V. Martinez Pillet, Vector magnetic fields of emerging solar flux. I. Properties at the site of emergence. Astron. Astrophys. **333**, 1053–1068 (1998)

B.W. Lites, G.B. Scharmer, T.E. Berger, A.M. Title, Three-dimensional structure of the active region photosphere as revealed by high angular resolution. Sol. Phys. **221**, 65–84 (2004)

B.W. Lites, M. Kubo, H. Socas-Navarro, T. Berger, Z. Frank, R. Shine, T. Tarbell, A. Title, K. Ichimoto, Y. Katsukawa, S. Tsuneta, Y. Suematsu, T. Shimizu, S. Nagata, The horizontal magnetic flux of the quiet-sun internetwork as observed with the Hinode spectro-polarimeter. Astrophys. J. **672**, 1237–1253 (2008). doi:10.1086/522922

B.W. Lites, R. Centeno, S.W. McIntosh, The solar cycle dependence of the weak internetwork flux. Publ. Astron. Soc. Jpn. **66**, 4 (2014). doi:10.1093/pasj/psu082

W.C. Livingston, Magnetic fields, convection and solar luminosity variability. Nature **297**, 208 (1982)

R. Manso Sainz, M.J. Martínez González, A. Asensio Ramos, Advection and dispersal of small magnetic elements in the very quiet Sun. Astron. Astrophys. **531**, 9 (2011). doi:10.1051/0004-6361/201117042

M.J. Martínez González, L.R. Bellot Rubio, Emergence of small-scale magnetic loops through the quiet solar atmosphere. Astrophys. J. **700**, 1391–1403 (2009). doi:10.1088/0004-637X/700/2/1391

M.J. Martínez González, A. Asensio Ramos, A. López Ariste, R. Manso Sainz, Near-IR internetwork spectropolarimetry at different heliocentric angles. Astron. Astrophys. **479**, 229–234 (2008). doi:10.1051/0004-6361:20078500

M.J. Martínez González, R. Manso Sainz, A. Asensio Ramos, L.R. Bellot Rubio, Small magnetic loops connecting the quiet surface and the hot outer atmosphere of the Sun. Astrophys. J. **714**, 94–97 (2010). doi:10.1088/2041-8205/714/1/L94

M.J. Martínez González, A. Asensio Ramos, R. Manso Sainz, E. Khomenko, V. Martínez Pillet, S.K. Solanki, A. López Ariste, W. Schmidt, P. Barthol, A. Gandorfer, Unnoticed magnetic field oscillations in the very quiet Sun revealed by SUNRISE/IMaX. Astrophys. J. Lett. **730**, 37 (2011a)

M.J. Martínez González, A. Asensio Ramos, R. Manso Sainz, E. Khomenko, V. Martínez Pillet, S.K. Solanki, A. López Ariste, W. Schmidt, P. Barthol, A. Gandorfer, Unnoticed magnetic field oscillations in the very quiet Sun revealed by SUNRISE/IMaX. Astrophys. J. **730**, 37 (2011b). doi:10.1088/2041-8205/730/2/L37

M.J. Martínez González, R. Manso Sainz, A. Asensio Ramos, E. Hijano, Dead calm areas in the very quiet Sun. Astrophys. J. **755**, 175 (2012a). doi:10.1088/0004-637X/755/2/175

M.J. Martínez González, L.R. Bellot Rubio, S.K. Solanki, V. Martínez Pillet, J.C. Del Toro Iniesta, P. Barthol, W. Schmidt, Resolving the internal magnetic structure of the solar network. Astrophys. J. Lett. **758**, 40 (2012b)

V. Martínez Pillet, Solar surface and atmospheric dynamics. The photosphere. Space Sci. Rev. **178**, 141–162 (2013). doi:10.1007/s11214-013-9967-8

V. Martínez Pillet, B.W. Lites, A. Skumanich, Active region magnetic fields. I. Plage fields. Astrophys. J. **474**, 810–842 (1997)

V. Martínez Pillet, J.C. Del Toro Iniesta, A. Álvarez-Herrero, V. Domingo, J.A. Bonet, L. González Fernández, A. López Jiménez, C. Pastor, J.L. Gasent Blesa, P. Mellado, J. Piqueras, B. Aparicio, M. Balaguer, E. Ballesteros, T. Belenguer, L.R. Bellot Rubio, T. Berkefeld, M. Collados, W. Deutsch, A. Feller, F. Girela, B. Grauf, R.L. Heredero, M. Herranz, J.M. Jerónimo, H. Laguna, R. Meller, M. Menéndez, R. Morales, D. Orozco Suárez, G. Ramos, M. Reina, J.L. Ramos, P. Rodríguez, A. Sánchez, N. Uribe-Patarroyo, P. Barthol, A. Gandorfer, M. Knölker, W. Schmidt, S.K. Solanki, S. Vargas Domínguez, The Imaging Magnetograph eXperiment (IMaX) for the Sunrise balloon-borne solar observatory. Sol. Phys. **268**, 57 (2011)

M. Meneguzzi, U. Frisch, A. Pouquet, Helical and nonhelical turbulent dynamos. Phys. Rev. Lett. **47**, 1060–1064 (1981). doi:10.1103/PhysRevLett.47.1060

N. Meunier, J. Zhao, Observations of photospheric dynamics and magnetic fields: From large-scale to small-scale flows. Space Sci. Rev. **144**, 127–149 (2009)

F. Meyer, H.U. Schmidt, Magnetically aligned flows in the solar atmosphere. Astron. J. Suppl. **73**, 71 (1968a)

F. Meyer, H.U. Schmidt, Magnetisch ausgerichtete Strömungen zwischen Sonnenflecken. Z. Angew. Math. Mech. **48**, 218–221 (1968b)

M.S. Miesch, W.H. Matthaeus, A. Brandenburg, A. Petrosyan, A. Pouquet, C. Cambon, F. Jenko, D. Uzdensky, J. Stone, S. Tobias, J. Toomre, M. Velli, Large-eddy simulations of magnetohydrodynamic turbulence in astrophysics and space physics. ArXiv e-prints (2015)

T. Moran, D. Deming, D.E. Jennings, G. McCabe, Solar magnetic field studies using the 12 micron emission lines. III. Simultaneous measurements at 12 and 1.6 microns. Astrophys. J. **533**, 1035–1042 (2000)

R. Muller, B. Mena, Motions around a decaying sunspot. Sol. Phys. **112**, 295–303 (1987)

R. Muller, T. Roudier, Variability of the quiet photospheric network. Sol. Phys. **94**, 33–47 (1984)

R. Muller, T. Roudier, Latitude and cycle variations of the photospheric network. Sol. Phys. **152**, 131–137 (1994)

S. Nagata, S. Tsuneta, Y. Suematsu, K. Ichimoto, Y. Katsukawa, T. Shimizu, T. Yokoyama, T.D. Tarbell, B.W. Lites, R.A. Shine, T.E. Berger, A.M. Title, L.R. Bellot Rubio, D. Orozco Suárez, Formation of solar magnetic flux tubes with kilogauss field strength induced by convective instability. Astrophys. J. Lett. **677**, 145–147 (2008)

G. Narayan, Transient downflows associated with the intensification of small-scale magnetic features and bright point formation. Astron. Astrophys. **529**, 79 (2011)

G. Narayan, G.B. Scharmer, Small-scale convection signatures associated with a strong plage solar magnetic field. Astron. Astrophys. **524**, 3 (2010)

P. Nisenson, A.A. van Ballegooijen, A.G. de Wijn, P. Sütterlin, Motions of isolated G-band bright points in the solar photosphere. Astrophys. J. **587**, 458–463 (2003)

Å. Nordlund, R.F. Stein, M. Asplund, Solar surface convection. Living Rev. Sol. Phys. **6**, 3 (2009). doi:10.12942/lrsp-2009-2, http://www.livingreviews.org/lrsp-2009-2

D. Orozco Suárez, Y. Katsukawa, On the distribution of quiet-sun magnetic fields at different heliocentric angles. Astrophys. J. **746**, 182 (2012). doi:10.1088/0004-637X/746/2/182

D. Orozco Suárez, L.R. Bellot Rubio, J.C. del Toro Iniesta, S. Tsuneta, B.W. Lites, K. Ichimoto, Y. Katsukawa, S. Nagata, T. Shimizu, R.A. Shine, Y. Suematsu, T.D. Tarbell, A.M. Title, Quiet-Sun internetwork magnetic fields from the inversion of Hinode measurements. Astrophys. J. **670**, 61–64 (2007). doi:10.1086/524139

D. Orozco Suárez, L.R. Bellot Rubio, J.C. del Toro Iniesta, S. Tsuneta, Magnetic field emergence in quiet Sun granules. Astron. Astrophys. **481**, 33 (2008)

D. Orozco Suárez, Y. Katsukawa, L.R. Bellot Rubio, The connection between internetwork magnetic elements and supergranular flows. Astrophys. J. **758**, 38 (2012). doi:10.1088/2041-8205/758/2/L38

A. Ortiz, S.K. Solanki, V. Domingo, M. Fligge, B. Sanahuja, On the intensity contrast of solar photospheric faculae and network elements. Astron. Astrophys. **388**, 1036–1047 (2002)

A. Ortiz, V. Domingo, B. Sanahuja, The intensity contrast of solar photospheric faculae and network elements. II. Evolution over the rising phase of solar cycle 23. Astron. Astrophys. **452**, 311–319 (2006)
E.N. Parker, The formation of sunspots from the solar toroidal field. Astrophys. J. **121**, 491 (1955)
E.N. Parker, Topological dissipation and the small-scale fields in turbulent gases. Astrophys. J. **174**, 499 (1972). doi:10.1086/151512
E.N. Parker, The reconnection rate of magnetic fields. Astrophys. J. **180**, 247–252 (1973). doi:10.1086/151959
E.N. Parker, Hydraulic concentration of magnetic fields in the solar photosphere. VI—Adiabatic cooling and concentration in downdrafts. Astrophys. J. **221**, 368–377 (1978)
C.E. Parnell, C.E. DeForest, H.J. Hagenaar, B.A. Johnston, D.A. Lamb, B.T. Welsch, A power-law distribution of solar magnetic fields over more than five decades in flux. Astrophys. J. **698**, 75–82 (2009)
V. Penza, B. Caccin, I. Ermolli, M. Centrone, Comparison of model calculations and photometric observations of bright "magnetic" regions. Astron. Astrophys. **413**, 1115–1123 (2004)
K. Petrovay, G. Szakaly, The origin of intranetwork fields: A small-scale solar dynamo. Astron. Astrophys. **274**, 543 (1993)
K. Petrovay, V. Martínez Pillet, L. van Driel-Gesztelyi, Making sense of sunspot decay—II. Deviations from the mean law and plage effects. Sol. Phys. **188**, 315–330 (1999)
J. Pietarila Graham, S. Danilovic, M. Schüssler, Turbulent magnetic fields in the quiet Sun: Implications of Hinode observations and small-scale dynamo simulations. Astrophys. J. **693**, 1728–1735 (2009). doi:10.1088/0004-637X/693/2/1728
J. Pietarila Graham, R. Cameron, M. Schüssler, Turbulent small-scale dynamo action in solar surface simulations. Astrophys. J. **714**, 1606–1616 (2010). doi:10.1088/0004-637X/714/2/1606
A. Pietarila, R. Cameron, S.K. Solanki, Expansion of magnetic flux concentrations: A comparison of Hinode SOT data and models. Astron. Astrophys. **518**, 50 (2010)
A. Pietarila, R.H. Cameron, S. Danilovic, S.K. Solanki, Transport of magnetic flux from the canopy to the internetwork. Astrophys. J. **729**, 136 (2011). doi:10.1088/0004-637X/729/2/136
G.W. Pneuman, S.K. Solanki, J.O. Stenflo, Structure and merging of solar magnetic fluxtubes. Astron. Astrophys. **154**, 231–242 (1986)
C. Quintero Noda, V. Martínez Pillet, J.M. Borrero, S.K. Solanki, Temporal relation between quiet-Sun transverse fields and the strong flows detected by IMaX/SUNRISE. Astron. Astrophys. **558**, 30 (2013). doi: 10.1051/0004-6361/201321719
C. Quintero Noda, J.M. Borrero, D. Orozco Suárez, B. Ruiz Cobo, High speed magnetized flows in the quiet Sun. Astron. Astrophys. **569**, 73 (2014). doi:10.1051/0004-6361/201424131
C. Quintero Noda, B. Ruiz Cobo, D. Orozco Suárez, Photospheric downward plasma motions in the quiet Sun. Astron. Astrophys. **566**, 139 (2014). doi:10.1051/0004-6361/201423461
D. Rabin, A true-field magnetogram in a solar plage region. Astrophys. J. Lett. **390**, 103–106 (1992a)
D. Rabin, Spatially extended measurements of magnetic field strength in solar plages. Astrophys. J. **391**, 832–844 (1992b)
M. Rempel, Numerical simulations of quiet Sun magnetism: On the contribution from a small-scale dynamo. Astrophys. J. **789**, 132 (2014). doi:10.1088/0004-637X/789/2/132
M. Rempel, R. Schlichenmaier, Sunspot modeling: From simplified models to radiative MHD simulations. Living Rev. Sol. Phys. **8**, 3 (2011). doi:10.12942/lrsp-2011-3, http://www.livingreviews.org/lrsp-2011-3
M. Rempel, M. Schüssler, M. Knölker, Radiative magnetohydrodynamic simulation of sunspot structure. Astrophys. J. **691**, 640–649 (2009). doi:10.1088/0004-637X/691/1/640
I.S. Requerey, J.C. Del Toro Iniesta, L.R. Bellot Rubio, J.A. Bonet, V. Martínez Pillet, S.K. Solanki, W. Schmidt, The history of a quiet-sun magnetic element revealed by IMaX/SUNRISE. Astrophys. J. **789**, 6 (2014)
R. Rezaei, O. Steiner, S. Wedemeyer-Böhm, R. Schlichenmaier, W. Schmidt, B.W. Lites, Hinode observations reveal boundary layers of magnetic elements in the solar photosphere. Astron. Astrophys. **476**, 33–36 (2007)
T.L. Riethmüller, S.K. Solanki, V. Martínez Pillet, J. Hirzberger, A. Feller, J.A. Bonet, N. Bello González, M. Franz, M. Schüssler, P. Barthol, T. Berkefeld, J.C. del Toro Iniesta, V. Domingo, A. Gandorfer, M. Knölker, W. Schmidt, Bright points in the quiet Sun as observed in the visible and near-UV by the balloon-borne observatory SUNRISE. Astrophys. J. Lett. **723**, 169–174 (2010)
T.L. Riethmüller, S.K. Solanki, S.V. Berdyugina, M. Schüssler, V. Martínez Pillet, A. Feller, A. Gandorfer, J. Hirzberger, Comparison of solar photospheric bright points between Sunrise observations and MHD simulations. Astron. Astrophys. **568**, 13 (2014)
D. Röhrbein, R. Cameron, M. Schüssler, Is there a non-monotonic relation between photospheric brightness and magnetic field strength in solar plage regions? Astron. Astrophys. **532**, 140 (2011)
P. Romano, F. Berrilli, S. Criscuoli, D. Del Moro, I. Ermolli, F. Giorgi, B. Viticchié, F. Zuccarello, A comparative analysis of photospheric bright points in an active region and in the quiet Sun. Sol. Phys. **280**, 407–416 (2012)

L.H.M. Rouppe van der Voort, V.H. Hansteen, M. Carlsson, A. Fossum, E. Marthinussen, M.J. van Noort, T.E. Berger, Solar magnetic elements at 0.1 arcsec resolution. II. Dynamical evolution. Astron. Astrophys. **435**, 327–337 (2005)

F. Rubio da Costa, S.K. Solanki, S. Danilovic, J. Hizberger, V. Martínez-Pillet, Centre-to-limb properties of small, photospheric quiet-Sun jets. Astron. Astrophys. **574**, 95 (2015). doi:10.1051/0004-6361/201424880

I. Rüedi, S.K. Solanki, D. Rabin, Infrared lines as probes of solar magnetic features. IV—Discovery of a siphon flow. Astron. Astrophys. **261**, 21–24 (1992a)

I. Rüedi, S.K. Solanki, W. Livingston, J.O. Stenflo, Infrared lines as probes of solar magnetic features. III—Strong and weak magnetic fields in plages. Astron. Astrophys. **263**, 323–338 (1992b)

R.J. Rutten, D. Kiselman, L. Rouppe van der Voort, B. Plez, Proxy magnetometry of the photosphere: Why are G-band bright points so bright?, in *Advanced Solar Polarimetry—Theory, Observation, and Instrumentation*, ed. by M. Sigwarth. Astronomical Society of the Pacific Conference Series, vol. 236 (2001), p. 445

A. Sainz Dalda, J. Martínez-Sykora, L. Bellot Rubio, A. Title, Study of single-lobed circular polarization profiles in the quiet Sun. Astrophys. J. **748**, 38 (2012). doi:10.1088/0004-637X/748/1/38

J. Sánchez Almeida, Inter-Network magnetic fields observed during the minimum of the solar cycle. Astron. Astrophys. **411**, 615–621 (2003). doi:10.1051/0004-6361:20031560

J. Sánchez Almeida, B.W. Lites, Physical properties of the solar magnetic photosphere under the MISMA hypothesis. II. Network and internetwork fields at the disk center. Astrophys. J. **532**, 1215–1229 (2000)

J. Sanchez Almeida, V. Martinez Pillet, The inclination of network magnetic fields. Astrophys. J. **424**, 1014 (1994)

J. Sánchez Almeida, A. Asensio Ramos, J. Trujillo Bueno, J. Cernicharo, G-band spectral synthesis in solar magnetic concentrations. Astrophys. J. **555**, 978–989 (2001)

J. Sánchez Almeida, J.A. Bonet, B. Viticchié, D. Del Moro, Magnetic bright points in the quiet Sun. Astrophys. J. Lett. **715**, 26–29 (2010). doi:10.1088/2041-8205/715/1/L26

G.B. Scharmer, K. Bjelksjo, T.K. Korhonen, B. Lindberg, B. Petterson, The 1-meter Swedish Solar Telescope, in *Society of Photo-Optical Instrumentation Engineers (SPIE) Conference Series*, ed. by S.L. Keil, S.V. Avakyan. SPIE Conf. Ser., vol. 4853 (2003), p. 341

G.B. Scharmer, J. de la Cruz Rodriguez, P. Sütterlin, V.M.J. Henriques, Opposite polarity field with convective downflow and its relation to magnetic spines in a sunspot penumbra. Astron. Astrophys. **553**, 63 (2013)

R.S. Schnerr, H.C. Spruit, The brightness of magnetic field concentrations in the quiet Sun. Astron. Astrophys. **532**, 136 (2011)

C.J. Schrijver, C. Zwaan, *Solar and Stellar Magnetic Activity* (Cambridge University Press, New York, 2000)

C.J. Schrijver, A.M. Title, A.A. van Ballegooijen, H.J. Hagenaar, R.A. Shine, Sustaining the quiet photospheric network: The balance of flux emergence, fragmentation, merging, and cancellation. Astrophys. J. **487**, 424–436 (1997)

C.J. Schrijver, A.M. Title, K.L. Harvey, N.R. Sheeley, Y.-M. Wang, G.H.J. van den Oord, R.A. Shine, T.D. Tarbell, N.E. Hurlburt, Large-scale coronal heating by the small-scale magnetic field of the Sun. Nature **394**, 152–154 (1998)

M. Schüssler, MHD models of solar photospheric magnetic flux concentrations, in *Small Scale Magnetic Flux Concentrations in the Solar Photosphere*, ed. by W. Deinzer, M. Knölker, H.H. Voigt (1986), p. 103

M. Schüssler, Theoretical aspects of small-scale photospheric magnetic fields, in *Solar Photosphere: Structure, Convection and Magnetic Fields*, ed. by J.O. Stenflo. IAU Symposium, vol. 138 (Kluwer/Springer, Dordrecht, 1990), p. 161

M. Schüssler, Solar Magneto-convection, in *Solar and astrophysical dynamos and magnetic activity*, ed. by A.G. Kosovichev, E. de Gouveia Dal Pino, Y. Yan. IAU Symposium, vol. 294 (Cambridge University Press, Cambridge, 2013), pp. 95–106. doi:10.1017/S1743921313002329

M. Schüssler, S.K. Solanki, Continuum intensity of magnetic flux concentrations—are magnetic elements bright points? Astron. Astrophys. **192**, 338–342 (1988)

M. Schüssler, A. Vögler, Strong horizontal photospheric magnetic field in a surface dynamo simulation. Astron. Astrophys. **481**, 5–8 (2008). doi:10.1051/0004-6361:20078998

M. Schüssler, S. Shelyag, S. Berdyugina, A. Vögler, S.K. Solanki, Why solar magnetic flux concentrations are bright in molecular bands. Astrophys. J. Lett. **597**, 173–176 (2003)

N. Shchukina, J. Trujillo Bueno, Determining the magnetization of the quiet Sun photosphere from the Hanle effect and surface dynamo simulations. Astrophys. J. Lett. **731**, 21 (2011). doi:10.1088/2041-8205/731/1/L21

S. Shelyag, M. Schüssler, S.K. Solanki, S.V. Berdyugina, A. Vögler, G-band spectral synthesis and diagnostics of simulated solar magneto-convection. Astron. Astrophys. **427**, 335–343 (2004)

S. Shelyag, M. Schüssler, S.K. Solanki, A. Vögler, Stokes diagnostics of simulated solar magneto-convection. Astron. Astrophys. **469**, 731–747 (2007)

T. Shimizu, B.W. Lites, Y. Katsukawa, K. Ichimoto, Y. Suematsu, S. Tsuneta, S. Nagata, M. Kubo, R.A. Shine, T.D. Tarbell, Frequent occurrence of high-speed local mass downflows on the solar surface. Astrophys. J. **680**, 1467–1476 (2008)
M. Sigwarth, Properties and origin of asymmetric and unusual Stokes V profiles observed in solar magnetic fields. Astrophys. J. **563**, 1031–1044 (2001)
M. Sigwarth, K.S. Balasubramaniam, M. Knölker, W. Schmidt, Dynamics of solar magnetic elements. Astron. Astrophys. **349**, 941–955 (1999)
S.K. Solanki, Velocities in solar magnetic fluxtubes. Astron. Astrophys. **168**, 311–329 (1986)
S.K. Solanki, The origin and the diagnostic capabilities of the Stokes V asymmetry observed in solar faculae and the network. Astron. Astrophys. **224**, 225–241 (1989)
S.K. Solanki, Smallscale solar magnetic fields—an overview. Space Sci. Rev. **63**, 1–188 (1993)
S.K. Solanki, Sunspots: An overview. Astron. Astrophys. Rev. **11**, 153–286 (2003)
S.K. Solanki, O. Steiner, How magnetic is the solar chromosphere? Astron. Astrophys. **234**, 519 (1990)
S.K. Solanki, J.O. Stenflo, Properties of solar magnetic fluxtubes as revealed by Fe I lines. Astron. Astrophys. **140**, 185–198 (1984)
S.K. Solanki, J.O. Stenflo, Models of solar magnetic fluxtubes—constraints imposed by Fe I and II lines. Astron. Astrophys. **148**, 123–132 (1985)
S.K. Solanki, D. Zufferey, H. Lin, I. Rüedi, J.R. Kuhn, Infrared lines as probes of solar magnetic features. XII. Magnetic flux tubes: evidence of convective collapse? Astron. Astrophys. **310**, 33–36 (1996)
S.K. Solanki, W. Finsterle, I. Rüedi, W. Livingston, Expansion of solar magnetic flux tubes large and small. Astron. Astrophys. **347**, 27 (1999)
S.K. Solanki, B. Inhester, M. Schüssler, The solar magnetic field. Rep. Prog. Phys. **69**, 563–668 (2006)
S.K. Solanki, P. Barthol, S. Danilovic, A. Feller, A. Gandorfer, J. Hirzberger, T.L. Riethmüller, M. Schüssler, J.A. Bonet, V. Martínez Pillet, J.C. del Toro Iniesta, V. Domingo, J. Palacios, M. Knölker, N. Bello González, T. Berkefeld, M. Franz, W. Schmidt, A.M. Title, Sunrise: Instrument, mission, data, and first results. Astrophys. J. Lett. **723**, 127 (2010)
S.K. Solanki, N.A. Krivova, J.D. Haigh, Solar irradiance variability and climate. Annu. Rev. Astron. Astrophys. **51**, 311–351 (2013). doi:10.1146/annurev-astro-082812-141007
H.C. Spruit, Pressure equilibrium and energy balance of small photospheric fluxtubes. Sol. Phys. **50**, 269 (1976)
H.C. Spruit, Magnetic flux tubes and transport of heat in the convection zone of the Sun, PhD thesis, University of Utrecht, The Netherlands (1977)
H.C. Spruit, Convective collapse of flux tubes. Sol. Phys. **61**, 363–378 (1979)
H.C. Spruit, Magnetic flux tubes. NASA Spec. Publ. **450**, 385 (1981)
H.C. Spruit, Theory of photospheric magnetic fields, in *Solar and Stellar Magnetic Fields: Origins and Coronal Effects*, ed. by J.O. Stenflo. IAU Symposium, vol. 102 (1983), p. 41
H.C. Spruit, E.G. Zweibel, Convective instability of thin flux tubes. Sol. Phys. **62**, 15–22 (1979)
J. Squire, A. Bhattacharjee, Generation of large-scale magnetic fields by small-scale dynamo in shear flows. ArXiv e-prints (2015)
S. Stangl, J. Hirzberger, On small scale magnetic structures in the solar photosphere. Astron. Astrophys. **432**, 319–329 (2005)
R.F. Stein, Solar surface magneto-convection. Living Rev. Sol. Phys. **9**, 4 (2012). doi:10.12942/lrsp-2012-4, http://www.livingreviews.org/lrsp-2012-4
O. Steiner, Photospheric magnetic field at small scales, in *Turbulence, Waves, and Instabilities in the Solar Plasma*, ed. by R. Erdelyi, K. Petrovay, B. Roberts, M. Aschwanden. NATO Advanced Research Workshop (2003), p. 117
O. Steiner, Radiative properties of magnetic elements. II. Center to limb variation of the appearance of photospheric faculae. Astron. Astrophys. **430**, 691–700 (2005)
O. Steiner, R. Rezaei, Recent advances in the exploration of the small-scale structure of the quiet Solar atmosphere: Vortex flows, the horizontal magnetic field, and the Stokes-V Line-ratio method, in *Fifth Hinode Science Meeting*, ed. by L. Golub, I. De Moortel, T. Shimizu. Astronomical Society of the Pacific Conference Series, vol. 456 (2012), p. 3
O. Steiner, U. Grossmann-Doerth, M. Schüssler, M. Sigwarth, The formation of extremely asymmetric Stokes V profiles, in *Astronomische Gesellschaft Meeting Abstracts*, ed. by R.E. Schielicke. Astronomische Gesellschaft Meeting Abstracts, vol. 15 (1999), p. 11
O. Steiner, P.H. Hauschildt, J. Bruls, Radiative properties of magnetic elements. I. Why are G-band bright points bright? Astron. Astrophys. **372**, 13 (2001)
O. Steiner, R. Rezaei, W. Schaffenberger, S. Wedemeyer-Böhm, The horizontal internetwork magnetic field: Numerical simulations in comparison to observations with Hinode. Astrophys. J. **680**, 85–88 (2008). doi:10.1086/589740
J.O. Stenflo, Magnetic-field structure of the photospheric network. Sol. Phys. **32**, 41–63 (1973)

J.O. Stenflo, Solar magnetic flux at small scales (invited review), in *Solar Magnetic Fields*, ed. by M. Schüssler, W. Schmidt (1994), p. 301
J.O. Stenflo, Distribution functions for magnetic fields on the quiet Sun. Astron. Astrophys. **517**, 37 (2010). doi:10.1051/0004-6361/200913972
J.O. Stenflo, Collapsed, uncollapsed, and hidden magnetic flux on the quiet Sun. Astron. Astrophys. **529**, 42 (2011)
J.O. Stenflo, Horizontal or vertical magnetic fields on the quiet Sun. Angular distributions and their height variations. Astron. Astrophys. **555**, 132 (2013). doi:10.1051/0004-6361/201321608
J.O. Stenflo, J.W. Harvey, Dependence of the properties of magnetic fluxtubes on area factor or amount of flux. Sol. Phys. **95**, 99–118 (1985)
J.O. Stenflo, S. Solanki, J.W. Harvey, J.W. Brault, Diagnostics of solar magnetic fluxtubes using a Fourier transform spectrometer. Astron. Astrophys. **131**, 333–346 (1984)
J.O. Stenflo, S.K. Solanki, J.W. Harvey, Center-to-limb variation of Stokes profiles and the diagnostics of solar magnetic fluxtubes. Astron. Astrophys. **171**, 305–316 (1987)
L.G. Stenholm, J.O. Stenflo, Multi-dimensional non-LTE radiative transfer in magnetic fluxtubes on the sun. Astron. Astrophys. **58**, 273 (1977)
A. Takeuchi, On the formation of intense solar photospheric flux tubes, in *IAU Joint Discussion*. IAU Joint Discussion, vol. 19 (1997), p. 51
I. Thaler, H.C. Spruit, Small-scale dynamos on the solar surface: Dependence on magnetic Prandtl number. Astron. Astrophys. **578**, 54 (2015). doi:10.1051/0004-6361/201423738
J.H. Thomas, B. Montesinos, Siphon flows in isolated magnetic flux tubes. III—The equilibrium path of the flux-tube arch. Astrophys. J. **359**, 550–559 (1990). doi:10.1086/169086
L.M. Thornton, C.E. Parnell, Small-scale flux emergence observed using Hinode/SOT. Sol. Phys. **269**, 13 (2011)
A.M. Title, T.D. Tarbell, K.P. Topka, S.H. Ferguson, R.A. Shine (SOUP Team), Statistical properties of solar granulation derived from the SOUP instrument on Spacelab 2. Astrophys. J. **336**, 475–494 (1989)
A.M. Title, K.P. Topka, T.D. Tarbell, W. Schmidt, C. Balke, G. Scharmer, On the differences between plage and quiet sun in the solar photosphere. Astrophys. J. **393**, 782–794 (1992)
S.M. Tobias, F. Cattaneo, S. Boldyrev, MHD dynamos and turbulence, in *Ten Chapters in Turbulence*, ed. by P.A. Davidson, Y. Kaneda, K.R. Sreenivasan (Cambridge University Press, Cambridge, 2013), pp. 351–404
K.P. Topka, T.D. Tarbell, A.M. Title, Properties of the smallest solar magnetic elements. I. Facular contrast near sun center. Astrophys. J. **396**, 351–363 (1992)
K.P. Topka, T.D. Tarbell, A.M. Title, Properties of the smallest solar magnetic elements. II. Observations versus hot wall models of faculae. Astrophys. J. **484**, 479–486 (1997)
S. Tsuneta, K. Ichimoto, Y. Katsukawa, S. Nagata, M. Otsubo, T. Shimizu, Y. Suematsu, M. Nakagiri, M. Noguchi, T. Tarbell, A. Title, R. Shine, W. Rosenberg, C. Hoffmann, B. Jurcevich, G. Kushner, M. Levay, B. Lites, D. Elmore, T. Matsushita, N. Kawaguchi, H. Saito, I. Mikami, L.D. Hill, J.K. Owens, The Solar Optical Telescope for the Hinode mission: An overview. Sol. Phys. **249**, 167 (2008)
H. Uitenbroek, A. Tritschler, The contrast of magnetic elements in synthetic CH- and CN-band images of solar magnetoconvection. Astrophys. J. **639**, 525–533 (2006)
H. Uitenbroek, A. Tritschler, Narrow-band imaging in the CN band at 388.33 nm. Astron. Astrophys. **462**, 1157–1163 (2007)
H. Uitenbroek, A. Tritschler, T. Rimmele, The discrepancy in G-band contrast: Where is the quiet Sun? Astrophys. J. **668**, 586–593 (2007)
R.K. Ulrich, L. Bertello, Solar-cycle dependence of the Sun's apparent radius in the neutral iron spectral line at 525 nm. Nature **377**, 214 (1995)
D. Utz, A. Hanslmeier, C. Möstl, R. Muller, A. Veronig, H. Muthsam, The size distribution of magnetic bright points derived from Hinode/SOT observations. Astron. Astrophys. **498**, 289–293 (2009)
D. Utz, A. Hanslmeier, R. Muller, A. Veronig, J. Rybák, H. Muthsam, Dynamics of isolated magnetic bright points derived from Hinode/SOT G-band observations. Astron. Astrophys. **511**, 39 (2010). doi:10.1051/0004-6361/200913085
D. Utz, J. Jurčák, A. Hanslmeier, R. Muller, A. Veronig, O. Kühner, Magnetic field strength distribution of magnetic bright points inferred from filtergrams and spectro-polarimetric data. Astron. Astrophys. **554**, 65 (2013)
D. Utz, J.C. del Toro Iniesta, L.R. Bellot Rubio, J. Jurčák, V. Martínez Pillet, S.K. Solanki, W. Schmidt, The formation and disintegration of magnetic bright points observed by Sunrise/IMaX. Astrophys. J. **796**, 79 (2014)
A.A. van Ballegooijen, P. Nisenson, R.W. Noyes, M.G. Löfdahl, R.F. Stein, Å. Nordlund, V. Krishnakumar, Dynamics of magnetic flux elements in the solar photosphere. Astrophys. J. **509**, 435–447 (1998)

P. Venkatakrishnan, Inhibition of convective collapse of solar magnetic flux tubes by radiative diffusion. Nature **322**, 156 (1986)
B. Viticchié, On the polarimetric signature of emerging magnetic loops in the quiet Sun. Astrophys. J. **747**, 36 (2012). doi:10.1088/2041-8205/747/2/L36
B. Viticchié, D. Del Moro, S. Criscuoli, F. Berrilli, Imaging spectropolarimetry with IBIS. II. On the fine structure of G-band bright features. Astrophys. J. **723**, 787–796 (2010)
B. Viticchié, J. Sánchez Almeida, D. Del Moro, F. Berrilli, Interpretation of HINODE SOT/SP asymmetric Stokes profiles observed in the quiet Sun network and internetwork. Astron. Astrophys. **526**, 60 (2011)
A. Vögler, M. Schüssler, A solar surface dynamo. Astron. Astrophys. **465**, 43–46 (2007). doi:10.1051/0004-6361:20077253
A. Vögler, S. Shelyag, M. Schüssler, F. Cattaneo, T. Emonet, T. Linde, Simulations of magneto-convection in the solar photosphere. Equations, methods, and results of the MURaM code. Astron. Astrophys. **429**, 335–351 (2005)
A.R. Webb, B. Roberts, Vertical motions in an intense magnetic flux tube. II—Convective instability. Sol. Phys. **59**, 249–274 (1978)
T. Wiegelmann, J.K. Thalmann, S.K. Solanki, The magnetic field in the solar atmosphere. Astron. Astrophys. Rev. **22**, 78 (2014)
E. Wiehr, A unique magnetic field range for nonspot solar magnetic regions. Astron. Astrophys. **69**, 279 (1978)
L. Yelles Chaouche, S.K. Solanki, M. Schüssler, Comparison of the thin flux tube approximation with 3D MHD simulations. Astron. Astrophys. **504**, 595 (2009)
K.L. Yeo, S.K. Solanki, N.A. Krivova, Intensity contrast of solar network and faculae. Astron. Astrophys. **550**, 95 (2013)
K.L. Yeo, N.A. Krivova, S.K. Solanki, Solar cycle variation in solar irradiance. Space Sci. Rev. **186**, 137–167 (2014). doi:10.1007/s11214-014-0061-7
V. Zakharov, A. Gandorfer, S.K. Solanki, M. Löfdahl, A comparative study of the contrast of solar magnetic elements in CN and CH. Astron. Astrophys. **437**, 43–46 (2005)
I. Zayer, S.K. Solanki, J.O. Stenflo, The internal magnetic field distribution and the diameters of solar magnetic elements. Astron. Astrophys. **211**, 463–475 (1989)
I. Zayer, J.O. Stenflo, C.U. Keller, S.K. Solanki, Dependence of the properties of solar magnetic flux tubes on filling factor. II—Results of an inversion approach. Astron. Astrophys. **239**, 356–366 (1990)
I.B. Zeldovich, A.A. Ruzmaikin, D.D. Sokolov, *Magnetic Fields in Astrophysics* (Gordon & Breach, New York, 1983)
H. Zhang, M. Song, Vector magnetogram and Dopplergram observation of magnetic flux emergence and its explanation. Sol. Phys. **138**, 69–92 (1992)
G. Zhou, J. Wang, C. Jin, Solar intranetwork magnetic elements: Flux distributions. Sol. Phys. **283**, 273 (2013)
H. Zirin, Weak solar fields and their connection to the solar cycle. Sol. Phys. **110**, 101–107 (1987)
H. Zirin, B. Popp, Observations of the 12 micron MG I lines in various solar features. Astrophys. J. **340**, 571–578 (1989)
C. Zwaan, The emergence of magnetic flux. Sol. Phys. **100**, 397–414 (1985)

Space Sci Rev (2017) 210:317–349
DOI 10.1007/s11214-016-0259-y

The Life Cycle of Active Region Magnetic Fields

M.C.M. Cheung[1,2] · L. van Driel-Gesztelyi[3,4,5] · V. Martínez Pillet[6] · M.J. Thompson[7]

Received: 17 December 2015 / Accepted: 27 April 2016 / Published online: 24 August 2016

Abstract We present a contemporary view of how solar active region magnetic fields are understood to be generated, transported and dispersed. Empirical trends of active region properties that guide model development are discussed. Physical principles considered important for active region evolution are introduced and advances in modeling are reviewed.

Keywords Sun · Active region · MHD · Magnetic fields · Sunspots

✉ M.C.M. Cheung
cheung@lmsal.com

L. van Driel-Gesztelyi
Lidia.vanDriel@obspm.fr

V. Martínez Pillet
vmpillet@nso.edu

M.J. Thompson
mjt@ucar.edu

1 Lockheed Martin Solar & Astrophysics Laboratory, 3251 Hanover Street, Palo Alto, CA 94304, USA

2 National Astronomical Observatory of Japan, 2-21-1 Osawa, Mitaka, Tokyo 181-8588, Japan

3 Mullard Space Science Laboratory, University College London, Holmbury St. Mary, Dorking, Surrey, RH5 6NT, UK

4 Konkoly Observatory, Research Centre for Astronomy and Earth Sciences, Hungarian Academy of Sciences, Budapest, Hungary

5 Observatoire de Paris, LESIA, CNRS, UPMC, Université Paris Diderot, 92195 Meudon Principal Cedex, France

6 National Solar Observatory, 3665 Discovery Drive, Boulder, CO 80303, USA

7 National Center for Atmospheric Research, P.O. Box 3000, Boulder, CO 80307-3000, USA

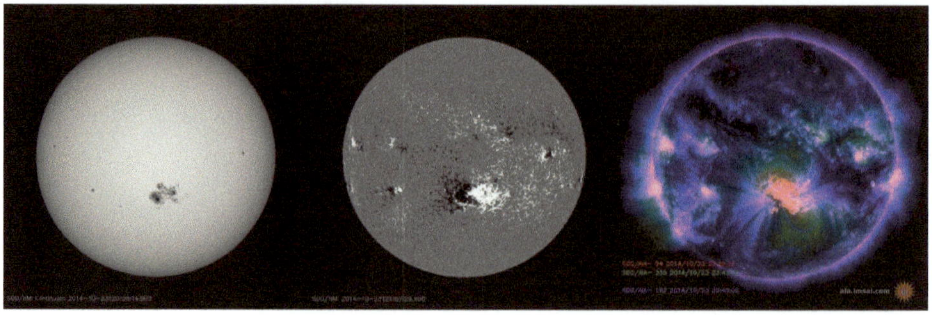

Fig. 1 Full-disk observations by NASA's Solar Dynamics Observatory (SDO) on 2014-10-23. The *left* and *middle panels* show a continuum image and line-of-sight magnetogram taken by the Helioseismic & Magnetic Imager (HMI), respectively. NOAA Active Region 12192, the largest sunspot group in cycle 24, appears as the dominant sunspot group and bipole in the southern hemisphere. The *right panel* shows a three-color composite EUV image (the 94, 335 and 193 Å channels of the Atmospheric Imaging Assembly (AIA) instrument) of the overlying corona

1 Introduction

Magnetic activity on the Sun drives a diverse range of dynamic phenomena observed to occur on the solar surface and in the overlying atmosphere. The global-scale distribution of magnetic fields on the solar surface gradually evolves under the action of differential rotation, meridional circulation, and buffeting by supergranular flows. These processes operate steadily to mold and to erode the Sun's magnetic terrain.

Before high-precision and high-resolution spectropolarimetry became available, the solar surface was roughly delineated into active regions and quiet Sun regions. Active regions were associated with magnetic complexes in and around sunspots, while the quiet Sun was thought to be devoid of magnetic fields. With improved resolution and sensitivity, we now know the quiet Sun to be far from quiet. With Zeeman observations, we know the not-so-quiet Sun is filled with magnetic fields down to the smallest resolved scales ($\approx$ 100 km, e.g. Lites et al. 1996; Domínguez Cerdeña et al. 2003; Khomenko et al. 2003; Harvey et al. 2007; Lites et al. 2008; Ishikawa et al. 2008; Ishikawa and Tsuneta 2009; Danilovic et al. 2010a,b). With the Hanle effect and realistic convection modeling, turbulent fields have been inferred to exist at unresolved scales (Trujillo Bueno et al. 2004). Although the properties and origin(s) of ubiquitous quiet Sun magnetism is still under debate (see the review paper by Borrero et al. in this volume), comparative studies of quiet Sun field observations and radiative magnetoconvection simulations (e.g. Vögler and Schüssler 2007; Schüssler and Vögler 2008) suggest much of the quiet Sun magnetic field to be generated by a distributed small-scale dynamo operating throughout the bulk of the convection zone (Pietarila Graham et al. 2009; Danilovic et al. 2010b; Rempel 2014; Hotta et al. 2015, 2016).

What about the origin and life cycle of active region magnetic fields? The definition of what constitutes a solar active region has evolved as our understanding of solar magnetism has progressed in the last century. As suggested by van Driel-Gesztelyi and Green (2015), a useful modern definition due to Kiepenheuer (1968) is *'[the] totality of all observable phenomena preceding, accompanying and following the birth of sunspots'*. Figure 1 shows a subset of such observations from NASA's Solar Dynamics Observatory. Active regions leave their mark on the photosphere in the form of sunspot umbrae and penumbrae, pores, and surrounding areas of small-scale field. The intense magnetic fields threading sunspots penetrate the overlying atmosphere and structure the solar corona.

Sunspot magnetic fields have been extensively studied since their detection via the Zeeman effect in the early 20th century (Hale 1908). How are magnetic fields on the Sun amplified, transported, and organized to form pores, sunspots and plages? How do the magnetic fields subsequently erode from sunspots, and where do they go? The life cycle of active regions is intimately connected to questions regarding the physical processes governing the global solar dynamo. The distribution of active region (old and new) magnetic fields on the solar surface shapes the global structure of the solar corona (Altschuler and Newkirk 1969; Schrijver and De Rosa 2003) and heliospheric current sheet (Riley et al. 2006), and determines the open flux of the interplanetary magnetic field (IMF; Wang et al. 2000; Lockwood et al. 2009; Jiang et al. 2011). Last but not least, non-potential (i.e. current-carrying) active regions spawn solar flares and coronal mass ejections (CMEs; Gosling 1993; Chen 2011; Webb and Howard 2012). So studying how active regions form, interact and decay is pertinent to our understanding of how the Sun's magnetism modulates space climate and space weather.

The physics of how the solar dynamo operates, how magnetoconvection modulates flux emergence, how active regions build up free energy to energize flares, and how they subsequently decay are all broad topics that deserve review papers in their own right. We do not aim for this paper to serve as a comprehensive reference for all topics associated with active regions. Rather, the aim is to synthesize the broad body of knowledge into a narrative summarizing our current understanding of the life cycle of active region magnetic fields. Our narrative begins with the subsurface origin of active regions.

2 The Subsurface Origin of Active Region Magnetic Fields

By all available observational accounts, sunspots form as large-scale flux bundles emerge from the solar interior onto the surface (see reviews by Zwaan 1978, 1985; Fan 2004, 2009; Lites 2009; Cheung and Isobe 2014; van Driel-Gesztelyi and Green 2015). At the scale of photospheric granulation ($L \sim 1$ Mm), the emerging flux reaches the surface as predominantly horizontal fields within the centers of granules (e.g. Cheung et al. 2007, 2008; Lites 2009; Ortiz et al. 2014). These are accompanied by opposite polarity patches anchored in the downflow lanes at the edges of granular cells. These telltale signs are evidence that granular-scale magnetic Ω-loops are transported to the surface from the solar interior by convective flows (and perhaps aided by magnetic buoyancy, see Sect. 5). Small-scale flux emergence events with these properties are not only found in developing active regions, but also in more quiet regions of the Sun (e.g. De Pontieu 2002; Centeno et al. 2007; Martínez González and Bellot Rubio 2009). In nascent active regions, ensembles of small-scale emergence events have a preferred orientation aligned with the final orientation connecting the leading and following polarity spots of the active region in a more mature state (Strous and Zwaan 1999; Pariat et al. 2004). Furthermore, opposite polarity patches stream in opposite horizontal directions. Like-polarity patches migrate in such a way that they coalesce into larger patches. When certain coherent patches have sufficient flux content ($\Phi \gtrsim 10^{19}$ Mx), pores (i.e. isolated umbral regions in the absence of penumbra) begin to appear. As the flux coalescence progresses, pores merge with adjacent pores and penumbral filaments begin to grow (e.g. Schlichenmaier et al. 2010). A detailed discussion of observations and theory concerning these 'phase transitions' is beyond the scope of this paper. Rather, we emphasize that an apparent underlying driver rooted in the solar interior appears to move emerged flux patches in a systematic, non-random fashion.

Extensive theoretical and numerical investigations of coherent magnetic flux bundles rising through the convection zone have been carried out since Parker (1955) first pointed out how magnetic buoyancy drives their rise toward the surface (see the review by Fan 2009). Statistical properties of active regions, including Hale's law (Hale and Nicholson 1925) and Joy's law (Hale et al. 1919), provide observational tests for theories of the subsurface progenitors of active regions. Detailed studies of the surface evolution of individual active regions offer further clues regarding their subsurface structure.

As a example, Toriumi et al. (2014) carried out numerical magnetohydrodynamics simulations of the rise of buoyant flux tubes to mimic the observed photospheric evolution of sunspots in NOAA AR 11158. This active region appeared on the photosphere as two pairs of magnetic bipoles, and evolved such that the following polarity of one bipole jousted against the leading polarity of the other bipole, resulting in a sheared polarity inversion line which subsequently produced an X-class flare and CME. Toriumi et al. (2014) carried out simulations with initial conditions that consist of (i) two adjacent, submerged flux tubes each with a buoyant segment, and (ii) a single submerged tube with two magnetically buoyant segments. Both initial conditions yielded a pair of bipoles emerging at the surface, but only the latter case resulted in a sheared polarity inversion line amenable for eruptive activity. This result gives support to that view that the drivers of solar eruptive events lie beneath the solar surface.

2.1 Helioseismic Constraints on Active Region Magnetic Fields

Helioseismology has taught us much about the properties of the solar interior, and offers the promise to teach us much about subsurface magnetic fields. So having adopted our working definition of active regions being the *'totality of observable phenomena preceding, accompanying and following the birth of sunspots'*, what observational constraints has helioseismology provided on the magnetic structure of active regions?

We first consider well-established helioseismology results that are relevant to, but not directly concerned with subsurface magnetic fields. Helioseismology has helped us learn about:

- The differential rotation profile $\Omega(r, \theta)$ throughout the bulk of the convection zone and in the upper radiative zone (Tomczyk et al. 1995; Thompson et al. 1996, 2003; Schou et al. 1998, see Fig. 2),
- The depth of the solar convection zone, indicated by the dashed line in Fig. 2 (Christensen-Dalsgaard et al. 1991),
- Patterns of meridional circulation (Giles 2000; Haber et al. 2002; Komm et al. 2013; Zhao et al. 2013), and
- Constraints on convective flows speeds (Hanasoge et al. 2012; Greer et al. 2015).

In the majority of global solar dynamo models, shearing by differential rotation, the so-called Ω-effect, plays a key role in the amplification of the mean azimuthal (toroidal) component of the magnetic field. In particular, certain mean-field (Krause and Rädler 1980) dynamo models require $\frac{\partial \Omega}{\partial r}$ to be positive or negative to give oscillatory solutions with equatorward propagating branches (mimicking the butterfly diagram, see Stix 1976). Helioseismic inversions of the rotation rate Ω as functions of radius and latitude have helped rule out the applicability of a number of such dynamo models.

The quasi-solid body rotation in the radiative core was an unexpected discovery (Thompson et al. 1996, 2003). This hints at the possible existence of a fossil field contained within the radiative core whose Maxwell stresses keep the core rotating as a solid body (MacGregor

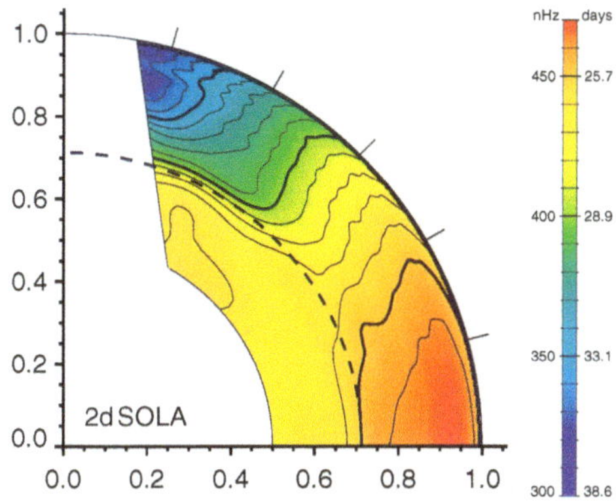

Fig. 2 Differential rotation profile of the solar interior from an inversion of global mode measurements by SoHO/MDI (figure adapted from Schou et al. 1998). The *bottom* of the convection zone is indicated by the *dashed curve*. Salient features include the conical rotation contours in the bulk of the convection zone, the solid body rotation in the radiative zone, the tachocline at low and high latitudes beneath the convection zone and the near-surface shear layer at low and mid-latitudes

and Charbonneau 1999). However, such a fossil field would presumably not undergo polarity reversals between cycles and so likely is not the birthplace of active region magnetic fields.

The tachocline, which is the region of transition from solid body rotation to radial isorotation profiles was recognized to contain large shear and so is a candidate for the operation of the Ω-effect (which, in some models, leads to strong toroidal fields that become buoyant unstable and rise toward the surface). The tachocline is prolate and at low latitudes lies substantially beneath the base of the convection zone. The region where rotational shear and convectively stable stratification coexist is thought to provide a suitable region for toroidal magnetic field to be generated and stored until it becomes buoyant and rises towards the surface.

According to flux-transport models, meridional circulation plays a key role in the working of the dynamo and in setting the length of the solar cycle. In such models the meridional flow advects flux from decaying active regions to the poles and causes the reversal of the poloidal component of the solar magnetic field in the polar regions. The simplest models envisaged a single cell in depth spanning from surface down to the deep convection zone. Time-distance helioseismic analysis of about 2 years of ground-based Global Oscillation Network Group (GONG) data by Jackiewicz et al. (2015) and 4 years of SDO/HMI data by Rajaguru and Antia (2015) support the existence of such a single meridional cell filling the entire convective zone. These studies dispute the reported detection of multiple cells structure in depth (Zhao et al. 2013) and in latitude (Schad et al. 2013; Zhao et al. 2013). The implications of multiple meridional cells for flux-transport dynamos have been explored by Hazra et al. (2014) and Belucz et al. (2015).

3 Active Regions and Ephemeral Regions

The magnetic pattern on the solar surface is continuously seeded by the emergence of magnetic loops over a range of spatial and temporal scales (Schrijver et al. 1997; Parnell 2001; Simon et al. 2001; Hagenaar et al. 2003; Meyer et al. 2011; Iida et al. 2012; Gošić et al.

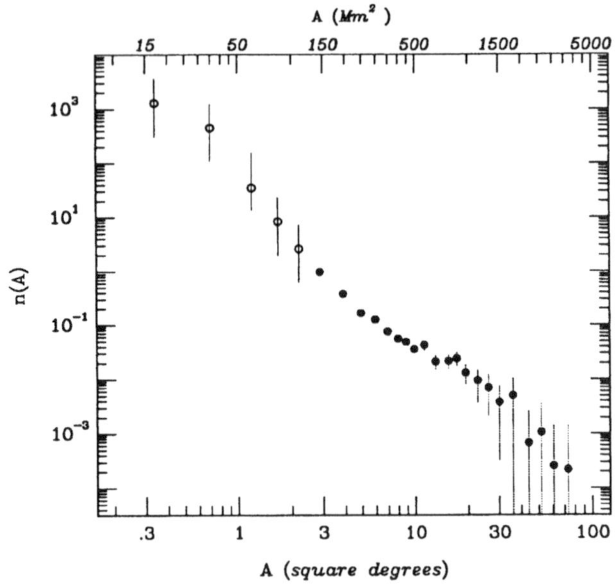

Fig. 3 Size distribution of 978 active regions (*solid dots*) and 9492 ephemeral regions (*open dots*) identified from Kitt Peak full-disk magnetograms (reproduced from Fig. 17, Chap. 8 of Harvey 1993). The measured quantity $n(A)$ corresponds to the number of bipolar regions with surface area A at the time of maximum development. For members in the sample with area $A > 3.5$ square degrees, the power-law shape of the size distribution was found to persist throughout Cycle 21 though the absolute appearance rates differed by a factor of 8 between cycle minimum and maximum

2014). To study where, when and how these loops appear is to get a glimpse of how the subsurface reservoir of dynamo-generated magnetic field feeds the surface. Some properties of the largest magnetic bipoles, the active regions are well-known. As a statistical ensemble, they yield Hale's Polarity Rule, Joy's Law and the butterfly diagram. What about their smaller counterparts? Are they simply smaller versions of the same phenomenon? Do they share a common origin? In summary, are ephemeral regions active regions?

This is one of the important questions addressed by K.L. Harvey in her PhD thesis. Harvey (1993) used daily Kitt Peak full-disk magnetograms taken throughout the duration of Cycle 21 (between 1975 through 1986) to examine the statistical properties of bipolar active regions and ephemeral regions (see also Harvey and Zwaan 1993). Figure 3 shows the size distribution histogram of magnetic bipoles corresponding to 978 active regions (solid dots) and 9492 ephemeral regions (open dots) identified over the entire data set. Their sample of bipolar regions included only ones that emerged and reached their maximum development on the visible hemisphere (although they correct for this observational limitation). The size distribution $n(A)$ for the entire sample, defined as the number of regions with area A per square degree per day, was found to follow a power-law. For active regions with $A \gtrsim 3.5$ square degrees, the absolute emergence rate varied in phase with the solar cycle with a modulation amplitude of 8 though the shape of the distribution remained unchanged. For smaller active regions with $2.5 \lesssim A \lesssim 3.5$ square degrees, the modulation amplitude is about 5. For ephemeral regions with $A \lesssim 2.5$ square degrees, the modulation amplitude is further diminished to a value of 2.

The emergence frequency is not the only property of magnetic bipoles that depends on the area. The distribution of orientation (tilt angles) of magnetic bipoles was also found to be size-dependent. Bipoles with $A \gtrsim 2.5$ square degrees had orientation distributions that are mostly consistent with Hale's polarity rule and Joy's law. However the bias in the preferred orientation diminishes with decreasing A and is virtually non-existent for $A \lesssim 2.5$ square degrees. Furthermore, the distribution of emergence latitudes follow the equatorward propagating branches of the butterfly diagram but the latitudinal spread of the distributions widen with decreasing A. Based on these findings, Harvey (1993) conclude that ephemeral regions are the smaller-scale manifestation of active regions over a continuous spectrum of magnetic bipolar regions spanning almost three orders of magnitude in size.

The seminal work by Harvey (1993) was extended by Hagenaar et al. (2003), who used full disk magnetogram data from the Michelson Doppler Imager (MDI) on board the Solar and Heliospheric Observatory (SoHO) to study the size distribution of ephemeral bipolar regions. Their study differs from Harvey (1993) in some important ways. Whereas Harvey (1993) manually tracked bipolar active regions, Hagenaar et al. (2003) used a fully automated algorithm. The algorithm (detailed in Hagenaar 2001) implements the following two steps. The first step involves the segmentation of individual magnetograms into discrete magnetic features with each magnetic feature consisting of contiguous pixels of like magnetic polarity (above some sensitivity threshold). The second step is the matching up of magnetic features in consecutive magnetograms (96 minute cadence) satisfying certain spatiotemporal criteria. This automated method only detects newly emerged regions and does not track the development of detected bipolar regions to maximum development and beyond. In lieu of active region area, the Hagenaar et al. (2003) study measures the size distribution in terms of magnetic flux content. Since the algorithm is designed to only detect new bipolar emergences that occur within 96 minute intervals, the detected emergence events are restricted to ephemeral regions (and small bipolar regions) with fluxes in the range $\Phi = 3 \times 10^{18}$–4×10^{20} Mx. The lower end of the range of fluxes is limited by the sensitivity of MDI magnetograms and the thresholding criteria used for feature detection. The upper end of the range presumably corresponds to an upper limit to the flux emergence rate of ephemeral regions and of small and large active regions in their infancy. Hagenaar et al. (2003) fitted decaying exponential profiles (with Φ as the independent variable) to bipolar emergence frequencies at different times between 1996 (minimum between Cycles 22 and 23) and 2001 (one year past the maximum of Cycle 23). They report a variation of the bipolar appearance rate in *antiphase* with the sunspot cycle with a modulation amplitude of at most 1.5.

Based on the results of Harvey (1993, see also Harvey and Zwaan 1993) and their own study of ephemeral regions, Hagenaar et al. (2003) produced the composite distribution of bipolar regions shown in Fig. 4. The magnetic bipole areas measured by Harvey (1993) were converted to magnetic flux content assuming a given area-to-flux ratio. The drop off in emergence rates at $\Phi < 10^{19}$ Mx is attributed to the sensitivity of the detection algorithm and source of data (MDI magnetograms). Disregarding that turn off, the bipole emergence rate is seen to be a smoothly decreasing function over four orders of magnitude in magnetic flux Φ.

From the size, tilt and orientation distributions from the Kitt Peak data set, Harvey (1993) concluded *'ephemeral regions are in the extrapolation of the small-scale end of a wide size spectrum of bipolar magnetic regions'*. At first glance, the findings by Hagenaar et al. (2003) would suggest a similar conclusion. However, Hagenaar et al. (2003) arrive at a markedly different interpretation. They propose that Harvey's findings and their own are *'compatible with an interpretation in which there are two distinct, but coupled origins of solar bipolar regions: (1) a globally operating cyclic dynamo and (2) a near-surface, small-scale dynamo'*.

It is perhaps timely to assess whether the remarkable suggestion offered by Hagenaar et al. (2003) holds up against more recent developments. As already mentioned in the introduction, there is now substantial empirical evidence that the solar surface is pervaded by mixed turbulent fields down to the smallest observable scales. Hinode spectropolarimetric observations of the Fe I 6302.5 Å photospheric line taken in quiet Sun regions near disk center show no significant variations of the turbulent field signal over the time period spanning 2006 to 2012 (Buehler et al. 2013).

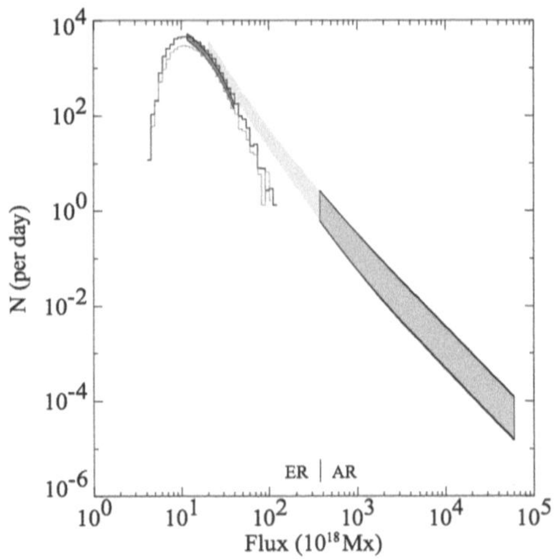

Fig. 4 Appearance rate of bipolar regions as a function of magnetic flux (reproduced from Fig. 11 of Hagenaar et al. 2003). The *curves* in the range $\Phi \lesssim 10^{20}$ Mx indicate distributions of ephemeral regions identified in SoHO/MDI magnetograms. The *black* and *gray curves* indicate distribution functions for data sets taken in Oct 1997 and Aug 2000, respectively. The drop off in emergence rates at the leftmost end of the flux range is attributed to limits in the detection sensitivity of the algorithm for the available data sets. The range of emergence rates reported by Harvey (1993) covering solar minimum to maximum (in cycle 21) is indicated by the diagonal *light* and *dark gray bands* in the *right half* of the plot. The *light gray band* indicates emergence rates for the ephemeral regions and active regions identified in that study

The terms 'local dynamo' and 'surface dynamo' are often used (sometimes interchangeably) in the literature as labels for the physical system responsible for generating the cycle-independent portion of the observed magnetic field. These terms carry the connotation that there is a clear segregation between the global cyclic dynamo and the mechanism generating small-scale magnetic flux. As numerical experiments of convective dynamos may suggest, segregation may well be an oversimplification. For instance, Rempel (2014) used the radiative MHD code MURaM to carry out a number of small-scale dynamo experiments in a Cartesian domain. The convective flows in the model are driven by radiative cooling at the surface as computed by solving the radiative transfer equation with realistic opacities and equation of state (the net vertical energy flux carried by convective motions below the surface and by radiation above corresponds to one solar luminosity). The Cartesian domain has a bottom boundary located at a few Mm below the surface and is open to inflows and outflows. Rempel (2014) examined the dependence of the surface magnetic flux densities on the choice of the bottom boundary condition. In the case where magnetic flux is prevented from entering the domain through upflows crossing the bottom boundary, the convective dynamo still operates but maintains unsigned flux at a level below what is reported from observations (~ 85 G). One possible resolution to this incompatibility with observed values of the unsigned flux is to allow inflows from the bottom boundary to advect magnetic flux into the domain. Rempel (2014) tested this scenario by enforcing that the r.m.s. field strength $B_{\rm rms}$ (as a function of depth) scale with the equipartition field strength ($B_{\rm eq} = (4\pi\varrho)^{1/2} v_{\rm rms}$) corresponding to the r.m.s. speed $v_{\rm rms}$ of convective flows. This setup led to surface field strengths in agreement with observations. If this case were representative of the solar interior, it would implicate feeding of flux from the deeper layers to the shallower layers as an important contributor to the surface magnetic flux budget. It would suggest that the convective dynamo operates over the bulk of the convection zone and is not localized near the solar surface (Fan and Fang 2014; Hotta et al. 2015, 2016).

4 The Small- and Large-Scale Dynamos as a Coupled System

In light of the aforementioned observational and theoretical developments, perhaps we should reframe the scientific discourse on the 'two dynamos' and the origin of active regions versus the origin of small-scale bipoles. If one takes the view that turbulent convection drives a distributed dynamo that operates over the bulk of the solar convection zone, one may expect magnetic fields to be generated and maintained over a continuous range of scales. At small scales (e.g. smaller than supergranulation) where the convective overturning times are short and the associated Rossby number ($Ro = \frac{P_{rot}}{\tau_{conv}}$, where P_{rot} and τ_{conv} are the rotation period and convective turnover time, respectively) is large, rotational effects such as the Coriolis force are small. In this regime, the dynamo-generated magnetic field patterns (at these scales) are expected to follow distribution functions with essentially random tilt angles and be largely insensitive to the phase of the solar cycle. This expectation is consistent with the properties of the smallest ephemeral regions. At larger scales (where $Ro < 1$), rotational effects such as shear and helical motions become important. Firstly, large-scale differential rotation is important for the generation of mean-field toroidal flux (Babcock 1961). Secondly, rotational motions generate helical magnetic fields, which in a turbulent medium exhibit an inverse cascade of magnetic helicity. This means the storage of magnetic twist migrates from the small scales to the large scales (Brandenburg 2001).

To summarize, both observations and theory indicate the lack of a clear separation of scales between the small and the large, between the convection-dominated and the rotation-dominated regimes. So perhaps we should ask the following questions: In a turbulent stellar interior like the Sun's convection zone where both the large and the small co-exist, what are the physical processes that govern the formation of magnetic bundles that become active regions? How does intermittency (i.e. the formation and destruction of spatially and temporally coherent structures) and order (i.e. systematic behavior followed by active regions) emerge (pun intended) out of this chaotic system? Do the small-scales impact the efficiency of magnetic flux generation at large (global) scales? The recognition that the global cyclic behavior of active regions and the details of magnetoconvection are intimately linked is pushing the limits of our understanding of the solar dynamo, and with that our understanding of the life cycle of active region magnetic fields. Recent advances have been enabled by numerical models of convective dynamos, a topic we will discuss in the following section.

5 The Cradle of Active Region Magnetic Fields

Parker (1955) first proposed the idea that sunspots form following the buoyant rise of magnetic flux tubes from the solar interior onto the surface. The underlying MHD concept is simple. Suppose a magnetic bundle is in pressure equilibrium with its surroundings. Due to the contribution of magnetic pressure $B^2/8\pi$ to the total pressure inside the tube, the interior gas pressure is reduced relative to the external gas pressure. Further assume the plasma in the interior and exterior are in thermal equilibrium. This requires a density deficit within the tube of order $\frac{\Delta\varrho}{\varrho} \sim -\frac{1}{\beta}$, where β is the ratio of gas to magnetic pressures (in the deep convection zone $\beta \gg 1$). Due to this density deficit, a flux tube has buoyancy and will rise toward the surface.

Parker assumed the existence of toroidal bands of magnetic field in the solar interior whose polarities switch from one sunspot cycle to the next. Such expectations motivate the search for certain mean-field dynamo models (Krause and Rädler 1980; Rädler 1980) with cyclic, wave-like solutions. While mean field dynamo (e.g. see Stix 1976; Moffatt 1978) solutions describe how the toroidal and poloidal components of the azimuthally-averaged field evolve, they do not provide a self-consistent picture for how active regions (and the associated Ω-type magnetic bundles) form. Nevertheless, the existence of a mean component of the toroidal field led researchers to study the linear (in)stability of thin magnetic flux tubes (Spruit and van Ballegooijen 1982; van Ballegooijen 1982; Ferriz-Mas and Schüssler 1993, 1995) subject to azimuthal perturbations.

The non-linear evolution of flux tube instabilities was studied by means of numerical simulations solving MHD equations that describe how coherent, slender flux tubes evolve under the influence of magnetic buoyancy, aerodynamic drag and the Lorentz and Coriolis forces (Spruit 1981; Choudhuri 1990; Longcope and Klapper 1997) in a rotating reference frame. A detailed review of developments within the so-called *thin flux tube approximation* is outside the scope of this paper. For that we refer the reader to Fan (2009). For our purpose we merely summarize some important results. First of all, the Coriolis force ($2\varrho\mathbf{\Omega} \times \mathbf{v}$) is found to deflect flux tubes rising from the bottom of the convection zone. This happens in two steps. First of all, the Coriolis force drives a retrograde flow in a rising flux tube (to conserve angular momentum). This motion then induces the Coriolis force to impart momentum to the flux tube in a direction parallel to the axis of rotation. Flux tubes with initial $B \lesssim 10^5$ G are deflected to latitudes higher than emergence latitudes of observed active regions on the Sun (Choudhuri and Gilman 1987; Choudhuri 1989). Since the sunspot cycle is approximately 11 years and sunspots groups appear over a large fraction of the cycle, the growth times of buoyancy instabilities are expected to be in the range of hundreds of days. For much shorter growth times, the underlying toroidal field would be prematurely depleted. For much longer growth times, buoyancy instabilities would not lead to the creation of sunspot groups. Schüssler et al. (1994) reported a threshold of $B \sim 10^5$ G to satisfy this requirement. For flux tubes initially located at latitudes within ± 70 deg of the equator in the overshoot layer, the $m = 1$ and $m = 2$ modes (m being the azimuthal number of the perturbation) have the highest growth rates. Furthermore, for flux tubes at these field strengths (and with flux content $\Phi \sim 10^{22}$ Mx), the thin flux tube model yields results that are consistent with empirically reported relationships between active region tilt and emergence latitude (see also D'Silva and Choudhuri 1993; Caligari et al. 1995), tilt and magnetic flux content (Fan et al. 1994; Fisher et al. 1995) and the scatter of active region tilt about Joy's law (Longcope and Fisher 1996).

Thin flux tube calculations have been used to study the development of asymmetries between the leading and following flanks of rising Ω-loops. Asymmetries can be characterized by a number of properties, such as the difference in the vertical orientation and the difference in field strengths between the two flanks. Caligari et al. (1998) compared model results from two types of initial conditions, those in (1) initial mechanical equilibrium ("MEQ", e.g. Moreno-Insertis et al. 1994; Caligari et al. 1995) in the overshoot layer, and those (2) with initial undulations and in thermal balance ("TBL", e.g. Fan et al. 1993) with the ambient stratification such that the crests are initially buoyant and already located in the convection zone. For initial field strengths of $B \sim 10^5$ G, the two types of initial conditions yield opposite results. For instance, MEQ (TBL) flux tubes have following flanks that are more (less) vertical than their leading counterparts. Caligari et al. (1998) consider the MEQ results to be more compatible with the observed proper motions of young active regions. The results from the two types of initial conditions are similar for tubes with weaker field ($B \sim 3 \times 10^4$ G). At these lower field strengths, both MEQ and TBL flux tubes approach the solar surface with the following flank more vertical and possessing stronger field strengths. In Sect. 5.1, we will discuss how flux tubes with 3×10^4 G initial strength are perhaps more consistent with observations when the interaction of the tube with convective flows is taken into account.

It is unclear the asymmetries in thin flux tube models are physically compatible with observational reports of relatively longer lifetimes of leading spots compared to following spots. For example, Meyer et al. (1977) assumed the following leg of an Ω-loop to be more inclined to the radial direction (opposite to the results of dynamical thin flux tube calculations) and invoked the interchange instability to explain the relatively shorter lifetimes of following spots. Rempel and Cheung (2014) perform a numerical MHD simulation of active region formation by emerging a magnetic loop with an imposed flow mimicking the retrograde flow driven by the Coriolis force. They found the leading polarity model spot to remain coherent longer than the following polarity spot. However, their simulations only capture the top 15 Mm of the convection zone and do not self-consistently treat the Coriolis effect. Further detailed investigations are necessary to clarify which physical effect may be responsible for morphological asymmetries in active regions.

Longcope and Klapper (1997) extended the thin tube model to treat flux tubes with magnetic twist. In their model, magnetic field lines within the tube are not assumed to be aligned within the tube axis. In addition to other quantities like field strength, mass density and momentum, the extended model solves for the magnetic pitch $q(l)$ as a function of position along the tube axis. Magnetic pitch is defined by the relation $B_\phi(l) = q(l) r B_l(l)$, where B_l and B_ϕ are, respectively, the longitudinal and transverse components of the magnetic field and r is the radial distance from the tube axis. By solving for $q(l)$, the Longcope and Klapper (1997) model captures important dynamical effects relevant to active region evolution. These include (1) the origin of active region twist (Longcope et al. 1998), (2) the relation between active region writhe and twist (Longcope et al. 1998), and (3) the driving mechanism for sunspot rotation (by magnetic torque) and injection of free energy into the corona (Longcope and Welsch 2000).

The apparent success of the thin flux tube models to reproduce well-known statistical properties of active regions on the Sun (and on other cool stars, see Schüssler et al. 1996; Granzer et al. 2000) opens up further questions about the origin of active region magnetic fields. For instance, how can flows near the bottom of the convection zone amplify magnetic fields to 100 kG? Depending on the model of the convection zone (e.g. 1D mixing length model or 3D dynamic model), the magnetic energy density associated with such fields is estimated to be 10–100 times the kinetic energy density of typical convective flows in the lower half of the convection zone. So Lorentz force feedback would likely prevent the amplification of magnetic fields to such strengths by convective flows (or, for that matter, the shear at the tachocline). One possible mechanism for intensification of magnetic fields to super-equipartition strengths was suggested by Moreno-Insertis et al. (1995). Working within the thin flux tube framework, they reported how Ω-loops from initial flux rings with $B = 10^4$ G fail to emerge because they experience dramatic expansion and weakening before reaching the surface. In the thin flux tube model, such a subsurface 'explosion' leads to a pressure gradient within the Ω-loop which sucks material out of the legs of the tube toward loop apex. Rempel and Schüssler (2001) modeled this process using 2D MHD simulations (see also Rempel 2001; Hotta et al. 2012, for corresponding models in 3D). As illustrated in Fig. 5, the rise of an Ω-loop through a super-adiabatically stratified layer (mimicking the convection zone) can lead to an abrupt expansion of the loop apex (even before it reaches the surface). The resulting pressure gradients drive upflows through the legs of the loop, which partially evacuates the anchored portions of the tube. The partial evacuation of the

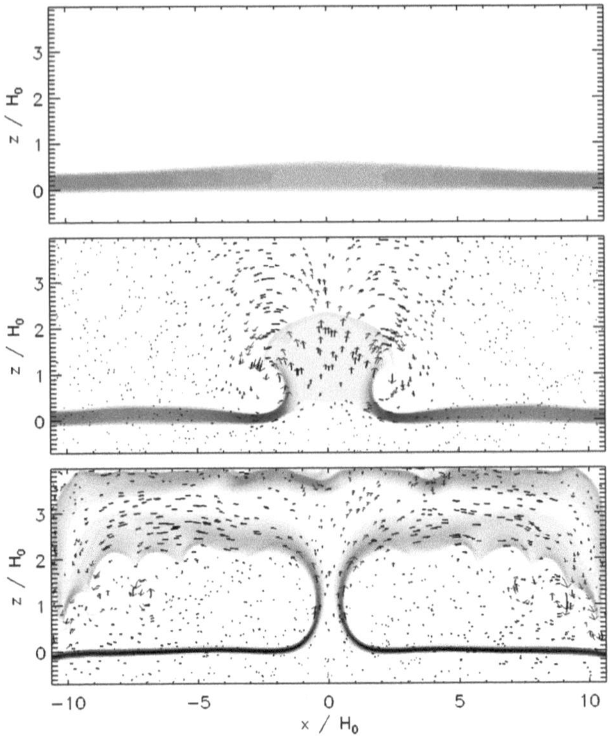

Fig. 5 Idealized 2D MHD simulation of the subsurface 'explosion' of an Ω-loop (from Rempel and Schüssler 2001). Following the explosion, the magnetic field in the horizontal portions of the loop is intensified

anchored portions leads to pressure deficit and subsequent compression. The combined action of the vertical compression and the stretching by outflows leads to an intensification of the magnetic field. Rempel and Schüssler (2001) suggest the repeated occurrence of such explosions may produce the superequipartition fields that are required by thin flux tube models for reproducing active region properties.

5.1 Magnetic Flux Tubes in a Convective Environment

Even as one grapples with the question of how 10^5 G fields may be generated inside the convection zone, one may also question whether such strong fields are really necessary, or just a requirement of overly simplified models. In particular, the thin flux tube models referenced above assume that the flux tubes do not interact with convective flows. They also assume that a solitary tube remains coherent on its journey through the turbulent convection zone. Does the picture change when convective turbulence is considered? The extent to which magnetoconvective effects are included varies widely between models. We first discuss models with the simplest treatment of convective turbulence before proceeding to review state-of-the-art models of 3D convective dynamos in Sun-like stars.

The consideration of the effects of convective turbulence on the life history of active region magnetic fields introduces both conceptual and technical challenges. When convective eddies interact with a rising flux tube, do they leave the flux tube intact or whitewash its identity either by enforcing magnetic reconnection at the scale of the eddies, or by turbulent diffusion? How does momentum transfer between a flux tube and convective eddies affect the former's trajectory (if it remains coherent)? A quantitative examination of the various

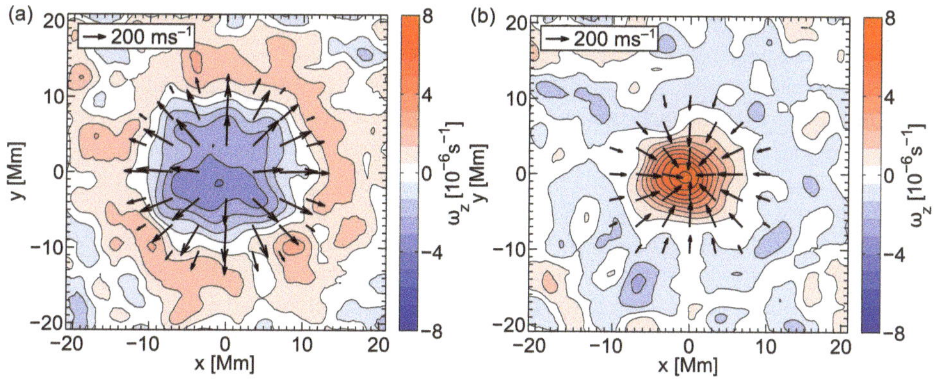

Fig. 6 Time-distance (TD) helioseismology flow maps averaged (**a**) divergent and (**b**) convergent regions (reproduced from Langfellner et al. 2015). These flow maps were computed by averaging over several thousand supergranules at a latitude of +40 deg and are representative of flows at depths between zero and a few Mm. The color-scale in *both panels* indicate the sign and magnitude of the average vertical component of flow vorticity (ω_z). This work shows that, at the supergranule-scale, diverging (converging) flows are correlated with negative (positive) vorticity in the northern hemisphere

effects relies on helioseismic measurements (or constraints) on the power spectrum of convective flows within the convection zone, which is a challenging technical feat in itself (e.g. compare results from Hanasoge et al. 2012; Greer et al. 2015).

The action of the Coriolis force on sufficiently large and long-lived flow structures results in net helical motion, which is characterized by a systematic correlation between horizontally diverging flows (associated with convective upflows) and vertical vorticity of one sign, and horizontally converging flows with vertical vorticity of the opposite sign (e.g. see Hathaway 1982; Rüdiger et al. 1999; Gizon and Duvall 2003). The importance of the Coriolis effect can be estimated by the Rossby number $Ro = (2\tau_c \boldsymbol{\Omega} \cdot \hat{g})^{-1}$, where τ_c is the lifetime (or correlation time) of turbulence, $\boldsymbol{\Omega}$ is the rotational velocity and $\hat{g}$ the unit vector for gravitational acceleration. For $Ro \lesssim 1$, the turbulent flows are rotationally constrained. Based on this simple analysis, surface granulation flows are too short lived to be rotationally constrained. However, larger scale, longer lived flows are expected to be modified by rotational effects. The action of the Coriolis is expected to generate clockwise motion in diverging flows in the northern hemisphere and anti-clockwise motion in the southern hemisphere. This expectation is supported by flow maps of the top few Mm of the convection zone from time-distance helioseismology inversions (Langfellner et al. 2015, see Fig. 6) and from surface flow maps derived from correlation tracking of photospheric magnetic patterns (Hathaway et al. 2013). As we shall see below, the helical nature of turbulence in the solar convection zone has important implications for the generation of twist in magnetic flux tubes.

The effect of angular momentum exchange between buoyant magnetic tubes and convective eddies was first considered by Choudhuri and D'Silva (1990) for axisymmetric thin flux tubes. They considered two scenarios. The first involves so-called 'giant cells', which the authors treat as a uniform, radially directed upflow that exert upward drag on rising tubes. In effect, this means a rising tube is no longer experiencing drag relative to a stationary background, but against a background with a net outflow velocity. For a range of field strengths, they found that an outflow with an amplitude of tens of m s^{-1} or more is sufficient to move flux tubes radially to $R = 0.9R_{\odot}$. At shallower depths, the flux tube expands so much that buoyancy and aerodynamic drag are negligible compared to the Coriolis force, which deflects tubes to latitudes beyond the Sun's activity belts.

In the second scenario, Choudhuri and D'Silva (1990) considered how interaction of the flux tube with small-scale turbulence can erode the tube's angular momentum. This effect is captured by adding a drag term to the azimuthal component of tube's equation of motion. For an initially stationary flux tube in a rotating frame, the tube gains angular momentum in a retrograde sense due to the Coriolis force. This in turn leads to a Coriolis force which deflects the tube poleward. Choudhuri and D'Silva (1990) suggested that the interaction between the flux tube and turbulent convection can reduce the angular momentum of the tube, which in turn may lead the flux tube (even those with initial field strengths of 10 kG, i.e. equipartition) to emerge within the Sun's activity belts. It should be noted that the effect of differential rotation in the solar interior is not considered in this early work. Instead, the effective drag of the background convection on the tube is computed assuming the background has zero angular velocity in the rotating reference frame. As can be seen in Fig. 2, except for the near-surface shear layer, $\frac{\partial \Omega}{\partial r} > 0$ within the bulk of the convection zone at latitudes below 30 deg. So the interaction between a rising flux tube and the super-rotating medium may be have a more pronounced effect than was reported by Choudhuri and D'Silva (1990).

Turbulent convective flows are expected to influence more than just the emergence latitude of rising magnetic tubes. The scatter of measured active region tilt angles far exceeds their latitudinal averages, being particularly large during the early emerge phase of active regions. Longcope and Fisher (1996) attribute the large scatter in tilt angles to stochastic interactions of convective flows with a rising flux tube. Using a mixing length model for the convection zone, they performed Monte Carlo simulations of thin flux tubes rising buoyantly while being buffeted by convective flows. The randomly generated convective flows modify the aerodynamic drag experienced by different sections of the tube and in doing so introduce scatter in the tilts of emerging Ω loops. When the initial field strength of the tube is 30 kG, the amplitude of the scatter in tilt angle generated by the Monte Carlo models compares well against observational results. Furthermore, ensemble averages reproduce the latitudinal dependence of the tilt angles.

The most sophisticated of thin flux tube models were carried out by Weber et al. (2011, 2013). Instead of parameterizing the effect of turbulent convective flows, they computed the aerodynamic drag force on the tube by using the 3D flow structure from a hydrodynamic global convection simulation. This non-magnetic convection model was tuned (in terms of latitudinal entropy gradient and the bottom and top boundaries and radial profiles of viscous and thermal diffusivities) to give solar-like differential rotation (see Fig. 2). Figures 7 and 8, respectively, show the shapes of the magnetic tubes (with flux $\Phi = 10^{22}$ Mx) at a time when their apices reach the top boundary ($r = 0.97 R_{\odot}$) of the simulation domain in the absence and presence of convective flows. In the absence of convection, the rise time of a flux tube with an initial field strength of $B = 15$ kG takes about half a sunspot cycle. Even a tube with initial $B = 40$ kG takes a couple of years to reach the top boundary. In contrast, the rise times of flux tubes in a convective environment is on the order of months irrespective of the initial field strength. In the Weber et al. (2013) study, they carried out a large parameter study by varying magnetic fluxes from $\Phi = 10^{20}$ to 10^{22} Mx, field strengths from 15 to 100 kG and by sampling different realizations of the convective flow pattern. They conclude that flux tubes with initial field strengths of $B \gtrsim 40$ kG produce active region properties that are most consistent with observed active region properties in terms of Joy's law, the scatter of tilt angles about the average, the most frequently occurring tilt angle and the rotation rate of emerging loops with respect to the ambient plasma.

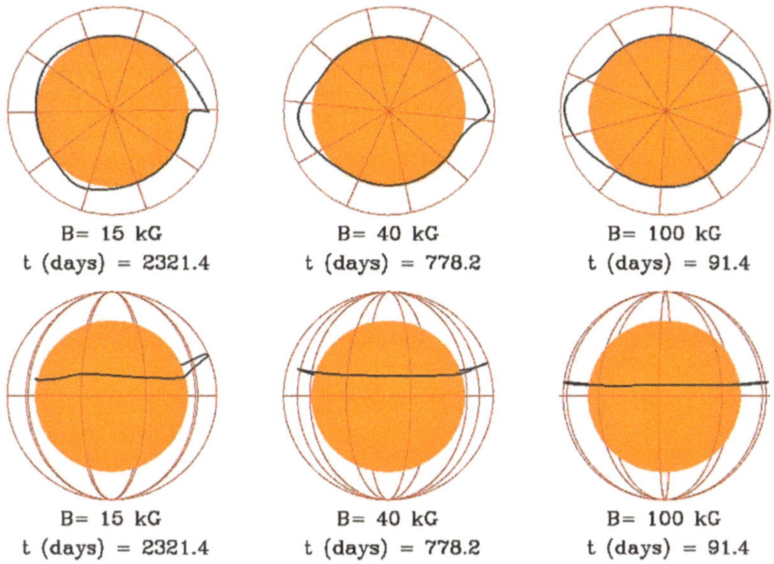

Fig. 7 Shapes of toroidal flux tubes (with initial strengths from $B = 15$ to $B = 100$ kG) at the time when the apices of the tubes reach the top boundary ($r = 0.97R_{\odot}$). This figure is reproduced from Weber et al. (2011) and shows flux tubes evolving in the absence of convective flows

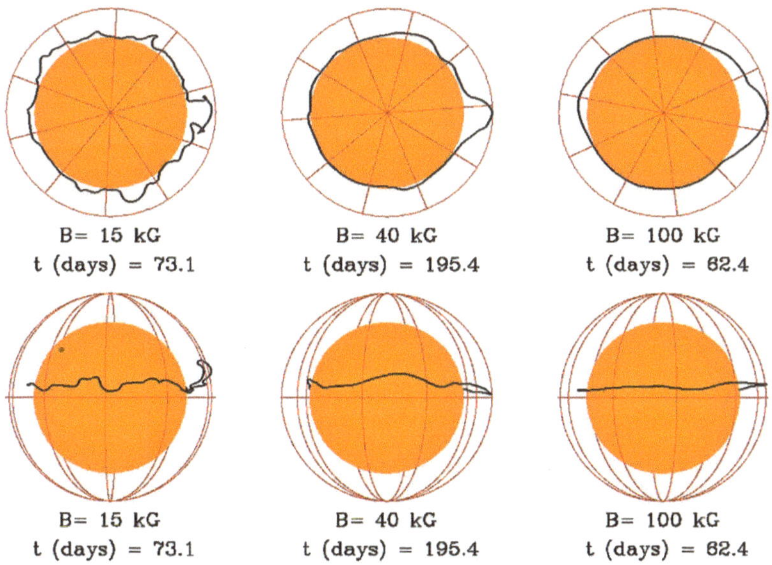

Fig. 8 Same as Fig. 7 but for flux tubes evolving against a 3D convection model

5.2 The Origin of Magnetic Twist and Chirality in Active Regions

Statistical tendencies for active regions to display signs (e.g. measurements of vertical current density) and symptoms (e.g. solar flares and eruptions) of magnetic twist in the solar atmosphere (Pevtsov et al. 1994, 2014; Wang 2013; Schmieder et al. 2014) offers clues that may help us rule out theories of the solar dynamo. An outstanding question is how active regions develop magnetic twist. Does the dynamo spawn flux concentrations that inherently have net twist? Or do the magnetic flux bundles develop twist as they rise through the convection zone or even as they erupt onto the surface? Finally, what is the relation between active region tilt and twist? First we briefly discuss ways by which magnetic twist is inferred from observations.

Once an active region has appeared at the solar surface, vector magnetograms can be used to probe the magnetic twist contained within the region. One commonly used quantity to characterize active region twist is the α parameter. Different definitions of α exist, and it is important to distinguish them for the sake of clarity. For any general magnetic field, the field-line torsion parameter can be locally defined as $\alpha_{\text{torsion}} = \mathbf{j} \cdot \mathbf{B} B^{-2}$, where $\mathbf{j} = \frac{c}{4\pi} \nabla \times \mathbf{B}$ is the current density. The sign of α_{torsion} reflects the handedness of neighboring field lines gyrating around the local magnetic field. The magnetic field is locally right-handed (left-handed) if $\alpha_{\text{torsion}} > 0$ (< 0). The amplitude of α_{torsion} relates to the amount of twisting of the field.

Measurements of all three components of $\mathbf{B}$ are typically accessible only at a single height. More precisely, photospheric vector magnetograms sample magnetic fields in a limited range of optical depths near $\tau = 0.1$. Due to the corrugation of the surfaces of constant optical depth (e.g. the Wilson depression), this impacts the accuracy of the horizontal field gradients needed for computing j_z, a problem that is especially pronounced at small spatial scales. In any case, only the $j_z B_z$ contribution to the dot product in the definition of α_{torsion} is accessible to observations. So a commonly used metric for the magnetic twist of active regions at the solar surface is $\alpha_z = j_z / B_z$ (e.g. Pevtsov et al. 1994; Leka et al. 1996; Leka and Skumanich 1999; Leka 1999). Here the dot product in the numerator has been replaced by $j_z B_z$ and B^2 in the denominator has been replaced by B_z^2, so one should recognize that α_z is *at best* a proxy for magnetic twist. The current helicity density H_c (e.g. Seehafer 1990) is related to field line torsion by the relation $H_c = \mathbf{j} \cdot \mathbf{B} = \alpha_{\text{torsion}} B^2$.

Another proxy for magnetic twist is the force-free parameter α_{FF}. If the magnetic field is force-free, i.e. $\mathbf{j} \parallel \mathbf{B}$, then α_{torsion} reduces to the force-free parameter. If the force-free field has the same value of α throughout the domain of interest, the field is called a linear force-free field and the associated twist parameter is α_{LFF}. Linear force-free fields are the solutions to the problem of minimizing the magnetic energy of the field subject to the imposed flux at the boundary and the constraint that magnetic helicity $\mathcal{H}_m$ be conserved (Woltjer 1958, also see discussion below for definition of $\mathcal{H}_m$). Observations reveal that α is usually not constant over an active region. α_{best} refers to the linear force-free field solution such that the discrepancy between the observed horizontal field (equivalently, j_z) with the model horizontal field is minimized (Pevtsov et al. 1995).

The tilt of an active region should be distinguished from the magnetic twist within the region. The tilt is generally taken to be indicative of the large-scale ($L \gtrsim d$, where d is the footpoint separation between preceding and following polarities) handedness of the flux bundle associated with an active region, while magnetic twist (especially those flavors of α using local derivative operators) measures the handedness of the magnetic field at small scales ($L \lesssim d$). Joy's law concerning active region tilts indicates a preference for right- and left-handed active regions in the northern and southern hemispheres, respectively. As discussed later, AR tilt is possibly indicative of the magnetic writhe of a flux tube. This hemispheric preference is maintained even as the magnetic polarity reverses from one cycle to the next. In contrast, vector magnetogram observations of active regions show a preference for left- and right-handed magnetic twist within active regions located in the northern and southern hemispheres, respectively. For example, Pevtsov et al. (1995) used the α_{best} metric to measure the average twist in individual active regions.

As discussed in Sect. 2.1, helioseismology has so far provided few direct probes of subsurface magnetic fields. However, knowledge about the flow structure in the solar interior offers some important clues. Local helioseismology reveals that, in the northern hemisphere of the Sun, divergent flows (associated with upwellings) are correlated with negative vertical vorticity ω_z and convergent flows (associated with downdrafts) are correlated with positive ω_z (Gizon and Duvall 2003; Langfellner et al. 2015, see Fig. 6). These results are consistent with the effect of the Coriolis force on the convective motions. In a local Cartesian frame in the northern hemisphere, $v_z \omega_z < 0$ for both upflows and downflows. Although the quantity $v_z \omega_z$ is only one component of the kinetic helicity $H_k = \mathbf{u} \cdot \nabla \times \mathbf{u}$, it nevertheless serves as a useful proxy for the sign of kinetic helicity in the turbulent solar convection. With respect to the sign of the mean kinetic helicity $\langle \mathbf{u} \cdot \nabla \times \mathbf{u} \rangle$, helioseismic results agree with theoretical (Rüdiger et al. 1999) and numerical calculations (Hathaway 1982; Ossendrijver et al. 2001) in that they suggest $\langle \mathbf{u} \cdot \nabla \times \mathbf{u} \rangle < 0$ in the northern hemisphere and $\langle \mathbf{u} \cdot \nabla \times \mathbf{u} \rangle > 0$ in the southern hemisphere.

Applying the model of Longcope and Klapper (1997) for twisted thin flux tubes (see brief discussion in Sect. 5), Longcope et al. (1998) considered how initially untwisted flux tubes rising from the bottom of the convection zone may develop net twist as a consequence of their interaction with turbulence with non-vanishing kinetic helicity. In their model, the generation of net twist (as measured in terms of the magnetic pitch $q(l)$ averaged over the tube) is a purely geometrical effect that can be explained in terms of magnetic helicity. As pointed out by Woltjer (1958) and Moffatt (1969), the magnetic helicity of a system, defined as

$$\mathcal{H}_m = \int_V \mathbf{A} \cdot \mathbf{B} \, dV, \tag{1}$$

where $\mathbf{B} = \nabla \times \mathbf{A}$, is conserved under ideal MHD evolution when no magnetic flux crosses the boundary ∂V. For a set of closed rings of magnetic flux, as in the case of a global toroidal twisted flux tube, the magnetic helicity can be decomposed into contributions attributed to the writhe (denoted Wr) of the axis of the flux ring and the total twist of field lines around the axis (denoted Tw, see Berger and Field 1984; Moffatt and Ricca 1992). In this case,

$$\mathcal{H}_m = \frac{\Phi^2}{2\pi}(\mathrm{Tw} + \mathrm{Wr}), \tag{2}$$

where Φ is the magnetic flux of the tube. Longcope and Klapper (1997) pointed out that a flux tube with zero initial twist and writhe (hence zero magnetic helicity) can rise such that it develops net writhe due to buffeting by turbulent flows with non-vanishing mean kinetic helicity $\langle \mathbf{u} \cdot \nabla \times \mathbf{u} \rangle$. In their scenario, the interaction with turbulence is in terms of the drag on different portions of the rising tube, so the evolution satisfies ideal MHD. The helical motion of the ambient flows leads to a net writhe of the flux tube. Since $\mathcal{H}_m$ must remain zero, magnetic twist of the same amplitude but opposite sign as writhe must result. The local generation of twist due to writhing of the tube is coined the Σ effect. Σ is the symbol used by Longcope et al. (1998) to denote the term that generates twist in time evolution equation of field line twist in the their thin flux tube model.

Based on theoretical expectations, Longcope et al. (1998) assumed $\langle \mathbf{u} \cdot \nabla \times \mathbf{u} \rangle$ to be negative (positive) in the northern (southern) hemisphere. As discussed earlier this expectation is consistent with helioseismic inferences (at least at supergranular scales, see Gizon and Duvall 2003; Langfellner et al. 2015). With this assumption, their model predicts $\mathrm{Wr} > 0$ (i.e. right-handed tilts of active regions) and $\mathrm{Tw} < 0$ (i.e. $\alpha_z < 0$) in the northern hemisphere. Longcope et al. (1998) argue that models relying on the α-effect (in the mean-field context, not to be confused with the magnetic twist parameter) would produce active regions with the same signs for Wr and Tw, which would be incompatible with empirical reports (e.g. Tian et al. 2001; Hagino and Sakurai 2004). This last claim deserves a bit of scrutiny. As demonstrated by Seehafer (1996), the α-effect produces magnetic helicity of opposite signs at the large and small scales. Consequently, the mean-field α-effect usually invoked to explain the tilts of active regions should generate small-scale twist of the opposite sign at sub-active region scales. So the tendency for $\mathrm{Tw} \times \mathrm{Wr} < 0$ does not, by itself, favor one picture over another.

6 Active Regions from Convective Dynamos

In Sect. 5, we introduced various physical mechanisms that are deemed important for the birth of active regions. These effects include magnetic buoyancy, magnetoconvection, helical flows induced by the Coriolis force, angular momentum conservation, and magnetic helicity conservation. Due to the complexity of the problem, these effects have mostly been studied in isolation. For example, the convective flow patterns used in the thin flux tube models of Weber et al. (2011, 2013, see Sect. 5.1) were purely hydrodynamic, so the work side-steps questions regarding the origin of toroidal flux tubes and the Lorentz force feedback of the magnetic field on the convection. The question of Lorentz force feedback is addressed in the 3D MHD models of flux tube rise by Fan (2008), Fan et al. (2013), Jouve and Brun (2009), Jouve et al. (2013) and Pinto and Brun (2013). The numerical models in these studies solve the MHD equations in a spherical domain without resorting to the thin flux tube approximation. However, the flux tubes in these models are introduced in an ad hoc fashion.

Recent global turbulent dynamo simulations are beginning to capture all these effects in a self-consistent manner. Some of these models are starting to reproduce certain aspects of solar-like behavior. In particular, some dynamo calculations produce large-scale buoyant flux bundles that rise to the surface. An example of turbulent dynamo calculations generating large-scale bipolar structures is the work of Jabbari et al. (2015). In order to be able to control the degree by which helical and non-helical motions drive the dynamo, they opted to not model convection, but instead imposed a forcing function in the momentum equation. They found that, by varying the latitudinal profile of the helical portion of the forcing function, they are able to generate bipolar active-region like structures in the mid-latitudes or in the polar regions. This result suggests the latitudinal shear profile in the Sun and Sun-like stars likely has an impact on the latitudes of activity belts.

The relation between helically-forced turbulence and rotationally affected thermal convection (as occurs on the Sun) is not straightforward. However, global convective dynamo models (with no externally prescribed forcing except for a chosen luminosity and background rotation rate) are also able to produce active-region scale flux concentrations. There are many unknowns about magnetoconvection in the high magnetic and dynamic Reynolds number, high density contrast regime of the solar convection zone. However, the solar luminosity is well measured and is not an adjustable parameter for a solar model. Since this paper is about solar active regions and is not intended to give an overview of convective dynamos, the following discussion will focus on models configured to transmit one solar luminosity ($L_\odot$).

The anelastic convective dynamo calculations by Nelson et al. (2011, 2013, 2014) for a model of a young Sun rotating at $\Omega = 3\Omega_\odot$ generate mean toroidal fields in the form of a pair of magnetic wreaths spanning the low to mid-latitudinal regions of the convection

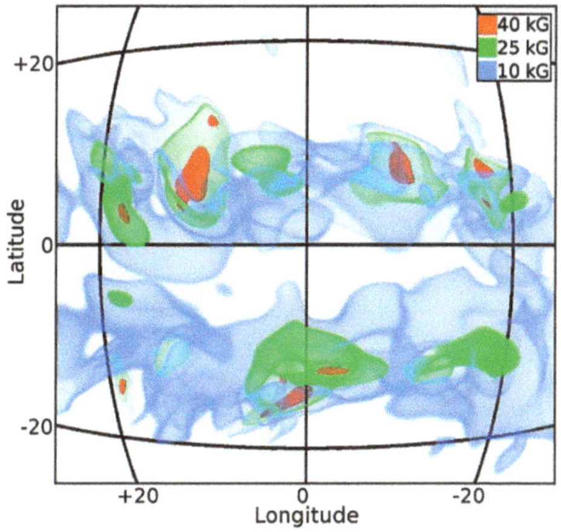

Fig. 9 Isosurface rendering of magnetic field strength in a convective dynamo simulation of a star rotating at $\Omega = 3\Omega_\odot$ by Nelson et al. (2013). Coherent toroidal magnetic wreaths are generated with field strengths beyond 10 kG. The magnetic wreathes contain cores that have superequipartition fields strengths reaching 45 kG

zone. (see Fig. 9). Convectively induced undulations of the wreaths spawn buoyant magnetic loops that rise toward the surface (see Fig. 10). The field strengths in the troughs of these undulations are roughly tens of kG, consistent with the field strengths of toroidal fields inferred from thin flux tube experiments (see Sect. 5).

The global convective dynamo model of Fan and Fang (2014, 2015) of a star rotating at $\Omega_\odot$ reproduces solar-like differential rotation, yields super-equipartition field strengths (also tens of kG) near the bottom of the convection zone and generates buoyant flux bundles that rise toward the surface. Statistical analysis shows that the bundles are Hale/anti-Hale at a rate of 2.4:1. Hale-like bundles have a mean tilt of 7.5 ± 1.6 deg.

6.1 Lorentz Force Feedback by Active Region Magnetic Fields

An intriguing finding from the Fan and Fang (2014, 2015) study is that a purely hydrodynamic convection zone model yields anti-solar (fast poles and slow equator) differential rotation. This hints at the (possible) importance of the back reaction of the Maxwell stresses on the differential rotation profile.

The recognition that Maxwell stresses may modulate differential rotation is not new. Soon after the discovery of the torsional oscillation (a butterfly-like equatorward propagating pattern of zonal flow residuals relative to the temporally averaged differential rotation profile, see Howard and Labonte 1980, and Chap. 9 of Howe 2009), the Lorentz force has been invoked as a possible driver of the oscillation (Schüssler 1981; Rempel 2007).[1] However the strength of the "torsional oscillation" (zonal flows) in terms of rotation speed less than one percent of differential rotation. So if the Fan and Fang (2014, 2015) model were representative of the solar interior, their result hints at a more important role for Maxwell stresses than previously assumed.

Recent global dynamo simulations of a model star rotating at $\Omega = \Omega_\odot$ (Hotta et al. 2016) suggest that Lorentz force feedback from turbulent fields generated at the small-scales may even impact the strength of the large-scale toroidal field. At a magnetic Reynolds number $R_m \approx 300$ (with constant explicit values of the magnetic diffusivity η and viscosity ν), the dynamo generates coherent toroidal magnetic wreaths. For the same numerical resolution but with implicit numerical diffusion set by a slope limiter scheme, the simulation becomes

[1] Drivers of a thermal nature have also been investigated by Spruit (2003) and Rempel (2007).

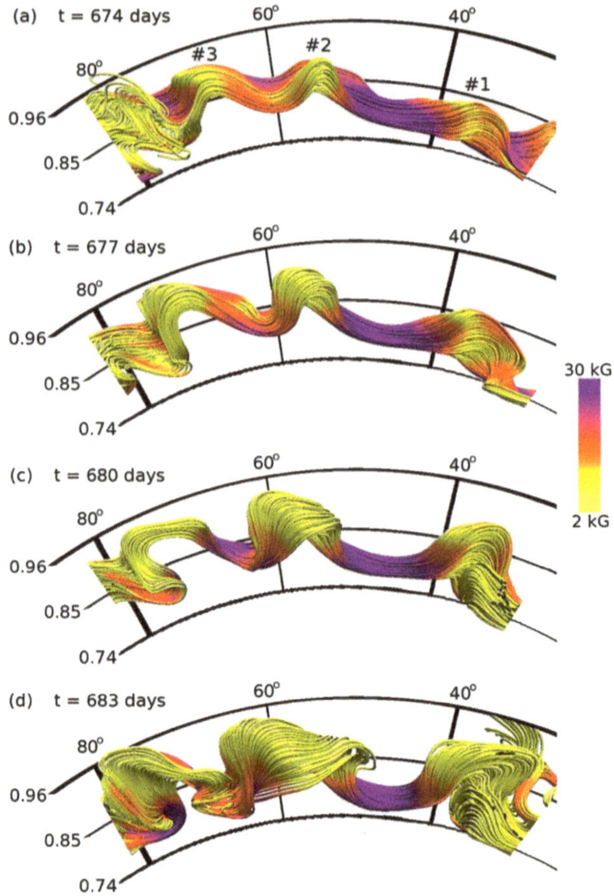

Fig. 10 Sequence of volume renderings of buoyant Ω-loops generated in the convective dynamo model of Nelson et al. (2014)

less diffusive, which increases the convective flow speeds at the small-scales. These stronger convective flows were found to have a destructive effect on the large-scale toroidal field. However, when the numerical resolution is increased even further such that $R_m \sim 2000$, the small-scale dynamo becomes efficient and the Lorentz-force feedback from the small-scale field dampens convective flows. This results again in a stronger large-scale toroidal field. In all three cases the large-scale toroidal field predominantly results from amplification by shear in the solar-like differential rotation profile and the poloidal field is amplified by convective turbulence. In short it is an $\alpha\Omega$ dynamo model. What this result demonstrates is that a large-scale field can persist (and as shown in Hotta et al. 2016, exhibit reversals) even in the presence of a small-scale dynamo if the latter is sufficiently efficient to have a strong Lorentz-force feedback.

Models where the interactions of the small and the large-scale fields are treated in a unified system are still in their infancy. There is at present no indication of convergence of the solutions with respect to magnetic and hydrodynamic Reynolds numbers. In particular, simulations have a tendency to operate in a regime where both Reynolds number are very similar, a regime that is known to favor turbulent dynamo action (see Tobias et al. 2013). The convective zone of the Sun is, however, in a very different regime where the dynamic Reynolds number overwhelmingly exceeds the magnetic one (in other words the magnetic

Prandtl number $Pr = \eta/\nu \ll 1$). Under these circumstances the nature of the turbulently driven dynamo action is not yet well understood (see Martínez Pillet 2013, and references therein; see also Thaler and Spruit 2015). On the other hand, all current numerical simulations of convective dynamos in the solar context are effectively large-eddy simulations where the smallest captured scales for both magnetic and velocity fields are far removed from the theoretical dissipation scales. It is unclear what the appropriate subgrid models for magnetic and viscous dissipation are and, given the appropriate models, whether they result in an effective Pr regime equivalent to that expected from atomic values of the diffusivities. However it is clear that a potential coupling between the two dynamos can be essential for advancing our understanding of the origin of active regions on the Sun.

7 Emergence and Active Life of AR Magnetic Fields

Having discussed the subsurface genesis and evolution of active region magnetic fields, we briefly turn to the next phase of their life cycle, namely emergence and development at and above the solar surface.

The evolutionary state of an AR is closely related to the level of magnetic activity it produces, i.e. the frequency and importance of flares and coronal mass ejections. For ARs that are flare productive, larger flares (i.e. above the C-class) are produced as long as high magnetic flux density/strong magnetic field concentrations are present, and the overall flare production rate decreases sharply with decreasing flux density, while the rate of CME production shows a much slower decrease with time, i.e. CMEs keep erupting from active regions well into their decay phase (van Driel-Gesztelyi and Green 2015). This is not surprising, as during the decay phase persistent flux cancellations along inversion lines occur (Green et al. 2011). This is taken as an observational signature of flux rope generation, either, through reconnection and submergence of sheared field lines (van Ballegooijen and Martens 1989) or via flux emergence (MacTaggart and Hood 2010). A prerequisite for sustained CME production is maintained magnetic shear or helicity in the active region field and sufficient amounts of flux cancellations. As CMEs consume magnetic free energy and carry away magnetic helicity, they exhaust this reservoir.

As an example, Cheung and DeRosa (2012) constructed time-dependent magnetofrictional coronal field models of AR 11158 by using SDO/HMI line-of-sight magnetograms. Magnetofriction assumes the plasma velocity in the induction equation to be parallel to the Lorentz force (Yang et al. 1986). By varying a shearing parameter that controls energy injection, they report that persistent shearing/twisting near the polarity inversion line is required to generate serial flux rope ejections. By using time sequences of *vector* magnetograms as well as surface flow maps from Doppler measurements and local correlation tracking, it is now possible to derive the full photospheric electric field of emerging and developing active regions. This allows one to quantify the fluxes of magnetic energy and relative helicity (Kazachenko et al. 2014, 2015) from the solar interior to the corona. Electric fields derived in this way have also been used to drive a magnetofrictional model of AR 11158 over a 5-day period (Fisher et al. 2015, see Fig. 11).

For an extensive review of active region evolution (during the phase when they are visible at and above the photosphere), we encourage the reader to consult van Driel-Gesztelyi and Green (2015). For discussions of the detailed processes related to magnetic flux emergence, we refer the reader to reviews by Schmieder et al. (2014) and Cheung and Isobe (2014).

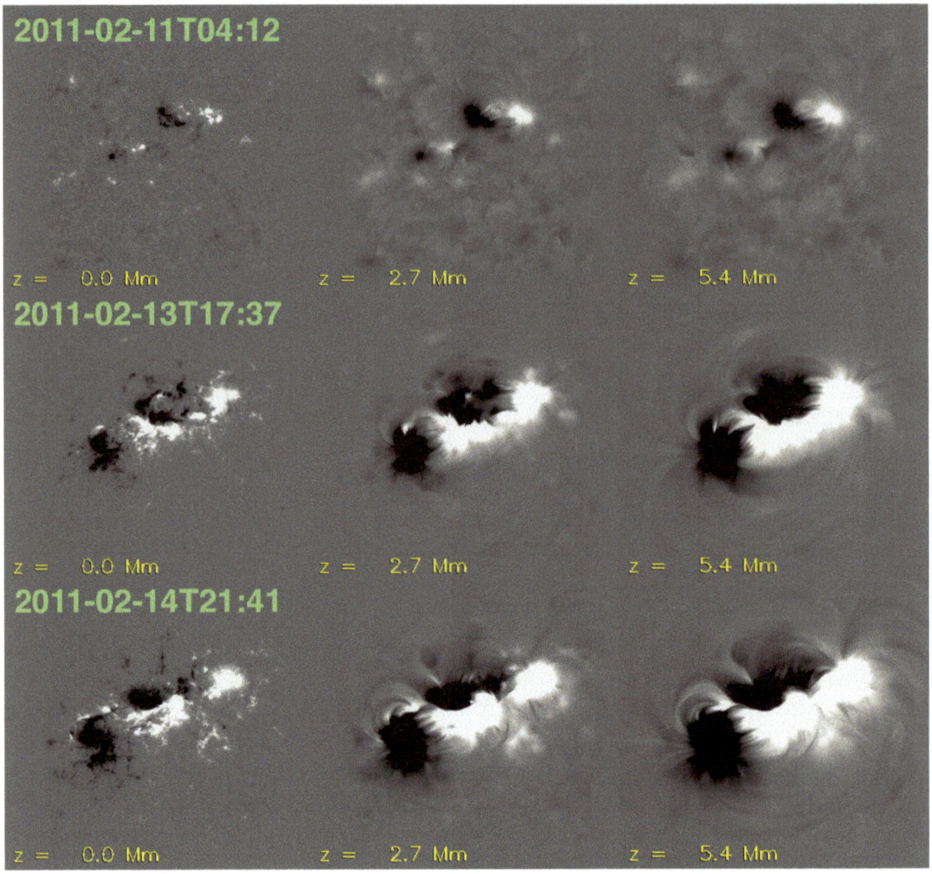

Fig. 11 Distribution of B_z at photospheric ($z = 0$ Mm), chromospheric ($z = 2.7$ Mm) and coronal heights ($z = 5.4$ Mm) of a data-driven model of NOAA AR 11158 (images produced from simulation data described in Fisher et al. 2015). The data-driven time dependent model retains information about the emergence history. Early in its evolution (see snapshot at 2011-02-11T04:12), the AR consists of two isolated bipolar regions. As the two bipolar regions grow, the following polarity of one (to the west) jousts against the leading polarity of the other and creates a sheared polarity inversion line

8 Ageing Active Regions

After the brief period of flux emergence, which lasts for up to five days, depending on the total flux to be emerged, starts the decay phase of active regions, which is at least an order of magnitude longer. How long it takes for the emerged flux to disperse to a quiet-sun magnetic flux density level while being removed by magnetic cancellation, depends on the (i) total magnetic flux (size) of the active region, (ii) level of magnetic activity outside and (iii) magnetic complexity within the active region. During high activity periods large active regions may become indistinguishable from the "background" magnetic field within four months, while during solar minima that may take as long as 10 months (Schrijver and Harvey 1994).

During the dispersion process the area of the active region has an approximate linear growth (van Driel-Gesztelyi et al. 2003), while the fields are more and more effectively shaped by differential rotation, rotating the polarity inversion line clockwise in the south, and counter-clockwise in the north hemisphere. However, the weighted polarity separation may not be affected by the dispersion (Li and Welsch 2008). Differential rotation shears the dispersing fields, i.e. injects magnetic helicity (DeVore 2000). However, the helicity injection is not a monotonous function of time. Once the magnetic inversion line has rotated to 45° with respect to the equator, the sign of helicity injection by differential rotation will change, and at any larger angle (i.e. more parallel to the equator) the accumulated helicity will be eroded and even reversed by this process. Démoulin et al. (2002) estimated a total magnetic helicity of 8.3×10^{42} Mx^2 produced by differential rotation during a five-month period in AR 7978, which would only supply enough helicity for 4 CMEs, while the AR 7978 produced 28 CMEs during this period (see Table 6 & 7 in van Driel-Gesztelyi and Green 2015). Démoulin et al. (2002) concluded that for AR 7978, which decayed in isolation, differential rotation was not the major source of magnetic helicity injection.

Removal of magnetic flux takes place by magnetic cancellation (Martin et al. 1985), which it is thought to reflect, in the majority of the cases, small-scale submergence of flux. There is an asymmetry between flux submergence and its counterpart, flux emergence, that is worth noticing. Flux emergence at all scales has a well-defined observational signature, namely fields parallel to the solar surface harbouring upflows. They have been detected in the large scale emergence of active regions (see Fig. 3 of Lites et al. 1998) and during the surfacing of small scale loops brought up by granules (see Figs. 5 and 6 of Guglielmino et al. 2010). The opposite, flux submergence detected as fields parallel to the solar surface in a downflow region is more scarce (but see Chae et al. 2004) and it has always a localized patchy nature. Submergence can be brought along by downward magnetic tension forces in high-curvature small-scale loops created in low-atmospheric magnetic reconnection (Cheung et al. 2008; Cameron et al. 2011). In cancellation processes opposite polarity flux concentrations approach each other, and an equal amount of flux disappears from each polarity. The main cancellation sites are along the magnetic inversion lines within the active region and along its external boundaries. In the latter case, submergence is expected to have the observed patchy nature as it will depend on the neighboring fields that are randomly encountered as the active region flux disperses. By contrast, along the active region neutral line, the systematic cancellation processes observed during the decay phase have not yet been unequivocally demonstrated to represent submergence episodes. Indeed the most detailed study of the transverse fields observed at an active region neutral line, concluded that these field lines display large scale upflows of a few hundreds meters per second (see Kuckein et al., 2012; see also the discussion in Mackay, 2015). It is clear that the nature of the cancellation processes at active region neutral lines urgently needs further investigation.

Magnetic complexity within the active region naturally increases the magnetic cancellation rate, resulting in the survival of the majority polarity locally. Such cancellation processes effectively wipe out complexity within an ageing active region, leading to a simplification of the magnetic structure, establishing a pair of opposite-polarity monopolar regions. Small-scale flux emergence episodes within either polarity may take place, but as long as the emerging flux is inferior to the flux of the ageing active region, complexity is only temporarily introduced. As such secondary flux emergence episodes are expected to bring up equal amounts of positive and negative fluxes (although see Wang and Shi 1993; Yardley et al. 2016, for examples when does not appear to be the case) and cancellation processes between the ageing and new flux remove equal amount of fluxes, these smaller-scale flux emergence episodes only temporarily increase the total flux present.

A number of detailed case studies indicate that active regions have flux loss rates in the range 5×10^{19}–5×10^{20} Mx day^{-1}. Green et al. (2011) reported a flux loss rate of 2×10^{20} Mx day^{-1}, corresponding to a relative loss rate of $\dot{\Phi}/\Phi = -0.1$ day^{-1}, where Φ is half of the total unsigned flux. At such a rate the active region would not last longer than 10 days and so is unlikely to contribute significantly to the polar flux budget. However, Wang et al. (1989, see their Fig. 3) measured the flux decay of an active region over 4 solar rotations and obtained a rate of 5×10^{19} Mx day^{-1}, a much reduced value. Dalda and Martinez Pillet (2008) measured a flux decay rate of 4×10^{20} Mx day^{-1} for a period of about 2.5 days that was followed by a plateau where the remaining flux evolved at a much slower pace. A similar curve, including a plateau region, was found by Sterling et al. (2010, see their Fig. 6). Tentatively, one would conclude that there is a flux decay phase of active regions that occurs at a pace of several 10^{20} Mx day^{-1} and that acts upon only a fraction of the total flux of the active region. For the case studied by Dalda and Martinez Pillet (2008), they estimated this to be about 50–70 % of the total active region flux. The remaining portion of the flux disappears at a much slower pace, perhaps at a rate compatible with the decay measured by the diffusion processes as obtained by Wang et al. (1989, basically one order of magnitude slower). It is worth noting that morphological changes in the spatial flux distribution of an active region (e.g. spots disintegrating into small-scale magnetic features) can complicate flux loss measurements. This due to the temperature-sensitivity of the Stokes profiles of some spectral lines. So one may detect apparent flux changes even though the flux has only been redistributed (flux "vanishing in situ"). In any case studies of the flux loss rates of a large sample of active regions is critically needed.

In some cases, unusual sunspot motions may occur, possibly owing to non-Ω-loop (e.g. U-loop, see van Driel-Gesztelyi et al. 2000) subsurface geometry, or multiple major flux emergence episodes in an active region, which drive large flux concentrations towards the internal inversion line, significantly increasing the cancellation rates there (Yardley et al. 2016). The presence of active regions in the vicinity of emerging flux, if they are favorably oriented, can lead to increased cancellation rates along the external boundaries and shortened active-region lifetimes. High decay rates have indeed been measured in complex active regions where not only small flux concentrations but spot-size magnetic fields cancel (e.g. 5×10^{20} Mx day^{-1} cancellation rate found in AR 11226 by Yardley et al. 2016; see also Sterling et al. 2010).

Magnetic flux cancellation along magnetic inversion lines do not only remove flux, but also change the overall magnetic topology of the ageing active region. Through these processes, magnetic shear is getting gradually converted into twist, forming a flux rope along the magnetic inversion line (van Ballegooijen and Martens 1989), and redistributing helicity (from small to large scales, see Mackay et al. 2014), concentrating it along the inversion line, which explains the gradient in force-free alpha found in active regions (Schmieder et al. 1996). Not all the cancelled flux is built in the forming flux rope, however, while the shear is transformed into twist. Green et al. (2011) very clearly show (see their Fig. 11) how cancellations in a five-fieldline sheared arcade would form a two-fieldline flux rope, whose footpoints had no opposite-polarity cancelling partners in the interior of the arcade. This suggests that the more sheared the arcade is the more flux can be built in the flux rope. At the opposite end of the scale, no flux rope forms in a potential arcade where each flux concentration has an opposite polarity with which to cancel. Savcheva et al. (2012) shows that most models are compatible with 60–70 % of the cancelled flux and 30–50 % of the total flux of a sheared active region being equal to the sum of toroidal (parallel to flux rope axis) and poloidal (transverse to flux rope axis) fluxes of flux ropes built up this way.

There exists a number of surface flux transport (SFT) models aiming to explain how active regions decay and how their magnetic flux disperses over the solar surface. SFT models treat the transport of radial magnetic flux on the solar surface due to advection by differential rotation and meridional circulation and by turbulent diffusion. For a comprehensive review, we refer the reader to papers in this volume by Cameron et al. (2016) and by Wang (2016) and the Living Reviews paper by Mackay and Yeates (2012). SFT modelers have had considerable success in applying models to explain the processes by which magnetic flux in the polar caps build up and subsequently reverse. The tilt of active regions is key to this. Due to Faraday's law (resulting in the conservation of magnetic flux), there are two ways to create net residual flux in a hemisphere. The first is for transport of flux across the equator. Due to Joy's law, trailing polarity spots are, on average, found at higher latitudes while leading polarities are found at lower latitudes. The flux from leading polarities is more likely to be transported across the equator, which results in residual flux from the trailing polarities to be transported toward the polar regions by meridional circulation and diffusion. The second way is to have tilted active regions emerge such that they straddle the equator.

SFT models are kinematic, which neglects the back reaction of the Lorentz force on flows. Furthermore, 2D SFT models do not take into account the 3D nature of active regions. For developing physical insight into how active regions may decay, one must resort to 3D radiative MHD simulations. Rempel and Cheung (2014) carried out a number of active region emergence and decay experiments by advecting semi-toroidal flux tubes through the bottom boundary of their Cartesian domain, which is located at 15.5 Mm below the photosphere. In their models the numerical resolution and the top boundary conditions were chosen such that the model active regions only developed naked spots. They examined the rate of flux dispersal for one of the model spots in the first two days during the decay phase and found it to be consistent with 2D diffusion with turbulent diffusivity $\eta_{\mathrm{turb}} = 350\,\mathrm{km}^2\,\mathrm{s}^{-1}$.

Rempel (2015) carried out 3D radiative MHD simulations of isolated sunspots (see Fig. 12). For an experiment where the top boundary condition enforces a potential magnetic field, the model spot has only very short, rudimentary penumbral filaments. It is almost 'naked'. For a numerical experiment with the same initial conditions but with a top boundary condition that forces the field to be more inclined to the vertical (so that the horizontal field is twice as strong compared to a potential field), the umbra is surrounded by a penumbra. For the naked spot, Rempel (2015) reports a flux decay rate of roughly $10^{21}\,\mathrm{Mx\,day}^{-1}$. In comparison, the sunspot with penumbra maintains its flux content over the same period of a few days. He analyzed the flow structure around the two spots and found that the filling factor of downflows in the periphery of the spot is suppressed in the case of the spot with a fully-fledged penumbra. This reduces the flux loss due to downward pumping of magnetic flux, and keeps the sunspot coherent for longer.

Consider a sunspot in a real active region on the Sun. The sunspot resides within an active region, which itself may be located in the vicinity of other active regions. Due to existing magnetic connectivities between the spot and surrounding polarities, the relaxation of the chromospheric and coronal field above the spot is not unconstrained and is unlikely to be in the lowest energy potential state. Instead the field will be closer to a non-linear force-free, current-carrying state. The results of Rempel (2015) gives some physical grounding to the idea that chromospheric and coronal complexity may have an impact on the underlying sunspot in terms of photospheric magnetic field and penumbral structure. And if the presence of a penumbra helps stabilize a spot from decay, it would suggest that the coronal field configuration may have some influence on the decay process of sunspots and active regions. These are speculations that can be tested by analysis of observations and by further numerical modeling.

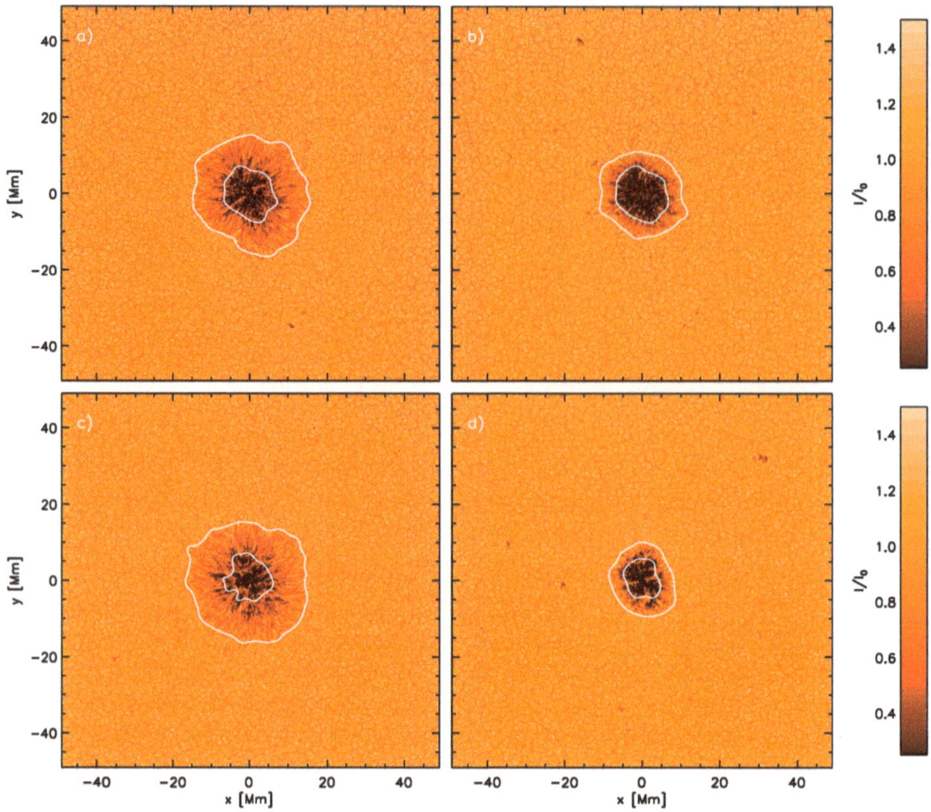

Fig. 12 3D radiative MHD simulations of sunspots with and without a penumbra (from Rempel 2015). Panels (**a**) and (**c**) show the brightness intensity of a model sunspot with a penumbra separated in time by 50 hr. This model was run with a top boundary condition that enforces the horizontal field to be more inclined than a potential field. Panels (**b**) and (**d**) show a corresponding model for a sunspot with rudimentary penumbra, which was run with a potential field top boundary condition. The latter decays at a rate of 10^{21} Mx day^{-1} whereas former is essentially stable

9 Closing (the Loop) Remarks

Statistical studies of the properties of active regions and ephemeral regions (e.g. frequency, latitude and orientation of emergence) suggest that the two features are drawn from a broad spectrum of magnetic structures generated within the solar interior. The existence of magnetic bipoles down to the scales of granulation provides evidence for the operation of a small-scale dynamo in addition to the large-scale dynamo. But the 'two' dynamos are not decoupled. The continuous distribution of bipole sizes motivates the search for a unified physical model to explain both active regions and smaller magnetic bipoles.

Helioseismology has so far not provided direct constraints on the strengths and geometries of active regions before they emerge. However, helioseismic characterization of flows in the solar interior gives indirect hints as to how active regions are generated and transported. For instance, the differential rotation profile in radius and latitude provides a constraint on convection and magnetoconvection models. In addition, the measurement of the kinetic helicity in the two hemispheres reveals how convective flows at supergranular scales introduce twist and writhe into rising flux tubes. There is empirical evidence that active regions exhibit opposite signs of small-scale twist and large-scale writhe (the latter in terms of tilt). Buffeting of the active region by convective flows introduces negative (positive) twist in the northern (southern) hemisphere and the flux bundle can as a result develop a writhe of the opposite sign. This is consistent with magnetic helicity conservation.

Both simple thin flux tube models and state-of-the-art global convective dynamo models suggest active regions begin their rise to the surface as superequipartition (tens of kG) strength magnetic bundles lying deep in the solar interior (bottom half of the convection zone).

The most recent global convective dynamo calculations (Nelson et al. 2011; Fan and Fang 2014; Hotta et al. 2016) yield toroidal magnetic wreaths amplified by differential rotation within the convection zone. Despite the apparent success of recent models, one should still be cautious about their fidelity to the Sun. Although convective dynamo models are starting to reproduce some solar-like properties, there remains a number of outstanding issues. In several models, different choices of diffusivities (viscous, thermal and magnetic) and numerical resolutions can yield qualitatively vastly different results (e.g. Nelson et al. 2013; Fan and Fang 2014; Hotta et al. 2016) in terms of the strength of the mean and maximum subsurface toroidal fields generated by the dynamo.

Furthermore, several hydrodynamic (Brown et al. 2008; Fan and Fang 2014) and magnetohydrodynamic (Fan and Fang 2014; Hotta et al. 2016) models are tuned so that a large fraction (about one third in the middle of the convection zone) of the outward energy flux is carried by radiative/thermal diffusion as opposed to enthalpy transport (i.e. positive correlations between temperature fluctuations and radial flows). The enhancement of thermal diffusion suppresses the otherwise large convective r.m.s. velocities, which in turn reduces the Reynolds stress transport of angular momentum to the poles. This artificial suppression of the Reynolds stress is generally recognized as a practical device to obtain solar-like differential rotation (slow poles and fast equator). For the time being this remains an ad hoc choice for a sub-grid model of energy transport. The implications of this sub-grid model for our understanding of solar magnetoconvection remain unclear.

The aforementioned generation of global solar dynamo models do not include the tachocline nor the near surface shear layer (see Fig. 2), which are the regions of strongest shear in the solar interior. Such models can be contrasted with earlier older models where the tachocline (e.g. Dikpati and Gilman 2001) or the near surface shear layer (Brandenburg 2005) act as primary agents of the Ω-effect and/or as the reservoir for field storage (e.g. Browning et al. 2006). Even without the tachocline and near surface shear layer the dynamo models are able to yield active region-like buoyant magnetic bundles. However they do not produce quasi-periodic reversals with solar-like butterfly diagrams. So the roles of the neglected shear layers remain unknown.

The buoyant flux bundles generated in the convective dynamo models do not rise through the surface to become compact active regions harboring sunspots. The primary reason is that the simulation domains do not capture the top twenty or so Mm (about half of the total pressure scale heights) of the convection zone. There are separate Cartesian MHD models of emerging active regions that capture those near-surface layers (Cheung et al. 2010; Stein and Nordlund 2012; Rempel and Cheung 2014), but the large-scale emerging fields in those models are imposed in an ad hoc fashion (by specifying certain bottom boundary conditions). The goal for future models is to connect this divide and to capture the formation, development and decay of active regions.

Surface flux transport models of active regions subject to differential rotation, turbulent diffusion and meridional circulation achieve considerable success in reproducing the migration of flux to the high latitudes to cause reversals of the large-scale poloidal field (refer to companion review papers in this volume by Cameron et al. and by Wang). How similar patterns of magnetic flux migration occur in the solar interior is less well known. The model of van Ballegooijen and Mackay (2007) treats the evolution of magnetic field in both the convection zone and corona in a simplified fashion. The coronal field evolves by magnetofriction (i.e. setting $\mathbf{v} \propto \mathbf{j} \times \mathbf{B}$ in the induction equation). The kinematically imposed flows in the model interior mimic (1) downward pumping by convection and (2) the ideal buoyant rise of a section of a toroidal flux tube. In this model, an Ω-loop is made to emerge to create an active region with flux expanding into the corona. Subsequent flux cancellation at the photospheric polarity inversion line leads to the formation of a flux rope that erupts due to loss of equilibrium. Associated with this is the submergence of flux back into the convection zone, which results in the repair of the original toroidal belt.

In this description of active region evolution, the final fate of the field lines is to submerge and return to their original state inside the Sun. But elsewhere in this paper we have described how some of the active region flux is known to be dispersed through the action of meridional circulation and diffusion, a critical process for the solar dynamo that leads to the reversal of the polar fields. This brings up the question of what fraction of the active region magnetic flux, if any, submerges and repairs the toroidal tube and what fraction is transported to the poles. One would also need to understand what distinguishes the two types of field lines: is the fraction of flux that participates in the diffusion process the one that somehow got disconnected from below and the fraction that submerges the one that is still tightly rooted to the deep interior? At any rate, it is unclear if the two processes co-exists during the decay phase of an active region and an effort must be made to understand if this is the case. Some indication for the co-existence of these two processes comes from the comparison of historical values of decay rates. Flux losses from recurrent active regions that are known to contribute to the polar reversal are hard to quantify as it is difficult to identify the amount of flux that belongs to the active region in each rotation.

Yeates and Muñoz Jaramillo (2013) extended the solar interior kinematic model of van Ballegooijen and Mackay (2007) to perform a data-driven 3D Babcock–Leighton flux transport model of a solar cycle (see also Miesch and Dikpati 2014). They prescribed subsurface flows designed to advect toroidal flux onto the surface. Idealized bipolar emergences were imposed at the times and locations (latitudes and longitudes) of large bipolar magnetic regions ($\Phi \geq 10^{22}$ Mx) spotted in Kitt Peak synoptic magnetograms during the years 1996 to 2008. The prescribed plasma expansion in the loops corresponds to an effective density contrast of 27 between the bottom of the convection zone and the photosphere. This is somewhat unrealistic since the actual density contrast in the Sun is 10^6. Nevertheless, the model is calibrated to yield bipolar magnetic regions at $r = R_\odot$ to have statistical properties similar to observed ones (e.g. tilts). The key advantage of this data-driven model is that it provides the 3D distribution of the subsurface magnetic field at different times of the cycle (see Fig. 13). This model includes downward turbulent pumping (as a prescribed background flow) and captures the toroidal belt repair mechanism proposed by van Ballegooijen and Mackay (2007). The model also includes differential rotation, which is the primary mechanism for toroidal field amplification. By virtue of being a kinematic model, it cannot be used to address questions regarding the structural stability of the toroidal belts. But recall that global convective dynamo calculations yield persistent toroidal magnetic wreathes pervading the mid- to low convection zone.

The range of measured loss rates of magnetic flux from active regions (5×10^{19}–5×10^{20} Mx day^{-1}) quoted in Sect. 8 are based on a few studies and need to be confirmed over a larger number of active regions. If the faster decay rates are associated with the cancellation at the neutral line, the importance of any potential flux repairing mechanism

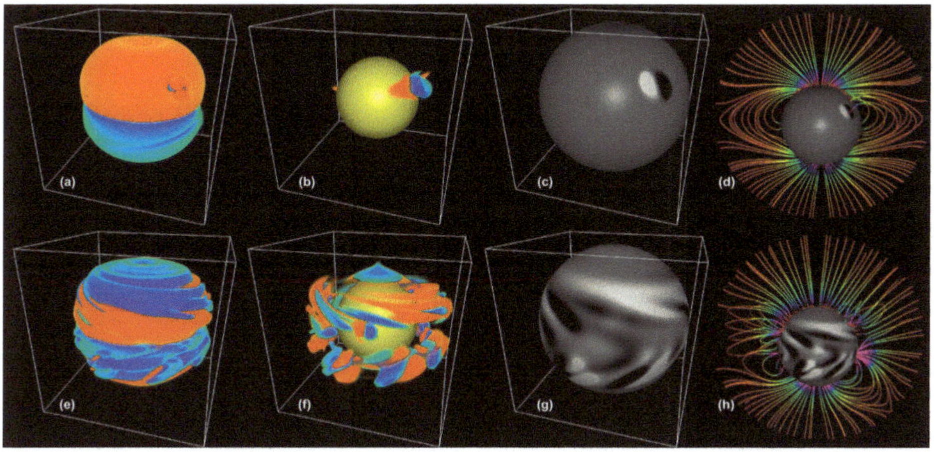

Fig. 13 3D data-driven Babcock–Leighton flux transport model by Yeates and Muñoz Jaramillo (2013). The *top* (*bottom*) *row* shows visualizations near solar minimum (maximum). Panels in the *first* and *second columns* (from the *left*) show 3D renderings of the toroidal and poloidal components of the subsurface magnetic field, respectively. Panels in the *third column* shows the radial field at $r = R_{\odot}$ (saturated at $B_r = \pm 25$ G). Panels in the *fourth column* show potential field source surface extrapolations

can be quantified. The flux that decays at a much slower pace would be the only one that remains available for surface transport processes to reverse the polar fields. Thus studying the flux history of a statistically significant number of active regions and the dynamics at the neutral line can prove crucial to understand the key players in the last phases of the evolution of an active region. Note that persistent submergence of active region flux and repairing of the toroidal tubes cannot continue beyond the corresponding solar cycle, thus raising the question about how the toroidal flux of a given cycle is finally destroyed. One option is the encounter of the two oppositely directed toroidal belts in each hemisphere near the equator where they would annihilate. Another important mechanism may be the unwinding of the toroidal field due to the reversal of the poloidal field. In any case, understanding which processes are key for closing the solar dynamo loop in its final stages requires first to detect the existence of any flux repairing mechanism and the dynamics of the magnetic fields at the neutral line of active regions. The answers to these questions may be key to closing the loop on the life cycle of active region magnetic fields.

Acknowledgements MCMC is supported by NASA's SDO/AIA contract (NNG04EA00C) and grant (NNX13AJ96G) to LMSAL. AIA is an instrument onboard the Solar Dynamics Observatory, a mission for NASA's Living With a Star program. Part of this paper was written while MCMC was a Visiting Associate Professor at the National Astronomical Observatory of Japan (NAOJ) in Tokyo. MCMC wishes to thank his NAOJ hosts, especially Professor Shin Toriumi, for their excellent hospitality. LvDG acknowledges the Hungarian Scientific Research Fund (OTKA) for support through grant no. K-109276. The National Solar Observatory is operated by AURA, Inc. under a cooperative agreement with the National Science Foundation. The National Center for Atmospheric Research is sponsored by the National Science Foundation. Finally, the authors wish to thank the anonymous referee for very constructive feedback.

References

M.D. Altschuler, G. Newkirk, Sol. Phys. **9**, 131 (1969)
H.W. Babcock, Astrophys. J. **133**, 572 (1961)

B. Belucz, M. Dikpati, E. Forgács-Dajka, Astrophys. J. **806**, 169 (2015)
M.A. Berger, G.B. Field, J. Fluid Mech. **147**, 133 (1984)
A. Brandenburg, Astrophys. J. **550**, 824 (2001)
A. Brandenburg, Astrophys. J. **625**, 539 (2005)
B.P. Brown, M.K. Browning, A.S. Brun, M.S. Miesch, J. Toomre, Astrophys. J. **689**, 1354 (2008)
M.K. Browning, M.S. Miesch, A.S. Brun, J. Toomre, Astrophys. J. Lett. **648**, L157 (2006)
D. Buehler, A. Lagg, S.K. Solanki, Astron. Astrophys. **555**, A33 (2013)
P. Caligari, F. Moreno-Insertis, M. Schüssler, Astrophys. J. **441**, 886 (1995)
P. Caligari, M. Schüssler, F. Moreno-Insertis, Astrophys. J. **502**, 481 (1998)
R. Cameron, A. Vögler, M. Schüssler, Astron. Astrophys. **533**, A86 (2011)
R.H. Cameron, M. Dikpati, A. Brandenburg, Space Sci. Rev. (2016). doi:10.1007/s11214-015-0230-3
R. Centeno, H. Socas-Navarro, B. Lites, M. Kubo, Z. Frank, R. Shine, T.D. Tarbell, A.M. Title, K. Ichimoto, S. Tsuneta, Y. Katsukawa, Y. Suematsu, T. Shimizu, S. Nagata, Astrophys. J. Lett. **666**, L137 (2007)
J. Chae, Y.-J. Moon, A.A. Pevtsov, Astrophys. J. Lett. **602**, L65 (2004)
P.F. Chen, Living Rev. Sol. Phys. **8** (2011)
M.C.M. Cheung, M.L. DeRosa, Astrophys. J. **757**, 147 (2012)
M.C.M. Cheung, H. Isobe, Living Rev. Sol. Phys. **11**, 3 (2014)
M.C.M. Cheung, M. Schüssler, F. Moreno-Insertis, Astron. Astrophys. **467**, 703 (2007)
M.C.M. Cheung, M. Schüssler, T.D. Tarbell, A.M. Title, Astrophys. J. **687**, 1373 (2008)
M.C.M. Cheung, M. Rempel, A.M. Title, M. Schüssler, Astrophys. J. **720**, 233 (2010)
A.R. Choudhuri, Sol. Phys. **123**, 217 (1989)
A.R. Choudhuri, Astron. Astrophys. **239**, 335 (1990)
A.R. Choudhuri, S. D'Silva, Astron. Astrophys. **239**, 326 (1990)
A.R. Choudhuri, P.A. Gilman, Astrophys. J. **316**, 788 (1987)
J. Christensen-Dalsgaard, D.O. Gough, M.J. Thompson, Astrophys. J. **378**, 413 (1991)
A.S. Dalda, V.M. Martinez Pillet, in *Subsurface and Atmospheric Influences on Solar Activity*, ed. by R. Howe, R.W. Komm, K.S. Balasubramaniam, G.J.D. Petrie. Astronomical Society of the Pacific Conference Series, vol. 383 (2008), p. 115
S. Danilovic, B. Beeck, A. Pietarila, M. Schüssler, S.K. Solanki, V. Martínez Pillet, J.A. Bonet, J.C. del Toro Iniesta, V. Domingo, P. Barthol, T. Berkefeld, A. Gandorfer, M. Knölker, W. Schmidt, A.M. Title, Astrophys. J. Lett. **723**, L149 (2010a)
S. Danilovic, M. Schüssler, S.K. Solanki, Astron. Astrophys. **513**, A1 (2010b)
B. De Pontieu, Astrophys. J. **569**, 474 (2002)
P. Démoulin, C.H. Mandrini, L. Van Driel-Gesztelyi, M.C. Lopez Fuentes, G. Aulanier, Sol. Phys. **207**, 87 (2002)
C.R. DeVore, Astrophys. J. **539**, 944 (2000)
M. Dikpati, P.A. Gilman, Astrophys. J. **559**, 428 (2001)
I. Domínguez Cerdeña, J. Sánchez Almeida, F. Kneer, Astron. Astrophys. **407**, 741 (2003)
S. D'Silva, A.R. Choudhuri, Astron. Astrophys. **272**, 621 (1993)
Y. Fan, Living Rev. Sol. Phys. **1** (2004)
Y. Fan, Astrophys. J. **676**, 680 (2008)
Y. Fan, Living Rev. Sol. Phys. **6** (2009)
Y. Fan, F. Fang, Astrophys. J. **789**, 35 (2014)
Y. Fan, F. Fang, ArXiv e-prints: 1512.08038 (2015)
Y. Fan, G.H. Fisher, E.E. Deluca, Astrophys. J. **405**, 390 (1993)
Y. Fan, G.H. Fisher, A.N. McClymont, Astrophys. J. **436**, 907 (1994)
Y. Fan, N. Featherstone, F. Fang, ArXiv e-prints: 1305.6370 (2013)
A. Ferriz-Mas, M. Schüssler, Geophys. Astrophys. Fluid Dyn. **72**, 209 (1993)
A. Ferriz-Mas, M. Schüssler, Geophys. Astrophys. Fluid Dyn. **81**, 233 (1995)
G.H. Fisher, Y. Fan, R.F. Howard, Astrophys. J. **438**, 463 (1995)
G.H. Fisher, W.P. Abbett, D.J. Bercik, M.D. Kazachenko, B.J. Lynch, B.T. Welsch, J.T. Hoeksema, K. Hayashi, Y. Liu, A.A. Norton, A.S. Dalda, X. Sun, M.L. DeRosa, M.C.M. Cheung, Space Weather **13**, 369 (2015)
P.M. Giles PhD thesis, Stanford University (2000)
L. Gizon, T.L. Duvall Jr., in *Local and Global Helioseismology: The Present and Future*, ed. by H. Sawaya-Lacoste. ESA Special Publication, vol. 517 (2003), pp. 43–52, GONG+ 2002
M. Gošić, L.R. Bellot Rubio, D. Orozco Suárez, Y. Katsukawa, J.C. del Toro Iniesta, Astrophys. J. **797**, 49 (2014)
J.T. Gosling, J. Geophys. Res. **98**, 18937 (1993)
T. Granzer, M. Schüssler, P. Caligari, K.G. Strassmeier, Astron. Astrophys. **355**, 1087 (2000)
L.M. Green, B. Kliem, A.J. Wallace, Astron. Astrophys. **526**, A2 (2011)

B.J. Greer, B.W. Hindman, N.A. Featherstone, J. Toomre, Astrophys. J. Lett. **803**, L17 (2015)
S.L. Guglielmino, L.R. Bellot Rubio, F. Zuccarello, G. Aulanier, S. Vargas Domínguez, S. Kamio, Astrophys. J. **724**, 1083 (2010)
D.A. Haber, B.W. Hindman, J. Toomre, R.S. Bogart, R.M. Larsen, F. Hill, Astrophys. J. **570**, 855 (2002)
H.J. Hagenaar, Astrophys. J. **555**, 448 (2001)
H.J. Hagenaar, C.J. Schrijver, A.M. Title, Astrophys. J. **584**, 1107 (2003)
M. Hagino, T. Sakurai, Publ. Astron. Soc. Jpn. **56**, 831 (2004)
G.E. Hale, Astrophys. J. **28**, 315 (1908)
G.E. Hale, S.B. Nicholson, Astrophys. J. **62**, 270 (1925)
G.E. Hale, F. Ellerman, S.B. Nicholson, A.H. Joy, Astrophys. J. **49**, 153 (1919)
S.M. Hanasoge, T.L. Duvall, K.R. Sreenivasan, Proc. Natl. Acad. Sci. **109**, 11928 (2012)
K.L. Harvey PhD thesis, Univ. Utrecht (1993)
K.L. Harvey, C. Zwaan, Sol. Phys. **148**, 85 (1993)
J.W. Harvey, D. Branston, C.J. Henney, C.U. Keller, Astrophys. J. Lett. **659**, L177 (2007)
D.H. Hathaway, Sol. Phys. **77**, 341 (1982)
D.H. Hathaway, L. Upton, O. Colegrove, Science **342**, 1217 (2013)
G. Hazra, B.B. Karak, A.R. Choudhuri, Astrophys. J. **782**, 93 (2014)
H. Hotta, M. Rempel, T. Yokoyama, Astrophys. J. Lett. **759**, L24 (2012)
H. Hotta, M. Rempel, T. Yokoyama, Astrophys. J. **803**, 42 (2015)
H. Hotta, M. Rempel, T. Yokoyama, Science **351**, 1427 (2016)
R. Howard, B.J. Labonte, Astrophys. J. Lett. **239**, L33 (1980)
R. Howe, Living Rev. Sol. Phys. **6** (2009)
Y. Iida, H.J. Hagenaar, T. Yokoyama, Astrophys. J. **752**, 149 (2012)
R. Ishikawa, S. Tsuneta, Astron. Astrophys. **495**, 607 (2009)
R. Ishikawa, S. Tsuneta, K. Ichimoto, H. Isobe, Y. Katsukawa, B.W. Lites, S. Nagata, T. Shimizu, R.A. Shine, Y. Suematsu, T.D. Tarbell, A.M. Title, Astron. Astrophys. **481**, L25 (2008)
S. Jabbari, A. Brandenburg, N. Kleeorin, D. Mitra, I. Rogachevskii, Astrophys. J. **805**, 166 (2015)
J. Jackiewicz, A. Serebryanskiy, S. Kholikov, Astrophys. J. **805**, 133 (2015)
J. Jiang, R.H. Cameron, D. Schmitt, M. Schüssler, Astron. Astrophys. **528**, A83 (2011)
L. Jouve, A.S. Brun, Astrophys. J. **701**, 1300 (2009)
L. Jouve, A.S. Brun, G. Aulanier, Astrophys. J. **762**, 4 (2013)
M.D. Kazachenko, G.H. Fisher, B.T. Welsch, Astrophys. J. **795**, 17 (2014)
M.D. Kazachenko, G.H. Fisher, B.T. Welsch, Y. Liu, X. Sun, Astrophys. J. **811**, 16 (2015)
E.V. Khomenko, M. Collados, S.K. Solanki, A. Lagg, J. Trujillo Bueno, Astron. Astrophys. **408**, 1115 (2003)
K.O. Kiepenheuer, in *Structure and Development of Solar Active Regions*, ed. by K.O. Kiepenheuer. Series IAU Symposium, vol. 35 (1968), p. 3
R. Komm, I. González Hernández, F. Hill, R. Bogart, M.C. Rabello-Soares, D. Haber, Sol. Phys. **287**, 85 (2013)
F. Krause, K. Rädler, *Mean-Field Magnetohydrodynamics and Dynamo Theory* (Pergamon, Elmsford, 1980)
C. Kuckein, V. Martínez Pillet, R. Centeno, Astron. Astrophys. **539**, A131 (2012)
J. Langfellner, L. Gizon, A.C. Birch, Astron. Astrophys. **581**, A67 (2015)
K.D. Leka, Sol. Phys. **188**, 21 (1999)
K.D. Leka, A. Skumanich, Sol. Phys. **188**, 3 (1999)
K.D. Leka, R.C. Canfield, A.N. McClymont, L. van Driel-Gesztelyi, Astrophys. J. **462**, 547 (1996)
Y. Li, B.T. Welsch, in *Subsurface and Atmospheric Influences on Solar Activity*, ed. by R. Howe, R.W. Komm, K.S. Balasubramaniam, G.J.D. Petrie. Astronomical Society of the Pacific Conference Series, vol. 383 (2008), p. 397
B.W. Lites, Space Sci. Rev. **144**, 197 (2009)
B.W. Lites, K.D. Leka, A. Skumanich, V. Martinez Pillet, T. Shimizu, Astrophys. J. **460**, 1019 (1996)
B.W. Lites, A. Skumanich, V. Martinez Pillet, Astron. Astrophys. **333**, 1053 (1998)
B.W. Lites, M. Kubo, H. Socas-Navarro, T. Berger, Z. Frank, R. Shine, T. Tarbell, A. Title, K. Ichimoto, Y. Katsukawa, S. Tsuneta, Y. Suematsu, T. Shimizu, S. Nagata, Astrophys. J. **672**, 1237 (2008)
M. Lockwood, A.P. Rouillard, I.D. Finch, Astrophys. J. **700**, 937 (2009)
D.W. Longcope, G.H. Fisher, Astrophys. J. **458**, 380 (1996)
D.W. Longcope, I. Klapper, Astrophys. J. **488**, 443 (1997)
D.W. Longcope, B.T. Welsch, Astrophys. J. **545**, 1089 (2000)
D.W. Longcope, G.H. Fisher, A.A. Pevtsov, Astrophys. J. **507**, 417 (1998)
K.B. MacGregor, P. Charbonneau, Astrophys. J. **519**, 911 (1999)
D.H. Mackay, in *Solar Prominences*, ed. by J.-C. Vial, O. Engvold. Astrophysics and Space Science Library, vol. 415 (2015), p. 355
D. Mackay, A. Yeates, Living Rev. Sol. Phys. **9** (2012)

D.H. Mackay, C.R. DeVore, S.K. Antiochos, Astrophys. J. **784**, 164 (2014)
D. MacTaggart, A.W. Hood, Astrophys. J. Lett. **716**, L219 (2010)
S.F. Martin, S.H.B. Livi, J. Wang, Aust. J. Phys. **38**, 929 (1985)
M.J. Martínez González, L.R. Bellot Rubio, Astrophys. J. **700**, 1391 (2009)
V. Martínez Pillet, Space Sci. Rev. **178**, 141 (2013)
F. Meyer, H.U. Schmidt, N.O. Weiss, Mon. Not. R. Astron. Soc. **179**, 741 (1977)
K.A. Meyer, D.H. Mackay, A.A. van Ballegooijen, C.E. Parnell, Sol. Phys. **272**, 29 (2011)
M.S. Miesch, M. Dikpati, Astrophys. J. Lett. **785**, L8 (2014)
H.K. Moffatt, J. Fluid Mech. **35**, 117 (1969)
H.K. Moffatt, Magnetic field generation in electrically conducting fluids (1978)
H.K. Moffatt, R.L. Ricca, in *Royal Society of London* (1992), pp. 411–429
F. Moreno-Insertis, P. Caligari, M. Schüessler, Sol. Phys. **153**, 449 (1994)
F. Moreno-Insertis, P. Caligari, M. Schüssler, Astrophys. J. **452**, 894 (1995)
N.J. Nelson, B.P. Brown, A.S. Brun, M.S. Miesch, J. Toomre, Astrophys. J. Lett. **739**, L38 (2011)
N.J. Nelson, B.P. Brown, A.S. Brun, M.S. Miesch, J. Toomre, Astrophys. J. **762**, 73 (2013)
N.J. Nelson, B.P. Brown, A. Sacha Brun, M.S. Miesch, J. Toomre, Sol. Phys. **289**, 441 (2014)
A. Ortiz, L.R. Bellot Rubio, V.H. Hansteen, J. de la Cruz Rodríguez, L. Rouppe van der Voort, Astrophys. J. **781**, 126 (2014)
M. Ossendrijver, M. Stix, A. Brandenburg, Astron. Astrophys. **376**, 713 (2001)
E. Pariat, G. Aulanier, B. Schmieder, M.K. Georgoulis, D.M. Rust, P.N. Bernasconi, Astrophys. J. **614**, 1099 (2004)
E.N. Parker, Astrophys. J. **121**, 491 (1955)
C.E. Parnell, Sol. Phys. **200**, 23 (2001)
A.A. Pevtsov, R.C. Canfield, T.R. Metcalf, Astrophys. J. Lett. **425**, L117 (1994)
A.A. Pevtsov, R.C. Canfield, T.R. Metcalf, Astrophys. J. Lett. **440**, L109 (1995)
A.A. Pevtsov, M.A. Berger, A. Nindos, A.A. Norton, L. van Driel-Gesztelyi, Space Sci. Rev. **186**, 285 (2014)
J. Pietarila Graham, S. Danilovic, M. Schüssler, Astrophys. J. **693**, 1728 (2009)
R.F. Pinto, A.S. Brun, Astrophys. J. **772**, 55 (2013)
K.-H. Rädler, Astron. Nachr. **301**, 101 (1980)
S.P. Rajaguru, H.M. Antia, Astrophys. J. **813**, 114 (2015)
M.D. Rempel, Struktur und Ursprung starker Magnetfelder am Boden der solaren Konvektionszone (Niedersächsische Staats-und Universitätsbibliothek) (2001)
M. Rempel, Astrophys. J. **655**, 651 (2007)
M. Rempel, Astrophys. J. **789**, 132 (2014)
M. Rempel, Astrophys. J. **814**, 125 (2015)
M. Rempel, M.C.M. Cheung, Astrophys. J. **785**, 90 (2014)
M. Rempel, M. Schüssler, Astrophys. J. Lett. **552**, L171 (2001)
P. Riley, J.A. Linker, Z. Mikić, R. Lionello, S.A. Ledvina, J.G. Luhmann, Astrophys. J. **653**, 1510 (2006)
G. Rüdiger, A. Brandenburg, V.V. Pipin, Astron. Nachr. **320**, 135 (1999)
A.S. Savcheva, L.M. Green, A.A. van Ballegooijen, E.E. DeLuca, Astrophys. J. **759**(2), 105 (2012)
A. Schad, J. Timmer, M. Roth, Astrophys. J. Lett. **778**, L38 (2013)
R. Schlichenmaier, N. Bello González, R. Rezaei, T.A. Waldmann, Astron. Nachr. **331**, 563 (2010)
B. Schmieder, P. Demoulin, G. Aulanier, L. Golub, Astrophys. J. **467**, 881 (1996)
B. Schmieder, V. Archontis, E. Pariat, Space Sci. Rev. **186**, 227 (2014)
J. Schou, H.M. Antia, S. Basu, R.S. Bogart, R.I. Bush, S.M. Chitre, J. Christensen-Dalsgaard, M.P. Di Mauro, W.A. Dziembowski, A. Eff-Darwich, D.O. Gough, D.A. Haber, J.T. Hoeksema, R. Howe, S.G. Korzennik, A.G. Kosovichev, R.M. Larsen, F.P. Pijpers, P.H. Scherrer, T. Sekii, T.D. Tarbell, A.M. Title, M.J. Thompson, J. Toomre, Astrophys. J. **505**, 390 (1998)
C.J. Schrijver, M.L. De Rosa, Sol. Phys. **212**, 165 (2003)
C.J. Schrijver, K.L. Harvey, Sol. Phys. **150**, 1 (1994)
C.J. Schrijver, A.M. Title, A.A. van Ballegooijen, H.J. Hagenaar, R.A. Shine, Astrophys. J. **487**, 424 (1997)
M. Schüssler, Astron. Astrophys. **94**, L17 (1981)
M. Schüssler, A. Vögler, Astron. Astrophys. **481**, L5 (2008)
M. Schüssler, P. Caligari, A. Ferriz-Mas, F. Moreno-Insertis, Astron. Astrophys. **281**, L69 (1994)
M. Schüssler, P. Caligari, A. Ferriz-Mas, S.K. Solanki, M. Stix, Astron. Astrophys. **314**, 503 (1996)
N. Seehafer, Sol. Phys. **125**, 219 (1990)
N. Seehafer, Phys. Rev. A **53**, 1283 (1996)
G.W. Simon, A.M. Title, N.O. Weiss, Astrophys. J. **561**, 427 (2001)
H.C. Spruit, Astron. Astrophys. **98**, 155 (1981)
H.C. Spruit, Sol. Phys. **213**, 1 (2003)
H.C. Spruit, A.A. van Ballegooijen, Astron. Astrophys. **106**, 58 (1982)

R.F. Stein, Å. Nordlund, Astrophys. J. Lett. **753**, L13 (2012)
A.C. Sterling, C. Chifor, H.E. Mason, R.L. Moore, P.R. Young, Astron. Astrophys. **521**, A49 (2010)
M. Stix, Astron. Astrophys. **47**, 243 (1976)
L.H. Strous, C. Zwaan, Astrophys. J. **527**, 435 (1999)
I. Thaler, H.C. Spruit, Astron. Astrophys. **578**, A54 (2015)
M.J. Thompson, J. Toomre, E.R. Anderson, H.M. Antia, G. Berthomieu, D. Burtonclay, S.M. Chitre, J. Christensen-Dalsgaard, T. Corbard, M. De Rosa, C.R. Genovese, D.O. Gough, D.A. Haber, J.W. Harvey, F. Hill, R. Howe, S.G. Korzennik, A.G. Kosovichev, J.W. Leibacher, F.P. Pijpers, J. Provost, E.J. Rhodes Jr., J. Schou, T. Sekii, P.B. Stark, P.R. Wilson, Science **272**, 1300 (1996)
M.J. Thompson, J. Christensen-Dalsgaard, M.S. Miesch, J. Toomre, Annu. Rev. Astron. Astrophys. **41**, 599 (2003)
L. Tian, S. Bao, H. Zhang, H. Wang, Astron. Astrophys. **374**, 294 (2001)
S.M. Tobias, F. Cattaneo, S. Boldyrev, in *Ten Chapters in Turbulence* (Cambridge University Press, Cambridge, 2013)
S. Tomczyk, J. Schou, M.J. Thompson, Astrophys. J. Lett. **448**, L57 (1995)
S. Toriumi, Y. Iida, K. Kusano, Y. Bamba, S. Imada, Sol. Phys. **289**, 3351 (2014)
J. Trujillo Bueno, N. Shchukina, A. Asensio Ramos, Nature **430**, 326 (2004)
A.A. van Ballegooijen, Astron. Astrophys. **113**, 99 (1982)
A.A. van Ballegooijen, D.H. Mackay, Astrophys. J. **659**, 1713 (2007)
A.A. van Ballegooijen, P.C.H. Martens, Astrophys. J. **343**, 971 (1989)
L. van Driel-Gesztelyi, L.M. Green, Living Rev. Sol. Phys. **8** (2015)
L. van Driel-Gesztelyi, J.-M. Malherbe, P. Démoulin, Astron. Astrophys. **364**, 845 (2000)
L. van Driel-Gesztelyi, P. Démoulin, C.H. Mandrini, L. Harra, J.A. Klimchuk, Astrophys. J. **586**, 579 (2003)
A. Vögler, M. Schüssler, Astron. Astrophys. **465**, L43 (2007)
Y.-M. Wang, Astrophys. J. Lett. **775**, L46 (2013)
Y.-M. Wang, Space Sci. Rev. (2016). doi:10.1007/s11214-016-0257-0
J. Wang, Z. Shi, Sol. Phys. **143**, 119 (1993)
Y.-M. Wang, A.G. Nash, N.R. Sheeley Jr., Science **245**, 712 (1989)
Y.-M. Wang, J. Lean, N.R. Sheeley Jr., Geophys. Res. Lett. **27**, 505 (2000)
D.F. Webb, T.A. Howard, Living Rev. Sol. Phys. **9** (2012)
M.A. Weber, Y. Fan, M.S. Miesch, Astrophys. J. **741**, 11 (2011)
M.A. Weber, Y. Fan, M.S. Miesch, Sol. Phys. **287**, 239 (2013)
L. Woltjer, Proc. Natl. Acad. Sci. USA **44**, 489 (1958)
W.H. Yang, P.A. Sturrock, S.K. Antiochos, Astrophys. J. **309**, 383 (1986)
S.L. Yardley, L.M. Green, D.R. Williams, L. van Driel-Gesztelyi, G. Valori, S. Dacie, ArXiv e-prints: 1606.08264 (2016)
A.R. Yeates, A. Muñoz-Jaramillo, Mon. Not. R. Astron. Soc. **436**, 3366 (2013)
J. Zhao, R.S. Bogart, A.G. Kosovichev, T.L. Duvall Jr., T. Hartlep, Astrophys. J. Lett. **774**, L29 (2013)
C. Zwaan, Sol. Phys. **60**, 213 (1978)
C. Zwaan, Sol. Phys. **100**, 397 (1985)

Space Sci Rev (2017) 210:351–365
DOI 10.1007/s11214-016-0257-0

Surface Flux Transport and the Evolution of the Sun's Polar Fields

Y.-M. Wang[1]

Received: 28 March 2016 / Accepted: 20 April 2016 / Published online: 26 April 2016

Abstract The evolution of the polar fields occupies a central place in flux transport (Babcock–Leighton) models of the solar cycle. We discuss the relationship between surface flux transport and polar field evolution, focusing on two main issues: the latitudinal profile of the meridional flow and the axial tilts of active regions. Recent helioseismic observations indicate that the poleward flow speed peaks at much lower latitudes than inferred from magnetic feature tracking, which includes the effect of supergranular diffusion and thus does not represent the actual bulk flow. Employing idealized simulations, we demonstrate that flow profiles that peak at mid latitudes give rise to overly strong and concentrated polar fields. We discuss the differences between magnetic and white-light measurements of tilt angles, noting the large uncertainties inherent in the sunspot group measurements and their tendency to underestimate the actual tilts. We find no clear evidence for systematic cycle-to-cycle variations in Joy's law during cycles 21–23. Finally, based on the observed evolution of the Sun's axial dipole component and polar fields up to the end of 2015, we predict that cycle 25 will be similar in amplitude to cycle 24.

Keywords Solar cycle · Solar dynamo · Solar polar fields · Magnetic flux transport on Sun

1 Introduction

The development of flux-transport dynamo models, starting with the work of Babcock (1961) and Leighton (1964, 1969), has been strongly influenced and guided by observations of processes at the solar surface. Of key significance were the first observations, after the invention of the Babcock magnetograph in the 1950s, of the polar fields and their reversal (Babcock 1959); Leighton's discovery of supergranular convection in the 1960s; and the detection and measurement of poleward-directed surface meridional flows from the late 1970s onward (see, e.g., Duvall 1979). The observationally driven, partly phenomenological nature of the flux transport model has been in contrast to the traditional $\alpha\omega$ mean-field dynamo models, where most of the important processes remain hidden beneath the surface.

✉ Y.-M. Wang
yi.wang@nrl.navy.mil

[1] Code 7682W, Space Science Division, Naval Research Laboratory, Washington, DC 20375, USA

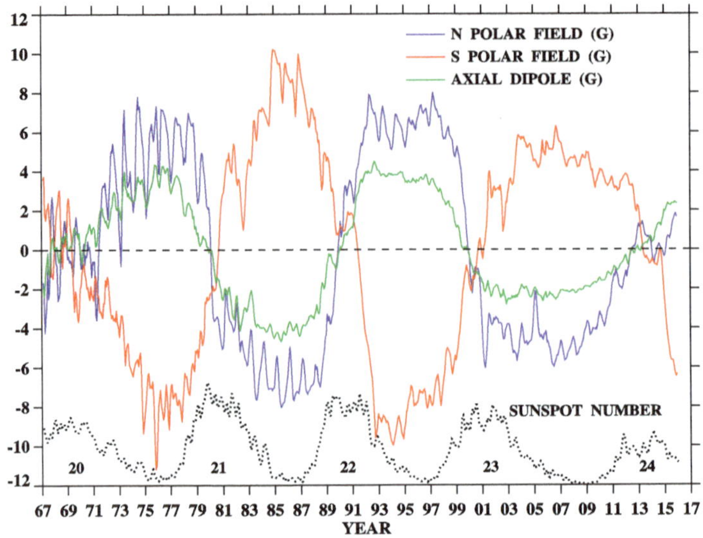

Fig. 1 Evolution of the Sun's polar fields during 1967–2015, derived from MWO and WSO observations. The curves show the mean flux density poleward of latitude $|L| = 60°$ in each hemisphere (*blue*: north pole; *red*: south pole). Also plotted is the Sun's axial dipole strength (*green*). The MWO and WSO measurements of the line-of-sight photospheric field have been deprojected by dividing by $\cos L$, multiplied by $(4.5 - 2.5 \sin^2 L)$ to correct for the saturation of the Fe I 525.0 nm line profile (see Ulrich 1992; Wang and Sheeley 1995), and averaged together during their period of overlap (1976–2012). For reference, the sunspot number (arbitrarily normalized and shifted) is shown by the *black dotted line*. All curves represent 3 Carrington rotation (CR) running means

The Sun's polar fields and their evolution have a central role in flux transport dynamos. Figure 1 shows the variation, from 1967 to the present, of the average photospheric field strength above latitude $L = \pm 60°$ in each hemisphere. The measurements are from the Mount Wilson Observatory (MWO: 1967–2012) and the Wilcox Solar Observatory (WSO: 1976–2015), with the two data sets being averaged during their period of overlap (see Wang et al. 2009). The polar fields reverse their sign rapidly at sunspot maximum and build up to their greatest strength during the declining phase of the cycle, with the variation sometimes showing a nonmonotonic behavior (ignoring fluctuations due to noise and the solar B0 angle). The polar field strengths remained in the range ∼6–9 G during the 1976, 1986, and 1996 sunspot minima, but fell to only ∼4–5 G during the 2008–2009 minimum. It may also be seen that the south polar field completed its reversal from positive to negative in 2014, and has strengthened rapidly since then. In contrast, the north polar field remains relatively weak, after having undergone a series of reversals: from negative to positive in 2012, from positive to negative at the beginning of 2014, and finally from negative to positive again at the beginning of 2015. The Sun's axial dipole component (also plotted in Fig. 1) shows a simpler, more monotonic variation, changing from negative to positive at the start of 2013.

The relationship between the polar fields, active regions, and surface flux transport is most easily seen by plotting the longitudinally averaged photospheric field as a function of latitude and time, as shown in Fig. 2 (top panel). As expected, the active regions form butterfly patterns as their latitudes of emergence migrate equatorward over each sunspot cycle. It is evident from the plot that the polar fields are formed by transporting magnetic flux from the active regions to the poles. This transport tends to occur in a series of discrete surges, due to the combination of the poleward meridional flow and the episodic nature

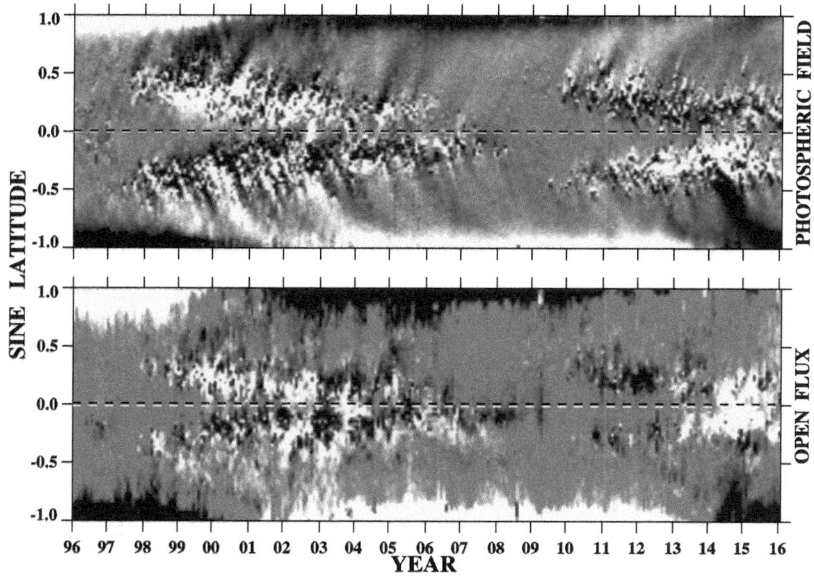

Fig. 2 Latitude–time plot of the photospheric field (*top panel*) and of open flux/coronal holes (*bottom panel*), constructed from photospheric field measurements from the National Solar Observatory (NSO: CR 1904–2046) and the Global Oscillation Network Group (GONG: CR 2047–2173). Gray scale for the radially oriented photospheric field ranges from $B_r < -4$ G (*black*) to $B_r > +4$ G (*white*); the GONG fields have been scaled upward by 1.65 to make them consist with the NSO fields. The longitudinally averaged distribution of open flux was determined by applying a PFSS extrapolation to the photospheric field maps; gray scale ranges from -0.5 G (*black*) to $+0.5$ G (*white*)

of active region emergence. Trailing-polarity surges are often followed by weaker leading-polarity surges, which causes the polar fields to fluctuate and sometimes undergo multiple reversals. As may be seen also from Fig. 1, a triple reversal occurred at the north pole during 2012–2015. A detailed description of the polar field evolution during 2010–2014, based on line-of-sight photospheric field measurements from the Helioseismic and Magnetic Imager on the *Solar Dynamics Observatory*, is given by Sun et al. (2015).

The bottom panel of Fig. 2 shows a latitude–time plot of the Sun's open flux (coronal holes), as determined by applying a potential-field source-surface (PFSS) extrapolation with source surface at $r = 2.5\ R_\odot$ to the NSO/GONG photospheric field. By early 2015, a large polar hole has formed in the south but not in the north, where the polar field remains weak and of mixed polarity. At this time, the negative-polarity open flux is concentrated toward the south pole, whereas most of the positive-polarity open flux is located near the equator. Later in 2015, however, positive-polarity open flux can be seen migrating toward the north pole, which suggests that a new polar hole will form there soon.

Solar cycle 24 reached its maximum in 2014, when the yearly mean sunspot number was one-third lower than in the previous maximum of 2000, making it the weakest cycle since cycle 16 (1923–1933). The weakness of the current cycle lends support to the idea that the polar field at solar minimum provides the seed for the toroidal flux and sunspots of the next cycle, as implicitly assumed in most models of the Babcock–Leighton type. Thus, based on the observed strength of the polar fields during the declining phase of cycle 23 compared with their strengths during the previous three cycles (see Fig. 1), Schatten (2005) and Svalgaard et al. (2005) predicted that cycle 24 would have a maximum yearly sunspot number R_{max} of 75–80, remarkably close to the value of 79 reached in 2014 (despite the uncertainties inherent in the polar field measurements). Given the success of the polar field predictor, a better understanding of the factors that determine long-term polar field evolution may improve our ability to forecast future cycles, as well as suggest explanations for past events such as the 1645–1715 Maunder Minimum. It is interesting to note, from Fig. 1, that the Sun's axial dipole strength in 2015 is already comparable to its strength during the 2008–2009 sunspot minimum. Since the post-maximum value of $|D_{ax}|$ may be regarded as a precursor of and proxy for the late-cycle polar field strength, this suggests that cycle 25 will be comparable in amplitude to cycle 24 (see Sect. 5).

In flux transport/Babcock–Leighton models, only a very small fraction of the flux emerging in active regions during a given sunspot cycle ends up contributing to the final polar fields; most of the flux is canceled/resubmerged at photospheric neutral lines. The amount of flux that reaches the poles depends both on the properties of the active region sources and on the latitudinal transport mechanisms, of which meridional flow is less well constrained by observations than supergranular convection. This overview/commentary is concerned only with the processes occurring at the solar surface; the subsurface aspects of the flux transport dynamo, including the return flow that drives the equatorward migration of sunspot activity, have been reviewed by Charbonneau (2010) and Cameron et al. (2016). For a recent comprehensive review of the Sun's polar fields and their evolution, see Petrie (2015).

2 Physical Basis of Polar Field Formation

The formation of the polar fields depends on three main factors (see, e.g., DeVore et al. 1984; Wang et al. 1989, 1991, 2002, 2005, 2009; Wang and Sheeley 1991; Dikpati and Charbonneau 1999; Mackay et al. 2002; Schrijver et al. 2002; Baumann et al. 2004; Dikpati and Gilman 2006; Dikpati et al. 2006, 2008, 2010; Choudhuri et al. 2007; Schrijver and Liu 2008; Yeates et al. 2008; Jiang et al. 2009, 2010, 2011a,b, 2013; Cameron and Schüssler 2010, 2012; Karak 2010; Muñoz-Jaramillo et al. 2010; Dikpati 2011; Dikpati and Anderson 2012; Mackay and Yeates 2012; Cameron et al. 2013; Hathaway and Upton 2014; Upton and Hathaway 2014a,b):

First, as shown by white-light and magnetic observations, bipolar magnetic regions (BMRs) tend to emerge at the photosphere with their axes slightly inclined relative to the east–west direction, such that the leading (westward) pole lies closer to the equator than the trailing pole. This axial inclination, which tends to increase with latitude ("Joy's law"), arises from the action of the Coriolis force on the toroidal flux as it rises to the surface (Caligari et al. 1995; D'Silva and Choudhuri 1993; Fan et al. 1994; Fisher et al. 1995; Wang and Sheeley 1991). This systematic tilt is the ultimate source of the Sun's axial dipole component and of the polar fields. The axial dipole moment of a BMR satisfying the Hale–Joy polarity rules is opposite to that associated with the polar field at the start of the given cycle, so the cumulative effect of the bipoles emerging during the cycle is to reverse the Sun's net axial dipole field, even if no flux were transported to the poles.

The second important factor in polar field formation is the transport of surface magnetic flux by turbulent convection at the photosphere. As first recognized by Leighton (1964), the continual random displacements of the flux elements in the nonsteady flow field associated with the supergranular convection cells (which have lifetimes on the order of a day and diameters of ~30 Mm) act to disperse or "diffuse" the flux across the solar surface. Although the supergranular flows are initially suppressed by the strong sunspot and plage fields in newly emerged active regions, the flux is eventually broken up and ends up as the main source of the polar fields. The preferential diffusion of the lower-latitude, leading-polarity flux across the equator leads to a net surplus of trailing-polarity flux in each hemisphere.

The third factor is the poleward transport of active region flux by the surface meridional flow over the course of each sunspot cycle. Because cross-equatorial diffusion produces an excess of trailing-polarity flux in each hemisphere, the meridional flow gives rise to polar fields dominated by the trailing polarity. By the end of the cycle, the flux becomes highly concentrated toward the poles, with the large-scale photospheric field approaching a quasi-equilibrium state in which the poleward flow just balances the tendency for the polar cap flux to diffuse back toward the equator.

It should be recognized that the formation of large-scale unipolar regions (including the polar fields near solar minimum) requires that at least some of the flux emerge in the form of BMRs that are much larger than the ephemeral regions studied by Hagenaar (2001) and Hagenaar et al. (2003, 2008), which have fluxes typically of order 10^{19} Mx and random or nearly random axial orientations. If all of the flux emerged in the form of randomly oriented ephemeral regions, no large-scale separation of the two polarities would occur, and the photospheric field would retain its salt-and-pepper nature even in the presence of supergranular convection, differential rotation, and meridional flow. Ephemeral regions can only contribute to the large-scale field if they retain some residual tendency to obey the Hale–Joy polarity rules; if such a tendency exists, it lies within the noise of present-day measurements.

Together, these factors determine the net flux that reaches the polar regions during the course of each sunspot cycle. The polar fields are built up from the successive contributions of individual active regions, the net flux inside each 30°-wide polar cap being roughly equal to the total flux in a medium-sized active region ($\sim 3 \times 10^{22}$ Mx). Especially critical in determining the final polar-field strength is the amount of leading-polarity flux that crosses the equator. The cross-equatorial transport involves a competition between the poleward-directed bulk flow and the equatorward diffusion of flux from the active region latitudes: the faster the flow speed at low latitudes, the smaller the amount of flux that manages to diffuse "upstream" into the other hemisphere, and the weaker the final polar fields. The leading-polarity flux that crosses the equator may either undergo cancellation with its opposite-hemisphere counterpart, or else migrate poleward to contribute directly to the polar field in that hemisphere.

In the following sections, we discuss in greater detail the role of the meridional flow and the axial tilts of BMRs.

3 Latitudinal Dependence of the Meridional Flow Speed

From measurements using helioseismology (Haber et al. 2002; González Hernández et al. 2006, 2011; Zhao et al. 2012, 2014), Doppler shifts (Hathaway 1996; Ulrich 2010), and magnetic feature tracking (Komm et al. 1993; Hathaway and Rightmire 2010, 2011; Rightmire-Upton et al. 2012), it is known that the surface and near-surface flow speeds have amplitudes in the range 10–20 $\mathrm{m\,s^{-1}}$ and are generally directed poleward. On average, the velocity goes through zero at the equator, although north–south asymmetries are often present. The flows are slower at mid latitudes near sunspot maximum, when their gradients also steepen near the equator; this solar cycle variation appears to be the result of residual flows (of order 5 $\mathrm{m\,s^{-1}}$) that converge toward the sunspot belts (see Zhao and Kosovichev 2004; Švanda et al. 2007; Hindman et al. 2009).

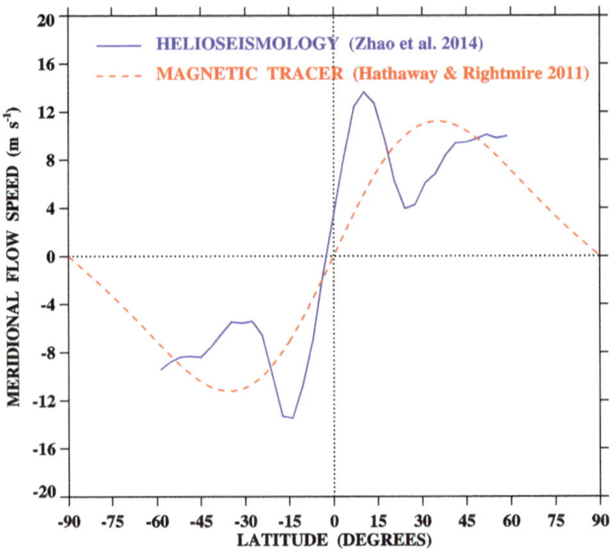

Fig. 3 Latitudinal variation of the surface meridional flow speed, as derived from helioseismology (Zhao et al. 2014) and magnetic feature tracking (Hathaway and Rightmire 2011). The flow speeds inferred by tracking magnetic flux elements include the contribution of supergranular diffusion, and therefore peak at higher latitudes than those derived from helioseismology. The helioseismic time–distance measurements, which are subject to large uncertainties at high latitudes, are here averaged over the 3-yr period 2010–2012. The mid-latitude dip in the helioseismic profile is attributed to the presence of residual flows that converge toward the active region latitudes near the surface and diverge from them at greater depths

The flow speeds derived from magnetic feature tracking (Komm et al. 1993; Hathaway and Rightmire 2010, 2011; Rightmire-Upton et al. 2012) differ systematically from those obtained from Doppler-shift measurements (Ulrich 2010) and from helioseismology (Zhao et al. 2012, 2014), being slower at low latitudes and faster at higher latitudes. As discussed by Dikpati et al. (2010) (see also Wang et al. 2009), the feature tracking measurements include the contribution of supergranular diffusion, which is directed away from the active region latitudes, thereby opposing the effect of the poleward meridional flow at lower latitudes but reinforcing it at higher latitudes. Because the rate at which magnetic flux is transported poleward is slower (faster) than the local flow speed at low (high) latitudes, the apparent "flow speed" inferred from magnetic tracers reaches its maximum at much higher latitudes than the actual flow speed. The helioseismic flow profiles recently derived by Zhao et al. (2014) (see Fig. 3) peak in the latitude range 10°–15°, in qualitative agreement with those employed in earlier flux-transport simulations by Wang et al. (2009).

To demonstrate the sensitivity of the polar fields to the flow speed profile, it is instructive to compare the results of two idealized flux-transport simulations. In the first simulation, the poleward flow speed is assigned the latitudinal dependence

$$v(L) = \pm 13\ \mathrm{m\,s^{-1}} \tanh(|L|/6^\circ) \cos^2 L, \tag{1}$$

so that the flow attains its peak speed of $12\ \mathrm{m\,s^{-1}}$ at $|L| = 13.2^\circ$. In the second simulation, the flow is taken to vary with latitude as

$$v(L) = \pm 12\ \mathrm{m\,s^{-1}} \sin(2|L|), \tag{2}$$

so that it peaks at $|L| = 45^\circ$, but otherwise has the same amplitude as in the first simulation. In both cases, the initial photospheric field consists of a large BMR centered at $L = +15^\circ$,

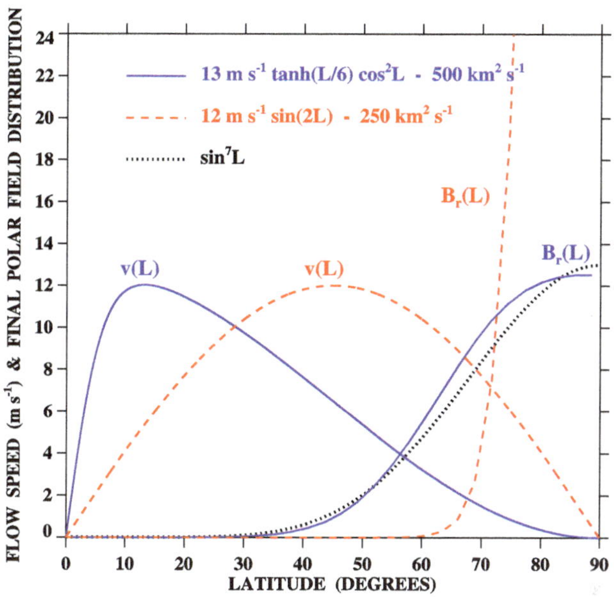

Fig. 4 Final polar-field distributions derived using two different meridional flow profiles $v(L)$, one peaking at $|L| = 13°$ (Eq. (1)), and the other peaking at $|L| = 45°$ (Eq. (2)). In both cases, the initial photospheric field consists of a large, tilted BMR centered at $L = +15°$, which is then evolved including the effects of meridional flow and supergranular diffusion (see also Figs. 5 and 6). The plot shows, as a function of latitude in the northern hemisphere, the adopted flow profiles and the resulting near-equilibrium distributions of the photospheric field (averaged over longitude and in arbitrary units). The flow profile that peaks at 45° gives much stronger polar fields than the one that peaks at 13°, even though the diffusion coefficient has been taken to be a factor of two smaller in the former case; the flux distribution also ends up far more concentrated toward the pole than the polar fields observed near solar minimum, which typically vary as $\sim \sin^7 L$ (*dotted curve*) (see Svalgaard et al. 1978; Petrie and Patrikeeva 2009; Wang et al. 2009)

with its leading pole located 4° equatorward of its trailing pole. The flux distribution is then evolved including the effects of flow and supergranular diffusion, until the system approaches a quasi-equilibrium configuration. (Note that differential rotation has no effect on the evolution of the longitudinally averaged or axisymmetric field component, which also does not depend on the longitudinal pole separation of the BMR.) As may be seen from Fig. 4, the flow profile that peaks at $|L| = 45°$ produces much stronger and narrowly concentrated polar fields than the profile that peaks at low latitudes. Indeed, to prevent the simulated polar field from "blowing up" further, we have reduced the diffusion coefficient κ to 250 km^2 s^{-1}, from the value of 500 km^2 s^{-1} adopted in the case where the profile peaks at 13°.

For the same two simulations, Fig. 5 displays the evolution of the photospheric field and PFSS-derived open field regions in latitude–time format, while Fig. 6 shows the time evolution of the north and south polar fields (averaged over 30°-wide polar caps) and of the axial dipole component. Comparing the top left and top right panels of Fig. 5, we see that far more leading-polarity (negative) flux diffuses across the equator in the simulation where the poleward flow peaks at latitude 45°, even though the diffusion coefficient has been reduced to one-half the value used in the simulation where the flow peaks at 13°. As a result, the northern hemisphere becomes heavily dominated by trailing-polarity flux, and very strong polar fields build up in both hemispheres. In addition, because the flow speed

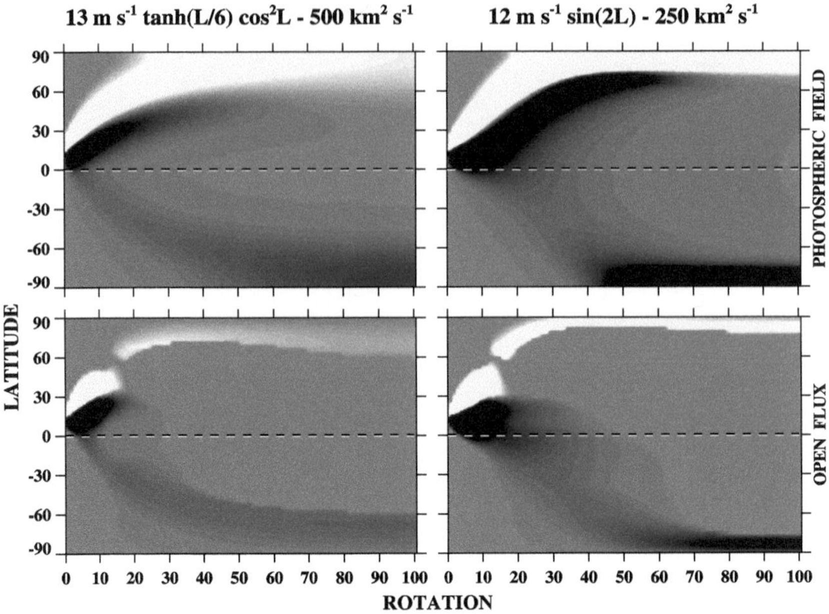

Fig. 5 Latitude–time plots of the photospheric field (*top panels*) and PFSS-derived open flux (*bottom panels*) for the two flow profiles shown in Fig. 4. *Left column*: case where the flow speed peaks at $|L| = 13.2°$ and $\kappa = 500\ \mathrm{km^2\,s^{-1}}$. *Right column*: case where the flow speed peaks at $|L| = 45°$ and $\kappa = 250\ \mathrm{km^2\,s^{-1}}$. Both simulations are initialized by depositing onto the photospheric grid a bipole centered at $L = +15°$, with its negative leading pole located 4° equatorward of its positive trailing pole; the flux is then transported over the solar surface for 100 27.3 day rotations (see also Fig. 6)

peaks at relatively high latitudes in this simulation, the flux becomes extremely concentrated toward the poles; the polar hole boundaries are thus confined to latitudes above ∼80° (Fig. 5, bottom right panel), instead of extending down to $|L| \sim 60°$ as typically observed near solar minimum. In contrast, in the case where the flow speed peaks at low latitudes (Fig. 5, left-hand panels), the poleward-directed flow strongly impedes the diffusion of leading-polarity flux across the equator, even though κ is as large as $500\ \mathrm{km^2\,s^{-1}}$. The net amount of trailing-polarity flux reaching the pole is thus relatively small; moreover, because the flow is very slow at high latitudes, the polar fields are less concentrated and the polar holes now extend down to latitude 60°, as observed.

As is evident from Fig. 5, the meridional flow transports flux of both polarities from the northern-hemisphere BMR to the nearby pole, so that the initial trailing-polarity surge is followed by a surge of the opposite polarity. This gives rise to the nonmonotonic variation of the north polar field seen in Fig. 6, which is particularly pronounced in the simulation where the flow speed peaks at latitude 45°. As also shown in Fig. 6, the axial dipole strength ends up twice as large in the latter case as in the case where the flow speed peaks at low latitudes.

It is clear from these idealized experiments that the evolution of the polar fields depends sensitively on the latitudinal variation of the meridional flow, and that a flow profile that peaks at mid to high latitudes will tend to produce polar fields that are too strong and concentrated. This problem can be mitigated to some extent by imposing a high-latitude cutoff in the flow field, as has been done by Cameron, Jiang, and their coworkers, following van Ballegooijen et al. (1998). If the flow profile peaks at low latitudes (as advocated here),

Fig. 6 Time evolution of the north and south polar fields (averaged over 30°-wide polar caps) and of the axial dipole strength for the two flow profiles plotted in Fig. 4 (see also Fig. 5). The field strengths are in arbitrary units

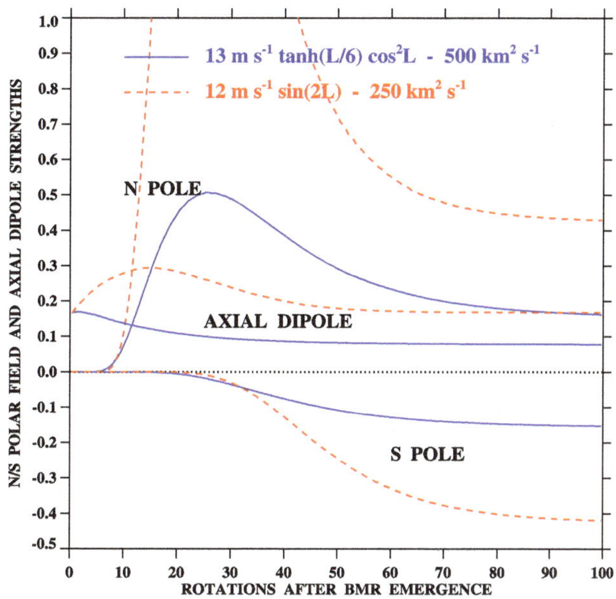

the polar field evolution becomes very sensitive to small changes in the flow speed, which controls the amount of flux that diffuses across the equator. Thus, as shown in Wang et al. (2009), a $\sim$15% increase in the poleward flow speed could account for the weakening of the polar fields during cycle 23 (see also Dikpati 2011). As discussed in the next section, systematic changes in the axial tilts of BMRs might also give rise to cycle-to-cycle variations in the polar fields.

4 Do the Axial Tilts of BMRs Undergo Long-Term Variations?

The polar fields have their ultimate source in the axial inclinations of active regions. The tilt angles and their latitudinal dependence have been measured using either white-light images of sunspot groups (Hale et al. 1919; Brunner 1930; Howard 1991a; Dasi-Espuig et al. 2010, 2013; Ivanov 2012; McClintock and Norton 2013; Baranyi 2015) or magnetograms (Wang and Sheeley 1989; Howard 1991b; Stenflo and Kosovichev 2012; Li and Ulrich 2012). In general, the tilt angles and Joy's law slopes derived from magnetograms tend to be larger than those inferred from white-light images.

Figure 7 shows magnetic and white-light determinations of the tilt angle as a function of latitude for cycles 21, 22, and 23. Here, the MWO magnetograph measurements (top panel) are from Li and Ulrich (2012); the white-light measurements are from the Debrecen Photoheliographic Data (DPD) sunspot-group catalog, which provides two sets of tilt angles, one based on the area-weighted positions of sunspot umbrae only ("U": middle panel), the other including umbrae, penumbrae, and pores ("UPP": bottom panel).

From Fig. 7, we note that the median tilt angles seem to have undergone relatively little long-term variation during cycles 21–23, whether the tilt angles are determined from magnetograms or from white-light images. This suggests that the weakening of the polar fields during cycle 23 was not caused by the emergence of active regions with unusually small

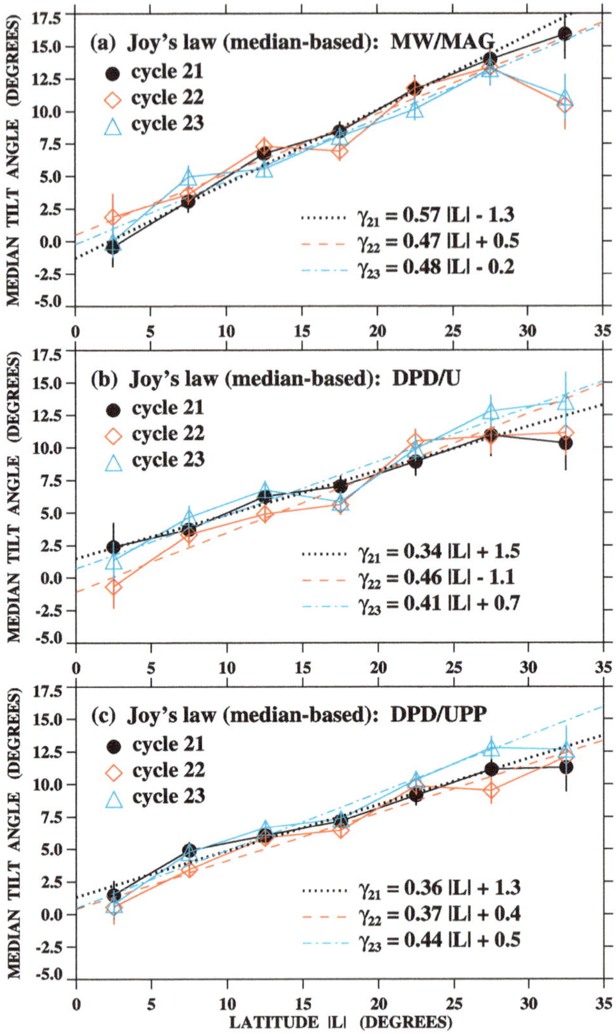

Fig. 7 Median-based Joy's law derived separately for cycle 21 (June 1976–August 1986), cycle 22 (September 1986–April 1996), and cycle 23 (May 1996–December 2008). (**a**) MWO magnetic measurements (Li and Ulrich 2012). (**b**) DPD/U white-light measurements (including dark umbrae only). (**c**) DPD/UPP white-light measurements (including umbrae, penumbrae, and faint pores)

axial inclinations. In addition, a comparison of the top and lower two panels of Fig. 7 confirms the tendency for magnetic measurements to yield larger tilt angles and steeper Joy's law slopes than white-light measurements. As shown by Baranyi (2015) and Wang et al. (2015), there are two main reasons for this discrepancy. First, many of the smaller white-light sunspot groups consist of sunspots of the same polarity (usually the leading polarity), and these groups tend to have small (pseudo-) tilts. In the case of the historical (1917–1985) MWO white-light database, as much as 43% of the sunspot groups have angular separations less than 2.5°, and 84% of these groups are unipolar. Second, the white-light measurements do not include the contribution of plage areas, which tend to have larger axial inclinations than sunspots. Physically, this is because the tilt imparted by the Coriolis force is proportional to the rise time of the buoyant loops, and this rise time is longer for the plage fields than for the stronger sunspot fields.

Figure 8 shows, for cycles 16–23, the median tilt angle of all active regions measured during each cycle, divided by their median latitude. The latitude-normalized tilt angles show

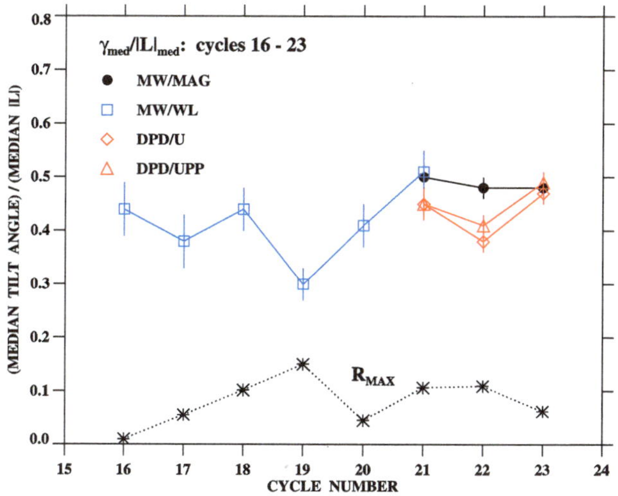

Fig. 8 Median tilt angles for cycles 16–23, divided by the median latitude of the active regions of each cycle. The data include MWO white-light measurements for cycles 16–21, as well as the MWO magnetic, DPD/U, and DPD/UPP measurements for cycles 21–23 (see Fig. 7). Normalizing the tilt angles by latitude reduces both the cycle-to-cycle variation and the spread among the different data sets. Also plotted is the (arbitrarily normalized and shifted) maximum sunspot number for each cycle. Except for the anomalously low value for cycle 19, the normalized tilt angles show no clear relationship to cycle amplitude

relatively little variation between cycles and among the different data sets, which include the earlier MWO white-light observations as well as the DPD/U, DPD/UPP, and Li and Ulrich (2012) measurements. An apparent exception is the data point for the very high-amplitude cycle 19, which was characterized by unusually small tilt angles in the southern hemisphere, according to the MWO white-light database (see McClintock and Norton 2013). If this outlier is ignored, there seems to be no systematic relationship between tilt angle and cycle amplitude. However, the uncertainties inherent in both the white-light and magnetic measurements are sufficiently large that the question of systematic long-term variations in the tilt angle remains open.

Jiang et al. (2015) have suggested that the weakening of the polar fields during cycle 23 was caused by the emergence of a few very low-latitude active regions with reversed north–south dipole moments. In their flux transport simulations, they adopted a meridional flow profile that peaks at mid latitudes, similar to that inferred by Hathaway and Rightmire (2010, 2011) from magnetic feature tracking. As discussed in the previous section, such profiles have the effect of maximizing the rate of cross-equatorial diffusion, and may thus lead to an overestimate of the contribution of near-equatorial active regions to the polar fields.

As a consequence of Joy's law and the equatorward progression of sunspot activity, the latitudinal centroids of the leading- and trailing-polarity flux of the active regions in each hemisphere tend to converge during the declining phase of each cycle. Based on photospheric field data from NSO and MWO, Petrie (2012) noted an unusually pronounced convergence of the polarity centroids after 2003 (especially in the southern hemisphere), and suggested that this effect might be related to the weakening of the polar fields. However, Fig. 1 shows that the Sun's axial dipole strength was already much weaker in 2001–2002 than it was during the corresponding period ($\sim$1–2 yr after the axial dipole reversed its sign) of the preceding cycles. It would thus appear that the active regions that emerge near sunspot maximum have a greater role in determining the final strengths of the axial dipole and polar fields than those that emerge late in the cycle (see also Sect. 5).

The current weak cycle provides an unprecedented opportunity to investigate the question of whether the axial tilts of BMRs vary systematically from cycle to cycle, and, in particular, to test the hypothesis that the tilt angles are larger in weak cycles than in strong ones (see Dasi-Espuig et al. 2010, 2013; Cameron et al. 2010; Muñoz-Jaramillo et al. 2013).

5 The Present State of the Polar Fields and the Amplitude of Cycle 25

As is evident from Figs. 1 and 2, the polar fields currently (at the beginning of 2016) show an unusually pronounced north–south asymmetry, with the net flux in the south polar cap being $\sim$3 times greater than that in the north polar cap. This dominance of the south polar field reflects the strong hemispheric asymmetry in the sunspot number during the recent maximum, when activity was heavily concentrated in the southern hemisphere. The opposite asymmetry occurred during the rising phase of cycle 24 (2010–2011), when northern hemisphere activity dominated. However, the mid-latitude BMRs emerging near the start of each cycle tend to undergo relatively little cross-equatorial diffusion, and are thus less efficient in supplying net flux to the polar regions than the lower-latitude active regions that emerge near sunspot maximum (when the rate of flux emergence is also greater). This difference may explain why the north polar field underwent multiple reversals starting as early as 2012, but remains much weaker than the south polar field, which reversed during 2013–2014.

When flux emergence is confined to one hemisphere, the evolution of the polar field in the opposite hemisphere lags that in the active hemisphere by roughly the time required for the leading-polarity flux to diffuse across the equator. This lag is clearly illustrated by the simulations plotted in Figs. 5–6, which were initiated by depositing a large BMR in the northern hemisphere. For both choices of the transport parameters, we see that the south polar field remains much weaker than the north polar field until as much as $\sim$4–5 yr after the BMR was deposited. Also, as remarked previously, the north polar field initially overshoots, with its strength rising rapidly to a peak as trailing-polarity flux arrives at the pole, then decreasing and leveling off as meridional flow carries some of the leading-polarity flux poleward.

Guided by these simulations, we may make a qualitative prediction about the evolution of the polar fields during the remainder of cycle 24: as sunspot activity continues to decline, the south polar field will weaken and the north polar field will strengthen. Their final strengths will be on the order of 4 G, approximately midway between their current values (see Fig. 1), with the south polar field remaining somewhat stronger than the north polar field.

Again from Fig. 1, we note that the Sun's axial dipole component $D_{\rm ax}$ undergoes a more regular solar-cycle variation than the polar fields themselves. In particular, during cycles 20–23, $D_{\rm ax}$ approached its final strength $|D_{\rm ax}^{\rm ssmin}|$ only $\sim$3 yr after it changed sign at sunspot maximum. We therefore assume that its current strength of $\sim$2 G is close to the value that it will have at the end of cycle 24. We may then obtain another estimate of $|B_{\rm pol}^{\rm ssmin}|$, the final polar-cap field strength, by using the fact that the longitudinally averaged photospheric field at solar minimum typically has the topknot form

$$B_r(L) \simeq B_r(L = 90^\circ) \sin^n L, \tag{3}$$

where L is latitude and $n \sim 7$ (see, e.g., Svalgaard et al. 1978; Petrie and Patrikeeva 2009; Wang et al. 2009). The corresponding axial dipole strength is

$$D_{\rm ax}^{\rm ssmin} \simeq \frac{3}{2} \int_{-90^\circ}^{90^\circ} B_r(L = 90^\circ) \sin^8 L \cos L \, dL \simeq \frac{1}{3} B_r(L = 90^\circ). \tag{4}$$

Setting $D_{ax}^{ssmin} \simeq 2$ G yields $B_r(L = 90°) \simeq 6$ G, so that the average field strength in each polar cap is

$$|B_{pol}^{ssmin}| \simeq \frac{\int_{60°}^{90°} |B_r(L = 90°)| \sin^7 L \cos L \, dL}{\int_{60°}^{90°} \cos L \, dL} \simeq 4 \text{ G}. \tag{5}$$

This agrees with the first estimate, derived by simply averaging the current strengths of the north and south polar fields.

The relative constancy of the Sun's axial dipole strength after sunspot maximum can be attributed not only to the decrease in the flux emergence rate, but also to the small axial inclinations of the remaining low-latitude active regions, which partially offsets the effect of the enhanced cross-equatorial diffusion in separating the two polarities. In addition, if flux emergence and meridional flow are neglected, the decay timescale for the axial dipole component is as long as $\tau_d \sim R_\odot^2/2\kappa \sim 15$ yr for a diffusion coefficient of $\kappa \sim 500$ km^2 s^{-1}. The equatorial dipole component decays on much shorter timescales because it is subject to rotational shearing, which hastens its diffusive annihilation by converting it into higher-order multipoles.

That the values of $|D_{ax}^{ssmin}|$ and $|B_{pol}^{ssmin}|$ inferred above are close to those measured during the 2008–2009 activity minimum suggests that cycle 25 will be similar in amplitude to cycle 24. This result is perhaps not surprising, given the close match between the amplitude of cycle 24 (as measured by R_{max}) and the axial dipole/polar field strength during the preceding minimum: flux emergence during cycle 24 has been just sufficient to reverse the sign of the initial, rather weak axial dipole, leaving its magnitude almost the same. However, even if the amplitude of a given cycle is assumed to be strongly correlated with the axial dipole strength during the previous minimum, large cycle-to-cycle variations in $|D_{ax}^{ssmin}|$ and $|B_{pol}^{ssmin}|$ (and thus in cycle amplitude) can still be introduced through systematic changes in the meridional flow field or in the axial tilts of BMRs.

6 Conclusions

The main points made in this review/commentary may be summarized as follows.

1. In surface flux-transport models, the evolution of the polar fields depends sensitively on the meridional speed and its variation with latitude.

2. Flow speed profiles derived from magnetic tracers include the contribution of supergranular diffusion, which is directed poleward above the sunspot latitudes and equatorward below them. As a result, the tracer-derived "flow speeds" peak at higher latitudes than the actual flow speeds.

3. Flow speed profiles that peak at mid or high latitudes tend to give polar fields that are too strong and concentrated compared with observations.

4. From the available white-light and magnetic data, it remains unclear whether Joy's law undergoes systematic long-term changes. White-light measurements of sunspot groups tend to underestimate the tilt angles of BMRs, while magnetic measurements as yet cover only three similar-amplitude cycles (21–23).

For the last four cycles, the Sun's axial dipole or polar field strength near sunspot minimum has been a good predictor of the amplitude of the following cycle. However, as cycle 23 showed, the amplitude of a given cycle is not necessarily a good indicator of the axial dipole or polar field strength during the following minimum. Fluctuations in the polar field strength from cycle to cycle could be caused by fluctuations (of unknown origin) in the surface meridional flow and/or in the axial tilts of BMRs. For example, a ~20% increase in the poleward flow speed from one cycle to the next may cause the polar field and axial dipole strength to decrease by as much as $\sim$50%. Relatively modest fluctuations in the flow amplitude, which remain barely above the uncertainty level of present-day helioseismic measurements, could thus have a major effect on the long-term evolution of the Sun's large-scale field.

In closing, we emphasize that comparison of the current weak cycle with the strong, well-documented cycles 21–23 may allow important constraints to be set on the relationship between cycle amplitude, surface flow speeds, and the axial tilts of BMRs.

Acknowledgements I am grateful to André Balogh for the opportunity to attend the workshop "Solar Magnetic Fields: From Measurements Towards Understanding," which was held in January 2015 at the International Space Science Institute in Bern. I also thank the referee for comments. This work was supported by the Chief of Naval Research.

References

H.D. Babcock, Astrophys. J. **130**, 364 (1959)
H.W. Babcock, Astrophys. J. **133**, 572 (1961)
T. Baranyi, Mon. Not. R. Astron. Soc. **447**, 1857 (2015)
I. Baumann, D. Schmitt, M. Schüssler, S.K. Solanki, Astron. Astrophys. **426**, 1075 (2004)
W. Brunner, Astron. Mitt. Eidgenöss. Sternwarte Zür. **124**, 67 (1930)
P. Caligari, F. Moreno-Insertis, M. Schüssler, Astrophys. J. **441**, 886 (1995)
R.H. Cameron, M. Dasi-Espuig, J. Jiang, E. Işik, D. Schmitt, M. Schüssler, Astron. Astrophys. **557**, A141 (2013)
R.H. Cameron, M. Dikpati, A. Brandenburg, Space Sci. Rev. (2016, this issue). doi:10.1007/s11214-015-0230-3
R.H. Cameron, J. Jiang, D. Schmitt, M. Schüssler, Astrophys. J. **719**, 264 (2010)
R.H. Cameron, M. Schüssler, Astrophys. J. **720**, 1030 (2010)
R.H. Cameron, M. Schüssler, Astron. Astrophys. **548**, A57 (2012)
P. Charbonneau, Living Rev. Sol. Phys. **7**, 3 (2010)
A.R. Choudhuri, P. Chatterjee, J. Jiang, Phys. Rev. Lett. **98**, 131103 (2007)
M. Dasi-Espuig, S.K. Solanki, N.A. Krivova, R. Cameron, T. Peñuela, Astron. Astrophys. **518**, A7 (2010)
M. Dasi-Espuig, S.K. Solanki, N.A. Krivova, R. Cameron, T. Peñuela, Astron. Astrophys. **556**, C3 (2013)
C.R. DeVore, N.R. Sheeley Jr., J.P. Boris, Sol. Phys. **92**, 1 (1984)
M. Dikpati, Astrophys. J. **733**, 90 (2011)
M. Dikpati, J.L. Anderson, Astrophys. J. **756**, 20 (2012)
M. Dikpati, P. Charbonneau, Astrophys. J. **518**, 508 (1999)
M. Dikpati, G. de Toma, P.A. Gilman, Geophys. Res. Lett. **33**, L05102 (2006)
M. Dikpati, G. de Toma, P.A. Gilman, Astrophys. J. **675**, 920 (2008)
M. Dikpati, P.A. Gilman, Astrophys. J. **649**, 498 (2006)
M. Dikpati, P.A. Gilman, R.K. Ulrich, Astrophys. J. **722**, 774 (2010)
S. D'Silva, A.R. Choudhuri, Astron. Astrophys. **272**, 621 (1993)
T.L. Duvall Jr., Sol. Phys. **63**, 3 (1979)
Y. Fan, G.H. Fisher, A.N. McClymont, Astrophys. J. **436**, 907 (1994)
G.H. Fisher, Y. Fan, R.F. Howard, Astrophys. J. **438**, 463 (1995)
I. González Hernández, R. Komm, F. Hill, R. Howe, T. Corbard, D.A. Haber, Astrophys. J. **638**, 576 (2006)
I. González Hernández, R. Komm, S. Kholikov, R. Howe, F. Hill, J. Phys. Conf. Ser. **271**, 012073 (2011)
D.A. Haber, B.W. Hindman, J. Toomre, R.S. Bogart, R.M. Larsen, F. Hill, Astrophys. J. **570**, 855 (2002)
H.J. Hagenaar, Astrophys. J. **555**, 448 (2001)
H.J. Hagenaar, M.L. DeRosa, C.J. Schrijver, Astrophys. J. **678**, 541 (2008)
H.J. Hagenaar, C.J. Schrijver, A.M. Title, Astrophys. J. **584**, 1107 (2003)
G.E. Hale, F. Ellerman, S.B. Nicholson, A.H. Joy, Astrophys. J. **49**, 153 (1919)
D.H. Hathaway, Astrophys. J. **460**, 1027 (1996)
D.H. Hathaway, L. Rightmire, Science **327**, 1350 (2010)
D.H. Hathaway, L. Rightmire, Astrophys. J. **729**, 80 (2011)
D.H. Hathaway, L. Upton, J. Geophys. Res. Space Phys. **119**, 3316 (2014)
B.W. Hindman, D.A. Haber, J. Toomre, Astrophys. J. **698**, 1749 (2009)
R.F. Howard, Sol. Phys. **136**, 251 (1991a)

R.F. Howard, Sol. Phys. **132**, 49 (1991b)
V.G. Ivanov, Geomagn. Aeron. **52**, 999 (2012)
J. Jiang, R.H. Cameron, D. Schmitt, E. Işik, Astron. Astrophys. **553**, A128 (2013)
J. Jiang, R. Cameron, D. Schmitt, M. Schüssler, Astrophys. J. Lett. **693**, L96 (2009)
J. Jiang, R.H. Cameron, D. Schmitt, M. Schüssler, Astron. Astrophys. **528**, A82 (2011a)
J. Jiang, R.H. Cameron, D. Schmitt, M. Schüssler, Astron. Astrophys. **528**, A83 (2011b)
J. Jiang, R.H. Cameron, M. Schüssler, Astrophys. J. Lett. **808**, L28 (2015)
J. Jiang, E. Işik, R.H. Cameron, D. Schmitt, M. Schüssler, Astrophys. J. **717**, 597 (2010)
B.B. Karak, Astrophys. J. **724**, 1021 (2010)
R.W. Komm, R.F. Howard, J.W. Harvey, Sol. Phys. **147**, 207 (1993)
R.B. Leighton, Astrophys. J. **140**, 1547 (1964)
R.B. Leighton, Astrophys. J. **156**, 1 (1969)
J. Li, R.K. Ulrich, Astrophys. J. **758**, 115 (2012)
D. Mackay, A. Yeates, Living Rev. Sol. Phys. **9**, 6 (2012)
D.H. Mackay, E.R. Priest, M. Lockwood, Sol. Phys. **207**, 291 (2002)
B.H. McClintock, A.A. Norton, Sol. Phys. **287**, 215 (2013)
A. Muñoz-Jaramillo, M. Dasi-Espuig, L.A. Balmaceda, E.E. DeLuca, Astrophys. J. Lett. **767**, L25 (2013)
A. Muñoz-Jaramillo, D. Nandy, P.C.H. Martens, A.R. Yeates, Astrophys. J. Lett. **720**, L20 (2010)
G.J.D. Petrie, Sol. Phys. **281**, 577 (2012)
G.J.D. Petrie, Living Rev. Sol. Phys. **12**, 5 (2015)
G.J.D. Petrie, I. Patrikeeva, Astrophys. J. **699**, 871 (2009)
L. Rightmire-Upton, D.H. Hathaway, K. Kosak, Astrophys. J. Lett. **761**, L14 (2012)
K. Schatten, Geophys. Res. Lett. **32**, L21106 (2005)
C.J. Schrijver, M.L. DeRosa, A.M. Title, Astrophys. J. **577**, 1006 (2002)
C.J. Schrijver, Y. Liu, Sol. Phys. **252**, 19 (2008)
J.O. Stenflo, A.G. Kosovichev, Astrophys. J. **745**, 129 (2012)
X. Sun, J.T. Hoeksema, Y. Liu, J. Zhao, Astrophys. J. **798**, 114 (2015)
L. Svalgaard, E.W. Cliver, Y. Kamide, Geophys. Res. Lett. **32**, L01104 (2005)
L. Svalgaard, T.L. Duvall Jr., P.H. Scherrer, Sol. Phys. **58**, 225 (1978)
M. Švanda, A.G. Kosovichev, J. Zhao, Astrophys. J. Lett. **670**, L69 (2007)
R.K. Ulrich, in *Cool Stars, Stellar Systems, and the Sun*, ed. by M.S. Giampapa, J.S. Bookbinder. ASP Conf. Ser., vol. 26 (Astronomical Society of the Pacific, San Francisco, 1992), p. 265
R.K. Ulrich, Astrophys. J. **725**, 658 (2010)
L. Upton, D.H. Hathaway, Astrophys. J. **780**, 5 (2014a)
L. Upton, D.H. Hathaway, Astrophys. J. **792**, 142 (2014b)
A.A. van Ballegooijen, N.P. Cartledge, E.R. Priest, Astrophys. J. **501**, 866 (1998)
Y.-M. Wang, R.C. Colaninno, T. Baranyi, J. Li, Astrophys. J. **798**, 50 (2015)
Y.-M. Wang, J. Lean, N.R. Sheeley Jr., Astrophys. J. Lett. **577**, L53 (2002)
Y.-M. Wang, J.L. Lean, N.R. Sheeley Jr., Astrophys. J. **625**, 522 (2005)
Y.-M. Wang, A.G. Nash, N.R. Sheeley Jr., Astrophys. J. **347**, 529 (1989)
Y.-M. Wang, E. Robbrecht, N.R. Sheeley Jr., Astrophys. J. **707**, 1372 (2009)
Y.-M. Wang, N.R. Sheeley Jr., Sol. Phys. **124**, 81 (1989)
Y.-M. Wang, N.R. Sheeley Jr., Astrophys. J. **375**, 761 (1991)
Y.-M. Wang, N.R. Sheeley Jr., Astrophys. J. Lett. **447**, L143 (1995)
Y.-M. Wang, N.R. Sheeley Jr., A.G. Nash, Astrophys. J. **383**, 431 (1991)
A.R. Yeates, D. Nandy, D.H. Mackay, Astrophys. J. **673**, 544 (2008)
J. Zhao, A.G. Kosovichev, Astrophys. J. **603**, 776 (2004)
J. Zhao, A.G. Kosovichev, R.S. Bogart, Astrophys. J. Lett. **789**, L7 (2014)
J. Zhao, K. Nagashima, R.S. Bogart, A.G. Kosovichev, T.L. Duvall Jr., Astrophys. J. Lett. **749**, L5 (2012)

Space Sci Rev (2017) 210:367–395
DOI 10.1007/s11214-015-0230-3

The Global Solar Dynamo

R.H. Cameron[1] **· M. Dikpati**[2] **· A. Brandenburg**[3,4,5]

Received: 17 August 2015 / Accepted: 14 December 2015 / Published online: 11 February 2016

Abstract A brief summary of the various observations and constraints that underlie solar dynamo research are presented. The arguments that indicate that the solar dynamo is an alpha-omega dynamo of the Babcock-Leighton type are then shortly reviewed. The main open questions that remain are concerned with the subsurface dynamics, including why sunspots emerge at preferred latitudes as seen in the familiar butterfly wings, why the cycle is about 11 years long, and why the sunspot groups emerge tilted with respect to the equator (Joy's law). Next, we turn to magnetic helicity, whose conservation property has been identified with the decline of large-scale magnetic fields found in direct numerical simulations at large magnetic Reynolds numbers. However, magnetic helicity fluxes through the solar surface can alleviate this problem and connect theory with observations, as will be discussed.

Keywords Solar dynamo · Sun

1 Introduction

The Sun's magnetic field is maintained by its interaction with plasma motions, i.e. dynamo action. From the 1950s to 1980 the most compelling question of whether dynamo action could in fact be responsible for the Sun's magnetic fields (a suggestion originally made by

✉ R.H. Cameron
cameron@mps.mpg.de

1 Max-Planck-Institut für Sonnensystemforschung, Justus-von-Liebig-Weg 3, 37077 Göttingen, Germany

2 National Center for Atmospheric Research 1, High Altitude Observatory, 3080 Center Green, Boulder, CO 80301, USA

3 Nordita, KTH Royal Institute of Technology and Stockholm University, 10691 Stockholm, Sweden

4 Department of Astronomy, Stockholm University, 10691 Stockholm, Sweden

5 JILA, Department of Astrophysical and Planetary Sciences, and Laboratory for Atmospheric and Space Physics, University of Colorado, Boulder, CO 80303, USA

Larmor 1919) was answered. That dynamo action is in principle possible for some prescribed flows in a body with uniform conductivity was established rigorously in the 1950's by Herzenberg (1958). In fact, a number of simple spatially periodic flows were later shown to act as dynamos (Roberts 1970). From 1955 to about 1980 the next step was taken, where it was shown that turbulence driven by convection in a rotating system can produce a magnetic field with scales comparable to the size of the system (Parker 1955a; Steenbeck et al. 1966). The key concept introduced here is helicity that is the breaking of the symmetry of the convecting system by the rotation Moffatt (1978), Krause and Rädler (1980). This answered the most compelling question of whether dynamo action could in fact be responsible for the Sun's magnetic fields, as suggested originally made by Larmor (1919).

The next step was to begin working out which motions are actually producing the Sun's magnetic field. Going in this direction substantially beyond what was presented by Parker (1955a), the work of Babcock (1961), Leighton (1964), and Leighton (1969) provide a phenomenological model for the solar dynamo with close contact to the observations. Their work has been put within the framework of mean-field models by Stix (1974). Hence the Babcock-Leighton model can be considered a special example of the mean-field model where the important dynamo effects are identified with what is seen on the Sun (such as the "rush to the poles" and the tilt of bipolar active regions).

The aim of dynamo theory today, in the context of the Sun, is to understand how the dynamo actually operates to produce the magnetic fields we see. The constraints that inform our understanding are thus the various different observations of solar magnetic fields. We will discuss some of the different types of observations that are available to us in Sect. 2. We will try to present some coverage of what time- and spatial-scales are covered by the observations but cannot hope to be comprehensive. Some of the observational constraints will then be presented in their distilled form in Sect. 3 (such as Hale's law, Joy's law, the Waldmeier effect, some properties of grand minima and maxima). A few implications for the dynamo that can easily be inferred from the observations will be discussed in Sect. 4.

Our coverage of the theoretical and modeling progress that has been made will be even more sparse. It is an exciting time when there are a large number of theoretical models backed by simulations to explain a number of aspects seen on the Sun. Again our choice of topics and models is necessarily biased. In Sect. 5 we will discuss some prominent results from Flux Transport Dynamo models, and in Sect. 6 we will discuss some results from detailed calculations of the evolution of the velocity and magnetic fields based on the MHD equations in a geometry resembling that of the Sun. We comment that because the parameters of the plasma inside the Sun are extreme, it is beyond the state of the art to hope to realistically model the solar dynamo at present. These simulations are therefore intended to give an insight into the physics that could possibly be occurring on the Sun.

2 What Do We Know from Observations?

2.1 What We Would Like to Know?

Our aim is to understand the solar dynamo. Because understanding requires synthesis, even a complete knowledge of the magnetic and velocity fields inside the Sun as a function of time would not necessarily be sufficient—however it would certainly be a much better start than what we have at present. What we have is incomplete in many ways. Firstly we only have reasonably consistent synoptic measurements based on seeing only one half of the Sun, and only for a few cycles. We know that there are Grand Minima for which we do not have this type of data, and in fact the last few cycles were part of a Grand Maximum so we do not have data representative of even the "typical" behavior of the Sun. Finally our only tool for probing the convection zone directly is helioseismology which comes with limitations in resolution due to finite-wavelength effects and in practice limitations due to the noise associated with granulation. Our (partial) blindness to the long-term and interior dynamics is a severe problem for understanding the solar dynamo.

2.2 What Observations Do We Have?

Since dynamo action occurs beneath the solar surface, the most directly relevant observations for dynamo theory are the time series of surface magnetic and velocity fields. For full-disk almost continuous space-based observations we have SDO/HMI (Scherrer et al. 2012) and SOHO/MDI (Scherrer et al. 1995) observations. These instruments allow, through inversion of the spectropolarimetric observations, the line-of-sight velocity and magnetic field at the solar surface to be inferred (SDO/HMI allows the full vector magnetic field to be determined). The tangential velocity can be inferred on scales larger than granules using Local Correlation Tracking techniques (e.g. Roudier et al. 1998).

These observations cover cycle 23 and so far the rising and maximum phase of cycle 24. This allows the study of the evolution of the magnetic field on timescales from minutes to a decade or more. Its main limitation is that only having two cycles means essentially we only have two data points (on solar-cycle timescales) and hence cannot make strong statements about the variation of activity level from cycle to cycle. Also while the resolution is adequate for some purposes, it is less than optimal near the poles.

The limited spatial resolution is partly compensated by high-resolution space missions such as Hinode (Lites et al. 2013) that can observe small magnetic flux concentrations near the poles (Tsuneta et al. 2008). The limited number of cycles covered is partly mitigated by observations which decrease in there temporal coverage and detail as we go further back in time. The most detailed of these are the magnetograms taken in synoptic programs which cover most of cycles 21 to 24. These include observations by KPNSO/VTT and SOLIS, the Mount Wilson observatory, the Wilcox solar observatory and for the later cycles GONG. These data allow us to follow the evolution of magnetic fields on periods of days to years.

Still in the era of photographic plates, we have only occasional magnetograms. In their place synoptic programs regularly recorded images of the Sun in white light (for examples see Howard et al. 1984, 1990) and in some important lines Ca II K (e.g. Bertello et al. 2010) at Mount Wilson and Kodaikanal. These types of programs have been undertaken for over 100 years.

Prior to photography, sunspots were drawn by hand, and we have systematic and continuous records of sunspot location and areas extending back to 1874 (Balmaceda et al. 2009), and less systematically to earlier times (e.g. Arlt et al. 2013; Diercke et al. 2015). Before this we have sunspot counts going back to before the Maunder Minimum (see the review by Clette et al. 2014).

Only sparse records of direct sunspot observations exist prior to this—there are for instance occasional reports extending back thousands of years. Instead we must rely on records of the interaction between the solar magnetic fields with the Earth's magnetic field—these exist in the form of records of auroral (e.g. Křivský and Pejml 1988) and more systematically in terms of the geomagnetic indices, such as the "*aa*" index. This index can be used to infer the interplanetary magnetic field near the earth, which in turn can be related to the open flux of the Sun because the field strength of the radial component of the interplanetary magnetic field largely depends only on the distance from the Sun (Smith and Balogh

1995). The "aa" index then gives us a record of global properties of the solar magnetic field that can be extended back to 1844 (Nevanlinna and Kataja 1993). Nature has also created records that can be used to infer solar activity, in particular cosmogenic nucleotides stored in ice cores, which extends our knowledge back to over 9600 years (Steinhilber et al. 2012).

2.3 Constraints

The dynamo problem is essentially concerned with plasma motions generating and sustaining magnetic fields. We therefore begin by *extremely briefly* outlining what is known about the motions themselves.

2.3.1 The Flows

Granulation and Supergranulation Heat is transported by convective motions in the outer 30 % of the Sun. The dominant scale of the convection near the surface (granulation) is well understood and depends mainly on the Sun's mass and luminosity (Stein and Nordlund 1989). The properties of flows at larger scales (both supergranulation and the lack of giant cells) are much more poorly understood theoretically (Lord et al. 2014) and observationally (compare Hanasoge et al., 2012 and Greer et al., 2015).

The interaction between the convection and rotation, especially at larger scales, drives global-scale flows such as differential rotation and meridional circulation. Improving our understanding of the large-scale convection is therefore a priority.

Rotation The total angular momentum of the Sun is a result of the angular momentum it had when it formed, and its evolution since then (e.g. due to magnetic breaking). The total angular momentum of the Sun, and in particular of the convection zone, is thus a basic parameter from the point of view of dynamo theory.

Differential Rotation—Latitudinal and Radial The Sun's differential rotation is well known as a function of both latitude and radius (Schou et al. 1998). The main properties are that latitudinal shear is much greater that the radial shear, the latter of which is localized at the tachocline at the base of the convection zone and in a near-surface shear layer. (For a detailed review see Howe 2009).

Torsional Oscillations The time dependence of the differential rotation is called "torsional oscillations" (Howard and Labonte 1980; Schou et al. 1998). These are clearly associated with magnetic activity, but are probably too weak to significantly influence the evolution of the Sun's magnetic field (they are however likely to be an important diagnostic).

Meridional Flow There is also a large-scale meridional flow, with a well-observed poleward component at the solar surface (Duvall 1979; Ulrich 2010). The subsurface structure of the flow is more controversial (compare Zhao et al. 2013; Schad et al. 2013; Jackiewicz et al. 2015). Given the important role of the subsurface meridional flow in transporting the field in the Flux Transport Dynamo model (discussed in Sect. 4.1) resolving this controversy should be seen as a priority.

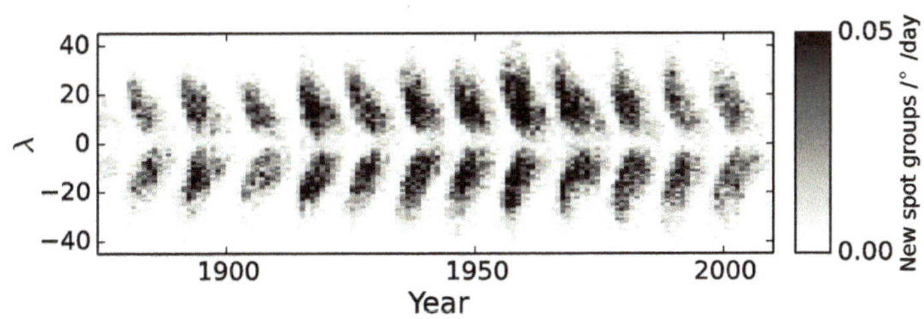

Fig. 1 A butterfly diagram: the number of sunspots appearing as a function of latitude λ and time (based on Royal Greenwich Observatory and USAF/NOAA SOON data)

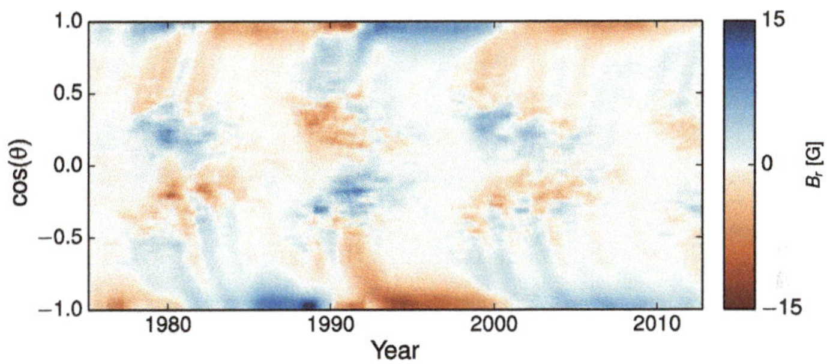

Fig. 2 A magnetic butterfly diagram: the longitudinally averaged radial component of the solar magnetic field as a function of latitude and time (based on Kitt Peak National Observatory Synoptic magnetograms)

2.3.2 Magnetic Field Evolution

Observations of the type described in Sect. 2.2 are the constraints we have for the solar dynamo. They contain a lot of information, some of which can be summarized in simple figures, and some of which can be distilled into "laws". Two figures which contain a lot of information are the butterfly diagram (Maunder 1904, see also Fig. 1) and the magnetic butterfly map (Fig. 2).

From the observations a number of properties (some of them "laws") have been described in the literature. A comprehensive solar dynamo model should be able consistent with these observational constraints. We begin with a number of constraints that are directly related to the magnetic activity.

11 Year Activity Cycles, 22 Year Magnetic Cycles A good proxy of magnetic activity is the number of sunspots, which varies in time with minima every 10 to 12 years. This was first noted by Schwabe (1849). Different cycles have different amplitudes (and modulation on longer periods might be present, e.g. Gleissberg 1939). The 11 year solar activity cycle corresponds to half of the 22 year magnetic cycle (Hale et al. 1919), with the dominant polarity of the leading sunspots in each hemisphere changing between each activity cycle.

Spörer's Law The emergence location of sunspots is observed to migrate equatorward during the cycle (Carrington 1858; Spörer 1879), beginning at about 35° and propagating to about 8° at the end of the cycle. The rate of propagation is similar for all cycles (Waldmeier 1955; Hathaway 2011).

Hales Law The magnetic nature of sunspots was discovered by Hale (1908). Sunspots typically appear in groups, with the leading and trailing spots (with respect to the solar rotation) having different polarities. The leading spots in each hemisphere mostly have the same polarity, and the polarity is opposite in the other hemisphere. The polarities of the leading and following spots switch between cycles (Hale et al. 1919).

Joy's Law As implied by Hales Law, sunspots often appear as bipolar pairs with the leading spots during one cycle and in one hemisphere having the same polarity. This is a statement about the east-west orientation of the sunspots. There is also a tendency for the leading spots to be slightly closer to the equator than the following spots. This tendency is much weaker than that of Hale's law, with the angle implied by the North-South separation compared to the East-West separation being about 7 degrees. This effect is known as Joy's law and was reported by Hale et al. (1919). There is some evidence that the strength of the effect depends on the strength of the cycle (Dasi-Espuig et al. 2010).

The effect is also much weaker in the sense that there is a lot of scatter in the North-South separation so that the effect is only robust when a large sample of sunspots is considered.

Waldmeier Effect The Waldmeier effect states that strong cycles peak earlier than weak cycles (Waldmeier 1941) (although the Waldmeier effect does not appear in all measures of the solar activity, cf. Dikpati et al. 2008; Cameron and Schüssler 2008). There is a closely related fact that strong cycles rise quickly, which Karak and Choudhuri (2011) call WE2, for the Waldmeier effect 2.

North-South Asymmetry Cycles are not symmetric (Spörer 1889a), and interestingly the asymmetric behavior can be coherent over many cycles (Carbonell et al. 1993).

Extended Cycle While the sunspot number has a period of around 11 years, the butterfly diagram indicates that the wings overlap so that sunspots corresponding to each cycle are present for about 13 years. Smaller than sunspots, ephemeral regions associated with a cycle have been shown to emerge about 5 years earlier, so the activity related to one cycle extends to about 18 years (Wilson et al. 1988).

Correlation Between Polar Fields, Open Flux and Strength of Next Cycle There is a strong correlation between the polar field at minimum (determined using polar faculae as a proxy) and the strength of the next cycle (Muñoz-Jaramillo et al. 2013); a stronger correlation exists between the Sun's open flux, determined using the minima of the *aa* index as a proxy, and the strength of the next cycle (Wang and Sheeley 2009). Cameron and Schüssler (2007) have suggested that this might be accounted for by the overlapping of cycles combined with the Waldmeier effect. A commonly claimed effect that the length of a minimum correlates positively with the weakness of the next cycle peak has been shown from sunspot data by Dikpati et al. (2010a) to be false for the most recent 12 cycles.'

Magnetic Fields at the Surface Are Advected by Surface Flows as if They Were Corks Outside of active regions, the radial component of the magnetic field is advected by the horizontal component of the velocity as if it were a passive tracer (DeVore et al. 1984). The details of the modeling and observations that support this were reviewed recently in Jiang et al. (2014).

Coronal Mass Ejections and Magnetic Helicity Fluxes The structure of the magnetic field in the solar atmosphere, with filaments and sigmoid shaped active regions as well as the various types of activity in the solar atmosphere, such as flares and coronal mass ejections also contain information related to the solar cycle. The interpretation of these structures in terms of the helicity generated by the dynamo is however not straightforward (Zirker et al. 1997). We return to the helicity in some detail in Sect. 5.

Grand Minima and Maxima The above "laws" and properties of the magnetic activity are mainly based on the last few hundred years of data. We however know that the solar dynamo does not always behave like this—there are extended periods of low activity including the Maunder minimum (Spörer 1889b). They occur on average every 300 years (Usoskin et al. 2007) and presumably represent a different state of the dynamo. We have very few observational constraints for this state of the dynamo and therefore do not discuss it further in this paper.

3 Synthesising the Observations and Theory

We now turn to the open problem of synthesizing the observations and the "laws" they embody with well known basic physics in order to gain an understanding of the solar dynamo.

3.1 The Omega and Alpha Effects

Cameron and Schüssler (2015) used the simplicity of the toroidal field implied by Hale's law to show that the surface magnetic field plays a key role in the solar dynamo. The argument begins by noting that Hale's law tells us that in each hemisphere, and during one cycle, the leading spots mainly (about 96 % for large active regions, Wang and Sheeley 1989) have the same magnetic polarity. This strong preference for the leading spots to have the same polarity indicates that the spots are coming from the emergence of toroidal flux that is all of the same polarity. The authors then considered the induction equation

$$\frac{\partial \boldsymbol{B}}{\partial t} = \nabla \times (\boldsymbol{u} \times \boldsymbol{B} - \eta \mu_0 \boldsymbol{J}), \tag{1}$$

where $\boldsymbol{u}$ and $\boldsymbol{B}$ are velocity and magnetic fields, t is time, $\boldsymbol{J} = \nabla \times \boldsymbol{B}/\mu_0$ is the current density, μ_0 is the magnetic permeability, and $\eta = 1/\sigma\mu_0$ is the magnetic diffusivity with σ being the conductivity. They then applied Stokes theorem to with a contour in a meridional plane and encompassing the convection zone in the northern hemisphere. This allowed them to demonstrate that the generation of net toroidal flux in each hemisphere is dominated by the winding up of the poloidal flux threading the poles of the photosphere at the poles by the latitudinal differential rotation.

The polar fields themselves are the remnants of flux that has crossed the equator (Durrant et al. 2004), which is dominated by either the emergence of tilted active regions across the equator (Cameron et al. 2013) or the advection of active region flux across the equator due to the random shuffling of the field lines due to the supergranular flows.

The simplicity of the toroidal magnetic field at solar maxima, and the poloidal magnetic field at solar minima, therefore indicates that the solar dynamo is of the Babcock-Leighton type. Explicitly it is an alpha-omega dynamo where the relevant poloidal field threads through the photosphere. The omega effect is simply the winding up of this poloidal

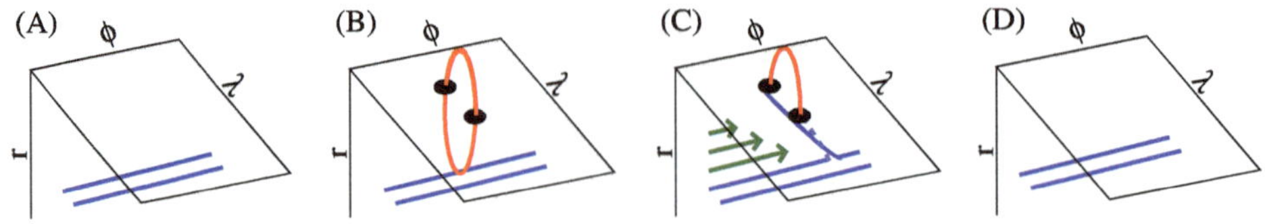

Fig. 3 Illustration of the cause of the equatorial propagation in the dynamo wave model. The model begins with toroidal field (shown in *blue*) situated away from the equator in Panel (**A**)—the equator is towards the top of the panel. In Panel (**B**) a sunspot group emerges from the toroidal flux. The rising tube is acted on by the Coriolis force, and by the convection, and poloidal field is produced (this is the alpha effect). In Panel (**C**) the poloidal field is sheared by radial differential rotation, producing positive and negative toroidal field. The *dashed blue line* represents toroidal field of the opposite sign to the original (*solid*) toroidal field. In Panel (**D**) the *dashed* poloidal field has canceled the existing toroidal field at high latitudes. The low lying newly created toroidal field leads to a new band of toroidal field with the same orientation as the existing field but at lower latitudes. The net action has thus been to move the toroidal field closer to the equator

field by the latitudinal differential rotation. The alpha effect is what produces the tilt of the active regions with respect to the equator (Joy's law).

We comment that the question of why we have a butterfly diagram, why sunspots obey Joy's law, why the cycle length is around 11 years, and why sunspots only emerge below about 40 degrees all remain open questions. They are difficult to answer because they are intimately related to the subsurface dynamics, which are mostly poorly understood theoretically and observationally. In the next two subsections we discuss some of the ideas which are in the literature.

3.2 Equatorial Migration of the Butterfly Wings

Spörer's law (discussed in Sect. 2.3.2) states that the latitude of emergence of sunspots propagates towards the equator as the cycle proceeds. The most straightforward (and probably correct) interpretation of this is that the underlying toroidal field is propagating towards the equator. There have been two main suggestions to explain the propagation of the toroidal field. The first is that of Parker (1955a), and explains the equatorward propagation in terms of a dynamo wave. The cause of the equatorial propagation in this model is explained in Fig. 3 in the case of a Babcock-Leighton dynamo. The essential idea is that radial differential shear causes toroidal flux to propagate latitudinally. The direction of propagation (equatorwards or polewards) depends on the sign of the alpha effect (which generates poloidal flux from toroidal flux) and the on whether the differential rotation rate increases or decreases with depth (Yoshimura 1975).

In terms of the Sun, the sign of the alpha effect is generally believed to be such that differential rotation has to increase with depth to obtain equatorward propagation. This is an issue on the Sun where the differential rotation decreases inwards near the base of the convection zone in the range of latitudes where sunspots form Brown and Morrow (1987). Thus if the solar dynamo is substantially located in the tachocline then this mechanism is excluded, the mechanism is however viable if the winding up of the toroidal field occurs in the near-surface shear layer.

The second mechanism for explaining the equatorward propagation is shown in Fig. 4 and is supposed to work at the base of the convection zone. The essential idea is that, if the meridional flow near the base of the tachocline is sufficiently strong, and if the latitudinal transport due to diffusion is sufficiently weak, then the simple advection of the toroidal flux by the meridional flow can produce the observed equatorial migration (Wang and Sheeley

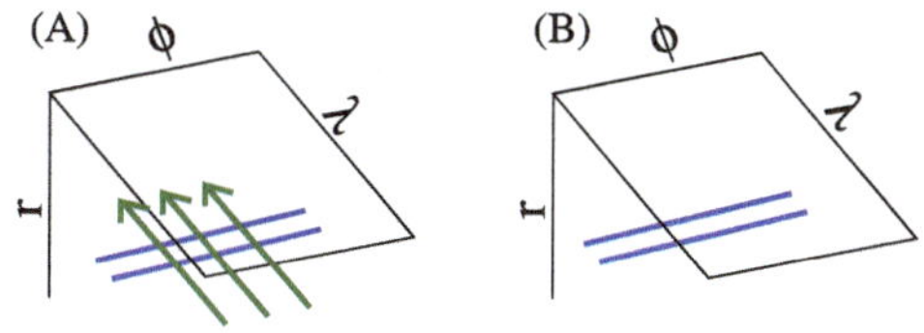

Fig. 4 Illustration of the cause of the equatorial propagation in the flux transport dynamo model—the toroidal flux (*blue lines*) is simply advected towards the equator by the meridional circulation (*green arrows*). The question are whether such a flow exists, where the toroidal field is stored and whether diffusive transport plays a role

1991; Choudhuri et al. 1995; Durney 1995; Dikpati and Charbonneau 1999; Küker et al. 2001).

Which, if either, of these two explanations accounts for the transport on the Sun remains an open question.

3.3 Emergence Latitudes

A second important question, which is probably closely related, is why sunspots overwhelmingly appear at latitudes below 40 degrees in latitude (see Fig. 1). An obvious explanation for this would be that the toroidal flux is concentrated at low latitudes, however this explanation is problematic for Babcock-Leighton type dynamos where the latitudinal differential shear occurs over essentially all latitudes. A second possible mechanism considers the instability which leads to flux emergence. This explanation assumes that the toroidal flux is stored in or near the tachocline. In this case flux tubes must have a field strength of about 10^5 G in order to have a substantial number of emergences at low latitude (Choudhuri and Gilman 1987). The argument is that if the tubes are weaker then their rise is more affected by the Coriolis force which preferentially causes them to rise with at a constant distance from the rotation axis (i.e. along cylinders). The value of 10^5 G is nicely consistent with the field strength required to produce Joy's law (D'Silva and Choudhuri 1993). Such flux tubes turn out to be much more unstable (Parker 1955b) at low latitudes than at high latitudes (Caligari et al. 1995), hence if the toroidal flux is stored in the tachocline then even if the 10^5 G loops form at high latitudes the loops are unable to escape to the surface.

This is an attractive possibility because the value of 10^5 G links several observational results. However we caution that this effort was focussed on the case where the storage is in the tachocline or near the base of the convection zone because this was the dominant paradigm at the time the work was carried out. Much less work has been put into considering why sunspots don't emerge at high latitudes in the case where the toroidal field is stored in the bulk of the convection zone. For this reason we regard the issue of the latitudinal range of the butterfly diagram as an open question.

3.4 Length of the Solar Cycle

It is somewhat sobering that after more than 160 years after the discovery by Schwabe (1849) that the level of solar activity varies with a period of about 11 years that we still don't have a good idea of why it is 11 years.

In principle the length of the solar cycle should reflect the latitude range over which activity appears (say from a latitude of 30 degrees at the start of a cycle to 8 degrees at the end of a cycle) and the rate at which it propagates towards the equator. Within the framework of flux transport dynamos, the rate at which activity propagates towards the equator and the period of the dynamo, is largely determined by the meridional flow circulation rate (Dikpati and Charbonneau 1999), however we have no clear basis in either observations or theory for understanding why the strength of the meridional flow at depth should be such that the dynamo period is 11 years. Within the alternative frame of a dynamo wave framework the period is set mainly by the magnitude of the alpha and omega effects, and the alpha effect in particular is not well understood.

To end this section on a positive light, it is clear that improving our understanding both what causes the equatorward propagation and what sets the latitudinal extent of the butterfly diagram, will give us a clearer handle on why the period of the solar cycle is 11 years.

3.5 The Alpha Effect

The above discussion indicates that the net toroidal field in each hemisphere is produced by the winding up of poloidal field by latitudinal differential rotation. In this regard it is only the field that threads through the surface which has an effect on the net flux. This winding up of the field is the omega effect of an alpha-omega dynamo. The details of how the alpha effect actually works in the Sun is less well understood—we observe the emergence of magnetic bipolar regions which are systematically tilted with respect to the equator, we observe their subsequent evolution, and we can infer (as in the previous section) that this is the field which is the poloidal field which gets wound up to produce the net toroidal flux in each hemisphere. What we do not observe are the processes which cause the field to emerge with a tilt.

It is relatively clear that the Coriolis force is implicated in the tilt; the uncertainty is whether the Coriolis for acts directly on the flows associated with the rise of the flux tube (e.g. D'Silva and Choudhuri 1993), or whether it acts on the convective flows which then interacts with the flux tube (Parker 1955a). This question is currently completely open.

Even without a proper understanding of the subsurface processes we can include flux emergence and Joy's law into idealized mean-field simulations. Most of the recent work along these lines has been in the context of the flux transport dynamo, and the next section will outline some results from these efforts.

4 Modeling Dynamo Action

4.1 Flux-Transport Dynamos

Two-dimensional "flux-transport" dynamos incorporate in some form all of the processes discussed above, including the idea of flux emergence from toroidal flux at the base of the convection zone producing tilted active regions at the surface. These models have been successful in simulating the most important features of the solar cycle, including the butterfly diagram, polar field reversals near solar cycle maximum, certain global coronal features, and certain asymmetries between North and South hemispheres.

These models were used to simulate and predict the timing and amplitude of solar cycle 24 (Dikpati et al. 2006, 2010b; Choudhuri et al. 2007; Nandy et al. 2011), with limited success. There are several possibilities for this. Some of the suggestions of what was not included (but needs to be) are changes in the global meridional circulation profile and speed (Belucz and Dikpati 2013), or localized inflow cells associated with active regions (Cameron

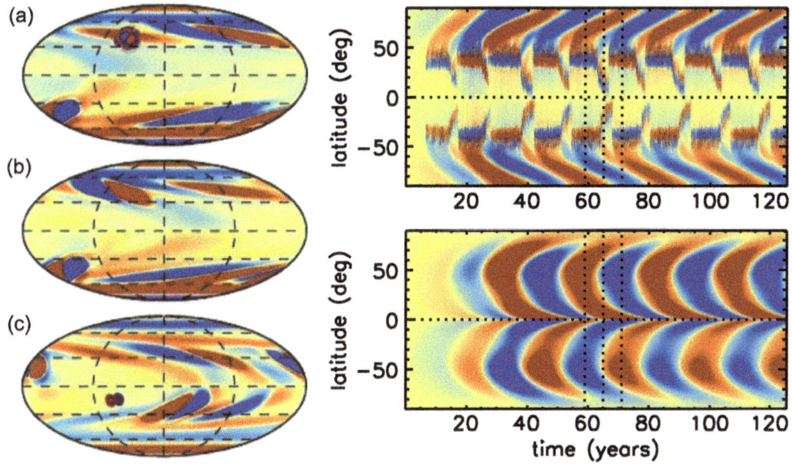

Fig. 5 *Left frame*: Eruption of tilted, bipolar spots and their dispersal by diffusion, meridional circulation and differential rotation; *right frame*: drift of trailing flux towards the poles appears in a series of streams, and cause polar reversal (*top right*) and toroidal field butterfly diagram (*bottom right*), showing equatorward migration and dynamo cycle-period governed by meridional circulation speed (frames adopted from Miesch and Dikpati (2014))

and Schüssler 2012; Shetye et al. 2015) or the scatter in tilt angles (Jiang et al. 2015). Unfortunately the results from helioseismology are too divergent to provide guidance on changes in the meridional flow (for example, compare the results in recent publications Ulrich 2010; Zhao et al. 2013; Schad et al. 2013; Jackiewicz et al. 2015).

One possibility, supported by the observations, is that the weak cycle 24 is the result of the actual values of the tilt angles in cycle 23 (Jiang et al. 2015). The essential point here is that poloidal source term in the Babcock-Leighton model is based on Joy's law, which is extremely noisy. The noise in Joy's law translates to noise in the alpha effect, and thus to the strength of the different cycles. For cycle 23 Jiang et al. (2015) used the observed tilt angles (Li and Ulrich 2012) and showed that alpha effect was indeed weak for cycle 23 (thus accounting for the weak cycle 24).

Current work includes extending the simulations to 3D, see, e.g., Miesch and Dikpati (2014). An example of such a simulation is shown in Fig. 5, which depicts the longitude-latitude pattern of emerged flux, and the patterns of surface mean poloidal fields and deep-seated toroidal fields that are created in a sequence of solar cycles.

However, perhaps the biggest challenge to the flux-transport dynamos model is whether the meridional circulation is single-celled or multiple-celled or is indeed not steady at all. Many solar cycle features can be reproduced well by a flux-transport dynamo model if the meridional circulation contains a single cell in each hemisphere, but models with two cells in depth do not (Belucz et al. 2015). There is currently no consensus from observational evidence and meridional circulation models on the dominant profile of meridional flow in depth and latitude within the convection zone and tachocline. At the photosphere there is clearly one primary, poleward cell or often along with that a weak reverse cell appears at high latitudes. If it is established that there are two cells in depth, then another paradigm shift in solar dynamo theory will be needed; however it seems to be too early to decide on that issue now.

4.2 Global Convective Dynamo Simulations

Attempts have also been made to simulate the convective motions and dynamo action in the Sun, with as few assumptions as possible—essentially from first principles. The first semi-successful attempts go back to the early 1980s when two different groups used the Cray 1 computer at the National Center for Atmospheric Research in Boulder (Colorado): Peter Gilman using a model of rotating convection (Gilman and Miller 1981) and Meneguzzi et al. (1981) using forced turbulence in a Cartesian domain. Both studies provided remarkable first steps into numerical studies of large-scale and small-scale dynamos, but they also demonstrated that simulating the Sun will be difficult.

Subsequently, Gilman (1983) obtained cyclic solutions that were however quite different from the Sun: instead of equatorward migration of magnetic activity, he obtained poleward migration. Furthermore, small-scale dynamo action was not well understood at the time and the original paper by Meneguzzi et al. (1981) did not even quote the now important reference to the paper by Kazantsev (1968). The role of helicity was also not clear in some of those first studies, see also Kida et al. (1991). This was partly because those early simulations did not have sufficient scale separation, i.e., the scale separation ratio k_f/k_1, where k_1 is the smallest wavenumber in the domain, was too small, as was noted subsequently (Haugen et al. 2004). In agreement with earlier work on rapidly rotating convection (Gilman 1977), the contours of angular velocity tend to lie on cylinders. This implies that the radial gradient of the local angular velocity, $\partial\Omega/\partial r$, is positive. Therefore, as expected predictions from mean-field dynamo theory, the dynamo simulations of Gilman (1983) and Glatzmaier (1985) produced poleward migration, and were thus unable to reproduce the solar butterfly diagram. (These simulations were subsequently applied to the geodynamo problem by Glatzmaier and Roberts 1995, where success was much clearer.)

From these simulations it is clear that reproducing the basic properties of the convection and the large-scale flows it drives are essential for a complete understanding of why a rotating middle-aged star such as the Sun should have an 11 year activity cycle and a butterfly diagram with equatorward propagation.

Computing power has increased dramatically since 1981, and simulations in more turbulent regimes became possible. This leads to flow patterns departing from otherwise nearly perfectly cylindrical contours (Miesch et al. 2000). Global simulations are now being conducted by many groups. Simulations with the anelastic spherical harmonics (ASH) code (Brun et al. 2004) work with a mean stratification close to that of mixing length theory, but at the solar rotation rate the resulting dynamo is statistically steady. Only at higher rotation rate, the solutions become time-dependent and cyclic (Brown et al. 2010). Simulations with the EULAG code (Ghizaru et al. 2010; Racine et al. 2011) also produce cyclic solutions, although the latitudinal migration of the mean magnetic field is weak. The pattern of meridional circulation in the simulations is, to date, mostly multicellular. This is in stark contrast to mean-field (flow) models, in which differential rotation is produced by the Λ effect and meridional circulation is dominated by one large cell (Brandenburg et al. 1992). As mentioned above, helioseismology does not yet provide a consistent answer to guide the theory or simulations.

A (surprising) key issue which has emerged is the convective power spectrum. Observations through correlation tracking (Rieutord et al. 2008; Hathaway 2012) and helioseismology (Hanasoge et al. 2012) suggest that the power in large-scale convective flows (giant cells) is very small. As alluded to in the introduction, the helioseismic evidence is contested (Greer et al. 2015), so the question remains open on the observational side. As already suggested by Spruit (1997), the structure of the large-scale convection is also unclear on the theoretical side—global simulations miss the physics of the radiating surface, and this may turn out to be crucial for producing simulations with realistic flow structures, which may be dominated by what is known as 'entropy rain'. This would be a fast small-scale downflow originating from the surface in such a way that the bulk of the convection zone is nearly isentropic, but with slightly stable stratification so as not to produce giant cell convection in the deeper parts (Brandenburg 2015). Clearly, simulations must eventually be able to reproduce the Sun, and reproducing the convective power spectrum and the large-scale flows is probably a precondition to accurately reproducing the solar dynamo. Knowing what those flows are is essential.

Currently the convective simulations are most useful in providing guidance on what is possible, and elucidating mechanisms. For example in simulations by Käpylä et al. (2012) using the PENCIL CODE,[1] there was pronounced equatorward migration, which was later identified as being due to a local negative radial Ω gradient (Warnecke et al. 2014). While this is a feature not expected to be present in the Sun, it does demonstrate that the dynamo wave in global simulations follows closely that expected from mean-field simulations (Parker 1955a; Yoshimura 1975). Global simulations have also produced evidence for strong ($\approx$40 kG) flux tubes in the bulk of the convection zone (Nelson et al. 2014; Fan and Fang 2014). This is interesting in view of understanding the overall magnetic flux concentration required to form active regions, although further amplification is needed; see Stein and Nordlund (2012) and Mitra et al. (2014) for possible mechanisms.

Both flux transport and full 3D dynamo models necessarily contain parametrizations of processes acting on scales smaller than spatially resolved. These parametrizations are all rooted in formulations of MHD turbulence, in which helicity of velocities, magnetic fields and electric currents play central roles. Therefore we consider helicity effects in detail in Sect. 5

5 The Roles of Magnetic Helicity

As revealed above, we are still struggling to understand the true nature of the solar dynamo and the relation between the large-scale dynamo in the Sun and in global simulations. In this regard we need to be sure that what we see in the simulations survives in the limit of large magnetic Reynolds numbers such as are found in the Sun. This has not been yet been fully confirmed. One example of this is that in the simulations the large-scale field magnetic field (as opposed to the small-field magnetic field) is often found to decrease with increasing values of R_{m}. In this section we show that this can be understood quantitatively in terms of magnetic helicity conservation, as will be discussed below.

Magnetic helicity is a quantity conserved by ideal MHD, and as such is conserved in the absence of microphysical diffusivity (i.e. it is not changed by the turbulent flow, or the associated turbulent magnetic diffusivity). In a closed volume, at low microphysical diffusivity of the sun, the amount of helicity is not expected to change on timescales of the solar cycle. This is remarkable and unprecedented in hydrodynamic turbulence. The only way that magnetic helicity can change and evolve (e.g., over the course of the 11 year cycle) is through magnetic helicity fluxes. They can be determined at the surface and this allows contact to be made between theory and observations. In this section we will therefore outline some of the different attempts to determine the helicity flux through the solar surface and what this tells us about the solar dynamo.

[1] https://github.com/pencil-code.

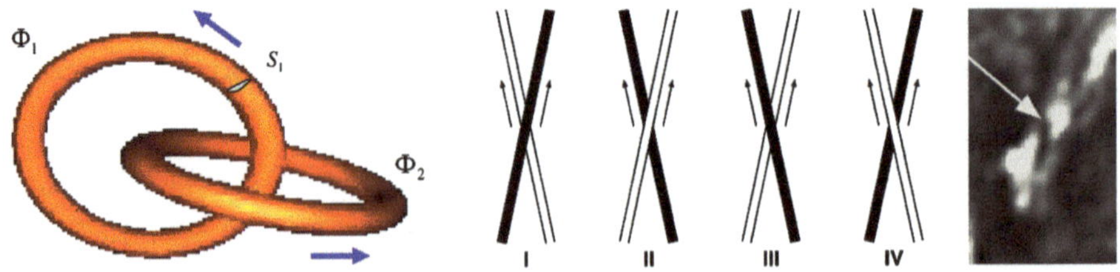

Fig. 6 Sketch of interlocked flux rings of positive helicity with two right-handed crossings (*left*), compared with two right-handed mutual crossings (*I* and *II*) and two left-handed mutual crossings (*III* and *IV*), as well as a high-resolution EUV image in 171 Å of a magnetic filament showing a crossing of type *III* in active region NOAA 8668 at 23° north on 1999 Aug 17 (right; courtesy of Jongchul Chae)

There are other types of helicity, including kinetic helicity, which also play important roles in dynamo theory. Most importantly kinetic helicity makes it relatively simple to generate a large-scale magnetic field (Moffatt 1969). We therefore begin by discussing the different types of helicity and their definitions.

5.1 Definitions

Mathematically, helicities are so-called pseudo-scalars, i.e., they are the dot product of a proper vector and an axial vector. (The latter changes its orientation when viewed in a mirror.)

Four such pseudoscalars are of particular interest: mean kinetic helicity density $\langle \boldsymbol{\omega} \cdot \boldsymbol{u} \rangle$, with $\boldsymbol{\omega} = \nabla \times \boldsymbol{u}$ being the vorticity, mean magnetic helicity density $\langle \boldsymbol{A} \cdot \boldsymbol{B} \rangle$, with $\boldsymbol{A}$ being magnetic vector potential such that $\boldsymbol{B} = \nabla \times \boldsymbol{A}$, mean current helicity density $\langle \boldsymbol{J} \cdot \boldsymbol{B} \rangle$, and finally the cross helicity, $\langle \boldsymbol{u} \cdot \boldsymbol{B} \rangle$; see Yokoi (2013) for a recent review, especially on cross helicity. All these helicities have topological interpretations that refer to the mutual linkage between interlocked structures, for example two flux rings in the case of magnetic helicity (Fig. 6, left), two vortex rings in the case of kinetic helicity, two current tubes in the case of current helicity, and a vortex tube with a magnetic flux tube in the case of cross helicity. This topological interpretation goes back to early work of Moffatt (1969) and is important for the existence of *qualitative* helicity indicators.

Qualitative Helicity Indicators Familiar to all of us is the case of kinetic helicity in the Earth's atmosphere: a glance at weather maps in England and Australia shows low pressure whirlpools of opposite orientation. This helicity is the direct result of the Coriolis force acting on large-scale flows. Likewise the Sun is rotating and has large-scale (supergranulation or larger) flows which feel the effect of the Coriolis force. Hence on the Sun we see the morphological or qualitative signatures of helicity in large-scale structures: Hα images reveal S-shaped structures in the south and N-shaped structures in the north (i.e., inverted S-shaped structures). They are referred to as sigmoidal structures and the importance of interpreting them was recognized by Sara Martin (1998a, 1998b) and many people after her (Canfield et al. 1999; Magara and Longcope 2001; Gibson et al. 2002). These helicity indicators can also be linked to mutual crossings of magnetic flux structures, which are best seen at EUV wavelengths where filaments appear occasionally in emission. This technique, which is due to Chae (2000), requires an additional assumption: the angle between two adjacent structures is an acute one (i.e., the invisible arrows of the field vectors point in roughly the same direction) rather than an obtuse one, as illustrated by the configurations I–IV in Fig. 6 from his paper. His study confirms that the magnetic field has negative helicity in the north (corresponding to crossings of types III and IV with N-shaped or so-called dextral filaments) and positive helicity in the south (corresponding to crossings of types I and II with S-shaped or so-called sinistral filaments). In Fig. 6 we also reproduce an EUV image from his paper with a filament of type III in the northern hemisphere, consistent with negative helicity.

Semiquantitative Indicators We turn now to semiquantitative indicators, by which we mean quantitative measures of something which is only qualitatively linked the helicity. As an important example, we can consider the product of what is known as the horizontal divergence and horizontal curl of the velocity, defined respectively as

$$\begin{aligned}(\operatorname{div}\boldsymbol{u})_{\rm h} &= u_{x,x} + u_{y,y},\\ (\operatorname{curl}\boldsymbol{u})_{\rm h} &= u_{y,x} - u_{x,y}.\end{aligned} \tag{2}$$

Here, commas denote partial differentiation. The horizontal curl is just the same as the z component of the usual curl. The product of $(\mathrm{div})_{\rm h}$ and $(\mathrm{curl})_{\rm h}$ has been determined for the Sun using local correlation tracking and local helioseismology by Langfellner et al. (2014). In fact, Rüdiger et al. (1999) have shown that the product of $(\mathrm{div})_{\rm h}$ and $(\mathrm{curl})_{\rm h}$ is a proxy of kinetic helicity. This is simply because one of the three terms in $\boldsymbol{\omega}\cdot\boldsymbol{u}$ is $u_z\,(u_{y,x} - u_{x,y})$, but, using the anelastic approximation, $\nabla\cdot(\rho\boldsymbol{u}) = 0$, where ρ is density, and the definition of the density scale height,

$$H_\rho = -(d\ln\rho/\mathrm{d}z)^{-1}, \tag{3}$$

we have $u_z = H_\rho\,\operatorname{div}\boldsymbol{u} \approx H_\rho(\operatorname{div}\boldsymbol{u})_{\rm h}$ and therefore $\langle\boldsymbol{\omega}\cdot\boldsymbol{u}\rangle \approx 3H_\rho\langle(\operatorname{div}\boldsymbol{u})_{\rm h}(\operatorname{curl}\boldsymbol{u})_{\rm h}\rangle$.

Quantitative Helicity Measures The measurement of current helicity density, $\boldsymbol{J}\cdot\boldsymbol{B}$, goes back to early work of Seehafer (1990), who determined B_z from circular polarization measurements, while B_x and B_y (giving J_z) were obtained from linear polarization. Such measurements are now obtained routinely from solar vector magnetograms. They all suggest that $\boldsymbol{J}\cdot\boldsymbol{B}$ is negative in the north and positive in the south. Typical values of $\mu_0\boldsymbol{J}\cdot\boldsymbol{B}$ are around $3\ \mathrm{G^2\,km^{-1}}$ (Zhang et al. 2014).

Regarding magnetic helicity, there is a notable complication in that $\langle\boldsymbol{A}\cdot\boldsymbol{B}\rangle$ is *gauge-dependent* and changes by adding an arbitrary gradient term to $\boldsymbol{A}$, which does not change $\boldsymbol{B}$. Exceptions are triply-periodic and infinite domains, for which $\langle\boldsymbol{A}\cdot\boldsymbol{B}\rangle$ turns out to be gauge-invariant. However, there exists a quantity called the relative magnetic helicity

$$H_{\rm rel} = \int_V (\boldsymbol{A}+\boldsymbol{A}_{\rm p})\cdot(\boldsymbol{B}-\boldsymbol{B}_{\rm p})\,\mathrm{d}V, \tag{4}$$

where $\boldsymbol{B}_{\rm p} = \nabla\times\boldsymbol{A}_{\rm p}$ is a potential field ($\nabla\times\boldsymbol{B}_{\rm p} = \boldsymbol{0}$) that satisfies $\boldsymbol{B}_{\rm p}|_{\rm surf}\cdot\hat{\boldsymbol{n}} = \boldsymbol{B}|_{\rm surf}\cdot\hat{\boldsymbol{n}}$ (Berger and Field 1984). It is gauge-independent, but it can only be determined over a finite volume. There is also a corresponding magnetic helicity flux $2\oint\boldsymbol{E}\times\boldsymbol{A}_{\rm p}$, where $\boldsymbol{E} = \boldsymbol{J}/\sigma - \boldsymbol{u}\times\boldsymbol{B}$ is the electric field. Simple examples of magnetic helicity and its flux have been presented by Berger and Ruzmaikin (2000) for theoretical models with rigid and differential rotation as well as with an assumed α effect. Quantitative measurements for the Sun's magnetic field are in the range of $10^{46}\ \mathrm{Mx^2/cycle}$ (DeVore 2000; Chae 2001; Welsch and Longcope 2003). This value is easily motivated by standard dynamo theory (Brandenburg 2009).

Let us finally comment on the cross helicity density $\boldsymbol{u}\cdot\boldsymbol{B}$, or its spatial average $\langle\boldsymbol{u}\cdot\boldsymbol{B}\rangle$. Just like $\langle\boldsymbol{A}\cdot\boldsymbol{B}\rangle$, it is a quantity that is conserved by the nonlinear interactions of the magnetohydrodynamic equations (Woltjer 1958), but it is often small, i.e., the *normalized* cross helicity $2\langle\boldsymbol{u}\cdot\boldsymbol{B}\rangle/\langle\boldsymbol{u}^2+\boldsymbol{B}^2\rangle$, which is the ratio of two conserved quantities (the latter being the total energy), is far away from its extrema of -1 and $+1$, if it was vanishing initially. An exception is a stratified layer with an aligned magnetic field. This also applies to the solar wind, where there can be regions where gravity $\boldsymbol{g}$ and the magnetic field are systematically aligned with each other. In those cases, a finite cross helicity can be driven away from zero. This can be understood by noting that there are two externally imposed vectors: gravity with a parallel (vertical) magnetic field $\boldsymbol{B}_0$. The latter is a pseudovector, giving rise to a pseudoscalar $\boldsymbol{g}\cdot\boldsymbol{B}_0$ that is odd in the magnetic field—just like $\langle\boldsymbol{u}\cdot\boldsymbol{B}\rangle$. Indeed, theoretical and numerical work by Rüdiger et al. (2011) showed that

$$\langle\boldsymbol{u}\cdot\boldsymbol{b}\rangle=(\tau/3)\left(u_{\mathrm{rms}}^2/c_{\mathrm{s}}^2\right)\boldsymbol{g}\cdot\boldsymbol{B}_0=-(\eta_{\mathrm{t}}/H_\rho)\,B_0, \tag{5}$$

where $\boldsymbol{b}=\boldsymbol{B}-\langle\boldsymbol{B}\rangle$ denotes the fluctuating magnetic field, τ is the correlation time of the turbulence, u_{rms} is its rms velocity, c_{s} is the sound speed, $\eta_{\mathrm{t}}=\tau u_{\mathrm{rms}}^2/3$ is the turbulent magnetic diffusivity, and $H_\rho=g/c_{\mathrm{s}}^2$ is the density scale height defined in Eq. (3) for an isothermal layer. Measurements for active regions are in the range 1–2 $\mathrm{G\,km\,s^{-1}}$ (Zhao et al. 2011; Rüdiger et al. 2012).

5.1.1 Large and Small Length Scales

The length-scale dependence of the different types of helicity can be investigated by looking at the spectra of magnetic energy and helicity.

Magnetic Helicity Spectra Given that the magnetic helicity is a conserved quantity, it should be zero if it was zero initially. However, because it is a signed quantity, "zero" can consist of a "mixture" of pluses and minuses. These two signs can be segregated spatially (typically into north and south) as well as spectrally (into large and small scales). This is discussed in more detail below when we talk about catastrophic quenching of a large-scale dynamo. For now it suffices to say that we can define a magnetic helicity spectrum $H_{\mathrm{M}}(k)$, where k is its wavenumber (inverse length scale), and whose integral gives the mean magnetic helicity density, i.e.,

$$\int_0^\infty H_{\mathrm{M}}(k)\,\mathrm{d}k=\langle\boldsymbol{A}\cdot\boldsymbol{B}\rangle. \tag{6}$$

If the turbulence or the magnetic field were homogeneous, it can be related to the Fourier transform of the two-point correlation function

$$M_{ij}(\boldsymbol{r})=\big\langle\boldsymbol{B}_i(\boldsymbol{x}+\boldsymbol{r})\boldsymbol{B}_j(\boldsymbol{x})\big\rangle, \tag{7}$$

which is independent of $\boldsymbol{x}$ owing to the assumption of homogeneity, and $r=|\boldsymbol{r}|$ is the separation. Its Fourier transform over $\boldsymbol{r}$ gives $\tilde{M}_{ij}(k)$, which, under isotropy, has the representation

$$4\pi k^2\tilde{M}_{ij}(k)=(\delta_{ij}-\hat{k}_i\hat{k}_j)2\mu_0E_{\mathrm{M}}(k)+\epsilon_{ijk}\mathrm{i}k_kH_{\mathrm{M}}(k), \tag{8}$$

where $\hat{\boldsymbol{k}}=\boldsymbol{k}/|\boldsymbol{k}|$ is the unit vector of $\boldsymbol{k}$ and $E_{\mathrm{M}}(k)$ is the magnetic energy spectrum with the normalization $\int E_{\mathrm{M}}(k)\,\mathrm{d}k=\langle\boldsymbol{B}^2\rangle/2\mu_0$. The dependence of $E_{\mathrm{M}}(k)$ and $H_{\mathrm{M}}(k)$ on the modulus $k=|\boldsymbol{k}|$ of the wavevector $\boldsymbol{k}$ is again an consequence of isotropy. This relation is slightly

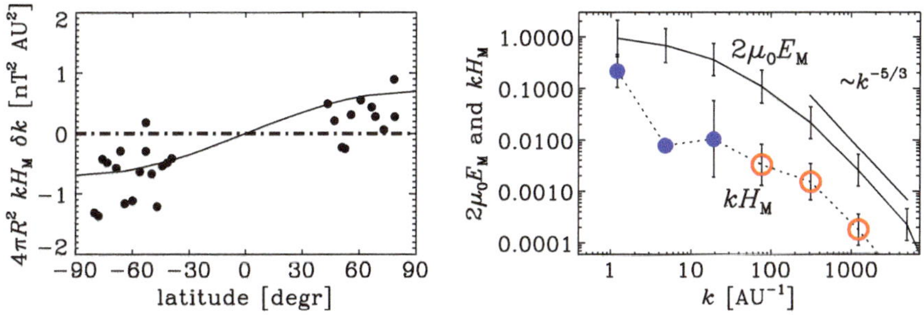

Fig. 7 Latitudinal dependence of spectral magnetic helicity for $k = 300\ \mathrm{AU}^{-1} \approx 2 \times 10^{-3}\ \mathrm{Mm}^{-1}$ (*left*) and the magnetic helicity spectrum for heliocentric distances above 2.8 AU for the northern hemisphere (*right*). *Filled blue symbols* denote negative values and *open red ones* positive values

modified when applied to the two-dimensional solar surface, but it allows us to obtain for the first time *spectra* of magnetic helicity, i.e., information about its composition from different length scales; see Zhang et al. (2014) for results that have confirmed that, in the southern hemisphere, the magnetic helicity is positive on wavenumbers of about 1 Mm^{-1}. This is in agreement with the results presented in Sect. 3.5.

Magnetic Helicity Spectra of the Solar Wind A similar technique to that presented above has been applied to the solar wind, where it is possible to obtain the magnetic field vector from *in situ* spacecraft measurements. The idea to compute the magnetic helicity spectrum by using Eq. (8) goes back to early work of Matthaeus et al. (1982), who used data from Voyager 1 and 2. To obtain measurements at positions $\boldsymbol{x}$ and $\boldsymbol{x}+\boldsymbol{r}$, one uses the Taylor hypothesis to relate the spatial separation $\boldsymbol{r}$ to a temporal separation t through $\boldsymbol{r} = \boldsymbol{r}_0 - \boldsymbol{u}_{\mathrm{W}} t$, where $\boldsymbol{u}_{\mathrm{W}}$ is the solar wind velocity of about 800 $\mathrm{km\,s}^{-1}$ and $\boldsymbol{r}_0$ is some reference position. However, since Voyager 1 and 2 flew close to the equatorial plane, the resulting magnetic helicity was expected to fluctuate around zero, which was indeed the case. This changed when data from *Ulysses* were used for such an analysis (Brandenburg et al. 2011b). In Fig. 7 we show the resulting spectrum, as well as the latitudinal dependence a specific k. We should point out that $H_{\mathrm{M}}(k)$ is here gauge-invariant because we are dealing with an infinite or periodic domain, which is automatically implied by the use of Fourier spectra. Note also that $k|H_{\mathrm{M}}(k)| \le 2\mu_0 E_{\mathrm{M}}(k)$, which is also known as the realizability condition.

The importance of magnetic helicity became particularly clear in connection with understanding the phenomenon of catastrophic quenching, which will be explained in the next section.

5.1.2 Magnetic Helicity Conservation

To understand the basics of dynamo action, including saturation mechanisms, it is often useful to work in idealized periodic geometries, and with turbulence either from simple forcing or driven by convection. While early work in this context at small values of R_{m} were promising (Brandenburg et al. 1990), subsequent studies at larger values of R_{m} showed that the field-aligned emf, $\langle \boldsymbol{u} \times \boldsymbol{b} \rangle_{\parallel} / B_{\parallel}$ approaches zero as $R_{\mathrm{m}} \to \infty$ (see Cattaneo and Hughes 1996).

This was understood as a consequence of magnetic helicity conservation (Gruzinov and Diamond 1996). Consider the equation for the fluctuation of the magnetic vector potential

$\boldsymbol{a} = \boldsymbol{A} - \langle \boldsymbol{A} \rangle$, where angle brackets denote volume averages and lower case symbols the fluctuations. The mean flow is assumed to vanish. Thus, we have

$$\frac{\partial \boldsymbol{a}}{\partial t} = \boldsymbol{u} \times \langle \boldsymbol{B} \rangle + \boldsymbol{u} \times \boldsymbol{b} - \eta \mu_0 \boldsymbol{j}. \tag{9}$$

We now derive the equation for the mean magnetic helicity density of the small-scale field as

$$\frac{\mathrm{d}}{\mathrm{d}t} \langle \boldsymbol{a} \cdot \boldsymbol{b} \rangle = -2 \langle \boldsymbol{u} \times \boldsymbol{b} \rangle \cdot \langle \boldsymbol{B} \rangle - 2\eta \mu_0 \langle \boldsymbol{j} \cdot \boldsymbol{b} \rangle. \tag{10}$$

In the steady state, we find that

$$\text{“}\alpha\text{”} \equiv \langle \boldsymbol{u} \times \boldsymbol{b} \rangle \cdot \langle \boldsymbol{B} \rangle / \langle \boldsymbol{B} \rangle^2 = -\eta \mu_0 \langle \boldsymbol{j} \cdot \boldsymbol{b} \rangle / \langle \boldsymbol{B} \rangle^2. \tag{11}$$

This relation for “α” is sometimes known as Keinigs relation (Keinigs 1983) and shows not only that “α” is positive when $\langle \boldsymbol{j} \cdot \boldsymbol{b} \rangle$ is negative (i.e., in the north), but also that “α” $\to 0$ when $\eta \to 0$, i.e., in the limit of large magnetic Reynolds numbers. This remarkable result seems like a disappointment to α effect theory, but it only means that no $\langle \boldsymbol{B} \rangle$, defined as a volume average (!), can be generated. This should then be no surprise, because the volume average of $\boldsymbol{B}$ is a conserved quantity for the periodic boundary conditions used in the study of Cattaneo and Hughes (1996) (which is the reason we put α in quotes).

When we consider the case where the averages are allowed to vary in space, the divergences of magnetic helicity flux will in general be non zero in the magnetic helicity equation.

Dynamical Quenching Consider planar averages, denoted by an overbar, magnetic helicity conservation yields instead

$$\alpha = \frac{\alpha_{\mathrm{K}} + R_{\mathrm{m}}[\mu_0 \overline{\boldsymbol{J}} \cdot \overline{\boldsymbol{B}} / B_{\mathrm{eq}}^2 - \nabla \cdot \overline{\boldsymbol{F}}_{\mathrm{f}} / 2B_{\mathrm{eq}}^2 - (\partial \alpha / \partial t)/(2\eta_{\mathrm{t}} k_{\mathrm{f}}^2)]}{1 + R_{\mathrm{m}} \overline{\boldsymbol{B}}^2 / B_{\mathrm{eq}}^2}. \tag{12}$$

Here, $\overline{\boldsymbol{F}}_{\mathrm{f}}$ is the magnetic helicity flux from the fluctuating magnetic field and $\alpha_{\mathrm{K}} = -(\tau/3)\langle \boldsymbol{\omega} \cdot \boldsymbol{u} \rangle$ is the kinetic α effect, which itself could depend on $\overline{\boldsymbol{B}}$, but this is here neglected. The main contribution to the quenching in Eq. (12) comes from the magnetic contribution α_{M} to the α effect, where $\alpha = \alpha_{\mathrm{K}} + \alpha_{\mathrm{M}}$ (Pouquet et al. 1976).

Equation (12) confirms first of all that α is catastrophically quenched (i.e., in an R_{m}-dependent fashion), when volume averages are used, i.e., when $\overline{\boldsymbol{J}} = \overline{\boldsymbol{F}}_{\mathrm{f}} = \boldsymbol{0}$ and the assumption of stationarity is made ($\partial \alpha / \partial t = 0$). In that case, we obtain

$$\alpha = \frac{\alpha_{\mathrm{K}}}{1 + R_{\mathrm{m}} \overline{\boldsymbol{B}}^2 / B_{\mathrm{eq}}^2} \quad \text{(for volume averages!).} \tag{13}$$

This equation was first motivated by Vainshtein and Cattaneo (1992) on the grounds that the energy ratio of small-scale to large-scale magnetic fields is proportional to R_{m}, i.e., $\langle b^2 \rangle / \langle \overline{\boldsymbol{B}}^2 \rangle \approx R_{\mathrm{m}}$ (e.g. Moffatt 1978; Krause and Rädler 1980), but this relation becomes invalid at large values of R_{m} where the rhs has to be replaced by $\ln R_{\mathrm{m}}$ (Kleeorin and Rogachevskii 1994). Furthermore, this relation assumes that the small-scale magnetic field is solely the result of tangling, so dynamo action is actually ignored in the old argument of Vainshtein and Cattaneo (1992).

It is now clear from Eq. (12) that catastrophic quenching is alleviated when there are mean currents ($\overline{\boldsymbol{J}} \neq \boldsymbol{0}$), which is already the case in triply-periodic helical turbulence, where, in fact, a super-equipartition field with $\langle\overline{\boldsymbol{B}}^2\rangle/\langle\boldsymbol{b}^2\rangle \approx k_\mathrm{f}/k_1$ can be generated, albeit only on a resistive time scale (Brandenburg 2001). Here, k_f/k_1 is the aforementioned scale separation ratio. Nevertheless, because of the long saturation time (which is determined by the microphysical diffusivity), such dynamos cannot be astrophysically relevant.

The ultimate rescue from catastrophic quenching seems to come from the presence of magnetic helicity flux divergences ($\nabla\cdot\overline{\boldsymbol{F}}_\mathrm{f} \neq 0$). Interestingly, what is primarily required is that the dynamo-generated field is no longer completely homogeneous as in α^2 dynamos, where $\overline{\boldsymbol{B}}$ is a Beltrami field with spatially constant $\overline{\boldsymbol{B}}^2$. For example, when there is shear, we can have $\alpha\Omega$-type dynamo action with finite $\nabla\cdot\overline{\boldsymbol{F}}_\mathrm{f}$ within an otherwise periodic (or shearing-periodic) domain where catastrophic quenching was indeed found to be alleviated (Hubbard and Brandenburg 2012). However, to demonstrate complete R_m independence is still difficult, and can only be expected for $R_\mathrm{m} \gtrsim 1000$ (Del Sordo et al. 2013).

At this point it is useful to return to the question of gauge-dependence. The evolution equation for the mean magnetic helicity density of the fluctuating field, $\overline{\boldsymbol{a}\cdot\boldsymbol{b}}$, can be written with a finite magnetic helicity flux divergence,

$$\frac{\partial}{\partial t}\overline{\boldsymbol{a}\cdot\boldsymbol{b}} = -2\overline{\boldsymbol{u}\times\boldsymbol{b}}\cdot\overline{\boldsymbol{B}} - 2\eta\mu_0\overline{\boldsymbol{j}\cdot\boldsymbol{b}} - \nabla\cdot\overline{\boldsymbol{F}}_\mathrm{f}. \tag{14}$$

On can now examine whether $\overline{\boldsymbol{a}\cdot\boldsymbol{b}}$, in the gauge under consideration, happens to be statistically stationary. In general, this does not need to be the case (for an example, see Fig. 2 of Brandenburg et al. 2002), but if it is, we can consider the overbars as denoting also an average over time, because then the left-hand side of Eq. (14) vanishes and we have

$$0 = -2\overline{\boldsymbol{u}\times\boldsymbol{b}}\cdot\overline{\boldsymbol{B}} - 2\eta\mu_0\overline{\boldsymbol{j}\cdot\boldsymbol{b}} - \nabla\cdot\overline{\boldsymbol{F}}_\mathrm{f}. \tag{15}$$

What is remarkable here is the fact that, at least in this special case ($\overline{\boldsymbol{a}\cdot\boldsymbol{b}}$ constant in time) the magnetic helicity flux divergence $\nabla\cdot\overline{\boldsymbol{F}}_\mathrm{f}$ is no longer gauge-dependent, i.e., it must be the same in all gauges. Moreover, unlike the aforementioned surface-integrated gauge-invariant magnetic helicity flux $2\oint\boldsymbol{E}\times\boldsymbol{A}_\mathrm{p}$, we can now make statements about its local dependence and its physical relation to mean flows and gradients of the magnetic helicity density. This has been done in several simulations which all confirm that an important part of the magnetic helicity flux is carried turbulent-diffusively like in Fickian diffusion (Mitra et al. 2010; Hubbard and Brandenburg 2010; Del Sordo et al. 2013).

5.1.3 Observational Clues

The Sign of the Helicity at Different Spatial Scales To make contact with solar magnetic helicity observations, we must ask about the scales on which helical fields are generated. If the large-scale field is really generated by an α effect, then both magnetic and current helicities of the large-scale field should have the same sign (Brandenburg 2001; Blackman and Brandenburg 2002) and should be positive in the north, where $\alpha > 0$. On large scales, the magnetic helicity $\langle\overline{\boldsymbol{A}}\cdot\overline{\boldsymbol{B}}\rangle$ obeys

$$\frac{\mathrm{d}}{\mathrm{d}t}\langle\overline{\boldsymbol{A}}\cdot\overline{\boldsymbol{B}}\rangle = 2\alpha\langle\overline{\boldsymbol{B}}^2\rangle - 2\eta_\mathrm{T}\mu_0\langle\overline{\boldsymbol{J}}\cdot\overline{\boldsymbol{B}}\rangle, \tag{16}$$

where we have assumed an isotropic α effect and an isotropic turbulent magnetic diffusivity η_t and $\eta_\mathrm{T} = \eta_\mathrm{t} + \eta$ is the total (turbulent plus microphysical) magnetic diffusivity. Note that, in the steady-state, we have

$$\mu_0\langle\overline{\boldsymbol{J}}\cdot\overline{\boldsymbol{B}}\rangle = (\alpha/\eta_\mathrm{T})\langle\overline{\boldsymbol{B}}^2\rangle, \tag{17}$$

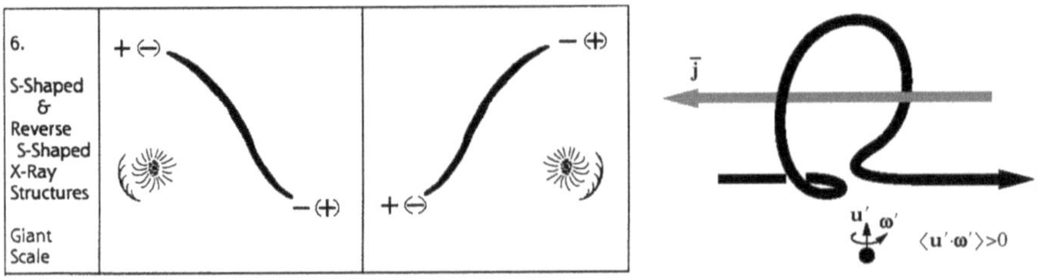

Fig. 8 Sketch of N-shaped and S-shaped sigmoidal structures in the northern and southern hemispheres (courtesy of Sara Martin) and sketch of an α loop for the southern hemisphere with positive kinetic helicity, taking S-shaped form when viewed from above (courtesy of Nobumitsu Yokoi)

i.e., the magnetic helicity of the large-scale field should indeed be positive in the north. At small scales, on the other hand, we expect the opposite sign. Only the latter has been observed directly. However, the N- and S-shaped structures in Hα images can indirectly be associated with large-scale fields resulting from a positive α affect in the north; see Fig. 8. Indeed, the barbs of filaments are an example, because right-handed (left-handed) barbs are found in filaments in which the purely axial threads (independent of the barb threads) have a slight but definite shape of a left-handed (right-handed) sigmoid. This way of interpreting the two signs of helicity within a single filament was discussed by Ruzmaikin et al. (2003); see Table 1 of Martin (2003) for details.

Helicity Reversals Within the Solar Wind The observation of the magnetic helicity spectrum in the solar wind poses some questions, because the sign is exactly opposite of what is observed at the solar surface. Theoretical support for this surprising result comes from simulations of dynamos with an extended outer layer that only supports turbulent diffusion, but no α effect. A magnetic helicity reversal was first recognized in the simulations of Warnecke et al. (2011, 2012), but such reversals were already present in early mean-field simulations of Brandenburg et al. (2009), which included the physics of turbulent-diffusive magnetic helicity fluxes.

Two related explanations have been proposed. Firstly, within the dynamo the effects of α and η_T nearly balance, which implies that both terms enter the magnetic helicity equation with opposite signs; see Eq. (16). However, within the wind, the α effect is basically absent, creating therefore an imbalance and thus a contribution of opposite sign. A related explanation assumes a steady state and invokes a turbulent-diffusive magnetic helicity flux obeying a Fickian diffusion law, i.e., $\overline{\boldsymbol{F}}_{\rm f} = -\kappa_{\rm f}\nabla\overline{\boldsymbol{a}\cdot\boldsymbol{b}}$, where $\kappa_{\rm f}$ is a turbulent diffusivity for magnetic helicity of the small-scale field. The difference from heat diffusion is that temperature is positive definite, but magnetic helicity is not. To transport positive magnetic helicity outward, we need a negative magnetic helicity gradient, which tends to drive it to (and even through) zero, which could explain the reversal. At present it is unclear which, if any, of these proposals is applicable. It is interesting to note, however, that similar reversals are seen also in the opposite orientations of coronal X-ray arcades in the northern and southern hemispheres; see Fig. 9. In essence, many filaments develop a sideways rolling motion that begins from the top down (Martin 2003) and evidence of this motion was found in Hα Doppler observations and in 304 Å images from SOHO (Panasenco and Martin 2008) and subsequently in 304 Å from SDO and STEREO (Panasenco et al. 2011). The most convincing evidence that the forces for this change come from the coronal environment is the correlation with coronal holes. Quiescent filaments without exception were found to roll from the top down away from adjacent coronal holes (Panasenco et al. 2013). The right-hand part of Fig. 9 shows the

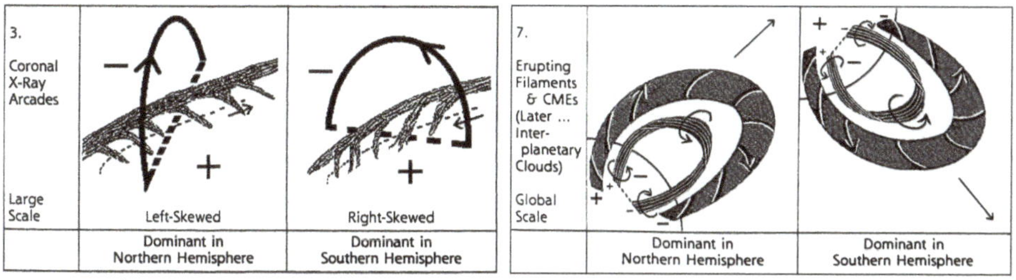

Fig. 9 Sketch of coronal X-ray arcades in the northern and southern hemispheres as well as their respective signs when they turn into interplanetary clouds (courtesy of Sara Martin)

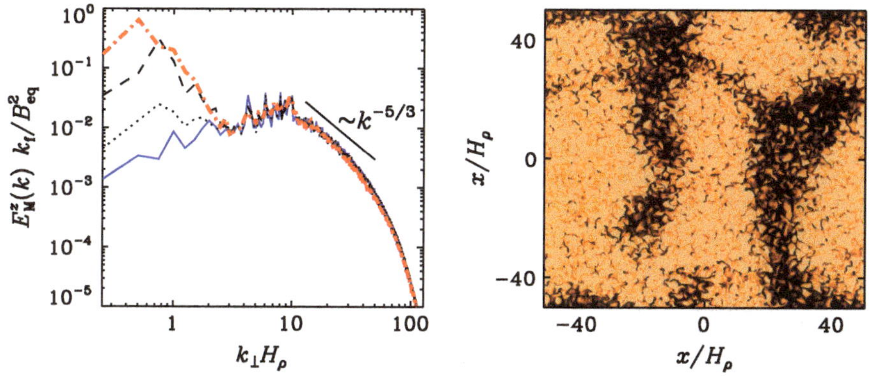

Fig. 10 *Left*: Normalized spectra of B_z from Run A40/1 of Brandenburg et al. (2014) at turbulent diffusive times $t\eta_t/H_\rho^2 \approx 0.2, 0.5, 1$, and 2.7 with $k_f H_\rho = 10$ and $k_1 H_\rho = 0.25$. *Right*: Magnetic field configuration at the upper surface near the end of the simulation

direction of motions in the filament that result in their becoming twisted during eruption. The general direction of the magnetic field is denoted by the polarity at the footpoints.

Magnetic Structures from Cross Helicity? In the presence of strong stratification, cross helicity is being generated if a large-scale magnetic field pierces the surface. This leads to a gradual evolution of the magnetic energy spectra of the vertical field, $E_M^z(k)$, showing a growth at small wavenumbers, akin to inverse transfer resulting from the α effect and approximate magnetic helicity conservation. This growth is associated with the development of magnetic structures; see Fig. 10.

The formation of such magnetic structures has also been associated with the possibility of a large-scale instability resulting from a negative contribution to the effective (turbulent) magnetic pressure. This instability is therefore referred to as negative effective magnetic pressure instability (NEMPI). Earlier work on NEMPI has shown a remarkably degree of predictive power of this theory in comparison with simulations (Brandenburg et al. 2011a; Kemel et al. 2013; Losada et al. 2013), but it is still unclear what drives the magnetic structures in simulations where the field strongly exceeds equipartition with the energy density of the turbulent motions (Mitra et al. 2014). Whether or not NEMPI or similar phenomena play a role in the formation of active regions or even sunspots is however an open question.

6 Conclusion

The question appears simple: Why does a star such as the Sun support have a global-scale magnetic field which reverses every 11 years and has equatorially propagating activity wings? We do not have anything like a complete answer to this question.

At the most fundamental level we do not we do not know the profiles of all the large-scale flows that are present. Beginning with convection, which is the driver for the global-scale motions and the magnetic field, we have to leave the reader with the following unanswered questions: Why is there so little power at scales larger than supergranulation in the solar photosphere (Lord et al. 2014)? Does this lack of power reflect a lack of power in these large scales at depth (compare the results in Hanasoge et al. 2012; Greer et al. 2015)? Is our fundamental picture of convection inapplicable to the solar convection zone (Spruit 1997; Brandenburg 2015)?

Moving to the flows on the global scale: What does the meridional circulation look like beneath the solar surface (Zhao et al. 2013; Schad et al. 2013; Jackiewicz et al. 2015)? What maintains the solar differential rotation? What maintains the meridional circulation? For the latter two questions we have some ideas (which give different answers), but they depend partly on the strength of the convective flows and will also depend on how the 'small-scale' turbulent dynamo, supposed to operate in the convection zone, modifies the flows.

These problems in our understanding, for example the power spectrum of the convective flows in the Sun, are not necessarily critical for understanding the solar dynamo if we take the observed flows as given. This simplifies the problem to asking how the (partially) observed flows produce the solar dynamo with all its observed features. Even this more limited problem is not solved. In this review we have briefly highlighted two important open questions. One of these questions is: Why do sunspots only appear at latitudes below about $\pm 40°$? The commonly given answer concerns the stability of toroidal flux located at the base of the convection zone, which for 10^5 G flux tubes are much more unstable at low latitudes than at high latitudes. Little consideration has however been given in this regard to the possibility that the toroidal flux is stored in the bulk of the convection zone. The second question we have highlighted is what causes the equatorial propagation of the activity belt towards the equator, and here we have competing answers. Both of these problems appear to be difficult, and their solution is likely to come from improvements in helioseismology and modeling efforts.

In order to end on a bright note, we think it is important to point out some of the recent progress which has occurred. For example we have Hale's law which tells us that the structure of the toroidal flux is quite simple, and at solar minimum the surface radial field is also simple (it is mainly near the poles and mainly of one sign in each hemisphere). This simplicity has enabled some rather strong conclusions to be drawn: the solar dynamo is an alpha-omega dynamo of the Babcock-Leighton type (Cameron and Schüssler 2015). Furthermore we have a number of ideas as to why the amount of activity varies from cycle to cycle (e.g. Jiang et al. 2015), and critically we are beginning to test these models against real data (Dikpati et al. 2006; Choudhuri et al. 2007; Dikpati et al. 2010b; Nandy et al. 2011). We thus have a long way to go but are making progress and have the right tools and perspective to make real progress in the coming years.

Acknowledgements This work was, in part, carried out in the context of Deutsche Forschungsgemeinschaft SFB 963 "Astrophysical Flow Instabilities and Turbulence" (Project A16).

This work was supported in part by the Swedish Research Council grants No. 621-2011-5076 and 2012-5797, as well as the Research Council of Norway under the FRINATEK grant 231444.

This work has been partially supported by NASA grant NNX08AQ34G. The National Center for Atmospheric Research is sponsored by the National Science Foundation.

NSO/Kitt Peak data used here are produced cooperatively by NSF/NOAO, NASA/GSFC, and NOAA/SEL. This work utilizes SOLIS data obtained by the NSO Integrated Synoptic Program (NISP), managed by the National Solar Observatory, which is operated by the Association of Universities for Research in Astronomy (AURA), Inc. under a cooperative agreement with the National Science Foundation. The synoptic magnetograms were downloaded from the NSO digital library, http://diglib.nso.edu/ftp.html

We acknowledge the support from ISSI Bern, for her participation in the ISSI workshop on.

Open access funding provided by Max Planck Society (or associated institution if applicable).

References

R. Arlt, R. Leussu, N. Giese, K. Mursula, I.G. Usoskin, Sunspot positions and sizes for 1825–1867 from the observations by Samuel Heinrich Schwabe. Mon. Not. R. Astron. Soc. **433**, 3165–3172 (2013). doi:10.1093/mnras/stt961

H.W. Babcock, The topology of the Sun's magnetic field and the 22-year cycle. Astrophys. J. **133**, 572 (1961). doi:10.1086/147060

L.A. Balmaceda, S.K. Solanki, N.A. Krivova, S. Foster, A homogeneous database of sunspot areas covering more than 130 years. J. Geophys. Res. Space Phys. **114**, 7104 (2009). doi:10.1029/2009JA014299

B. Belucz, M. Dikpati, Role of asymmetric meridional circulation in producing North-South asymmetry in a solar cycle dynamo model. Astrophys. J. **779**, 4 (2013). doi:10.1088/0004-637X/779/1/4

B. Belucz, M. Dikpati, E. Forgács-Dajka, A Babcock-Leighton solar dynamo model with multi-cellular meridional circulation in advection- and diffusion-dominated regimes. Astrophys. J. **806**, 169 (2015). doi:10.1088/0004-637X/806/2/169

M.A. Berger, G.B. Field, The topological properties of magnetic helicity. J. Fluid Mech. **147**, 133–148 (1984). doi:10.1017/S0022112084002019

M.A. Berger, A. Ruzmaikin, Rate of helicity production by solar rotation. J. Geophys. Res. **105**, 10481–10490 (2000). doi:10.1029/1999JA900392

L. Bertello, R.K. Ulrich, J.E. Boyden, The Mount Wilson Ca II K plage index time series. Sol. Phys. **264**, 31–44 (2010). doi:10.1007/s11207-010-9570-z

E.G. Blackman, A. Brandenburg, Dynamic nonlinearity in large-scale dynamos with shear. Astrophys. J. **579**, 359–373 (2002). doi:10.1086/342705

A. Brandenburg, The inverse cascade and nonlinear alpha-effect in simulations of isotropic helical hydromagnetic turbulence. Astrophys. J. **550**, 824–840 (2001). doi:10.1086/319783

A. Brandenburg, The critical role of magnetic helicity in astrophysical large-scale dynamos. Plasma Phys. Control. Fusion **51**(12), 124043 (2009). doi:10.1088/0741-3335/51/12/124043

A. Brandenburg, Stellar mixing length theory with entropy rain (2015). arXiv:1504.03189

A. Brandenburg, A. Nordlund, P. Pulkkinen, R.F. Stein, I. Tuominen, 3-D simulation of turbulent cyclonic magneto-convection. Astron. Astrophys. **232**, 277–291 (1990)

A. Brandenburg, D. Moss, I. Tuominen, Stratification and thermodynamics in mean-field dynamos. Astron. Astrophys. **265**, 328–344 (1992)

A. Brandenburg, W. Dobler, K. Subramanian, Magnetic helicity in stellar dynamos: new numerical experiments. Astron. Nachr. **323**, 99–122 (2002). doi:10.1002/1521-3994(200207)323:2<99::AID-ASNA99>3.0.CO;2-B

A. Brandenburg, S. Candelaresi, P. Chatterjee, Small-scale magnetic helicity losses from a mean-field dynamo. Mon. Not. R. Astron. Soc. **398**, 1414–1422 (2009). doi:10.1111/j.1365-2966.2009.15188.x

A. Brandenburg, K. Kemel, N. Kleeorin, D. Mitra, I. Rogachevskii, Detection of negative effective magnetic pressure instability in turbulence simulations. Astrophys. J. Lett. **740**, 50 (2011a). doi:10.1088/2041-8205/740/2/L50

A. Brandenburg, K. Subramanian, A. Balogh, M.L. Goldstein, Scale dependence of magnetic helicity in the solar wind. Astrophys. J. **734**, 9 (2011b). doi:10.1088/0004-637X/734/1/9

A. Brandenburg, O. Gressel, S. Jabbari, N. Kleeorin, I. Rogachevskii, Mean-field and direct numerical simulations of magnetic flux concentrations from vertical field. Astron. Astrophys. **562**, 53 (2014). doi:10.1051/0004-6361/201322681

T.M. Brown, C.A. Morrow, Depth and latitude dependence of solar rotation. Astrophys. J. Lett. **314**, 21–26 (1987). doi:10.1086/184843

B.P. Brown, M.K. Browning, A.S. Brun, M.S. Miesch, J. Toomre, Persistent magnetic wreaths in a rapidly rotating sun. Astrophys. J. **711**, 424–438 (2010). doi:10.1088/0004-637X/711/1/424

A.S. Brun, M.S. Miesch, J. Toomre, Global-scale turbulent convection and magnetic dynamo action in the solar envelope. Astrophys. J. **614**, 1073–1098 (2004). doi:10.1086/423835

P. Caligari, F. Moreno-Insertis, M. Schussler, Emerging flux tubes in the solar convection zone. 1: asymmetry, tilt, and emergence latitude. Astrophys. J. **441**, 886–902 (1995). doi:10.1086/175410

R. Cameron, M. Schüssler, Solar cycle prediction using precursors and flux transport models. Astrophys. J. **659**, 801–811 (2007). doi:10.1086/512049

R. Cameron, M. Schüssler, A robust correlation between growth rate and amplitude of solar cycles: consequences for prediction methods. Astrophys. J. **685**, 1291–1296 (2008). doi:10.1086/591079

R.H. Cameron, M. Schüssler, Are the strengths of solar cycles determined by converging flows towards the activity belts? Astron. Astrophys. **548**, 57 (2012). doi:10.1051/0004-6361/201219914

R. Cameron, M. Schüssler, The crucial role of surface magnetic fields for the solar dynamo. Science **347**, 1333–1335 (2015). doi:10.1126/science.1261470

R.H. Cameron, M. Dasi-Espuig, J. Jiang, E. Işık, D. Schmitt, M. Schüssler, Limits to solar cycle predictability: cross-equatorial flux plumes. Astron. Astrophys. **557**, 141 (2013). doi:10.1051/0004-6361/201321981

R.C. Canfield, H.S. Hudson, D.E. McKenzie, Sigmoidal morphology and eruptive solar activity. Geophys. Res. Lett. **26**, 627–630 (1999). doi:10.1029/1999GL900105

M. Carbonell, R. Oliver, J.L. Ballester, On the asymmetry of solar activity. Astron. Astrophys. **274**, 497 (1993)

R.C. Carrington, On the distribution of the solar spots in latitudes since the beginning of the year 1854, with a map. Mon. Not. R. Astron. Soc. **19**, 1–3 (1858)

F. Cattaneo, D.W. Hughes, Nonlinear saturation of the turbulent α effect. Phys. Rev. E **54**, 4532 (1996). doi:10.1103/PhysRevE.54.R4532

J. Chae, The magnetic helicity sign of filament chirality. Astrophys. J. Lett. **540**, 115–118 (2000). doi: 10.1086/312880

J. Chae, Observational determination of the rate of magnetic helicity transport through the solar surface via the horizontal motion of field line footpoints. Astrophys. J. Lett. **560**, 95–98 (2001). doi:10.1086/324173

A.R. Choudhuri, P.A. Gilman, The influence of the Coriolis force on flux tubes rising through the solar convection zone. Astrophys. J. **316**, 788–800 (1987). doi:10.1086/165243

A.R. Choudhuri, M. Schussler, M. Dikpati, The solar dynamo with meridional circulation. Astron. Astrophys. **303**, 29 (1995)

A.R. Choudhuri, P. Chatterjee, J. Jiang, Predicting solar cycle 24 with a solar dynamo model. Phys. Rev. Lett. **98**(13), 131103 (2007). doi:10.1103/PhysRevLett.98.131103

F. Clette, L. Svalgaard, J.M. Vaquero, E.W. Cliver, Revisiting the sunspot number. A 400-year perspective on the solar cycle. Space Sci. Rev. **186**, 35–103 (2014). doi:10.1007/s11214-014-0074-2

M. Dasi-Espuig, S.K. Solanki, N.A. Krivova, R. Cameron, T. Peñuela, Sunspot group tilt angles and the strength of the solar cycle. Astron. Astrophys. **518**, 7 (2010). doi:10.1051/0004-6361/201014301

F. Del Sordo, G. Guerrero, A. Brandenburg, Turbulent dynamos with advective magnetic helicity flux. Mon. Not. R. Astron. Soc. **429**, 1686–1694 (2013). doi:10.1093/mnras/sts398

C.R. DeVore, Magnetic helicity generation by solar differential rotation. Astrophys. J. **539**, 944–953 (2000). doi:10.1086/309274

C.R. DeVore, J.P. Boris, N.R. Sheeley Jr., The concentration of the large-scale solar magnetic field by a meridional surface flow. Sol. Phys. **92**, 1–14 (1984). doi:10.1007/BF00157230

A. Diercke, R. Arlt, C. Denker, Digitization of sunspot drawings by Spörer made in 1861–1894. Astron. Nachr. **336**, 53 (2015). doi:10.1002/asna.201412138

M. Dikpati, P. Charbonneau, A Babcock-Leighton flux transport dynamo with solar-like differential rotation. Astrophys. J. **518**, 508–520 (1999). doi:10.1086/307269

M. Dikpati, G. de Toma, P.A. Gilman, Predicting the strength of solar cycle 24 using a flux-transport dynamo-based tool. Geophys. Res. Lett. **33**, 5102 (2006). doi:10.1029/2005GL025221

M. Dikpati, P.A. Gilman, G. de Toma, The Waldmeier effect: an artifact of the definition of wolf sunspot number? Astrophys. J. Lett. **673**, 99–101 (2008). doi:10.1086/527360

M. Dikpati, P.A. Gilman, R.P. Kane, Length of a minimum as predictor of next solar cycle's strength. Geophys. Res. Lett. **37**, 6104 (2010a). doi:10.1029/2009GL042280

M. Dikpati, P.A. Gilman, G. de Toma, R.K. Ulrich, Impact of changes in the Sun's conveyor-belt on recent solar cycles. Geophys. Res. Lett. **37**, 14107 (2010b). doi:10.1029/2010GL044143

S. D'Silva, A.R. Choudhuri, A theoretical model for tilts of bipolar magnetic regions. Astron. Astrophys. **272**, 621 (1993)

B.R. Durney, On a Babcock-Leighton dynamo model with a deep-seated generating layer for the toroidal magnetic field. Sol. Phys. **160**, 213–235 (1995). doi:10.1007/BF00732805

C.J. Durrant, J.P.R. Turner, P.R. Wilson, The mechanism involved in the reversals of the Sun's polar magnetic fields. Sol. Phys. **222**, 345–362 (2004). doi:10.1023/B:SOLA.0000043577.33961.82

T.L. Duvall Jr., Large-scale solar velocity fields. Sol. Phys. **63**, 3–15 (1979). doi:10.1007/BF00155690

Y. Fan, F. Fang, A simulation of convective dynamo in the solar convective envelope: maintenance of the solar-like differential rotation and emerging flux. Astrophys. J. **789**, 35 (2014). doi:10.1088/0004-637X/789/1/35

M. Ghizaru, P. Charbonneau, P.K. Smolarkiewicz, Magnetic cycles in global large-eddy simulations of solar convection. Astrophys. J. Lett. **715**, 133–137 (2010). doi:10.1088/2041-8205/715/2/L133

S.E. Gibson, L. Fletcher, G. Del Zanna, C.D. Pike, H.E. Mason, C.H. Mandrini, P. Démoulin, H. Gilbert, J. Burkepile, T. Holzer, D. Alexander, Y. Liu, N. Nitta, J. Qiu, B. Schmieder, B.J. Thompson, The structure and evolution of a sigmoidal active region. Astrophys. J. **574**, 1021–1038 (2002). doi:10.1086/341090

P.A. Gilman, Nonlinear dynamics of Boussinesq convection in a deep rotating spherical shell. I. Geophys. Astrophys. Fluid Dyn. **8**, 93–135 (1977). doi:10.1080/03091927708240373

P.A. Gilman, Dynamically consistent nonlinear dynamos driven by convection in a rotating spherical shell. II—dynamos with cycles and strong feedbacks. Astrophys. J. Suppl. Ser. **53**, 243–268 (1983). doi:10.1086/190891

P.A. Gilman, J. Miller, Dynamically consistent nonlinear dynamos driven by convection in a rotating spherical shell. Astrophys. J. Suppl. Ser. **46**, 211–238 (1981). doi:10.1086/190743

G.A. Glatzmaier, Numerical simulations of stellar convective dynamos. II—field propagation in the convection zone. Astrophys. J. **291**, 300–307 (1985). doi:10.1086/163069

G.A. Glatzmaier, P.H. Roberts, A three-dimensional convective dynamo solution with rotating and finitely conducting inner core and mantle. Phys. Earth Planet. Inter. **91**, 63–75 (1995). doi:10.1016/0031-9201(95)03049-3

W. Gleissberg, A long-periodic fluctuation of the sun-spot numbers. Observatory **62**, 158–159 (1939)

B.J. Greer, B.W. Hindman, N.A. Featherstone, J. Toomre, Helioseismic imaging of fast convective flows throughout the near-surface shear layer. Astrophys. J. Lett. **803**, 17 (2015). doi:10.1088/2041-8205/803/2/L17

A.V. Gruzinov, P.H. Diamond, Nonlinear mean field electrodynamics of turbulent dynamos. Phys. Plasmas **3**, 1853–1857 (1996). doi:10.1063/1.871981

G.E. Hale, On the probable existence of a magnetic field in Sun-spots. Astrophys. J. **28**, 315 (1908). doi:10.1086/141602

G.E. Hale, F. Ellerman, S.B. Nicholson, A.H. Joy, The magnetic polarity of Sun-spots. Astrophys. J. **49**, 153 (1919). doi:10.1086/142452

S.M. Hanasoge, T.L. Duvall, K.R. Sreenivasan, Anomalously weak solar convection. Proc. Natl. Acad. Sci. **109**, 11928–11932 (2012). doi:10.1073/pnas.1206570109

D.H. Hathaway, A standard law for the equatorward drift of the sunspot zones. Sol. Phys. **273**, 221–230 (2011). doi:10.1007/s11207-011-9837-z

D.H. Hathaway, Supergranules as probes of the Sun's meridional circulation. Astrophys. J. **760**, 84 (2012). doi:10.1088/0004-637X/760/1/84

N.E. Haugen, A. Brandenburg, W. Dobler, Simulations of nonhelical hydromagnetic turbulence. Phys. Rev. E **70**(1), 016308 (2004). doi:10.1103/PhysRevE.70.016308

A. Herzenberg, Geomagnetic dynamos. Philos. Trans. R. Soc. Lond. Ser. A **250**, 543–583 (1958). doi:10.1098/rsta.1958.0007

R. Howard, B.J. Labonte, The sun is observed to be a torsional oscillator with a period of 11 years. Astrophys. J. Lett. **239**, 33–36 (1980). doi:10.1086/183286

R. Howard, P.I. Gilman, P.A. Gilman, Rotation of the sun measured from Mount Wilson white-light images. Astrophys. J. **283**, 373–384 (1984). doi:10.1086/162315

R.F. Howard, K.R. Sivaraman, S.S. Gupta, P.I. Gilman, Sunspot motions from a study of Kodaikanal and Mount Wilson observations, in *Basic Plasma Processes on the Sun*, ed. by E.R. Priest, V. Krishan. IAU Symposium, vol. 142 (1990), pp. 107–111

R. Howe, Solar interior rotation and its variation. Living Rev. Sol. Phys. **6**, 1 (2009). doi:10.12942/lrsp-2009-1

A. Hubbard, A. Brandenburg, Magnetic helicity fluxes in an $\alpha 2$ dynamo embedded in a halo. Geophys. Astrophys. Fluid Dyn. **104**, 577–590 (2010). doi:10.1080/03091929.2010.506438

A. Hubbard, A. Brandenburg, Catastrophic quenching in $\alpha\Omega$ dynamos revisited. Astrophys. J. **748**, 51 (2012). doi:10.1088/0004-637X/748/1/51

J. Jackiewicz, A. Serebryanskiy, S. Kholikov, Meridional flow in the solar convection zone. II. Helioseismic inversions of GONG data. Astrophys. J. **805**, 133 (2015). doi:10.1088/0004-637X/805/2/133

J. Jiang, D.H. Hathaway, R.H. Cameron, S.K. Solanki, L. Gizon, L. Upton, Magnetic flux transport at the solar surface. Space Sci. Rev. **186**, 491–523 (2014). doi:10.1007/s11214-014-0083-1

J. Jiang, R.H. Cameron, M. Schüssler, The cause of the weak solar cycle 24. Astrophys. J. Lett. **808**, 28 (2015). doi:10.1088/2041-8205/808/1/L28

P.J. Käpylä, M.J. Mantere, A. Brandenburg, Cyclic magnetic activity due to turbulent convection in spherical wedge geometry. Astrophys. J. Lett. **755**, 22 (2012). doi:10.1088/2041-8205/755/1/L22

B.B. Karak, A.R. Choudhuri, The Waldmeier effect and the flux transport solar dynamo. Mon. Not. R. Astron. Soc. Lett. **410**, 1503–1512 (2011). doi:10.1111/j.1365-2966.2010.17531.x

A.P. Kazantsev, Enhancement of a magnetic field by a conducting fluid. Sov. Phys. JETP **26**, 1031 (1968)

R.K. Keinigs, A new interpretation of the alpha effect. Phys. Fluids **26**, 2558–2560 (1983). doi:10.1063/1.864446

K. Kemel, A. Brandenburg, N. Kleeorin, D. Mitra, I. Rogachevskii, Active region formation through the negative effective magnetic pressure instability. Sol. Phys. **287**, 293–313 (2013). doi:10.1007/s11207-012-0031-8

S. Kida, S. Yanase, J. Mizushima, Statistical properties of MHD turbulence and turbulent dynamo. Phys. Fluids A **3**, 457–465 (1991). doi:10.1063/1.858102

N. Kleeorin, I. Rogachevskii, Effective Ampère force in developed magnetohydrodynamic turbulence. Phys. Rev. E **50**, 2716–2730 (1994). doi:10.1103/PhysRevE.50.2716

F. Krause, K.-H. Rädler, *Mean-Field Magnetohydrodynamics and Dynamo Theory* (Akademie-Verlag, Berlin, 1980)

L. Křivský, K. Pejml, *Solar Activity, Aurorae and Climate in Central Europe in the Last 1000 Years* (Astronomical Institute of the Czechoslokak Academy of Sciences, Prague, 1988), p. 75

M. Küker, G. Rüdiger, M. Schultz, Circulation-dominated solar shell dynamo models with positive alpha-effect. Astron. Astrophys. **374**, 301–308 (2001). doi:10.1051/0004-6361:20010686

J. Langfellner, L. Gizon, A.C. Birch, Time-distance helioseismology: a new averaging scheme for measuring flow vorticity. Astron. Astrophys. **570**, 90 (2014). doi:10.1051/0004-6361/201424201

J. Larmor, How could a rotating body such as the Sun become a magnet? Rep. Br. Assoc. Adv. Sci. **87th Meeting**, 159–160 (1919)

R.B. Leighton, Transport of magnetic fields on the Sun. Astrophys. J. **140**, 1547 (1964). doi:10.1086/148058

R.B. Leighton, A magneto-kinematic model of the solar cycle. Astrophys. J. **156**, 1 (1969). doi:10.1086/149943

J. Li, R.K. Ulrich, Long-term measurements of sunspot magnetic tilt angles. Astrophys. J. **758**, 115 (2012). doi:10.1088/0004-637X/758/2/115

B.W. Lites, D.L. Akin, G. Card, T. Cruz, D.W. Duncan, C.G. Edwards, D.F. Elmore, C. Hoffmann, Y. Katsukawa, N. Katz, M. Kubo, K. Ichimoto, T. Shimizu, R.A. Shine, K.V. Streander, A. Suematsu, T.D. Tarbell, A.M. Title, S. Tsuneta, The Hinode spectro-polarimeter. Sol. Phys. **283**, 579–599 (2013). doi:10.1007/s11207-012-0206-3

J.W. Lord, R.H. Cameron, M.P. Rast, M. Rempel, T. Roudier, The role of subsurface flows in solar surface convection: modeling the spectrum of supergranular and larger scale flows. Astrophys. J. **793**, 24 (2014). doi:10.1088/0004-637X/793/1/24

I.R. Losada, A. Brandenburg, N. Kleeorin, I. Rogachevskii, Competition of rotation and stratification in flux concentrations. Astron. Astrophys. **556**, 83 (2013). doi:10.1051/0004-6361/201220939

T. Magara, D.W. Longcope, Sigmoid structure of an emerging flux tube. Astrophys. J. Lett. **559**, 55–59 (2001). doi:10.1086/323635

S.F. Martin, Conditions for the formation and maintenance of filaments (invited review). Sol. Phys. **182**, 107–137 (1998a). doi:10.1023/A:1005026814076

S.F. Martin, filament chirality: a link between fine-scale and global patterns (review), in *IAU Colloq. 167: New Perspectives on Solar Prominences*, ed. by D.F. Webb, B. Schmieder, D.M. Rust. Astronomical Society of the Pacific Conference Series, vol. 150 (1998b), p. 419

S.F. Martin, Signs of helicity in solar prominences and related features. Adv. Space Res. **32**, 1883–1893 (2003). doi:10.1016/S0273-1177(03)90622-3

W.H. Matthaeus, M.L. Goldstein, C. Smith, Evaluation of magnetic helicity in homogeneous turbulence. Phys. Rev. Lett. **48**, 1256–1259 (1982). doi:10.1103/PhysRevLett.48.1256

E.W. Maunder, Note on the distribution of sun-spots in heliographic latitude, 1874-1902. Mon. Not. R. Astron. Soc. Lett. **64**, 747–761 (1904)

M. Meneguzzi, U. Frisch, A. Pouquet, Helical and nonhelical turbulent dynamos. Phys. Rev. Lett. **47**, 1060–1064 (1981). doi:10.1103/PhysRevLett.47.1060

M.S. Miesch, M. Dikpati, A three-dimensional Babcock-Leighton solar dynamo model. Astrophys. J. Lett. **785**, 8 (2014). doi:10.1088/2041-8205/785/1/L8

M.S. Miesch, J.R. Elliott, J. Toomre, T.L. Clune, G.A. Glatzmaier, P.A. Gilman, Three-dimensional spherical simulations of solar convection. I. Differential rotation and pattern evolution achieved with laminar and turbulent states. Astrophys. J. **532**, 593–615 (2000). doi:10.1086/308555

D. Mitra, S. Candelaresi, P. Chatterjee, R. Tavakol, A. Brandenburg, equatorial magnetic helicity flux in simulations with different gauges. Astron. Nachr. **331**, 130 (2010). doi:10.1002/asna.200911308
D. Mitra, A. Brandenburg, N. Kleeorin, I. Rogachevskii, Intense bipolar structures from stratified helical dynamos. Mon. Not. R. Astron. Soc. Lett. **445**, 761–769 (2014). doi:10.1093/mnras/stu1755
H.K. Moffatt, The degree of knottedness of tangled vortex lines. J. Fluid Mech. **35**, 117–129 (1969). doi:10.1017/S0022112069000991
H.K. Moffatt, *Magnetic Field Generation in Electrically Conducting Fluids* (Cambridge University Press, Cambridge, 1978)
A. Muñoz-Jaramillo, M. Dasi-Espuig, L.A. Balmaceda, E.E. DeLuca, Solar cycle propagation, memory, and prediction: insights from a century of magnetic proxies. Astrophys. J. Lett. **767**, 25 (2013). doi:10.1088/2041-8205/767/2/L25
D. Nandy, A. Muñoz-Jaramillo, P.C.H. Martens, The unusual minimum of sunspot cycle 23 caused by meridional plasma flow variations. Nature **471**, 80–82 (2011). doi:10.1038/nature09786
N.J. Nelson, B.P. Brown, A. Sacha Brun, M.S. Miesch, J. Toomre, Buoyant magnetic loops generated by global convective dynamo action. Sol. Phys. **289**, 441–458 (2014). doi:10.1007/s11207-012-0221-4
H. Nevanlinna, E. Kataja, An extension of the geomagnetic activity index series aa for two solar cycles (1844–1868). Geophys. Res. Lett. **20**, 2703–2706 (1993). doi:10.1029/93GL03001
O. Panasenco, S.F. Martin, Topological analyses of symmetric eruptive prominences, in *Subsurface and Atmospheric Influences on Solar Activity*, ed. by R. Howe, R.W. Komm, K.S. Balasubramaniam, G.J.D. Petrie. Astronomical Society of the Pacific Conference Series, vol. 383 (2008), p. 243
O. Panasenco, S. Martin, A.D. Joshi, N. Srivastava, Rolling motion in erupting prominences observed by STEREO. J. Atmos. Sol.-Terr. Phys. **73**, 1129–1137 (2011). doi:10.1016/j.jastp.2010.09.010
O. Panasenco, S.F. Martin, M. Velli, A. Vourlidas, Origins of rolling, twisting, and non-radial propagation of eruptive solar events. Sol. Phys. **287**, 391–413 (2013). doi:10.1007/s11207-012-0194-3
E.N. Parker, Hydromagnetic dynamo models. Astrophys. J. **122**, 293 (1955a). doi:10.1086/146087
E.N. Parker, The formation of sunspots from the solar toroidal field. Astrophys. J. **121**, 491 (1955b). doi:10.1086/146010
A. Pouquet, U. Frisch, J. Leorat, Strong MHD helical turbulence and the nonlinear dynamo effect. J. Fluid Mech. **77**, 321–354 (1976). doi:10.1017/S0022112076002140
É. Racine, P. Charbonneau, M. Ghizaru, A. Bouchat, P.K. Smolarkiewicz, On the mode of dynamo action in a global large-eddy simulation of solar convection. Astrophys. J. **735**, 46 (2011). doi:10.1088/0004-637X/735/1/46
M. Rieutord, N. Meunier, T. Roudier, S. Rondi, F. Beigbeder, L. Parès, Solar supergranulation revealed by granule tracking. Astron. Astrophys. **479**, 17–20 (2008). doi:10.1051/0004-6361:20079077
G.O. Roberts, Spatially periodic dynamos. Philos. Trans. R. Soc. Lond. Ser. A **266**, 535–558 (1970). doi:10.1098/rsta.1970.0011
T. Roudier, J.M. Malherbe, J. Vigneau, B. Pfeiffer, Solar mesogranule lifetime measurements. Astron. Astrophys. **330**, 1136–1144 (1998)
G. Rüdiger, A. Brandenburg, V.V. Pipin, A helicity proxy from horizontal solar flow patterns. Astron. Nachr. **320**, 135 (1999)
G. Rüdiger, L.L. Kitchatinov, A. Brandenburg, Cross helicity and turbulent magnetic diffusivity in the solar convection zone. Sol. Phys. **269**, 3–12 (2011). doi:10.1007/s11207-010-9683-4
G. Rüdiger, M. Küker, R.S. Schnerr, Cross helicity at the solar surface by simulations and observations. Astron. Astrophys. **546**, 23 (2012). doi:10.1051/0004-6361/201219268
A. Ruzmaikin, S. Martin, Q. Hu, Signs of magnetic helicity in interplanetary coronal mass ejections and associated prominences: case study. J. Geophys. Res. **108**, 1096 (2003). doi:10.1029/2002JA009588
A. Schad, J. Timmer, M. Roth, Global helioseismic evidence for a deeply penetrating solar meridional flow consisting of multiple flow cells. Astrophys. J. Lett. **778**, 38 (2013). doi:10.1088/2041-8205/778/2/L38
P.H. Scherrer, R.S. Bogart, R.I. Bush, J.T. Hoeksema, A.G. Kosovichev, J. Schou, W. Rosenberg, L. Springer, T.D. Tarbell, A. Title, C.J. Wolfson, I. Zayer, MDI Engineering Team, The solar oscillations investigation—Michelson Doppler imager. Sol. Phys. **162**, 129–188 (1995). doi:10.1007/BF00733429
P.H. Scherrer, J. Schou, R.I. Bush, A.G. Kosovichev, R.S. Bogart, J.T. Hoeksema, Y. Liu, T.L. Duvall, J. Zhao, A.M. Title, C.J. Schrijver, T.D. Tarbell, S. Tomczyk, The Helioseismic and Magnetic Imager (HMI) investigation for the Solar Dynamics Observatory (SDO). Sol. Phys. **275**, 207–227 (2012). doi:10.1007/s11207-011-9834-2
J. Schou, H.M. Antia, S. Basu, R.S. Bogart, R.I. Bush, S.M. Chitre, J. Christensen-Dalsgaard, M.P. Di Mauro, W.A. Dziembowski, A. Eff-Darwich, D.O. Gough, D.A. Haber, J.T. Hoeksema, R. Howe, S.G. Korzennik, A.G. Kosovichev, R.M. Larsen, F.P. Pijpers, P.H. Scherrer, T. Sekii, T.D. Tarbell, A.M. Title, M.J. Thompson, J. Toomre, Helioseismic studies of differential rotation in the solar envelope by the solar oscillations investigation using the Michelson Doppler imager. Astrophys. J. **505**, 390–417 (1998). doi:10.1086/306146

M. Schwabe, Sonnen-und Saturn-Beobachtungen im Jahre 1848, von Herrn Hofrath Schwabe in Dessau. Astron. Nachr. **28**, 302 (1849)

N. Seehafer, Electric current helicity in the solar atmosphere. Sol. Phys. **125**, 219–232 (1990). doi:10.1007/BF00158402

J. Shetye, D. Tripathi, M. Dikpati, Observations and modeling of North–South asymmetries using a flux transport dynamo. Astrophys. J. **799**, 220 (2015). doi:10.1088/0004-637X/799/2/220

E.J. Smith, A. Balogh, Ulysses observations of the radial magnetic field. Geophys. Res. Lett. **22**, 3317–3320 (1995). doi:10.1029/95GL02826

G. Spörer, Beobachtung der Sonnenflecken etc. Astron. Nachr. **96**, 23 (1879). doi:10.1002/asna.18790960205

G. Spörer, Mémoires et observations. Sur les différences que présentent l'hémisphère nord et l'hémisphère sud du soleil. Bull. Astron., Sér. I **6**, 60–63 (1889a)

G. Spörer, Über die Periodicität der Sonnenflecken seit dem Jahr 1618. Nova Acta-Kais. Leopold.-Carol. Dtsch. Akad. Nat.forsch. **53**(2), 281–324 (1889b)

H. Spruit, Convection in stellar envelopes: a changing paradigm. Mem. Soc. Astron. Ital. **68**, 397 (1997)

M. Steenbeck, F. Krause, K.-H. Rädler, Berechnung der mittleren Lorentz-Feldstärke $\overline{\mathbf{v} \times \mathbf{B}}$ für ein elektrisch leitendes Medium in turbulenter, durch Coriolis-Kräfte beeinflußter Bewegung. Z. Naturforsch. Teil A **21**, 369 (1966)

R.F. Stein, A. Nordlund, Topology of convection beneath the solar surface. Astrophys. J. Lett. **342**, 95–98 (1989). doi:10.1086/185493

R.F. Stein, Å. Nordlund, On the formation of active regions. Astrophys. J. Lett. **753**, 13 (2012)

F. Steinhilber, J.A. Abreu, J. Beer, I. Brunner, M. Christl, H. Fischer, U. Heikkila, P.W. Kubik, M. Mann, K.G. McCracken, H. Miller, H. Miyahara, H. Oerter, F. Wilhelms, 9,400 years of cosmic radiation and solar activity from ice cores and tree rings. Proc. Natl. Acad. Sci. **109**, 5967–5971 (2012). doi:10.1073/pnas.1118965109

M. Stix, Comments on the solar dynamo. Astron. Astrophys. **37**, 121–133 (1974)

S. Tsuneta, K. Ichimoto, Y. Katsukawa, B.W. Lites, K. Matsuzaki, S. Nagata, D. Orozco Suárez, T. Shimizu, M. Shimojo, R.A. Shine, Y. Suematsu, T.K. Suzuki, T.D. Tarbell, A.M. Title, The magnetic landscape of the Sun's polar region. Astrophys. J. **688**, 1374–1381 (2008). doi:10.1086/592226

R.K. Ulrich, Solar meridional circulation from Doppler shifts of the Fe I line at 5250 Å as measured by the 150-foot Solar Tower telescope at the M. Wilson observatory. Astrophys. J. **725**, 658–669 (2010). doi:10.1088/0004-637X/725/1/658

I.G. Usoskin, S.K. Solanki, G.A. Kovaltsov, Grand minima and maxima of solar activity: new observational constraints. Astron. Astrophys. **471**, 301–309 (2007). doi:10.1051/0004-6361:20077704

S.I. Vainshtein, F. Cattaneo, Nonlinear restrictions on dynamo action. Astrophys. J. **393**, 165–171 (1992). doi:10.1086/171494

M. Waldmeier, *Ergebnisse und Probleme der Sonnenforschung* (Becker & Erler Kom.-Ges., Leipzig, 1941)

M. Waldmeier, *Ergebnisse und Probleme der Sonnenforschung* (Geest & Portig, Leipzig, 1955)

Y.-M. Wang, N.R. Sheeley Jr., Average properties of bipolar magnetic regions during sunspot cycle 21. Sol. Phys. **124**, 81–100 (1989). doi:10.1007/BF00146521

Y.-M. Wang, N.R. Sheeley Jr., Magnetic flux transport and the Sun's dipole moment—new twists to the Babcock-Leighton model. Astrophys. J. **375**, 761–770 (1991). doi:10.1086/170240

Y.-M. Wang, N.R. Sheeley, Understanding the geomagnetic precursor of the solar cycle. Astrophys. J. Lett. **694**, 11–15 (2009). doi:10.1088/0004-637X/694/1/L11

J. Warnecke, A. Brandenburg, D. Mitra, Dynamo-driven plasmoid ejections above a spherical surface. Astron. Astrophys. **534**, 11 (2011). doi:10.1051/0004-6361/201117023

J. Warnecke, A. Brandenburg, D. Mitra, Magnetic twist: a source and property of space weather. J. Space Weather Space Clim. **2**(27), 11 (2012). doi:10.1051/swsc/2012011

J. Warnecke, P.J. Käpylä, M.J. Käpylä, A. Brandenburg, On the cause of solar-like equatorward migration in global convective dynamo simulations. Astrophys. J. Lett. **796**, 12 (2014). doi:10.1088/2041-8205/796/1/L12

B.T. Welsch, D.W. Longcope, Magnetic helicity injection by horizontal flows in the quiet Sun. I. Mutual-helicity flux. Astrophys. J. **588**, 620–629 (2003). doi:10.1086/368408

P.R. Wilson, R.C. Altrocki, K.L. Harvey, S.F. Martin, H.B. Snodgrass, The extended solar activity cycle. Nature **333**, 748–750 (1988). doi:10.1038/333748a0

L. Woltjer, On hydromagnetic equilibrium. Proc. Natl. Acad. Sci. **44**, 833–841 (1958). doi:10.1073/pnas.44.9.833

N. Yokoi, Cross helicity and related dynamo. Geophys. Astrophys. Fluid Dyn. **107**, 114–184 (2013). doi:10.1080/03091929.2012.754022

H. Yoshimura, Solar-cycle dynamo wave propagation. Astrophys. J. **201**, 740–748 (1975). doi:10.1086/153940

H. Zhang, A. Brandenburg, D.D. Sokoloff, Magnetic helicity and energy spectra of a solar active region. Astrophys. J. Lett. **784**, 45 (2014). doi:10.1088/2041-8205/784/2/L45

M.Y. Zhao, X.F. Wang, H.Q. Zhang, The correlation between the magnetic and velocity fields on the full solar disk. Sol. Phys. **270**, 23–33 (2011). doi:10.1007/s11207-011-9742-5

J. Zhao, R.S. Bogart, A.G. Kosovichev, T.L. Duvall Jr., T. Hartlep, Detection of equatorward meridional flow and evidence of double-cell meridional circulation inside the Sun. Astrophys. J. Lett. **774**, 29 (2013). doi:10.1088/2041-8205/774/2/L29

J.B. Zirker, S.F. Martin, K. Harvey, V. Gaizauskas, Global magnetic patterns of chirality. Sol. Phys. **175**, 27–44 (1997). doi:10.1023/A:1004946303735

Space Sci Rev (2017) 210:397–426
DOI 10.1007/s11214-015-0208-1

Prospects of Solar Magnetometry—From Ground and in Space

Lucia Kleint[1] · Achim Gandorfer[2]

Received: 12 June 2015 / Accepted: 8 October 2015 / Published online: 17 December 2015

Abstract In this review we present an overview of observing facilities for solar research, which are planned or will come to operation in near future. We concentrate on facilities, which harbor specific potential for solar magnetometry. We describe the challenges and science goals of future magnetic measurements, the status of magnetic field measurements at different major solar observatories, and provide an outlook on possible upgrades of future instrumentation.

Keywords Sun: magnetic fields · Instrumentation: high angular resolution · Instrumentation: polarimeters · Space vehicles: instruments · Telescopes · Sun: helioseismology · Sun: photosphere

1 Introduction: Complementary Worlds—The Advantages and Drawbacks of Ground-Based and Space-Borne Instruments

To nighttime astronomers it usually sounds as a paradox that solar magnetic measurements are photon-starved. Detecting four polarization states (I, Q, U, V) with high enough spatial resolution (sub-arcsecond) in a relatively large field-of-view (several dozen arcsec), in a short time (less than a second), plus in sufficient wavelengths in and around a spectral line poses strict limitations on the instrumentation. Night-time telescopes simply increase their aperture in order to collect more photons, with the currently largest aperture of 10.4 m at the Gran Telescopio Canarias (Grantecan) on La Palma, Spain. This is also desirable for solar telescopes, but the technical requirements are more complicated owing to the heat management and the required corrections for seeing variations by adaptive optics (AO).

✉ A. Gandorfer
gandorfer@mps.mpg.de

L. Kleint
lucia.kleint@fhnw.ch

1 Fachhochschule Nordwestschweiz, Bahnhofstrasse 6, 5210 Windisch, Switzerland

2 Max-Planck-Institut für Sonnensystemforschung, Justus-von-Liebig-Weg 3, 37077 Göttingen, Germany

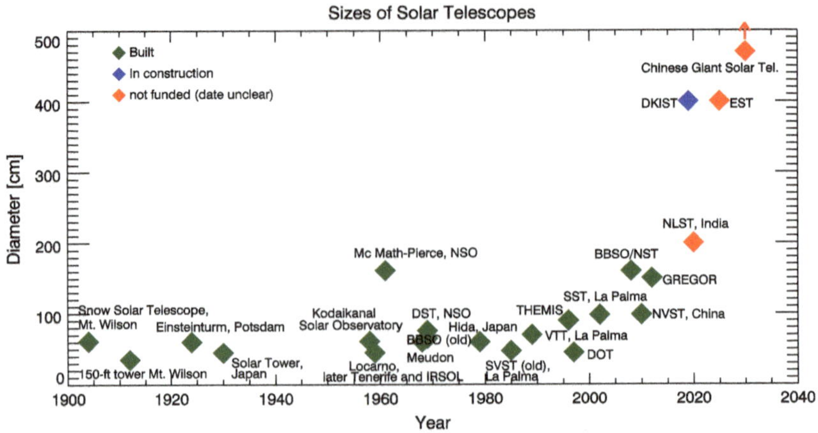

Fig. 1 The effective aperture size of selected telescopes shown for the year that the telescope was inaugurated. The DKIST will be an impressive jump in size (4 m) compared to the currently largest BBSO/NST (1.6 m)

The construction of the world's largest solar telescope, the 4 m DKIST, led by the National Solar Observatory, is currently underway with planned first-light in 2019. Compared to the current largest solar telescope, the 1.6 m New Solar Telescope (NST) at the Big Bear Solar Observatory, the photon collecting area will increase by a factor of more than six. A selection of effective aperture size of solar telescopes is shown in Fig. 1. Only recently, with the exception of the McMath-Pierce telescope, the aperture sizes surpassed 1 m.

1.1 The Need for Large Apertures

A large telescope serves two main purposes: to increase the number of collected photons and to increase the spatial resolution. Both obviously can be traded for another, depending on the science question that is being investigated.

There are several science questions that currently cannot be answered because of the limited available resolutions and sensitivities—temporal, spatial, and polarimetric. For example, the need for a large number of photons is evident when searching for horizontal magnetic fields, which are observed in linear polarization. These small-scale fields are important for a better understanding of the surface magnetism, e.g. whether they are created by local dynamo processes. The answer to the long-standing question of whether small-scale magnetic fields are more horizontal or vertical seems to currently depend on the method of analysis, and especially how the noise in Q and U is dealt with (e.g., Lites et al. 2008; Asensio Ramos 2009; Stenflo 2010; Borrero and Kobel 2011; Orozco Suárez and Bellot Rubio 2012; Steiner and Rezaei 2012; Martínez Pillet 2013). Because the signals of linear polarization (Q, U) are much smaller than the circular polarization (V) in the photospheric quiet Sun, there is a bias to detect line-of-sight fields more easily. The solution would require a higher polarimetric sensitivity, which can be reached by a larger number of collected photons. In the chromosphere, the problem becomes even more complicated because the magnetic fields are generally weaker leading to even lower polarization signals, and because the selection of spectral lines is more limited, also in terms of Landé factors, again leading to lower polarization signals. Simulations of the chromospheric 8542 Å line have shown that a noise level below $\sim 10^{-3.5}$ I_c, I_c being the continuum intensity, is required to detect quiet

Sun magnetic fields (de la Cruz Rodríguez et al. 2012). Currently, this level is rarely reached, except for observations with the Zurich Imaging Polarimeter (ZIMPOL, Povel 1995; Gandorfer and Povel 1997; Gandorfer et al. 2004; Kleint et al. 2011a). ZIMPOL modulates the polarization with frequencies of several kHz, which eliminates seeing-induced crosstalk and allows to reach polarimetric sensitivities of up to 10^{-5} I_c. However, this is done at the expense of spatial resolution, which generally is $> 1''$ pixel^{-1}, and requires averaging over a large part of the spectrograph slit.

Another argument to aim for large numbers of photons are coronal measurements. The free energy stored in coronal magnetic fields is believed to drive solar flares and coronal mass ejections. But so far, it cannot be measured with the desired temporal and spatial resolution. The coronal intensity I_{cor} is about 10^{-5}–10^{-6} I_c and its polarimetric signal is even smaller: 10^{-3}–10^{-4} I_{cor}. Currently, it takes 70 minutes of exposure time at the 0.4 m Solar-C Haleakala telescope with a huge pixel size of $20'' \times 20''$ to obtain the desired sensitivity of 1 G for coronal measurements (Lin et al. 2004).

Scattering polarization is another prime candidate for photon-starved observations. Because of its low amplitudes, usually only small fractions of a percent of the continuum intensity, it is notoriously difficult to observe (Stenflo and Keller 1997). With integration times of several minutes and averaging over most of the length of the spectrograph slit (few dozen of arcsec), one obtains a puzzling result: it seems as if the field strength of the turbulent magnetic field depends on the spectral line that was observed. For example, measurements in the atomic Sr I line consistently give higher magnetic field values on the order of 100 G (Trujillo Bueno et al. 2006), than molecular lines, such as CN ($\sim$80 G) or C_2 ($\sim$10 G) (Shapiro et al. 2011; Kleint et al. 2011b). An explanation based on modeling was proposed by Trujillo Bueno (2003), suggesting that Sr I may be formed in intergranular lanes, where the magnetic field is stronger and C_2 in granules, where the field is weaker. But for a conclusive explanation, one would need Hanle imaging without having to average over large spatial scales. Current solar telescopes are unable to collect a sufficient number of photons for the required polarimetric sensitivity and spatial resolution on time scales before the granulation changes.

Flare observations would mostly benefit from an increased temporal resolution, which again is only feasible with more photons and faster instrumentation. A scan of a spectral line with full polarimetry takes about 15 seconds at the Dunn Solar Telescope with the IBIS instrument. The evolution and motion of flaring plasma is clearly noticeable during this time. The Stokes vector and possible changes of the magnetic fields are then hard to interpret.

On the other hand, increased spatial resolution will also prove highly interesting to for example compare the Sun with state-of-the-art 3D radiative MHD simulations whose resolution reaches few km (for details see Rempel and Schlichenmaier 2011), a factor of $\sim$5 better than current observations. Open questions about the origin of the Evershed flow, the overturning convection and small-scale features, such as umbral-dots and penumbral filament structure may then be resolved.

In summary, one can identify many science topics, which will greatly benefit from a 4 m telescope, e.g.:

- Magnetic field measurements throughout the solar atmosphere (including the corona)
- Comparisons with high-resolution simulations, including fast polarimetry at maximum resolution
- Turbulent magnetic fields and studies of the solar dynamo, which require Hanle imaging
- Flares, small-scale dynamics and trigger mechanisms. Space weather.

1.2 Advantages and Disadvantages of Ground- and Space-Based Telescopes

There are many advantages of ground-based observations, one being the possibility to upgrade and to repair instruments. The flexibility of the instrumentation and the wavelength selection, e.g. by simply replacing a prefilter, are also important. The "unlimited" telemetry, which possibly is only limited by data transfer rates and the size of data storage, is another factor when comparing e.g. the SST (several TB during a good day) with SDO, the solar space mission with the currently highest data rates, which requires a dedicated groundstation to download its $\sim$1.5 TB/day (Pesnell et al. 2012). Another advantage of ground-based observations is the possibility of real-time target changes, especially important when targeting flares or rapidly changing features, for example. Additionally, the technical possibilities to launch a 4 m telescope currently do not exist and the costs are generally much lower for ground-based observations.

There are more or less obvious reasons for launching telescopes into space. The most obvious one is the absorptance of wide portions of the solar spectrum by the Earth´s atmosphere; especially in the UV, observations of the Sun (e.g. the transition region) are restricted to space-borne instruments. Less obvious, but of increasing interest is the capability of space telescopes to observe the Sun from a different vantage point, offering complementary views on our star. The Stereo mission has impressively demonstrated the advantages of stereoscopic viewing (Harrison et al. 2009; Aschwanden 2011). Of particular interest are observations of the solar polar regions, which can be obtained from the ecliptic with only marginal quality (up to 7 degrees, thanks to the inclination of the solar rotation axis). Views from inclined orbits my greatly improve on this shortcoming. Solar Orbiter is the first mission carrying telescopes to an orbit, which is inclined with respect to the ecliptic by up to 34 degrees. But even "normal" observations, which could *per se* be done from ground, are worth being done from space: The fact that most solar observatories are located in low to mid geographic latitudes excludes a 24 hour view on our target, a severe drawback considering the dynamic timescales of solar activity. For a limited time of up to several days, stratospheric ballooning in high geographic latitudes essentially offers uninterrupted observations, as demonstrated by the Flare Genesis Experiment (Bernasconi et al. 2000) and by Sunrise (Barthol et al. 2011; Solanki et al. 2010). But only space guarantees uninterrupted viewing for weeks, months, or even years, with constantly high data quality and without the effects of seeing.

This comes, however, to a high prize, not only literally. Space instruments generally lack the flexibility to adjust for new observing strategies (especially the choice of observed spectral bands), and are notoriously short of telemetry, not accounting for potential loss of data due to radiation induced upsets. The need to build an instrument under severe restrictions in terms of mass, volume, power allocation, and often in combination with harsh radiation and varying thermal environment makes it necessary to agree on technical compromises, which often limit the performance of the instrument. Only thanks to the unrivaled stability of the observing conditions, these instruments can—for specific observations—outperform their ground-based counterparts. A drawback for space-based missions is their long lead time and often their electronics, such as cameras, are no longer state-of-the-art by the time they launch. Because space instruments generally have a fixed configuration and often are simpler in design than their ground-based counterparts in order to be fail-safe, data pipelines for space-based instruments are better developed and more stable. The importance of this issue has been recognized and there currently is an effort in the DKIST project to develop such pipelines for ground-based projects as well. In general, ground- and space-based observations ideally complement each other, and this will always be the case. The question is not, which route—ground or space—to take. The key point is how to make optimum use of the complementary aspects of both worlds.

2 Current and Future Projects in Ground-Based Solar Physics

In this section, we review a selection of current and future optical telescopes from all over the world and their measurements of magnetic fields. Access to observing time of most European telescopes (and to the DST) is possible through the SOLARNET consortium, a European FP7 Capacities Programme. Access to US-based publicly funded telescopes (DST and later DKIST) is possible through an open proposal process.

2.1 GREGOR

Design and Instrumentation The 1.5 m GREGOR telescope on Tenerife, Spain, is the currently largest European solar telescope (e.g., Schmidt et al. 2012, and further papers in the special issue of AN Vol. 333, Issue 9). Its design started in 1998/1999 and the commissioning in 2012. The early science phase is ongoing from 2014–2015 with access mostly limited to the consortium, which is led by the German Kiepenheuer Institute. The first open call for proposals from SOLARNET was released in March 2015, but is restricted to EU and associated countries.

The design consists of a Gregory telescope with three on-axis mirrors and an image derotator further down in the image path, which is planned to be installed in November 2015. The large primary mirror suffered from several fabrication problems with the originally planned silicon carbide (Cesic) material and is now made from light-weighted Zerodur with active cooling. High-order adaptive optics, which allow diffraction-limited observations at $0.08''$ at 500 nm for good seeing ($r_0 \geq 10$) are integrated into the path. Three (later four) post focus instruments allow to observe a very large spectral range from 350 nm to the near Infrared (several micron) and have a field-of-view of up to $150''$. The instruments include a broadband imager (no polarimetry), the GREGOR Fabry Perot Interferometer (GFPI, Puschmann et al. 2013), a dual collimated Fabry Perot system, and the GREGOR Infrared Spectrograph (GRIS, Collados et al. 2012), which can be combined with the ZIMPOL system (Ramelli et al. 2014). A stellar spectrograph is planned to be installed in the future.

Science The main science goals include high-resolution photospheric observations to investigate the structure and dynamics of sunspots and of granulation, small-scale magnetic field studies to investigate the presence of a local dynamo, and the largely unexplored chromospheric magnetic field. Figure 2 shows some first results of the fine structure of magnetic features obtained from an inversion of GRIS data of a sunspot with lightbridges (AR 12049). For another example, see Lagg et al. (2015, this issue). GRIS has already shown a good performance in the infrared.

Future Proposals for the first open observing season in 2016 will be solicited at the end of 2015. It is also planned to replace the M2 mirror, currently made of Cesic, by a version made of Zerodur in the next couple of years, which is expected to improve the contrast and thus the use of the AO. Polarimetric capabilities are a possible upgrade for the stellar spectrograph.

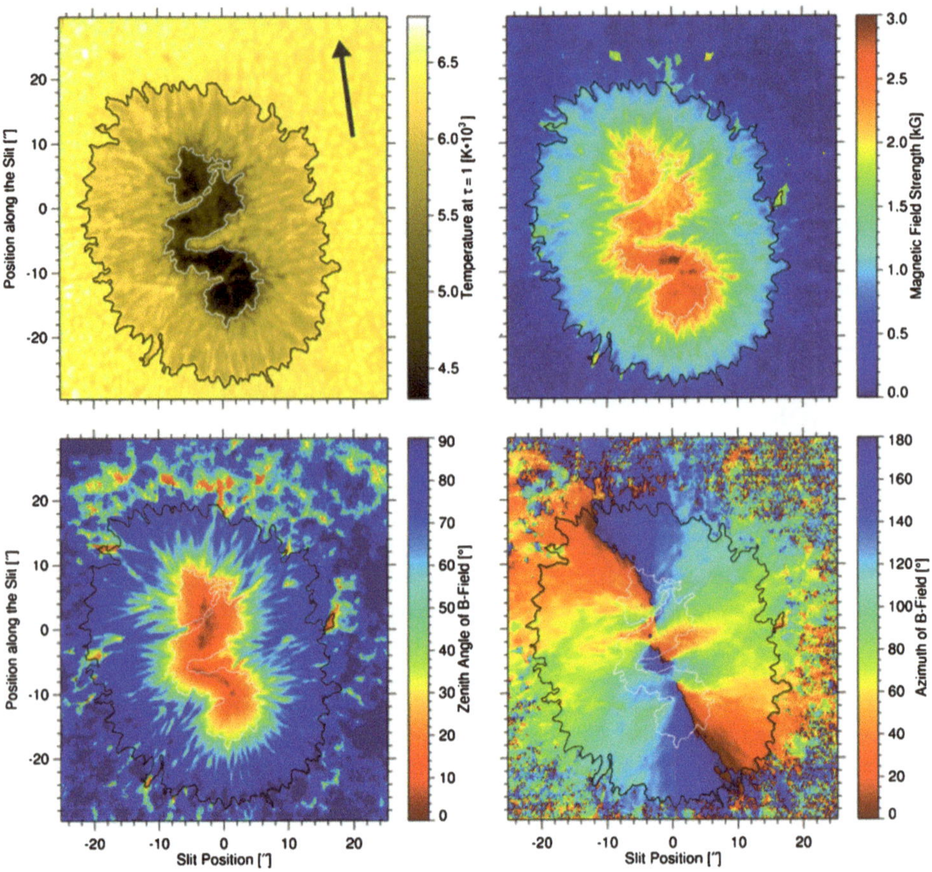

Fig. 2 Inversion of a spectropolarimetric raster scan of a sunspot in AR 12049, observed at 1.568 μm with GRIS at GREGOR in May 2014. Clockwise from top left: temperature, magnetic field strength, azimuth and inclination of the magnetic field. Courtesy of GREGOR under the leadership of the Kiepenheuer-Institut für Sonnenphysik (KIS) with the German partners AIP, IAG and MPS, the Spanish IAC and with contributions by the Czech ASCR

2.2 SST

Design and Instrumentation The Swedish 1-m Solar Telescope (SST, Scharmer et al. 2003) on La Palma, Spain, has produced some of the highest-resolution magnetograms to date, both in the chromosphere and in the photosphere. Its design consists of a fused silica lens with a clear aperture of 97 cm, a turret with two 1.4 m flat mirrors, and an evacuated tube to minimize air turbulence caused by heating. Chromatic aberration induced by the front lens is corrected by the Schupmann system, which consists of a negative lens (being passed twice by the light) and a mirror, and creates an achromatic image at the secondary focus. With only 6 mirrors in the beam before the optical table, the SST's throughput is very high. An Echelle-Littrow spectrograph (called TRIPPEL) is available for spectroscopy, simultaneously for three spectral windows with a resolution of R $\sim$ 200000 (Kiselman et al. 2011). For polarimetry, SST's main instrument is CRISP, a telecentric dual Fabry Perot system with a spectral resolution of $\sim$60 mÅ at 6302 Å, and an image scale of $0.059''$ pixel^{-1} (Scharmer

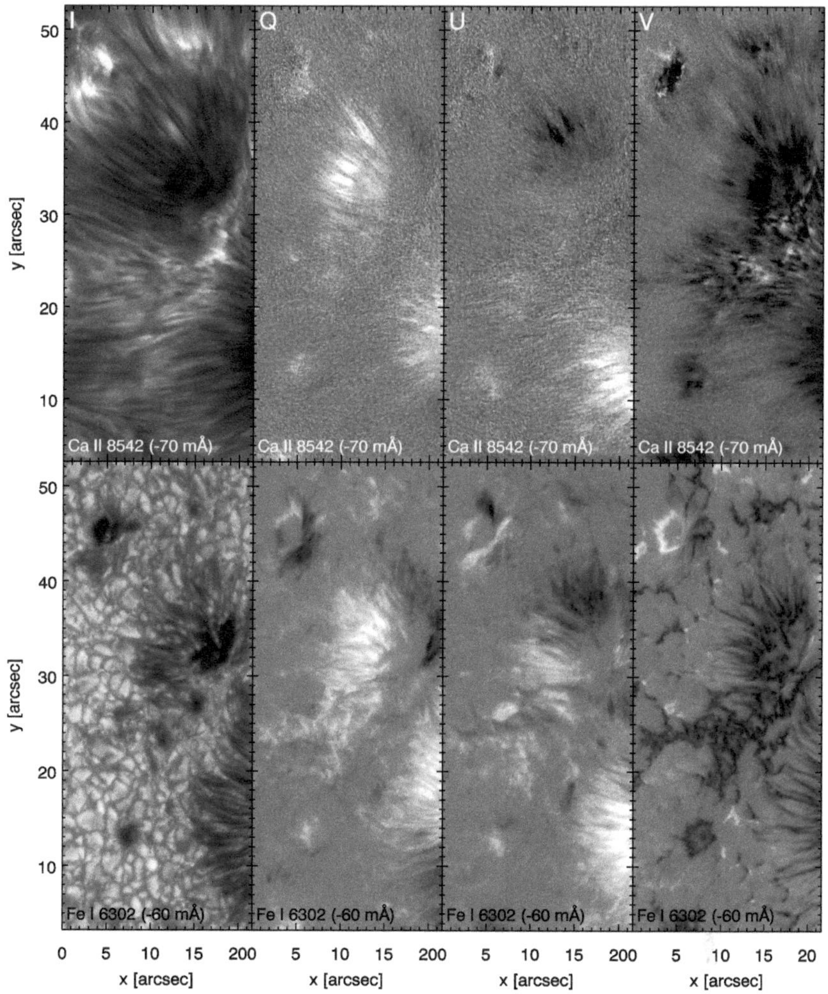

Fig. 3 AR11793 taken on 2013-07-19, 13:34 UT in the photosphere (*bottom row*) and chromosphere (*top row*) in intensity (I), linear polarization states (Q, U) and circular polarization (V), taken with SST/CRISP showing the very high spatial resolution that can be obtained. Courtesy of Jaime de la Cruz Rodríguez (Institute for Solar Physics, Stockholm University)

et al. 2008; de la Cruz Rodríguez et al. 2015). The incoming beam is split and the short wavelengths are sent to separate fast cameras, while CRISP records the red part of the light. The polarization modulation was recently upgraded to ferroelectric liquid crystals, which change modulation state in less than 1 ms. The modulation is thus limited by the camera exposure time and readout, giving a speed of ~28 ms per state, i.e. 112 ms for a full polarization cycle. After the prefilter for the Fabry Perot, a part of the light is split off to a separate camera, serving as a broad-band reference for image reconstruction. The remainder of the light passes through the two Fabry Perot interferometers and the two orthogonal beams after the polarizing beamsplitter are recorded with two separate cameras. The SST group has pioneered the use of the multi-object multi-frame blind deconvolution (MOMFBD) technique (van Noort et al. 2005), which is applied on the frames already before the demodulation.

Coupled with exceptionally good seeing, it produces high spatial resolution, as illustrated in Fig. 3.

Science Recent scientific results with polarimetry include the discovery of opposite polarities in the penumbra (Scharmer et al. 2013) and a study of the effects of umbral flashes and running penumbral waves on the chromospheric magnetic field through inversions of the chromospheric Ca II 8542 Å line (de la Cruz Rodríguez et al. 2013). Chromospheric inversions were also used to demonstrate that fibrils often, but not always trace the magnetic field lines (de la Cruz Rodríguez and Socas-Navarro 2011).

Future The next planned upgrade at the SST is the CHROMIS instrument, a Fabry Perot Interferometer to observe the Ca II H (3934 Å) and K (3968 Å) lines. While its first version is planned without polarimetry, an upgrade would be possible later and it would be interesting to investigate the rather strong scattering polarization of these lines. It would be possible to upgrade TRIPPEL for polarimetry, enabling for example rasters of He 10830 to derive chromospheric magnetic fields, but there are no immediate plans.

2.3 THEMIS

Design and Instrumentation The THEMIS telescope on Tenerife, Spain, is a 90 cm Ritchey-Chrétien reflector with an alt-az mount. Belonging to the French Centre National de la Recherche Scientifique (CNRS), THEMIS was operated by France & Italy until 2009, and by France since then, but with a lower level of funding. Its specialty is being virtually polarization-free due to its symmetric design and the polarimeter being placed next to the primary focus, enabling high-accuracy spectropolarimetry. Seeing in and around the telescope is minimized by a helium-filled tube and a specially constructed dome whose small opening coincides and co-rotates with the telescope aperture. A highly complex system of transfer optics and vertically oriented spectrographs feeds light to the instruments. Its design with many optical elements however leads to light loss, especially towards the blue wavelengths. Long-slit multi-line observations are possible with the MTR (MulTi-Raies) instrument (López Ariste et al. 2000), which can observe up to 6 different wavelength regions, recorded by 12 cameras in a dual-beam setup. THEMIS pioneered several new instrument concepts: The MSDP (Multi Channel Subtractive Double Pass, Mein 2002), allowed to observe two spectral lines simultaneously, with several windows per spectral line and two orthogonal polarization states recorded simultaneously through an elaborate double-pass through a grating. It is no longer available at THEMIS, but a new version is operating at the Meudon Observatory. TUNIS (Tunable Universal Narrowband Imaging Spectrograph) (López Ariste et al. 2010, 2011) was developed after MSDP's idea and expanded through a so-called Hadamard mask, encoding full spectral information in the images. After 63 measurements shifting the Hadamard mask to predefined positions, the cube of x, y, λ can be reconstructed, making it a fast instrument compared to regular scanning spectrographs, for example.

Science Due to its low instrumental polarization, THEMIS is well-suited for high-precision polarimetric studies. Studies have focused on magnetic fields in prominences (e.g., López Ariste et al. 2006; Schmieder et al. 2014) and the second solar spectrum (Faurobert and Arnaud 2003; Faurobert et al. 2009). López Ariste et al. (2012) investigated the scattering polarization of Na in the exosphere of Mercury, which could be used in the future to study its magnetic field.

Future There are no observing campaigns in 2015 and 2016 to allow refurbishing THEMIS' full optical path to enhance the "polarization-free" feature of the telescope. It is also planned to install an AO system to improve its spatial resolution. THEMIS plans to resume observations in 2017 and stay operational until the EST comes online.

2.4 BBSO/NST

Design and Instrumentation The Big Bear Solar Observatory (BBSO) is located in Big Bear, California on a pier inside a lake and is operated by the New Jersey Institute of Technology. Its location results in less turbulent air motions, which is very beneficial for the seeing. The 1.6 m New Solar Telescope (NST) was inaugurated in 2009 (Goode et al. 2010; Cao et al. 2010; Goode and Cao 2012), with the high order adaptive system following in 2013 (Varsik et al. 2014). It is currently the world's largest solar telescope and has provided some of the highest resolution images to date. Currently, there are no calls for proposals, but some data and overviews are available on their website http://www.bbso.njit.edu/ and they plan to accept observing proposals in the near future. NST's design is an off-axis Gregorian system with a wavelength range of 390 nm–5 µm. Its instrumentation consists of fast imaging (e.g. Broadband Filter Imagers—BFI—for G-band and TiO and the Visible Imaging Spectrometer—VIS—for Hα and Na D2), spectrographs (e.g. Fast Imaging Solar Spectrograph—FISS—used for He 10830 Å and Ca II 8542 Å and the Cryogenic Infrared Spectrograph—Cyra—for observations in the infrared 1–5 µm), and polarimetry. Polarimetric observations are carried out with the Near InfraRed Imaging Spectropolarimeter (NIRIS, Cao et al. 2012). NIRIS is a dual Fabry Perot system in a telecentric configuration with an 85$''$ FOV with a wavelength coverage from 1.0 to 1.7 microns. As of 2015, the instrument is in routine operations and acquires full-Stokes images in the near infrared line pair at Fe I 1564.85 nm and 1565.29 nm.

Science During the past few years, NST studies have focused on fast imaging (no polarimetry yet in the following science highlights). For example, they studied rather rare three-ribbon flares (Wang et al. 2014), sunspot oscillations (Yurchyshyn et al. 2015), flux emergence coupled with a jet (Zeng et al. 2013; Vargas Domínguez et al. 2014), the eruption of a flux rope (Wang et al. 2015), and the connection between small-scale events in the photosphere and subsequent coronal emission (Ji et al. 2012). From observations of a C-flare, Zeng et al. (2014) concluded that the He 1083 nm triplet is formed primarily by photoionization of chromospheric plasma followed by radiative recombination. A first result from NIRIS is shown in Fig. 4.

Future Spectroscopy and spectropolarimetry in the He I 1083.0 nm are the next goal of NIRIS to investigate chromospheric magnetic fields in high resolution. Benefiting from an existing Lyot filter, the NST explored the scientific potential of such observations. First-light for this NIRIS upgrade is expected in 2016. VIS is also being upgraded to dual-FPIs with a diameter of 100 mm each to provide spectroscopic/polarimetric measurements of the lines in a wavelength range of 550–860 nm.

2.5 NSO/DST

The National Solar Observatory (NSO) operates several telescopes, one of them being the 76 cm Dunn Solar Telescope (DST) at Sacramento Peak in New Mexico (e.g. Zirker 1998). Inaugurated in 1969, it was the world's leading telescope for many years.

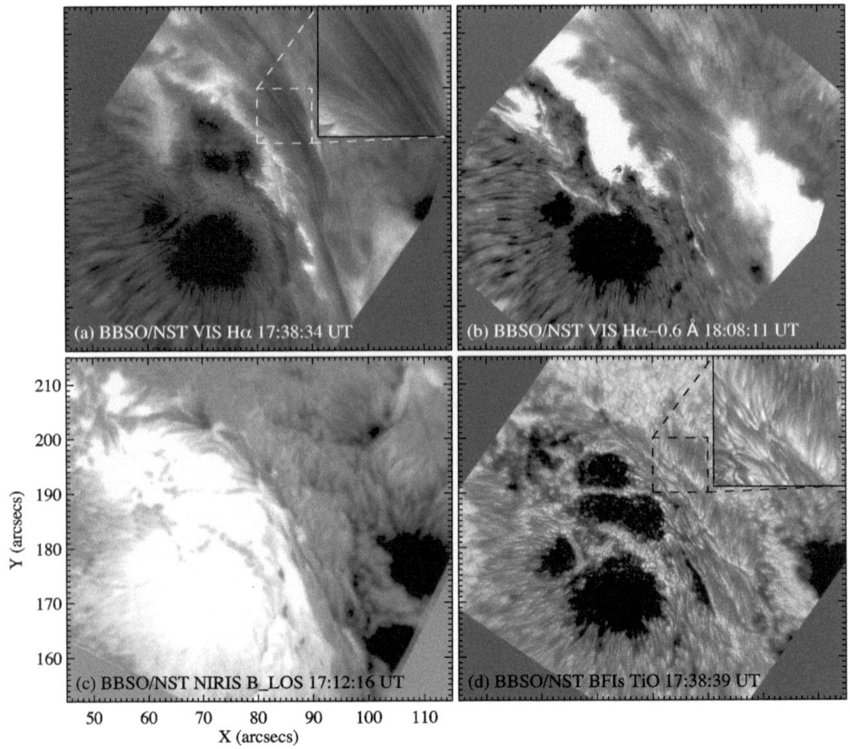

Fig. 4 Observation from June 22, 2015 showing NOAA 12371, which produced a long-duration M6.5 flare at 17:39 UT (start)—18:23 UT (peak)—after midnight (decay). Panels (**a**) and (**d**) show the speckle-reconstructed pre-flare images in Hα and TiO, respectively with a plate scale of 0.034″ pixel^{-1}. The pre-flare magnetogram from the Fe I 1565 nm line is visible in panel (**c**) with a plate scale of 0.083″/pixel. Panel (**b**) shows a speckle-reconstructed image of the blue wing of Hα during the flare peak with two flare ribbons and connecting loops. Courtesy of BBSO, New Jersey Institute of Technology

Design and Instrumentation The design of the DST is very peculiar, with a turret on top of a 41.5 m tower and a 250-ton optical system suspended on a liquid mercury bearing to counteract the solar image rotation caused by the alt-az mount.

With its long history, instruments at the DST were constantly upgraded and are now some of the most versatile in the world. For polarimetric measurements, the Diffraction Limited Spectropolarimeter (DLSP, Sankarasubramanian et al. 2004), the Interferometric Bidimensional Spectropolarimeter (IBIS, Cavallini 2006), the Facility Infrared Spectropolarimeter (FIRS, Jaeggli et al. 2010), or the Spectro-Polarimeter for Infrared and Optical Regions (SPINOR, Socas-Navarro et al. 2006) can be used, covering the range from visible to infrared and from spectroscopy to imaging. IBIS is currently one of the few instruments capable of imaging polarimetry in the chromospheric Ca II 8542 Å line. The ASP and DLSP helped in the development of the highly successful SP instrument onboard Hinode. The DST led the development of solar adaptive optics, and was the first telescope to be equipped with high-order adaptive optics (Rimmele et al. 2004), enabling diffraction limited studies of small-scale features.

Science The DST enabled many discoveries, whose list would exceed the scope of this paper. Some examples are sunspot and penumbral oscillations (Beckers and Schultz 1972), the first study of the subsurface structure of sunspots (Thomas et al. 1982), the confirmation that the penumbral magnetic field is composed of two components using the ASP (Lites et al. 1993), the discovery of small-scale short-lived (∼minutes) horizontal magnetic fields in the internetwork (Lites et al. 1996) and the first magnetic map of a prominence (Casini et al. 2003). Using the multi-line capabilities of SPINOR, the magnetic field of the quiet Sun (Socas-Navarro et al. 2008), of spicules (Socas-Navarro and Elmore 2005) and its 3D structure in sunspots (Socas-Navarro 2005) could be studied. FIRS allowed to confirm through spectropolarimetric observations in the He 10830 Å line that the superpenumbral fibrils trace the magnetic field (Schad et al. 2013). More recently, polarimetric observations by IBIS allowed to study the inclination change of the magnetic field during the formation of a penumbra (Romano et al. 2014), and a coordinated observing campaign with Hinode and IRIS was led from the DST, resulting in the "best-observed X-flare" (Kleint et al. 2015). In recent years, the DST carried out 3 cycles of service mode observations. While common in night-time astronomy and for satellite missions, they were a first in solar ground-based observations and a preparation for significantly more efficient DKIST observing modes.

Future NSO plans to cease operations of the DST by the end of 2017 while preparing for first light with their new 4-m DKIST on Maui, but negotiations with a consortium of universities and institutes that could operate the facility after 2017 are ongoing.

2.6 NLST

The 2 m National Large Solar Telescope (NLST) is a proposed project in India (Hasan 2012). The design studies and site selection are complete, but the construction is not funded (yet). The proposal is awaiting formal clearance from the government of India. The fabrication will take about 3.5 years and it is expected to take place from late 2016–2020. The project is led by the Indian Institute of Astrophysics.

Design and Instrumentation The NLST's Gregorian design is similar to the GREGOR telescope, but with fewer mirrors, leading to a higher throughput. It is designed to observe from 380–2500 nm. The planned first-light instruments include a narrow band imager and a spectropolarimeter, whose design has not been finalized yet. After four years of site surveys, Hanle near the border to Tibet was selected as primary site. At an altitude of 4500 m, it has very low water vapor and favorable weather.

Currently, India is commissioning the 50 cm off-axis MAST telescope. The integration of its 19 actuator AO and of two $LiNbO_3$ Fabry Perot interferometers to observe the photospheric 6173 Å line and the chromospheric 8542 Å line is ongoing (Raja Bayanna et al. 2014). The post-focus instruments include broad band and tunable Fabry-Perot narrow band imaging instruments; a high resolution spectropolarimeter and an echelle spectrograph for night time astronomy.

2.7 NVST

The 98.5 cm New Vacuum Solar Telescope (NVST) at Fuxian Lake is a Chinese telescope, which has recently been upgraded for high-resolution observations (Yang et al. 2014; Liu et al. 2014a).

Design and Instrumentation As a vacuum telescope, it contains an entrance window, which is followed by a modified Gregorian design. It is capable of observing from 300 to 2500 nm. A new high-order adaptive optics system with 151 actuators was developed in 2015 and great care is taken to reduce the local seeing, with the location of the telescope (lakeshore), by cooling the building's roof with a shallow pool of water, with a wind screen, and with the vacuum inside the telescope tube. There are two optical paths after the AO, one leading to three broadband channels (H-α, TiO and G-band), which are imaged by three separate cameras, and the other to a spectrograph system, which is being upgraded to polarimetric capabilities. A polarimeter for observations of the Fe I 5324 Å line and the line pair at 5247–5250 Å has been developed and calibrations are ongoing and a near-IR magnetograph for the 1.56 μm line is planned.

Future An ambitious Chinese project is the Chinese Giant Solar Telescope, currently in the planning phase with no final design yet (Liu et al. 2012, 2014b). Current ideas include a ring of mirrors on a 8 m diameter aperture, which would have the light-gathering power equivalent to a 5-m telescope. The telescope is planned to work mostly in the infrared (0.3–15 μm), because the seeing variations are lower. One goal is to measure the magnetic field in lines near 10 μm, e.g. 10.4 μm. After 5 years of site surveys, two candidate sites have been selected (one lake and one mountain site), with no final decision expected in the near future.

2.8 EST

The European Solar Telescope (EST) is a joint project of several European institutions with the goal of building a 4 m telescope on La Palma in the Canary Islands (Collados et al. 2013). Its conceptual design study was finished successfully in 2011 and it is currently in the detailed design stage (2013–2017). The construction phase might take place earliest from 2018–2023, but its funding, estimated at 150 M€, is not approved yet.

Design and Instrumentation EST's design consists of an on-axis Gregorian on an alt-az mount, similar to GREGOR. The instruments are not defined yet, but will include a broad band imager, a narrow-band tunable imager and a grating spectrograph. The advantage of the EST compared to the DKIST is that the design is favorable for polarimetric observations (though at the expense of coronal observations) and one of EST's requirements is high-resolution spectropolarimetry simultaneously in the photosphere and in the chromosphere. The telescope Mueller matrix is unity for all wavelengths and independent of elevation and azimuth and the calibration optics are located on axis, which is the ideal configuration for polarimetry. Another feature of EST is to include multi-conjugate adaptive optics (MCAO) in the beam, which so far has not been done before, even though first MCAO tests are ongoing. Distortions of the incoming wavefront through atmospheric turbulence occur at multiple heights. Current AO systems include one deformable mirror, which allows for a good correction within the so-called isoplanatic patch. In other words, at its edges, away from the AO lockpoint, the FOV is more variable and possibly blurry. MCAO consists of multiple deformable mirrors and would allow to correct for turbulence occurring at multiple heights, giving a more uniform FOV. Because the EST is on nearly the opposite side of the world than the DKIST, a near-continuous coverage of a target would be possible.

2.9 DKIST

The 4 m Daniel K. Inouye Solar Telescope (DKIST, formerly ATST) led by NSO will be the world's largest solar telescope when it is commissioned in 2019. The four meter aperture is

Fig. 5 *Left*: DKIST telescope enclosure with details labeled. *Right*: DKIST construction progress on the summit of Haleakalā (Hawaii), April 30, 2015. Courtesy of NSO/DKIST

a factor of more than 6 in light gathering power compared to the currently largest telescope, the NST. With a wavelength range of 380–5000 nm, and possibly up to 28000 nm (28 μm) for second generation instruments, the DKIST will provide unprecedented magnetic field measurements from the photosphere to the corona up to 1.5 solar radii.

Design and Instrumentation The DKIST is currently under construction on Haleakala on Maui HI (see Fig. 5). The enclosure, which was shipped from Spain to Hawaii, will be installed in 2015. Its design allows for thermal control and dust mitigation, especially important for coronal observations where any scattering through dust on the mirrors needs to be avoided. The Telescope Mount Assembly will be installed in 2016–2017 and the optics system installation and integration is planned for 2017–2018. After the instrument commissioning, first light is planned for the middle of 2019. The off-axis Gregory setup was chosen in view of the low straylight for coronal observations. However, the off-axis design results in significant polarization, variable over the day, and will need to be calibrated carefully.

There will be five first-generation instruments, four of which are capable of polarimetric measurements. Contrary to many other telescopes, most of them can be operated simultaneously (with the exception of the coronal instrument Cryo-NIRSP) by splitting spectral ranges off the incoming beam. The Visible Spectropolarimeter (VISP), led by the High Altitude Observatory, consists of an Echelle spectrograph from 380–900 nm, and will record three spectral ranges on different cameras simultaneously. The Visible Tunable Filter (VTF), led by the Kiepenheuer Institute, is a dual Fabry Perot with coatings that permit observations from 550–860 nm. The Diffraction Limited Near Infrared Spectro-Polarimeter (DL-NIRSP), led by the University of Hawaii, consists of a fiber-fed multi-slit spectrograph and a selection of cameras that cover the range of 900–2300 nm. Additionally to these polarimetric instruments, the Visible Broadband Imager (VBI), led by NSO, will record the intensity in selected filters from 390–860 nm and the images will be speckle-reconstructed directly after their acquisition to improve the spatial resolution. The fourth polarimetric instrument is the Cryogenic Near Infrared Spectro-Polarimeter (Cryo-NIRSP), led by the University of Hawaii, which is the only instrument that does not allow beam-sharing and which will not utilize the adaptive optics. Its wavelength range 1000–5000 nm is optimized for diffraction-limited coronal observations. Polarimetry is possible up to 4000 nm. For a more complete overview of the instruments, see Elmore et al. (2014).

Science The DKIST will be the largest advance in ground-based solar observations in several decades. It is expected that its high spatial and temporal resolution and its polarimetric capabilities will lead to ground-breaking discoveries, especially for chromospheric magnetic fields, turbulent magnetic fields and local dynamo mechanisms, coronal magnetism, and the photospheric fine-structure, including that of sunspots.

3 Next Generation Instrumentation

Even with the significant advances of DKIST, some desired observations will still not be possible. For example, Hanle imaging to investigate turbulent magnetic fields, or extremely fast flare polarimetry are outside the realm of any currently planned DKIST instrument. While a 4 m telescope will provide the necessary light gathering power and the spatial resolution, possibly with MCAO, it would be desirable to record the cube $[x, y, \lambda, Stokes]$ quasi-simultaneously, or at least faster than any seeing-variation (~hundred Hz). This requires a sophisticated instrument design, plus very fast detectors that can modulate and read-out faster than ~100 Hz.

3.1 Spatial and Spectral data Simultaneously

For the instrument design, there are several possibilities to obtain $[x, y, \lambda]$ simultaneously, with two examples depicted in Fig. 6. One option is a fiber array (integral field unit) that

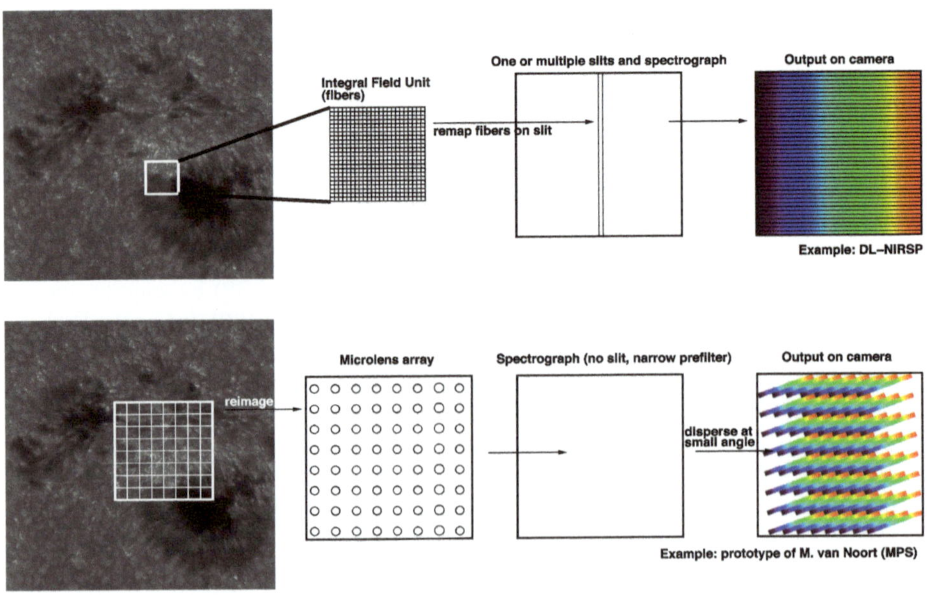

Fig. 6 Two principles of how to record $[x, y, \lambda]$ simultaneously. Both are currently being used in prototype instruments. Figure adapted from M. van Noorts and T. Schads versions

subdivides the focal plane into parts that are then fed into a spectrograph and re-arranged onto the detector. An example for this type of instrument is the DL-NIRSP of DKIST, which is planned to have a mode with 19200 fibers that are fed into 5 slits. A slight disadvantage is the rather limited FOV (for good spatial resolution) and thus a requirement to scan to obtain a picture of e.g. a full sunspot.

Another option is to re-image the focal plane with a microlens array, in principle shrinking the pixels. They are then dispersed by a spectrograph at a small rotation angle, so that they can fit on a detector. A prefilter needs to be used to avoid an overlap of spectra of different image elements. FOV and spectral range can be traded for another. A prototype of such an instrument is currently being tested by M. van Noort at the Max Planck Institute for Solar System Research.

A third option would be to employ an image slicer that separates the FOV into equal slices that are re-imaged by mirrors onto a spectrograph. Within one of these slices, the spatial information is retained, which is not the case for principles (a) and (b). Such an option is currently being explored for the EST (Calcines et al. 2013, 2014).

3.2 Fast Solar Polarimeter

A very promising project for extremely fast modulation and readout is the *Fast Solar Polarimeter (FSP)*, led by A. Feller at the Max Planck Institute for Solar System Research (Feller et al. 2014).

The FSP overcomes several drawbacks of ZIMPOL, while reaching similar polarimetric sensitivity. ZIMPOL's pixels are not square (due to the masked rows used for charge shifting), and rather large, which is not ideal for high-resolution images. The overall ZIMPOL throughput is also relatively low, even with the microlenses that focus light from the masked rows to the unmasked rows. Also, ZIMPOL's overhead is rather large (above 100 %) for short exposures, which is not ideal for fast changing solar features, such as flares. The FSP tackles these problems by a different design, and still reaches frame rates high enough to sufficiently suppress seeing-induced crosstalk. It has a specially designed pnCCD sensor, which can read out at speeds up to 400 fps. This is achieved by shifting the charges of each half of the CCD to one side (split frame transfer), and by parallel readout of all sensor columns. The polarization modulation is achieved with two ferro-electric liquid crystals, which operate at the same frame rates, reaching a duty cycle of > 95 %. A polarimetric accuracy of 10^{-4} I_c at sub-arcsec resolution can be achieved by summing images below solar evolution time scales ($\sim$ minutes). Alternatively, about 1 restored set of Stokes images can be obtained per second with a typical S/N of order 300–500 by combining some 400 single frames. This allows to apply image restoration with a cadence down to about 1 s, which to our knowledge is not possible so far with any other ground-based solar polarimeter. The first prototype of the FSP with a 256×256 pixel pnCCD was tested at the VTT (see Fig. 7). Currently, a 1024×1024 pnCCD is in development, which will be used in a dual beam setup. First observations are planned for 2016. The FSP may be the instrument, which will allow to clarify the question of the inconsistent field strengths derived from scattering polarization observations (cf Sect. 1.1), or to enable fast polarimetry during flares.

4 Future Space Initiatives and Projects with Relevance to Solar Magnetometry

In this section we present a brief overview of space missions with relevance to solar magnetometry. We restrict the list to space missions, which have not yet been launched, but rather

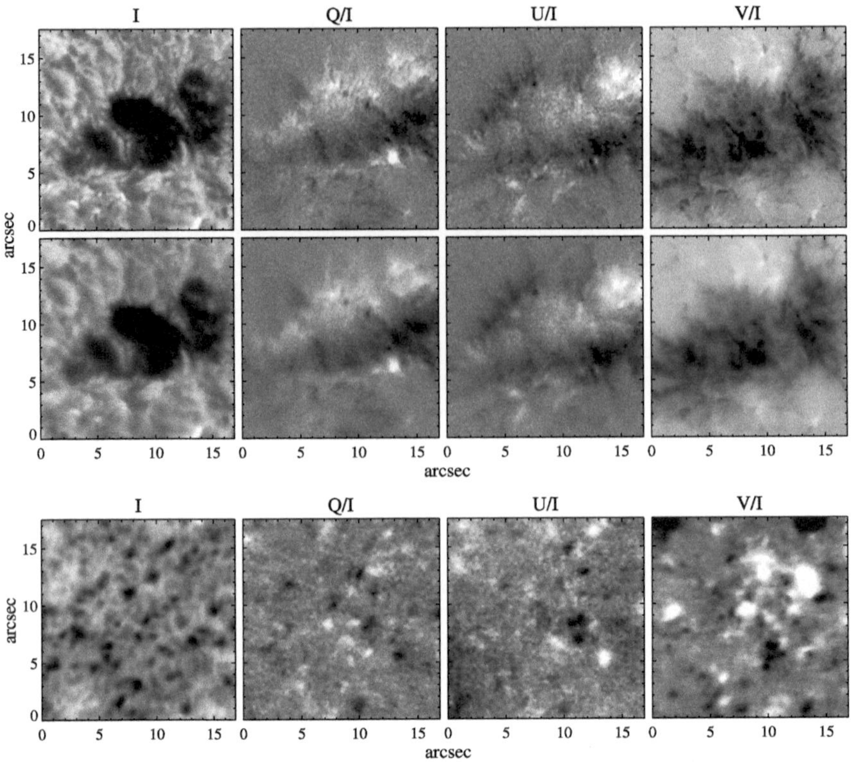

Fig. 7 Example results obtained with the FSP prototype at the TESOS filtergraph on the VTT. *Upper panel*: Part of a small active region recorded in Fe I 630.2 nm. The *four columns* represent the full state of polarization, i.e. Stokes $I, Q/I, U/I$, and V/I with grayscales $I : \langle I \rangle \pm 40\ \%$, $Q/I, U/I : \pm 8\ \%$, $V/I \pm 15\ \%$. *Each image* of the *upper row* is a result of a MOMFBD restoration involving 640 individual frames, spanning 1.6 s. For comparison, the *images* in the *second row* have just been averaged, which significantly reduces spatial resolution. *Lower panel*: Small-scale and very faint magnetic fields of the quiet Sun, recorded in the same Fe I line, with a very high polarimetric sensitivity of 0.02 %. A 3×3 pixels binning has been applied to further increase S/N. *Grayscales*: $I : \langle I \rangle \pm 10\ \%$, $Q/I, U/I : \pm 0.25\ \%$, $V/I : \pm 0.5\ \%$. Courtesy A. Feller

are in the preparation or even planning phase. Details of some of these missions are also mentioned in Lagg et al. (2015, this issue).

4.1 CLASP—Chromosheric Lyman-alpha SpectroPolarimeter

The observation and interpretation of linear polarization caused by scattering in the solar atmosphere, and its modification by magnetic fields via the Hanle effect, has been identified as a diagnostic tool for solar magnetic fields already early (see Stenflo and Stenholm 1976). From a technological as well as from interpretational aspects, the field of scattering polarimetry has been boosted by the advent of highly sensitive polarimeters, mainly ZIMPOL in its various evolutionary generations. Ground based observations of the second solar spectrum, both with ZIMPOL and at the THEMIS telescope, stimulated the development of new theoretical concepts and numerical tools, which brought us into a situation, where we can finally use scattering polarization as a diagnostic tool for solar magnetometry.

The first attempts to measure the scattering polarization in the Lyman alpha line of hydrogen and in this way retrieve information on the physical conditions in the upper chromosphere and transition region was undertaken by Stenflo and collaborators in the late 1970s. A small polarimeter for measuring linear polarization around 121 nm (Stenflo et al. 1976) was installed on the russian Intercosmos 16 satellite. Although the experiment failed due to contamination issues, it was able to set an upper limit to the polarization degree, however with a large uncertainty (Stenflo et al. 1980). Stenflo et al. (1980) claim that "The average polarization of the Lyman alpha solar limb was found to be less than 1 %. It is indicated that future improved VUV polarization measurements may be a diagnostic tool for chromospheric and coronal magnetic fields and for the three-dimensional geometry of the emitting structures."

Although the essential usefulness of UV polarimetry is widely accepted, and the development of UV polarization technology is explicitly recognized in the ASTRONET Infrastructure Roadmap as a "perceived gap" (see p. 73 of the Astronet Infrastructure Roadmap 2008), until today there has been no further instrument operating in this wavelength regime. Lyman alpha polarimeters have been proposed twice as strawman instruments for ESA missions, first for the COMPASS mission proposed in response to the call for the M2 slot (Fineschi et al. 2007), later as part of the SOLMEX mission proposed in 2008 in response to ESAs M3 call (Peter et al. 2012). It is obvious that—given the explorative character of such an instrument—a demonstration of the principle including the successful interpretation of the harvested data is mandatory.

To this end, by initiative of the Japanese Space Agency JAXA, together with NASA and with European contributions, the CLASP mission was established. CLASP stands for Chromospheric Lyman-alpha SpectroPolarimeter and is an explorative sounding rocket experiment with the aim of demonstrating the capabilities of UV spectropolarimetry from space (c.f. Kano et al. 2012; Kubo et al. 2014, and references therein). CLASP consists of a 279 mm diameter Cassegrain telescope feeding a dual-beam spectrograph equipped with a rotating half-wave plate modulator for the measurement of Stokes I, Q, and U in and close to the Lyman alpha line in a wavelength range from 121.1–122.1 nm.

The target is the chromosphere and transition region as seen on disk radially from the solar limb. The spatial resolution is $1.5''$ (set by the slit width) times $2.9''$ (along the slit) in a field of view of $1.5''$ times $400''$. With a spectral resolution of 0.01 nm, in 5 min observing time, a noise level of 0.1 % can be achieved when further binning the data by averaging along the slit. A context imager with an angular resolution of $2.2''$ images a field of view of $550'' \times 550''$ in order to identify the structures, which are observed by the spectropolarimeter.

The requirements on polarimetric sensitivity and accuracy are derived from models of the expected polarization signals (c.f. Ishikawa et al. 2014b, and references therein), based on reported estimates of the field strength in chromospheric structures (Trujillo Bueno et al. 2011). For 5–50 G, the calculated polarization profiles indicate that a sensitivity of 0.1–0.2 % is necessary. While in the line core a sensitivity of 0.1 % in the linear polarization is aimed for, the sensitivity in the wings of the line can be decreased to 0.5 % due to the larger polarization signals there. More recent 3-D simulations of scattering polarization and the Hanle effect in MHD chromospheric models have been undertaken by Štěpán (2015) and Štěpán et al. (2015).

The spectropolarimeter employs a spherical constant-line-spaced grating with 3000 grooves per mm and uses the plus and minus first diffraction orders for dual beam polarimetry. Two camera mirrors feed two cameras. Two reflective polarization analyzers are employed in front of the cameras. This setup maximizes the photon efficiency and allows for high sensitivity polarimetry. To this end, all polarimetric error sources must be controlled by a thorough calibration concept and a rigorous polarimetric error budget tracking (Ishikawa et al. 2014a). The linear polarization is modulated by a rotating wave plate made from MgF_2

with a retardance of half a wave, rotating at a period of 4.8 seconds. With sixteen exposures per round, the optimum exposure time of 0.3 seconds per frame is achieved.

In order to cope with the demanding false-light suppression requirements and for thermal reasons, the primary mirror acts as a cold mirror. The coating of this mirror has high reflectivity only around 121 nm (54 % as measured on test samples), but is transparent in the visible, thus transmitting most of the incoming solar energy to a thermally insulated light trap.

The international effort is led by Japan, which provides the experiment structure, telescope optics, polarimeter optics, slitjaw optics, and the waveplate rotation mechanism. The sounding rocket, flight operations, CCD cameras, and avionics are under responsibility of NASA. Spain contributes by modeling of the Hanle effect, while the modeling tools for the chromosphere are developed under Norwegian responsibility. France contributes to the mission by providing the diffraction grating.

At the time of writing, CLASP was scheduled for launch in September 2015. During the manuscript review process, CLASP has been launched, but no results have been announced yet. The total mission observing time was 5 min. If the mission is successful, CLASP will pave the way for the application of UV polarimetry from space.

4.2 Sunrise

Although not being real space missions according to classical definition, stratospheric balloons offer a valid alternative for a low cost access to a near space like environment (for a recent review see Gaskin et al. 2014, and references therein). Since the technological aspects of balloon borne observatories are much closer to those of space instruments than to those of ground-based facilities, we list Sunrise here.

Sunrise is the largest and most complex UV/Vis solar observatory up to date, which could escape the disturbing influences of the terrestrial atmosphere. Based on a 1-m aperture optical telescope and two scientific post focus instruments in the near UV and the visible range of the solar spectrum, it was designed for highest resolution observations of solar surface magnetic fields. Sunrise is an international mission led by the Max-Planck-Institute of Solar System Research (MPS) in Göttingen, Germany, together with the German Kiepenheuer Institute of solar physics in Freiburg, a Spanish consortium under leadership of the IAC (until 2012) and the IAA (since 2013), and the High Altitude Observatory in Boulder, USA. Also involved is the Lockheed Martin Solar and Astrophysics laboratory in Palo Alto. The ballooning aspects are under responsibility of NASA's Columbia Ballooning Facility.

In two 6-day flights from ESRANGE space center near Kiruna (Sweden) to Northern Canada in summer 2009 and in summer 2013, respectively, Sunrise could demonstrate its unique potential. Despite of some technical issues in both flights the magnetograms recorded of the extremely quiet Sun in 2009 and of active regions in 2013 represent some of the highest resolution seeing free magnetic field and Doppler maps of the photosphere. The co-spatial brightness maps of the photosphere and low chromosphere in the near UV (down to 220 nm, which is not accessible from the ground) are the highest contrast images of the solar surface ever recorded (Hirzberger et al. 2010), thanks to the absence of seeing and by exploiting the contrast transfer capabilities of a 1-m aperture telescope in the near UV. In this wavelength range, the sensitivity of the brightness to temperature fluctuations is very pronounced because of the steepness of the Planck curve; this helped in mapping the temperature structure and thus identifying magnetic bright points with high sensitivity (Riethmüller et al. 2010), while the magnetic field and the flows were directly measured in the visible. A number of scientific insights into the physics of small-scale solar surface magnetism resulted from these observations (see e.g. Solanki et al. 2011, and references therein).

Sunrise contains an instrument suite, which consists of a near UV filtergraph (SuFI, Gandorfer et al. 2011) and an imaging magnetograph (IMaX, Martínez Pillet et al. 2011), which are both fed with light from a 1 m aperture diameter Gregory telescope by a light distribution unit (ISLiD, Gandorfer et al. 2011). This unit also provides the high resolution images for a correlation tracker and wave front sensor (CWS, Berkefeld et al. 2011), which reduces residual pointing errors of the telescope by a fast tip tilt mirror inside the light distribution unit, and keeps control of the telescope alignment. The IMaX magnetograph is based on two key technologies: wavelength selection is done by a tunable solid state etalon, and the polarization analysis makes use of two nematic liquid crystals. Both technologies have proven to be applicable in near-space conditions and represent a big step forward in our efforts to build compact and lightweight polarimeters for space applications. The Sunrise experience thus directly influences the development of the *Polarimetric and Helioseismic Imager* onboard *Solar Orbiter* (see below) and is considered a necessary and successful predecessor.

Ballooning offers unique opportunities for testing new instrumental concepts for their usage in space. Most solar instruments have reached a technological complexity, which cannot be transferred to space in one step. This applies in particular to new detector architectures and the complex electronics systems associated with them. Special newly developed sensors like the one of the Fast Solar Polarimeter are today in a prototype status for ground-based demonstration of the concept. It is only natural to expand these efforts to use such sensors in Sunrise, and maybe—after successful qualification—some day in a new space observatory. Even without being affected by seeing, very high resolution polarimeters in space suffer from the internal vibrations of the spacecraft, which cannot be completely avoided. This becomes more and more the dominant cost driver in the development of satellite platforms for solar observations. Fast detectors greatly relax the requirements on the residual pointing errors and thus will some day enable polarimetry at extremely high resolution from space. The jitter spectrum of a stratospheric balloon is typically much more demanding and thus represents a worst case validation. Therefore, if Sunrise will fly again, it would be desirable to have a prototype of a near-space FSP-type sensor on board.

4.3 Solar Orbiter and Its Polarimetric and Helioseismic Imager PHI

More than twenty years after the launch of SoHO, a new European solar mission will be launched to space in 2018, Solar Orbiter, a collaborative mission of ESA and NASA.

Solar Orbiter is conceived and designed to clarify the magnetic coupling from the photosphere throughout the solar atmosphere into the inner heliosphere, and aims at providing answers to questions like the following: How and where do the solar wind plasma and magnetic field originate in the corona? How do solar transients drive heliospheric variability? How do solar eruptions produce energetic particle radiation that fills the heliosphere? How does the solar dynamo work and drive the connections between the Sun and the heliosphere?

Carrying a suite of remote sensing instruments and an in-situ analysis package, Solar Orbiter is an integrated and unique approach to heliophysics, since it combines aspects of a space solar observatory with the characteristics of an encounter mission. Solar Orbiter is not particularly focussed on magnetometry, but the mission will offer unique opportunities to study surface fields and to probe the solar dynamo (Löptien et al. 2014).

Solar Orbiter's science can be addressed thanks to a unique orbit design: During a 3.5 year transfer orbit the spacecraft undergoes several gravity assist maneuvers (GAM) at Venus and Earth, which help the spacecraft to lose orbital energy, and thus allow *Solar Orbiter* to come close to the Sun. After the second GAM at Venus, *Solar Orbiter* begins its operational phase. From then on its orbit is in a three-to-two resonance with Venus, such that after each third orbit the inclination of the orbital plane with respect to the ecliptical plane can be increased by a further gravity assist. This particular feature gives *Solar Orbiter* access to the high latitude regions of the Sun for the first time.

While the in-situ instrument suite will be operational over the full orbit, the remote sensing instruments will be used in three distinct science phases per orbit, the perihelion passage, and the phases of maximum and minimum solar latitude. During the perihelion passages *Solar Orbiter* can follow the evolution of surface structures and solar features not only from close-by, but in addition under practically unchanged geometrical viewing conditions for several days, thanks to a corotating vantage point.

The *Solar Orbiter* Instrumentation can be grouped in three major packages, each consisting of several instruments:

- Field Package: Radio and Plasma Wave Analyzer and Magnetometer.
- Particle Package: Energetic Particle Detector and Solar Wind Plasma Analyser.
- Solar remote sensing instrumentation: Visible-light Imager and Magnetograph, Extreme Ultraviolet Spectrometer, EUV Imager, Coronagraph, and Spectrometer/Telescope for Imaging X-rays, Heliospheric Imager.

From all instruments onboard, the visible light imager and magnetograph, called "Polarimetric and Helioseismic Imager PHI" is of highest relevance for solar surface magnetometry. Together with the other high resolution instruments it observes the same target region on the solar surface with an identical angular sampling of 0.5 arcsec per pixel. By combining its observations with the other instruments on Solar Orbiter, PHI will address the magnetic coupling between the different atmospheric layers. Extrapolations of the magnetic field observed by PHI into the Sun's upper atmosphere and heliosphere will provide the information needed for other optical and in-situ instruments to analyze and understand the data recorded by them in a proper physical context.

The instruments on Solar Orbiter are very challenging to build. No space magnetograph has ever flown in such a difficult environment, with the spacecraft following a strongly elliptical trajectory, leading to significant thermal changes in the course of an orbit. The technical description below follows closely the more complete instrument descriptions by Gandorfer et al. (2011) and Solanki et al. (2015). For the sake of completeness and readability, we present a shortened version here. PHI makes use of the Doppler- and Zeeman-effects in a single spectral line of neutral iron at 617.3 nm. The physical information is decoded from two-dimensional intensity maps at six wavelength points within this line, while four polarization states at each wavelength point are measured. In order to obtain the abovementioned observables, PHI is a

- diffraction limited,
- wavelength tunable,
- quasi-monochromatic,
- polarization sensitive imager with two telescopes,

which (alternatively) feed a common filtergraph and focal plane array (for illustration see Fig. 8): The High Resolution Telescope (HRT) provides a restricted FOV of 16.8 arcmin squared and achieves a spatial resolution that, near the closest perihelion pass, will correspond to about 200 km on the Sun. It is designed as a decentered Ritchey-Chrétien telescope with a pupil of 140 mm diameter. The all-refractive Full Disk Telescope (FDT), with a FOV of 2.1° in diameter and a pixel size corresponding to 730 km (at 0.28 AU), provides a complete view of the full solar disk during all orbital phases. These two telescopes are used alternatively and their selection is made by a feed selection mechanism.

Both telescopes are protected from the intense solar flux by special heat-rejecting entrance windows, which are part of the heat-shield assembly of the spacecraft. These multilayer filters have more than 80 % transmittance in a narrow notch around the science wavelength, while effectively blocking the remaining parts of the spectrum from 200 nm to the far infrared by reflection. Only a fraction of the total energy is absorbed in the window, which acts a passive thermal element by emitting part of the thermal radiation to cold space; emission of infrared radiation into the instrument cavity is minimized by a low emissivity coating on the backside of the window (acting at the same time as an anti-reflection coating for the science wavelength). Thus the heat load into the instruments can be substantially reduced, while preserving the high photometric and polarimetric accuracy of PHI.

The filtergraph unit FG uses two key technologies with heritage from IMaX onboard Sunrise: A $LiNbO_3$ solid state etalon in a telecentric configuration selects a passband of 100 mÅ width. Applying a high voltage across the crystal allows changing the refractive index of the material and its thickness, and thus tuning the passband in wavelength across the spectral line. Gensemer and Farrant (2014) report on the fabrication technology for these etalons, whose absolute thickness, approximately 250 μm, is controlled to < 10 nm, maintaining a thickness uniformity of < 1 nm over the 60 mm aperture. A 3 Å wide prefilter acts as an order sorter for the Fabry-Perot channel spectrum. The polarimetric analysis is performed by two Polarization Modulation Packages (PMP) in each of the telescopes. Each PMP consists of two nematic liquid crystal retarders, followed by a linear polarizer as an analyzer. The liquid crystal variable retarders have been successfully qualified for the use in Solar Orbiter (Alvarez-Herrero et al. 2011). Both, the FG and the PMPs, are thermally insulated from the Optics Unit and actively temperature stabilized.

The opto-mechanical arrangement is designed to operate in a wide temperature range. To this end the optics unit structure consists of a combination of AlBeMet (an aluminum-beryllium alloy) and low expansion carbon-fibre reinforced plastic. An internal image stabilization system in the HRT channel system acts on the active secondary mirror, which greatly reduces residual pointing error by the spacecraft to levels compatible with high resolution polarimetry.

The error signal for the piezo-driven mirror support is derived from a correlation tracker camera inside the HRT. For details on the Image stabilization system we refer to Carmona et al. (2014). The focal plane assembly is built around a 2048 by 2048 pixel Active Pixel Sensor (APS), which is specially designed and manufactured for this instrument. It delivers 10 frames per second, which are read out in synchronism with the switching of the polarization modulators.

Besides all the technological complexity of the cutting-edge hardware, the most critical aspect of the instrument lies, however, in the data reduction strategy: The extremely limited telemetry rate and the large amount of scientific information retrieved from the PHI instrument demand a sophisticated on-board data reduction and necessitate to employ a non-linear, least-square, inversion technique, which numerically solves the radiative transfer equation on board. This inversion is based on the Milne-Eddington approximation and is provided by the highly adaptable onboard s/w, two powerful reprogrammable FPGAs (Fiethe et al. 2012), and a large non-volatile data storage unit.

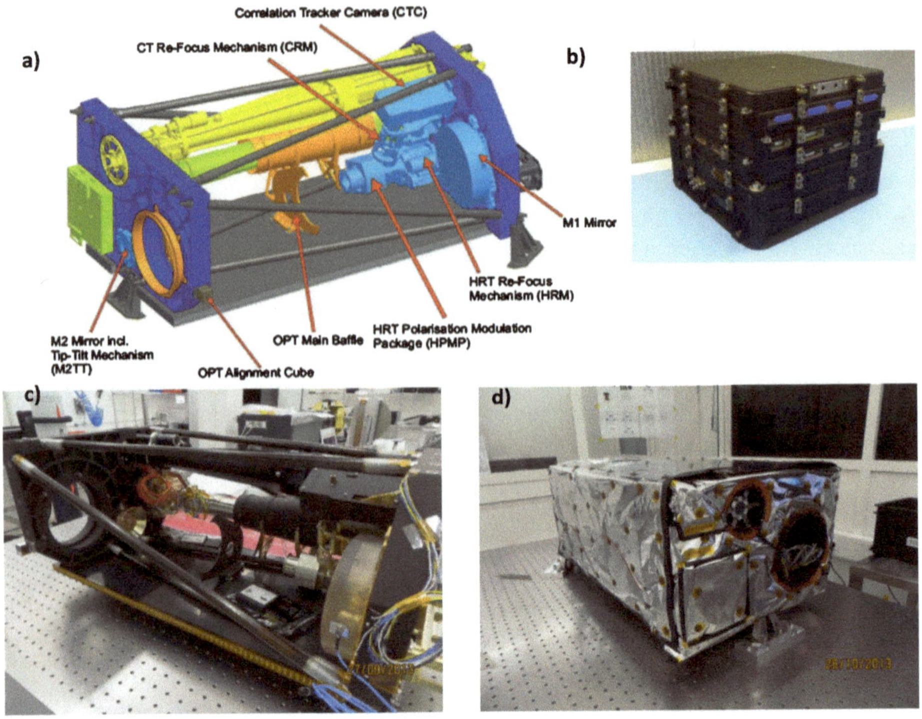

Fig. 8 Images of the PHI instrument: (**a**) CAD representation of the PHI Optics Unit with FDT tube (*yellow*) and HRT subsystems; (**b**) structural thermal model (STM) of the electronics unit; (**c**) STM of the optics unit without multi-layer insulation (MLI); (**d**) same, but now with mounted MLI. From Solanki et al. (2015)

4.4 Solar-C

Solar-C (Watanabe 2014) is an international vision for the ultimate future solar space observatory, initiated by Japan in the wake of the highly successful Hinode mission.

Solar-C is a JAXA led solar space observatory with the overarching science goal of "understanding the Sun's magnetized atmosphere from bottom to top by

- understanding the dynamic structuring and mass and energy loading of the solar atmosphere
- understanding the basic plasma processes at work throughout the solar atmosphere
- understand the causes of solar activity, which affects our natural and technical environment" (Solanki and the EPIC Consortium 2015).

To this end, Solar-C represents an extremely powerful space observatory in a geosynchronous orbit, with a payload that significantly improves our capabilities in imaging and spectropolarimetry in the UV, visible, and near infrared with respect to what is available today or foreseen in the near future. The mission concept and strawman instrumentation has been conceived and designed by the efforts of the ISAS/JAXA Solar-C working group, and a summary of the mission has been published by Watanabe (2014). A strong European Contribution to Solar-C is envisaged. A first proposal to ESA in response to the call for the M4 mission (Solanki and the EPIC Consortium 2015) was very highly ranked but was not selected in this round.

Solar-C aims at probing the different temperature regimes in the solar atmosphere with unprecedented angular resolution and at simultaneously covering the entire solar atmosphere from the solar surface to the outer corona. Three telescopes, working in distinct spectral regimes, are required for this task:

- SUVIT, a 1.4 m aperture UV/VIS/NIR telescope for imaging and spectropolarimetry with an order of magnitude increase in photon collecting area over the largest solar telescope in space. With its instrumentation suite, which will be described below, this telescopes covers seamlessly the solar photosphere and chromosphere.
- EUVST, a 30 cm aperture VUV telescope feeding a spectrometer for imaging spectroscopy covering line formation temperatures ranging from the chromosphere to the hottest parts of the corona. Over this extended temperature range, EUVST will have a resolution and effective collecting area an order of magnitude better than available today. EUVST is based on the LEMUR instrument described by Teriaca et al. (2012).
- HCI, a next generation high resolution extreme-ultraviolet imager with 32 cm aperture operating at high cadence in multiple spectral lines sampling the chromosphere and corona.

The telescope, which is of most relevance to magnetometry, is the Solar Ultraviolet and Visible Telescope SUVIT (Suematsu et al. 2014b). SUVIT will for the first time measure chromospheric magnetic fields from space and is expected to obtain photospheric vector fields at very high resolution with the highest sensitivity ever for subarcsec resolution.

SUVIT is a telescope with an aperture of 1.4 m, providing an angular resolution of $0.1''$ in combination with spectropolarimetry in wavelengths ranging from 280 nm to 1080 nm.

With its aperture, SUVIT would have at least 10 times higher sensitivity than any other current or planned space mission. The increase in spatial resolving power and light gathering power directly results in an order of magnitude improvement in sensitivity of magnetic flux, electric current and magnetic and kinetic energy. Another major advantage of SUVIT is that it samples a large range of line formation heights almost seamlessly. In order to achieve this, the telescope feeds several highly dedicated post-focus instruments via a common optical interface unit. This interface unit provides each instrument with a collimated beam with the required spectral bands; it also contains the common tip-tilt correction mirror for stabilizing image motion caused by the residual pointing error of the S/C. A similar strategy—although for a much narrower spectral range—has been successfully implemented in the SOT of Hinode (Suematsu et al. 2008). The post-focus instrumentation consists of a tunable filtergraph (FG) with polarimetric sensitivity for the spectral range from 525 nm to 1083 nm. The baseline design employs a Lyot filter equipped with liquid crystals as variable retardation plates, which allows tuning in wavelength without any moving or rotating parts. The classical echelle-type grating spectropolarimeter (SP) covers the same wavelength range and also possesses vector magnetometric capabilities in a variety of spectral lines. In order to achieve the key science goals of retrieving the magnetic field information over the full atmospheric height, nearly simultaneous observations in several lines are mandatory. This can be achieved in two different ways: Either the grating is moved and the spectral bands are recorded in series with a single camera, or—in order to obtain strictly simultaneous recordings—three cameras are employed; in both options interference filters act as order sorters for the echelle grating. While the three camera solution—to first sight—seems more complex, it should not be forgotten, that it spares two mechanisms, which is always beneficial in terms of reliability and microvibration control. The final decision on which option to follow must be done by a careful assessment during the study phase of the mission. Photospheric and chromospheric magnetograms with simultaneous two-dimensional coverage can be achieved only by an integral field unit (IFU). Such a unit is foreseen for the SP as well, it maps a 9″ squared patch of the field of view onto three entrance slits of the SP, located parallel to the "normal" entrance slit. It can be based on an array of optical fibers (Lin 2012) or on micro image slicers (Suematsu et al. 2014a).

The capabilities of high-resolution imaging in the near UV for providing new insights into the photosphere and chromosphere have been established by the SUFI imager onboard the Sunrise balloon-borne stratospheric observatory. IRIS has unveiled the upper chromosphere in the light of the Mg II h& k lines, uncovering the dynamic and energetic phenomena by high resolution UV spectroscopy. In the wake of these achievements, the Ultraviolet Imager and Spectropolarimeter (UBIS) will further advance those observations and provide UV filtergrams at much higher spatial resolution. In addition, expanding the capabilities of IRIS, UBIS aims at spectropolarimetry in the Mg II h & k lines, allowing for magnetometry in the upper chromosphere. To this end, UBIS consists of two sub-units, a re-imager with filter wheels, and an optional spectropolarimeter behind it. The spectropolarimeter should be considered in this stage as an optional instrument, since the potential of spectropolarimetry in the Mg II lines still needs to be thoroughly demonstrated. Recent significant progress in forward modelling suggests that the chromospheric magnetic field can be reliably diagnosed with these lines (Belluzzi and Trujillo Bueno 2012).

5 Outlook

We are at an exciting stage in solar magnetic measurements and the next major missions and telescopes are within reach. By 2020, we expect our newest ground- and space-based telescopes, DKIST and Solar Orbiter, to provide unprecedented data.

The 4 m ground-based DKIST, currently in construction and scheduled for first light in 2019, will improve our current spatial resolution (diffraction limit) by a factor of 2.5. This will for example allow to compare features of sizes of fractions of arcseconds in sunspots, such as umbral dots, or penumbral grains to the most advanced simulations to investigate details of sunspot dynamics, formation and decay. The polarimetric sensitivity of better than 10^{-4} I_c, coupled with the large aperture will hopefully allow us to solve the mystery of the orientation of small-scale magnetic fields in the quiet Sun and thus to investigate the local dynamo processes. Employing fast polarimetry will enable us to study the magnetic structure of rapidly changing features, such as flares, filament eruptions, or dynamic flows.

After Solar Orbiter's launch, currently scheduled for 2018, and its cruise phase, we will for the first time have a polarimeter on a significantly inclined orbit around the Sun allowing us to study its polar regions. We will not only be able to study the mechanism of flux transport towards the poles in more detail, but also combine the observations with a whole suite of 9 additional instruments on Solar Orbiter, giving us a more complete picture of the different solar layers.

Plans for the far future include the EST, a 4 m class ground-based telescope in Europe, with a design specifically optimized for polarimetric measurements and Solar-C, the successor of the Japanese Hinode satellite, that will hopefully enable us to study chromospheric polarimetry systematically and from space for the first time, shedding light on a highly dynamic and yet not well explored solar layer.

Acknowledgements We thank the experts of the telescopes and instruments for their advice and information, in particular, Wenda Cao, Gianna Cauzzi, Manolo Collados, Jaime de la Cruz Rodriguez, Alex Feller, Bernard Gelly, S.S. Hasan, Bruce Lites, Chang Liu, Arturo Lopez Ariste, Zhong Liu, Valentin Martinez Pillet, Rolf Schlichenmaier, Alexandra Tritschler, Michiel van Noort, and Haimin Wang.

References

A. Alvarez-Herrero, N. Uribe-Patarroyo, P. García Parejo, J. Vargas, R.L. Heredero, R. Restrepo, V. Martínez-Pillet, J.C. del Toro Iniesta, A. López, S. Fineschi, G. Capobianco, M. Georges, M. López, G. Boer, I. Manolis, Imaging polarimeters based on liquid crystal variable retarders: an emergent technology for space instrumentation. Adv. Space Res. (2011). doi:10.1117/12.892732

M.J. Aschwanden, Solar stereoscopy and tomography. Living Rev. Sol. Phys. **8**, 5 (2011). doi:10.12942/lrsp-2011-5

A. Asensio Ramos, Evidence for quasi-isotropic magnetic fields from Hinode quiet-sun observations. Astrophys. J. **701**, 1032–1043 (2009). doi:10.1088/0004-637X/701/2/1032

P. Barthol, A. Gandorfer, S.K. Solanki, M. Schüssler, B. Chares, W. Curdt, W. Deutsch, A. Feller, D. Germerott, B. Grauf, K. Heerlein, J. Hirzberger, M. Kolleck, R. Meller, R. Müller, T.L. Riethmüller, G. Tomasch, M. Knölker, B.W. Lites, G. Card, D. Elmore, J. Fox, A. Lecinski, P. Nelson, R. Summers, A. Watt, V. Martínez Pillet, J.A. Bonet, W. Schmidt, T. Berkefeld, A.M. Title, V. Domingo, J.L. Gasent Blesa, J.C. del Toro Iniesta, A. López Jiménez, A. Álvarez-Herrero, L. Sabau-Graziati, C. Widani, P. Haberler, K. Härtel, D. Kampf, T. Levin, I. Pérez Grande, A. Sanz-Andrés, E. Schmidt, The Sunrise Mission. Sol. Phys. **268**, 1–34 (2011). doi:10.1007/s11207-010-9662-9

J.M. Beckers, R.B. Schultz, Oscillatory motions in sunspots. Sol. Phys. **27**, 61–70 (1972). doi:10.1007/BF00151770

L. Belluzzi, J. Trujillo Bueno, The polarization of the solar Mg II h and k lines. Astrophys. J. Lett. **750**, 11 (2012). doi:10.1088/2041-8205/750/1/L11

T. Berkefeld, W. Schmidt, D. Soltau, A. Bell, H.P. Doerr, B. Feger, R. Friedlein, K. Gerber, F. Heidecke, T. Kentischer, O.v.D. Lühe, M. Sigwarth, E. Wälde, P. Barthol, W. Deutsch, A. Gandorfer, D. Germerott, B. Grauf, R. Meller, A. Álvarez-Herrero, M. Knölker, V. Martínez Pillet, S.K. Solanki, A.M. Title, The wave-front correction system for the Sunrise Balloon-Borne Solar Observatory. Sol. Phys. **268**, 103–123 (2011). doi:10.1007/s11207-010-9676-3

P.N. Bernasconi, D.M. Rust, H.A. Eaton, G.A. Murphy, Balloon-borne telescope for high-resolution solar imaging and polarimetry, in *Airborne Telescope Systems*, ed. by R.K. Melugin, H.-P. Röser Society of Photo-Optical Instrumentation Engineers (SPIE) Conference Series, vol. 4014 (2000), pp. 214–225

J.M. Borrero, P. Kobel, Inferring the magnetic field vector in the quiet Sun. I. Photon noise and selection criteria. Astron. Astrophys. **527**, 29 (2011). doi:10.1051/0004-6361/201015634

A. Calcines, R.L. López, M. Collados, MuSICa: the Multi-Slit Image Slicer for the EST spectrograph. J. Astron. Instrum. **2**, 50009 (2013). doi:10.1142/S2251171713500098

A. Calcines, R.L. López, M. Collados, N. Vega Reyes, *MuSICa Image Slicer Prototype at 1.5-m Gregor Solar Telescope*. Society of Photo-Optical Instrumentation Engineers (SPIE) Conference Series, vol. 9147 (2014), p. 3. doi:10.1117/12.2053577

W. Cao, N. Gorceix, R. Coulter, K. Ahn, T.R. Rimmele, P.R. Goode, Scientific instrumentation for the 1.6 m New Solar Telescope in Big Bear. Astron. Nachr. **331**, 636 (2010). doi:10.1002/asna.201011390

W. Cao, P.R. Goode, K. Ahn, N. Gorceix, W. Schmidt, H. Lin, NIRIS: the second generation near-infrared imaging spectro-polarimeter for the 1.6 meter New Solar Telescope, in *Second ATST-EAST Meeting: Magnetic Fields from the Photosphere to the Corona*, ed. by T.R. Rimmele, A. Tritschler, F. Wöger, M. Collados Vera, H. Socas-Navarro, R. Schlichenmaier, M. Carlsson, T. Berger, A. Cadavid, P.R. Gilbert, P.R. Goode, M. Knölker Astronomical Society of the Pacific Conference Series, vol. 463 (2012), p. 291

M. Carmona, J.M. Gómez, D. Roma, A. Casas, M. López, J. Bosch, A. Herms, J. Sabater, R. Volkmer, F. Heidecke, T. Maue, E. Nakai, W. Schmidt, System model of an image stabilization system. Proc. SPIE (2014). doi:10.1117/12.2054867.

R. Casini, A. López Ariste, S. Tomczyk, B.W. Lites, Magnetic maps of prominences from full stokes analysis of the He I D3 line. Astrophys. J. Lett. **598**, 67–70 (2003). doi:10.1086/380496

F. Cavallini, IBIS: a new post-focus instrument for solar imaging spectroscopy. Sol. Phys. **236**, 415–439 (2006). doi:10.1007/s11207-006-0103-8

M. Collados, R. López, E. Páez, E. Hernández, M. Reyes, A. Calcines, E. Ballesteros, J.J. Díaz, C. Denker, A. Lagg, R. Schlichenmaier, W. Schmidt, S.K. Solanki, K.G. Strassmeier, O. von der Lühe, R. Volkmer, GRIS: the GREGOR infrared spectrograph. Astron. Nachr. **333**, 872 (2012). doi:10.1002/asna.201211738

M. Collados, F. Bettonvil, L. Cavaller, I. Ermolli, B. Gelly, A. Pérez, H. Socas-Navarro, D. Soltau, R. Volkmer (EST Team), The European solar telescope. Mem. Soc. Astron. Ital. **84**, 379 (2013)

J. de la Cruz Rodríguez, H. Socas-Navarro, Are solar chromospheric fibrils tracing the magnetic field? Astron. Astrophys. **527**, 8 (2011). doi:10.1051/0004-6361/201016018

J. de la Cruz Rodríguez, H. Socas-Navarro, M. Carlsson, J. Leenaarts, Non-local thermodynamic equilibrium inversions from a 3D magnetohydrodynamic chromospheric model. Astron. Astrophys. **543**, 34 (2012). doi:10.1051/0004-6361/201218825

J. de la Cruz Rodríguez, L. Rouppe van der Voort, H. Socas-Navarro, M. van Noort, Physical properties of a sunspot chromosphere with umbral flashes. Astron. Astrophys. **556**, 115 (2013). doi:10.1051/0004-6361/201321629

J. de la Cruz Rodríguez, M.G. Löfdahl, P. Sütterlin, T. Hillberg, L. Rouppe van der Voort, CRISPRED: a data pipeline for the CRISP imaging spectropolarimeter. Astron. Astrophys. **573**, 40 (2015). doi:10.1051/0004-6361/201424319

D.F. Elmore, T. Rimmele, R. Casini, S. Hegwer, J. Kuhn, H. Lin, J.P. McMullin, K. Reardon, W. Schmidt, A. Tritschler, F. Wöger, *The Daniel K. Inouye Solar Telescope First Light Instruments and Critical Science Plan*. Society of Photo-Optical Instrumentation Engineers (SPIE) Conference Series, vol. 9147 (2014), p. 7. doi:10.1117/12.2057038

M. Faurobert, J. Arnaud, Center-to-limb variation of scattering polarization in molecular solar lines: Observations and modeling. Astron. Astrophys. **412**, 555–565 (2003). doi:10.1051/0004-6361:20031468

M. Faurobert, M. Derouich, V. Bommier, J. Arnaud, Hanle effect in the solar Ba II D2 line: a diagnostic tool for chromospheric weak magnetic fields. Astron. Astrophys. **493**, 201–206 (2009). doi:10.1051/0004-6361:200810474

A. Feller, F.A. Iglesias, K. Nagaraju, S.K. Solanki, S. Ihle, Fast solar polarimeter: description and first results, in *Solar Polarization 7*, ed. by K.N. Nagendra, J.O. Stenflo, Q. Qu, M. Sampoorna Astronomical Society of the Pacific Conference Series, vol. 489 (2014), p. 271

B. Fiethe, F. Bubenhagen, T. Lange, H. Michalik, H. Michel, J. Woch, J. Hirzberger, Adaptive hardware by dynamic reconfiguration for the solar orbiter Phi instrument, in *2012 NASA/ESA Conference on Adaptive Hardware and Systems (AHS)*, 2012, pp. 31–37. doi:10.1109/AHS.2012.6268666

S. Fineschi, S. Solanki, C. Team, COMPASS: coronal magnetism, plasma and activity studies from space: a formation flying mission to measure the solar magnetism. Proposal to ESA Cosmic Vision (2007)

A.M. Gandorfer, H.P. Povel, First observations with a new imaging polarimeter. Astron. Astrophys. **328**, 381–389 (1997)

A.M. Gandorfer, H.P. Povel, P. Steiner, F. Aebersold, U. Egger, A. Feller, D. Gisler, S. Hagenbuch, J.O. Stenflo, Solar polarimetry in the near UV with the Zurich Imaging Polarimeter ZIMPOL II. Astron. Astrophys. **422**, 703–708 (2004). doi:10.1051/0004-6361:20040254

A. Gandorfer, B. Grauf, P. Barthol, T.L. Riethmüller, S.K. Solanki, B. Chares, W. Deutsch, S. Ebert, A. Feller, D. Germerott, K. Heerlein, J. Heinrichs, D. Hirche, J. Hirzberger, M. Kolleck, R. Meller, R. Müller, R. Schäfer, G. Tomasch, M. Knölker, V. Martínez Pillet, J.A. Bonet, W. Schmidt, T. Berkefeld, B. Feger, F. Heidecke, D. Soltau, A. Tischenberg, A. Fischer, A. Title, H. Anwand, E. Schmidt, The filter imager SuFI and the image stabilization and light distribution system ISLiD of the Sunrise Balloon-Borne Observatory: instrument description. Sol. Phys. **268**, 35–55 (2011). doi:10.1007/s11207-010-9636-y

J.A. Gaskin, I.S. Smith, W.V. Jones, Introduction to the special issue on scientific balloon capabilities and instrumentation. J. Astron. Instrum. **3**, 3001 (2014). doi:10.1142/S2251171714030019

S.D. Gensemer, D. Farrant, Fabrication and metrology of lithium niobate narrowband optical filters for the solar orbiter. Adv. Opt. Technol. **3**, 309–314 (2014). doi:10.1515/aot-2014-0016

P.R. Goode, W. Cao, The 1.6 M off-axis New Solar Telescope (NST) in Big Bear, in *Second ATST-EAST Meeting: Magnetic Fields from the Photosphere to the Corona*, ed. by T.R. Rimmele, A. Tritschler, F. Wöger, M. Collados Vera, H. Socas-Navarro, R. Schlichenmaier, M. Carlsson, T. Berger, A. Cadavid, P.R. Gilbert, P.R. Goode, M. Knölker Astronomical Society of the Pacific Conference Series, vol. 463 (2012), p. 357

P.R. Goode, R. Coulter, N. Gorceix, V. Yurchyshyn, W. Cao, The NST: first results and some lessons for ATST and EST. Astron. Nachr. **331**, 620 (2010). doi:10.1002/asna.201011387

R.A. Harrison, J.A. Davies, A.P. Rouillard, C.J. Davis, C.J. Eyles, D. Bewsher, S.R. Crothers, R.A. Howard, N.R. Sheeley, A. Vourlidas, D.F. Webb, D.S. Brown, G.D. Dorrian, Two years of the STEREO heliospheric imagers. Invited Review. Sol. Phys. **256**, 219–237 (2009). doi:10.1007/s11207-009-9352-7

S.S. Hasan, India's National Large Solar Telescope, in *Second ATST-EAST Meeting: Magnetic Fields from the Photosphere to the Corona*, ed. by T.R. Rimmele, A. Tritschler, F. Wöger, M. Collados Vera, H. Socas-Navarro, R. Schlichenmaier, M. Carlsson, T. Berger, A. Cadavid, P.R. Gilbert, P.R. Goode, M. Knölker Astronomical Society of the Pacific Conference Series, vol. 463 (2012), p. 395

J. Hirzberger, A. Feller, T.L. Riethmüller, M. Schüssler, J.M. Borrero, N. Afram, Y.C. Unruh, S.V. Berdyugina, A. Gandorfer, S.K. Solanki, P. Barthol, J.A. Bonet, V. Martínez Pillet, T. Berkefeld, M. Knölker, W. Schmidt, A.M. Title, Quiet-sun intensity contrasts in the near-ultraviolet as measured from SUNRISE. Astrophys. J. Lett. **723**, 154–158 (2010). doi:10.1088/2041-8205/723/2/L154

R. Ishikawa, A. Asensio Ramos, L. Belluzzi, R. Manso Sainz, J. Štěpán, J. Trujillo Bueno, M. Goto, S. Tsuneta, On the inversion of the scattering polarization and the Hanle effect signals in the hydrogen Lyα line. Astrophys. J. **787**, 159 (2014a). doi:10.1088/0004-637X/787/2/159

R. Ishikawa, N. Narukage, M. Kubo, S. Ishikawa, R. Kano, S. Tsuneta, Strategy for realizing high-precision VUV spectro-polarimeter. Sol. Phys. **289**, 4727–4747 (2014b). doi:10.1007/s11207-014-0583-x

S.A. Jaeggli, H. Lin, D.L. Mickey, J.R. Kuhn, S.L. Hegwer, T.R. Rimmele, M.J. Penn, FIRS: a new instrument for photospheric and chromospheric studies at the DST. Mem. Soc. Astron. Ital. **81**, 763 (2010)

H. Ji, W. Cao, P.R. Goode, Observation of ultrafine channels of solar corona heating. Astrophys. J. Lett. **750**, 25 (2012). doi:10.1088/2041-8205/750/1/L25

R. Kano, T. Bando, N. Narukage, R. Ishikawa, S. Tsuneta, Y. Katsukawa, M. Kubo, S.-n. Ishikawa, H. Hara, T. Shimizu, Y. Suematsu, K. Ichimoto, T. Sakao, M. Goto, Y. Kato, S. Imada, K. Kobayashi, T. Holloway, A. Winebarger, J. Cirtain, B. De Pontieu, R. Casini, J. Trujillo Bueno, J. Štepán, R. Manso Sainz, L. Belluzzi, A. Asensio Ramos, F. Auchère, M. Carlsson, *Chromospheric Lyman-Alpha Spectro-Polarimeter (CLASP)*. Society of Photo-Optical Instrumentation Engineers (SPIE) Conference Series, vol. 8443 (2012), p. 4. doi:10.1117/12.925991

D. Kiselman, T.M.D. Pereira, B. Gustafsson, M. Asplund, J. Meléndez, K. Langhans, Is the solar spectrum latitude-dependent? An investigation with SST/TRIPPEL. Astron. Astrophys. **535**, 14 (2011). doi:10.1051/0004-6361/201117553

L. Kleint, A. Feller, D. Gisler, Imaging spectropolarimetry with two $LiNbO_3$ Fabry Pérot interferometers and a spectrograph. Astron. Astrophys. **529**, 78 (2011a). doi:10.1051/0004-6361/201015870

L. Kleint, A.I. Shapiro, S.V. Berdyugina, M. Bianda, Solar turbulent magnetic fields: non-LTE modeling of the Hanle effect in the C_2 molecule. Astron. Astrophys. **536**, 47 (2011b). doi:10.1051/0004-6361/201015857

L. Kleint, M. Battaglia, K. Reardon, A. Sainz Dalda, P.R. Young, S. Krucker, The fast filament eruption leading to the X-flare on March 29, 2014. ArXiv e-prints (2015)

M. Kubo, R. Kano, K. Kobayashi, T. Bando, N. Narukage, R. Ishikawa, S. Tsuneta, Y. Katsukawa, S. Ishikawa, Y. Suematsu, H. Hara, T. Shimizu, T. Sakao, K. Ichimoto, M. Goto, T. Holloway, A. Winebarger, J. Cirtain, B. De Pontieu, R. Casini, F. Auchère, J. Trujillo Bueno, R. Manso Sainz, L. Belluzzi, A. Asensio Ramos, J. Štěpán, M. Carlsson, A sounding rocket experiment for the Chromospheric Lyman-alpha Spectro-polarimeter (CLASP), in *Solar Polarization 7*, ed. by K.N. Nagendra, J.O. Stenflo, Q. Qu, M. Sampoorna Astronomical Society of the Pacific Conference Series, vol. 489 (2014), p. 307

A. Lagg et al., Space Sci. Rev. (2015, this issue)

H. Lin, J.R. Kuhn, R. Coulter, Coronal magnetic field measurements. Astrophys. J. Lett. **613**, 177–180 (2004). doi:10.1086/425217

H. Lin, *SPIES: The Spectropolarimetric Imager for the Energetic Sun*. Society of Photo-Optical Instrumentation Engineers (SPIE) Conference Series, vol. 8446 (2012), p. 1. doi:10.1117/12.926830

B.W. Lites, D.F. Elmore, P. Seagraves, A.P. Skumanich, Stokes profile analysis and vector magnetic fields. VI. Fine scale structure of a sunspot. Astrophys. J. **418**, 928 (1993). doi:10.1086/173450

B.W. Lites, K.D. Leka, A. Skumanich, V. Martinez Pillet, T. Shimizu, Small-scale horizontal magnetic fields in the solar photosphere. Astrophys. J. **460**, 1019 (1996). doi:10.1086/177028

B.W. Lites, M. Kubo, H. Socas-Navarro, T. Berger, Z. Frank, R. Shine, T. Tarbell, A. Title, K. Ichimoto, Y. Katsukawa, S. Tsuneta, Y. Suematsu, T. Shimizu, S. Nagata, The horizontal magnetic flux of the quiet-sun internetwork as observed with the Hinode spectro-polarimeter. Astrophys. J. **672**, 1237–1253 (2008). doi:10.1086/522922

Z. Liu, Y. Deng, Z. Jin, H. Ji, *Introduction to the Chinese giant solar telescope*. Society of Photo-Optical Instrumentation Engineers (SPIE) Conference Series, vol. 8444 (2012), p. 5. doi:10.1117/12.926033

Z. Liu, J. Xu, B.-Z. Gu, S. Wang, J.-Q. You, L.-X. Shen, R.-W. Lu, Z.-Y. Jin, L.-F. Chen, K. Lou, Z. Li, G.-Q. Liu, Z. Xu, C.-H. Rao, Q.-Q. Hu, R.-F. Li, H.-W. Fu, F. Wang, M.-X. Bao, M.-C. Wu, B.-R. Zhang, New vacuum solar telescope and observations with high resolution. Res. Astron. Astrophys. **14**, 705–718 (2014a). doi:10.1088/1674-4527/14/6/009

Z. Liu, Z. Jin, S. Yuan, J. Lin, Y. Deng, H. Ji, Y. Yan, *The progress of Chinese giant solar telescope*. Society of Photo-Optical Instrumentation Engineers (SPIE) Conference Series, vol. 9145 (2014b), p. 26. doi:10.1117/12.2055382

A. López Ariste, C. Le Men, B. Gelly, Double-pass spectroimaging with spectral multiplexing: TUNIS. Contrib. Astron. Obs. Skaln. Pleso **41**, 99–105 (2011)

A. López Ariste, J. Rayrole, M. Semel, First results from THEMIS spectropolarimetric mode. Astron. Astrophys. Suppl. Ser. **142**, 137–148 (2000). doi:10.1051/aas:2000144

A. López Ariste, G. Aulanier, B. Schmieder, A. Sainz Dalda, First observation of bald patches in a filament channel and at a BARB endpoint. Astron. Astrophys. **456**, 725–735 (2006). doi:10.1051/0004-6361:20064923

A. López Ariste, C. Le Men, B. Gelly, A. Asensio Ramos, Double-pass spectro-imaging: TUNIS. Astron. Nachr. **331**, 658 (2010). doi:10.1002/asna.201011396

A. López Ariste, F. Leblanc, R. Casini, R. Manso Sainz, B. Gelly, C. Le Men, Resonance scattering polarization in the magnetosphere of Mercury. Icarus **220**, 1104–1111 (2012). doi:10.1016/j.icarus.2012.07.001

B. Löptien, A.C. Birch, L. Gizon, J. Schou, T. Appourchaux, J. Blanco Rodríguez, P.S. Cally, C. Dominguez-Tagle, A. Gandorfer, F. Hill, J. Hirzberger, P.H. Scherrer, S.K. Solanki, Helioseismology with solar orbiter. Space Sci. Rev. (2014). doi:10.1007/s11214-014-0065-3

V. Martínez Pillet, Solar surface and atmospheric dynamics. The photosphere. Space Sci. Rev. **178**, 141–162 (2013). doi:10.1007/s11214-013-9967-8

V. Martínez Pillet, J.C. del Toro Iniesta, A. Álvarez-Herrero, V. Domingo, J.A. Bonet, L. González Fernández, A. López Jiménez, C. Pastor, J.L. Gasent Blesa, P. Mellado, J. Piqueras, B. Aparicio, M. Balaguer, E. Ballesteros, T. Belenguer, L.R. Bellot Rubio, T. Berkefeld, M. Collados, W. Deutsch, A. Feller, F. Girela, B. Grauf, R.L. Heredero, M. Herranz, J.M. Jerónimo, H. Laguna, R. Meller, M. Menéndez, R. Morales, D. Orozco Suárez, G. Ramos, M. Reina, J.L. Ramos, P. Rodríguez, A. Sánchez, N. Uribe-Patarroyo, P. Barthol, A. Gandorfer, M. Knoelker, W. Schmidt, S.K. Solanki, S. Vargas Domínguez, The Imaging Magnetograph eXperiment (IMaX) for the Sunrise Balloon-Borne Solar Observatory. Sol. Phys. **268**, 57–102 (2011). doi:10.1007/s11207-010-9644-y

P. Mein, The MSDP of THEMIS: capabilities, first results and prospects. Astron. Astrophys. **381**, 271–278 (2002). doi:10.1051/0004-6361:20011418

D. Orozco Suárez, L.R. Bellot Rubio, Analysis of quiet-sun internetwork magnetic fields based on linear polarization signals. Astrophys. J. **751**, 2 (2012). doi:10.1088/0004-637X/751/1/2

W.D. Pesnell, B.J. Thompson, P.C. Chamberlin, The Solar Dynamics Observatory (SDO). Sol. Phys. **275**, 3–15 (2012). doi:10.1007/s11207-011-9841-3

H. Peter, L. Abbo, V. Andretta, F. Auchère, A. Bemporad, F. Berrilli, V. Bommier, A. Braukhane, R. Casini, W. Curdt, J. Davila, H. Dittus, S. Fineschi, A. Fludra, A. Gandorfer, D. Griffin, B. Inhester, A. Lagg, E.L. Degl'Innocenti, V. Maiwald, R.M. Sainz, V.M. Pillet, S. Matthews, D. Moses, S. Parenti, A. Pietarila, D. Quantius, N.-E. Raouafi, J. Raymond, P. Rochus, O. Romberg, M. Schlotterer, U. Schühle, S. Solanki, D. Spadaro, L. Teriaca, S. Tomczyk, J.T. Bueno, J.-C. Vial, Solar magnetism eXplorer (SolmeX). Exploring the magnetic field in the upper atmosphere of our closest star. Exp. Astron. **33**, 271–303 (2012). doi:10.1007/s10686-011-9271-0

H.-P. Povel, Imaging Stokes polarimetry with piezoelastic modulators and charge-coupled-device image sensors. Opt. Eng. **34**, 1870–1878 (1995). doi:10.1117/12.200596

K.G. Puschmann, C. Denker, H. Balthasar, R.E. Louis, E. Popow, M. Woche, C. Beck, T. Seelemann, R. Volkmer, GREGOR Fabry-Pérot interferometer and its companion the blue imaging solar spectrometer. Opt. Eng. **52**(8), 081606 (2013). doi:10.1117/1.OE.52.8.081606

A. Raja Bayanna, S.K. Mathew, P. Venkatakrishnan, N. Srivastava, Narrow-band imaging system for the multi-application solar telescope at Udaipur Solar Observatory: characterization of lithium niobate etalons. Sol. Phys. **289**, 4007–4019 (2014). doi:10.1007/s11207-014-0557-z

R. Ramelli, D. Gisler, M. Bianda, N. Bello González, S. Berdyugina, D. Soltau, *First Successful Deployment of the Zimpol-3 System at the Gregor Telescope*. Society of Photo-Optical Instrumentation Engineers (SPIE) Conference Series, vol. 9147 (2014), p. 3. doi:10.1117/12.2056934

M. Rempel, R. Schlichenmaier, Sunspot modeling: from simplified models to radiative MHD simulations. Living Rev. Sol. Phys. **8**, 3 (2011). doi:10.12942/lrsp-2011-3

T.L. Riethmüller, S.K. Solanki, V. Martínez Pillet, J. Hirzberger, A. Feller, J.A. Bonet, N. Bello González, M. Franz, M. Schüssler, P. Barthol, T. Berkefeld, J.C. del Toro Iniesta, V. Domingo, A. Gandorfer, M. Knölker, W. Schmidt, Bright points in the quiet sun as observed in the visible and near-UV by the Balloon-Borne Observatory SUNRISE. Astrophys. J. Lett. **723**, 169–174 (2010). doi:10.1088/2041-8205/723/2/L169

T.R. Rimmele, K. Richards, S. Hegwer, S. Fletcher, S. Gregory, G. Moretto, L.V. Didkovsky, C.J. Denker, A. Dolgushin, P.R. Goode, M. Langlois, J. Marino, W. Marquette, First results from the NSO/NJIT Solar adaptive optics system, in *Telescopes and Instrumentation for Solar Astrophysics*, ed. by S. Fineschi, M.A. Gummin Society of Photo-Optical Instrumentation Engineers (SPIE) Conference Series, vol. 5171 (2004), pp. 179–186. doi:10.1117/12.508513

P. Romano, S.L. Guglielmino, A. Cristaldi, I. Ermolli, M. Falco, F. Zuccarello, Evolution of the magnetic field inclination in a forming penumbra. Astrophys. J. **784**, 10 (2014). doi:10.1088/0004-637X/784/1/10

K. Sankarasubramanian, C. Gullixson, S. Hegwer, T.R. Rimmele, S. Gregory, T. Spence, S. Fletcher, K. Richards, E. Rousset, B. Lites, D. Elmore, K. Streander, M. Sigwarth, The diffraction limited spectro-polarimeter: a new instrument for high-resolution solar polarimetry, in *Telescopes and Instrumentation for Solar Astrophysics*, ed. by S. Fineschi, M.A. Gummin Society of Photo-Optical Instrumentation Engineers (SPIE) Conference Series, vol. 5171 (2004), pp. 207–218. doi:10.1117/12.508790

T.A. Schad, M.J. Penn, H. Lin, He I vector magnetometry of field-aligned superpenumbral fibrils. Astrophys. J. **768**, 111 (2013). doi:10.1088/0004-637X/768/2/111

G.B. Scharmer, K. Bjelksjo, T.K. Korhonen, B. Lindberg, B. Petterson, The 1-meter Swedish Solar Telescope, in *Innovative Telescopes and Instrumentation for Solar Astrophysics*, ed. by S.L. Keil, S.V. Avakyan

Society of Photo-Optical Instrumentation Engineers (SPIE) Conference Series, vol. 4853 (2003), pp. 341–350
G.B. Scharmer, G. Narayan, T. Hillberg, J. de la Cruz Rodríguez, M.G. Löfdahl, D. Kiselman, P. Sütterlin, M. van Noort, A. Lagg, CRISP spectropolarimetric imaging of penumbral fine structure. Astrophys. J. Lett. **689**, 69–72 (2008). doi:10.1086/595744
G.B. Scharmer, J. de la Cruz Rodriguez, P. Sütterlin, V.M.J. Henriques, Opposite polarity field with convective downflow and its relation to magnetic spines in a sunspot penumbra. Astron. Astrophys. **553**, 63 (2013). doi:10.1051/0004-6361/201220899
W. Schmidt, O. von der Lühe, R. Volkmer, C. Denker, S.K. Solanki, H. Balthasar, N. Bello Gonzalez, T. Berkefeld, M. Collados, A. Fischer, C. Halbgewachs, F. Heidecke, A. Hofmann, F. Kneer, A. Lagg, H. Nicklas, E. Popow, K.G. Puschmann, D. Schmidt, M. Sigwarth, M. Sobotka, D. Soltau, J. Staude, K.G. Strassmeier, T.A. Waldmann, The 1.5 meter solar telescope GREGOR. Astron. Nachr. **333**, 796 (2012). doi:10.1002/asna.201211725
B. Schmieder, H. Tian, T. Kucera, A. López Ariste, N. Mein, P. Mein, K. Dalmasse, L. Golub, Open questions on prominences from coordinated observations by IRIS, Hinode, SDO/AIA, THEMIS, and the Meudon/MSDP. Astron. Astrophys. **569**, 85 (2014). doi:10.1051/0004-6361/201423922
A.I. Shapiro, D.M. Fluri, S.V. Berdyugina, M. Bianda, R. Ramelli, NLTE modeling of Stokes vector center-to-limb variations in the CN violet system. Astron. Astrophys. **529**, 139 (2011). doi:10.1051/0004-6361/200811299
H. Socas-Navarro, The three-dimensional structure of a sunspot magnetic field. Astrophys. J. Lett. **631**, 167–170 (2005). doi:10.1086/497334
H. Socas-Navarro, D. Elmore, Physical properties of spicules from simultaneous spectropolarimetric observations of He I and Ca II lines. Astrophys. J. Lett. **619**, 195–198 (2005). doi:10.1086/428399
H. Socas-Navarro, D. Elmore, A. Pietarila, A. Darnell, B.W. Lites, S. Tomczyk, S. Hegwer, Spinor: visible and infrared spectro-polarimetry at the National Solar Observatory. Sol. Phys. **235**, 55–73 (2006). doi:10.1007/s11207-006-0020-x
H. Socas-Navarro, J.M. Borrero, A. Asensio Ramos, M. Collados, I. Domínguez Cerdeña, E.V. Khomenko, M.J. Martínez González, V. Martínez Pillet, B. Ruiz Cobo, J. Sánchez Almeida, Multiline spectropolarimetry of the Quiet Sun at 5250 and 6302 Å. Astrophys. J. **674**, 596–606 (2008). doi:10.1086/521418
S. Solanki (The EPIC Consortium), EPIC: European participation in Solar-C. A proposal in response to the call for the M4 Mission of ESA (2015)
S.K. Solanki, P. Barthol, S. Danilovic, A. Feller, A. Gandorfer, J. Hirzberger, T.L. Riethmüller, M. Schüssler, J.A. Bonet, V. Martínez Pillet, J.C. del Toro Iniesta, V. Domingo, J. Palacios, M. Knölker, N. Bello González, T. Berkefeld, M. Franz, W. Schmidt, A.M. Title, SUNRISE: instrument, mission, data, and first results. Astrophys. J. Lett. **723**, 127–133 (2010). doi:10.1088/2041-8205/723/2/L127
S.K. Solanki, P. Barthol, S. Danilovic, A. Feller, A. Gandorfer, J. Hirzberger, A. Lagg, T.L. Riethmüller, M. Schüssler, T. Wiegelmann, J.A. Bonet, V.M. Pillet, E. Khomenko, J.C. del Toro Iniesta, V. Domingo, J. Palacios, M. Knölker, N.B. González, J.M. Borrero, T. Berkefeld, M. Franz, M. Roth, W. Schmidt, O. Steiner, A.M. Title, The Sun at high resolution: first results from the Sunrise Mission, in *IAU Symposium*, ed. by D. Prasad Choudhary, K.G. Strassmeier IAU Symposium, vol. 273 (2011), pp. 226–232. doi:10.1017/S1743921311015298
S.K. Solanki, J.C. del Toro Iniesta, J. Woch, A. Gandorfer, J. Hirzberger, W. Schmidt, T. Appourchaux, A. Alvarez-Herrero, *The Polarimetric and Helioseismic Imager for Solar Orbiter: So/Phi*. IAU Symposium, vol. 305 (2015), pp. 108–113. doi:10.1017/S1743921315004615
O. Steiner, R. Rezaei, Recent advances in the exploration of the small-scale structure of the quiet solar atmosphere: vortex flows, the horizontal magnetic field, and the stokes- V line-ratio method, in *Fifth Hinode Science Meeting*, ed. by L. Golub, I. De Moortel, T. Shimizu Astronomical Society of the Pacific Conference Series, vol. 456 (2012), p. 3
J.O. Stenflo, Distribution functions for magnetic fields on the quiet Sun. Astron. Astrophys. **517**, 37 (2010). doi:10.1051/0004-6361/200913972
J.O. Stenflo, C.U. Keller, The second solar spectrum. A new window for diagnostics of the Sun. Astron. Astrophys. **321**, 927–934 (1997)
J.O. Stenflo, L. Stenholm, Resonance-line polarization. II—Calculations of linear polarization in solar UV emission lines. Astron. Astrophys. **46**, 69–79 (1976)
J.O. Stenflo, H. Biverot, L. Stenmark, Ultraviolet polarimeter to record resonance-line polarization in the solar spectrum around 130–150 nm. Appl. Opt. **15**, 1188–1198 (1976). doi:10.1364/AO.15.001188
J.O. Stenflo, D. Dravins, N. Wihlborg, A. Bruns, V.K. Prokofev, I.A. Zhitnik, H. Biverot, L. Stenmark, Search for spectral line polarization in the solar vacuum ultraviolet. Sol. Phys. **66**, 13–19 (1980). doi:10.1007/BF00150514
Y. Suematsu, S. Tsuneta, K. Ichimoto, T. Shimizu, M. Otsubo, Y. Katsukawa, M. Nakagiri, M. Noguchi, T. Tamura, Y. Kato, H. Hara, M. Kubo, I. Mikami, H. Saito, T. Matsushita, N. Kawaguchi, T. Nakaoji, K.

Nagae, S. Shimada, N. Takeyama, T. Yamamuro, The solar optical telescope of Solar-B (Hinode): The optical telescope assembly. Sol. Phys. **249**, 197–220 (2008). doi:10.1007/s11207-008-9129-4

Y. Suematsu, T. Sukegawa, Y. Okura, T. Nakayasu, Y. Enokida, M. Koyama, K. Saito, S. Ozaki, S. Tsuneta, *Development of micro image slicer of integral field unit for spaceborne solar spectrograph*. Society of Photo-Optical Instrumentation Engineers (SPIE) Conference Series, vol. 9151 (2014a), p. 1. doi:10.1117/12.2054569

Y. Suematsu, Y. Katsukawa, H. Hara, R. Kano, T. Shimizu, K. Ichimoto, *Large aperture solar optical telescope and instruments for the Solar-c Mission*. Society of Photo-Optical Instrumentation Engineers (SPIE) Conference Series, vol. 9143 (2014b), p. 1. doi:10.1117/12.2054373

L. Teriaca, V. Andretta, F. Auchère, C.M. Brown, E. Buchlin, G. Cauzzi, J.L. Culhane, W. Curdt, J.M. Davila, G. Del Zanna, G.A. Doschek, S. Fineschi, A. Fludra, P.T. Gallagher, L. Green, L.K. Harra, S. Imada, D. Innes, B. Kliem, C. Korendyke, J.T. Mariska, V. Martínez-Pillet, S. Parenti, S. Patsourakos, H. Peter, L. Poletto, R.J. Rutten, U. Schühle, M. Siemer, T. Shimizu, H. Socas-Navarro, S.K. Solanki, D. Spadaro, J. Trujillo-Bueno, S. Tsuneta, S.V. Dominguez, J.-C. Vial, R. Walsh, H.P. Warren, T. Wiegelmann, B. Winter, P. Young, LEMUR: large European module for solar ultraviolet research. European contribution to JAXA's Solar-C mission. Exp. Astron. **34**, 273–309 (2012). doi:10.1007/s10686-011-9274-x

J.H. Thomas, L.E. Cram, A.H. Nye, Five-minute oscillations as a subsurface probe of sunspot structure. Nature **297**, 485–487 (1982). doi:10.1038/297485a0

J. Trujillo Bueno, New diagnostic windows on the weak magnetism of the solar atmosphere, in *Solar Polarization*, ed. by J. Trujillo-Bueno, J. Sanchez Almeida Astronomical Society of the Pacific Conference Series, vol. 307 (2003), p. 407

J. Trujillo Bueno, A. Asensio Ramos, N. Shchukina, in *The Hanle Effect in Atomic and Molecular Lines: A New Look at the Sun's Hidden Magnetism*, ed. by R. Casini, B.W. Lites Astronomical Society of the Pacific Conference Series, vol. 358 (2006), p. 269

J. Trujillo Bueno, J. Štěpán, R. Casini, The Hanle effect of the hydrogen Lyα line for probing the magnetism of the solar transition region. Astrophys. J. Lett. **738**, 11 (2011). doi:10.1088/2041-8205/738/1/L11

J. Štěpán, *Three-Dimensional Simulations of Scattering Polarization and the Hanle Effect in MHD Chromospheric Models*. IAU Symposium, vol. 305 (2015), pp. 360–367. doi:10.1017/S1743921315005050

J. Štěpán, J. Trujillo Bueno, J. Leenaarts, M. Carlsson, Three-dimensional radiative transfer simulations of the scattering polarization of the hydrogen Lyα line in a magnetohydrodynamic model of the chromosphere/corona transition region. Astrophys. J. **803**, 65 (2015). doi:10.1088/0004-637X/803/2/65

M. van Noort, L. Rouppe van der Voort, M.G. Löfdahl, Solar image restoration by use of multi-frame blind de-convolution with multiple objects and phase diversity. Sol. Phys. **228**, 191–215 (2005). doi:10.1007/s11207-005-5782-z

S. Vargas Domínguez, A. Kosovichev, V. Yurchyshyn, Multi-wavelength high-resolution observations of a small-scale emerging magnetic flux event and the chromospheric and coronal response. Astrophys. J. **794**, 140 (2014). doi:10.1088/0004-637X/794/2/140

J. Varsik, C. Plymate, P. Goode, A. Kosovichev, W. Cao, R. Coulter, K. Ahn, N. Gorceix, S. Shumko, *Control and operation of the 1.6 M New Solar Telescope in Big Bear*. Society of Photo-Optical Instrumentation Engineers (SPIE) Conference Series, vol. 9147 (2014), p. 5. doi:10.1117/12.2056688

H. Wang, C. Liu, N. Deng, Z. Zeng, Y. Xu, J. Jing, W. Cao, Study of two successive three-ribbon solar flares on 2012 July 6. Astrophys. J. Lett. **781**, 23 (2014). doi:10.1088/2041-8205/781/1/L23

H. Wang, W. Cao, C. Liu, Y. Xu, R. Liu, Z. Zeng, J. Chae, H. Ji, Witnessing magnetic twist with high-resolution observation from the 1.6-m New Solar Telescope. Nat. Commun. **6**, 7008 (2015). doi:10.1038/ncomms8008

T. Watanabe, *The Solar-c Mission*. Society of Photo-Optical Instrumentation Engineers (SPIE) Conference Series, vol. 9143 (2014), p. 1. doi:10.1117/12.2055366

S. Yang, J. Zhang, Z. Liu, Y. Xiang, New vacuum solar telescope observations of a flux rope tracked by a filament activation. Astrophys. J. Lett. **784**, 36 (2014). doi:10.1088/2041-8205/784/2/L36

V. Yurchyshyn, V. Abramenko, A. Kilcik, Dynamics in sunspot umbra as seen in New Solar Telescope and interface region imaging spectrograph data. Astrophys. J. **798**, 136 (2015). doi:10.1088/0004-637X/798/2/136

Z. Zeng, W. Cao, H. Ji, Observation of magnetic reconnection driven by granular scale advection. Astrophys. J. Lett. **769**, 33 (2013). doi:10.1088/2041-8205/769/2/L33

Z. Zeng, J. Qiu, W. Cao, P.G. Judge, A flare observed in coronal, transition region, and Helium I 10830 Å emissions. Astrophys. J. **793**, 87 (2014). doi:10.1088/0004-637X/793/2/87

J.B. Zirker, The Sacramento Peak Observatory. Sol. Phys. **182**, 1–19 (1998). doi:10.1023/A:1005048409308

FSC
www.fsc.org
MIX
Papier | Fördert gute Waldnutzung
FSC® C083411

Kolibri 360 GmbH
Ferdinand-Jühlke-Straße 7
99095 Erfurt, Deutschland
produktsicherheit@kolibri360.de